The Composting Handbook

The Composting Handbook

A how-to and why manual for farm, municipal, institutional and commercial composters

Editor

Robert Rynk

Associate Editors

Ginny Black
Johannes Biala
Jean Bonhotal
Leslie Cooperband
Jane Gilbert
Mary Schwarz

ELSEVIER

ACADEMIC PRESS
An imprint of Elsevier

Academic Press is an imprint of Elsevier
125 London Wall, London EC2Y 5AS, United Kingdom
525 B Street, Suite 1650, San Diego, CA 92101, United States
50 Hampshire Street, 5th Floor, Cambridge, MA 02139, United States
The Boulevard, Langford Lane, Kidlington, Oxford OX5 1GB, United Kingdom

Notices
Knowledge and best practice in this field are constantly changing. As new research and
experience broaden our understanding, changes in research methods, professional practices,
or medical treatment may become necessary.

Practitioners and researchers must always rely on their own experience and knowledge in
evaluating and using any information, methods, compounds, or experiments described
herein. In using such information or methods they should be mindful of their own safety and
the safety of others, including parties for whom they have a professional responsibility.

To the fullest extent of the law, neither the Publisher nor the authors, contributors, or editors,
assume any liability for any injury and/or damage to persons or property as a matter of
products liability, negligence or otherwise, or from any use or operation of any methods,
products, instructions, or ideas contained in the material herein.

Library of Congress Cataloging-in-Publication Data
A catalog record for this book is available from the Library of Congress

British Library Cataloguing-in-Publication Data
A catalogue record for this book is available from the British Library

ISBN: 978-0-323-85602-7

For information on all Academic Press publications visit our
website at https://www.elsevier.com/books-and-journals

About the covers. The large background photo on the front cover was provided by Karin Grobe.
It shows a windrow composting facility handling discarded salad from a neighboring packing
factory. The inset photos depict various composting methods and compost uses. From top to
bottom, the sources are: Robert Rynk, Monica Ozores-Hampton, Cary Oshins, Judy Puddester,
and Ji Li. The back cover shows selected posters promoting International Compost Awareness
Week, including poster sponsored by the Composting Research and Education Foundation
in the U.S., the Compost Council of Canada, and the European Composting Network.

Publisher: Charlotte Cockle
Acquisitions Editor: Nancy Maragioglio
Editorial Project Manager: Lindsay Lawrence
Production Project Manager: Paul Prasad Chandramohan
Cover Designer: Matthew Limbert

Typeset by TNQ Technologies

This book is dedicated to Jerry Goldstein, and the Goldstein family, who fed a fledgling group of composters with knowledge and positivity, and nurtured us into an important industry.

Contents

CHAPTER 6 **Forced aeration composting, aerated static pile, and similar methods................................ 197**
Frederick Michel, Tim O'Neill and Robert Rynk

CHAPTER 7 **Contained and in-vessel composting methods and methods summary.................................... 271**
Frederick Michel, Tim O'Neill and Robert Rynk

CHAPTER 8   Composting animal mortalities 307

Jean Bonhotal, Mary Schwarz and Robert Rynk

CHAPTER 15 Compost characteristics and quality 737

Richard Stehouwer, Leslie Cooperband and Robert Rynk

                    Monica Ozores-Hampton, Johannes Biala,
                    Gregory Evanylo and Britt Faucette

CHAPTER 19 Composting economics 913
Craig S. Coker and Mark King

Authors and Contributors

John Aber
Professor Emeritus, University of New Hampshire, Durham, NH, United States

Ron Alexander
R. Alexander Associates, Inc., Apex, NC, United States

Susan Antler
Compost Council of Canada, Toronto, ON, Canada

Johannes Biala
Centre for Recycling of Organic Waste & Nutrients, The University of Queensland, Gatton, QLD, Australia

Ginny Black
Compost Research and Education Foundation (CREF), Raleigh, NC, United States

Anna F. Bokowa
Environmental Odour Consulting Corporation, Oakville, ON, Canada

Jean Bonhotal
Cornell Waste Management Institute, Cornell University, Ithaca, NY, United States

Nellie J. Brown
Workplace Health & Safety Program, ILR, Cornell University, Buffalo, NY, United States

Sally Brown
School of Forest Resources, University of Washington, Seattle, WA, United States

Michael Bryant-Brown
Green Mountain Technologies, NE Bainbridge Island, WA, United States

Van Calvez
Green Mountain Technologies, NE Bainbridge Island, WA, United States

Andrew Carpenter
Northern Tilth, LLC., Belfast, ME, United States

Craig S. Coker
Coker Composting and Consulting, Troutville, VA, United States

Leslie Cooperband
Praire Fruits Farm and Creamery, Champagne, IL, United States

Matthew Cotton
Integrated Waste Management Consulting, Richmond, CA, United States

Jeffrey A. Creque
Rangeland and Agroecosystem Management, Carbon Cycle Institute, Petaluma, CA, United States

Gregory Evanylo
School of Plant and Environmental Sciences, Virginia Polytechnic Institute and State University, Blacksburg, VA, United States

Britt Faucette
Filtrexx International, Decatur, GA, United States

Frank Franciosi
U.S. Composting Council, Raleigh, NC, United States

Jeff Gage
Green Mountain Technologies, Bainbridge Island, WA, United States

Scott Gamble
Organic Waste Specialist, Professional Engineer, Edmonton, AB, Canada

Jane Gilbert
Carbon Clarity, Rushden, Northamptonshire, United Kingdom

Thomas Halbach
Extension Professor Emeritus, University of Minnesota, Minneapolis, MN, United States

James Hardin
Associate Professor, SUNY Cobleskill, Cobleskill, NY, United States

Harry A. Hoitink
Professor Emeritus, Ohio State University, Wooster, OH, United States

Harold Keener
Professor Emeritus, Ohio State University, Wooster, OH, United States

Mark King
Organics Management Specialist, Maine Dept. of Environmental Protection, Bangor ME, United States

Nanci Koerting
Environmental Compliance, Grant County Mulch, Boonsboro, MD, United States

Nancy J. Lampen
Associates for Human Resource Development, Pittsford, NY, United States

Tera Lewandowski
Growing Media, The Scotts Miracle-Gro Company, Marysville, OH, United States

Ji Li
Professor, Department of Ecology and Ecological Engineering, China Agricultural University, Beijing, China

Dan Lilkas-Rain
Growing Media, Town of Bethlehem, Bethlehem, NY, United States

Lorrie Loder-Rossiter
Revinu, Inc., Fleming Island, FL, United States

Pierce Louis
Dirt Hugger, Dallesport, WA, United States

Frederick Michel
Ohio State University, Wooster, OH, United States

Robert Michitsch
University of Wisconsin—Stevens Point, Stevens Point, WI, United States

Deborah A. Neher
University of Vermont, Burlington, VT, United States

Hilary Nichols
U.S. Composting Council, Raleigh, NC, United States

Tim O'Neill
Engineered Compost Systems, Seattle, WA, United States

Cary Oshins
U.S. Composting Council, Raleigh, NC, United States

Monica Ozores-Hampton
TerraNutri, LLC, Miami Beach, FL, United States

John Paul
Transform Compost Systems, Abbotsford, BC, Canada

Tom L. Richard
Pennsylvania State University, State College, PA, United States

Jonathan M. Rivin
Materials Evaluation Specialist, Oregon Department of Environmental Quality, Portland, OR, United States

Nancy Roe
Agricultural Consultant, Tucson, AZ, United States

Robert Rynk
Professor Emeritus, SUNY Cobleskill, Cobleskill, NY, United States

Mary Schwarz
Cornell Waste Management Institute, Cornell University, Ithaca, NY, United States

Ronda Sherman
North Carolina State University, Raleigh, NC, United States

Stefanie Siebert
European Compost Network, Bochum, Germany

Matthew Smith
USDA National Agroforestry Center, Lincoln, NE, United States

Richard Stehouwer
Pennsylvania State University, State College, PA, United States

Dan Sullivan
Department of Crop and Soil Science, Oregon State University, Corvallis, OR, United States

Rod Tyler
Green Horizons Environmental, Medina, OH, United States

Rudy Wentz
Agricultural Equine Industry Economics, Formerly with the State University of New York, Cobleskill, NY, United States

Holly Wescott
Heart Beet Gardens, Ashfield, MA, United States

Steven Wisbaum
CV Compost, Charlotte, VT, United States

Jeff Ziegenbein
Regional Compost Operations, Inland Empire Utilities Agency, Chino, CA, United States

Preface

The Composting Handbook began as an effort to produce a second edition of the *On-Farm Composting Handbook (OFCH)*, which was originally published in 1992 by NRAES[1]. This book is not, however, that anticipated second edition. Instead, it should be considered the OFCH's sequel; a much-expanded update that encompasses all applications of composting beyond the backyard—farm, commercial, institutional, and municipal. *The Composting Handbook* adopts the practical attitude of the original book, and it still retains a farm flavor. Indeed, readers who are familiar with the OFCH will recognize the connection. For me, there is some sadness in losing "on-farm" from the title. Personally, I continue to believe that farms offer the best situations for composting. Among other positives, the decentralized, and often remote, locations of farms pose many advantages. On-farm composting should be liberally encouraged, for almost all feedstocks.

But this book takes a much wider swipe and a deeper dive in its topics, applications, and geography. Unintentionally, the OFCH found readership among nonfarm and international readers. *The Composting Handbook* intentionally aims to serve nonfarm and international readers.

I submit that this book was written for the composting community *by* the composting community. Please look at the list of authors and contributors. I expect that you will be impressed by the length of the list and the talent represented. Also note the diversity of positions, perspectives, and locations of the contributors. Collectively, we are teachers, researchers, engineers, compost producers, facility managers, extension specialists, technical consultants, business consultants, compost users, public servants, nonprofit professionals, vendors of equipment and systems, soil scientists, horticulturists, and farmers. In addition, we authors benefitted from the good work of past and current colleagues. We borrowed knowledge from many published journal articles, guidelines, professional reports, presentations, books, and even graduate student thesis. I think that some of the more valuable elements of this book are in its references pages. At the end of every chapter, we itemized cited references, as required, but we also listed references that we authors relied on for wisdom and references that we consider potentially valuable resources for readers.

The Composting Handbook has a large and somewhat risky ambition. It seeks to serve nearly everyone in the composting community. It presents useful advice for new and prospective composters, as well as composting veterans. It targets composting practitioners, but professionals in any role within the composting industry should

---

[1]*The On-Farm Composting Handbook* was originally published by The Northeast Regional Agricultural Engineering Service (NRAES), a program of 13 land-grant universities in the northeastern US Electronic copies are available through Cornel University's e-commons system at https://ecommons.cornell.edu/handle/1813/67142.

find it useful. While the emphasis is on compost production, there is ample advice for compost users too. The risk in trying to serve everyone is that no one will be satisfied. However, after reading and rereading the chapters, I am confident that we succeeded. Everyone will find some information of value. Most readers will find a lot of it.

The book also aims at a broad target geographically. The book is written with the expectation that it will be used internationally. It rightly draws from international publications and practices, regardless of the audience. But also, we consciously included international examples, photographs, rules, and regulations. I hope that we "internationalized" the book well enough. I apologize to those mostly non-English speaking citizens who I suspect we neglected.

Despite our efforts, there is no denying that the book has a US bias. This bias is an almost unavoidable consequence of the American genesis of the book. Conditions and regulations in the US steal much of the spotlight. The dollar values and spellings are also US (although a few "odours" and "fertilisers" might have slipped by).

Some balance is restored by the fact that metric (SI) measurement units are primary, and US (imperial) units are secondary. Having to acknowledge two sets of units throughout the book is annoying, and sometimes confusing, but it is currently necessary. Conversions between metric and US units are generally approximate throughout the book. For example, the difference between a US short ton (2000 lbs) and a metric ton, or tonne (1000 kg), is occasionally ignored. Either or both of these units are normally referred to as simply a "ton," except where the situation requires more specificity. We can be approximate because composting is so forgiving. It usually permits approximations. The composting sage, Peter Moon of O_2Compost, famously says, *"There are no decimal points in composting."* Peter is being poetic, of course, but his point is well taken—composting is such a robust process that it does not deserve a high degree of precision. Therefore, if you find a case where a metric dimension is coarsely converted to a round number in US units, please forgive us and move on. Sometimes one meter can equal three feet. Sometimes 1123 kg can be a ton.

The information presented in *The Composting Handbook*—the principles and practices herein—are current to the time of its writing, circa 2021. The contents of this book will gradually become outdated. Some topics will soon become outdated. Therefore, we anticipate that you will continue to follow progress in the related scientific disciplines and in the practices within the composting industry. Please consult research journals and popular industry media, like *BioCycle*. Please join professional organizations associated with composting, and participate in their conferences and educational activities. Accurate knowledge will always be the best tool for success. Because regulations, best practices, and science will inevitably change, this book should not be regarded as a set of rules, or a surrogate book of regulations (the OFCH was sporadically used in that fashion). Instead think of *The Composting Handbook* a collection of guidelines, to be considered and applied according to the situation at hand, along with other information currently available.

Robert Rynk, Principal Editor

Acknowledgments and appreciations

This book was a long time coming. In 2005, Marty Sailus, then manager of PALS[1] (NRAES' successor) asked Leslie Cooperband and I to develop the second edition of the On-Farm Composting Handbook (OFCH). Leslie and I sat at her farm table in Illinois and mapped out a plan, including the contents, potential contributors, and timeline for the new edition. To save embarrassment, I will not reveal the timeline. Within a year, we even had a few authors complete a draft of several chapters. But then Leslie's farm enterprise began to flourish, and I started my job at SUNY Cobleskill. As it turns out, it is really hard to produce an ambitious book while running a farm or tending to a demanding day job. Although I attempted to keep up the momentum, the effort languished. In the meantime, PALS perished and composting science and practice evolved. As the project revived, we needed to find new content for the book, a new partner, and a new publisher. In the end, the editors and authors stepped up to revitalize the content. The Composting Research and Education Foundation (CREF) stepped forward to help produce the book. And Elsevier stepped in to publish it.

Given the book's history, this section is devoted to acknowledgments, appreciations, and a few apologies. In general, I would like to express my sincere and gratitude to everyone who contributed to the book and moved it forward. Specifically:

- Thank you to my patient co-editors, who put in many hours, and enlivened the work when it was floundering.
- Thank you to the book's authors and contributors for sharing their knowledge. Extra thanks (if that is a thing) to the long-suffering authors who have been with the project since the early years.
- Thank you, and sincere apologies, to Marty Sailus for his painstaking yet doomed efforts to motivate me to hurry the book toward its completion. Good try, Marty.
- Thank you to editorial assistant and counselor, Judy Puddester, for her patience, help, advice, and wagging forefinger (yes, her *forefinger*).
- My apologies to the many people who I naively told, over the years, that the book is nearly done and will be ready "next year." Most of the time, I truly believed it myself.
- Extra special thank yous to the New York State Energy Research and Development Authority (NYSERDA) and Cornell University for financially supporting the project in its early stages. That early support was essential.

---

[1]Plant And Life Sciences (PALS) Publishing was a program of the Dept. of Horticulture at Cornell University. PALS was the successor to NRAES the publisher of the original OFCH. PALS ceased operation in 2018. Its publications are still available through Cornell University. https://www.cornellstore.com/pals-publishing

- An extra special thank you to the Kevin Tritz Memorial Fund (KTMF) for supporting the book in the later stages. Like Kevin himself, KTMF is dedicated to advancing the compost industry through sound scientific research and education. Kevin was a strong advocate for composting and a former president of the US Composting Council. Kevin would have been been "all in" with the book's development. http://www.mncompostingcouncil.org/kevin-tritz-memorial-fund.html
- Thank you to the original authors of the OFCH who laid the foundation for *The Composting Handbook*: Maarten van de Kamp, George B. Willson, Mark E. Singley, Tom L. Richard, John J. Kolega, Francis (Frank) Gouin, Lucien Laliberty, Jr., David Kay, Dennis W. Murphy, Harry A. J. Hoitink, and William F. Brinton. I am aware that Maarten, George, Mark, John, and Frank have passed on. They may have applied some divine guidance to the new book. Certainly, Maarten is chatting about it to others in the afterlife.

Why compost?

1

Authors: **Robert Rynk[1], Leslie Cooperband[2]**

[1]*SUNY Cobleskill, Cobleskill, NY, United States;* [2]*Praire Fruits Farm and Creamery, Champagne, IL, United States*

Contributors: **Cary Oshins[3], Holly Wescott[4], Jean Bonhotal[5], Mary Schwarz[5], Ronda Sherman[6], Sally Brown[7]**

[3]*U.S. Composting Council, Raleigh, NC, United States;* [4]*Heart Beet Gardens, Ashfield, MA, United States;* [5]*Cornell Waste Management Institute, Cornell University, Ithaca, NY, United States;* [6]*North Carolina State University, Raleigh, NC, United States;* [7]*School of Forest Resources, University of Washington, Seattle, WA, United States*

1. Introduction

"Composting" is the *aerobic*, or oxygen-requiring, decomposition of organic materials by *microorganisms* under controlled conditions. It has been practiced for eons, on wide-ranging scales of operation, from backyard piles to huge automated systems contained in warehouse-like buildings. Similarly, composts, made from a long list of feedstocks, have long been appreciated and used to the benefit of farmers, gardeners, and landscapers in a variety of applications (Platt et al., 2014). It is a means to conserve resources, preserve the environment, and create value (U.S. PIRG 2019). It yields useful natural products from less useful, and often wasted, organic ingredients. In so doing, composting opens opportunities for:

- farmers to improve handling and enhance the value of manure and crop residues,
- municipalities and other public entities to better use the organic materials that generally fall under their purview, including leaves, yard trimmings, biosolids[1], and solid waste,
- industrial, commercial, and residential food waste generators and collectors to dispose of food in an environmentally benign manner through recycling,
- operators of anaerobic digestors to create added-value outlets for digester effluents,
- generators of organic materials to become recyclers of organic materials,

---

[1] *The word "biosolids" is used throughout this book as a term for the solids, or sludge, from wastewater treatment. The term biosolids implies that the material is suitable for land application (i.e., low concentrations of contaminants and treated to reduce pathogens and vector attraction).*

The Composting Handbook. https://doi.org/10.1016/B978-0-323-85602-7.00001-7

- farmers, ranchers, horticulturalists, gardeners, and other plant growers to enrich their soils and improve their methods and products,
- environmental managers to help offset the climate impacts of greenhouse gases by sequestering carbon in soils, and,
- businesses, farms, and small entrepreneurs to earn money through composting services and the production and sale of compost.

The ingredients, or "feedstocks", for composting are organic residuals—manure, leaves, yard trimmings, wood and paper products, food residuals, biosolids, and variety of other organic materials (Chapter 4). These residuals inherently possess utility and value. They contain resources in the form of organic matter, energy, nutrients, minerals, and microorganisms that can benefit soils, crops, the landscape, livestock, people, the atmosphere, and the environment. Composting retains the nature of these resources and transforms them into soil-building products. The value of the composted products nearly always exceeds that of the original feedstocks.

Composting is only one of many alternatives for managing organic residuals. Many types of residuals can be applied to farmland directly as a source of organic matter and nutrients, as they have been used traditionally. Some residuals can be applied directly in the landscape as mulch. They can be processed for energy production or livestock feed. Other biological conversion options include vermicomposting and anaerobic digestion, neither of which is covered in detail in this book. Although vermicomposting shares many features with conventional composting, the processes are different enough to merit separate coverage (see Box 1.1).

So, with these options for recycling organic residuals, the question that needs asking is: "Why compost?"

Box 1.1 Vermicomposting

Author: Rhonda Sherman.

Vermicomposting is a controlled biological process that relies on earthworms and microorganisms to decompose and stabilize organic materials. The earthworms ingest organic particles and resident microorganisms to obtain sustenance. The digested particles that they excrete are called "castings." The resulting product, a mixture of castings and otherwise-decomposed feedstocks, is called vermicompost or vermicast.

It is crucial to understand the differences between thermophilic composting and vermicomposting. They are two separate methods of managing organic materials and have diverse outcomes. Vermicomposting takes place at lower, mesophilic, temperatures. Vermicomposting worms thrive at temperatures in the range of roughly 15−28°C (about 60−80 F) and do not tolerate temperatures above 35°C (95 F). Temperatures are kept low by using shallow windrows, beds, and bins and by adding feedstocks gradually, as needed by the worms. The absence of high temperatures means that the vermicast is not heat-sanitized, although pathogens numbers are greatly reduced by the worms' digestion and the competing microorganisms (Sawati and Hait, 2018). Precomposting of feedstocks is encouraged to destroy pathogens and seeds with more certainty, make the worm feed homogenous, decompose potential toxins, reduce ammonia levels, and to reduce the weight, volume, and heat in feedstocks.

Earthworms can process food residuals, animal manures, crop wastes, paper products, industrial organic byproducts, and brewery wastes. They can also consume biosolids and sludge produced from paper and pulp mills, and milk processing plants. Manure produced by cows, pigs, horses, goats, llamas, alpacas, sheep, and rabbits is commonly vermicomposted. Chicken manure is an exception because it is too high in ammonia.

Box 1.1 Vermicomposting—cont'd

Vermicomposting does not require specialized training or equipment. Many farmers utilize unused animal housing for their vermicomposting operations. Worm beds or windrows can be set directly on the ground, in pits or trenches, or in bins. Bins are typically made of lumber, concrete blocks, bricks, poured concrete, or clay tile. Discarded items may also be repurposed as worm bins. For larger operations, many choose to build or buy continuous flow-through digesters that automate feeding and harvesting of vermicompost.

Although vermicomposting does not necessitate advanced schooling, there are several key points of knowledge that are required to be successful. Many people initially focus on bins and equipment and miss the point that earthworms are doing the majority of the work, not machines. Therefore, it is essential to master earthworm animal husbandry skills on a small scale before expanding an operation. Vermicomposters must use the appropriate earthworm species and provide their needs to help them thrive.

There are more than 9000 species of earthworms, but only seven species are suitable for vermicomposting. Most earthworms live underground and consume soil. Earthworms used for vermicomposting live on top of the ground in decaying organic materials such as manure and leaf piles. Most people use one species of earthworm for vermicomposting. It has many common names, such as red wiggler, red worm, brandling worm, tiger worm, or California red worm, but these are all nicknames for the species *Eisenia fetida*.

Vermicast is a premium product, generally superior to conventional compost. It is rich in nutrients that are readily available to plants. It also contains massive numbers and varieties of microorganisms that benefit soil and plant health. Because vermicast is not subject to thermophilic temperatures, it has higher numbers and greater varieties of microorganisms compared to conventional compost. Plant growth hormones and humic and fulvic acids in vermicast increase plant growth and crop yields. Phenolic compounds repel insects that eat plants. Vermicast and its liquid extracts have been found to suppress plant diseases (see Chapter 17). Because of its beneficial effects on soil and plants, vermicast sells for a premium price, in the range of $250 to $1300 per cubic meter ($200 to $1000 per cubic yard).

Only a small amount of vermicast can provide the benefits to plants described earlier. Mixing 10% to 20% volume of vermicast with soil significantly increases plant growth and yields and suppresses pests and diseases. Vermicast is commonly added to soils in gardens, vineyards, golf courses, nurseries, farm fields, lawns, and in potted plants. Some thermophilic compost manufacturers and soil blenders mix vermicompost into their products to provide added benefits.

Thousands of people all over the world are vermicomposting to manage organic waste and generate part or all of their income from gate fees or sales of vermicast or earthworms. *The Worm Farmer's Handbook* (Sherman, 2018) provides detailed descriptions of two dozen vermicomposting operations taking place at farms, institutions, municipalities, schools, and businesses. The book also provides information about earthworm biology, vermicomposting methods, and management practices. Additional resources, including videos, are available at the website, https://composting.ces.ncsu.edu/.

Composting performs two fundamental functions—it converts difficult materials into a valuable commodity and an easily handled material. A composter benefits from both functions, regardless of whether composting is practiced primarily to manage organic residuals or primarily to produce compost (Fig. 1.1). For instance, a landscaping business might compost yard trimmings from the surrounding

FIGURE 1.1

Composting offers the potential to generate two revenue streams: services for recycling waste materials and sales of compost products.

Source: Brendon Mallia.

community to obtain the resulting compost for its own use but, in doing so, also collect fees for recycling the yard trimmings.

The first of these functions reflects a waste management goal, which has been the primary driver for composting for several decades. The second function is manufacturing, and it is becoming more prevalent as growing numbers of users recognize the value of compost. In fact, many composting experts and practitioners have embraced the manufacturing model as the new paradigm for composting (USCC, 2020). As a manufacturing process, the emphasis of composting is producing specific compost products with specific characteristics for particular uses or markets. Like any manufacturing enterprise, composters have to ensure that the revenues and benefits outweigh the costs and drawbacks.

Practitioners have developed their own notions about composting and compost that seem true, perhaps even obvious. However, some of these notions do not hold true for every situation. The prominent facts and fictions of composting and compost are addressed in Section 5 of this chapter.

2. Benefits and drawbacks of composting

Composting offers a variety of economic and environmental benefits. When considering the benefits of composting, it is important to distinguish between *composting*, the "process" and *compost*, the "product."

Although the benefits are numerous, composting can be a major undertaking, with associated costs and drawbacks (Table 1.1). One cannot simply dump leaves on a hillside or pile manure behind the barn and expect to have compost several weeks later. A successful composting operation deserves the same planning and commitment given to other functions, like crop production or landscape maintenance. Although it is often integral to the other operations, composting should be viewed as an enterprise in its own right. Like any enterprise, a composter needs to consider the labor, physical infrastructure, financial resources, and time available to compost properly.

Table 1.1 Summary of the benefits and drawbacks of on-farm composting.

Economic benefits of composting and compost	Environmental benefits of composting and compost	Drawbacks
Revenue from processing or "gate" fees	Improved soil health and plant vigor	Upfront and sustained investments in time and money
Revenue from compost sales	Retention of soil nutrients	Land requirement (and possibly building space)
Production of a useable product; reduced costs of substitute inputs	Water conservation	Odor and other nuisance complaints
Increase in crop yields and plant production and quality	Plant disease suppression; reduction in pesticide use.	Management in unfavorable weather
Generation of an animal bedding substitute	Erosion control	Diversion of manure and crop residues from crops if compost is sold off-farm
Destruction of weed seeds; reduce herbicide costs	Destruction of human, animal, and plant pathogens	Potential loss of nitrogen and generation of methane under anaerobic conditions
Reduction in waste disposal costs	Decomposition of hormones, antibiotics, and pesticides	Slow release of plant nutrients in finished product
Reduction in handling costs	Treatment of animal mortalities	Variable levels of plant-available nitrogen
Expansion of outlets for organic residuals	Lower environmental impacts from compost versus raw feedstocks	Zoning risk of being considered a commercial enterprise (rather than a farm)
	Reduction of greenhouse gas emissions	Need for environmental permits and adherence to regulations

3. Economic benefits of composting

There are direct and indirect economic benefits to composting. Direct benefits result in revenue from processing organic residuals and/or selling finished products derived from composting (Chapter 18). Indirect benefits arise from cost savings associated with purchasing, handling, storing, transporting, and dispersing the organic residuals in other ways. Together these benefits can increase the efficiency of a manufacturing or farming system by reducing handling and transportation costs and/or capturing revenue from resources that might otherwise be wasted.

3.1 Revenue from processing or "gate" fees

Many municipalities, institutions, commercial enterprises, and industries that generate organic "wastes" do not have the capacity to process those materials properly, let alone compost them. This situation creates an opportunity for compost-ers to collect processing fees by composting organic residuals generated by someone else. The fee collected for accepting waste materials is commonly referred to as a tipping fee or gate fee. This book uses the term "gate" fee.

"Other people's residuals" can be primary feedstocks for a composting operation, or they can supplement existing feedstocks. In the latter case, the off-site feedstocks can improve a composting mix. For instance, on-farm composters often need to mix manures with relatively dry materials that are good sources of carbon. Examples are leaves, newspaper, cardboard, sawdust, bark, and wood shavings. Conversely, municipal composters frequently need to mix their high carbon feedstocks with high nitrogen materials such as manures from farms or food waste from curbside collection programs.

Compost feedstocks generated by others must be considered with a measure of caution. First, gate fees can be difficult to capture. Alternative uses for organic residuals often exist, and the competition for the generator's dollar can be strong. Second, some of these "unwanted" residuals are unwanted for a reason. They can be difficult to handle or have the potential to create nuisances (e.g., odors, trash).

Composting off-site wastes might lead to extra processing at the composting site, odor problems and odor control measures, resistance from neighbors, and more restrictive environmental regulations. For instance, adding just one additional feed-stock to a mix can double or triple materials handling efforts prior to composting. The impact on the quality and value of the compost product also must be considered since the feedstocks influence the compost's market value and use, and the concen-tration of contaminants (such as plastic, glass, or heavy metals) reduce the compost market value.

3.2 Saleable product

One of the most attractive features of composting is that there is a market for the product (Fig. 1.2). Potential buyers include home gardeners, landscapers, developers and construction contractors, public transportation (e.g., highway) agencies and their contractors, vegetable farmers, turf growers, vineyards, operators of golf

FIGURE 1.2

Compost has value; at least as much value as topsoil. (Note: prices are circa 2006 in New Zealand dollars).

courses, and ornamental crop growers (Chapter 18). The price of compost varies considerably depending on the quality of the finished product and the markets available and their level of awareness and appreciation of the value of using compost. The price range is large and variable. The current price of bulk compost typically ranges roughly from $10 per cubic meter (or cubic yard), for high volume users that are not discriminating about compost qualities, to over $70 per cubic meter for premium markets interested in specific compost characteristics. As an example of the extreme high value of compost, aggregated, aged leaf compost used to filter out pollutants in storm water sells for over $300 per cubic meter. At the other extreme, some municipalities allow citizens to take a limited amount of compost for free.

3.3 Useable product

Compost that is not sold still has economic value for the composter, if the composter has good uses for the compost (Chapter 16). The monetary benefits can be realized through better crop yields or plant quality, or more directly as savings in the cost of purchased compost, topsoil, mulch, and fertilizer. Many composters have become composters simply to produce compost for their own use.

3.4 Animal bedding substitute

Compost has been used for poultry litter and bedding in livestock barns. Research and experience have shown that compost is generally a safe and effective bedding

FIGURE 1.3

Windrow composting of manure solids for the production of recycled livestock bedding.

Source: M. Schwarz.

material. Increasingly, the solid manure fraction, mechanically separated from liquid manure, is recycled as bedding after a short period of composting, which kills pathogens and drives off moisture (Fig. 1.3). Bedding is also recovered from the effluent of anaerobic digesters after separation and subsequent composting.

3.5 Destruction of weed seeds, pathogens, and reduced pesticide costs

The high temperature achieved during composting effectively destroys weed seeds contained in grass, leaves, yard trimmings, manure, crop residues, and other vegetation. Applying compost versus the raw feedstocks (e.g., manure or leaves) greatly decreases weeds, reducing, if not eliminating, the need to apply herbicides. Crop and plant producers can achieve savings in both chemical costs and application costs (labor and fuel). Similarly, the destruction of plant pathogens during composting, the disease suppressive nature of compost and the increased plant vigor can reduce the disease and insect pressure on crops, leading to savings in application of other crop chemicals.

3.6 Reduced disposal costs

Composting can be a cheaper alternative to waste collection and disposal, especially where the associated fees are high or the disposal site is a great distance away. In such cases, the generator can compost the material "on-site" or contract with a local composter to process the material. The on-site approach is practical on a small scale and for larger waste generators that have the land base, equipment, and skill set to carry out large-scale composting. Examples include municipal public works agencies, farms, vineyards, nurseries, greenhouses, food processing companies, brewers, and lumber yards. Numerous institutions, including schools and hospitals compost food waste in on-site systems. Specific on-farm examples include composting of poultry and livestock mortalities and composting of grass seed straw in locations where field burning has been prohibited.

3.7 Reduced handling costs

Composting reduces handling costs by making the organic residuals easier to manage. Compost is typically drier and substantially reduced in weight and volume compared to the raw feedstocks. The volume reduction typically falls in the range of 50% to 75%, depending on the feedstocks. The reduced weight and volume can significantly lower the cost to load, unload, store, and transport materials. In some cases, the savings can offset the compost production costs.

3.8 Expanded outlets for organic residuals

In addition to decreasing the volume and weight of raw feedstocks, composting also changes their character. The compost has a low rate of decomposition, little or no odor and less moisture. As a result, the compost is acceptable for a broader spectrum of uses than the raw feedstocks. Although some other organic residuals can be directly land-applied in their raw state, certain conditions can limit the practice. For example, wood and paper residuals are limited by nitrogen immobilization. Direct land application of leaves is discouraged by their low bulk density and weed seeds. Biosolids are constrained by community resistance and harvest waiting periods. Composting these residuals opens up other options.

3.9 Improve manure management

Farms have reported reductions in manure handling expenses due to composting, including savings in the cost of transportation, labor, storage, and fly control chemicals (Chapter 18). Compost is easier to handle than manure and stores well without odors or fly problems. Spills of compost during road transport are a much smaller concern compared to raw manure. Because of its storage qualities, compost can be applied at convenient times of the year. This advantage minimizes runoff and nitrogen loss in the field and reduces the need for extended storage of raw manure. Also, composting manure with a large amount of bedding lowers the carbon/nitrogen ratio to acceptable levels for land application (minimizing the potential for nitrogen immobilization when mixed with the soil). In organic farming applications, composting manure can effectively shorten the waiting period between land application and crop harvest.

On farms with insufficient or no cropland for land spreading, direct application of manure to cropland is constrained or simply not an option. In some cases, many years of manure and fertilizer applications have saturated the soil with nutrients, particularly phosphorus, creating conditions for nonpoint source pollution of surface and ground waters. In some regions with excessive soil nutrient loads, regulations have either curtailed or eliminated the application of manure to cropland. Composting makes it easier to export nutrients off the farm or to fields farther from the barn. In general, composting expands the outlets for manure, increasing the distance at which it can be economically transported and the number of neighboring farmers, and other users, willing to accept it.

4. Environmental benefits of composting

Both composting and compost offer environmental benefits. The benefits from composting derive from the diversion of feedstocks from less desirable alternatives (e.g., landfills) from the conversion of waste and from the ability of the process to destroy pathogens and decompose worrisome organic compounds (e.g., antibiotics). The use of compost supplies a wide range of environmental benefits resulting from improved plant growth and healthier soils (Bell and Platt, 2014; Gilbert et al., 2020; Soils for Salmon, 2020).

4.1 Soil health and plant vigor

Compost is an excellent soil conditioner. Compost adds decomposed organic matter, which improves soil structure, improves soil water balance, and increases the soil nutrient reserves, particularly the cation exchange capacity. The organic matter benefits of compost translate to reduced potential for soil erosion, runoff, and the subsequent nutrient and sediment losses. The soil improving effects of compost encourage extensive root growth and generally increases plant vigor. The more vigorous plants are better able to withstand stresses such as drought, insects, and disease (Fig. 1.4).

4.2 Nutrient retention

Compost contains both major plant nutrients (N, P, K) and minor nutrients or trace elements. Applying compost to farm fields, landscapes, and gardens reduces the need for commercial fertilizers. Compared to commercial fertilizers and raw manure, the nutrients in compost are less water-soluble, especially nitrogen (N). As a result, the nutrients are less likely to be lost through leaching and runoff. Regular applications of compost increase the reserve of nutrients stored in the soil. In addition, the soil-building qualities of compost increase the soils ability to hold soluble nutrients in place. Replacing fertilizers with compost reduces the negative environmental impacts associated with manufacturing those fertilizers as well as reducing the potential to overload soils with nutrients that could then be lost to the environment.

FIGURE 1.4

The soil-improving benefits of compost typically result in vigorous plants with healthy root systems.

Source: R. Alexander.

4.3 Water conservation

Compost application to soils adds organic matter that improves soil water holding capacity and water balance in the soil. The resulting benefits include increased efficiency in water usage, increases drought tolerance, and decreases reliance on irrigation. The water conserving benefits depend on the existing soil texture. The potential to conserve moisture is greatest in sandy soils. In clayey soils, the compost may have little or no effect on available moisture holding capacity. However, compost still improves the structure and aggregation of clay soils, creating greater soil porosity and greater range of pore sizes. This effect makes soils more amenable to increased water infiltration, thereby reducing runoff. Increased water infiltration means that plant roots can access more water and that plants are better able to extract water under drought conditions. The improved water holding capacity through application of compost is further discussed in Chapter 15.

4.4 Plant disease suppression

Generally, compost has been found to reduce soil-borne plant diseases. The effect is not consistent among all composts and all diseases, but it is well documented. If composts can be used as alternatives to fungicides, fumigants, and other pesticides, this benefit reduces or eliminates the environmental impact of such pesticides (alteration of soil properties, loss to atmosphere or water bodies) and the human health issues associated with worker application of such pesticides. The disease-suppressing qualities of compost are discussed in greater detail in Chapter 17.

4.5 Erosion control

Compost can be used as a surface mulch on slopes and bare ground to reduce erosion from rainfall and runoff. In addition, erosion control techniques have been developed that use compost installed in berms or within filter "socks" (compost filled mesh cylinders). In these instances, the compost serves to slow or eliminate runoff, remove sediment and filter contaminants. A wider choice of application equipment for compost has increased the uses for compost in erosion control and other environmental applications. This topic is discussed in greater detail in Chapter 16.

4.6 Pathogen destruction

The high temperatures generated during composting, plus the vigorous biological activity, destroy pathogenic organisms that can infect plants, livestock, and humans. Thus, compared to direct land application of fresh organic residuals, composting reduces the risks of spreading disease to plants and animals. The effects of composting on pathogen destruction have been studied for a large variety of pathogens including pathogenic viruses (e.g., avian influenza), bacteria (e.g., *Salmonella, e-coli*), fungi and protozoa (e.g., *Giardia* and *Cryptosporidium parvum*). While a few plant pathogens (e.g., tobacco mosaic virus) survive in lower numbers, the studies have shown that composting effectively sanitizes the compost of nearly all pathogens of concern (Ryckeboer et al., 2002). To date, little is known about composting's effect on the destruction of prions, the agents responsible for livestock diseases like BSE (cattle), scrapie (sheep), and chronic wasting (deer). Current and future research in this area should provide guidance about the efficacy of composting for prion destruction (see Chapter 8). Chapters 3 and 8 provide more information about the fate of pathogens during composting.

4.7 Destruction of hormones, antibiotics, and pesticide residues

Recent studies evaluating water quality of rivers and streams flowing have discovered the presence of hormones, antibiotics, and pesticides in waterways. Many of these xenobiotic compounds are human introduced, nonnaturally occurring compounds, and from farms where hormones and antibiotics are fed routinely to livestock or used in livestock or crop production. Pesticides, particularly herbicides like atrazine, tend to persist in the environment and make their way to surface or ground waters. Other studies associated similar xenobiotic compounds with municipal wastewater treatment facilities. Composting of biosolids, manures, yard trimmings, and crop residues that often contain these synthetic organic compounds is an effective means to degrade them, so they are less susceptible to be lost to the environment. Most pesticides readily decompose to safe levels during composting, with the notable exception of a few persistent compounds, like the herbicides clopyralid, aminopyralid, aminocyclopyrachlor, and picloram (see Chapter 15).

4.8 Treatment of animal mortalities

Although not universally accepted, composting is widely practiced to treat carcasses from animals that have died routinely on farms, along roads, on beaches, and in catastrophic situations that result in mass mortalities, such as disease outbreaks in poultry barns. In many cases, composting is the preferred method of treatment. The resulting compost is typically used on-site.

The emergence and growth of animal mortality composting is largely driven by the reduction of local rendering facilities. However, regulations restricting burial and incineration have also been a factor. Composting of dead animals is not approved in all states, provinces, and countries, particularly outside of the United States. Even within the United States, there are restrictions. Chapter 8 covers this topic in detail.

4.9 Low risk of environmental impacts from compost use

When compost is used or land-applied, it presents a low overall risk of air, soil, and water pollution. The environmental impacts of using compost are much lower than the raw feedstocks from which it is made, and products that it might replace, like chemical fertilizer.

Composting greatly reduces the risk of odors and airborne contaminants associated with land application and storage of manure, biosolids, and other odor-prone raw materials. At the time of application, mature compost does not emit strong odors. Composting requires a moderate moisture content (between 45% and 65% by weight) and relatively high carbon to nitrogen ratio (>20:1). These moderate moisture conditions greatly reduce the degree of nitrogen loss, either dissolved in water or as a gas, thus reducing the risk of odors.

As noted above, composting converts water-soluble nutrients, including most nitrogen and some phosphorus compounds, to more stable forms, making them less likely to be dissolved in runoff and leachate from fields and storages. The increased portability of compost, compared to raw feedstocks, evens out soil nutrient imbalances at the landscape or watershed scale that may have developed from long-term high-rate applications of manures, fertilizers, or other nutrient-rich soil amendments. The relative homogeneity and stability of compost compared to raw feedstocks also increases the windows of opportunity for timing of land application.

The lower impacts of compost use must be weighed against the potential environmental impacts of converting the feedstocks into compost. Nevertheless, if composters follow best management practices in making compost, there is likely to be a net environmental benefit.

4.10 Reduction of greenhouse gas emissions

Recent studies comparing gas emissions from static manure piles and actively composted (high temperature composting) manure piles show significant reductions in emissions of greenhouse gases like methane and nitrous oxide from composting manure piles (Brooksbank, 2018). Diversion of organic wastes such as food wastes and industrial organic wastes from landfills also has the potential to reduce fugitive greenhouse gas emissions.

Additional reductions in greenhouse gases can result from the use of compost in agricultural and horticultural settings. Since compost use reduces the need for synthetic fertilizers, fungicides, and herbicides, there is potential for reduced greenhouse gas emissions associated with input manufacturing. At present, the impact of composting on greenhouse gases and global warming is still being determined because methane and nitrous oxide gases can be emitted from composting piles. However, current information suggests that composting overall has a positive impact, especially when taking into account compost use as a soil amendment and nutrient supplement (Box 1.2).

Box 1.2 Climate, carbon, and composting

Author: Sally Brown.

Composting can impact carbon emissions and carbon storage in multiple ways. Much of the focus on carbon accounting for composting has been on avoided greenhouse gas emissions when organics are diverted from a landfill and composted under controlled conditions. There has also been work to characterize greenhouse gas emissions during the composting process. What is also critical to consider is the impact of compost use both in terms of carbon accounting and in the range of ecosystem benefits that compost provides when used to improve soils.

Avoided emissions When highly putrescible organics such as food waste are landfilled, they have sufficient moisture and nutrients to start degrading quickly. Most of the landfills in the developed world are "sanitary landfills." That means that waste is deposited in lined landfill cells and compacted so that oxygen is limited. Methane (CH_4) gas is formed when microbes decompose the organic waste where oxygen is limited. While valuable as a fuel, when CH_4 is released to the atmosphere it has 23 times ($23\times$) the impact of carbon dioxide (CO_2) over a 100-year time frame. When the cell is full, it is closed, and methane gas is collected to be flared or used as natural gas. It typically takes two to five years between the time a cell starts receiving waste and cell closure. However, methane generation often starts within months after organic waste is put into a cell. Because food waste is so putrescible, it creates methane even faster. Much of the methane generated by the decomposing material is released to the atmosphere before gas capture is functional.

The US EPA WARM (Waste Reduction Model) recently revised its estimates of methane release from food waste in landfills. According to the WARM model, about 1 ton of carbon dioxide equivalent (CO_2e) is released from landfilled food waste in the form of methane. That estimate is on a wet weight basis, which means that taking food waste out of landfills and to compost piles yields huge greenhouse gas savings.

Not all materials generate as much methane as food waste. For example, woody materials are relatively inert in landfills. Yard trimmings decompose moderately well to slowly in landfills, depending on the mix of wood and fresh plant debris. Animal manures are rarely landfilled, but they generate abundant methane when stored in moist piles or lagoons prior to land application. These materials make excellent compost and can also provide carbon credits when composted instead of improperly stored.

Emissions during the composting process Composting is a predominately aerobic process. Access to sufficient oxygen lets the range of microorganisms involved rapidly decompose the composting feedstocks. This decomposition releases a lot of CO_2, heat, and transforms the organics into a stable soil amendment. However, composting is not entirely aerobic. Micro sites even within a well-aerated compost pile can be anaerobic. A well-aerated compost pile generates fewer odors and reaches sufficient temperature to kill all pathogens and weed seeds. Good aeration also limits the release of greenhouse gases during composting.

There are three types of gas that can be emitted during composting that qualify as carbon emissions—carbon dioxide, methane, and nitrous oxide (N_2O).

CO_2—Carbon dioxide is produced from oxidizing carbon compounds. As we eat and digest the carbon in our food, we breath out carbon dioxide. The same thing happens as microbes eat the materials in a compost pile. Although CO_2 is a primary culprit in carbon emissions from burning fossil fuels, it is not a factor in carbon emissions during composting. The CO_2 released from burning fossil fuels is from the long-term carbon cycle. The CO_2 released from decomposition is considered "biogenetic." It comes from the short-term carbon cycle, the annual cycle of growth and decay of living things. However, composting does generate some nonbiogenic CO_2 emissions through equipment use, transportation, electricity consumption, and any other input that consumes fossil fuel.

Box 1.2 Climate, carbon, and composting—cont'd

CH_4—Methane is produced by specialized microorganisms that decompose organics in cases where oxygen is limited. It only lasts in the atmosphere for 12 years but is a highly potent greenhouse gas. It can be produced during composting when oxygen is scarce. Methane is most commonly detected when the composting process first starts and feedstocks are cool and wet. As pile heats up, moisture evaporates and allows for greater air flow and oxygen penetration. Emissions from composting are best controlled by making sure piles stay aerobic throughout the process.

N_2O—Nitrous oxide is a highly potent greenhouse gas with $296\times$ the impact of CO_2 over a 100-year time frame. It can be formed as ammonia converts to nitrate. However, it is most commonly formed as nitrate is transformed back to nitrogen gas in the absence of oxygen, a mostly biological process called denitrification. In the initial stages of composting, most of the nitrogen is present as organic nitrogen. As the carbon is mineralized, the nitrogen is converted first to ammonia and then to nitrate. Nitrous oxide can be detected during composting. It typically is seen after a portion of the nitrogen in the feedstocks has been converted from organic nitrogen. Although the quantities detected are generally low, it merits concern because of the high potency of this gas. Sufficient aeration is one tool to limit formation of N_2O. Selecting feedstocks with a sufficiently high carbon to nitrogen ratio (e.g., greater than 30:1) also discourages N_2O formation.

There are cases where composting can emit significant quantities of CH_4 and N_2O; for instance, with quickly degrading and/or nitrogen-rich feedstocks that are poorly aerated. However, these cases are the exception, not the rule. In addition, the CO_2e is almost always much lower than would be emitted if these same materials were landfilled or improperly stored.

Carbon sequestration from using compost Studies have shown that adding organic matter to soils can increase the quantity of carbon stored in the soil. By increasing soil carbon, the compost is effectively taking CO_2 out of the atmosphere and storing it in the soil. The EPA WARM model gives a credit of 0.2 tons of CO_2e per ton of feedstock composted for soil carbon sequestration. The actual amount of carbon stored varies, depending on how disturbed the receiving soil is, whether the soil is cultivated and the climate where the compost is used. When the compost is used to replace fertilizers, there is an additional benefit as fertilizers require a great deal of fossil energy to produce.

Table 1.2 provides estimates of the net CO_2e emissions from landfilling and composting food scraps, yard trimmings and a 50/50 mix of each. On balance, composting is the superior CO_2e option when food is a prominent feedstock. Landfilling and composting perform nearly the same with 100%-yard trimmings, owing to the assumption that the carbon in the woody yard trimmings remains sequestered within the landfill.

Table 1.2 Estimated emissions from landfill and composting for food scraps, yard waste, and a 50:50 blend of the two materials. Values are tons of CO_2e per wet ton of feedstock.

Material	Fugitive emissions		Carbon sequestration		Balance	
	Landfill	Compost	Landfill	Compost	Landfill	Compost
Food scraps	0.72	0.05	−0.06	−0.22	**0.65**	**−0.17**
Yard trimmings	0.29	0.06	−0.49	−0.22	**−0.20**	**−0.15**
50:50 blend	0.50	0.05	−0.28	−0.22	**0.23**	**−0.16**

Data from Brown, 2016. Greenhouse gas accounting for landfill diversion of food scraps and yard waste. Compost Sci. 24 (1), 11–19 and U.S. EPA, 2014. Solid Waste Management and Greenhouse Gasses. Documentation for Greenhouse Gas Emission and Energy Factors Used in the Waste Reduction Model (WARM). http://epa.gov/epawaste/conserve/tools/warm/SWMGHGreport. Html.

Continued

Box 1.2 Climate, carbon, and composting—cont'd

Counting carbon credits and demerits is an inexact exercise. It involves broad assumptions, imperfect data, and generalized conditions. A well-run composting operation will minimize emissions from composting. Composting feedstocks that would otherwise produce CH_4 will provide significant emissions reductions. If the compost is used as a soil amendment, soil carbon storage will provide additional benefits, and possibly the largest advantage. The magnitude of the soil-sequestration benefit will vary based on how healthy the receiving soil is already. There is no argument, however, about the general environmental benefits of using compost. Adding compost to soils is an excellent way to improve soil health. Healthy soils are the basis for a broad number of ecosystem services including improved rainwater infiltration, reduced soil erosion, moisture and nutrient retention, disease suppression, and carbon sequestration.

References cited

Brown, S., 2016. Greenhouse gas accounting for landfill diversion of food scraps and yard waste. Compost Sci. 24 (1), 11−19.

U.S. EPA, 2014. Solid Waste Management and Greenhouse Gasses. Documentation for Greenhouse Gas Emission and Energy Factors Used in the Waste Reduction Model (WARM). http://epa.gov/epawaste/conserve/tools/warm/SWMGHGreport. Html.

---

5. The drawbacks

Although composting harnesses natural biological processes, nevertheless, it can pose challenges. Composting deals with a diverse collection of actively decomposing organic materials that can be wet, dry, bulky, particulate, potentially odorous, and attractive to pests. Materials handling, space, and equipment are major elements of composting. In addition, composting is normally practiced outdoors, and year-round, at the perils of bad and changing weather. Trouble-free operation and the production of quality compost require investment, resources, effort, and diligence.

5.1 Time and money

Like any other manufacturing system, composting requires equipment, labor, and management (Chapter 19). The initial investment for a composting operation can be very low, if existing equipment and facilities are used. This approach is fine where the volume of material is relatively small. However, most medium to large-scale operations have found that only adapting existing equipment requires too much labor and restricts composting process management and quality control. Many composters have found it necessary to purchase special composting equipment, develop a separate infrastructure, and hire one or more employees dedicated to the composting operation. When the sale of compost becomes a major objective, the composting operation can become a business in itself (Chapter 2).

The investment needed is inherently tied to the scale of operation. With special equipment, currently it could cost as little as $20,000 to start a small-scale on-site composting operation. Investments in the $200,000 to $750,000 range are typical for moderately sized facilities (Platt et al., 2014). Large operations may require investments of a few million dollars. As the operation grows, so do the maintenance and labor costs.

5.2 Land and site

The composting site, storage for raw material, and storage for finished compost can occupy a considerable area of land and sometimes building space (Fig. 1.5). It is usually necessary to create a designated area for composting, and make modifications to the site, including compacting the surface or building a permanent engineered composting pad. Additionally, locating the composting area in relation to competing on-site activities, access to roads, and neighbors is often an exercise in compromise.

5.3 Odor

Controlling and minimizing odors from feedstocks and decomposing organic materials is perhaps the most challenging management task in composting. Although the end products of the composting process are not odorous, the feedstocks being composted can emit offensive odors before and during the process. Until they begin to compost, biologically unstable materials like manure, biosolids, grass clippings, food processing residuals, postconsumer food wastes, and livestock mortalities can produce odors, especially if they have been stockpiled prior to composting. Odors

FIGURE 1.5

Aerial view of compost facility in Upstate New York, showing the space and facilities often necessary to carryout composting on a large scale.

Source: O₂Compost OCRRA.

can also be generated from the process itself, especially in the early stages when the materials are actively decomposing and the oxygen demand is high. Odor emissions become odor problems when unfavorable weather conditions carry the smells of site to neighbors (Chapter 12).

Sensitivity to odors is essential and requires careful planning for managing the receiving of feedstocks and composting operation to minimize odor generation. Some sites, because of their location, may require expensive odor control measures. With most feedstocks, the odors emitted during composting are periodic and short lived. Composting can still represent an improvement over conventional methods of handling biosolids, manures, and other "rapidly degrading" organic residuals.

5.4 Weather

Cold weather can slow the composting process by lowering the temperature of the composting material. It can also cause other problems like freezing materials and equipment. The effects of rain and snow are potentially more serious. Precipitation creates the need for runoff controls. Heavy precipitation adds water to the composting mix; snow and mud limit access to the composting site. It is possible that a heavy snow fall could interrupt the operation until spring. If this occurs, an alternative method to store or dispose of the feedstocks is necessary. In semi-arid and arid climates, it can be difficult to maintain adequate moisture in compost piles. Extreme weather conditions (too wet, too cold, or too dry) often can lead to investments in enclosures for composting (roofs, buildings, compost covers), which add to the overall costs of operation. Fortunately, even without a building, composters have management options, like adjusting pile size, to minimize the effects of unfavorable weather (Chapter 14).

5.5 Diversion of manure and crop residues from cropland

Composting manure on farms, and then selling it as compost, diverts the nutrients, organic matter, and soil-building qualities of that manure from cropland. This situation also holds true for crop residues that are composted rather than returned to the land. Buying commercial fertilizers to make up for the lost nutrients may not make good economic or agronomic sense, assuming that the soil needs additional nutrients for crop production. Farm composters have the option to use compost to satisfy the needs of the farm's soils first and sell the remainder.

5.6 Potential loss of nitrogen

Composts made from nitrogen-rich feedstocks can contain less than half the nitrogen of the fresh feedstocks. The nitrogen loss can be substantial if the initial carbon to nitrogen ratio of the composting mix is low (e.g., < 20:1). However, whether composting conserves or squanders nitrogen depends on how the feedstocks are otherwise handled and how composting is practiced (see Chapters 3 and 11).

5.7 Slow release of nutrients

The nutrients in compost are mostly in a complex organic form and must be mineralized in the soil before they become available to plants (Chapters 15 and 16). Typically, only 5% to 20% of the total nitrogen in compost is available in the first cropping season, depending on the climate and compost characteristics. This slow release of nutrients is generally an environmental benefit, but for crop producers who depend on compost for fertility, it can be a short-term drawback. Compared to the raw manure or biosolids, initial applications of compost must be greater to achieve the same nitrogen fertilization level. Adding enough compost to satisfy 100% of the crop's nitrogen needs in a given year is usually impractical. In most cases, it is not economical to apply the high rates of compost needed to meet the nitrogen needs of the crop. In addition, such high application rates likely exceed the phosphorus needs of the crop and lead to build up of phosphorus in the soil, especially with manure- or biosolids-based composts. If it is possible to apply composts to soils annually at low to moderate rates over several years, the soil's organic nitrogen reserve can eventually meet 50%–80% of the nitrogen demands of most crops.

5.8 Uncertain availability of nitrogen

In addition to the slow release of nitrogen from compost, the rate at which nitrogen mineralizes is uncertain, even from a given type of compost. This uncertainty makes it difficult for users to plan how much supplemental N to apply. The availability (i.e., mineralization) of N depends on several factors including the soil type, moisture, and temperature as well as the plants grown. The characteristics of the specific batch of compost applied also affect the mineralization rate. For example, a dairy manure-yard trimmings compost may have a C:N ratio of 15:1 for one batch and 25:1 for another batch. Assuming that these composts are applied to the same soil type with similar weather conditions, the compost with higher C:N ratio may take longer to mineralize N than the compost with lower C:N ratio.

Because of the inherent variability associated with soils, climate, and compost batches, it is useful to have each batch of compost tested by a laboratory before land applying so you know the C:N ratio and the concentrations of both nitrate and ammonium forms of mineral N. Conducting a "presidedress N" type soil test (to assess the status of available N) after the compost has been applied can be used to determine how much supplemental N you will need to add to meet the plants N needs. If the time and resources are not available to test the soil or compost, N availability can be estimated using average values, obtained from either: (1) field tests of the mineralization of specific compost products over time; (2) the compost supplier; or (3) published mineralization values for the type of compost used (e.g., biosolids compost, poultry manure compost, yard trimmings, yard waste trimmings-food scraps compost, etc.).

5.9 Risk of losing farm or nonprofit classification

It is possible to be too successful. If a farm sells a large amount of compost or handles off-farm wastes for a fee, neighbors and local regulators may contend that the operation is a commercial enterprise, rather than an agricultural activity. A farm could conceivably lose its standing as an agricultural entity in regard to zoning, permitting, or environmental regulations. This possibility should be carefully considered before establishing or expanding a composting operation. A farm-based composter should determine at what point and under what conditions a farm composting operation becomes a commercial enterprise in her/his state, province, and local community. The same situation may exist for public or nonprofit organizations (e.g., colleges) that earn revenue from composting services or the sale of compost.

6. Facts and fiction of composting and compost

Composting is a robust process that is practiced by many people at a remarkable range of scales. Its simplicity and versatility underlie composting's broad appeal. However, its many applications and many practitioners make composting vulnerable to misinterpretations, generalizations, myths, and fictions. This section addresses some of the common fictions, and facts, of composting and compost before moving on to the science and practice discussed in the ensuing chapters.

6.1 Composting is an art; composting is a science

Fact: Both statements are true. The science of composting includes—biology, chemistry, physics, engineering and even economics, psychology, and meteorology. Composting is most successful when operators and compost users understand and apply the underlying scientific principles. However, because science cannot predict, explain, and cover all of the elements, art remains an essential part of composting. The art of composting combines scientific knowledge with an operator's or manager's experience, judgment, intuition, business acumen, and ethics (Chapters 2−19).

6.2 Composting is a natural process

Both fiction and fact: While composting harnesses the natural process of decomposition, it takes more than nature to produce compost, at least on the scale and time-frame practiced by compost manufacturers. Nature rarely builds piles 3 m (10 ft) high and 30m (100 ft) long and turns, grinds, and screens the material. On the other hand, it is a fact that a pile of leaves left undisturbed will eventually turn into compost through the forces of nature alone. Composting at the backyard level comes closer to nature's approach (Chapters 5−7 and 9).

6.3 Composting is aerobic

Both fiction and fact: It is a fact that composting is an aerobic process. At the very least, the process has to be aerobic in its final stages—when the product of decomposition oxidized to produce usable and mature compost. However, anaerobic decomposition is part of the overall composting process, and it would be misleading to claim that composting is *entirely* aerobic. In the transformation of raw feedstocks to compost, some of the chemistry and biology takes place anaerobically. Even in a well-aerated compost system, anaerobic pockets exist within piles and within clumps of solid and liquid materials. Furthermore, anaerobic conditions can be advantageous in breaking down specific organic compounds. The defining feature of composting is that it is *predominately* aerobic, particularly in its final steps— enough to contain and complete the anaerobic processes that inevitably occur within the system (Chapter 3).

6.4 Composting does not create odors, when properly done

Fiction: There is always a risk of odors with composting, simply because large quantities of actively decomposing organic materials are amassed in one place. The "smells" generated from composting become odor problems when high-enough concentrations of odorous compounds reach the noses of someone who finds them objectionable. Whether or not this happens depends on many factors including, but not limited to, the feedstocks, the effectiveness of aeration, volume of material, location and conditions of the composting site, weather conditions, and the expectations and sensitivities of neighbors (Chapter 12).

6.5 Composting is thermophilic

Both fact and fiction: While the primary stages of composting are usually and preferably thermophilic, composting readily and successfully takes place at lower temperatures. Thermophilic means that at least a large part of the process occurs at high temperatures (>40°C, 105 F) due to the heat generated. High temperatures speed the process, drive off moisture, and destroy more pathogens and weed seeds. Some feedstocks (e.g., biosolids, food scraps, and yard scraps) are required to reach high temperatures by regulations. Absent regulatory requirements, composting takes place at lower temperatures and good compost can be made. Hordes of organisms can do the job whether it is cold, cool, warm, or hot (Chapter 3).

6.6 Composting is impractical in cold climates

Fiction: Indeed, many successful composting facilities are located in climates where the temperatures drop below freezing for long periods. Piles and windrows are typically large enough to insulate the decomposer organisms within from the frigid temperatures without. As the protected organisms continue to decompose the organic feedstocks, they also continue to generate heat that warms their surroundings. Winter conditions make the situation more challenging for the human operators, creating extra work and possibly interrupting their manipulation of the process, but composting process continues nonetheless (Chapter 14).

6.7 Meats, fats, and oils should not be composted

Fiction: Backyard composters are routinely discouraged from composting meat and oily materials because these materials are problematic on a small scale (e.g., attract pests, generate odors). However, on a large scale, flesh from meat and fish and materials with oil readily decompose. In fact, composting is now used as means to treat mixed food residuals, fish, meat wastes from butcher shops, road kill, and poultry and livestock carcasses (Chapter 8).

6.8 Compost can be made in a week

Fiction: Claims have been made that this process or that system can produce compost within a week's time. Such claims are, at best, exaggerations. Under ideal conditions, a week of composting may produce a well-decomposed material that might not resemble the original feedstocks. This material may be usable for direct land application and soil incorporation during a time when plants are not growing actively (e.g., winter, fallow ground). However, it cannot be considered compost, and it is not suitable for general use. Compost must be "mature" enough to ensure that intermediate compounds and further decomposition do not injure plants. Acquiring this level of maturity inherently takes time—more than two months in most cases. How quickly finished compost can be produced depends on the feedstocks, composting methods and processing, intensity of management, marketing factors and, especially, the intended use for the product. Typically, compost is manufactured over periods that range from six months to over one year (Chapters 3 and 5–7).

6.9 Composting and compost are safe and environmentally beneficial

Both fact and fiction: This statement is mostly factual, but the word "completely" renders it fictional. Composting and compost present many environmental benefits. However, in the production of compost, there are opportunities to do environmental harm if accepted practices are ignored. Composting and compost presents almost no threat of harm to compost users and neighbors of composting facilities. But the threat is not zero. Some people are allergic to particular biological elements associated with compost and composting facilities. Composting feedstocks and compost can harbor biological constituents that affect susceptible people. Like any manufacturing process that involves equipment and materials handling, composting has its share of safety risks and demands adherence to safety rules. Workers would be at the great risk and should be screened for allergies and immune suppression issues (Chapters 10 and 13).

6.10 Compost is compost; all composts are created equal

Fiction: Some people have the perception that the composting process converts all feedstocks to a finished product called compost and that all composts possess similar characteristics, regardless of the ingredients. In truth, compost encompasses a variety of products that vary with the feedstocks, the composting methods, and how long they are processed. The finished composts retain many of the traits inherent in the feedstocks used to make them. The phrase "you are what you eat," applies well to composts (Chapters 15 and 18).

6.11 Compost sells itself

Fiction: While many compost users appreciate the value and benefits of compost, most potential users need to be convinced that compost is worth the purchase price. Most users also need to be instructed about how to use it. Although customers may know how to grow plants, they may not have experience using compost, and need to be convinced of its benefits. A marketing effort is necessary to sell compost (Chapters 2, 18 and 19).

6.12 Compost suppresses plant diseases

Both fact and fiction: Compost can suppress plant diseases but not all compost and not always. There is a large and growing body of literature and grower experiences that suggest that compost products can suppress soil-borne diseases such a root rots or nematode diseases. Particular qualities are required of compost to suppress particular diseases. Most foliar diseases are not suppressed by compost. Compost teas are more commonly used to protect against foliar diseases (Chapter 17).

6.13 Compost is "organic"

Both fact and fiction: The answer depends on how one uses the word "organic." In the general sense, compost is an organic material as it is made from and composed of organic substances—carbon-based compounds that are or were once alive. However, with the growth of organic agriculture and the organic foods industry, the word "organic" has taken on explicit definitions. Some materials, like biosolids, food scraps, or yard trimmings, which are organic in composition are excluded from certified organic agriculture use because they may also contain undesirable components. Regulations, guidelines and certification authorities may determine whether a given compost product can be marketed as a product for particular "organic" uses. Moreover, a given compost product can contain a substantial amount of nonorganic mineral and/or inert components, sometimes in proportions exceeding that of the organic fraction (Chapters 4 and 15).

6.14 Compost is a fertilizer

Both fact and fiction: Like fertilizer, compost contains chemicals that nourish plant growth. However, compared to conventional fertilizers, the plant nutrients in compost are relatively low and vary in their concentration and availability. Therefore, the fertilizer value of compost is usually de-emphasized in favor of compost's soil-building properties. In addition, fertilizers are defined and regulated by government authorities. In most jurisdictions, a compost product that carries nutrient claims must be registered as a fertilizer. Most compost producers choose to avoid this procedure. Nevertheless, some producers have officially registered their products as fertilizers and sell them as such. In addition, many people rely on compost as a primary source of fertility for their plants and crops, generally applying compost in large amounts or annually over several years (Chapters 15, 16 and 18).

6.15 Compost contains "heavy metals"

Fact: All compost products contain heavy metals, just as all soils and food product do. The fact that heavy metals are present in compost is not important. What is important is the concentration of these elements and their relative availability for plant uptake. Customarily known to soil scientists as "trace elements," heavy metals are a specific group of chemical elements with a relatively high molecular weight. Many are commonly recognized as familiar metals (e.g., iron, zinc, copper, lead, and cadmium). High concentration of some elements in this group represent health risks and are regulated. In some jurisdictions, regulations and compost standards require the trace element contents of compost to be tested and maintained below standards levels, at least for composts made from certain feedstocks (e.g., biosolids, MSW). At low concentrations, some heavy metals (e.g., molybdenum, copper, zinc, and selenium) are essential nutrients to plants and animals. It is rare that a compost product contains metal concentrations high enough to cause concern (Chapters 4 and 15).

6.16 Composting makes nutrients more available to plants

Fiction—mostly, with a hint of fact: Some compost users have assumed that because composting involves decomposition, plant nutrients become more available as the raw feedstocks, like manure, break down. However, in general, the reverse is true. Composting stabilizes many nutrients, making them less available as they are incorporated in the organic compounds of the compost. The degree of stabilization depends on the particular nutrient and its balance with other nutrients. Nitrogen, especially, is usually less available to plants after composting manure, food, and green vegetation. On the other hand, after composting, nitrogen and other nutrients are less likely to be lost to the environment. One instance where composting can increase the availability of nutrients is where the feedstocks start with a high carbon to nitrogen ratio (e.g., wood residues, manure with bedding). Also, composts contain organic compounds that "chelate" or bind trace nutrients, like iron, making them more readily available for plant uptake particularly in slightly alkaline soils (Chapters 15 and 16).

6.17 Compost is humus

Fiction: Compost is often referred to as "humus" because many people believe that the composting process converts raw organic materials into this very stable organic component of soils. However, compost and soil humus are not the same. Soil humus is a combination of stable organic compounds that develop over a long time period as a result of repeated cycles of organic matter decomposition and biological metabolism. While the chemical constituents of finished composts contain humic compounds, they are not identical to soil humus. It is fair to say that compost is on its way to becoming humus, but it is not there yet (Chapters 3 and 15).

References

Cited references

Bell, B., Platt, B., 2014. Building Healthy Soils With Compost to Protect Watersheds. Institute for Local Self-Reliance, Washington, DC. https://ilsr.org/wp-content/uploads/2013/05/Compost-Builds-Healthy-Soils-ILSR-5-08-13-2.pdf.

Brooksbank, K., 2018. Composting to avoid methane production. Department of Primary Industries and Regional Development, Government of Western Australia. https://www.agric.wa.gov.au/climate-change/composting-avoid-methane-production.

Gilbert, J., Ricci-Jürgensen, M., Ramola, A., 2020. Benefits of Compost and Anaerobic Digestate When Applied to Soil. ISWA — International Solid Waste Association. https://www.iswa.org/media/publications/iswa-soils-project/#c8146.

Platt, B., Goldstein, N., Coker, C., 2014. State of Composting in the U.S.; What, Why, Where and How. Institute for Local Self-Reliance, Washington, D.C. https://ilsr.org/state-composting-us-what-why/.

Ryckeboer, J., Cops, S., Coosemans, J., 2002. The fate of plant pathogens and seeds during anaerobic digestion and aerobic composting of source separated household wastes. Compost Sci. Util. 10 (3), 204—216. https://doi.org/10.1080/1065657X.2002.10702082.

Sherman, R., 2018. The Worm Farmer's Handbook. Chelsea Green Publishing, White River Junction, VT.

Soils for Salmon, 2020. https://www.soilsforsalmon.org/.

Swati, A., Hait, S., 2018. A comprehensive review of the fate of pathogens during vermicomposting of organic wastes. J. Environ. Qual. 47 (1), 16—29. https://doi.org/10.2134/jeq2017.07.0265.

USCC (U.S. Composting Council), 2020. USCC Vision and Mission Statements. https://www.compostingcouncil.org/.

U.S. EPA, 2014. Solid Waste Management and Greenhouse Gasses. Documentation for Greenhouse Gas Emission and Energy Factors Used in the Waste Reduction Model (WARM). http://epa.gov/epawaste/conserve/tools/warm/SWMGHGreport. Html.

U.S. PIRG, 2019. Composting in America. A Path to Eliminate Waste, Revitalize Soil and Tackle Global Warming. United States Public Interest Research Group. https://uspirg.org/reports/usp/composting-america.

Consulted and suggested references

Most of the references that were consulted for this chapter can be found in the corresponding chapters identified in the foregoing text. Additional references and resources that are particularly relevant to this chapter are listed below.

Alexander, R., 2001. Field Guide to Compost Use. U.S. Composting Council. http://www.compostingcouncil.org.

Amlinger, F., Peyr, S., Geszti, J., Dreher, P., Weinfurtner, K., Nortcliff, S., 2007. Beneficial Effects of Compost Application on Fertility and Productivity of Soils Literature Study. Federal Ministry for Agriculture and Forestry, Environment and Water Management, Austria.

BioCycle, 2020. https://www.biocycle.net/.

Brown, S., 2016. Greenhouse gas accounting for landfill diversion of food scraps and yard waste. Compost Sci. 24 (1), 11−19.

Cornell Waste Management Institute, 2020. http://cwmi.css.cornell.edu/composting.htm.

Ingham, E., 2020. Food Web & Soil Health. Natural Resources Conservation Service, United States Department of Agriculture. Institute for Local Self-Reliance (ILSR). https://www.nrcs.usda.gov/wps/portal/nrcs/main/soils/health/. https://ilsr.org/composting/.

Rynk, R. (Ed.), 1992. On-Farm Composting Handbook. Northeast Regional Agricultural Engineering Service, Ithaca, NY.

Schwarz, M., Bonhotal, J., 2018. Carbon footprint of a University compost facility: case study of Cornell farm services. Compost Sci. Util. 0 (0), 1−22. https://doi.org/10.1080/1065657X.2018.1438934.

Stofella, P.J., Kahn, B.A. (Eds.), 2001. Compost Utilization in Horticultural Cropping Systems. Lewis Publishers, Boca Raton, FL.

Sweeney, J., 2019. Community-scale Composting Systems. Chelsea Green Publishing, White River Junction, VT.

WRAP (Waste and Resources Action Programme), 2020. In: http://www.wrap.org.uk/collections-and-reprocessing/organics/guidance/guidance-on-compost-and-composting.

Enterprise planning

Authors: Susan Antler[1], Leslie Cooperband[2], Craig S. Coker[3]

[1]*Compost Council of Canada, Toronto, ON, Canada;* [2]*Praire Fruits Farm and Creamery, Champagne, IL, United States;* [3]*Coker Composting and Consulting, Troutville, VA, United States*

Contributors: Mary Schwarz[4], Robert Rynk[5]

[4]*Cornell Waste Management Institute, Cornell University, Ithaca, NY, United States;* [5]*SUNY Cobleskill, Cobleskill, NY, United States*

1. Introduction

Turning your passion for composting into a viable enterprise requires planning, and more. Passion is not enough. You must take the time up front to establish goals, develop strategic objectives, evaluate strengths and weaknesses, and think through the many essential elements of a composting enterprise, determining how each aspect will be managed and how collectively they will function. From this process, the business plan for your composting enterprise will develop and a document will emerge that will become your guide to track progress. As your business evolves, so too will your plan, allowing you to take your business through the many stages of growth and development.

While there are common elements in the management of any enterprise, each initiative is unique, embodying the values and vision of the enterprise, incorporating specific strategies and decisions to realize success. Fortunately, there are many composting enterprises already in existence that can serve as case studies or models for your enterprise (Fig. 2.1). Studying both the successes and lessons learned of existing composting enterprises offers lifetimes of invaluable experience from which to learn. The best time to do this studying and planning is at the start of the enterprise.

This chapter is intended to be a beginning guide to planning a composting enterprise or the future of a current enterprise. It is most relevant to those readers for which composting is expected to be a stand-alone venture. Nevertheless, the steps and tasks involved in planning an enterprise also apply well to operations that are a component of a larger enterprise, such as a farm or even a public agency. Furthermore, many composting ventures that begin as a part of a larger organization eventually evolve into a distinct enterprise. This chapter provides a compost-flavored initiation to the framework of enterprise planning. Many of the concepts presented here are covered in more detail in "Building a Sustainable Business" (DiGiacomo et al., 2003). Current information on composting enterprise development is regularly

The Composting Handbook. https://doi.org/10.1016/B978-0-323-85602-7.00006-6

FIGURE 2.1

Four different composting enterprises: (A) Bulk sale of compost to public; (B) Revenues for composting services; (C) Compost production for self-use on an organic farm; (D) Composting as a means to recycle autumn leaves for a municipality.

published in bulletins and newsletters from organics recycling organizations. Publications like BioCycle (2021) and Composting News (2021) frequently provide information about enterprise development and management. BioCycle publishes a regular series of articles under the category of "Business and Finance" (Coker, 2020).

2. Starting a composting enterprise

The reasons for starting a composting enterprise vary (see Boxes 2.1−2.4). A livestock farmer might be motivated primarily to improve manure and nutrient management. For a municipality, the main driver may be managing leaves, yard trimmings, and food generated within the community to reduce transportation and/or disposal costs. When the reason for composting is primarily to make and sell compost, a product manufacturing mindset is needed. In fact, it is almost universally important to view composting as a manufacturing operation, with material and labor inputs, investment as well as product quality requirements. Regardless of what the motivation is, there are costs involved in converting feedstocks, into compost and then distributing the compost to customers, or for use on-site.

Box 2.1 Composting enterprise options: composting to produce compost *for sale*

Rationale Compost is a valuable product that can be sold for profit. Prices vary greatly, typically from roughly $5 to well over $50 per cubic yard depending on the markets, quantity, and compost quality (see Chapter 19).

Considerations
- High profit potential.
- Requires a product manufacturing approach to production and distribution.
- Requires an extensive marketing effort and marketing skills.
- Attention to product quality is key and product testing will be required to verify quality specifications.
- Distribution network necessary (unless sold from an on-site outlet like a greenhouse or farm stand).
- Transportation necessary.
- May require multiple products.
- Successive enterprises may become separate from the central business.
- Business may be viewed as a commercial rather than agricultural or public enterprise.
- Requires knowledge about compost qualities and uses.

Box 2.2 Composting *off-site* feedstocks as a revenue source via *gate fees*

Rationale An enterprise may provide composting services, supplying the land, labor, and equipment for recycling organic residuals generated elsewhere. A fee (i.e., gate fee) is collected for this service. The enterprise can provide an outlet for the compost if it is associated with a municipality, landscape company, nursery, or other type of farm.

Considerations
- Feedstocks that generate gate fees may require more processing (e.g., grinding) and/or a greater degree of management.
- When they pay a fee, generators are less diligent about the quality of the feedstocks delivered to the facility.
- Off-site feedstocks may be contaminated with trash.
- Sale of the compost produced may be restricted by regulations.
- The enterprise can be the outlet for the compost if it is associated with a municipality, landscape company, nursery, or other type of farm.
- If any portion of the compost is sold, attention to compost quality and marketing is essential.
- May bring objections from neighbors and scrutiny from regulators.
- May shift a farm to a commercial category regarding regulations, zoning, and land use.
- Application of the compost produced must accommodate agronomic practices and nutrient management plans of the farm.

It is valuable to go through the exercise of identifying the internal and external factors related to your composting goals and situation (Fig. 2.2). External factors are those conditions that are largely, although not entirely, out of your control. External factors include:

- regulations related to siting and permitting a composting operation, acceptable feedstocks and marketing compost products (e.g., fertilizer and soil amendment rules),

---

Box 2.3 Composting on-site feedstocks to make compost for self-use

Rationale Composting of organic residuals generated on-site can provide a useful purpose for those residuals (as opposed to landfilling). Also, composting can reduce cost associated with alternative options, such as direct land application. Composting makes most organic residuals easier to handle, in large part because it reduces volume and homogenizes amendments relative to the raw feedstocks (e.g., food waste, leaves, manure, bedding and crop residuals, straw).

Considerations
- Reduces the C:N ratio of crop residues and highly bedded manure, thus avoiding N immobilization in the soil.
- Compost is easier to handle and spreads more uniformly than the raw feedstocks.
- Allows wet feedstocks, like some manures, to be stored and handled as a solid.
- Odors can be generated from the composting operation as well as the raw feedstocks.
- May need different land application equipment, better suited to granular materials.
- Greatly reduces flies, pathogens, and weeds seeds in vegetation, manure, and other raw organic materials.
- Better disease suppressive qualities than direct land application of the raw feedstocks.
- Nutrients are less available to plants in the short term.
- Compost does not have to be fully mature, especially if applied after harvest or to fallow land.
- Composting period can be relatively short, stopped after high temperatures destroy pathogens and weed seeds.

Box 2.4 Composting *off-site* feedstocks to make compost for *self-use*

Rationale Composting makes it possible to use organic residuals, generated by others, to become sources of soil amendments, mulch, organic matter, and nutrients. This approach can avoid or reduce the cost of purchased compost and other organic inputs for organic farmers in particular. It can be beneficial to all crop producers, nurseries, greenhouses, landscapers, and landscape managers. It can also generate some revenues from gate fees.

Considerations
- Off-site feedstocks can improve composting recipe when added to feedstocks generated on-site.
- Once composted, many off-site organic residuals qualify as inputs for organic agriculture.
- There is more control of the compost quality, compared to purchased compost.
- Composting time can be shortened if the compost does not need to be highly mature.
- Possible to collect a fee for accepting and composting off-farm materials.
- May bring objections from neighbors and scrutiny from regulators.
- Must accommodate agronomic practices and nutrient management plans.
- Some portion of the compost produced can be sold.

- local competitors including existing composting facilities and other competing options (e.g., peat bogs, mushroom farms, landfills),
- the potential customer base, both customers who need to manage their organic residuals and customers who might use compost products,
- proximity to markets (for both waste management and compost products), which is generally limited to 80–120 km (50–75 miles),
- distance to, and types of neighboring land uses, and,
- quality and capacities of nearby road networks to handle truck traffic.

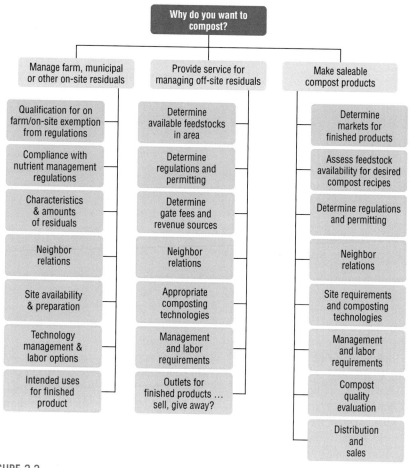

FIGURE 2.2

Decision tree: reasons for composting; things to consider for planning each of these enterprise options.

Internal factors, which you have some degree of control over, can include:

- the composting site location,
- the organic residuals or feedstocks to be composted,
- the selected production process and equipment,
- the financing support structure for the enterprise, in terms of the means of acquiring funds to support the venture as well as the system to be used to track revenues and expenditures,
- the distribution network to service customers,
- the management team and workforce, and,
- the marketing plan, including the research and development efforts to identify and target customers as well as to determine, develop, and provide the right products and services for their needs.

Take time to assess the various trends that could impact your composting enterprise, such as: encroaching development; changes in public policies and regulations, landfill gate fees (one of your primary competitors); availability of feedstocks (e.g., food processing factories, supermarkets, and landscapers); emerging markets, technologies, and regulations; potential competitive changes and potential shifts in consumer attitudes and behaviors.

For help, you can tap into many resources. Start with trade association publications, industry directories, internet searches, and accessing government documents. Conferences and visits to composting facilities outside of your geographic trading area can also provide insight. Meetings with the regional and local regulators offer the opportunity to streamline your information search and begin positive relationships with the regulatory community. A number of resources for the composting industry are listed in Appendix A).

3. Assessing your resources

Before beginning the planning process, take stock of existing resources that can be also put to work in a composting operation. The general categories include operations, human resources, and finances.

Physical resources include land, buildings, other structures, equipment, and feedstocks. These tangible assets are used to build a composting enterprise. Production systems include the source, timing, and condition of feedstocks. For example, if you are a livestock farmer, your principal feedstock will be manure, generated on a daily basis. You need to assess how the animals are raised, what kind of bedding material is used, how frequently the barns are cleaned out, and in what manner manure is removed (e.g., scraped off the barn floor or flushed out with water). If you are a municipality, your feedstocks might be autumn leaves that arrive almost all at once or regular loads of wastewater biosolids. Management systems would include an evaluation of how current enterprises are run (a farm, a greenhouse, a yard debris collection facility, and a college) and how organic residues are currently handled. For example, in the case of a commercial greenhouse, how are discarded potting mixes stored? What other kinds of organic residues are generated and how are they currently managed?

Human resource needs are covered in greater detail later in the chapter. At this stage, identify who in the current work force would provide the labor needed for composting, what their current skill set is, and what kind of training they might require.

As with any enterprise, finances will need to be considered. Will you need to use your savings to start your enterprise? Will it be a cash-generating operation? What will future financial needs be? What are the risks involved? Managing these and other financial and business strategies will be essential for the sustainability of your composting enterprise.

4. SWOT analysis

Once you have taken stock of enterprise resources, a good way to evaluate them is through a SWOT analysis (Kuligowski, 2020). SWOT stands for "Strengths,

Weaknesses, Opportunities and Threats." A SWOT analysis is a useful technique for identifying and evaluating a strategy—business, professional, or personal. It applies to a variety of strategies or decisions from identifying markets for products/services to selecting a career path. Although a SWOT analysis seems self-evident on the surface, it is usually a surprisingly enlightening exercise.

To conduct a SWOT analysis, you list the strengths and weakness that you or your business can bring to a particular situation and the opportunities and threats that the situation poses to you or your business. Some key questions that you might ask in a SWOT analysis include:

4.1 Strengths

- What can you offer that competitors cannot?
- What do you do best?
- What do you do better or for less cost than competitors?
- What do you do well?
- What do other people recognize as your strengths?
- What have you been successful doing (and why)?
- What resources do you have access to or can obtain easily or inexpensively?
- What management and labor skills do you possess?

4.2 Weaknesses

- What knowledge or skills do you lack?
- What do you do poorly?
- What can competitors provide that you cannot?
- What have you been unsuccessful doing (and why)?
- What resources do you lack (and that competitors have)?
- What can you improve?
- What have other people mentioned as weaknesses or complained about?

4.3 Opportunities

- What particular opportunities are at hand?
- How do your strengths match against a particular venture?
- What other advantages are at hand?
- What are the trends and do they work in your favor?
- Is the timing suitable to your situation?

4.4 Threats

- What barriers exist for you?
- How much investment is required?
- What are the risks of failure?
- What could impact the quality of the compost products that you intend to produce?

- How long will it take to become established or generate income?
- How do your weaknesses match against a particular venture?
- What trends work against you?
- Do your competitors have an advantage?

Example 2.1 gives an example of a SWOT analysis for a hypothetical composting enterprise, using a worksheet that divides the SWOT components between internal and external factors. SWOT worksheets and tools are available from numerous books, articles, and internet sources that focus on project management or decision-making topics (Caramela, 2020; Nediger, 2020).

Example 2.1 SWOT analysis for a hypothetical composting enterprise

Situation: A local commercial horse stables proposes to make compost for sale to local garden centers.

Strengths	Weaknesses
INTERNAL FACTORS	**INTERNAL FACTORS**
Ample supply of free horse manure Locally produced product Already know local merchants Ample land Most production equipment available No feedstock amendments needed No grinding equipment needed (lower cost)	No screening equipment Cannot produce bagged product Cannot produce multiple products Limited marketing experience No horticultural expertise
EXTERNAL FACTORS	**EXTERNAL FACTORS**
Three large and four small garden centers within 20 miles; all sell bulk products No large compost producers within 30 miles Low permit requirements	Local garden centers and gardeners have little experience using compost Free mushroom soil available 100 miles away Regulations prevent fertilizer claims
Opportunities	Threats
INTERNAL FACTORS	**INTERNAL FACTORS**
Eliminate current manure trucking expense Revenue to offset horse stable expenses Possible business enterprise if successful Compost manure from neighboring farms to increase product volume Compost manure from neighboring farms for fee to offset composting expenses Can supply compost to horse stable clients thereby strengthen the stable business Diversify business	Distraction from horse stable business May need to hire an employee Odors upset neighbors and community May need to purchase specialized equipment if can't maintain production rate Possible biosecurity breach from incoming truck movement to other farms/facilities Eventually may need improved site and separate building
EXTERNAL FACTORS	**EXTERNAL FACTORS**
Large retail merchants looking to buy local	Regulations may require full permit in future Town may produce and give away leaf compost in future

5. Defining success—start with the end in mind

The best way to get the planning process started is to begin with the end in mind. How will you define (and measure) success? What is it that you want to do? What is your primary motivation for composting? Do you want to compost to better manage on-farm residuals? Are you a public agency mainly interested in providing a residuals management service to the community or region? Are you planning to develop a profitable business around the sale of compost products?

If you are composting to make saleable products, you might define success as:

- developing and producing consistent quality composts for several valuable markets in the region,
- generating more revenues through product sales than through gate fees,
- increasing the use of compost and compost products in several regional markets, and,
- convincing the public transportation agency to specify compost in highway construction projects and erosion control projects.

If you are composting primarily to provide an organic residuals management service, you may define success as:

- recovering the costs of operating the composting facility through gate fees,
- eliminating the need for landfilling valuable organic materials,
- reducing greenhouse gas emissions (compared to landfilling those materials),
- eliminating pathogen concerns inherent in raw feedstocks, and,
- creating products that can either be sold or given back to the communities that provide the feedstocks (in the case of municipalities composting yard trimmings).

If you are planning a composting enterprise to improve manure and nutrient management, you may define success as:

- reducing manure handling costs,
- reducing the need for land for land-spreading raw manure,
- improving the use of manure nitrogen and phosphorus in crop production and reducing or eliminating fertilizer costs,
- eliminating the need for storage lagoons or high-cost storage infrastructure,
- reducing odor emissions and improving air quality by reducing emissions of ammonia and greenhouse gases, and,
- improving neighbor relations.

5.1 Vision and mission

Acknowledging your motivations for composting and then understanding how to define success helps create the vision from which the business plan can be developed. A *vision statement* identifies the future that the enterprise will create. For example, an enterprise vision statement might read: "To supply the local community with natural high-quality compost products while sustaining the family farm and its environmental quality." How this vision will be achieved is described by the mission statement.

The mission statement is the guidepost that identifies what the enterprise is and does, its operating values and purpose. It helps to steer the enterprise toward its long-term destination as well as help in the day-to-day choices and actions. For example, the mission statement based on the vision statement above might read: "The enterprise will provide a variety of compost products that meet customer demands for soil amendments and exceed their expectations for product quality."

For both the vision and mission statements, it's important to dream big. They set the path, direction, and impact of the enterprise on society. They should boldly declare what it is that will be done and what the enterprise is to become. These statements deserve considerable thought. Develop an operating philosophy, purpose, and scope that distinguish your enterprise from others. One way to start is by filling in the blanks on statements such as:

- Our composting enterprise will be known for …… and recognized as ……,
- The purpose of our composting enterprise will be to ……,
- Our composting operation believes in … and values … and is committed to ….,
- We will serve our customers by …… providing solutions for ……,
- Our composting enterprise does … (*what*) … and…. (*for whom*) … and … (*wants to become*)….

Examples of both vision and mission statements are found below
Vision statement examples:

- To be an environmentally responsible producer of healthy food and related products.
- To reduce our farm's ecological footprint through recycling of organic residuals.
- To be the region's best source for high quality compost.
- To provide a recycling service to local businesses and farms with the widespread reputation of being an asset to the community.
- To reduce the cost of managing the County's organic wastes while protecting the environment and providing citizens with high quality compost at a low cost.

Mission statement examples:

- We will produce excellent compost products at our facility that will exceed customer expectations of quality and environmental respect.
- We will recycle and reuse all of the organic residuals generated on our farm.
- We will collect and compost all of our region's organic residuals with the acceptance of our neighbors and the community.
- We will farm responsibly and profitably, retaining open space, environmental quality, and the rural character of the community.

5.2 Goal setting

The next step is to focus on the establishment of specific goals and to develop the working details of the business plan. Goals zero in on success, identifying specific and measurable achievements. Goals sharpen your focus, direct and prioritize your efforts and resources. They steer your enterprise in the direction of your vision.

Goals must be clear, concise, prioritized, and capable of being monitored and associated with specific timeframes. Short-term goals target a time horizon of one to five years while medium-term goals span between five and ten years and long-term goals look beyond 10 years. Goals can be both quantitative and qualitative. Examples of quantitative goals include: Our composting enterprise will realize $100,000 of sales and $45,000 of profit by year 2025; or, we will reduce our operating costs by 33% in 5 years. An example of a qualitative goal is, "We will be recognized as an environmental champion by receiving the local sustainable business award at each of our facilities."

Devote time to setting goals for each of the operating functions of your composting enterprise. Whether procuring feedstocks, handling materials, managing the composting process, marketing, or handling finances, each function's specific goal needs to be clearly aligned with the overall enterprise direction and definition of success.

6. Scoping out availability of feedstocks and markets for compost

Since many composting enterprises generate revenue "on the front end" from gate fees for residuals processing, market assessment includes gaining an understanding of feedstock availability and procurement. This assessment involves evaluation of the types of organic residuals available, the overall quantity and quality of feedstocks, their source, proximity to your facility, and seasonal factors that can impact ongoing supply. It is important to understand how these organic residuals are currently being managed—whether they are being landfilled, used for animal feed, diverted to another composting facility, etc.—to determine the ease of access and the fees that you might be able to charge to manage them. This information can be acquired through detailed surveys and interviews of industrial, commercial, and institutional sources of organic wastes. Although it can be difficult to obtain accurate information, a persistent effort usually yields worthwhile information.

If the primary reason for composting is to generate revenues by producing saleable products, become knowledgeable about the market(s) that you are targeting. An assessment of the local or regional markets for composts and compost products is a prerequisite for product distribution (Chapter 18). Even if the sale of compost products is not your primary driver, knowledge of the markets helps define future options. Many facilities that begin composting for other reasons shift toward selling compost because of the revenue opportunities.

For marketing finished products, understand the quality and other product characteristics that your potential customer base requires. You should become familiar with the purchasing and usage habits of the customers who will be buying your compost products. Seasonality will be a significant factor in market sales, one that will impact your inventory management system as well as storage requirements. Please see the *Compost Marketing* chapter for more details about compost markets and marketing.

7. Determine compost facility regulatory requirements

Operating plans must recognize and reflect on the requirements imposed by regulations that directly impact composting operations, compost product sales, and business operation. Regulatory requirements strongly influence decisions about infrastructure, technology, and overall costs.

Regulations that affect compost facility siting, design, operations, and compost quality are usually established and administered by state, provincial, or regional environmental agencies rather than local or federal government (although local government zoning laws can affect siting). Almost all state and provincial agencies require permits be issued for composting facilities that accept regulated wastes from off-site sources. In some cases, selected materials (e.g., yard trimmings, food processing residuals) are exempted from permits and bring a reduced level of review, such as a registration process. Many state and provincial authorities adopt federal regulations developed for the beneficial use of biosolids as appropriate performance and product quality criteria for composting all feedstocks.

These environmental regulations stipulate minimum acceptable setback requirements from various potentially sensitive features, e.g., streams, wells, wetlands, off-property residences, etc. Design-related regulations address minimum acceptable materials for construction of composting pads or protective lining requirements for storm water ponds. Operations-based regulations primarily address material volume limits, maximum pile size, process monitoring, nuisance conditions (e.g., time of operation), and record-keeping.

Local governments impact composting facilities primarily through land use or zoning regulations. Few localities explicitly mention composting as a land use category. Thus, land use/zoning committees or officers usually determine the classification. Many municipalities allow composting facilities in areas zoned for industry or manufacturing. On-farm composting is generally considered an agricultural land use and routinely allowed on agriculturally zoned land. Frequently, commercial composting is deemed agricultural as well because of the agricultural nature of composting and compost. However, an unreceptive local government may challenge the agricultural classification of even an on-farm operation if the composter accepts off-farm feedstocks (especially for a gate fee) or even if the farm sells compost to the public. Some communities have argued that composting is strictly a commercial enterprise. In such cases, the decisions are ultimately decided in court. Court decisions have generally favored composters, but the added expense of court fees, attorneys, expert witnesses, down and lost time, and clients can be problematic. Part of enterprise planning is determining where and how composting will be accepted by a local community.

Local regulations can also affect composting operations via codes and infrastructure requirements. A community may have requirements for access roads, water, fire suppression, and wetlands. Farms are typically exempted from these requirements. Some local authorities have established regulations specifically for composting, often to limit the number and size of composting facilities in the community.

However, in some cases, courts have determined these local regulations to be exclusionary and unlawful.

Finally, in most jurisdictions, regulations exist that affect the sale of compost products. In the U.S., these regulations are administered at the state level. In other countries, compost products are regulated at the national level, with and without additional state/provincial regulations. Compost generally is sold as a soil amendment, not a fertilizer, although it can be deemed a fertilizer if specific conditions are met. Similarly, some jurisdictions have requirements for products labeled as soil amendments. The requirements may depend on the amount and packaging of the products. Through the enterprise planning process, a prospective compost producer should understand the full complement of regulations governing the sale of compost products within their sales region.

8. Planning human resource needs of a compost enterprise

Many composting enterprises begin as one-person operations, with the manager also serving as the equipment operator and marketer. For municipalities and farms, labor is often drawn part-time from other areas of the organization in between other tasks. This model, however common, is short-lived. Nearly all composting operations, on-farm and otherwise, expand to the point of requiring at least one dedicated employee. Many commercial operations grow to require a team of staff specialized by function such as management, equipment operation, or marketing. Even when making compost for one's own use, staff must be allocated and trained to manage the composting operation.

To begin shaping your human resource strategy, determine tasks required to deliver the composting services and products. For each task or function, list all the duties and responsibilities expected to get the job done. For larger operations, determine and diagram the interrelationships between job functions and oversight for those functions. Estimate the time involved for the various tasks of each job function and account for seasonal workload factors.

An organizational chart depicts the operating dynamics of the business, the interactions and back-up support within and between the various business functions. Consider developing an organizational chart for future conditions, even if you're the one doing most of the work at first. This exercise will help you with your overall business planning.

Fig. 2.3 depicts an organizational chart for a typical composting operation that produces and sells compost on a moderate scale (e.g., 5000 tons of compost per year). Smaller-scale composting operations often evolve to this level of operations once they begin to sell compost and become successful at it. As an enterprise grows, it will eventually need a workforce with other skills—procurement people who understand waste management markets, if charging gate fees for off-site materials; operations people who are skilled at materials handling and equipment maintenance; salespeople who understand the importance of compost products in horticultural and landscape markets; and office people to handle the paperwork and manage the finances/bookkeeping. Additionally, job descriptions will be needed to detail

FIGURE 2.3

Organizational chart for a typical moderate-scale (e.g., produces 5000 tons of compost per year) commercial composting facility that sells compost.

the tasks, responsibilities, and expectations involved with each function and to identify the knowledge and skills required. A job description helps assess the compensation for each position. A good way to get started is to seek information from other composting facilities to find out how they operate and what workload and skill sets are involved.

Be mindful of the government and legal requirements involved with being an employer. Understand what is involved in hiring, retaining, and firing employees as well as to maintain good staff relations. This type of information can be crucial to protect you and your business from liability.

Hiring the right people for each task is not easy and should never be underestimated, both in terms of time and effort required. There are a number of hiring options that can be pursued to help with workload demands including contract service providers, custom operators, full- and part-time employees, family, interns, volunteers, and seasonal laborers. As your operation is growing, some jobs may be under contract. For example, it is common for composters to subcontract for grinding or screening services. Wages and benefits need to be carefully determined to ensure staff retention. Industry standards and local business associations can provide the baseline against which you can then decide your compensation strategy.

Training, along with developing and maintaining excellent communication among your team, pays dividends in terms of productivity and job satisfaction. Even with a single employee, it is important to establish a training program so that each staff member understands his/her job responsibilities and expectations. Well-trained staff members are more motivated to take on responsibilities and meet productivity expectations. At a minimum, employees should receive the appropriate occupational health and safety training. Many safety courses exist on this topic, delivered by both private training companies and public agencies. Other relevant training programs include: compost operator training, diesel engine and equipment repair, welding, and marketing and sales.

Training programs covering these topics are routinely offered in many regions by public agencies, universities, and professional associations connected with composting, recycling, solid waste management, wastewater treatment, and food processing and distribution. Many one-day classes are attached to professional conferences. Numerous multi-day composting training workshops are available. Check with state and regional associations, agencies that regulate composting, and national organizations that represent the composting industry (see Appendix A). Such education programs often lead to, or are a minimum requirement of, operator and manager certifications.

9. Production planning—the business of manufacturing

Compost operations focused on producing saleable compost and production of compost-derived products is similar to any manufacturing initiative. With labor and energy, inputs are processed into outputs, including products and by-products. Manufacturing elements that need to be clarified before setting up your own operation include: market assessment, feedstock procurement for developing compost recipes based on market needs, process set-up, production scheduling, equipment needs, and space requirements. Before setting up your production process, invest the time and effort to determine how other composting operations manage these elements.

Table 2.1 lists some of the considerations involved in production planning. Develop a production schedule, from feedstock procurement to product distribution (or use). The schedule should cover the manufacturing actions that will take place, the processing timeframes, and the equipment and labor resources required for each stage. Understanding the capacity of your operation and its limitations throughout the year allow you to assess your business potential and anticipate potential bottlenecks in the process.

There are various options for acquiring the funding needed for equipment and infrastructure. Outright purchase, leasing, sharing or renting are viable options, each with their advantages and disadvantages. The available funding affects decisions about what technologies to employ to process and compost feedstocks and further process and store finished products.

Develop a chart or table, similar to Table 2.2, that identifies the potential composting technologies you are considering, and the equipment needs for each technology. You should note how often, and through which phases of the composting process, you will use the equipment, comparing equipment purchase costs versus rental costs. For example, a composter who is creating a refined product may purchase a windrow turner to operate more efficiently, but rent a screen, as needed, to refine the finished product.

Successful entrepreneurs plan for scenarios that represent start up conditions and full production capacity. Anticipate constraints you might encounter (and when) as your enterprise grows. This approach allows you to forecast ongoing investments and infrastructure requirements to support your growth.

Table 2.1 Example compost production-planning summary.

Manufacturing step	Manufacturing actions needed	Processing timeframes	Possible equipment & infrastructure considerations	Labor resource considerations
Feedstock procurement	Move on-site wastes to composting facility. Accept off-site wastes	On-site wastes: daily to weekly, depending on size. Off-site wastes: usually daily or delivered on a schedule	On-site wastes: loaders, trucks. Off-site wastes: truck scales, office, receiving bay. Both: sampling equipment	On-site wastes: laborer. Off-site wastes: weigh-master, office personnel
Feedstock mixing	Mix feedstocks and bulking agents together according to recipe	Dependent on quantities of waste; usually daily or on a schedule	Loaders, mixers, hardened surface, concrete walls	Loader and equipment operators; laborers
Active composting	Manage primary decomposition process to produce stable product	Dependent on composting approach; generally, 30–90 days	Loaders, windrow turners, blowers, distribution pipe, hardened composting pad, adequate space, process monitoring tools and equipment	Loader and equipment operators; laborers
Curing	Manage secondary decomposition to produce mature products	Generally 30–60 days	Loaders, hardened surfaces, adequate space, sampling equipment	Loader operators. Laborers
Product screening	Screen large particles out to recycle back to mixing step	Depends on scale of production. Several hours per day to several consecutive days twice yearly.	Screening equipment, loaders, hardened surfaces	Loader and equipment operators; laborers
Product refinement	Blend screened compost with sand/soils/etc., for specialty products; bag compost and/or specialty products	Dependent on market factors and strategies; can vary with seasons; can be several hours per day	Specialty soil blending equipment, bagging equipment, hardened surfaces, loaders, palletizing equipment	Loader and equipment operators; laborers
Product storage	Manage product inventories for seasonal sales	Dependent on market and seasonal factors; plan on 120–180 days	Storage space, hardened surfaces, loaders, sampling equipment	Loader operators; laborers
Product distribution	Distribute product(s) to wholesale and/or retail customers (both on-site pickup and off-site delivery)	Seasonal— maximum in spring and fall, less in summer and winter	Loaders, truck scales, delivery trucks	Loader operators; truck dispatcher; product salesman/broker
Record keeping and reporting	Process quality control monitoring; regulatory compliance reporting	Daily to weekly depending on parameter to be monitored	Thermometers, moisture meters, oxygen meters, sampling equipment, data recording devices, weather station	Trained staff; office personnel

Table 2.2 Example of equipment needs for two composting technologies.[a]

Composting technology	Equipment to be used daily/ weekly	Estimated purchase costs	Equipment to be used every 3–4 months	Estimated total annual rental costs
Open air windrow turning	Tractor	$0 (existing)	Chipper	$4700
	Bucket loader	$0 (existing)		
	Compost turner	Tractor-pulled turner: $58,000	Screen	$5200
		Self-propelled turner: $280,000 (used)		
Open-air aerated-static pile	Bucket loader	$0 (existing)	Grinder	$13,500
	Mixer	$35,000 (used)		
	Blowers (5)	$1500	Screen	$4800
	PVC pipe network	$300		

[a] The costs listed are rough estimates, provided as examples. See Chapter 19 for additional equipment and cost information. Seek actual cost estimates from equipment vendors.

10. Financial strategy—the business of business

Your financial strategy is an offshoot of your marketing and overall business or enterprise plan. Keep tight control of what is happening on the financial side of the composting operation, even if you don't plan to generate revenue from product sales. You still need to account for costs, avoided costs, and actual income sources. There are many choices to make regarding how to establish and organize your financial management system.

Seek external advice from both legal and accounting counsel to set-up your enterprise optimally from tax, liability, and profitability perspectives. There are many business development organizations, usually set-up by government agencies, which exist specifically to help you set-up your business. An internet search or phone call to your local chamber of commerce, and organizations dedicated to small business (e.g., small business development centers) or extension service can steer you in the direction of these usually free resources.

If you develop the composting enterprise as a separate business, the various organizational options are: sole proprietorship, partnership, corporation, limited liability corporation, and cooperative. Each alternative has advantages and disadvantages from a variety of perspectives including ownership rights, financial protection, taxes, and growth opportunities. The option selected needs to be suited to your circumstances and reflect your long-term vision for your enterprise.

Managing the many risks associated with your financial and business strategies is essential for short- and long-term sustainability. Some risks will be minimal while others could effectively wipe you out (see contingency planning sidebar in Chapter 12). Risk takes on many forms: personal, production, market conditions, personnel, and financial. Your job is to minimize these situations through your own internal management systems and by transferring this risk externally. Internal risk management options include: business and income diversification, storage, leasing options, and cash/credit reserves. External risk management options include: insurance, production, and marketing contracts such as annual guarantees of tonnage deliveries or purchases. Try to predict financing requirements through the life of your business. Consider requirements for start-up, annual operation, capital expenditures, and long-term growth investments.

From an internal perspective, savings and/or financing from family might get you started. Investigate the opportunity for government financing support, including low-interest loans or loan guarantees. Externally, banks and other lending institutions can become sources for financing and credit (Goldstein, 2020). Operating and capital leases versus outright, immediate ownership of equipment and/or land can help reduce short-term costs and enable you to build equity for the long-term. The key to any of these options is balancing the business control that you retain, or have to relinquish, with the associated risk to your financial and business future.

To determine financial viability, analyze the various factors that might impact your business' financial health. Projecting your financial situation translates your goals and expectations into dollars. Developing financial statements allows you to record how and whether you were able to attain them (Box 2.5).

Your financial plan should be based on assumptions and reflect the strategies you have developed for your marketing programs, operations, and human resources support. To develop budget projections, you need to forecast sales, estimate operating expenses, and account for capital expenditures and financing. In projecting sales, reflect both the dollars as well as the units involved in attaining this sales base. Include both direct and fixed costs in your projections. Direct costs would include labor, materials, and the cost of operating equipment. Fixed or overhead costs include building, rent, insurance, depreciation, cost of capital, and marketing programs.

Far too often, plans are developed on a best-case scenario only and don't reflect caution or contingencies in the event of changes versus expectations. Crunch the numbers based on a variety of scenarios. At a minimum, assume scenarios that represent the best case, worst case, and anticipated case. This approach ensures that you are aware of the risks involved in your business and where the opportunity for cost controls can be included, if business does not progress according to your original expectations.

Another critical aspect of your financial management strategy is cash flow. As you will quickly find out, it doesn't matter how much profit is recorded on your financial statements unless you also have cash on-hand to pay your bills and manage your obligations. As important as realizing sales is the collection of the money owed for these sales. Careful attention to the prompt issuance of invoices and the timing of the payments or collection of these amounts is critical. The monies collected will be

---

Box 2.5 Key financial statements

Financial statements tell the story of your business from several perspectives: revenue and income, assets and liabilities, and cash flow. They can be monitored across a variety of time periods: sometimes weekly but mostly monthly, quarterly, and annually. As your composting business develops, comparisons between time periods will become ever more important, allowing you to assess your progress and anticipate future-forward issues.

Two statements of critical importance for you to be familiar with are: *the income statement* and *the balance sheet.*

The income statement This statement records the operations of your business within a specified time. Sales are subtracted by expenses and ultimately reflect the net income from your efforts.

Included in the income statement are:
- revenue or sales,
- cost of goods sold (these are the variable costs involved in the production of your composting services and products),
- expenses (these are fixed costs that are incurred to operate your business, including marketing, sales, administration, and building), and,
- net income (the result of your sales less your expenses). It is this figure that then gets inserted into your balance sheet.

The balance sheet The balance sheet records the financial position of your business, documenting your assets and liabilities. It consists of:
- assets (the resources owned or owing to your company),
- liabilities (the items and amounts that your company owes to others), and,
- owner's equity (the amount remaining once the liabilities are subtracted from the assets).

Both assets and liabilities can be divided into "current" (within 12 months) and "long term" (beyond 1 year). Current assets are those items that can translate into cash within the 12-month period while current liabilities are what is owed or needs to be paid within the same timeframe.

---

offset by cash outflows in a variety of areas including costs of production, overhead expenses such as rent, insurance and taxes, as well as repayments of loans.

The seasonality of your composting operation will also be a factor that needs to be included in your cash flow calculations. If your operation is focused on leaf and yard trimmings, the winter period will be light in terms of gate fee revenues. At the same time, the sales from your finished compost will generally be strongest in the spring and fall time periods.

Because of the significant potential for swings in the timing of revenue periods for a composting operation and the fact that many operating costs are constant every month, you will need to take great care of your cash flow situation to ensure that you can pay your bills throughout the year.

Time must be taken on a regular basis to analyze your financial situation and to plan your future. You must be disciplined in both recording your transactions and managing the financial health of your business—or all your hard work and investment could be severely jeopardized.

Annual budgets and financial projections allow you to set the stage for your annual operating plan and program activities. As you go through the year, regular reviews of achievements allow you to track your progress, accommodate adjustments, and work toward future goals.

Financial planning and management of the enterprise involves many tasks, financial tools, and options, far too many to cover in this chapter in the depth and breadth that they deserve. The references listed at the end of this chapter contain a great deal more information. Also, accountants, business attorneys, and small business associations are sources of much information, much of it free, including publications on many "getting-started in business" topics (Box 2.5).

11. Enterprise planning—case study

The following case study provides insight into the enterprise planning process for a real composting enterprise on a real dairy farm. The name of the farm and its owner are fictional, however, to protect their privacy. Monetary values are reported in US dollars and adjusted to 2020 prices.

The HW Dairy, located in Wisconsin, has been producing and selling compost in addition to milk for over 10 years. For Will Holmes and family, the decision to make and sell compost was primarily driven by the need to reduce the cost of land and land application of manure. Twenty years ago, the farm had 300 cows and 120 ha (300 acres). The farm eventually expanded its herd size to 600 cows without proportionally increasing its land base. Based on phosphorus limitations, over 600 ha (1500 acres) would have been needed to accommodate manure from 600 cows. Holmes explains, "we wanted to find a way to continue dairying without having to farm so much land." They determined composting to be the most cost-effective option for achieving this goal (see Table 2.3).

Table 2.3 Enterprise Summary for HW Dairy Composting Enterprise.

Mission	To remove as many nutrients as possible from the waste stream at HW. Dairy off the farm in an efficient manner.
Strategy	Separate the liquids from the solids in the dairy barn and compost the solids.
Goal	Create a compost product that is economical to produce and is genuinely good for the customers at a price that generates a profit.
Marketing strategy	We will sell only in bulk to separate us from the poor quality, cheap bagged product that is mostly sand.
Target markets	Landscapers, organic producers, gardeners, and garden stores. 80%–90% wholesale in large truck lots.
Marketing approach	Print advertising (65%); Gimmick promotion (10%); Personal sales (15%); Trade show (10%).

Because of the number of cows on their dairy (over 500 cows or 1000 animal units), HW Dairy meets the USDA-USEPA[1] definition of a "concentrated animal feeding operation" or CAFO. As such, the farm is required to submit a nutrient management plan annually and demonstrate how composting fits into its nutrient management plan. The Holmes also had to construct a permitted composting facility because the CAFO designation precluded any on-farm exemptions from composting regulations in Wisconsin.

Presently, HW Dairy composts all of the solid portion of the manure, leaving only the lagoon liquids for on-farm use. Solids from the liquid/solid separator are hauled to a two-acre asphalt pad, where the solids are stored prior to composting. Holmes uses a 4-meter-wide (14 ft) self-propelled windrow turner to turn the compost weekly.

The dairy's cows generate about 9200 cubic meters (12,000 cubic yards) of solids annually. After composting, the volume shrinks to about 3800 cubic meters (5000 cubic yards), which has significantly reduced costs associated with hauling manure. Holmes estimates that his dairy spends only about 10% of the cost that it would otherwise take to haul and land apply all the manure. This estimate does not include the time savings and equipment costs avoided by not having to apply raw manure to the 600 ha.

Although not an initial goal that guided their decision to compost manure on the farm, marketing the finished product now is a major part of the HW Dairy's success. Compost sales have been generating a profit since 2012. Tested regularly for chemical and biological quality, finished compost is sold for $40 per cubic meter ($30 per cubic yard) in bulk. Sales are divided between organic farmers (40%), landscape contractors (30%–40%), and individual homeowners (20%–30%), mostly within 80-kilometer (50 mile) radius of the farm. Holmes does most of the marketing himself. Also, the HW Dairy has a "free for neighbors" policy for the compost. The farm managers think this along with sound lagoon management and reduced odors associated with *not* land applying manure solids is a major reason why they do not get complaints about odor that so often plague dairies with residential neighbors.

Holmes' decision to adopt composting has definitely paid off. He has increased the herd size and avoided increased land requirements while being environmentally responsible with the nutrients produced. The result is a dairy that is profitable and geared toward what Holmes enjoys—being a dairyman.

Below is a sample enterprise planning budget from the HW Dairy.

Capital Budget, adjusted to 2020 dollars

Compost facility—1.2 ha (3 acres) of asphalt	$281,000
Compost turner, self-propelled, used	$101,000
End loader use	$40,000

[1] USDA is the US Department of Agriculture; USEPA is the US Environmental Protection Agency.

Delivery truck- new	$54,000
Total	$476,000
Operating budget	
Income	
Compost sales- 4600 m³ (6000 cubic yards)	$138,000
Compost freight	$13,000
Total	$151,000
Expenses	
Depreciation	$32,600
Repairs	$1300
Fuel	$3400
Labor	$53,600
Advertising	$5000
Interest	$33,900
Total	$126,400

The savings in manure handling costs occur as follows:

- pumping and irrigating the liquid is about 10% of the cost of hauling solid or slurry manure.
- hauling the separated solids requires less weight so the equipment needed is cheaper.
- hauling the solids to an adjacent site is less costly than to a field much farther away.
- the land application of unsold compost is cheaper than raw manure because the volume is reduced by over 50%.
- an added benefit is reduced odor compared to nonseparated manure.

There is a perception that farms that practice composting are better for the environment than those who do not. This perception also can be of value to the dairy.

References

Cited references

BioCycle, 2021. BioCycle Connect e-Newsletter; Business+Finance. JG Press, Emmaus, PA. https://www.biocycle.net/category/businessfinance/.

Caramela, S., 2020. 10 SWOT Analysis Tools for Small Businesses. Business News Daily. https://www.businessnewsdaily.com/6828-swot-analysis-tools.html.

Composting News, 2021. McEntee Media, Inc. http://compostingnews.com/.

DiGiacomo, G., Morse, D.E., King, R., 2003. Building a Sustainable Business: *Minnesota Institute for* Sustainable Agriculture Research. Also, USDA Sustainable Agriculture Research and Education (SARE) Network. Available from: http://www.misa.umn.edu/Publications/BuildingaSustainableBusiness/.

Goldstein, N., 2020. Financing composting infrastructure. BioCycle. https://www.biocycle.net/financing-composting-infrastructure-development/.

Kuligowski, K., 2020. Why You Need a SWOT Analysis for Your Business. Business.com. https://www.business.com/articles/swot-analysis-for-small-business-planning/.

Nediger, m., 2020. 20+ SWOT Analysis Templates, Examples & Best Practices. Venngage, Inc. https://venngage.com/blog/swot-analysis-templates/.

Consulted and suggested references

Aich, A., Ghosh, S.K., 2016. Application of SWOT Analysis for the selection of technology for processing and disposal of MSW. Proc. Environ. Sci. 35 (2016), 209−228.

Business Queensland, 2020. SWOT Analysis. Queensland Government, AU. https://www.business.qld.gov.au/starting-business/planning/market-customer-research/swot-analysis.

Coker, C., 2020. Composter tool for triple bottom line assessment; composting business management series Part V. BioCycle. https://www.biocycle.net/composting-business-management-composter-tool-triple-bottom-line-assessment/.

Cooperband, L., 2002. The Art and Science of Composting Science of Composting. Center for Integrated Agricultural Systems, University of Wisconsin-Madison.

Earthsquad, 2020. Compost: A Business Model for Small Communities. https://www.earthsquad.global/compost-a-business-model-for-small-communities/.

Gregory, A., 2020. The Pros and Cons of Starting a Composting Business. The Balance Small Business Website. Dotdash Publishing. https://www.thebalancesmb.com/the-pros-and-cons-of-starting-a-composting-business-2951463.

Grubinger, V.P., 1999. Sustainable Vegetable Production From Start up to Market. Northeast Regional Agricultural Engineering Service, Ithaca, NY.

Grudens-Schuck, N., Knoblauch, W., Green, J., Saylor, M., 1988. Farming Alternatives: A Guide to Evaluating the Feasibility of New Farm-Based Enterprises. Northeast Regional Agricultural Engineering Service, Ithaca, NY.

Lindsay, M., 2020. Lessons in community composting: business focus and structure. BioCycle. https://www.biocycle.net/lessons-community-composting-business-focus-structure/.

Raviv, M., 2013. SWOT analysis of the use of composts as growing media components. Acta Hortic. 1013, 191−202.

Ryndin, R.,E., Tuuguu, E., 2015. Composting Manual for Cold Climate Countries. ACF (Action Contre le Faim), Ulaanbaatar, Mongolia. http://www.susana.org/en/resources/library/details/2389. (Accessed 29 March 2017).

Solus Group, 2019. Human Resource Planning for Commercial Compost Providers. Solus Group, St. Louis, MO. https://solusgrp.com/blog/human-resource-planning-for-commercial-compost-providers.html.

Truic, 2020. Starting a Compost Business. TRuiC How to Start an LLC Website. https://howtostartanllc.com/business-ideas/compost-business.

The composting process

Authors: Cary Oshins[1], Frederick Michel[2]

[1]U.S. Composting Council, Raleigh, NC, United States; [2]Ohio State University, Wooster, OH, United States

Contributors: Pierce Louis[3], Tom L. Richard[4], Robert Rynk[5]

[3]Dirt Hugger, Dallesport, WA, United States; [4]Pennsylvania State University, State College, PA, United States; [5]SUNY Cobleskill, Cobleskill, NY, United States

1. Introduction

Composting is the aerobic, or oxygen-requiring, decomposition of organic materials by microorganisms under controlled conditions. During composting, microorganisms use oxygen (O_2) while consuming organic matter present in the raw materials, commonly called feedstocks (Fig. 3.1). Active composting generates considerable heat while carbon dioxide (CO_2) and water vapor are released into the air (Fig. 3.2). Water losses can amount to more than half the weight of the original feedstocks. Composting thus reduces both the volume and mass of the feedstocks while transforming them into compost, a valuable soil conditioner.

The feedstocks provide the nutrients (e.g., carbon and nitrogen) and energy for the microorganisms. The microorganisms come with the feedstocks and from the surrounding environment, or may be added by recycling finished compost or with inoculants. The volume and mass of the finished compost can be one half to less than one quarter the starting volume.

Composting is most rapid when conditions that encourage the growth of the decomposer microorganisms are established and maintained (Table 3.1). The most important conditions include:

- Organic materials appropriately mixed and sized to provide the nutrients needed for microbial, activity, and growth, including a balanced supply of carbon and nitrogen (C:N ratio),
- Enough moisture to permit biological activity without hindering aeration,
- Oxygen at levels that support aerobic decomposition, and,
- Temperatures that encourage vigorous microbial activity.

The composting process is robust, and many aspects of it are inexact. The process occurs over a wide range of conditions and with many feedstocks. The speed of composting and the qualities of the finished compost are largely determined by the

The Composting Handbook. https://doi.org/10.1016/B978-0-323-85602-7.00008-X

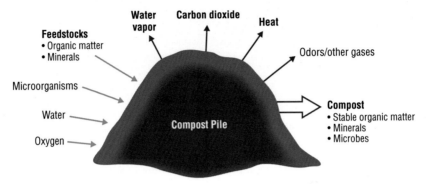

FIGURE 3.1

The composting process.

FIGURE 3.2

A composting pile generates heat, moisture, and CO_2, which are released to the pile's surroundings.

selection and mixing of feedstocks and influenced by the process conditions maintained.

The principles of composting have been well known for years, but continue to be scrutinized and fine-tuned by research and practice. Along the way, accepted theories,

Table 3.1 Favorable conditions for composting.

Condition	Reasonable range[a]	Preferred range
Moisture content	40%–65%	50%–60%
Carbon to nitrogen (C:N) ratio	20:1–60:1[b]	25:1–40:1
Oxygen, minimum concentration within the interior pore spaces	Greater than 5%	Greater than 10%
Temperature[c]	45–70°C (113–160°F)	50–65°C (120–150°F)
Ph	5.5–9.0	6.5–8.0
Particle size	3–50 mm (1/8 to 2 in.)	Depends on feedstocks and use for compost
Bulk density	Less than 700 kg/m^3 (1200 lbs/yd^3)	400–600 kg/m^3 (700–1000 lbs/yd^3)

[a] Generally for rapid composting. Composting can still be successful outside of these ranges.
[b] Some feedstocks can be composted successfully even at C:N ratios greater than 60:1, although the composting rate is slow and the time period is long.
[c] Temperatures as low as below 45°C are conducive to rapid composting, but sanitization specifications require temperatures to be held at 55°C or above for a period of time (e.g., three days). See Section 4.10.

principles, and practices have been summarized in influential references, including but not limited to Howard (1935), Waksman (1939), Rodale (1971), Gouleke (1972), Gray et al. (1973), Jeris and Regan (1973), Poincelot (1975), Willson (1980), Haug (1980 and 1993), Epstein (1997), BioCycle magazine (and its predecessor Compost Science), and numerous conference proceedings (e.g., Gasser, 1985; Hoitink and Keener, 1993). Drawing generously from these and more recent sources, this chapter presents the current knowledge about the composting process and the factors that influence it.

2. What happens during composting?

Composting begins as soon as moist organic materials are piled together. Decomposer microorganisms (bacteria and fungi) are already present on the surfaces of every particle and plenty of oxygen is available. Almost immediately, the microorganisms start increasing in numbers. They use the readily available sugars, starches, and proteins as food, providing the energy and compounds they need for growth and metabolism. As the process continues and the easily degradable organic substances are depleted, the composition of the microbial populations shifts toward species that are better adapted to metabolize the more difficult organic compounds.

Organic materials used in composting contain a significant amount of stored energy. The original source of this energy was the transformation of solar energy to chemical energy during photosynthesis in plants. By breaking the chemical bonds, microorganisms obtain energy for growth and synthesis of nutrients from the organic materials. During this process, some of the chemical energy is transformed to heat.

The composting feedstocks themselves act as insulation, slowing the dissipation of heat. When heat is generated faster than it is lost, the temperature of the pile rises. The temperatures can continue to rise until microorganisms are inhibited or die. The overall process is complex and dynamic where different sets of microorganisms thrive, die, and become food for others as they attack various components of the feedstocks over a range of temperatures.

In aerobic systems, energy production is proportional to the amount of oxygen consumed. Rapidly decomposing feedstocks consume a great deal of oxygen and produce abundant heat. Since the release of heat is directly related to the microbial activity, temperature is usually a good indicator of how well the process is working. An increase in temperature as a result of microbial activity is typically noticeable within a few hours of forming a pile or windrow. The temperature of composting material characteristically follows a pattern of rapid increase to 50–71°C (120–160°F), which can be maintained for many weeks. As active composting slows, temperatures gradually drop toward 40°C (100°F) and finally to near ambient air temperature. The activity, and thus the temperature, may fall due to poor aeration, low moisture, or lack of nutrients and then recover when these deficiencies are corrected by agitation, resupply of oxygen, or the addition of water. This characteristic pattern of temperature rises and fall over time reflects changes in the rate and type of decomposition taking place as composting proceeds and also reflects the "stability" of the organic matter in the compost (Fig. 3.3).

During composting, a wide range of temperature, moisture, and oxygen conditions can exist within the pile, even at a given point in time. The surface is near ambient temperature, oxygen, and humidity while the interior could be hot, oxygen depleted, and very moist. Beneath the surface, the conditions vary within the composting mass because the pile is not homogeneous and some areas possess or receive more moisture, nutrients, and air than other locations. This variation of conditions within the composting mass is reduced by mixing and moderated by containment, such as composting within vessels or under covers. The conditions change as the feedstocks decompose and as the pile is managed and manipulated. As the initial oxygen present in the pile is consumed, aerobic decomposition slows and may eventually changeover to mostly anaerobic if the oxygen is not replenished.

Microorganisms obtain their oxygen from air via *aeration*—ventilation for the composting piles. Aeration is required to recharge the continually depleting oxygen supply and to vent carbon dioxide and other gasses that accumulate in the pore space. Aeration is provided either by *forced aeration* or by passive air exchange (natural convection and diffusion). Mechanical mixing of the composting materials, or turning, temporarily supplies fresh air and oxygen; but this oxygen is quickly consumed and must be replenished by passive or forced air movement.

Through the active composting period, falling temperatures indicate decreased microbial activity. The microorganisms may become less active because the nutrients they need become depleted or unavailable, because they do not have enough moisture, or because they are not getting the oxygen they need. Oxygen flow may be insufficient because the pores between the particles are too small (the bulk density

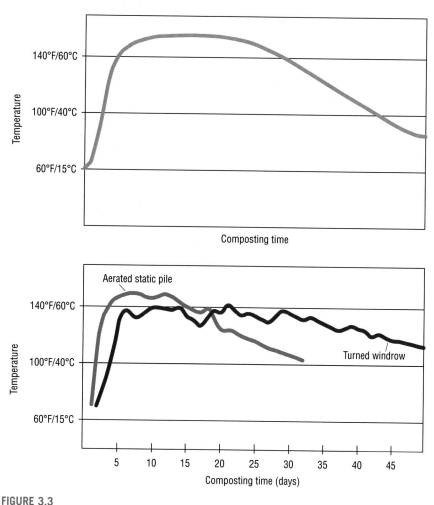

FIGURE 3.3

Temperature-time patterns for composting—generalized (top) and typical (bottom).

is too high), because the pores are filled with water, or because the pile is too big. Under these conditions, decomposition continues anaerobically and generates unwanted odorous compounds. On the other hand, if oxygen is available and the microbial activity is intense, the temperature can rise well above 60°C (140°F). As temperatures climb above 65–70°C (150–160°F), many microorganisms begin to die or become dormant (Strom, 1985). Cooling the pile by forced aeration or reducing the pile size helps to keep the temperature from reaching these damaging levels. Turning the pile is not a reliable cooling method during the vigorous phases of composting. Turning generally stimulates composting in the active phase. After an initial drop following turning, temperatures tend to recover, and often increase beyond the previous level. In the later stages, turning may have a lasting cooling effect—an indication that the active stage is ending.

A curing or maturation stage follows the active composting stage. In the curing pile, the compost continues to decompose but at a much slower pace and typically at lower temperatures. The rate of oxygen consumption decreases to the point where natural convection and diffusion usually meet the oxygen demand without forced aeration. However, curing piles are sometimes turned and/or aerated with blowers to ensure that aerobic conditions and moderate temperatures are maintained throughout the pile and odorous conditions do not occur.

The composting process does not stop at a particular point. Material continues to break down slowly. The nutrients from the original feedstocks become bound within the microbial community and its byproducts and continue to be slowly recycled. Compost is judged to be "done" by process indicators like temperature decline, oxygen demand, carbon dioxide evolution, and ammonia/nitrate concentrations. Compost manufacturers also use characteristics related to the compost's use and handling, such as texture, color, moisture, and odor. The degree of doneness is called its maturity. Mature compost is stable; that is, it has a low level of biological activity and will not cause harm to growing plants due to continued decomposition. However, the compost's usefulness can be limited by other factors unrelated do the compost's stability, such as soluble salts (Chapter 15).

3. Changes in the materials during composting

During composting the microorganisms transform organic feedstocks into compost (Fig. 3.4). This transformation changes the nature of the materials. The feedstocks

(A)   **(B)**

FIGURE 3.4

Composting changes the nature of the materials from a collection of diverse feedstocks into a uniform easily handled product.

begin as a diverse mixture of particles and compounds, many of which are easily degraded and potentially odorous, others highly resistant to decomposition, and many on a spectrum between those two extremes. By the time composting is complete, the mix of compounds becomes more uniform and significantly more stable. Little or no trace of the original feedstocks is discernible, with the common exception of wood particles (and inert components). The material becomes dark brown to black in color. The particles reduce in size and become consistent and soil-like in texture. In the process, the amount of humus-like compounds increases, the carbon:nitrogen (C:N) ratio decreases, the bulk density increases, and the nutrient exchange capacity of the material increases.

Composting leads to both a large weight reduction, typically on the order of 40% to 50%, and a substantial shrinkage in volume, of one-half to up to 90% of the initial volume. Part of this volume reduction represents the loss of mass, as CO_2 and water escape to the atmosphere. However, much of it occurs as loose, bulky raw feedstocks break down into crumbly, fine-textured compost. Where a particular feedstock mixture falls in this spectrum depends upon the specific feedstocks and, to a lesser extent, on the methods of composting and processing. High moisture, readily degradable materials like vegetative food residuals, and loose bulky feedstocks like leaves, exhibit the greatest loss in volume. A volume reduction of 75% to 90% is not uncommon for leaves and mixed food scraps. Typical agricultural materials, like manure, exhibit a more moderate shrinkage, generally finishing at 30% to 70% of the original volume (Larney et al., 2008).

Fig. 3.5 shows dry mass and volume reductions of several different feedstock mixes after 100 days of turned-windrow composting (Breitenbeck and Schellinger, 2004). The largest percentage volume losses are associated with bulky vegetative materials. From these results, the authors generated the following regression formula ($r = 0.72$) relating volume loss (ΔV) to initial dry bulk density (dBD_i).

$$\Delta V = 0.80 - (2.0 \times dBD_i) \tag{3.1}$$

The C:N ratio gradually falls during composting because of the loss of carbon as CO_2 from the starting materials. The amount of carbon lost during composting usually exceeds the nitrogen loss. However, if the starting C:N ratio is low, less than 15:1, the nitrogen losses may be large enough to cause little change in the C:N ratio.

Some loss of nitrogen from the composting pile is inevitable. Significant amounts of the nitrogen in the feedstocks can also be lost as ammonia and, to a lesser extent, as nitrous oxides. Nitrogen loss increases with lower starting C:N ratio, higher temperatures, higher pH, and more frequent turning. Thus, composters with a goal of nitrogen conservation should manage their system accordingly (Chapter 14).

Most other nutrients that are released from the feedstocks during decomposition remain in the compost. They become part of the bodies of new microorganisms or

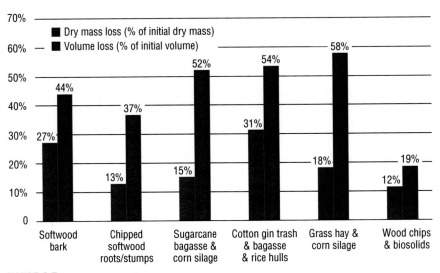

FIGURE 3.5

Reductions in dry mass and volume of several feedstock mixes after 100 days of turned-windrow composting. Mix ratios: 100% shredded softwood bark; 100% shredded softwood root balls and stumps; equal volumes of sugarcane bagasse and spoiled corn silage; equal volumes of cotton gin trash, bagasse, and rice hulls; two volumes of grass hay plus one volume of spoiled corn silage; and four volumes of wood chips plus one volume of biosolids.

Based on data from Breitenbeck and Schellinger (2004).

are incorporated into stable forms of organic matter. As carbon is lost, nutrients become concentrated in the finished compost. However, in areas of high rainfall, potassium, and some phosphorous, can be lost by runoff and leaching.

Microorganisms decompose organic materials progressively, attacking simple sugars and starch first, then proteins, and eventually the larger complex polymers like cellulose and hemicellulose. Microorganisms break down raw materials into simpler compounds, then reform some of them into new complex compounds, while modifying other compounds such as lignin to form humic and fulvic acids. The final product has a low rate of microbial activity but is still rich in microorganisms and the remains of microorganisms.

Some organic compounds present initially in the feedstocks pass through the composting process with little or no change. Lignin, found in woody materials, is difficult to break down in the typical time span of a composting pile. Lignin and other biologically resistant substances are concentrated in the compost. They are partially responsible for compost's characteristic qualities.

4. Factors affecting the composting process

The compost pile is initially made up of a mixture of solid particles of feedstocks with pores spaces between the particles filled with water, air, and other gases. One can envision a microscopic situation in which each solid particle is surrounded by a liquid film, with gases occupying the pore space between the particles (Fig. 3.6). Most of the aerobic decomposition of composting occurs within this liquid film, especially near the surface of the particles where the nutrients are concentrated. Oxygen moves readily as a gas through the air-filled pore spaces but diffuses very slowly through the liquid and solid portions of the particles (Van Ginkel et al., 2002). A population of aerobic microorganisms builds up in the liquid layer surrounding the surface of particles. The microorganisms use the available oxygen dissolved in the liquid film as they decompose organic compounds near the particle's surface (Bemal et al., 1998). Meanwhile microorganisms also are active in the oxygen-starved interior of the particle. The particle shrinks as the composting microorganisms work their way inward.

Factors that affect how quickly the composting process transforms these feedstock particles into compost include: oxygen, aeration; moisture; nutrients (e.g., carbon and nitrogen); feedstock degradability, porosity, free air space (FAS), permeability, structure, bulk density, and particle size and shape; pH; temperature; and time. Many of these factors are interrelated and all impact the rate of decomposition.

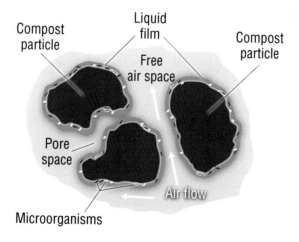

FIGURE 3.6

Depiction of particle environment within composting pile.

4.1 Oxygen

Aerobic composting consumes large amounts of oxygen. For example, one cubic foot of yard trimmings may ultimately require all of the oxygen in 10,000 cubic feet of air for the conversion of organic carbon to CO_2. During the initial days of composting, readily degradable components of the feedstocks are rapidly metabolized. Therefore, the need for oxygen and the production of heat, water vapor, and other gases are greatest at early stages. After peaking, typically early in the process, the composting rate and oxygen consumption decrease as the process ages to a relatively steady level near the end, as depicted in Fig. 3.7. A similar pattern can be expected after a pile is turned. The graph is a very generalized illustration of the pattern of microbial oxygen demand. The shape of the curve and time scale depends greatly on the feedstocks and other conditions.

If oxygen is not replaced at the rate it is consumed, the composting process becomes anoxic or anaerobic. Anoxic means that the oxygen levels are low and limit aerobic respiration. Anaerobic technically refers to the absence of oxygen, although the term is generally used to describe both anoxic and anaerobic conditions. In either case, the rate of decomposition slows, and odorous compounds accumulate (Box 3.1).

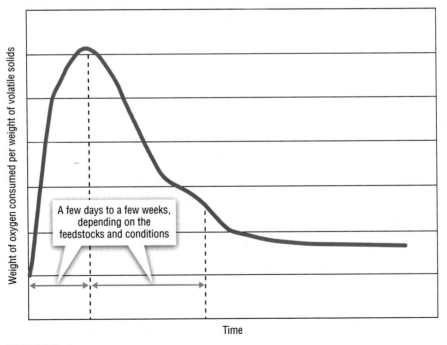

FIGURE 3.7

General pattern of oxygen consumption (i.e., uptake rate) during composting.

Box 3.1 Why the stink?

Author: Tom L. Richard

Although some plants and animals produce odors as a repellant, to the best of our knowledge microorganisms are not intentionally using this strategy to gain privacy from prying eyes. Instead, odors are an inherent byproduct of their metabolism, especially the anaerobic respiration and fermentation processes that provide them with energy. Microorganisms liberate the energy needed for metabolism and growth by combining electron donors (carbohydrates, proteins, fats, etc.) with electron acceptors, a process commonly referred to as respiration. A compound that accepts electrons takes on a negative charge and is thus said to be "reduced." In aerobic respiration, oxygen serves as the electron acceptor. Oxygen is the most energy-efficient electron acceptor, resulting in the most "complete" decomposition and providing a competitive advantage to aerobic organisms. Nitrate (NO_3^-) is a close second and thus readily consumed under low oxygen conditions (i.e., anoxic).

Under anaerobic conditions, sulfate (SO_4^{-2}), and carbon dioxide (CO_2) and minerals such as ferric iron (Fe^{+3}) and manganese (Mn) also serve as electron acceptors, but with significantly lower amounts of energy released. The reduction of sulfate produces the gas hydrogen sulfide (H_2S) and leads to other odorous sulfide gases including dimethyl sulfide and dimethyl disulfide. Other anaerobic reactions recycle part of the carbon liberated during anaerobic degradation as an electron acceptor, producing a biogas that includes both CO_2 and methane (CH_4). Energy can also be liberated anaerobically through fermentation, which uses organic molecules as both electron donors and electron acceptors. Fermentations can produce a variety of compounds (including hydrogen gas and alcohols), but of particular interest, here is the production of organic acids, many of which are very odorous and toxic to plants.

Because they do not liberate all the energy from the feedstocks, anaerobic respiration and fermentation leave useable energy in the products of decomposition (e.g., methane, acetate). Thus, aerobic organisms can break down the products of these anaerobic processes to extract more energy. Also, anaerobic respiration and fermentation yield products like methane and ethanol, which can be harvested from these processes as alternative fuels.

Although respiration reactions produce and consume the largest quantities of odors, ammonia (NH_3) can be odorous and forms under both aerobic and anaerobic conditions. Urea is another common source of ammonia. Also, protein degradation produces ammonia as the amino acids are broken down (i.e., deamination). Ammonia is particularly volatile under alkaline (high pH) conditions, when it readily escapes from the composting pile.

Respiration and fermentation reactions using the example of glucose ($C_6H_{12}O_6$):

Aerobic respiration (oxygen serves as the electron acceptor to form only carbon dioxide and water):

$$C_6H_{12}O_6 + 6O_2 \rightarrow 6CO_2 + 6H_2O$$

Anaerobic respiration (sulfate is the electron acceptor, forming odorous hydrogen sulfide):

$$C_6H_{12}O_6 + 3SO_4^{-2} + 6H^+ \rightarrow 6CO_2 + 6H_2O + 3H_2S$$

Fermentation (glucose decomposes without an external electron acceptor to form odorous butyric acid and hydrogen gas):

$$C_6H_{12}O_6 \rightarrow CH_3CH_2CH_2COOH_6 + 2CO_2 + 2H_2$$

Fermentation (glucose decomposes without an external electron acceptor to form ethanol):

$$C_6H_{12}O_6 \rightarrow 2C_2H_6O + 2CO_2$$

Anaerobic decomposition involves different biochemical reactions and a different set of microorganisms. Anaerobic processes are less efficient than aerobic processes, leaving behind a variety of "intermediate" compounds that can further decompose if oxygen is present. Little heat is generated to evaporate water from the materials. Anaerobic processes produce a variety of volatile intermediate compounds, including methane, organic acids, hydrogen sulfide, and other substances. Many of these compounds have strong odors, and some are toxic to plants (phytotoxic). Under anaerobic conditions, these compounds accumulate and can be released to the environment in noticeable concentrations, especially when the pile is turned, aerated, or otherwise disturbed. If oxygen becomes available or the compounds migrate to an area in the compost pile where oxygen is available, they can oxidize to inoffensive compounds like carbon dioxide and water. An adequate supply of oxygen gives the more efficient aerobic organisms a competitive advantage over the anaerobes. Thus, maintaining aerobic conditions is important in avoiding the offensive odors associated with anaerobic decomposition.

Although composting is called "aerobic," it is more accurate to say that composting is *predominantly* aerobic, rather than exclusively aerobic. The availability and distribution of oxygen varies within different sections of any given pile or vessel. Also, the average level of oxygen varies among different composting methods and operations. But even in well-aerated systems, part of the organic matter likely decomposes anaerobically. The critical factor is that enough oxygen is present such that most of the anaerobic compounds eventually decompose aerobically before leaving the pile. In composting, aerobic conditions should prevail—enough to degrade odorous compounds to a tolerable level and enough to render the compost largely free of phytotoxic compounds.

In the early stages of the composting process, the rate of oxygen consumption can outpace the rate at which it can be replenished. Therefore, the oxygen levels can be low at the start of composting and then recover as the decomposition rate and oxygen demand decline. Correspondingly, the carbon dioxide concentration shows the reverse pattern (Fig. 3.8).

The numerical oxygen concentration that divides aerobic and anaerobic states is both indistinct and debated. The minimum oxygen concentration traditionally recommended for aerobic composting is 5%. This value generally refers to the concentration of oxygen in pore spaces within an active pile, windrow, or vessel, although supporting research measured it in the exhaust from laboratory reactors (Schulze, 1962). In comparison, fresh air contains about 21% oxygen.

Although the 5% threshold is still widely observed, the targets for minimum oxygen concentration have increased in recent years. Recommendations for minimum oxygen concentrations now favor 10%, and some recommendations are as high as 18%. This large range is due to several factors including differing philosophies about composting and compost, differing risks and sensitivities to odors, the cost of supplying oxygen and differences in the methods and time constraints of composting. Some composters believe that the quality of the compost suffers whenever anaerobic conditions exist and thus try to maintain oxygen concentrations above 10% (or CO_2 concentrations below 11%). In other cases, a higher oxygen concentration threshold, in the 10% to 18% range is maintained for process and odor control. These cases might include situations where the consequence of odor is severe or compost must be produced quickly and the composting method accommodates rapid

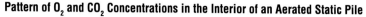

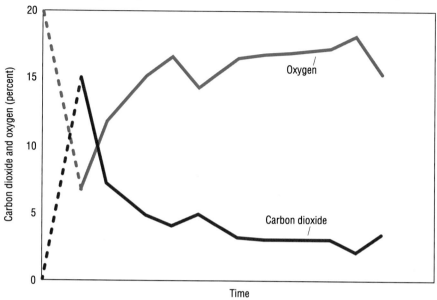

FIGURE 3.8

Example of how oxygen and CO₂ concentrations mirror each other, in this case in a aerated static pile of biosolids and woodchips. The dotted lines crudely represent the abrupt drop in oxygen, and rise in carbon dioxide, that occurs when feedstocks are first piled. The corresponding pattern for a passively aerated pile or windrow would typically show oxygen levels hovering below 5% for a few weeks and rising much more gradually, with carbon dioxide concentrations almost the exact inverse of oxygen.

Adapted from Epstein (1997).

decomposition. Conversely, lower oxygen concentrations are tolerated where odors are not a large concern, and a slower composting method is tolerated.

Maintaining a high concentration of oxygen during the active phases of composting is a challenge. Passively aerated piles, including windrows, can show near zero concentrations of oxygen in the core of the pile, especially in the early stages of composting. Higher oxygen concentrations can be achieved only by limiting pile size and using feedstocks with sufficient FAS for effective convection. With forced aeration composting, oxygen concentration can be maintained more easily because of the control afforded by the fans. However, increasing the rate of forced aeration has drawbacks including energy consumption and potentially excessive loss of heat and moisture (see following section). Ultimately, the most practical oxygen management strategy is dictated by the individual site characteristics, including pile density, moisture, feedstocks, cost of aeration, wind and topographic conditions, and distance to neighbors.

Some compost operators monitor oxygen as well as pile temperatures and odors to learn how these factors relate in their particular system. Various sensors and probes can be used to measure the oxygen concentration in the pore space within a pile. Oxygen can be monitored by using a sensor that reads oxygen directly, or by using a carbon dioxide meter. Carbon dioxide (CO_2) is produced in direct proportion to the oxygen (O_2) being consumed. These devices are described in Chapter 14. Oxygen or carbon dioxide sensors are probably the second most common piece of monitoring equipment at composting operations (thermometers are overwhelmingly first).

4.2 Aeration

Oxygen is supplied to the composting organisms with fresh air.[1] The air is delivered via aeration—either forced or passive.

With forced aeration, mechanical blowers or fans, with a network of pipes, push or pull air through the composting materials (Fig. 3.9). As the air moves through the pile, it delivers oxygen and removes water, heat, carbon dioxide, and other gasses produced within the pile, including odorous gases. The air also cools the pile, primarily by evaporating water. With positive aeration fans push air into and through the pile. Positive aeration is the more energy-efficient direction. The advantage of negative aeration, even though it requires more energy, is that the exhaust air from the pile is concentrated in the outlet of the fan and can be treated for odors.

Passive aeration relies on natural forces to exchange air between the interior of the composting mass and the surrounding environment (Fig. 3.10). The principal force in passive systems is thermal convection—the movement of warm gas upwards due to a decrease in their density as they warm. In rising out of the pile, the exiting gas creates a partial vacuum and cool fresh air flows in to replace it (aka the "chimney effect"). The rate of thermal convection depends on the temperature difference between the gases inside the composting mass and the ambient air plus the resistance to air flow within the material (i.e., permeability). Convection increases with high temperatures within the pile, cool ambient temperatures outside, and an open porous

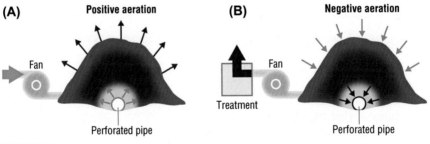

FIGURE 3.9

Forced aeration of a composting pile by positive (A) and negative (B) modes.

---

[1] Although attempts have been made to directly supply pure molecular oxygen, they have not yet been economically successful.

Modes of passive aeration

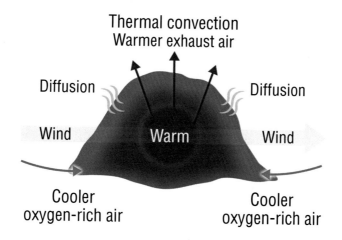

FIGURE 3.10

Driving forces for passive aeration. Passive air flow is primarily caused thermal convection. Molecule diffusion and wind-driven air movement are less consequential.

matrix of materials. Passive aeration also occurs by molecular diffusion, the migration of molecules from areas of high concentration to low concentration due to random molecular movement. Oxygen diffuses into composting materials because there is an imbalance—more oxygen outside than inside. Similarly, carbon dioxide has tendency to diffuse outward. As oxygen levels decrease and carbon dioxide levels increase in a pile, the concentration difference widens and movement due to diffusion increases. Diffusion is a relatively slow means of providing oxygen, especially within wet materials. Finally, in open outdoor windrows and piles, wind can assist in forcing gases into and out of the composting mass.

In addition to providing oxygen, aeration removes heat, water vapor, and other gases from within the composting materials. In fact, the required rate of aeration for heat removal (i.e., temperature control) can be 10 times greater than that for supplying oxygen. Therefore, temperature control, instead of oxygen supply, usually dictates how much, and how frequently, forced aeration is required. As noted earlier, the need for oxygen and the generation of heat, water vapor, and other gases are greatest early in the process. Therefore, the required rate of aeration also is greatest at the start and progressively declines as the feedstocks decompose.

4.3 Turning

"Turning" is the common term for the agitation or mixing of materials in compost piles, bins, vessels and, especially, windrows (Fig. 3.11). Turning is performed

FIGURE 3.11

Tractor-powered windrow turner in action.

intentionally to stimulate composting by homogenizing the composting materials. It occurs incidentally when materials are moved from one location to another, sometimes between two distinct stages of composting. Turning invigorates the composting process, but it is not a reliable means for aerating or cooling piles (Box 3.2).

Turning blends the composting materials and moves them to and from hotter and cooler areas. Turning overcomes the stagnation that can occur in static composting systems because materials are not completely homogeneous and passive aeration is not uniform. It can eliminate localized limitations due to too little or too much moisture, lack of particular nutrients, air channels, or other process factors. Turning breaks apart particles, mixes feedstocks, and promotes more even decomposition and pathogen destruction. The overall effect of turning is to stimulate composting.

The effects of turning on the composting process have been demonstrated by several illumining studies, including but not limited to Buckner (2002), Michel et al. (1996) and Michel and Tirado (2010). Such studies have shown that higher turning frequencies increase the rate of decomposition, pathogen destruction, and compost bulk density. Generally, more frequent turning decreases compost particle size and the time required to reach full maturity and also reduces the potential phytotoxicity of composts. Beyond a certain point, however, turning has only a modest effect on increasing the composting rate. Daily turnings usually do not decrease the time required to stabilize compost, compared to biweekly turning or even turning every two weeks. Studies have also shown that turning frequency has minimal *long-term* effects on oxygen concentrations or temperatures within windrows or on final compost properties (Chapter 5). Windrow turning frequency appears to have less of an impact on the composting process than other variables such as feedstock composition (e.g., C:N ratio), moisture content, and pile size.

Box 3.2 Compost turning: falsehoods, facts, and functions

- Turning provides aeration: False! The oxygen introduced by mechanical agitation (turning) may be consumed in a matter of minutes in an active pile. Passive or forced aeration is needed to sustain oxygen levels after turning.
- Turning "fluffs" the pile and restores pore space: False! (Mostly) Depending on the feedstocks and type of turning equipment, turning tends to reduce particle size, which increases bulk density and reduces total pore space. Turning can temporarily increase the porosity of materials that are already dense (e.g., cattle manure) but materials eventually resettle and compact.
- Turning cools the pile: False! (Mostly) During the active phase of composting, turning reduces temperature only temporarily. Generally, it is followed by an increase in temperature beyond the preturning point by exposing fresh surfaces for decomposition and breaking up anaerobic pockets. During the later phase of composting, when heat generation slows, turning can have a permanent cooling effect.
- Turning speeds decomposition: Fact! Turning has an invigoration effect on the composting process, even when forced aeration is applied. It releases trapped gases, exposes fresh surfaces, breaks apart clumps, and redistributes moisture, nutrients, and microorganisms. Because turning performs these functions, the composting process tends to surge forward after turning. Each turning advances and accelerates the process in such surges. However, as the turnings increase in frequency, each successive turning has a smaller effect. Each composter must decide when the positive effects of turning are outweighed by the costs.

The advantages of turning can extend beyond the composting process to the compost product. Turning with a specialized windrow turner typically shreds and homogenizes the material, thus producing a compost product that is more uniform in texture and moisture. These effects can not only improve screening efficiency but also the compost's appeal to its users. Therefore, even if the composting process does not call for it, many composters continue to turn windrows to enhance product quality.

While turning has a stimulating effect on the process, excessive turning can have some undesired consequences. First, turning can lead to excessive drying and particle size reduction. Second, turning releases ammonia (NH_3) and other gaseous decomposition products that have accumulated in the internal void spaces of the windrow. Some of the products released by turning are odorous. Turning has also been shown to increase nitrogen loss from low C:N ratio feedstocks, like many types of animal manures (Yang et al., 2019). Thus, turning can have detrimental effects by increasing nitrogen loss (principally as NH_3) and odor emissions. Turning windrows and piles more than necessary increases the labor and fuel costs required to produce stable compost and wears out equipment faster.

Finally, and back on the plus side, turning provides access to the interior of a pile or windrow, making it easier to observe and evaluate the progress of decomposition. It provides the opportunity and a means to evenly add water and/or nutrients to the composting materials. In short, turning is an aid to process management.

4.4 **Moisture**

After the feedstocks themselves, moisture is perhaps the most important factor in the decomposition of organic materials. Without enough moisture, virtually nothing happens. Composting materials should be maintained at a moisture content between 40% and 65%, depending on the feedstocks and stage of composting. The preferred range is narrower, between 50% and 60%. Too little moisture slows biological activity, even to a halt, while too much moisture interferes with aeration and oxygen transfer. In theory, biological activity is optimal when the materials are saturated and oxygen is plentiful. While this situation might be achieved in liquid composting systems by bubbling in air and continuously mixing, the costs would be very high. Thus, composting solid feedstocks at roughly 40% to 65% moisture, balances the microorganisms' need for abundant moisture and the ability to economically supply them with oxygen.

Water is essential to biological activity. The moisture that envelopes particles in the compost pile supports the metabolic processes of the microorganisms. Water provides the medium for chemical reactions, dissolves and transports nutrients, and allows the microorganisms to move about.

However, aerobic microorganisms also need oxygen. The oxygen in the air between the particles must dissolve into the water around each particle and then move by diffusion to the microorganisms. As the moisture content increases, the film of liquid surrounding the solid particles grows thicker and oxygen and carbon dioxide diffuse more slowly into and out of the film (Fig. 3.12). Because oxygen diffuses 10,000 times slower in water than in air, a thicker moisture layer results in more anaerobic activity. As the lower diagram of Fig. 3.12 illustrates, a thick water film limits the penetration of oxygen to the solid substrate. With limited oxygen, the resulting anaerobic compounds break through to the pore space before completely decomposing. The odors may be decomposed within the pile or they may be carried out of the pile with the exhaust or liberated at turning.

In addition, at high moisture levels, generally greater than 65%, water displaces much of the air in the pore spaces of the composting materials. The water in the pores reduces air movement into and through the pores and further encourages anaerobic conditions. The abundant moisture also adds to the weight of the materials, which causes more settling, compaction, and loss of pore space within the pile. As a result, air movement through the pores is impeded. In short, too much moisture interferes with aeration and leads to anaerobic conditions.

On the other hand, insufficient moisture negatively affects the decomposer organisms. Biological activity essentially ceases below 15% moisture content, but experience has shown that the composting process becomes noticeably slower as the moisture content drops below 40%. For rapid composting, 50% moisture is usually a better lower limit, depending on the feedstocks and phase of composting. Composters often allow moisture to drop below 50% during the later phase of composting to create a drier product that is more easily screened.

Since the moisture content tends to decrease as composting proceeds, the moisture content of the blended feedstocks should start above 50%. For many compost mixtures, materials that are too dry are blended with materials that are too wet to achieve the target moisture content. With some dry materials, such as leaves, water

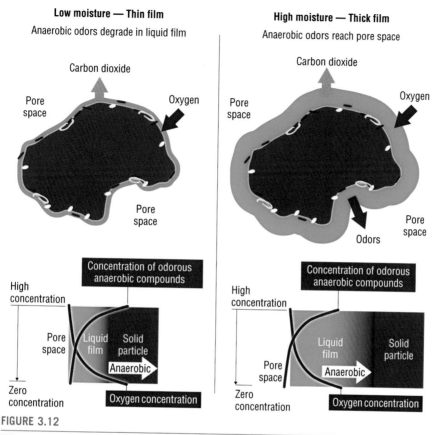

Low moisture — Thin film
Anaerobic odors degrade in liquid film

High moisture — Thick film
Anaerobic odors reach pore space

FIGURE 3.12

Effect of high moisture on oxygen transport and diffusion.

Based on presentation by Tom Richard and Cary Oshins.

may be added directly. As a source of moisture for dry feedstocks, some composters are able to obtain liquid wastes, such as wet manure, food processing residuals, and even out-of-date beverages.

During composting, moisture levels are constantly changing. Water may be added as rain or snow, released from cells as they degrade, or intentionally added with a watering system. Moisture decreases as respiration transforms liquid water to water vapor and as water evaporates from the pile. Usually more water evaporates than is added by microbial respiration and precipitation, so the overall moisture content tends to decline as composting proceeds. However, the moisture balance depends on heat production within the pile, seasonal and short-term weather patterns, and the pile size and exposure. Moisture levels should be monitored and maintained such that materials are thoroughly wetted without being waterlogged or dripping excessive water.

The 40%−65% moisture content range is a general recommendation that works well in practice for most feedstocks. The acceptable moisture limits depend on the particle size, porosity, absorbency, and ash or mineral content of the feedstocks.

Highly porous materials with large particles can be wetter than densely packed materials with small particles. A mixture with highly absorbent materials may need to be maintained well above 50% moisture to support rapid composting. Feedstocks that contain a high amount of inert (nonbiodegradable) material compost well at moisture contents below 50% because it is only the noninert fraction that is decomposing. Examples include animal manure with a lot of soil and a municipal waste with plastic. In such cases, the "ash-free" moisture content can be used as a guide (Chapter 4). An ash-free moisture content in the range of 50% to 60% generally works well for composting.

4.5 Nutrients and the carbon to nitrogen (C:N) ratio

Carbon (C), nitrogen (N), phosphorus (P), potassium (K), and sulfur (S) are the primary nutrients required by the microorganisms involved in composting. N, P, and K are also the primary nutrients for plants, so their concentrations can also influence the value of the compost as a fertilizer. Manures, biosolids, plant residues, and food residuals contain ample quantities of most nutrients to nourish the microorganisms. The nutrients that mainly influence the composting process are carbon (C) or nitrogen (N), either in excessive or insufficient amount. A balanced supply of C and N (i.e., a good C:N ratio) usually ensures that the other required nutrients are present in adequate amounts. However, the lack and abundance of P and S have occasionally been found to substantially affect the process.

4.5.1 C:N ratio

Carbon is the basic building block of virtually all organic compounds, so it is the major element in all compost feedstocks. Microorganisms use carbon compounds for both energy and growth, just as humans do. Nitrogen is an essential element for the production of protein and for cellular reproduction. In general, living organisms, from bacteria to humans, contain 10 to 15 units of carbon for each unit of nitrogen, by weight. However, because organisms continually respire and lose carbon as CO_2, they need to consume about 25 times more C than N. It is, therefore, important to provide C and N in appropriate proportions at the start of the process. The ratio of carbon to nitrogen is referred to as the C:N ratio. The C:N ratio expresses the elemental weight of C present relative to the elemental weight of N present. Typical C and N data for common feedstocks are listed in Appendix B. A feedstock with a C:N ratio that is too low or too high can be blended with other feedstocks to better balance these nutrients.

Individual feedstocks or feedstock blends with a C:N ratio of 25:1 to 40:1 are considered ideal for active composting. In this range, there is usually enough nitrogen to promote the growth and reproduction of microorganisms and sufficient carbon available to sustain the process while also conserving nitrogen. However, initial C:N ratios from 20:1 up to 60:1 consistently result in successful composting.

With C:N ratios below 20:1, composting is rapid but the microorganisms use the available C without converting all of the available N into cellular compounds. The excess N is converted into nonorganic N compounds like ammonia or nitrous oxide, which can volatilize (i.e., evaporate). In this situation, as much as 50% of the total initial N can then be lost to the atmosphere. Establishing a starting C:N ratio above 30:1 does much to retain nitrogen (Fig. 3.13).

The consequence of a high initial C:N ratio (above 40:1) is a slowing of the process because there is not enough N to feed enough organisms to decompose the available carbon. As the C:N ratio increases, the rate of composting generally decreases. However, for most applications, composting proceeds well even with feedstock mixes with C:N ratios up to 60:1. In fact, it can be a good management practice to establish an initial C:N ratio that is higher than the "ideal" for the microorganisms. While the higher ratio may reduce the rate of composting, the slower pace can make the process more manageable and forgiving. In addition, feedstocks with high C:N ratios typically include a good proportion of dry and bulky materials like wood chips, leaves, and sawdust. These materials tend to increase the porosity of the mix, improving aeration and further reducing the risk of odors.

The C:N ratio is an important but imperfect guide for balancing these two nutrients. The types of compounds that contain the carbon atoms must also be considered. Different carbon-containing compounds biologically decompose at different rates (Fig. 3.14). For example, chopped straw decomposes and releases its C to the microorganisms more easily than sawdust. This effect occurs because the straw is mostly cellulose, whereas the carbon compounds in sawdust and woody materials are largely bound by lignin, which is highly resistant to biological break down (Chapter 4). Similarly, the C in the simple sugars of fruit wastes

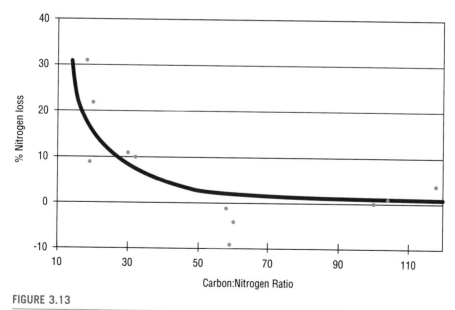

FIGURE 3.13

Effect of C:N ratio on nitrogen retention.

Adapted from data presented by Larsen and McCartney (2000).

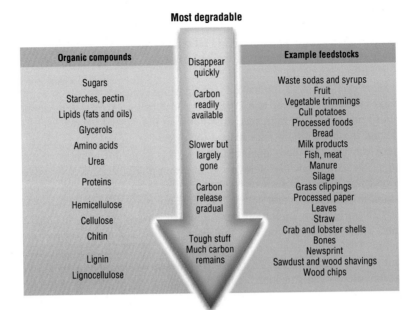

Most degradable

Organic compounds		Example feedstocks
Sugars	Disappear quickly	Waste sodas and syrups
Starches, pectin		Fruit
Lipids (fats and oils)	Carbon readily available	Vegetable trimmings
Glycerols		Cull potatoes
Amino acids		Processed foods
Urea	Slower but largely gone	Bread
Proteins		Milk products
	Carbon release gradual	Fish, meat
Hemicellulose		Manure
Cellulose		Silage
Chitin		Grass clippings
	Tough stuff Much carbon remains	Processed paper
Lignin		Leaves
Lignocellulose		Straw
		Crab and lobster shells
		Bones
		Newsprint
		Sawdust and wood shavings
		Wood chips

Least degradable

FIGURE 3.14

Relative degradability and carbon-availability of organic compounds.

and the starches of grains is more quickly consumed than the cellulose-carbon in straw. The biodegradability of organic carbon compounds, and their contribution to the composting process, can be estimated by the lignin content. Although it is an uncommon practice, a biodegradability factor can be used to roughly estimate the amount of "available" C and the C:N ratio can be adjusted accordingly. The N in nearly all feedstocks can be considered completely biodegradable during composting.

Finally, providing a mix of feedstocks with an appropriate C:N ratio has a larger impact, beyond the availability of nutrients. Feedstocks that have a high N concentration (i.e., low C:N ratio), like wet grass, unbedded manures or food residuals, tend to be wet, have small particles, and decompose quickly. Feedstocks that have a low N concentration (i.e., high C:N ratio), like branches, bark, or straw, tend to be dry, contain larger particles, and decompose more slowly. Thus, mixing feedstocks to obtain the right ratio of C to N also creates a mix that is more balanced with respect to moisture, particle size, bulk density, porosity, and decomposition rate. In short, a mix with a good C:N ratio is good in many respects. This larger effect of balancing the feedstocks may be more important to how well the composting proceeds than the actual ratio of C to N itself.

Table 3.2 General targets for feedstock starting nutrient ratio (weight of carbon ÷ weight of other nutrient).

Nutrient ratio	Target range
Carbon to nitrogen (C:N)	20:1 to 60:1
Carbon to phosphorus (C:P)	120:1 to 250:1
Carbon to potassium (C:K)	100:1 to 150:1
Carbon to sulfur (C:S)	Approximately or greater than 100:1

4.5.2 Other nutrient ratios: P, K, and S

Although C and N are usually the principal nutrients in the composting process, in some situations, the lack of other nutrients also can affect the process (Table 3.2). If a compost pile that appears satisfactory in terms of C:N and moisture is not progressing as expected, other nutrient ratios may indicate a limiting nutrient.

Several researchers have identified feedstocks in which the addition of P considerably stimulated the composting process. Those feedstocks tend to be dominated by paper products and by-products (e.g., paper mill residuals, municipal solid waste). A recommended ratio of C to P falls in the range of 120:1 to 240:1 (Brown et al., 1998).

Potassium has not been implicated as a limiting nutrient for the composting process. Nevertheless, a C to K ratio of between 100 and 150:1 appears to be a good target, based on microbial biomass composition and carbon conversion.

The ratio of carbon to sulfur (C:S) also is worth noting—not because of a lack of S, but rather its abundance. Organic matter generally has a C:S ratio of roughly 100:1. When the C:S ratio is appreciably lower than 100:1, there is an excess of S relative to the amount of C that the microorganism can use. Just as excess nitrogen forms ammonia, excess S can lead to the evolution of volatile sulfur compounds, many of which are odorous (Miller, 1993). Maintaining a C:S ratio near or above 100:1 might limit the generation of odorous compounds. However, currently there is little guidance from research or actual operations on this practice.

4.6 Physical factors: particle size, structure, porosity, free air space, and permeability

In addition to providing the chemical compounds for decomposition, the feedstocks also provide the physical foundation for the process. Several characteristics of the feedstocks collectively determine the physical arrangement of particles as they pile up. These characteristics primarily affect the composting process by their influence on aeration. In addition, particle size also affects the microorganisms' access to the organic matter, nutrients and moisture. The physical properties of the pile can be adjusted by the selection of the individual feedstocks and by preprocessing. Preprocessing might include grinding to reduce the size of the particles, mixing to increase homogeneity and screening to change particle size distribution. Feedstocks added to adjust these properties are referred to as amendments or bulking agents.

A compost pile is a collection of solid particles that physically support one another to stack together in a pile. The 3-dimensional matrix of solid particles creates a vast network of voids, or pore spaces. The pores are filled with air, other gases, and liquids, primarily water. The network of pores provides multiple pathways for air to move into and through the compost pile and for gaseous products of decomposition to move through and out of the pile. The pores serve as the pile's ventilation system. The arrangement and size of pores is determined by size and size distribution of particles, the rigidity of the particles (i.e., structure), and the amount of water present. The matrix of pore spaces can be described by several closely related measures including porosity, FAS, and permeability.

4.6.1 Particle size and shape

Microbial decomposition occurs at a particle's surface. As the size of a particle decreases, it has a greater surface area relative to its volume. Therefore, the rate of aerobic decomposition increases with smaller particle size—that is, within limits. Smaller particles pack more tightly together and effectively reduce the size and continuity of pore spaces and generally constrain the flow of gases into and out of a pile. The size distribution and shape of particles also affects aeration. Feedstocks with a broad range of particle sizes and shapes restrict aeration because the small particles fill the voids created by the large ones. Similarly, particles with a uniform shape roughly equal in all dimensions, form larger and more stable pores than long thin or flat particles, while the latter provide greater surface area for decomposition (Fig. 3.15).

The condition of a particle's surface is likewise important. Consider two particles of equal volume and mass, one smooth-surfaced and one rough and jagged. The rough one has considerably more surface area than the smooth one, and therefore decomposes faster since more microorganisms can inhabit the larger surface area.

Size also affects the ability of a solid particle to absorb moisture, and therefore decompose. Small particles are more readily wetted. This relationship is particularly true for materials that are "anisotropic" (i.e., their qualities differ in different directions). For example, a straw stem, which is waxy along its length, takes on water primarily through its ends. Cutting the straw exposes more ends, allows better moisture penetration, and faster decomposition. The same situation applies to some other materials like wood and coated-cardboard.

Optimal particle size is therefore a compromise between available surface area and ease of aeration. The optimal size depends on the specific feedstocks, the method of composting, and the intended market for the product. Good composting results are usually obtained with a variety of starting particles ranging in size from 3 to 50 mm (1/8 to 2 in.).

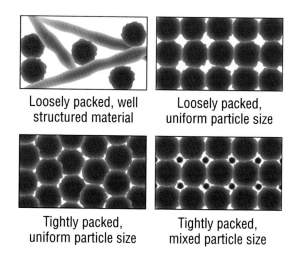

Loosely packed, well
structured material

Loosely packed,
uniform particle size

Tightly packed,
uniform particle size

Tightly packed,
mixed particle size

FIGURE 3.15

Effect of particle size and shape on porosity and bulk density.

Source: Richard (1996).

4.6.2 Structure

Structure refers to the rigidity of the particles—that is, their ability to resist settling and compaction. Good structure prevents the loss of porosity in the moist environment of the compost pile, allowing gasses to circulate into and out of the pile. Larger composting piles, especially those that are not regularly turned, require better structure to resist settling and maintain air channels during the entire composting period.

Good structure is provided by feedstocks with large rigid particles that decompose slowly. Ground or chipped wood is the most common amendment used to supply structure. Corn cobs, nut shells, and even inert amendments like chipped tires can be used to supply structure. However, any material must be present in sufficient proportions to substantially support the pile. Increasing moisture levels reduces the rigidity of most materials, although woody feedstocks are less affected than other amendments like straw, cornstalks, and leaves. Amendments added purposely to mainly provide structure are often called "bulking agents."

4.6.3 Porosity

Porosity refers to the nonsolid portion of the composting mass, and includes the space occupied by both the liquid and gases. It does not, however, include the small pore spaces within solid particles. Porosity is determined by the size, gradation, and type of the particles of the materials, and the continuity of the air spaces. It increases with particles that are larger, stiffer (e.g., woodier), more uniform in shape, and when most of the particles are similar in size. Porosity is expressed as percentage of the total volume. A porosity of 60% means that 60% of the volume is occupied by air and water (plus other liquids and gases) while the remaining 40% of the volume is solid particles.

4.6.4 Free air space and permeability

FAS is the portion of pore spaces not occupied by liquids or solids. It is sometimes referred to as the "air-filled porosity," although gases other than air also occupy that space. Like porosity, FAS is expressed as a percentage of the total volume. An FAS of 40% means that 40% of the volume is gases, the remaining 60% is filled with solids and liquids. FAS is the porosity minus the volume of liquids present (Fig. 3.16).

Compared to porosity, FAS is a better indicator of a pile's ability to channel air because it accounts for the presence of water. The FAS essentially determines how easily air can move through the composting mass. In general, the resistance to airflow decreases as FAS increases. Recommended values for FAS range widely from 30% to 60%, although most research has generally favored values between 30% and 40%. In part, the wide range is due to different techniques for measuring FAS (Alburquerque et al., 2008). Currently, field measurements are the most practical means of measuring FAS (see Chapter 4).

If the FAS is low, there is much resistance to aeration, both passive and forced. If the FAS is too high, then the pile temperatures may remain low; possibly because the heat generated by the microbes quickly dissipates. Compression from the overlying material reduces the porosity and FAS near the base of the pile (Fig. 3.17). Porosity and FAS generally diminish during composting as the materials decompose and particles are reduced in size.

FAS is a useful indicator of how well a pile aerates, but it can possibly be deceptive, at least in theory. A feedstock with few large pores can have the same FAS value as another feedstock with many more small pores. On one hand, large pores can be widely dispersed, leaving pockets of anaerobic solids with air flowing around them.

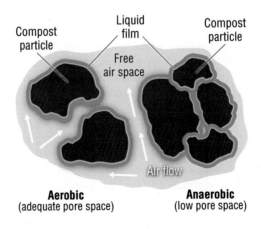

FIGURE 3.16

Conceptual depiction of free air space.

Source: C. Oshins.

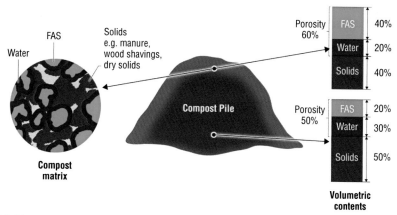

FIGURE 3.17

Relationship of free air space (FAS) to pile depth (hypothetical values).

Adapted from McCartney and Chen (2001).

On the other hand, a feedstock with a great number of small pores may have a high porosity and FAS yet the movement of air is greatly restricted by the small and tortuous pore spaces.

A parameter known as permeability has been used to directly characterize the ease of aeration. Permeability is determined by measuring the pressure required to push air through a sample or section of the composting mass. It is related to pressure drop in a forced aeration system. Researchers have used several methods to measure permeability, such as pycnometers (Ruggieri et al., 2009). However, no instrument has yet emerged for field use. Also, standard recommendations for permeability are lacking. Fortunately, permeability has shown to correlate closely with FAS and bulk density, which are easily determined.

4.7 Bulk density

Bulk density is the mass or weight per unit volume of an accumulation of particulate materials.[2] For instance, bulk density applies to a pile of wood chips while a single wood chip is characterized by particle density or simply density. In the realm of composting, bulk density is commonly expressed in kilograms per cubic meter (kg/m^3) and pounds per cubic yard (lbs/yd^3).

---

[2] Density is technically a measure of *mass* per unit volume. However, when using the US system of units, density is commonly expressed in units of *weight* (lbs, tons) rather than mass (slugs). The quantity given by weight per unit volume is more correctly termed "specific weight."

Bulk density is perhaps the best single index of how well a mix of feedstocks aerates because it is influenced by several relevant factors—moisture content, porosity, particle size, and the density of the individual particles. It increases with higher moisture and with lower porosity. Therefore, it also reflects the FAS in a composting material. As the straight lines in Fig. 3.18 suggest, bulk density correlates very well with porosity and FAS. In short, bulk density provides an overall indication for the physical and aeration conditions of a composting mass. Furthermore, it is easily measured in the field. Thus, bulk density serves as a good gauge for combining feedstocks and managing the composting process.

As a rule of thumb for most composting feedstocks, the starting bulk density of a feedstock mix should be lower than 700 kg/m^3 (1200 lbs/yd^3), and preferably below 600 kg/m^3 (roughly 1000 lbs/yd^3). Above 600 kg/m^3, aeration becomes increasingly difficult, and beyond 700 kg/m^3 it nearly stops. If the bulk density is very low (below 400 kg/m^3 or about 700 lbs/yd^3), the material may be so porous that heat escapes quickly and the temperatures fail to reach the desired level. Alternatively, the low temperature might be due to abundance slowly decomposing large, woody, or dry

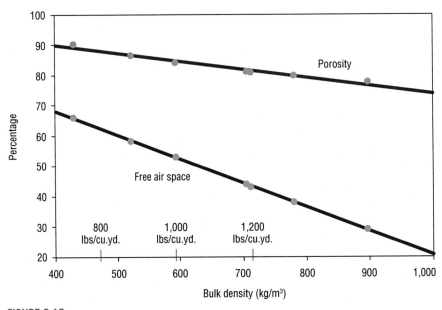

FIGURE 3.18

Correlation of bulk density with free air space (FAS) and porosity for paper-mill deinking sludge.

Adapted from Day and Shaw (2001).

feedstocks responsible for the low bulk density. In either case, this problem is rare and can be overcome by building larger piles.

One problem with using bulk density as a management criterion is that it differs throughout a pile or windrow. Because of compression, bulk density increases from the top to the base of a pile. Thus, bigger (taller) piles tend to have a greater overall bulk density and are more difficult to effectively aerate. When the bulk density of a standing pile is measured, samples must be taken at different depths in the pile or at a depth known to represent the average condition. A large difference in bulk density from top to bottom can be corrected by turning, at least temporarily. Methods used to measure bulk density should account for changes in bulk density as samples are removed from the pile. A standard method for field measurement of bulk density is presented in Chapter 4.

4.8 Pile size

As a pile grows in size, conditions outside the pile exert less influence on conditions inside the pile. A larger pile loses proportionally less heat and moisture to the ambient air, and gains less moisture from precipitation because there is less surface area per unit of pile volume. As a result, larger piles are typically warmer than smaller windrows or piles. The path to oxygen-rich ambient air is longer and more tortuous so natural air currents are less effective at supplying the pile with needed oxygen and cooling. Similarly, CO_2, water vapor, and other gases have a more difficult time migrating outward. In addition, as a pile grows in height, the material above presses more weight on the material below. On average, the pile becomes increasingly dense due to compaction. There is a large difference in bulk density between the compacted material at the base of the pile and the loose material near the surface.

Passive aeration is at the mercy of these effects of increasing pile size. It is simply more difficult for a large pile to naturally "breathe" and breathe evenly throughout. Thus, as a pile gains size, it becomes less aerobic throughout, hotter and the composting process slows. Forced aeration or pipes laid beneath the compost can improve conditions in a larger pile, but only up to a certain point. In piles taller than 3 m (10 ft), the densification resulting from settling and compaction tends to cause the air flow to channelize and be less effective. The optimum pile size depends on many factors—the weather, feedstocks, stage of composting, moisture, composting method, etc. Finding the right pile size becomes a balance among these factors plus equipment capabilities and the effective use of the land area. In many cases, the equipment forming the piles determines the maximum height. It is never advisable to drive on a pile, as this causes increased compaction and reduced airflow, thus slowing decomposition. As a general rule, the pile becomes too large when the equipment has to drive onto the pile to reach the top.

4.9 pH

The broad spectrum of microorganisms involved in composting generally allows the process to proceed over the wide pH range typically found in composting feedstocks. The preferred pH is in the range of 6.5–8.0, but the microbial diversity and the natural buffering capacity of the process makes it possible to work within a wider pH range (Box 3.3). Composting may proceed effectively at pH levels between 5.5 and 9. However, it is likely to be slower at 5.5 or 9 than it is at a pH near neutral (pH of 7). Extremely low pH (<5.5) can develop in the first weeks of composting and hinder the process. Also, high pH at the finish can encourage ammonia volatilization and impair compost quality for some uses. While adjusting the pH is rarely necessary, and often inadvisable, there are exceptional cases when additives that raise or lower the pH might be helpful.

Two groups of molecules generally control the pH during composting; organic acids, which are acidic, and ammonium (NH_4^+), which is alkaline. When oxygen is scarce, particularly when composting is just beginning, carbon compounds cannot be completely oxidized to CO_2. They accumulate as organic acids, such as acetate, propionate, butyrate, and others (many of which are odorous). The pH drops, potentially to levels between 4 and 5. Fortunately, these acids oxidize when oxygen is available, allowing the pH to return to neutral pH (Sunberg, 2005; Michel and Reddy, 1998). Ammonium, and subsequently ammonia, form more gradually as urea and proteins decompose. When ammonium is generated, the pH tends to rise to levels between 8 and 9 (i.e., the pKa range of ammonium[3]). This is why the pH of compost is commonly in this range.

The pH tends to change as decomposition proceeds (Fig. 3.19). The pattern and extent of the pH change varies with the materials and the process conditions. During the early stages of composting, rapidly decomposing feedstocks form organic acids that lower the pH. Usually, the pH recovers within a few days as the organic acids aerobically decompose. If aeration is poor, the organic acids can persist and the pH remains low until enough oxygen becomes available and the acids breakdown. In many cases, depending on the feedstocks, the organic acids decompose almost as quickly as they form and the pH drop is unnoticeable, or it does not occur. In the later stages of composting, the pH either remains stable or slightly rises or slightly falls due to the formation of ammonium and its subsequent conversion to nitrate or protein.

---

[3] In simple terms, the pKa value indicates how strongly a chemical compound acts as an acid at a given temperature (i.e., donates H^+ ions or protons). Stronger acids have lower pKa values. Without diving into the details of acid/base chemistry, if the pH of a solution is higher than a compound's pKa, the compound donates H^+ ions to the solution, and thus lowers the pH of the solution. This continues until the pH equals the pKa (equilibrium). If the pH is lower than the pKa, the compounds accepts H^+ ions from the solution, raising the solution's pH. Thus, the pKa indicates the level to which a particular compound potentially raises or lowers pH if it is present in high enough concentrations.

Box 3.3 pH by the numbers

pH is a measure of the acidity or alkalinity of a substance. Substances with a pH less than 7 are considered acids. Substances with a pH greater than 7 are alkaline, also called "bases."

Technically, pH is a measure of the concentration of hydrogen ions (H^+) that are present when a substance is in a solution of water. The presence of many H^+ ions create acidic conditions. More specifically, pH is the negative logarithm of the H^+ concentration, written as:

$$pH = - \log[H^+] \tag{3.2}$$

For all but the most extreme acids and bases, H^+ concentrations fall in the range of 1×10^0 (or 1.0) moles per liter to 1×10^{-14} (or 0.00000000000001) moles per liter. So, all but the most extreme acids and bases have pH values ranging from 0 to 14. A substance with a low pH has a higher concentration of H^+ ions than a substance with a high pH. The negative sign in Eq. (3.2) is to blame for this apparent contradiction. Moreover, because pH is based on a logarithm, each whole number change represents a 10-fold different. A pH of 5 is 10 times more acidic than a pH of 6 (i.e., the H^+ concentration is 10 times greater). The alkalinity is 10 times greater at pH 9 versus pH 8.

A pH of 7 is neutral, neither acidic nor alkaline. In pure water, a few molecules disassociate into H^+ and OH- ions while the majority of molecules remain together as H_2O. The "normal" concentration of H^+ ions in pure water is 1×10^{-7}, which is balance with an equal concentration of 1×10^{-7} hydroxide ions (OH^-). Thus, the pH of pure water is 7, neutral.

Anything that increases the H^+ ion concentration in water is an acid and has a pH less than 7. Anything that reduces the H^+ concentration is alkaline. For example, when dissolved in water, an alkaline substance might contribute OH^- ions that combine consume H^+ ions to form water molecules. Alkaline substances have pH values greater than 7. The further that a substance veers from pH 7, the more acidic or alkaline it is. A buffer is a substance that counteracts the effect of acids and bases by either absorbing or releasing H^+ ions. Buffers stabilize pH against changing conditions.

Feedstock mixes that are rich in easily degradable carbohydrates, such as food residuals, can quickly generate an abundance of organic acids that overwhelm the process and drive the pH to excessively low levels (<5.5). As the organic acids accumulate, a sour odor develops in the pile and the escape of volatile organic compounds (VOCs) may lead to odor problems. Research by Sunberg et al. (2005) suggests that keeping the temperatures below 40 C (105 F) affords mesophilic microorganisms the opportunity to degrade the organic acids. Conversely, if temperatures rise above 40 C, the ensuing thermophilic microorganisms are less capable of degrading organic acids, either because of the low pH or the lack of suitable enzymes. In addition, the higher temperatures promote greater volatility of the organic acids and the release of VOCs.

One way to prevent, or correct, this situation is to increase aeration (Fig. 3.20). The effects of the enhanced aeration are to increase the oxygen concentrations to oxidize the organic acids and to better cool the pile and maintain low temperatures that favor mesophilic organisms. With forced aeration, the fan aeration rate can be increased and/or the fan on-time can be extended. With passive aeration, windrows and piles can be made less dense by using more bulking agent, and/or kept smaller until pH rises above 5.5.

Fundamentally, excessively low pH points to an imbalance in the feedstock mix; either the mix contains too high a proportion of quickly degradable feedstocks (e.g., heavy in food residuals), the C:N ratio is too low, and/or the mix lacks FAS and

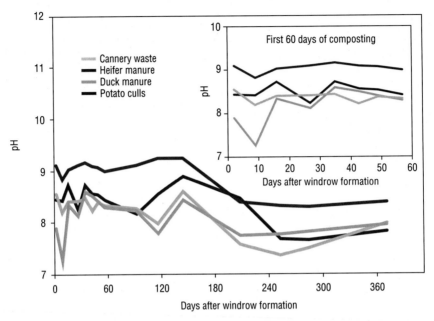

FIGURE 3.19

Example of change in pH over time for four different feedstocks. These particular feedstocks are uncommon, and the pH values are uncharacteristically high. Nevertheless, they are varied, and display different patterns of pH evolution through a long period of composting.

Courtesy of Leslie Cooperband, unpublished data from a University of Wisconsin study of the effects of selected feedstock.

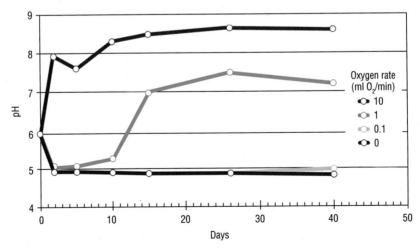

FIGURE 3.20

Effects of oxygen on composting pH. Lower amounts of oxygen result in organic acid formation that reduces compost pH.

Adapted from Michel and Reddy (1998).

structure. Before addressing possible aeration shortcomings, the feedstock recipe should be reexamined. If the feedstocks recipe is not changed, adding lime or wood ash to the feedstocks can immediately raise the low pH. If such additives are used to raise the pH, they should be used in small quantities and should be thoroughly mixed with the other feedstocks.

As composting proceeds, conversion of organic nitrogen to ammonium (NH_4^+) tends to raise the pH. As long as the ammonium persists in the compost, the pH remains elevated. However, over time the ammonium normally converts to gaseous ammonia (NH_3). The ammonia gas either volatilizes or forms new biomass protein. When either reaction occurs, H^+ ions are liberated and the pH decreases, in effect, reversing the pH-elevating effect of ammonium. The pH increase typically occurs a few days into the process, lasts for several weeks, and drops slowly as the concentration of ammonium in the compost declines. Similarly, H^+ ions are released and pH goes down (acidifies) as ammonium is oxidized to form nitrate during the curing phase. These effects are more evident with feedstocks that have high concentrations of nitrogen.

Ammonia volatilization greatly increases at high pH (above 7.5–8.0), which leads to nitrogen losses and reduces the nitrogen content of the finished compost. If desired, ammonia loss can be limited by including more acidic feedstocks or adding gypsum to keep the pH below 8.0. Even better, by providing a moderate to high C:N ratio (>30:1), ammonia is conserved as it is converted to microbial protein. Similarly, volatilization of hydrogen sulfide (H_2S) increases as the pH decreases below 7. Maintaining the pH levels near or above 7, limits the emission of this odorous gas.

In the end, composting tends to yield a product with a stable pH that is usually neutral to slightly alkaline, between 7 and 8.5. In many cases, composting results in a slight increase in pH compared to that of the feedstocks. Some feedstocks, like manure, that start with a high pH may produce compost with a slightly lower pH.

4.10 Temperature

Temperature is both a cause and an effect within the composting process. It is a cause in that it influences the rate of biochemical reactions and the organisms involved. It is an effect in that the biological activity produces heat, which in turn changes the temperature. Because of this relationship, temperature is a reliable indicator of the composting process.

Temperature has a profound effect on the composition and activity of microorganisms and the chemical reactions that take place. Within limits, biochemical reactions occur faster at higher temperatures, thus accelerating decomposition. Faster decomposition increases heat production. Higher temperatures increase oxygen demand, sanitize the materials, and promote more evaporation of volatile compounds and water. At lower temperatures, the composting process occurs more slowly. However, nutrients are conserved, organic matter breaks down more slowly, and less odors and volatile compounds are emitted.

As a matter of convenience, scientists have subdivided and given names to the ranges of temperatures within which certain microorganisms thrive. Composting principally takes place within the two ranges known as mesophilic, approximately 20–45°C (68–113°F), and thermophilic, 45–75°C (113–167°F). The range below 20°C (68°F) is called *psychrophilic*. The boundaries between the ranges are not exact, nor important. The relevant point is that different species of microorganism thrive in different temperature regimes.

Composting can take place effectively at mesophilic temperatures, as it typically does in backyard composting piles. In fact, some evidence suggests that, for the same degree of compost maturity, more carbon and nitrogen are conserved when composting takes place under mesophilic temperatures compared to thermophilic (Adler and Sikora, 2005). Mesophilic microorganisms are better able to perform certain biochemical tasks, like degrading organic acids and converting ammonia to nitrate.

Nevertheless, large-scale composting is generally managed with the goal of attaining and maintaining temperatures in the thermophilic range for a prolonged time period. In addition to faster decomposition, thermophilic temperatures are desirable because they destroy more pathogens, weed seeds and fly larvae present in the composting materials than mesophilic temperatures alone. Research has shown that human and animal pathogens are reduced by more than 99.999% after three days of composting at 55°C (131°F). This temperature should destroy most plant pathogens as well. Therefore, various national, state, and provincial composting regulations set the critical temperature for killing human pathogens during composting to be 55°C (131°F) or higher, when maintained for a specified time period. This pathogen-destroying process is generally termed "sanitization." In the United States, temperature/time sanitization requirements are known by the acronym PFRP (Process to Further Reduce Pathogens) (Box 3.4).

An effective temperature target for destroying weed seeds is 60°C (140°F). The critical temperature depends on the weed species, the time of exposure, and moisture (dry conditions tend to protect seeds). Most weed seeds cannot survive the 55°C threshold level for even a few hours. The vast majority of seeds are destroyed within an hour between 60°C and 70°C (158°F) (Dahlquist et al., 2007). However, with a few especially hardy weed species, a percentage of seeds can remain viable even after weeks of composting at these high temperatures.

60°C (140°F) is often cited as the optimal composting temperature for microbial diversity and general rate of decomposition. However, the optimal temperature is likely a moving target that changes with the phase of composting, specific feedstocks, and desired final product qualities. For example, high temperatures, near 75°C (167°F), have been reported to enhance the decomposition of feedstocks of fats, oils, or waxes. Experience and research indicate that temperatures in the range of 50–70°C range (roughly 120 to 160°F) generally work well. As a rule of thumb, to speed the process and promote decomposition and evaporation, one should

Box 3.4 Sanitization and the "Process to Further Reduce Pathogens (PFRP)"

A key function of composting is "sanitization" (aka "pasteurization")—producing compost that is effectively devoid of pathogens, organisms that cause disease. Human pathogens are of greatest concern, but the destruction of animal and plant pathogens is also important. The goal is to diminish pathogen numbers to a level small enough that the risk of passing on a disease is very low, *acceptably* very low. Acceptably low results might require the destruction 99.9% to 99.999% of a particular pathogen.

Composting kills pathogens primarily with heat. The harsh environment of a composting pile, including competition and predation by other organisms, also destroy pathogens, but high temperature is the most certain and effective killer. The US EPA considers 55°C (131°F) the threshold temperature for killing pathogens via composting. Other countries set the bar a bit higher, at 60°C, and even up to 65°C (149°F) for certain feedstocks. However, simply reaching the lethal temperature threshold is not enough. To have its desired effect, lethal temperatures must be attained throughout the composting mass, and maintained for a sufficient length of time. Therefore, composting regulations that govern sanitization typically specify a minimum temperature *and* a minimum time of exposure at that temperature. Some regulations also specify a minimum moisture content to maintain during the sanitization period because dry materials interfere with heat transfer. Regulations also address the likelihood that composting piles and windrows are not uniformly hot; cool spots can endure at the margins.

Different countries, provinces, and states have varying rules and regulations regarding how composters accomplish and document sanitization. Although the specific requirements vary, the temperature/time model is almost universally followed. The reader should research the specific regulations in his/her own political jurisdictions.

Sanitization requirements for composting are epitomized by the set of rules that underlies regulations in the United States. These rules are known by the acronym PFRP. The PFPR rules were established by the US EPA as one of several options for the treatment of wastewater biosolids. In actuality, composting is only one of the "processes" specified to "further reduce" pathogens. However, the composting part of these rules have since become the basis for regulations and/or practices regarding composting of many feedstocks. When composters say that they are meeting PFRP, they mean that they are meeting the requirements for composting under the broader PFRP rules.

The US EPA's PFRP criteria depend on the method of composting.
- For composting by either the aerated static pile or an in-vessel methods, the temperature must be maintained at or above 55°C or for three days.
- For composting by the turned windrow method, the temperature must be maintained at 55°C or higher for 15 days or longer. Also, the windrow must be turned a minimum of five times during the period that the temperature remains at 55°C or higher.

Turned windrows merit a longer time is because they tend to be smaller in size. A large portion of the material is at or near the exposed exterior surface and remains cool even if the interior is hot. Turning the windrow five times ensures that all of the material in the windrow spends time within the hot core. In contrast, aerated static piles and in-vessel systems are considered to be large enough or well-insulated enough to experience high temperatures throughout.

In the United States, implementation and oversight of the PFRP rules are the responsibility of individual states. State environmental agencies can impose stricter criteria and additional requirements, such as minimum temperature monitoring and record keeping provisions. On a national level, the PFRP rules apply only to biosolids and septage. In many states, composters who do not handle these feedstocks are not compelled to meet PFRP rules, though it is a good idea to do so anyway. However, most states also require PFRP compliance with higher-risk feedstocks like municipal solid wastes and postconsumer food waste. A few states have expanded the PFRP burden to nearly all feedstocks, including yard trimmings. Fortunately, it is easy for most composting piles and windrows to reach 55°C, even 65°C, and stay at that level for several weeks.

operate at the higher end of this temperature range. To slow things down, conserve nutrients and limit odors, one should maintain lower temperatures, that is, *after* the compost is sanitized of pathogens and weed seeds, if and as necessary.

The temperatures of a compost pile are determined by both the amount of heat generated *and* the amount of heat lost to the surrounding environment. When heat generation exceeds heat loss, the temperatures increase. When heat loss prevails, the temperatures decline. Organic matter decomposition by microorganisms inherently releases large amounts of energy as heat. The self-insulating qualities of the composting materials lead to an accumulation of heat, which initially raises the temperature. At the same time, the materials continuously lose heat as water evaporates and as air movement carries away the water vapor and other warm gases. Aeration accelerates heat loss and, therefore, is used to maintain temperatures in the desired range. Cold weather and small piles also promote heat loss.

Heat accumulation can push temperatures well above 60°C (140°F). When this occurs, microorganisms begin to suffer the effects of high temperature, and the composting process slows. The temperature can continue to rise above 70°C (about 160°F) because of heat generated by ongoing microbial activity and the insulating qualities of the composting materials. At temperatures approaching 80°C (175°F), many microorganisms die or become dormant (e.g., form heat-resistant spores). The process effectively stops and does not recover until the population of microorganisms recovers. To prevent this situation, temperature should be monitored. When the temperature approaches 160°F (71°C), heat loss should be accelerated by increasing aeration or by reducing pile size. Turning does not reliably lower temperatures. If a pile is set back by high temperature, the recovery may be quickened by remixing the pile, either with the parts of the pile that are not as hot, like the outer edges, or, in extreme cases, with material from other more active batches.

Maintaining adequate moisture is a key to limiting temperature rises in a pile. Most of the heat loss in composting occurs by the evaporation of water. Thus, moisture serves as a temperature buffer. The pile should not be allowed to dry below 40% moisture. Very low moisture, below 40%, increases the chance of damaging high temperatures as well as spontaneous combustion (Chapter 14).

4.11 Time

The length of time required to transform raw feedstocks into usable compost depends upon many factors including the feedstocks themselves, temperature, moisture, frequency of turning, effectiveness of aeration, and the intended use of the compost. Proper moisture content, a suitably low C:N ratio, regular agitation plus adequate aeration ensure the shortest possible composting period. Conditions which slow the process include lack of moisture, excessive moisture, a high C:N ratio, insufficient aeration, large particles, and a high percentage of feedstocks that are resistant to biodegradation, such as woody materials. The required composting period also depends on the intended use of the compost. It can be shortened if the compost does not need to be fully mature. For instance, if the compost is to be applied to cropland well before the growing season, it can cure in the field.

In general, the entire decomposition and stabilization of materials, from feedstock to usable compost product, is typically accomplished in three months to a year. Some compost producers prefer to manipulate the process as little possible and wait as long as two years to harvest compost. Under tightly controlled conditions, it is possible to produce useable compost within one month but a minimum period of 10 weeks is more reasonable. Typical composting times for several common applications are given in Table 3.3.

Although some in-vessel composting systems may claim less than one week to produce compost, an additional composting phase, lasting four to eight weeks, is required before the compost is suitable for general use. Also, a given process might achieve stabilization quickly by drying the materials to a low moisture content,

Table 3.3 Typical composting times for selected methods and feedstocks.

Composting method	Example feedstocks	Primary composting period		Curing period
		Range	Typical	
Passive composting (little or no turning)[a]	Leaves	1½–3 years	1½ years	—
	Well-bedded manure	6 months to 2 years	1 year	—
Windrow—infrequent turning[b]	Leaves	6 months to 1 year	9 months	4 months
	Manure + amendments[e]	4–8 months	6 months	1–2 months
Windrow –frequent turning[c]	Manure, food scraps + amendments	1–4 months	2 months	1–2 months
Passively aerated windrow	Well-bedded manure	10–12 weeks	—	1–2 months
	Fish scraps + peat moss	8–10 weeks	—	1–2 months
Aerated static pile	Biosolids, manure food scraps + amendments	3–5 weeks	1 month	1–2 months
Agitated bed	Biosolids, manure food scraps + amendments	2–4 weeks	3 weeks	1–2 months
Rotating drums	Biosolids + MSW manures	3–8 days	3 days	2 months[d]

[a] Little or no turning = static piles that may be moved with a bucket loader once or twice.
[b] Infrequent turner—typified by turning with a bucket loader or excavator every other month on average.
[c] Frequent turner—typified by turning with a windrow turner every 2–4 weeks on average.
[d] These methods involve a second stage of composting in windrows or aerated piles.
[e] Amendments—feedstocks used to balance moisture, bulk density, or other parameters, e.g., wood chips added to biosolids or yard debris added to food scraps.

which inhibits biological activity. This is fine if the end use for the compost does not dictate biologically stable compost. Partially stabilized compost, however, is not suitable for most uses. It is also important to recognize that as the dried material regains moisture, biological activity resumes. Odors, pathogen regrowth, and other problems can then develop if the material is not stabilized.

5. Curing

Curing is an extremely important but too-often neglected stage of composting during which compost matures. Curing occurs at lower, mesophilic temperatures. The oxygen consumption, heat generation, and moisture evaporation are much lower than in the active composting stage. There is no precise point at which curing should begin (or end). Conceivably, the curing stage begins when a windrow or pile no longer reheats after turning. With forced aeration, curing begins after the pile temperature shows a steady decrease and approaches mesophilic levels (40°C or 105°F), and air is no longer required for cooling, assuming there is adequate moisture.

Curing furthers the aerobic decomposition of resistant compounds, organic acids, large particles, and clumps of material that remain after active composting. Some fungi and actinomycetes that can breakdown cellulosic compounds are only active in the mesophilic temperature regimes of curing. As a result, the C:N ratio decreases, the exchange capacity increases, and the concentration of humic compounds increase. Some desired changes take place only at low temperatures or within well-decomposed organic matter, which is not present during thermophilic composting. One example is nitrification, or the conversion of ammonia to nitrate-nitrogen. Nitrification only becomes noticeable during the curing stage. Another is the recolonization of the pile by soil organisms, which can give the compost disease-suppressing qualities. The development of humus-like compounds is also believed to occur more readily at these conditions.

The importance of curing, and its duration, increases if the active composting stage is either shortened or poorly managed. A long curing period provides a safety net that helps to overcome the shortcomings of the composting method and also reduces the chance that immature compost will be used. Intermediary compounds produced during thermophilic decomposition, some of which are phytotoxic,[4] are fully broken-down during curing. Immature compost continues to consume oxygen and thereby reduces the availability of oxygen to the plant roots. It can also contain high levels of organic acids and ammonium, a high C:N ratio, and other characteristics which can be damaging when the compost is used for certain horticultural or agricultural applications. In principle, the duration of curing should be determined

---

[4] "Phytotoxic" means harmful to plants.

by the desired maturity level of the compost. The prevailing rule of thumb recommends a minimum curing time of one month.

Because curing continues the aerobic decomposition process, adequate aeration remains a necessity. The continued need for oxygen limits the size and moisture content of the curing piles. Compost, or pockets of compost, that becomes anaerobic within the curing piles develop some of the same detrimental qualities found in immature compost. However, because curing is less biologically active than the thermophilic phase, oxygen demand is much less, so curing piles can be larger than their predecessors without becoming anaerobic. If anaerobic conditions do occur in curing piles, the resulting anaerobic products can be degraded relatively quickly when the cured compost is exposed to air. It is a good practice to move compost to smaller curing or storage piles, and allow them to aerate, before using or selling the compost.

Shaffer (2010) presents an excellent explanation of the importance of curing and its role in producing compost that is fit for its intended use.

6. When is it done?

The question of when the compost process is finished vexes many novice composters. The process is dynamic and complex. As noted above, temperature alone is not sufficient because a drop in temperature may only indicate the transition from the thermophilic phase to curing, or it may indicate one of the key parameters, such as oxygen or moisture, has become limiting. Part of the challenge is that the answer depends on the intended use for the compost. Some applications require a more-finished compost; others can use immature compost. Two terms used to describe a compost's state of completion are stability and maturity (see Chapter 15).

Stability refers to the biological activity of the compost. As the available food for microbial growth and development diminishes, populations decline. This condition is accompanied by a decrease in respiration, reducing how much oxygen is used and carbon dioxide is given off. Either of these gases can be used as a measure of stability. However, respiration can also decrease because other growth conditions become unfavorable. For example, a pile that gets too dry may appear stable as measured by CO_2 release, but will become more active when remoistened. Using this compost could lead to nitrogen immobilization or oxygen depletion in the soil that can inhibit or damage plant growth.

Maturity is a related but more complex and less well-defined concept that generally describes the degree of doneness of compost. Since rapidly decomposing organic matter produces compounds that can be phytotoxic, maturity is typically interpreted as the ability of compost to support plant growth. Mature compost presents little risk of plant damage in almost any situation. Immature composts may damage some plants in some situations. Because of the problems mentioned above with stability, mature compost is stable but stable compost may not be mature. Moreover, it is still possible for a mature compost to cause harm to some plants for reasons other than the degree of decomposition, such as a high soluble salt concentration or the presence of persistent herbicides.

Table 3.4 Characteristics of compost maturity levels.

Maturity level	Traits	End use examples
Immature	May produce odors Significant potential to inhibit plant growth	Amend fallow soil Add organic matter to depleted soils
	Significant potential to impact nitrogen availability in soil	Feedstock for compost Mulch for weed control in orchards
Mature	Unlikely to produce odors Limited potential to inhibit plant growth Minimal impact on soil nitrogen	General field use Vineyards, row crops Garden or landscape bed amendment
Very mature	No malodors No potential to inhibit plant growth	Container plant mixes Turfgrass topdressing

Adapted from CalRecycle. 2003.

No single test parameter is sufficient to measure maturity, though many have been proposed (Chapter 15). Most tests for maturity include a measure of stability. The surest way to tell if compost has reached adequate maturity is to grow plants in it. This is called a bioassay, a procedure that follows testing protocols that include optimizing moisture, seed selection, lighting, and other growth factors, and compares the results to a known control. As noted, not all uses require mature compost (Table 3.4).

7. Composting microbiology[5]

The composting process occurs through the work of microorganisms, mainly bacteria and fungi. A single teaspoon of active compost can contain 10 billion bacteria. Tending the compost pile has been called the "care and feeding of the microherd." Just as a farmer provides the right environment (food, water, and shelter) for a herd of livestock to thrive, so a composter provides the right conditions for the microbial "herd" to do their work breaking down and stabilizing the feedstocks.

The microorganisms that are at work in a compost pile are extremely diverse, with perhaps 1000's of species at work at various times in the process. Some microbes are fairly specialized in the food and conditions to which they are adapted while others can use many different substrates and survive under a broad range of

---

[5] The book, *Microbes at Work: From Wastes to Resources* (Insam et al., 2010), presents much more information about the types and functions of microorganisms in composting.

conditions. The compost pile usually has a variety of local environments that change as the process advances, so the mix of microbial communities is always shifting.

Bacteria and fungi are the primary decomposing organisms (Figs. 3.21–3.23). Bacteria are by far the most diverse organisms on earth having evolved over four billion years. They are small (0.5 to 3 microns), simple single-celled organisms that exist in a wide variety of forms. They are often subdivided based on their genetic sequences or into groups along functional lines, including aerobes, anaerobes, actinomycetes, and nitrogen fixers. Fungi are less diverse and include yeasts and molds, some of which produce mushrooms. They are larger than bacteria (10 to 50 microns). Yeasts are unicellular. Molds form long multicellular filaments called mycelia. As a group, fungi are more tolerant of dry conditions and lower nitrogen levels but less tolerant of anaerobic situations and high temperatures. Actinomycetes are a type of bacteria that form multicellular filaments like fungal mycelia. These networks of cells allow both actinomycetes and molds to "forage" farther for nutrients than single-celled bacteria can access from their surroundings. However, turning and other forms of agitation can disrupt and distribute the networks of mycelia and filaments of actinomycetes.

Bacteria clearly dominate the composting realm. Their numbers are typically about 100 times greater than fungi. Bacteria flourish particularly well when the conditions are good for decomposition—abundant moisture, easily degradable feedstocks, and thermophilic temperatures. Overall numbers are highest at the beginning of the process and decline as organic matter becomes more stable. Fungi, and actinomycetes, gain some ground near the end of the process when the tougher organic compounds remain.

New molecular techniques in microbial identification and analysis have greatly increased our knowledge of the microbiology of composting. For example, Fig. 3.24 shows the relative numbers of bacterial and fungal classes found in 10 different compost products by sequencing conserved parts of the DNA that makes up their

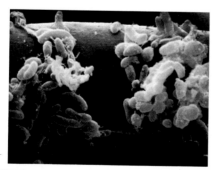
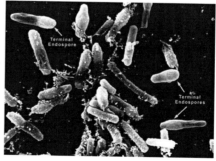

FIGURE 3.21

Thermophilic *Bacilli*, the genus of bacteria most commonly isolated by culturing from active thermophilic composts. Note the characteristic spores.

Permission granted by Fred Michel and the USCC.

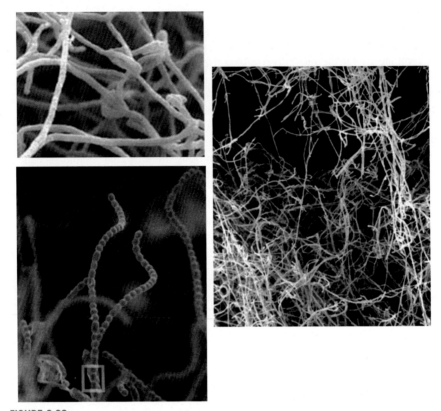

FIGURE 3.22

Actinomycetes—a type of bacteria found in mature composts that forms a network of threads or mycelia.

Courtesy of Fred Michel and the USCC.

genomes and comparing the sequences to databases of similar genes (Michel, 2020). Genome sequencing has greatly expanded and simplified our ability to identify microorganisms in composts compared to the microbial culturing techniques used in the past. However culturing and other methods can provide complimentary information about the specific capabilities and processes that microorganisms carry out.

These molecular methods of identification have revealed that the diversity of microorganisms at both the phylum and species level is significantly greater than previously understood. This diversity is what gives the composting process the ability to adapt to changing conditions. It also holds out the promise of discovering strains of microorganisms that will be particularly suited to advancing work in disease control, bioremediation, enzyme discovery, and other biotechnologies.

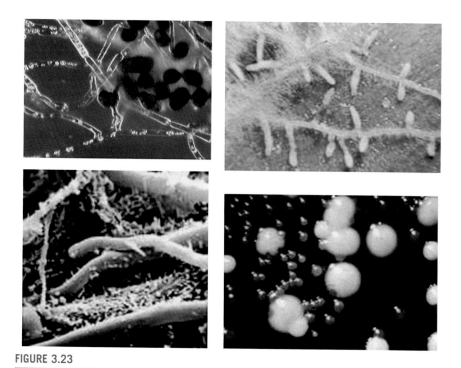

FIGURE 3.23

Fungi—molds and yeasts (bottom right photo). Fungi are more temperature sensitive than bacteria.

Courtesy of Fred Michel and the USCC.

In addition to bacteria and fungi, other organisms participate in the composting process but to a much lesser extent and primarily at the later stages. They are more prominent in backyard composting situations and other applications where temperatures are largely mesophilic. These organisms include protozoa, algae, nematodes, and macroorganisms including worms, mites, millipedes, insects, and similar creatures (CWMI, 1996). Some of these organisms, like protozoa, prey on bacteria and fungi, helping to incorporate the biomass into the compost. Although they do not play a large role, algae can colonize moist surfaces of pile, use nutrients from the pile, and add biomass through photosynthesis. Larger organisms, such as worms and insects, produce both biological and physical degradation when temperatures remain at mesophilic levels, mostly at the end of the process. The high temperatures attained during the early and middle phases of composting are lethal to these organisms. Earthworms, fly larvae, and some species of beetles are able to transform raw feedstocks into a valuable compost-like product, but these processes are different from conventional composting and occur at mesophilic temperatures. For the most part, composting relies on the services of microorganisms—principally bacteria and fungi—to manufacture compost.

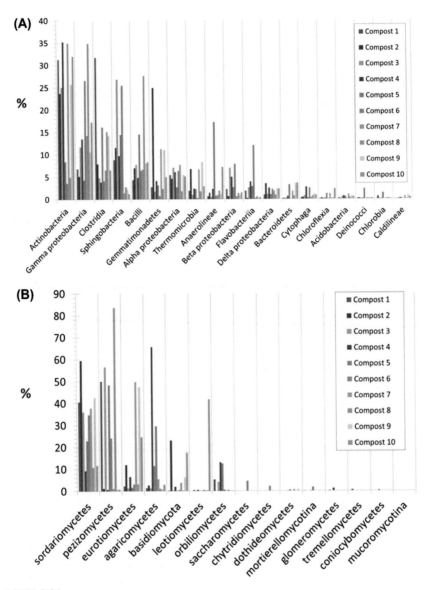

FIGURE 3.24

Prominent classes, of bacteria (A) and fungi (B) and their relative abundance in 10 different mature composts, identified using ribosomal 16S DNA and intergenic spacer region (ITS) amplification and sequencing.

Permission granted by Fred Michel (Michel, 2020).

7.1 Microbial functions

There are a number of important processes that microorganisms perform in the compost process. These include:

- Decomposition: breakdown of plant and animal remains into simpler parts,
- Mineralization: conversion of organic compounds into inorganic ions (e.g., nitrate, phosphate, ammonium, carbon dioxide, hydrogen sulfide, etc.),
- Immobilization: incorporation of water-soluble inorganic and organic molecules into more complex molecules and microbial cells,
- Humification: conversion of lignin, proteins, and other organic fractions into humic substances (large molecules resistant to further decay).

As the composting process proceeds, there is a succession of microorganisms based on food availability and temperature. When the pile is first formed, simple compounds such as proteins, sugars, and starches are consumed by mesophilic bacteria. Heat generated in the breakdown process is trapped in the pile and the temperature rises, allowing thermophilic bacteria to increase at the expense of the mesophiles. The thermophiles go to work on the proteins and fats, which are also readily degradable. This period is the time of greatest oxygen demand. As oxygen levels decrease, or are locally depleted, facultative bacteria switch from aerobic to anaerobic respiration. Those species that require oxygen die or become dormant while the anaerobes increase. If and when oxygen becomes more available, the reverse occurs.

Meanwhile, other thermophilic bacteria work on more resistant compounds, such as hemicellulose, pectin, cellulose, and chitin. Cellulosic compounds are the building blocks of plants; chitin is a structural component in the cells of fungi, insects, and crustaceans. Finally, as the proteins and other readily degraded food sources are exhausted and temperatures drop, the fungal activity increases relative to the bacteria. Fungi work with bacteria, especially actinomycetes, to complete the breakdown of cellulose and begin to work on the most resistant compound, lignin. This stage is also when ammonia is converted to nitrate (nitrification) and free-living (as opposed to symbiotic) nitrogen-fixing bacteria can proliferate, as readily available nitrogen compounds are depleted.

The concentration of the various classes of microorganisms in mature compost depends on the initial feedstocks and the management of the process, and can vary by many orders of magnitude.

7.2 Microbial sources

The microorganisms needed for composting are found throughout the natural environment. They are present as spores or dormant cells on virtually all the feedstocks, as well as in the soil and air. Some manufacturers and some experienced composters promote the use of an inoculant at the start of composting, adding specific bacteria and fungi that help the process proceed faster or with fewer odors or to produce compost with specific qualities. There is little scientific evidence to support this practice, as these additions are usually overwhelmed by the naturally occurring microorganisms. Inoculation may be most effective for feedstocks that have been treated to reduce their microbial numbers; for example, cooked or otherwise preserved foods. In any case, the best inoculant may be a small amount of finished or immature compost made from similar feedstocks. Unlike an inoculant made from a standard recipe, this compost already contains organisms specifically adapted to the conditions and feedstocks of the composting system at hand.

There has been some research to support the use of certain inoculants during the curing phase, particularly to impart a specific disease-suppressive ability to the compost. However, this requires very tight control and generally has not proven to be as consistent as needed to be economically viable. See Chapter 17 for more information on using compost for disease suppression.

References

Cited references

Adler, P.R., Sikora, L.J., 2005. Mesophilic Composting of arctic char manure. Compost Sci. Util. 13 (1), 34—42. https://doi.org/10.1080/1065657X.2005.10702215.

Alburquerque, J.A., McCartney, D., Yu, S., Brown, L., Leonard, J.J., 2008. Air space in composting research: a literature review. Compost Sci. Util. 16 (3), 159—170. https://doi.org/10.1080/1065657X.2008.10702374.

Böhm, R., 2007. Pathogenic agents. In: Diaz, et al. (Eds.), Compost Science and Technology. Elsevier Science. ISBN 10 0080439608/ISBN 13:9780080439600.

Breitenbeck, G., Schellinger, D., 2004. Calculating the reduction in material mass and volume during composting. Compost Sci. Util. 12 (4), 365—371.

Brown, K.H., Bouwkamp, J.C., Gouin, F.R., 1998. The influence of C:P ratio on the biological degradation of municipal solid waste. Compost Sci. Util. 6 (1), 53—58. https://doi.org/10.1080/1065657X.1998.10701909.

Buckner, S.C., 2002. Effects of turning frequency and mixture composition on process conditions and odor concentrations during grass composting. In: Michel, F.C., Rynk, R., Hoitink, H.A.J. (Eds.), 2002. International Compost Symposium Proceedings. The JG Press, Inc., Emmaus, PA, pp. 251—280. www.jgpress.com.

CalRecycle, 2003. The Importance of Compost Maturity. CIWMB Pub. #443-03-007. https://www2.calrecycle.ca.gov.

Dahlquist, R.M., Prather, T.S., Stapleton, J.J., 2007. Time and temperature requirements for weed seed thermal death. Weed Sci. 55, 619—625.

Day, M., Shaw, K., 2001. Biological, chemical, and physical processes in composting. In: Stofella, P.J., Kahn, B.A. (Eds.), Compost Utilization in Horticultural Cropping Systems. Lewis Publishers, Boca Raton, FL.

Epstein, E., 1997. The Science of Composting. Technomic Publishing Co., Inc. Bassel, Switzerland.

Francis, J., Larney, F.J., Olson, A.F., Carcamo, A.A., Chang, C., 2000. Physical changes during active and passive composting of beef feedlot manure in winter and summer. Bioresour. Technol. 75 (2), 139–148.

Gasser, J.K.R. (Ed.), 1985. Composting of Agricultural and Other Wastes. Elsevier Applied Science Publishers, London and New York. x + 320 pp. ISBN 0-85334-357-8. Biomass, 9, 318.

Gouleke, C.G., 1972. Composting: A Study of the Process and its Principles. Rodale Press, Inc. Emmaus, PA.

Haug, R.T., 1980. Com Post Engineering, Principles and Practices. Ann Arbor Science Publishers Inc, M ichagen.

Haug, R.T., 1993. The Practical Handbook of Compost Engineering. CRC Press, Boca Raton, FL.

Hoitink, H.A.J., Keener, H. (Eds.), 1993. Science and engineering of composting. Proceedings of an International Composting Symposium, March, 1992. Renaissance Publications, Columbus, OH; Worthington, OH.

Insam, H., Franke-Whittle, I., Goberna, M., 2010. Microbes at Work: From Wastes to Resources. Springer Science & Business Media, p. 329. https://doi.org/10.1007/978-3-642-04043-6. Springer-Verlag Berlin Heidelberg 2010. Print ISBN 978-3-642-04042-9 Online ISBN 978-3-642-04043-6.

Jeris, J.S., Regan, R.W., 1973. Controlling environmental parameters for optimal composting. Parts I, II, III. Compost Sci. 14 (1), 10–15, 14(2) 8-15; 14(3) 16-22.

Journal of Compost Science and Utilization. (Quarterly Journal) JG Press, Emmaus, PA.

Larney, F.J., Olson, A.F., Miller, J.J., Demaere, P.R., Zvomuya, F., McAllister, T.A., 2008. Physical and chemical changes during composting of wood chip-bedded and straw-bedded beef cattle feedlot manure. J. Environ. Qual. 37 (2), 725–735.

Larsen, K.L., McCartney, D.M., 2000. Effect of C:N ratio on microbial activity and N retention: bench-scale study using pulp and paper biosolids. Compost Sci. Util. 8 (2), 147–159.

McCartney, D.M., Chen, H., 2001. Using a biocell to measure effect of compressive settlement on free air space and microbial activity in windrow composting. Compost Sci. Util. 9 (4), 285–304.

Michel, F., 2020. Unpublished data from the author at. The Ohio State University.

Michel Jr., F.C., Forney, L.J., Huang, A.J.-F., Drew, S., Czuprenski, M., Lindeberg, J.D., Reddy, C.A., 1996. Effects of turning frequency, leaves to grass ratio, and windrow vs. pile configuration on the composting of yard trimmings. Compost Sci. Util. 4 (1), 26–43.

Michel, F.C., Quensen, J., Reddy, C.A., 2001. Bioremediation of a PCB-contaminated soil via composting. Compos. Sci. Util. 9 (4), 274–284.

Michel Jr., F.C., Reddy, C.A., 1998. Effect of oxygenation level on yard trimmings composting rate, odor production and compost quality in bench scale reactors. Compost Sci. Util. 6 (4), 6–14.

Michel, F., Tirado, S., 2010. Effects of turning frequency, windrow size and season on the production of dairy manure/sawdust composts. Compost Sci. Util. 18, 70–80. https://doi.org/10.1080/1065657X.2010.10736938.

Paul, J., 2017a. Killing potential pathogens in compost – cold and calculated. Trans. Compost. Syst. Blog. http://www.transformcompostsystems.com/blog/2017/11/02/killing-potential-pathogens-in-compost-cold-and-calculated/.

Paul, J., 2017b. Pathogen kill – more than just high temperatures. Trans. Compost. Syst. Blog. http://www.transformcompostsystems.com/blog/2017/09/04/pathogen-kill-more-than-just-high-temperatures/.

Poincelot, R.P., 1975. The Biochemistry of Composting. The Connecticut Agricultural Experiment Station, New Haven, CT.

Richard, T., 1996. Inadequate Porosity. Cornell Waste Management Institute. http://compost.css.cornell.edu/odors/inadeq.porosity.html.

Rodale, J.I. (Ed.), 1971. The Complete Book of Composting. Rodale Books, Inc. Emmaus, PA.

Ruggieri, L.T.G., Artola, A., Sánchez, A., 2009. Air filled porosity measurements by air pycnometry in the composting process: a review and a correlation analysis. Bioresour. Technol. 100, 2655–2666.

Shaffer, B., 2010. Curing compost: an antidote for thermal processing. Acres, U.S.A. 40 (11). November 2010.

Siebert, S., Auweele, W.V., 2018. European Quality Assurance Scheme for Compost and Digestate. European Compost Network ECN e.V., Germany.

Strom, P.F., 1985. Effect of temperature on bacterial species diversity in thermophilic solid - waste composting. Appl. Environ. Microbiol. 50 (4), 899–905.

Sunberg, C., 2005. Improving Compost Process Efficiency by Controlling Aeration, Temperature and pH. Doctoral thesis. Swedish University of Agricultural Sciences, Uppsala. ISSN 1652-6880, ISBN 91-576-6902-3.

Van Ginkel, J.T., Van Haneghem, I.A., Raats, P.A.C., 2002. Physical properties of composting material: gas permeability, oxygen diffusion coefficient and thermal conductivity. Biosyst. Eng. 81 (1), 113–125.

Willson, G.B., 1980. Manual for Composting Sewage Sludge by the Aerated Static Pile Method. USEPA, Cincinnati, OH.

Yang, X., Liu, E., Zhu, X., Wang, H., Liu, H., Liu, X., Dong, W., 2019. Impact of composting methods on nitrogen retention and losses during dairy manure composting. Int. J. Environ. Res. Publ. Health 16 (18), 3324. https://doi.org/10.3390/ijerph16183324.

Consulted and suggested references

Agnew, J.M., Leonard, J.J., 2003. The physical properties of compost. Compost Sci. Util. 11 (3), 238–264.

Antunes, L., Martins, L., Pereira, R., et al., 2016. Microbial community structure and dynamics in thermophilic composting viewed through metagenomics and metatranscriptomics. Sci. Rep. 6, 38915. https://doi.org/10.1038/srep38915.

Bell, R.G., Pos, J., Lyon, R.J., 1973. Production of compost from soft wood lumber mill wastes. Compost Sci. 14 (2), 5–7.

Bernai, M.P., Paredes, C., Sanchez-Mondero, M.A., Cegarra, J., 1998. Maturity and stability parameters of composts prepared with a wide range of organic wastes. Bioresour. Technol. 63, 91–99.

Biddlestone, A.J., Gray, K.R., 1973. Composting - application to municipal and farm wastes. Chem. Eng. Lond. 270, 76–79.

Blanc, M., Marilley, L., Beffa, T., Aragno, M., 1999. Thermophilic bacterial communities in hot composts as revealed by most probable number counts and molecular (16S rDNA) methods. Fed. Eur. Microbiol. Soc. Microbiol. Ecol. 28 (2), 141–149. https://doi.org/10.1111/j.1574-6941.1999.tb00569.x.

CWMI (Cornell Waste Management Institute), 1996. Invertebrates of the Compost Pile. Cornell University, Ithaca, NY. http://compost.css.cornell.edu/invertebrates.html.

Das, K., Keener, H.M., 1997. Moisture effect on compaction and permeability in composts. J. Environ. Eng. 123 (3), 275–281.

Das, K., Keener, H.M., 1996. Process control based on dynamic properties in composting: moisture and compaction considerations. In: de Bertodli, M., et al. (Eds.), The Science of Composting, Proceedings of the European Commission International Symposium, September, 1995. Blackie Academic & Professional, London, U.K, pp. 116–125.

Das, K.C., Tollner, E., 2003. Comparison between composting of untreated and anaerobically treated paper mill sludges. Trans. Am. Soc. Agric. Eng. 46 (2), 475–481.

Day, D.L., Krzymien, M., Shaw, K., Zaremba, W.R., Wilson, C., Botden, C., Thomas, B., 1998. An investigation of the chemical and physical changes occurring during commercial composting. Compost Sci. Util. 6 (2), 44–66.

The science of composting. In: de Betoldi, M., Sequi, P., Lemmes, B., Papi, T. (Eds.), 1996. Proceedings of the European Commission International Symposium, September, 1995. Blackie Academic & Professional, London, U.K.

Diaz, L.F., de Bertoldi, M., Bidlingmaier, W., 2007. Compost Science and Technology. Elsevier Science. ISBN 10 0080439608/ISBN 13:9780080439600.

Finstein, M.S., Miller, F.C., MacGregor, F.C., Psarianos, K.M., 1985. The Rutgers Strategy for Composting: Process Design and Control. Rutgers University, New Brunswick, NJ.

Finstein, M.S., Miller, F.C., Strom, P.F., MacGregor, S.T., Psarianos, K.M., 1983. Composting Ecosystem Management for Waste Treatment. Bio/Technology, June edition, pp. 3 4 7–353.

Girard, D., August 7, 2018. Particle sizing for composting. Biocycle 59 (7).

Golueke, C.G., Card, B.J., McGauhey, P.H., 1954. A critical evaluation of inoculums in composting. Appl. Microbiol. 2, 45–53.

Gray, K.R., Biddlestone, A.J., 1974. Decomposition of urban waste. In: Dickson, C.H., Pugh, G.J.F. (Eds.), Biology of Plant Litter Decomposition, vol. 2. Academic Press, London, pp. 743–755.

Gray, K.R., Biddlestone, A.J., Clark, R., 1973. A review of composting Part 3. Process Biochem. Oct. 1973.

Gray, K.R., Sherman, K., Biddlestone, A.J., 1971. A review of composting Parts 1 and 2. Process Biochem. June 1971 and Oct. 1971.

Green, S.J., Michel Jr., F.C., Hadar, Y., Minz, D., April 1, 2004. Similarity of bacterial communities in sawdust- and straw-amended cow manure composts. FEMS Microbiol. Lett. 233 (1), 115–123. https://doi.org/10.1016/j.femsle.2004.01.049. PMID: 15043877.

Hewings, G., 2007. Design and Management of Composting Systems. Doctoral Dissertation. Cardiff School of Engineering, Cardiff University. Published by ProQuest LLC.

Howard, A., 1935. The manufacture of humus by the Indore process. J. Roy. Soc. Arts 82, 25.

Hyatt, G.W., Richard, T.L. (Eds.), 1992. Biomass Bioenergy 3 (3–4), 121–299.

Jain, M.S., Daga, M., Kalamdhad, A.S., A.S, 2019. Variation in the key indicators during composting of municipal solid organic wastes. Sustain. Environ. Res. 29, 9. https://doi.org/10.1186/s42834-019-0012-9.

Keener, H.M., Ekinci, K., Michel, F.C., 2005. Composting process optimization - using on/off controls. Compost Sci. Util. 13 (4), 288–299.

Keener, H.M., Hansen, R.C., Elwell, D.L., 1997. Airflow through compost: design and cost implications, 13 (3), 377–384.

Keener, H.M., Marugg, C., Hansen, R.C., Hoitink, H.A.J., 1993. In: Hoitink, H.A.J., Keener, H.M. (Eds.), Optimizing the Efficiency of the Composting Process. Science & Engineering of Composting: Design, Environmental & Microbial & Utilization Aspects. Renaissance Publications, Worthington, USA, pp. 59–94.

Lopez-Real, J., Baptisa, M., 1996. A preliminary study of three manure composting systems and their influence on process parameters and methane emissions. Compost Sci. Util. 4 (3), 71–82.

Lynch, J.M., 1993. Substrate availability in the production of composts. p.24-35. In: Science and Engineering of Composting. In: Hoitink, H.A.J., Keener, H. (Eds.), Proceedings of an International Composting Symposium, March, 1992. Renaissance Publications, Columbus, OH. Wothington, OH.

MacGregor, S.T., Miller, F.C., Psarianos, K.M., Finstein, M.S., 1981. Composting process control based on interaction between microbial heat output and temperature. Appl. Environ. Microbiol. 41 (6), 1321−1330.

Martin, J.P., Wang, Y., 1944. Utilization of plant residues for the production of artificial manures. J. Am. Soc. Agron. 36, 373−385.

Mason, I.G., 2007. An Evaluation of Substrate Degradation Patterns in the Composting Process. Part 1: Profiles at Constant Temperature. Waste Management.

Mason, I.G., Milke, M.W., 2005. Physical modelling of the composting environment: a review. Part 1: React. Syst. Waste Manag. 25 (5), 481−500.

Mathur, S.P., Patni, N.K., Lévesque, M.P., 1990. Static pile, passive aeration composting of manure slurries using peat as a bulking agent. Biol. Waste 34 (4), 323−333. https://doi.org/10.1016/0269-7483(90)90033-O. ISSN 0269-7483.

McCartney, D.M., Chen, H., 1999. Physical modeling of composting using biological load cells: effect of compressive settlement on free air space and microbial activity. P. 37. In: Warman, P., Munro-Warman, T.R. (Eds.), Abstracts of the International Composting Symposium. September, 1999. Coastal BioAgresearch, Ltd., Truro, Nova Scotia.

Michel, F.C., Reddy, C.A., Forney, L.J., 1993. Yard waste composting: studies using different mixes of leaves and grass in a laboratory scale system. Compos. Sci. Util. 1, 85−96.

Michel, F.C., Reddy, C.A., Forney, L.J., 1995. Microbial-degradation and humification of the lawn care pesticide 2,4-dichlorophenoxyacetic acid during the composting of yard trimmings. Appl. Environ. Microbiol. 61 (7), 2566−2571.

Miller, F.C., 1993. Minimizing odor generation. p.219-241. In: science and engineering of composting. In: Hoitink, H.A.J., Keener, H. (Eds.), Proceedings of an International Composting Symposium, March, 1992. Renaissance Publications, Columbus, OH; Worthington, OH.

Natural Resources Conservation Service (NRCS), 1992. National Engineering Handbook Part 651: Agricultural Waste Management Field Handbook. United States Dept. of Agriculture (USDA), Washington, DC. http://www.ftw.nrcs.usda.gov/awmfh.html.

Naylor, L.M., 1996. Composting. In: Girovich, M.J. (Ed.), Biosolids Treatment and Management Processes for Beneficial Use. Marcel Dekker, NY.

Pepe, O., Ventorino, V., Blaiotta, G., 2013. Dynamic of functional microbial groups during mesophilic composting of agro-industrial wastes and free-living (N2)-fixing bacteria application. Waste Manag. 33 (7), 1616−1625. https://doi.org/10.1016/j.wasman.2013.03.025. Epub 2013 May 3. PMID: 23647951.

Peters, S., Koschinsky, S., Schwieger, F., Tebbe, C.C., 2000. Succession of microbial communities during hot composting as detected by PCR-single-strand-conformation polymorphism-based genetic profiles of small-subunit rRNA genes. Appl. Environ. Microbiol. 66 (3), 930−936. https://doi.org/10.1128/aem.66.3.930-936.2000.

Puyuelo, B., Gea, T., Sánchez, A., 2014. GHG emissions during the high-rate production of compost using standard and advanced aeration strategies. Chemosphere 109, 64−70. https://doi.org/10.1016/j.chemosphere.2014.02.060.

Randle, P., Flegg, P.B., 1978. Oxygen measurements in a mushroom composting stack. Sci. Hortic. 8, 315−323.

Richard, T.L., 1992. Municipal solid waste composting: physical and biological processing. Biomass Bioenergy 3 (3−4), 163−180.

Richard, T.L., Hamelers, H.V.M., Veeken, A., Silva, T., 2002. Moisture relationships in composting processes. Compost Sci. Util. 10 (4), 286–302.

Robinzon, R., Kimm el, E., Avnimelech, Y., 2000. Energy and mass balances of windrow composting system. Trans. Am. Soc. Agric. Eng. 43 (5), 1253–1259.

Robinzon, R., Kimmel, E., Krasovitski, B., Avnimelech, Y., 1999. Estimation of bulk parameters of a composting process in windrows. J. Agric. Eng. Res. 73, 113–121.

Sartaj, M.L., Fernandes, L., Patni, N.K., 1997. Performance of forced, passive, and natural aeration methods for composting manure slurries. Trans. Am. Soc. Agric. Eng. 40 (2), 457–463.

Schulze, K.L., 1962. Continuous thermophilic composting. Appl. Microbiol. 10, 108–122.

Singley, M., Higgins, A.J., Franklin-Rosengaus, M., 1982. Sludge Composting and Utilization — A Design and Operating Manual. Rutgers University, New Brunswick, NJ.

Soares, M.A.R., Quina, M.J., 2013. Prediction of free air space in initial composting mixtures by a statistical design approach. J. Environ. Manag. 128 (2013), 75–82.

Stentiford, E.I., 1981. Proceedings of the International Conference on Composting of Solid Waste and Slurries. University of Leeds, Leeds. U.K.

Stentiford, E.I., 1996. Composting control: principles and practice.p.49-59. In: the Science of Composting. In: de Bertodli, M., et al. (Eds.), Proceedings of the European Commission International Symposium, September, 1995. Blackie Academic & Professional, London, U.K.

Suler, D.J., Finstein, M.S., 1977. Effect of temperature, aeration and moisture on CO_2 formation in bench-scale, continuously thermophilic composting of solid waste. Appl. Environ. Microbiol. 33 (2), 345–350.

Sundberg, C., Jönsson, H., 2008. Higher pH and faster decomposition in biowaste composting by increased aeration. Waste Manag. 28 (3), 518–526. https://doi.org/10.1016/j.wasman.2007.01.011. ISSN 0956-053X.

Tchobanoglous, G., Theisen, H., Vigil, S.A., 1993. Integrated Solid Waste Management: Engineering Principles and Management Issues. McGraw-Hill, New York, USA.

USEPA (U.S. Environmental Protection Agency), 1985. Composting of Municipal Wastewater Sludges. EPA/625/4-85-016. Cincinnati, OH.

Vandecasteele, B., Willekens, K., Steel, H., D'Hose, T., Van Waes, C., Bert, W., 2017. Feedstock mixture composition as key factor for C/P ratio and phosphorus availability in composts: role of biodegradation potential, biochar amendment and calcium content. Waste Biomass Valor. 8, 2553–2567. https://doi.org/10.1007/s12649-016-9762-3.

Waksman, S.A., Cordon, T.C., Hulpoi, N., 1939. Influence of temperature upon the microbiological population and decomposition processes in composts of stable manure. Soil Sci. 47, 83–113.

Weppen, P., 2001. Process calorimetry on composting of municipal organic wastes. Biomass Bioenergy 21, 289–299.

Wiley, J.S., 1962. Pathogen survival in composting municipal wastes. J. Water Pollut. Control Fed. 34, 80–90.

Compost feedstocks

Authors: **Robert Rynk**[1], **Mary Schwarz**[2]

[1]*SUNY Cobleskill, Cobleskill, NY, United States;* [2]*Cornell Waste Management Institute, Cornell University, Ithaca, NY, United States*

Contributors: **Tom L. Richard**[3], **Matthew Cotton**[4], **Thomas Halbach**[5], **Stefanie Siebert**[6]

[3]*Pennsylvania State University, State College, PA, United States;* [4]*Integrated Waste Management Consulting, Richmond, CA, United States;* [5]*Extension Professor Emeritus, University of Minnesota, Minneapolis, MN, United States;* [6]*European Compost Network, Bochum, Germany*

1. Introduction

Feedstocks are the raw ingredients for composting. They are organic materials, usually solid, and usually in an active state of decomposition. More than any other factor, the feedstocks determine the character of the compost and the temperament of the composting process. Important process and product-related properties of feedstocks include moisture, carbon and nitrogen content, physical characteristics, and level of contamination and value.

Overall, the list of potential ingredients is large and diverse (Table 4.1). Nearly any organic material can be converted to compost. Common feedstocks include manures; leaves, and yard trimmings collected from homes, parks, and other landscapes; food waste from residential and commercial sources (grocery stores, universities, schools, restaurants, hotels, hospitals, outdoor markets, fairs, and festivals and warehouses); biosolids from wastewater treatment (i.e., treated sewage sludge), digestate from anaerobic digesters; food processing residuals; industrial byproducts like paper mill sludge; shredded wood; and mixed municipal solid waste (MSW) (Fig. 4.1). Efforts to limit greenhouse gas emissions are diverting increasing amounts of food waste to composters (Frischmann, 2018). This trend is expected to continue as long as methane escapes from landfills and climate change is taken seriously.

Despite the fact that the feedstocks change drastically through the process, they carry through many of their original qualities to the compost, including organic matter, nutrients, minerals, salts, particle size, and contaminants. Feedstocks with relatively high levels of nutrients and salts tend to produce compost with relatively high levels of nutrients and salts. Feedstocks littered with inert contaminants tend

The Composting Handbook. https://doi.org/10.1016/B978-0-323-85602-7.00005-4

Table 4.1 Common and uncommon feedstocks for composting.

Alcohol stillage	Fats, oils and grease (e.g., from traps)	Poultry manure
Anchovies	Filter press cake (apple, olive, etc.)	Rabbit manure
Animal mortalities	Fish kills	Rice hulls
Apple pomace	Fish processing wastes	Salsa processing waste
Aquatic weeds	Flower seed screenings	Sausage casings
Bamboo, stems and leaves	Fruit, culled and spoiled	Sea urchin
Banana pulp	Garlic culls and plant residues	Seaweed
Biosolids (wastewater sludge)	Gelatin production waste	Septage (septic take solids)
Blood (from abattoirs)	Glucose solution	Shark carcasses
Bread and breadcrumbs	Grain (burnt, spoiled and wasted)	Sheep manure
Brewer's grains	Grape pomace	Shrimp waste
Brush (chips, shredded, etc.)	Grass clippings	Silage, spoiled and feed refusals
Butcher wastes (meat, bones, fat)	Gypsum and dry wall	Spent hops
Cannabis, seized and legal residues	Hay, spoiled and wasted	Straw
Carcasses from road kill	Horse manure, stables and racetrack	Sugarcane bagasse
Cardboard (corrugated, coated, paperboard, etc.)	Ice cream manufacturing waste	Sugarcane press mud (filter cake)
Cattle manure	Leather	Sunflower shells
Citrus fruit rinds and pulp	Lees (fruit fermentation sediment)	Swine manure
Coal ash	Lime	Tannery wastes (e.g., hides, sludges)
Cocoa shells	Mixed food waste (residential, grocery, school, restaurant, etc.)	Tare dirt (soil from beets, potatoes)
Coconut coir	Mollusk shells	Tea
Coffee grounds	Newspaper	Telephone books
Compost (recycled)	Night soil (human waste)	Tobacco
Contaminated soil (petroleum, munitions, etc.)	Olive processing wastes	Tomato paste processing waste
Cork	Onion culls	Tree bark
Corn cobs and stover	Organic textiles and textile residues	Trout manure
Cotton gin trash (leaves, stems, etc.)	Palm oil processing wastes	Vanilla bean residues
Cotton mattresses	Paper	Vegetables—culled and spoiled
Crab and lobster shells	Paper mill sludge	Vitamin residuals
Cranberry plant residue		Walnut shells
Currency, shredded	Pasta (wet, dry and boxed)	Waste beverages, in bulk tankers
Deciduous tree leaves	Paunch manure	Waste beverages, in containers
Diatomaceous earth (w/filter cake)	Peanut shells	Water hyacinth
Pharmaceutical gel caps	Peat moss	Whey (from cheese and yogurt)
Dissolved air flotation (DAF) sludge	Wood chips	Wood ash
Dog food	Pine needles (and other conifers)	Wood sawdust and shavings
Drilling mud	Potato culls	Yard trimmings, mixed
Elephant manure	Potato peels	Zoo manure and bedding

FIGURE 4.1

Examples of common and diverse composting feedstocks: (A) mixed yard trimmings, (B) supermarket wastes, (C) dairy cattle manure, (D) horse stable wastes, (E) straw, and (F) food processing waste from tomato products.

to produce compost with more of these contaminants, although processing can improve the result with some organic contaminants. The market value of a compost product depends on the feedstocks from which it was made and a very low level of contamination.

With respect to the process, the feedstocks determine how fast and how far decomposition proceeds, the character of odors generated, the need for moisture, rate of aeration, the temperature rise, and equipment needs. The degradability of a given feedstock dictates how it decomposes within a compost pile, and hence the release of its carbon, nutrients, and odor, and the maintenance of its physical structure. Quickly decomposing feedstocks are more likely to produce odors and rapidly rising temperatures than feedstocks that decompose more slowly. Consequently, the former feedstocks require more aeration and attention than the latter, all other things being equal.

In addition, the feedstocks often dictate what regulations and level of oversight are imposed on the composting operation. Regulations in many jurisdictions name specific materials for which requirements are relaxed and/or specific materials that carry special regulatory provisions. Agricultural by-products are often generated and managed on site tend to be exempt from many regulatory burdens. The regulations tighten with more challenging materials, like food wastes and biosolids. The highest degree of regulation and scrutiny come with feedstocks that have been traditionally viewed as waste materials, like wastewater biosolids and MSW.

Composting can be successful with a single feedstock, such as leaves or bedded livestock manure. However, more often, several different feedstocks are combined, either to improve the process or because multiple feedstocks are readily available. The combination and proportions of feedstocks used in a composting mix are commonly referred to as a "recipe." Sometimes the goal is to compost a given "primary" feedstock, such as a food processing residual, livestock manure, or biosolids. Then, other feedstocks are added to improve the composting characteristics of the primary feedstock. A feedstock added to a recipe to deliberately complement the processing of another feedstock is called an "amendment."

2. Feedstock value

Typically, feedstocks are unwanted by-products, derived from other processes or activities. In the best situations, feedstocks have a positive value for the composter; that is, composters are paid a fee to accept and process feedstocks into compost. Such a processing fee is often called a "tipping" fee or "gate" fee. In other cases, a desired feedstock may need to be purchased in order to produce compost with particular qualities or to improve the composting process. The latter situation usually relates to amendments, like wood chips, when suitable amendments are locally scarce.

Feedstocks that bring a gate fee are tempting. However, such fees can be a mixed blessing. Many feedstocks earn a fee because they are troublesome to handle. Some decompose quickly and quickly turn odorous (e.g., food waste). Others are odorous from the start (e.g., biosolids, fish waste). Some fee-bearing feedstocks need extra processing or carry contamination (e.g., brush, packaged food waste). In addition, just earning a processing fee may shift the composting facility into a more restrictive regulatory category.

On balance, accepting a feedstock for a fee is usually a plus for composters. Indeed, many composters earn more revenue from processing fees than the sale of compost. Whether or not it is practical, profitable and desirable to do so depends on: the fee itself; the costs to obtain the feedstock (e.g., transport); the cost to handle and process the feedstock (e.g., covered storage); and the characteristics of the feedstock including moisture, nitrogen content, physical characteristics like structure and bulk density, level of contamination, and odor risk. The last thing a composter should do is risk the existence of the composting site because of a nuisance-prone feedstock with an enticing gate fee.

3. Feedstock characteristics

The previous chapter discussed the important factors for the composting process. Those factors that can be affected by feedstocks are summarized in Table 4.2. Ideally, the feedstocks should be chosen and mixed in the right proportion to produce characteristics within the ranges listed in Table 4.2. However, it is not necessary, and often not possible, to achieve these values. Composting is a robust and forgiving process. It occurs over a broad range of conditions that can vary quite far from the ideal.

The allowable deviation from the ideal depends on the time available to complete composting, the effect on the compost product, and the consequences when the process strays from the ideal—most prominently the impacts of odors. For rapid composting, or for feedstocks and situations with a high risk of odors, it is prudent to stay close to the ranges in Table 4.2. Moisture content and the carbon-to-nitrogen (C:N) ratio are the feedstock characteristics of greatest concern to the process, and together commonly determine the recipe of the mix. Particle size, bulk density, and pH are also important process-related characteristics that should be considered when selecting feedstocks. The potential for odor should be another prime consideration in selecting feedstocks.

The characteristics of feedstocks not only have an effect on compost process they also have an impact on the finished compost's value. In fact, nearly every attribute of

Table 4.2 Optimal feedstock characteristics for rapid composting.

Condition	Acceptable	Ideal
Moisture content	40%–65%	50%–60% by weight
C:N ratio of combined feedstocks	20:1– 60:1	25–40:1
Feedstock particle size	<5 cm (2 in.)	Variable
Bulk density	<700 kg/m^3 (1200 lbs/yd^3)	400–600 kg/m^3 (700–1000 lbs/yd^3)
pH	5.5–9.0	6.5–8.0

a feedstock contributes to the character of the compost. Organic matter/ash content, macro- and micronutrient levels, soluble salts, and contamination of the feedstock all influence the utility of the resulting compost.

3.1 Moisture

Moisture content is usually a principal factor in developing a composting recipe (see following section). Feedstocks that are dry decompose very slowly and must be either mixed with wet feedstocks or watered. Although the recommended lower limit for moisture content is commonly 40%, a better target for rapid decomposition is 50%. As the moisture content dips below 35%, not only does microbial activity slow down but also dust becomes a concern. Conversely, wet feedstocks compost poorly because excessive moisture increases density and inhibits aeration. The recommended upper limit is typically in the 60%–65% moisture range. The high end of this range works well for feedstocks that are inherently bulky and have particles with a rigid structure. The primary examples are wood by-products. Conversely, feedstocks that are dense or contain a relatively high proportion of soil or other inert components call for a lower level of moisture. Feedstocks that are too wet for composting can be dewatered mechanically (common for some types of livestock manure and biosolids), drained of excess water, and/or dried by the sun and wind. The common remedy for wet feedstocks is to mix them with dry feedstocks.

3.2 C:N ratio

The relative proportions of carbon to nitrogen (C:N ratio) in the feedstocks influences the composting process and the qualities of the compost. The C:N ratio affects the process speed and nitrogen conservation, as the previous chapter explains. During composting the loss of carbon usually far outpaces the loss of nitrogen so the C:N ratio of the compost is substantially less than that of the feedstocks. Still, the starting C:N ratio is a factor in the C:N ratio of the final product. Feedstocks with a high C:N ratio are likely to produce compost with a high C:N ratio, especially if the carbon is bound in resistant compounds, like wood. For general horticultural and agricultural use, it is desirable for the C:N ratio of the compost to be less than 20:1, and preferably below 18:1. Using compost with a C:N ratio greater than 20:1 can lead to N immobilization. On the other hand, compost with a high C:N ratio is good for most mulch and erosion control applications.

The concentration of N in a feedstock largely determines its C:N ratio because N is the lesser quantity. A feedstock with 1% N typically has a C:N ratio half that of a feedstock with 0.5% N. Most raw organic materials have carbon concentrations in the 40% to 50% vicinity. Therefore, feedstocks with low C:N ratios tend to be abundant in N while feedstocks with high C:N ratios can be considered low nitrogen materials.

3.3 Particle size

As described in the previous chapter, particle size has a conflicting influence on the composting process—small particles decompose more quickly but also inhibit aeration. The effect of particle size depends on a material's degradability. Easily degradable materials, like food products, quickly lose their particle integrity within a composting pile. Poorly degradable materials, like wood, hardly decompose regardless of the particle size. Particle size is most relevant to moderately resistant feedstocks such as straw, leaves, vegetation, and paper. Thus, the general recommendation for particle size is fairly broad—0.3 to 5 cm (1/8—2 in.). In general, feedstocks with particles larger than 5 cm (2 in.) benefit from size reduction. Even sawdust particles survive the composting process with little change and become part of the compost. The size of woody particles therefore affects the texture of the compost. Processing operations, such as screening and grinding, offer some control of the compost particle size and texture.

3.4 Bulk density

Bulk density, the mass or weight per unit volume, is a good general index of the quality of a feedstock for composting, with $600 \, kg/m^3$ ($1000 \, lbs/yd^3$) being a good target. Bulk density reflects the feedstock's moisture content, porosity, free air space (FAS), and aeration capabilities (see Chapter 3). The bulk density (like porosity and FAS) of a mixture of feedstocks cannot be predicted with accuracy from individual ingredients because particles of different feedstocks intermingle.[1] For instance, the small particles of a dense feedstock fill the voids spaces within a bulky one. Hence, one cannot simply take a weighted average of two feedstocks to calculate the bulk density of their mixture. Still, in sufficient proportions, feedstocks with a low bulk density generally lower the bulk density of a mix and thus improve aeration.

The bulk density of a given feedstock can be highly variable, depending on moisture and its tendency to settle and compact. The compacted material at the bottom of a pile is denser than the material near the top. In addition to the overbearing weight, small particles tend to settle toward the base. The moist material at the center has a higher bulk density than the drier material near the surface. Hence, accurately quantifying the bulk density requires good sampling technique. If the moisture content of a particular feedstock varies considerably, it can be characterized by its dry bulk density (wet bulk density x dry matter content).

3.5 pH

As described earlier, the effect of feedstock pH on the composting process is usually minor, although composting is generally most effective if the mixture of feedstocks

---

[1] Researchers have used statistical methods to predict the FAS of a feedstock mix based on the characteristics of individual feedstocks. See Soares et al. (2013).

has a near-neutral pH. Some quickly degrading feedstocks that are rich in carbohydrates (e.g., potato culls and food waste) can sharply lower the pH of the pile early in the composting cycle. This effect, which can acerbate odors, is preventable with strong aeration. The pH of the starting materials does influence the product. The pH of the compost tends to be close to that of the feedstocks. The final pH may be either slightly higher or slightly lower, depending on the process conditions. For those feedstocks with pH near the extremes, amendments can be added to adjust pH upward or downward. For example, wood ash or lime can be added to raise the pH of acid feedstocks (Wang et al., 2017). Similarly, feedstocks with high pH (e.g., some manures and lime-treated biosolids) can be amended with acidic feedstocks or sulfur products typically used to lower soil pH (Ekinci et al., 2000). More often, however, high pH is managed by adjusting the pH of the resulting composts. Again, intentionally manipulating the pH of feedstocks is rarely necessary.

3.6 Organic matter, volatile solids, and ash

The amount of organic matter in the feedstocks plays a part in how much organic matter is in the compost. Generally speaking, feedstocks with more organic matter yields compost with more organic matter. For example, a compost made from manure usually will be higher in organic matter than one made from leaf and yard waste. This relationship is especially true with poorly degradable feedstocks like wood. Highly degradable feedstocks, like food materials, can defy this generality as they thoroughly decompose to carbon dioxide. Also, how the compost is processed also plays a part. Turned compost generally has less organic matter than static pile compost made with the same ingredients, especially if that compost is on a soil pad where soil can be incorporated into the mix during turning.

Organic matter content is nearly the same as another characteristic called "volatile solids" (VS). The difference is that VS also includes nonorganic compounds that easily vaporize, like ammonia. Usually, one value can be taken as a reasonable approximation of the other. Ash is the reverse of VS (or organic matter). Ash content, in percent, is 100% minus the VS content in percent (or organic matter).

3.7 Nutrients

The nutrient concentrations in the feedstocks determine the nutrients of the compost. The levels of nutrients such as nitrogen (N), phosphorus (P), potassium (K), sodium (Na), calcium (Ca), magnesium (Mg), sulfur (S), and zinc (Zn) in the compost product can have an effect on its quality and use. Because nutrients are mostly conserved through the composting process, as carbon is lost, the remaining nutrients are concentrated in the compost. Nitrogen is a possible exception but only if conditions like a low C:N ratio and high pH encourage large losses of N. Minor losses of S, P, and K can also occur but the general relationship holds—the nutrients in the feedstocks determine the nutrient levels in the compost and the concentrations tend to increase from start to finish.

3.8 Soluble salts

Soluble salts are ions of minerals that easily dissolve in water. They are not harmful in themselves. In fact, many salts are essential nutrients for plants and microorganisms (e.g., nitrate). However, high concentrations of salts can damage plants, so it is desirable to limit the concentration in the compost. Salts in compost come from the feedstocks. Hence, it is generally advisable to avoid feedstocks that carry relatively high concentrations of salts. Unfortunately, livestock and poultry manures fall into this category, along with many sources of food residuals. Allowable salt concentrations in the compost depend on the intended use of the compost. Salt concentrations are affected by precipitation and leaching. Composting in the open, without a cover, or at least curing in the open can reduce compost salt level. If salts are high, blending of compost products and mixing with low salt content feedstock dilutes or manages the salt. Specific quantitative recommendations for maximum salt concentrations in feedstocks have not emerged. The common approach is to avoid "salty" feedstocks if necessary and when possible and otherwise control salt levels by blending the compost product or managing its use.

3.9 Odor risk

Feedstocks that have a strong odor or turn rancid quickly require special handling. Odor risk is closely related to degradability. There are two related odor risks. First, odor-prone feedstocks can impose their distinct odors to the composting site when they are delivered. Second, these feedstocks require more aeration and process control to minimize odors during the active composting stage. In locations that are vulnerable to odor complaints, strong-smelling materials such as fish processing waste, food, poultry manure, and biosolids are best avoided or composted inside a building or vessel.

Balch et al. (2019) rate the odour potential of numerous feedstocks. Overall, the feedstocks judged to have the highest odour potentials are those that are wet, quickly degradable (putrescible), and/or rich in N, S, protein or fats. Chapter 12 discusses odor management principles in detail.

4. Feedstock contaminants

Possible contaminants in composting feedstocks include physical contaminants that detract from the appearance and value of the compost, chemicals that might compromise its use, and trace elements (e.g., heavy metals) and biological materials that present health concerns at elevated concentrations.

4.1 Physical contaminants

Physical contaminants are unwanted inert (i.e., inorganic) items that arrive with the feedstocks. Inert contaminants pass through the composting process and become concentrated and more visible in the compost. Common examples include: rocks and metal hardware scraped up with manure; bottles, synthetic fiber, tennis balls, glass, parts from vehicles, and many other items collected in with leaves and yard trimmings; staples in cardboard boxes; labels attached to fruits and vegetables; rubber bands and containers in food waste; and plastics in almost any feedstock (Fig. 4.2).

The most troublesome contaminant is plastic. Plastic is the scourge of nearly all composters. The rise of food waste composting has only worsened composters' contamination woes. Plastic comes in film, foam, hard, and composite forms. It is ubiquitous and difficult to thoroughly separate from compost by mechanical means. Feedstocks that are heavily contaminated with plastic and other inert items are costly to process. The best approach is to replace or separate plastic items at the source. Compostable plastic products and plastic substitutes offer a potential source-based solution but also have their pitfalls (see Section 8.11).

A more recent type of plastic contamination is microplastics. Microplastics are plastic particles less than 5 mm in size. They primarily find their way into compost from feedstocks that contain items that are lined or coated with plastic, including paper beverage cartons, drinking cups, frozen food packages, plastic-lined paper bags, and some paper plates (Brinton et al., 2018). Small plastic particles remain behind after the organic fraction of these items decompose. Microplastics have been found in compost and soils amended with compost. They pose potential concerns to the soil environmental and that of adjacent surface waters (Weithmann et al., 2018; Watteau et al., 2018). At present, there are no regulations or guidelines regarding microplastics in composting or compost. The best prevention is to avoid specific plastic-treated items that carry these contaminants or the associated feedstocks (e.g., food waste, biosolids, MSW, and some digestates). There is evidence

FIGURE 4.2

A variety of physical contaminants removed with screen overs after screening compost. The contaminants arrive with the feedstocks.

Source C. Oshins.

that particular composting practices, such as very high temperatures and/or inoculants, can help degrade microplastics through the composting process (Chen et al., 2020), but the issue is still developing.

4.2 Chemical contaminants

Possible chemical contaminants in composts primarily include pesticide residues and medications. Pesticide residues may accompany hay, straw, greenhouse residues, and other crop residues; grass weeds, leaves, and general yard trimmings; or manure. Antibiotics and other medications pass through the manure when animals are treated for illness. Although not every chemical compound has been studied, research results suggest that most pesticides and medications generally decompose during composting. However, there are a few notable exceptions.

Concerns about contamination of compost with pesticides include the effect of herbicide residues on plants; human exposure from eating crops; children ingesting the compost directly; and general environmental and ecosystem effects. The occurrence, fate, and degradation of pesticides during composting have been widely studied. Investigations of pesticide residues in composting feedstocks and finished compost detected few pesticides and those that were found occurred at low concentrations (Buyuksonmez et al., 1999, 2000; Lemmon and Pylypiw, 1992; Strom, 2000; Vandervoot et al., 1997). A few specific herbicides defy this positive generality. Herbicides known as "pyridine" or "pyrimidine carboxylic acids" can persist in high enough concentrations within compost to damage plants when the compost is used (see Chapter 15).

As much as 75% of the antibiotics consumed by livestock in feed are excreted unchanged. In addition, medications other than antibiotics are also used in the livestock industry. Steroids and other hormones, nonsteroidal anti-inflammatory drugs (NSAIDs), ectoparasitics (fly, tick, and lice medications), and anthelmintics (wormers) are commonly used. In human waste, hormones, tranquilizers and stimulants, analgesics, caffeine and vaccines might occur. When contaminated manure and effluent are used to fertilize agricultural soils, loads of up to kilograms per hectare may be reached. These antibiotics and medications are of concern because they may increase the incidence of antibiotic resistance among a wide variety of pathogenic bacteria as well as having potential deleterious effects on soil bacterial populations. Research examining the fate of antibiotics in composted manure indicates that chlortetracycline, monensin, tylosin, and oxytetracycline are rapidly reduced during the composting process (Arikan, 2007, 2009; Dolliver et al., 2008; Donoho, 1984; Storteboom et al., 2007).

Steroids such as estrogen, estadiol, progesterone, and testosterone (largely excreted in human and animal urine) degrade slowly in manure, soil, and water. Normal wastewater treatment plant conditions do little to degrade these compounds (Khan et al., 2008; Pauwels et al., 2008). Composting and composts rich in microorganisms, including bacteria, actinomycetes, and fungi can degrade or transform these compounds into less toxic substances and/or lock up pollutants within the organic matrix, thereby reducing pollutant bioavailability (Puglisi et al., 2007).

NSAIDs are quickly metabolized in the body and are generally not excreted in the feces. However, vultures in South Asia were poisoned by diclofenac (an NSAID) after scavenging on livestock treated with the drug shortly before death. Composting these carcasses would not only have degraded the diclofenac, but also removed them from access by the vulture. Concern has been raised that the use of antiparasitic agents and anthelmintics in livestock may adversely affect harmless or beneficial organisms, which breed in or feed on dung and play a vital role in the processes in the decay of manure on pastures. There has been little to no research done on the fate of these compounds in composting, but it appears that they do not have much effect on the decomposers in the manure, so composting can occur (McKellar, 1997). Finally, there is concern, about the use of the euthanasia drug, sodium pentobarbital, when composting mortalities. Sodium pentobarbital has been shown to degrade during the composting process so that by the time composting is finished (within six months), very low concentrations of the drug remain (see sidebar Chapter 6).

Other possible chemical contaminants in feedstocks include residues from cleaning and disinfecting products and personal care products. These items might be present in some levels in MSW, septage, and biosolids. Nothing in these categories of chemicals has been found to impair the composting process or compost products. Some composts derived from MSW have contained boron at phytotoxic levels.

At the time of this writing, there is heightened apprehension about a group of widely used industrial compounds known by the acronym PFAS (per- and polyfluoroalkyl substances). PFAS includes the more notorious compounds: Perfluorooctanoic acid (PFOA) and perfluorooctanesulfonic acid (PFOS). Some of these chemicals have been linked to some health issues in humans. Consequently, they are currently raising questions and spurring changes to drinking water, environmental and consumer product standards. Many industries have phased out the use of the PFOA and PFOS.

The problem for composters is that PFAS have been widely used, over decades, in numerous products, including textiles (e.g., carpets, furniture, and clothing), cookware, food service products (e.g., paper plates), and firefighting foam. They tend to be persistent chemicals and water-soluble. Therefore, residues of PFAS compounds are prevalent in the environment and in wastes, including composting feedstocks. Food waste, biosolids, and paper mill residuals earn the most attention, but small concentrations can even be found in leaves and yard trimmings. In Europe and elsewhere, compost standards include limits on fluorine concentration, partly in response to current PFAS concerns. Such standards may especially affect the development and use of biodegradable food service products.

The present situation, circa 2020, does not require composters to do anything out of the ordinary regarding feedstocks and PFAS, other than remain aware of developments. The situation is fluid. It may never significantly affect composters. For more and updated information about PFAS and composting, monitor the pages of BioCycle Connects (see Beecher and Brown, 2018; Coker, 2000a). Also, the USCC maintains a PFAS web page at (https://www.compostingcouncil.org/page/pfas).

4.3 Trace elements

Trace elements, including heavy metals, are regulated because high concentrations of some elements pose health and/or environmental risks. Prime examples include lead, copper, zinc, arsenic, and cadmium. All naturally occurring materials contain trace elements. Some trace elements are necessary nutrients for plants and humans. With these elements, it is the concentrations that matter, not their presence per se. The concentrations of trace elements in typical composting feedstocks are usually well below the standards established (internationally, the standards differ greatly among countries, see Chapter 15). Wastewater biosolids and MSW are the feedstock most likely to challenge the limits. However, copper or zinc levels also can be high in manure where copper and zinc sulfate footbaths are used, or if copper is provided as a dietary supplement for livestock.

4.4 Biological contaminants

The primary biological "contaminants" of concern are pathogenic organisms that are inherent to the feedstock. Since the composting process effectively destroys animal and plant pathogens, the concern is small. Nevertheless, feedstocks that contain more pathogens merit more attention to temperature management and handling to avoid cross-contamination. For human pathogens, biosolids, MSW, and postconsumer food waste pose the greatest concern. Manure and plant debris carry livestock and plant pathogens and should be handled and composted at thermophilic temperatures. When the compost process is rushed in the primary phase, pathogen regrowth or reestablishment can occur. Another possible biological contaminant is weed seeds. Feedstocks that carry abundant weed seeds should be either avoided or composted at thermophilic temperatures to kill the seeds.

In most jurisdictions, regulations require specific time-temperature regimes for sanitization (e.g., PFRP) when composting biosolids, MSW, or postconsumer food waste (see Box 3.4 in Chapter 3). In some jurisdictions, these sanitization requirements apply to all feedstocks.

5. Biodegradability

The degradability of a given feedstock governs how quickly and completely the feedstock decomposes within a compost pile. Hence, degradability also determines the availability of the feedstock's carbon and the maintenance of its physical structure. In composting, it is the biodegradability that counts because organic compounds are primarily decomposed by biological agents (i.e., microorganisms and enzymes).

Organic materials greatly differ in their speed and extent of decomposition. Compare, for example, a pile of wasted fruit to a pile of green wood chips. The fruit pile promptly decomposes, slumps in a wet puddle, and releases odor while the wood chip pile stands unchanged for months or years. Fig. 3.13 in Chapter 3 presents a hierarchy of biodegradability with various feedstocks included.

Ignoring the effects of moisture and particle size, the biodegradability of a feedstock depends on its chemical make-up. Sugars and starches decompose quickly because of their simple molecular structure. Fewer enzymes and fewer steps are involved in their decomposition. It takes longer to disassemble compounds of cellulose and hemicellulose, which are chains of simple sugars. Lignins are large and complex organic molecules that defy decomposition (Box 4.1). It follows that feedstocks dominated by sugars and starches decompose quickly while those with abundant lignin hardly decompose at all. The concentration of lignin is a fairly reliable predictor of how thoroughly a feedstock will decompose via composting. Eq. (4.1) is one formula for estimating biodegradation based on lignin content (Richard, 1996, after Chandler et al., 1980).

$$\text{Biodegradable fraction (\% of volatile solids)} = 0.83 - (0.028 \times \text{lignin\% of volatile solids})$$
$$(4.1)$$

Box 4.1 The effect of lignin on biodegradability

Author: Tom L. Richard

Plant cell walls are composed of four important constituents: cellulose, hemicelluloses, pectins, and lignin. Lignin is particularly difficult to biodegrade and reduces the bioavailability of the other cell wall constituents. Knowing a bit about each of these constituents is helpful in understanding the vastly different rates at which different plant materials decompose. Cellulose is made of long chains of glucose molecules, linked one to another. The simplicity of cellulose's structure, using repeated identical bonds, means that only a small number of enzymes are required to degrade this material. Hemicelluloses and pectins are branched polymers of several sugar monomers including xylose, arabinose, galactose, mannose, glucose, and acidic sugars. Hemicelluloses and pectins surround and link cellulose, enhancing the strength and flexibility of the cell wall. Hemicelluloses can also cross-link with lignin, creating a complex web of bonds that provides structural strength to the plant, but also inhibits microbial degradation.

Lignin is a complex polymer of phenylpropane units, which are cross-linked to each other with a variety of chemical bonds. This complexity has thus far proven as resistant to detailed chemical characterization as it is to microbial degradation. Nonetheless, some organisms, particularly fungi, have developed the necessary enzymes to break lignin apart. Actinobacteria (aka actinomycetes) can also decompose lignin, but typically degrade less than 20% of the total lignin present. Lignin degradation is primarily an aerobic process, and in an anaerobic environment, lignin can persist for very long periods.

Because lignin is the most resistant component of the plant cell wall, it largely determines the biodegradability of a given type of plant material. With higher proportions of lignin, less of the total carbon is available to microorganisms ("bioavailable"). The effect of lignin on the bioavailability of other cell wall components is thought to be largely a physical restriction, with lignin molecules either reducing the surface area available for enzymes to bind and degrade other components of the cell wall or sequestering those enzymes in an inactive form.

Although significant lignin degradation is possible during aerobic composting, a number of factors are likely to affect the decomposition rate. Conditions that favor the growth of fungi, including adequate nitrogen, moisture, and temperature, all appear to be important in encouraging lignin decomposition, as does the composition of the plant material itself. Lignin is present in all plants, but it is especially abundant in trees and other woody species. Without lignin, our wooden houses would crumble to the ground.

Adapted from the Science and Engineering section of Cornell Waste Management Institute website, http://compost.css.cornell.edu/calc/lignin.html (Richard, 1996).

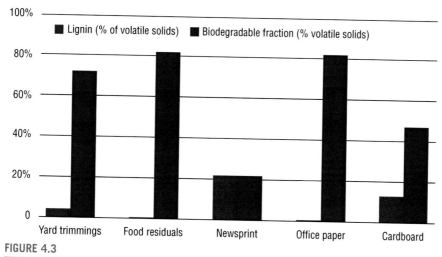

FIGURE 4.3

Estimated degradability of selected feedstocks. Lignin is expressed in percent of volatile solids concentration. Biodegradable fraction is the percentage of dry volatile solids.

Based on data presented by Tchobanoglous et al. (1993).

Fig. 4.3 provides examples of the predicted biodegradability of selected feedstocks based on their lignin content. Degradability can also be tested in the laboratory, or on a pilot scale, if desired, by monitoring carbon dioxide evolution and/or loss of mass over time (Mason, 2009).

A measure of biodegradability can be applied to a feedstock's total carbon in order to estimate its "available" carbon. This approach has been proposed by several researchers including Kayhanian and Tchobanoglous (1992) and Puyuelo et al. (2011). It would be useful for assessing things like oxygen demand, heat generation, and effective C:N ratio. It is somewhat remarkable that using an "available C:N ratio" has not taken hold as common practice, although the recommended C:N benchmarks would also need to be revisited. Presently, judging the bioavailability of carbon in a feedstock mix is left to the experience and intuition of the operator, and remains part of the art of composting.

6. Combining feedstocks—amendments and recipes

In few cases does a given feedstock, in its available condition, possess all the characteristics required for efficient composting. Therefore, it is often necessary to blend several ingredients, in suitable proportions, to achieve a mix with the desired overall characteristics (Fig. 4.4). The mixture of ingredients, and their relative proportions,

FIGURE 4.4

Combining feedstocks for better composting.

is referred to as a *recipe*. A common recipe involving food waste is typically dominated by yard trimmings or/and wood chips. For farms, a composting recipe is often a blend of manure and bedding materials, and possibly crop residues. Sometimes waste products from nearby lumber operations, such as untreated sawdust or bark, are used.

In many cases, composters deal with a primary feedstock, like food, manure, or biosolids. Other feedstocks are then added, either to improve composting of the primary ingredient or simply because they are available for composting.

Ingredients that are deliberately added to a recipe to improve the composting characteristics of a primary feedstock are referred to as amendments, bulking agents, or carbon sources. An amendment is added to adjust any characteristic of the mix, including moisture content, texture, bulk density, or C:N ratio. A bulking agent improves the physical characteristics so that the feedstocks have sufficient pore space and structure for aeration; that is, they stand in a pile without slumping or collapsing. Carbon sources are added to raise the C:N ratio. The three terms are often used interchangeably because common amendments, like wood chips and straw, also add bulk and carbon. "Amendment" is the more general term and is used in this handbook to describe any ingredient added to improve the qualities of a primary feedstock.

In many cases, the primary feedstock to be composted is wet and high in nitrogen, like manure or biosolids. Therefore, dry bulky amendments with a high C:N ratio are often in great demand. Since amendments must often be obtained from outside sources, cost and availability are important factors. For composting to remain economical, the amendments must be inexpensive. Fortunately, many free or inexpensive materials are suitable and available for composting.

6.1 Approaches to developing recipes

A composting recipe is like a cooking recipe. It prescribes the ingredients and their proportions for successfully making the desired product at an acceptable cost. As in cooking, the recipe affects both the process and the product.

A composting recipe can result from any of the following strategies:

- Look and feel—relying on senses to combine feedstocks in a way that produces a mix that looks and feels right for composting. The *feel* typically relates to the moisture level while the *look* may simply answer the question, "Does the pile stand up well on its own?" For instance, more manure or water is added if the mix feels dry. Dry amendments are added until the mix stands in a pile without slumping. This approach to recipe making requires experience and good composting instincts.
- Trial and error—a common approach that is typically partnered with the look and feel strategy. Again, this tactic requires some experience and understanding the composting process. There is the risk that a trial that results in error will cause odor or product quality problems. The key to success is maintaining an awareness of the moisture content, C and N balance, and the other aeration characteristics of the feedstock mix.
- Tried and true—learning from and copying successful recipes from other composters. Some recipes have even been "standardized." For example, a standard recipe for composting biosolids by the aerated static pile method is mixing three volumes of wood chips to one volume of biosolids. The common backyard recipe of three volumes of "browns" to one volume of "greens" is generally sound advice. In this vernacular, the "browns" are low-nitrogen feedstocks, which tend to be dry and often brownish in color. The greens are high nitrogen feedstocks, which tend to be moist and typified by green vegetation. Although brown in color, manure is very "green" by the definition used here. The browns to greens volume ratio of 3:1 usually leads to a safe and workable recipe in large scale composting and the backyard. However, it can result in an unnecessarily costly mix if brown amendments must be purchased.
- Target bulk density—mixing feedstocks to yield a desired bulk density. Because bulk density is a good general index for the physical and moisture character of feedstocks, it is also a reasonable criterion for combining feedstocks. Bulk density cannot be accurately predicted from the densities and proportions of individual feedstocks, so finding the proportions to achieve the target is a trial-and-error exercise. Once the feedstocks mixed, the bulk density easily can be measured directly in the field (Box 4.2). As noted earlier, a good bulk density target falls in the range of 400–600 kg/m^3 (about 700–1000 lbs/ yd^3).
- Mathematically balancing moisture—calculations based on only the moisture content of individual feedstocks. With many feedstocks, moisture is the pivotal factor in successful composting. In such cases, it is often sufficient to determine recipes based on the moisture contents of the individual feedstocks. This task can be done mathematically, and easily, using the formulas presented in Table 4.3.

Box 4.2 Procedure for field measurement of Bulk Density and Free Air Space (FAS)

The following procedures for field testing of bulk density and free-airspace (FAS) are recommended by the Composting Research and Education Foundation (CREF) and used in the CREF's Compost Operators Training Course (Oshins, 2019). The bulk density test is completed first and is used as the basis for determining FAS.

Bulk Density Bulk density is calculated by dividing the mass or weight of a collection of particulate material by the volume that the collection occupies. It can be determined by filling a container of known volume and weight with the material to be tested and then weighing the filled container. The bulk density equals the filled container weight minus the empty container weight divided by the container volume.

An inconvenient feature of bulk density is that it is inconsistent. It changes over time as particles settle, lose or gain moisture. It also changes a great deal within the cross section of a pile, so getting a good composite sample is important.

Furthermore, when sampling a pile of material, the inherent density is disturbed. The sample tends to be less dense than the material in its pile. Therefore, some standard repeatable means of compaction is necessary to obtain a consistently representative result. A simple and commonly used procedure for determining bulk density on-site is described below.

1. Tools needed

   - A garden shovel,
   - A 20-liter or 5-gallon bucket or similar standard container, and,
   - A scale able to weigh at least 25 kgs or 50 lbs (e.g., a luggage scale).

2. Confirm and record the volume of the bucket. If necessary, place a mark on the bucket wall at the location that corresponds to 20 liters or 5 gallons. A simple way to find this location is to place the bucket on the scale and add water until the scale reads 20 kg plus the bucket weight or 42 lbs plus the bucket weight. Otherwise, you can pour in four 5-liter containers of water or five 1-gallon jugs.

3. Using the garden shovel obtain representative samples from different locations of the pile. Add the samples to the bucket until the bucket is about *one-third* full (Figure 4.5A). Sample the pile from top to bottom and from the surface to the core (review the sampling tips in Chapter 14).

FIGURE 4.5

Field measurement of bulk density (A) and free air space (FAS) (B) using a 5-gallon pail, a shovel, a luggage scale, and water.

Box 4.2 Procedure for field measurement of Bulk Density and Free Air Space (FAS)—cont'd

4. Lift the bucket to so the base is about 15 cm (6 inches) above the ground and drop the bucket onto a firm flat surface. Repeat this procedure for a total of 10 drops.
5. Add additional representative samples to the bucket until it is about *two-thirds* full.
6. Again lift the bucket to 15 cm (6 inches) and drop it to the ground 10 times.
7. Fill the bucket with additional samples to the fill line (e.g., 20-liter or 5-gallon mark).
8. Once again, drop the bucket to the ground 10 times from a height of 15 cm (6 inches).
9. Add more material to the bucket to the 20-liter or 5-gallon mark and weigh bucket and its contents.

Using this procedure, the estimated bulk density is:

$$\text{Bulk Density} = \frac{\text{Filled container weight} - \text{Empty Container Weight}}{\text{Container Volume}} \qquad (4.2)$$

Eq. (4.2) gives an answer in kgs per liter or lbs per gallon. The bulk density in more useful units can be calculated directly using Eqs. (4.3) and (4.4). (There are fifty 20-liter buckets in a cubic meter and forty 5-gallon buckets in a cubic yard).

$$\text{Bulk Density} \left(\text{kg/m}^3\right) = \text{kgs/liter from Eq. } 4.2 \times 50 \qquad (4.3)$$

$$\text{Bulk Density} \left(\text{lbs/yd}^3\right) = \text{lbs/gallon from Eq. } 4.2 \times 40 \qquad (4.4)$$

Free Air Space FAS is a measure of the amount of open space within the material through which oxygen is provided for aerobic bacteria to do their work. It is the ratio of gas-filled volume to total sample volume. It can be measured in the field in much the same way that bulk density is measured. After following steps 1 − 9 above for measuring bulk density, water is added to the bucket to displace the gas-filled void-space between particles. The weight of water added, and its corresponding volume, determines the FAS.

10. Follow steps 1 − 9 above for bulk density.
11. Add water to the filled bucket dually to allow the air bubbles to percolate to the surface. Continue just until the water covers the surface of the material in the bucket (Figure 4.5B). Weigh the bucket again.
12. The difference between the weights of the bucket before and after adding water equals the weight of water added to it.

Weight of water added = Weight of bucket after adding water − weight of bucket before

13. Volume of water added is the volume of the pore space displaced. It is equal to the weight of water added divided by the density of water. In the case of kilograms and liters, the weight and volume of water are numerically equivalent.

$$\text{Volume of gas-filled pore space (liters)} = \text{Weight of water added (kg)} / 1 \text{ kg/L} \qquad (4.5)$$

$$\text{Volume of gas-filled pore space (gallons)} = \text{Weight of water added (lbs)} / 8.34 \text{ lbs/gal} \qquad (4.6)$$

The estimated FAS is the volume of the gas-filled pore space divided by the volume of the contents in the bucket (e.g., 20 liters or 5 gallons). Multiplying by 100 expresses the answer in its common percentage form. of the water added to fill all of the air spaces, divided by the weight of water that would fill the volume of the bucket used. For example, 40 lbs of water will fill a 5-gallon bucket. The resulting number is the percentage of free air space in the sample.

$$\text{FAS}(\%) = 100\% \times \frac{\text{Volume of gas-filled pore space}}{\text{Volume of bucket contents}} \qquad (4.7)$$

- Mathematically balancing moisture and C:N ratio—calculating feedstock mixes based on chemical analysis of individual feedstocks. Moisture content and C:N ratio have been the main variables customarily used to develop composting recipes. A recipe for multiple feedstocks can be worked out mathematically, if the moisture, carbon and nitrogen contents of the individual ingredients are known. Table 4.3 presents the related formulas, and Section 6.3 explains the procedures. Computer spreadsheet and dedicated recipe-building programs make the calculations tolerable for even math-phobic composters. By this method, a well-balanced recipe can be developed on paper, and without the cost and effort of trial and error. Because it includes several key factors, this method of recipe development is preferred in situations where process control and speed are important. Thus, it is discussed in more detail in the following sections.

6.2 The prominence of moisture and C:N ratio

Moisture and C:N ratio are two primary characteristics commonly used to determine composting recipes. In part, these characteristics are emphasized because they are measurable and mathematically predictable. However, they are also key factors in the composting process. If the moisture content and C:N ratio are within the recommended ranges, composting will likely go well.

As Table 4.2 indicates, when compost feedstocks are combined, the resulting mixture ideally should fall in a moderate moisture range (50%–65%) and possess a ratio of carbon to nitrogen in the vicinity of 25:1 to 40:1. One can visualize a feedstock's character by plotting these properties on a chart such as Fig. 4.6. The center of the chart represents an excellent balance. The surrounding quadrants delineate conditions for which a feedstock is either too wet, too dry, nitrogen poor, and/or overabundant in nitrogen. Again, it is the nitrogen content of a feedstock that primarily determines its C:N ratio.

Fig. 4.7 maps typical moisture contents and C:N ratios for selected example feedstocks. For example, livestock and poultry manures exhibit the full range of moisture contents but in most cases, manure is wet and high in nitrogen. Livestock bedding and poultry litter materials (e.g., wood shavings) bring down the moisture content and raise the C:N ratio. Assorted feedstocks that come from the landscape or gardens are likely to be balanced if they are well blended. Individually, green materials (e.g., grass clippings, fresh vegetation) tend to be moist and moderately high in nitrogen while brown materials (e.g., autumn leaves, hay/straw, wood trimmings) are usually dry and lack nitrogen. Feedstocks from the food stream are diverse but those most commonly composted (e.g., market produce, kitchen scraps) tend to be very wet and nitrogen-rich. It is vital to understand that specific materials with each type of feedstock tagged in Fig. 4.7 vary greatly and can differ from the positions in the figure.

Charts like these can help a composter visualize how a particular feedstock can be amended or altered to improve its composting performance; that is, move it

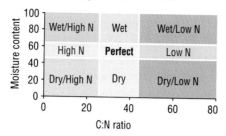

FIGURE 4.6

Generic feedstock moisture and C:N ratio map.

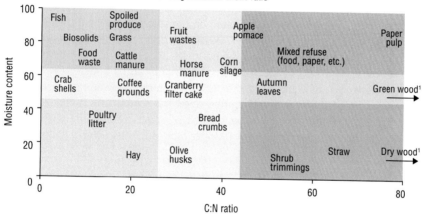

1. The C:N ratio of wood is "off-the-chart." ranging from 250 to over 500

FIGURE 4.7

Typical moisture and C:N ratio positions for selected food feedstocks.

toward the center of the chart. Appendix B lists typical values for moisture, C:N ratio and other characteristics of numerous feedstocks common to composting. It is important to recognize that the C:N ratio and, especially, the moisture content of a specific material can vary greatly depending on its source, management and handling. A laboratory analysis of specific candidate feedstocks from the actual sources is recommended.

6.3 Recipe math

This section outlines the mathematical procedure for developing a composting feedstock recipe based on moisture content and C:N ratio (total C and N). To some readers, the formulas and procedure may be daunting, especially readers without algebra experience. Fortunately, there are ways to follow this recipe-making option without doing the math. Numerous computer spreadsheets and recipe "calculators" are available, either free of charge or for a small fee. An internet search will reveal numerous alternatives. Many of these recipe programs include an embedded inventory of potential feedstocks and an estimate of their C, N, and moisture contents. Some programs go a little further and also consider feedstock costs (Calisti et al., 2020). Appendix C includes a link to the recipe calculator used by CREF in its composting operator training program.

Mathematically developing a composting recipe is a balancing act because both the C:N ratio and the moisture content need to be within acceptable ranges. Usually, one of these attributes takes priority, and an appropriate recipe is determined. Then, if necessary, the proportions are adjusted to bring the second attribute in line without excessively changing the first. Sometimes an acceptable balance cannot be achieved, and a different set of ingredients must be considered.

With wet feedstocks, the moisture content is particularly critical because high moisture content leads to anaerobic conditions, odors, and slow decomposition. The consequences of a poor C:N ratio are less damaging. It is usually better to develop an initial composting recipe based on moisture content and then adjust it, if necessary, to achieve an acceptable C:N ratio.

When working with primarily dry feedstocks, feedstocks can be first proportioned on the basis of C:N ratio. If necessary, water can be added as needed. In this case, the weight of water needed to achieve a target moisture content can be calculated using Eq. (4.8), recognizing that water has a moisture content of 100%. Eq. (4.9) calculates the corresponding volume of water by dividing the required weight by the density of water, about 1 kg per liter or 8.3 lbs per US gallon.

$$\text{Weight of water needed} = \frac{(\text{fw} \times \text{M}) - (\text{fw} \times \text{mf})}{(100 - \text{M})} \tag{4.8}$$

fw is the feedstock weight.
mf is the feedstock moisture content in percent.
M is the target moisture content in percent.

$$\text{Volume of water needed} = \text{Weight of water/Density of water} \tag{4.9}$$

The formulas for calculating a composting recipe are given in Table 4.3. For each ingredient, the moisture content, the concentration of nitrogen and either the carbon concentration or the C:N ratio must be known. If it is necessary to convert from weight to volume or vice versa, the bulk density of the feedstock is also required. Note that carbon and nitrogen concentrations are expressed and calculated on a dry weight basis.

If the literature or test results do not report the carbon concentration but the percentage of organic matter or ash is known, the carbon content can be roughly estimated by the following equation.

$$\% \text{ Carbon} = \frac{\% \text{ Organic matter}}{1.8} = \frac{100 - \% \text{ Ash}}{1.8} \tag{4.10}$$

Eq. (4.10) was proposed in the 1951 by researchers in New Zealand (Golueke, 1972) and it remains valid. It assumes that carbon comprises slightly more than half (56%) of organic matter by weight (i.e., what is not ash). The factor in the denominator (i.e., 1.8) is an average based on many measurements. In individual samples, this factor has since been determined to be as low as 1.6 and as high as 2.2.

Table 4.3 Recipe Math—formulas for calculating composting recipes based on moisture content and C:N ratio.

Formulas for an individual ingredient

Moisture content = % Moisture content/100
Weight of water = total weight × moisture content
Dry weight = total weight – weight of water
 = total weight × (1 – moisture content) = total weight × solids content
Nitrogen content = dry weight × (%N /100)
%carbon = %N × C : N ratio
Carbon content = dry weight × (%C/100)
 = N content × C : N ratio

General formulas for a mix of materials

$$\text{Moisture content} = \frac{\text{weight of water in feedstock a + water in b + water in c} + \dots}{\text{total weight of all feedstocks}} = \frac{(a \times m_a) + (b \times m_b) + (c \times m_c) + \dots}{a+b+c}$$ (Eq. 4.11)

$$\text{C : N ratio} = \frac{\text{weight od C in feedstock a + weight of C in b + weight of C in c} + \dots}{\text{weight of N in a + weight of N in b + weight of N in c} + \dots}$$

$$= \frac{[\%C_a \times a \times (1 - m_a)] + [\%C_b \times b \times (1 - m_b)] + [\%C_c \times c \times (1 - m_c)] + \dots}{[\%N_a \times a \times (1 - m_a)] + [\%N_b \times b \times (1 - m_b)] + [\%N_c \times ac \times (1 - m_c)]}$$ (Eq. 4.12)

where: a = total weight of feedstock a
b = total weight of feedstock b
c = total weight of feedstock c

$m_a, m_b, m_c, \dots$ = moisture content of feedstocks a, b, c, ...
$\%N_a, N_b, N_c,$ = % nitrogen of feedstocks a, b, c, ... (% of dry weight)
$\%C_a, C_b, C_c, \dots$ = % carbon of feedstocks a, b, c, ... (% of dry weight)

Shortcut formulas for only two feedstocks (for example, leaves plus grass clippings)

Required amount of feedstock a per kg (or lb) of feedstock b, based on the desired *moisture content* $a = \frac{m_b - M}{M - m_a}$ (Eq. 4.13)

Then check the C:N ratio using the general formula.

Required amount of feedstock a per kg (or lb) of feedstock b, based on the desired C:N ratio $a = \frac{\%N_b}{\%N_a} \times \frac{(R - R_b)}{(R_a - R)} \times \frac{(1 - m_b)}{(1 - m_a)}$ (Eq. 4.14)

Then check the moisture content using the general formula.

where: a = kgs (or lbs) of feedstock a per kg (or lb) of feedstock b
M = desired mix moisture content
m_a = moisture content of feedstock a (for example, amendment)
m_b = moisture content of feedstock b (for example, manure)
R = desired C:N ratio of the mix
R_a = C:N ratio of feedstock a
R_b = C:N ratio of feedstock b

The word "weight" is used to represent both mass and weight units. It is technically incorrect but pragmatic, as long as practitioners remain on a given planet. Also, the formulas for C, N and C:N ratio are for total C and N on a dry weight basis (not available C and N).

To develop a recipe mathematically, it is necessary to assume the *relative* proportions of the feedstocks being considered. If there is a primary feedstock, it is best to assume that it has a unit of 1 (e.g., 1 ton, 1 metric ton, 1 cubic yard, 1 cubic meter). The proportions of the other ingredients are relative to that ingredient. It is easiest to work in weight or mass units and then convert the final weight proportions to volume (volume = weight or mass ÷ bulk density). Making the calculations with volume-based proportions is extra work, although easily done by a computer.

The sample calculations in Example 4.1 describe a procedure for calculating the recipe proportions, moisture content, and C:N ratio. With only two ingredients, such

---

Example 4.1 Sample calculations for composting recipe based on moisture content and C:N ratio

On a weekly basis, a farm composter currently handles 20 cubic meters of separated solids from dairy manure and mixes it with eight cubic meters of horse manure. The composter has been offered organic waste from the local market. On average, every week the market generates about five cubic meters of organic waste, which is mostly food with some paper products. The composter plans to add more horse manure to accommodate the moist market waste.

The question is: How much more horse manure should the composter add to maintain a reasonable recipe? As a start, the composter assumes that increasing the horse manure to 10 cubic yards might suffice. The average feedstocks' analysis is listed below.

Feedstocks	Moisture (% wet wt)	Nitrogen (% dry wt)	Carbon (% dry wt)	C:N ratio	Bulk density kg/m^3 (lb/yd^3)
Dairy manure solids (DMSs)	70	1	55	55	550 (927)
Horse manure (HM)	50	1.5	48	32	770 (1289)
Market organics (MO)	78	2.8	52	19	800 (1348)

Given the above analysis and the assumed weekly volumes, the amount of each feedstock available in terms of wet weight, dry weight and water weight are listed below. "Wet weight" is the actual weight (or mass) of the feedstock. "Water weight" is the amount of water in the feedstock. "Dry weight" is the weight (or mass) of everything except the water.

Feedstocks	Weekly volume	Wet weight[a]	Water weight[b]	Dry weight[c]
DMS	20 m^3	11,000 kg	7700 kg	3300 kg
	26.2 yd^3	24,287 lb	17,004 lb	7286 lb
HM	10 m^3	7700 kg	4235 kg	3465 kg
	13.1 yd^3	17,004 lb	9667 lb	7652 lb
MO	5 m^3	4150 kg	3120 kg	880 kg
	6.6 yd^3	9170 lb	6887 lb	1942 lb
Totals	33 m^3	22,700 kg	15,055 kg	7645 kg
	43.3 yd^3	50,121 lb	33,240 lb	16,880 lb

[a] Wet weight = Volume × Bulk density.
[b] Water weight = Wet weight × Moisture content.
[c] Dry weight = Wet weight − water weight.

Step 1: Determine the combined moisture of the feedstocks:
From Table 4.3, the formula for moisture content is:

$$\text{Moisture content} = \frac{\text{weight of water in ingredient a} + \text{water in b} + \text{water in c} + \ldots}{\text{Total weight of all ingredient}}$$

$$\text{In kilograms: M} = \frac{7700 + 4235 + 3120}{22,700} = \frac{15,055 \text{ kg}}{22,700 \text{ kg}} = 0.663 = 66.3\%$$

$$\text{In pounds: M} = \frac{17004 + 9667 + 6887}{50,121} = \frac{33,240 \text{ lb}}{50,121 \text{ lb}} = 0.663 = 66.3\%$$

The resulting moisture content, 66% is slightly higher than desired.
Step 2: Check the resulting C:N ratio:
From Table 4.3, the formula for C:N ratio is:

$$\text{C:N ratio} = \frac{\text{weight of C in ingredient a} + \text{weight of C in b} + \text{weight of C in c} + \ldots}{\text{weight of N in a} + \text{weight of N in b} + \text{weight of N in c} + \ldots}$$

The weight of each nutrient is calculated by multiplying the dry weight of the feedstock by its corresponding nutrient content which is demonstrated in the following table.

Feedstocks	Dry weight[a]	Carbon (C) content[a]	Carbon weight[b]	Nitrogen (N) content[a]	Nitrogen weight[b]
DMS	3300 kg	0.55	1815 kg	0.01	33 kg
	7286 lb		4007 lb		73 lb
HM	3465 kg	0.48	1663 kg	0.015	52 kg
	7652 lb		3673 lb		115 lb
MO	880 kg	0.52	458 kg	0.028	25 kg
	1942 lb		1010 lb		54 lb
Totals	7645 kg		3936 kg		110 kg
	16,880 lb		8690 lb		242 lb

[a] Carbon content = %Carbon ÷ 100%; Nitrogen content = %Nitrogen ÷ 100%.
[b] Nutrient weight = Dry weight x Nutrient content.

Filling in the values from the above table into the formula:

$$\text{In kilograms : C:N} = \frac{1815 + 1663 + 458}{33 + 52 + 25} = \frac{3936 \text{ kg C}}{110 \text{ kg N}} = 35.8 = 36{:}1 \text{ C:N}$$

$$\text{In pounds : C:N} = \frac{4007 + 3673 + 1010}{73 + 115 + 54} = \frac{8689 \text{ lb C}}{242 \text{ lb N}} = 35.9 = 36{:}1 \text{ C:N}$$

The resulting C:N ratio, 36:1, is excellent. The addition of the market waste will lower the overall C:N ratio and improve it relative to the current recipe using only dairy manure solids and horse manure.

Continued

Example 4.1 Sample calculations for composting recipe based on moisture content and C:N ratio—cont'd

Conclusion The composter would be wise to use more horse manure (the drier feedstock) to bring down the moisture content below 65%. Doing so does not harm the overall C:N ratio because the horse manure has a good C:N ratio in itself, 32:1. Repeating these calculations assuming a weekly horse manure volume of 14 cubic meters (18.3 cubic yards) yields a combined moisture of 65% and content and C:N ratio of 35:1.

All of these calculations are made quickly and easily by a computer spreadsheet program. The following image depicts a spreadsheet layout for this example problem. After inputting the data, a composter only has to try different amounts in the cells of the "Volume" column (orange background) to ascertain the resulting moisture contents and C:N ratios (yellow background) for various combinations of the feedstocks.

Feedstock	Volume (m³)	Bulk density (kg/m³)	Wet wt. (kg)	Moisture content (%/100)	Water wt. (kg)	Dry wt. (kg)	C (% dry wt.)	N (% dry wt.)	C:N Ratio	Wt. of C (kg)	Wt. of N (kg)
DMS	20	550	11,000	0.7	7700	3300	55	1	55	1815	33
HM	10	770	7700	0.55	4235	3465	48	1.5	32	1663	52
MO	5	800	4000	0.78	3120	880	52	2.8	19	458	25
Total and results	35		22,700	66.3%	15,055	7645			36	3936	110

Feedstock	Volume (yd³)	Bulk density (lb/yd³)	Wet wt. (lb)	Moisture content (%/100)	Water wt. (lb)	Dry wt. (lb)	C (% dry wt.)	N (% dry wt.)	C:N Ratio	Wt. of C (lb)	Wt. of N (lb)
DMS	26.2	927	24,287	0.7	17,001	7286	55	1	55	4007	73
HM	13.1	1298	17,004	0.55	9352	7652	48	1.5	32	3673	115
MO	6.55	1348	8829	0.78	6887	1942	52	2.8	19	1010	54
Total and results	46		50,121	66.3%	33,240	16,880			36	8690	242

as manure plus an amendment, the amendment proportion can be calculated directly from the desired moisture content or the desired C:N ratio, as shown in the example. However, if three or more ingredients are used, the recipes must be calculated by trial and error using the general formulas in Table 4.3. In this case, the proportions of the ingredients are first assumed and then the corresponding C:N ratio and moisture content are calculated. If either the C:N ratio or moisture content is unacceptable, proportions are adjusted and calculations are repeated until an acceptable C:N ratio and moisture content are obtained. Appendix D is a presentation-based tutorial for calculating composting recipes, including additional examples.

A computer spreadsheet program is tailor-made for repetitive calculations of this type. Once the characteristics of the candidate ingredients are entered into the spreadsheet, the relative proportions are easily altered to find the best recipe.

7. Determining feedstock characteristics

It is frequently helpful and sometimes necessary to determine the physical and chemical characteristics of specific composting feedstocks, current and prospective (Table 4.4). Knowing the characteristics helps in developing recipes, indicates a material's suitability for composting, infers the plant nutrient content of the compost, and identifies suspected contaminants. Values for some physical and chemical characteristics of many feedstocks can be estimated from experience or found in literature (as in Appendix B). Simple "in-house" tests can determine some qualities, like moisture content and bulk density. However, getting specific and reliable values for the concentration of nitrogen and other constituents requires the specialized equipment and skills of a professional laboratory (private or public). Testing for biological organisms (e.g., pathogens) and chemical contaminants also requires laboratory services. Professional laboratories possess the required knowledge of the methodologies, analytical equipment, technical skills, and quality control. Many labs that serve composting facilities offer analytical "packages" that bundle useful tests together for one price. When having feedstocks analyzed for compost recipes, it is advisable to secure a laboratory that conducts analysis specifically for composting feedstocks and compost.

The extent and frequency of laboratory analysis for feedstocks depend on the purposes for the analysis. The most common purpose is to provide information for developing composting recipes. In this case, a lab analysis should determine, at a minimum, the moisture (or solids) content, carbon concentration, nitrogen concentration, and organic matter content (or VS or ash). Laboratories usually break down the total amount of nitrogen by its primary chemical components—organic nitrogen and ammonia. Depending on the political jurisdiction and its regulations, some feedstocks, like municipal biosolids or MSW, may require testing for prescribed potential contaminants such as certain trace elements and biological constituents.

Table 4.4 Feedstock characteristics often or sometimes worthy of lab analysis.

Relevant to composting process and recipe	Indicators of compost product qualities and use	Environmental indicators (if required or otherwise warranted) [a]
Bulk density	Organic matter (or ash)	Heavy metals
Moisture content	Total nitrogen	Inert contaminants
pH	Organic nitrogen	Pesticide residues
Carbon content	Phosphorus	
Total nitrogen	Potassium	
Ammonia nitrogen	Calcium	
Phosphorus[a]	Magnesium	
Sulfur[a]	Sodium and chloride	
Organic matter (or ash)[a]	pH	

[a] Rarely pertinent for evaluating feedstocks.

ANALYTICAL CHEMISTS
and
BACTERIOLOGISTS
Approved by State of California

SOIL CONTROL LAB

42 HANGAR WAY
WATSONVILLE
CALIFORNIA
95076
USA

TEL: 831-724-5422
FAX: 831-724-3188
www.compostlab.com

CODE: FS-compost
Account #: 6070677-2/6-9383
Group: Jul.16 D #63
Reporting Date: July 29, 2016

Robert Rynk- CEST
Cobleskill, NY 12043

Feedstock Analysis

Date Received: 22 Jul. 16
Sample Identification: Old wood grindings BP
Sample ID #: 6070677 - 2/6

Nutrients-Primary + Secondary	Units	as Received	Dry Weight
Total Nitrogen (N):	%	0.54	0.77
Organic Nitrogen (Org.-N):	%	0.54	0.77
Ammonia (NH4-N):	%	0.0025	0.0036
Nitrate (NO3-N):	%	0.00034	0.00049
Phosphorus (as P2O5):	%	0.10	0.15
Potassium (as K2O):	%	0.28	0.40
Calcium (Ca):	%	0.72	1.0
Magnesium (Mg):	%	0.043	0.061
Sulfate (SO4):	%	0.016	0.023
C/N Ratio	Ratio	61	61
AgIndex	Ratio	150	150
Carbonates (as CaCO3)	lbs/ton	5.9	8.4
Moisture	%	29.7	0
Organic Matter:	%	66.0	93.9
Ash:	%	4.3	6.1
pH value	units	6.45	NA
Salts			
Sodium (Na):	%	0.0017	0.0024
Chloride (Cl):	%	0.0044	0.0062
Electrical Conductivity (EC5):	mmhos/cm	NA	0.62
Void Space	% v/v	NA	55.6
Bulk Density	g/cc	0.29	0.21
Void Space (> 4mm fraction):	% v/v	NA	70.8
Volume (> 4mm fraction):	% v/v	NA	100.0
Volume (< 4mm fraction):	% v/v	NA	15.3
Voids left	% v/v	NA	55.6
Size			
Greater than 4 mm fraction:	% w/w	NA	85.3
Less than 4 mm fraction:	% w/w	NA	14.7
*Material Cost ($ per unit)	$		NA
*Availability (1=least to 5=most)	Rating		NA

*=Information provided by client for formulation purpose.

Analyst: A.S.

FIGURE 4.8

Example of a typical laboratory analysis for a composting feedstock, in this case aged ground wood.

Courtesy of Soil Control Lab.

Typically, laboratory results also include analysis of pH, electrical conductivity (a measure of salinity), and concentrations of other nutrients such as phosphorus, potassium, calcium, magnesium, and sodium. Although not essential to developing a recipe, these characteristics can identify potential problems with composting or the compost products. Sulfur is not typically included in the suite of nutrients analyzed by laboratories. However, in some situations, a composter may wish to request it because the relative level of sulfur indicates the odor character of a given feedstock. Other tests that can be requested, as needed, include particle size distribution, feedstock degradability, lignin content, presence of pesticide or pharmaceuticals, and more.

Fig. 4.8 is a copy of a typical feedstock analysis report, in this case for ground wood. The report presents the values in two columns, "as Received" and "dry weight." The "as Received" column reports concentrations based on the actual or wet weight of the sample. Only moisture content, bulk density, and C:N ratio should be read from the "as Received" column. All of the other characteristics are regularly based on the "dry weight" concentrations. Figure 4.9 is an example of an abbreviated feedstock analysis, which this particular lab offered upon the request of this client. It is sufficient for the purpose of developing feedstock recipes.

Whether tests are conducted in-house or by a laboratory, getting a good sample is an essential first step, followed by properly handling and preserving the sample. Sampling is especially important to accurately characterizing a feedstock. Most organic materials are notoriously heterogeneous; that is, they are inconsistent in their texture and composition. Furthermore, bulk materials tend to stratify, with the fines and moisture migrating toward the base of piles. A random sample from a pile of bedded livestock manure could yield a shovel full of mostly sawdust or a shovel full of mostly manure. The same situation could occur with a pile of mixed grass, leaves, and brush.

Chapter 14 offers more detailed information about sampling techniques, professional laboratory analysis, and in-house testing procedures.

8. Common feedstocks for composting

The list of organic materials suitable for composting is nearly endless. With the trend in managing organic residuals in a more sustainable way, more and different feedstocks are increasingly common. In particular, a variety of food residuals are being diverted for composting, either directly or after anaerobic digestion (AD).

This section summarizes the qualities of general categories of "the usual" composting feedstocks. Specific feedstocks within these categories are profiled in more detail in Appendix E. Table 4.5 very generally presents composting characteristics of common feedstocks. For each feedstock, listed there are exceptions to the characterizations presented in the table due to specific circumstances. The descriptions presented here are based on: numerous research publications; several composting manuals, guidebooks, and handbooks (e.g., Paul and Geesing, 2015); popular articles in *BioCycle* magazine, *BioCycle Connects*, and other periodicals; and the

13611 B Street ☐Omaha, Nebraska 68144-3693 ☐(402) 334-7770 ☐FAX (402) 334-9121 ☐www.midwestlabs.com

Lab# 8941198	**Report of Analysis**			Report Number: 21-209-4012

Account: 61xxx	Cary Oshins US Composting Council		₥ Account Manager 402-xxx-xxxx
Date Sampled:	2021-07-20		
Date Received:	2021-07-21		Feedstock Analysis
Sample ID:	Spoiled Silage		

		Analysis (as rec'd)	Analysis (dry weight)	Total content, lbs per ton (as rec'd)
NUTRIENTS				
Nitrogen				
Total Nitrogen	%	1.52	2.23	30.4
Organic Nitrogen	%	none	----	----
Major and Secondary Nutrients				
Micronutrients				
OTHER PROPERTIES				
Moisture	%	31.80		
Total Solids	%	68.20		1364.0
C:N Ratio		19 : 1		
Total Carbon	%	29.38	43.08	
pH		none		

FIGURE 4.9

Example of an abbreviated feedstock laboratory analysis with enough information for developing feedstock recipes.

Source: Midwest Laboratories, In.

experience of the authors and our colleagues. Many of these sources are listed in the "Consulted and suggested reading" section of this chapter.

Composting feedstocks are subject to substantially different regulations among different countries, states, provinces, and other political jurisdictions. In addition, various feedstocks are usually regulated differently within a given jurisdiction. It is important for facility managers and operators to understand, and follow, the regulations and guidelines that apply in their specific locations, especially in regard to safety and sanitization. For example, regulations that govern composting of animal by-products exist both across the European Union and in individual countries (see Box 4.3). With tiered levels of restrictions, these regulations apply to many common composting feedstocks, including manure, food wastes, and animal flesh.

> ## Box 4.3 Composting regulated animal by-products
>
> To borrow from the current vernacular, "animal by-products" are *literally* anything that arises from animals—manure, stomach contents (e.g., paunch manure), whole carcasses, partial carcasses, organs, hides, hooves, blood, bones, butcher waste, discarded meat, food waste that might contain meat (e.g., catering waste), and numerous other animal discards. They are of particular concern because of their *potential* to spread disease to humans or animals. As such, several countries thoroughly regulate the handling, treatment, and disposal of animal by-products.[3] These regulations impact composting (and AD) in ways that can dictate what feedstocks can be composted (e.g., no farm animal mortalities) and process sanitization requirements (e.g., temperatures/times). As the foregoing list of by-products suggests, many common composting feedstocks fall under these regulations where they exist including all animal manures and many, if not most, sources of food waste. Fortunately, the regulations do not require great changes to composting practices for most affected composters, especially for low-risk feedstocks like manures. Usually, the composting process easily meets the temperature and time mandates. The biggest burden may be related to sanitizing equipment and documentation.
>
> The European Union has the most wide-ranging animal by-products regulations ((EC) No 1069/2009 and (EU) No 142/2011). Animal by-products are categorized according to their relative risk. Category 1 includes the highest risk materials like brain and spinal tissue that might carry prions that cause transmissible spongiform encephalopathies diseases like BSE (aka Mad Cow). Pet animals are also in this category. Category 1 materials cannot be composted. Category 2 and 3 materials can be composted following specified practices. Category 2 includes by-products like manure, paunch manure, and fish. Category 3 covers food waste and low-risk materials from slaughtering animals (e.g., hooves, feathers). The foregoing summary is an extreme simplification of the European regulation. The European Compost Network has a "Good Practice Guide" that explains and helps composters navigate these animal by-products regulations (Amlinger and Blytt, 2013).
>
> ---
>
> [3] Current animal by-products regulations likely arose in response to catastrophic outbreaks of animal diseases in the 1990s and early 2000s, including foot-and-mouth, BSE (Mad Cow), and various swine and bird flus.

8.1 Yard trimmings

Also called green waste and yard waste, the category of yard trimmings covers a broad swath of materials including deciduous leaves, pine needles, grass clippings, unused fruits and vegetables, shrubs, tree branches, garden vegetation, aquatic plants, and other vegetative materials. A common thread among yard trimmings is that they are generated from maintenance of yards, gardens, parks, and other public and private landscapes. A stronger commonality is that they are largely vegetative in nature, and usually a varying mix of herbaceous and of woody vegetation. Depending on the amount and thickness of the woody materials, yard trimmings may require grinding before composting. Depending on the compost use, screening might be necessary.

It is more than a pun to proclaim that yard trimmings are the low hanging frui*ts* of composting feedstocks. They are easily composted with little oversight in residential backyards and other points of generation. In fact, on-site composting is the best approach to managing yard trimmings. Yard trimmings tend to amass in a somewhat balanced collection of moist green vegetation, dry brown vegetation, leaves, flowers, stalks, stems, branches green wood, and old wood. The mix is often, if not usually, ready for composting "as is." Most yard trimmings compost at a moderate rate—fast enough to produce compost within a normal composting cycle, yet slow enough to present a low risk of odors, again with important exceptions. One advantage for commercial and farm composters is that yard trimmings are typically someone's burden. These materials are offered up as composting feedstocks by property owners, public works departments, commercial landscape companies, and refuse haulers. Normally, they are a source of gate fees for commercial and municipal compost producers. At worst, they are available free, except possibly for the cost of trucking. Regulations for composting yard trimmings are relatively lenient, compared to other waste materials. In many jurisdictions, yard trimmings composters are not compelled to meet time/temperature requirements (e.g., PFRP). Nevertheless, it is prudent to achieve pathogen-killing temperatures because of the possible presence of pet wastes, weed seeds, and pesticide residuals. Fortunately, high temperatures are easily reached and maintained in commercial, municipal, and farm-scale yard trimmings piles and windrows.

Yard trimmings are not without drawbacks. They may contain soil, which does not impede composting but can accelerate wear on machine parts. If soil needs to be removed prior to grinding, the feedstocks can be run through a course screen to separate the soil particles. Yard trimmings often carry pesticides, weed seeds, plastics, and other possible contaminants. Weed seeds and most pesticides are effectively negated through the composting heat and biochemistry. Pyridine herbicides remain a potential problem with grass clippings (see Chapter 15). The degree of contamination with plastics and other inert objects depends on how yard trimmings are collected. Plastic bags are a large problem where property owners and landscapers use plastic bags to dispose of leaves, grass, and other yard trimmings. Paper bags should be encouraged as an alternative. Contaminants are prevalent in yard trimmings that are piled near the roadside for bulk collection. Such loads often include plastic shopping bags, food wrappers, tennis balls, hardware, car parts, and anything else that litters roadsides. Although most items can be removed by screening, loads of bulk yard trimmings should be inspected for large stiff items that could damage machinery.

Usually, yard trimmings arrive at the composting site as a mix of constituents that proportionally takes on the character of those constituents. That mix is very much dependent on the local region and its climate. Yard trimmings might be relatively uniform through the year or it can change drastically in amount and character through the seasons, especially in regions with cold winters. In the spring and early summer, moist green vegetation tends to prevail. The mix becomes dryer and browner starting in late summer and continuing through autumn. It may stop almost entirely in the winter. Generally, yard trimmings are balanced and diverse enough to compost well without adding other feedstocks. Dry vegetation and woody materials add good structure and lower the bulk density of mixed yard trimmings. Whether or not additional water must be added depends on the season, regional climate, and recent weather. Also, yard trimmings can serve as amendments for wet feedstocks. For instance, food residuals are often composted together with yard trimmings. However, when grass clippings are abundant, additional dry feedstocks may be needed.

Two particular types of yard trimmings are distinctive—deciduous leaves and grass clippings. In temperate climates, deciduous leaves standout because they tend to arrive over a relatively short period of time in the autumn. The composting site must be able to accommodate the huge autumn volume. Given the timing and annual cycle of autumn leaves, composters can adopt a relaxed schedule, allowing leaves to compost over the winter and into the spring and even summer. Leaves and partially composted leaves are also a good feedstock to compost with other materials, like grass clippings. Grass clippings are notable because they are a concentrated source of nitrogen, which contributes to their rapid decomposition. Without added amendments to balance the C:N ratio and promote aeration, grass clippings can generate odors.

8.2 Wood

Wood is generally available to composters as sawdust, shavings, chips, and large items including tree trunks, limbs, branches, pallets, and discarded lumber. These large items must be chipped, ground, or shredded at the composting site, if size reduction does not occur at the source.

Wood products are primarily valued in composting as amendments to absorb moisture and bulking agents to add structure and improve aeration. However, wood also brings a lot of organic matter to the process and compost products. Wood products have three overriding composting characteristics: (1) very low nitrogen concentrations (less than 0.1%) and thus very high C:N ratios (e.g., 400:1); (2) an inherent stiffness that gives most wood particles good structure; and, (3) a very slow rate of degradation, owing to the high lignin content.

When the particles of wood are very small, as with sawdust, the structure disappears and the rate of degradation increases but even small wood particles remain resistant to biological decomposition. The wood particles degrade at the edges, shrink in size, and change color but otherwise retain their basic characteristics and lend those characteristics to the compost. Large particles of wood remaining in the compost can be screened out and recycled for additional composting. Because wood resists degradation, only a fraction of the wood's carbon is released during composting. Therefore, the C:N ratio of a feedstock mix that includes wood is effectively overstated.

There is some variability among wood materials due to tree species and the specific condition of the wood. Generally, hardwood species degrade faster than softwood species, probably because softwoods tend to have more lignin. Wood with a larger proportion of sapwood, the peripheral living portion of the tree, has more nitrogen and moisture compared to wood that has relatively less sapwood. For example, freshly cut branches have proportionally more sapwood than large limbs, trunks, and dead branches. Dead trees and branches have no sapwood. Freshly cut, or "green," wood is relatively moist (50% to 60%) and then loses moisture after cutting. By the time they reach the composting pile, most woody materials are dry (20% to 30%). Wood waste from pallets, lumber, and other kiln-dried products is typically very dry, in the range of 10% to 20%.

8.3 Paper

Almost all paper is made from wood; wood that is pulped to very fine fibers and further processed to varying degrees. True to its wood origin, paper contains very little nitrogen. It can also be deficient in phosphorus. Consumer-based paper products that find their way to composting piles include newsprint, corrugated cardboard, and highly processed products like napkins, tissue, paper cups and plates, office paper and magazines. Discarded paper goods are generally extremely dry (less than 10%), although some paper items absorb additional moisture during use and transport. After paper takes on moisture, it starts to disintegrate into its fibrous constituents. Highly processed products, like office paper, are chemically stripped of lignin and, unlike wood, decompose quickly. Lightly processed paper products, like newsprint, are close to wood in composition and the fibers retain the biological qualities of raw wood. Corrugated cardboard, which has some structure built in by design, takes longer to physically collapse. It can provide structure to a compost pile for several weeks.

In addition, treatment of the water from paper manufacturing generates a solid residue or "sludge" that can be composted. Paper mill sludge is moist to wet (e.g., 60% to 80%), depending on the degree of dewatering.

8.4 Food waste

Food waste, also called food residuals, comes from residential collections, grocery stores, universities, schools, restaurants, hotels, hospitals, outdoor markets, fairs and festivals, separated organics, and warehouses. Mixed food waste from these sources can include both pre- and post-plate waste, vegetables, fruits, meat, milk, condiment containers, soiled food wrappers, food within wrappers, bottles and cans, plates, napkins, and other paper products as well as coffee grounds. Some may even include compostable service ware products such as utensils and clamshell packaging. One advantage of taking this type of waste is to help local schools and businesses manage their organic waste and possibly generate a farm to table (or store) relationship. However, one should be prepared for the "seasonality" of mixed food waste from these sources. Secondary schools produce waste only 10 months of the year and food waste type depends on what is abundant at that time of year. One week's load may be mostly pineapple tops, while other loads may have only discarded pizza dough and cardboard.

Because there is such a range, it is difficult to characterize food waste. Sources should be evaluated on an individual basis. By itself, and very generally, the food component of food waste is moist to wet (60% to 80%), moderately high in nitrogen (1% to 3%) and decomposes quickly. Again, these characteristics are very dependent on the actual composition. Paper products within the food lower the moisture and nitrogen concentrations and give the food some structure for aeration. Paper and cardboard are usually not a problem for composting but the plastic and other inert items must be removed. The best place to separate these contaminants from the food is at the source, which merits strong communication with the generator and may require placing restrictions on the contents of the food waste loads. Increasingly, depackaging machines are being used to separate the food from wrappers and containers.

Taking on food waste raises the management bar for composters, at least compared to yard trimmings, the feedstock that is usually composted with food. In addition to the potential contamination, food decomposes quickly and is attractive to flies, rodents, and other animals. It is often unsightly and can become odorous, even before it arrives at the composting site. Therefore, some level of containment is necessary for food waste composting. The containment can be in outdoor piles if the food is mixed with abundant yard trimmings or other amendments. The food should remain within the envelop of the pile, and the pile should not be turned until the food decomposes to the point at which it is no longer a nuisance. Alternatively, the pile can be turned and then covered with amendment or finished compost. Otherwise, food waste can be initially composted in rotating drums, aerated containers, or other in-vessel systems until the food essentially disappears. The main advantage of adding food waste to yard trimmings or manure composting enterprise is that the composter can charge a fee for handling the food waste. However, it also tends to increase the regulatory burden.

8.5 Biosolids

Biosolids is the label used in the United States for treated sewage sludge that can be recycled by land application and composting. Sewage sludge is the accumulation of microbial biomass and other organic solids generated at wastewater treatment plants. Biosolids usually derives from the anaerobic or aerobic digestion of the sludge. Biosolids has a long history as composting feedstock. Fairly well-established procedures and recipes have been developed for composting biosolids. Generally, biosolids is nitrogen-rich (2%–7%) and very wet (70%–85% moisture). The nitrogen and moisture content depend on the dewatering and treatment processes at the wastewater plant. Composting biosolids requires two to four volumes of dry amendment per volume of biosolids. Typically, the amendment used is wood chips, which is commonly screened from the compost and reused for additional batches of biosolids.

Biosolids is a good feedstock for making nutrient-rich compost but it comes with "baggage." Biosolids is a highly regulated material. Composting it involves operational and site permits, process monitoring, and product analysis. It is mandatory for composters to demonstrate pathogen destruction, either by maintaining records of process temperatures or by product testing (or both). Depending on the initial quality of the biosolids, regulations governing compost use might apply due to heavy metal concentrations. Also, biosolids presents a moderate to high risk of odors. A well-buffered site, process controls and/or containment are necessary to control odors. For these reasons, biosolids has a public image problem and a composter may meet stiff resistance from the community. For the same reasons, there is a good opportunity to collect fees for composting biosolids.

8.6 Livestock and poultry manure

Manure generated from livestock and poultry farms can be tricky to characterize because it arises in different forms, depending on the animal species, husbandry practices, manure collection approaches, climate, and processing. The various sectors of livestock and poultry industries tend to follow similar manure management practices, as a group. Still, there are large difference between farms in any given sector.

As excreted by the animal, manure is wet and has a relatively high nitrogen content. As-excreted moisture content ranges from about 75% for poultry, sheep, and horses to 90% for cattle and swine. Nitrogen concentrations start at about 2.5% for horses and can surpass 6% for poultry and swine. Ultimately, the consistency and nutrient content of manure depends on the husbandry practices.

One primary factor is the amount of bedding or litter used in the animal rearing areas ("litter" is bedding for poultry). Bedding/litter is typically dry fibrous by-products like straw, sawdust, wood shavings, nut shells, rice hulls, chopped corn stalks, shredded paper and the like—the same types of materials that make good composting amendments. Bedding can dilute the nitrogen concentration by one-third to one-half or more. Moisture contents of bedded manure range from 50% to about 80%. Where bedding or litter is used generously, the manure tends to be solid and might be compostable without additional amendments. Where bedding is used sparingly, manure tends to have a wet porridge-like consistency and is often referred to as a "slurry." The moisture content of slurry is in the 80% to 90% range. Slurry can be composted by creating a trench down the center of a prebuilt windrow with periodic dams to slow the movement of liquid so it can be absorbed.

Manure consistency is also determined by how and how often manure is collected. Horse stalls tend to be cleaned regularly by stable hands with manure forks. Horse manure is typically 50% to 70% moist due to the large amount of bedding being removed. Manure that accumulates over time beneath the animals for weeks or months is usually liberally bedded and tends to be solid, in the 60% to 70% moisture range. This form of manure is commonly called a "bedded pack" on livestock farms and simply "litter" by poultry producers. It is commonly gathered-up with skid-steer or tractor buckets. Their composting characteristics depend on how much bedding is applied, the climate, and whether the bedded pack is located under a roof or not.

At the other extreme, some livestock farms remove manure from barns by flushing the floors with water. Flushed manure is truly liquid and not suitable for composting in its liquid form, except as a supplemental moisture source. However, liquid manure is frequently squeezed and screened to separate the solids from the liquid. The solids portion, which is dominated by wasted feed and bedding, can have a surprisingly high C:N ratio (e.g., over 30:1). Moisture content typically ranges from 70% to 85%, depending on the method of separation and also the climate. These solids are good composting feedstocks. They are often partially composted on the farm to sanitize them before being recycled as bedding.

The weather greatly influences the characteristics of manure, especially where animals are raised and fed outdoors. An abundance of sun and wind reduces moisture while an abundance of rain adds water and washes away nutrients and salts. Loss of ammonia nitrogen is increased by sun, wind, and warm temperatures; cold temperatures conserve it. In addition, manure from dry lot dairy farms, beef feedlots and other animals confined to unpaved feedlots, and paddocks tend to include a fair amount of soil that is scraped up with the manure. The soil increases the manure's density and ash level, which consequently lowers the organic matter.

While manure is a natural composting feedstock, it is not without a few potential concerns. Many manures have relatively high levels of salt, pH, and ammonia, which can be a detriment in some situations There is a "silver lining" in the fact that the high salt levels are primarily attributed to high levels of nutrients. Depending on the source, manure can carry unwanted debris like medicine containers, trash, and scrap metal. However, it is generally free of the plastic contamination common to yard trimmings and food waste. Manures carry weed seeds, parasites, and likely contain some traces of antibiotics and other medicines. These biological contaminants are reliably killed and/or decomposed during thermophilic composting.

Some types of manure are also possible sources of the persistence pyridine herbicides like clopyralid, aminopyralid, and picloram. These herbicides may be used on animal feed and forage crops, especially grass hay but also on sugarbeets, wheat, and mint. It is worth noting that damaging levels of these herbicides have been found in cattle and horse manure.

8.7 Agricultural crop and processing residue

The production of farm crops, and value-added products, generates a number of organic by-products that are not necessarily applied back to crop land. Prominent examples include straw, hay, spoiled silage, wasted feed, corn stover, corn cobs, bagasse, culled fruit and vegetables, rice hulls, nut shells and pomace from pressing olives, apples, grapes, and other crops. Some of these materials may come from locations beyond the farm fields, including packing houses, storages, markets, and processing facilities.

As a category, the composting characteristics of agricultural residues cannot be generalized because they are specific to the materials at hand. One group of residues is consistently dry, low in nitrogen, and decompose moderately well (e.g., straw, corn stalks, bagasse). They make good amendments for composting with wet feedstocks like most manures. Some residues in this group also have a particularly strong structure and make good bulking agents (e.g., nut shells, rice hulls). Another group is generally wet, nitrogenous, and degrades quickly (e.g., unused fruits, vegetable culls, spoiled silage). These materials need to be combined with drier feedstocks for composting.

In general, agricultural crop residues are clean feedstocks that decompose well and present little odor risk. However, it is important that agricultural residues are composted at thermophilic temperatures. Most agricultural by-products carry plant pathogens, weed seeds and, in some cases, pesticide residues. Each material should be evaluated separately to determine its applicability as a compost feedstock.

8.8 Food processing waste

A wide variety of by-products from food processing make good feedstocks for composting. Some examples of food processing by-products that have been composted include potato peels, wasted pasta, filter press cake from cranberries and other fruits, spent brewers' grain, almond hulls, tomato paste, corn cobs, olive mill pomace, spent coffee grounds, gelatin waste, out-of-date beverages, and seafood processing waste (see below). These by-products differ greatly from one another; for instance, ranging from dry solids (e.g., peanut shells) to completely liquid (e.g., cheese whey). Therefore, each type must be evaluated on its own. Meat processing and butchering wastes carry special restrictions. Composting of butcher waste should follow the same procedures as composting animal mortalities, which are discussed in Chapter 8.

Many food processing by-products still have food value and can be attractive to flies, rodents, birds, and other animals. Such by-products must be composted in a contained manner, as previously described for food waste. The odor risk depends entirely of the specific material and ranges from low to moderate (e.g., pomace) to potentially high (e.g., gelatin waste). Compared to commercial and residential food wastes, food processing products are relatively free of contaminants. However, some by-products may contain source-specific contaminants like machinery and cleaning solutions used at the processing plant and poorly degradable additives such as filtering and pressing aids (e.g., diatomaceous earth, rice hulls). Because a feedstock comes from a single generator, contamination is relatively easy to correct, compared to consumer and commercial food wastes. A major advantage of food processing by-products is the opportunity to receive a tipping fee.

8.9 Fish and seafood processing waste

Fish and seafood wastes includes whole fish, racks, frames, heads, tails, crustacean and mollusk shells, and ground-up mixed fish waste (called "gurry" in some regions). Some packing operations may even generate associated by-products like wasted bread crumbs. Fish and seafood wastes have a long history of composting. Few other recycling alternatives can economically take advantage of their rich nutrient content and the nuisance factors that fish and seafood wastes present. These materials usually come with a tipping fee for the composter.

Most fish wastes are wet to very wet (70% to 90%). Almost all fish wastes are high in nitrogen (between 7% and 15%), giving them a very low C:N ratio. Lobster, crab, shrimp, and mollusk shells, with an average of about 50% moisture, are drier than fish waste. The shells retain enough meat and nitrogen to still give them a low C:N ratio. The shells provide good structure to a compost pile. The crustacean shells decompose during composting; mollusk shells do not decompose at all, although they may become more brittle with thermophilic temperatures. Fish flesh decomposes very rapidly and quickly generates putrid amine-related odors. Bones are slower to decompose but generally disappear during composting.

Fish and seafood waste must be handled in manner that minimizes odor and discourages birds, flies, and other pests. A composting recipe that includes fish and seafood requires a large amount of chunky, dry amendment to reduce odor, and absorb moisture. As with other food wastes, fish and seafood items need to be isolated from the surrounding environment until they are well-decomposed.

8.10 Anaerobic digestate

AD is a popular treatment option for manure, sewage sludge and, increasingly, food waste. It recovers biogas energy from these organic feedstocks but leaves behind most of the volume and nutrients of the feedstocks in the "digestate," the partially stabilized effluent of digestion. Regardless of the feedstock, digestate can be composted to increase its value and broaden its potential uses. Municipal biosolids, which has a long history of composting, is the digestate of anaerobic digestors at the wastewater treatment plant.

Digestate reflects the characteristics of the feedstocks from which it is made. Relative to the raw feedstock mix, organic matter, carbon and solids concentrations decrease but the volume diminishes only slightly, becoming slightly more liquid. Nearly all of the plant nutrients entering with the feedstock remain in the digestate, although much of nitrogen changes from organic to ammonia nitrogen. With the conservation of nitrogen and the loss of carbon, the C:N ratio decreases slightly. AD reduces pathogens and the viability of weeds but does not eliminate them unless the digestion process is thermophilic, which is uncommon. The digestate is somewhat more biologically stable than the raw feedstock, which means it is slower to decompose. Still, digestates typically retain enough degradable organic matter to decompose well and generate heat during composting. One advantage of digestate is that it tends to be less odorous than the raw feedstocks. Many of the volatile odorous compounds formed, like hydrogen sulfide, leave with the biogas.

Digestate that results from "dry" AD systems are moderately wet with moisture contents in the range of 60% to 80%. Most AD systems, however, process dilute wet feedstocks and produce a digestate in the range of 90% to 95% moisture. This digestate can serve as a source of moisture for composters in arid climates. However, usually digestate is separated into liquid and solid fractions. The solid fraction, at 75% to 80% moisture, is composted. In this case, most of the nitrogen in the digestate is carried in the liquid fraction, thereby by-passing the compost.

Much of the nitrogen in either fraction of digestates is in the ammonia form, which readily volatilizes. Therefore, high N digestates are best combined with abundant low-N and low-pH feedstocks, preferably those with available carbon (e.g., straw rather than wood). Also, turning and other modes of agitation should be minimized in the first few weeks of composting to avoid exposing the interior of the pile, where the ammonia is concentrated, to the outside environment.

8.11 Compostable plastics

With compostable plastics, more food wastes can be captured and composted without the effort of separating plastic contaminants or sacrifices to compost quality; at least that is the hope. Over the past decade, many technological and strategic advances have made compostable plastics a reality for the composting industry. However, their widespread acceptance remains frustrated by several challenges (Box 4.4).

Box 4.4 The promise and perils of compostable plastics

Author: Matthew Cotton

Plastic products that thoroughly decompose during composting offer the promise that food residuals, and other organics, can be widely composted without the worry and inconvenience of physical contamination. This promise has yet to be fully realized, even though great strides have been made in creating compostable plastic products. Industry can make plastics that decompose under standardized composting conditions. A number of products have been certified to do so. However, peripheral factors, beyond the chemistry and design of the plastic products create confusion and hinder the potential of compostable plastics (EEA, 2020). Such factors include sporadic and inappropriate use, lack of identification, and variability of composting practices.

The promise of *compostable* materials is that they may function just like plastics in their primary use, especially in food service applications—like plates, cups, and bags—thus making the separation of food for composting more productive. However, there are several challenges to the composter who would accept and process these materials. And in the end, replacing single use disposables with single-use compostables does not necessarily reduce their environmental impact.

Although there are still questions to be addressed about how quickly compostable plastics break down in a well-managed compost system, there are larger, fundamental questions which offer perhaps greater challenges:

Identification Compostable plastics look nearly identical to their noncompostable counterparts. Regular plastics are a contaminant that most composters strive to avoid, and spend considerable time and money trying to remove before they compromise finished products. It can be extremely difficult for composters to separate the compostable from the noncompostable materials. Depackaging machines, which some composters use as part of preprocessing can remove significant amounts of contaminants in a food waste stream, but the technology does not differentiate between compostable and noncompostable. The same is true of a back-end screen. Some have hoped to overcome the identification problem with a single bold choice of color or design, like a green stripe or a consistent color. Unfortunately, many of the large national food manufacturers have invested in branded color schemes that (in their minds) conflict with a green stripe, and there's nothing stopping traditional plastics manufacturers from making copycat items. This widespread confusion leads to the majority of compostable plastics collected with food being removed and disposed of, wasting the opportunity for them to be composted.

User confusion Compostable products are evolving at a time when compost infrastructure is not consistently developed in the United States. Once a Starbucks user learns that their coffee cup, lid, and stirrer or straw goes in the compost bin, they may think ALL coffee cups, lids, etc., go in the compost bin. Some composters readily accept them, and some avoid them as much as they can. This confusion leads to noncompostable items being put in compost bins and vice versa.

Organic certification Getting compost approved for use by certified organic farms is an important market for some composters. Even some composters who sell to conventional growers go through the effort to get their products approved for organic use as a marketing tool. Organic certification is a powerful imprimatur. The US Department of Agriculture's National Organic Program currently views compostable plastics as synthetic and while the use of compostable plastics in compost is not expressly prohibited, many certifiers will not approve a compost that contains compostable plastic, unless it can be demonstrated that the plastic materials are removed.

Compost residence time How long a compost process takes, from raw feedstocks to finished product, varies considerably among facilities, but in general, compost residence times are shrinking as facilities handle more and more feedstocks and become more proficient in managing the process. The lab tests developed to ascertain "compostablilty" assume disintegration in 12 weeks and complete biodegradation in 180 days. Many composters make compost in much faster time frames, leaving compostable materials inadequate time to break down.

> **Box 4.4 The promise and perils of compostable plastics—cont'd**
>
> Therefore, if "compostable" plastics are included in the compost process, a composter should make sure the products, at a minimum, meet the specifications for "compostable" products. Products that meet these standards can be identified by the "Compostable" Logo stamped on the products. In addition, it is wise to test a compostable product on a small scale in the composting operation to see if it disintegrates and/or biodegrades well enough within in the required time and under your conditions. For example, if compost is to be used or sold in 90 days and the plastic takes 180 days to completely disintegrate and biodegrade, that particular product may be problematic. Any composters considering accepting compostable plastics should seek as much information as possible and evaluate manufacturer claims with skepticism considering all of the possible implications of accepting these materials.

There are several types of plastics resins that are designed to be "biodegradable." Polylactic acid (PLA), made by fermentation of agricultural materials, is probably the most recognizable and most common. Others are derived from starches (e.g., polyhydroxyalkanoate or PHA) and cellulose (e.g., cellophane). These and other resins are used to manufacture a variety of biodegradable products, such as bags, wrappers, food utensils, containers, drinking cups, and films for agricultural mulch. Although the resins used and a product's function might suggest how well a given product decomposes, these features are not good enough predictors. Testing is still necessary.

People often use the terms *degradable*, *biodegradable*, and *compostable* plastics[4] interchangeably, but they are not the same. Under the conditions, and in the duration of, a composting pile, the "degradable" and "biodegradable" plastics may not completely degrade. Most operations sell their compost after 6 to 12 months, so "compostable" plastics must reach significant levels of biodegradation in these time frames. Otherwise, they are noticeable in the finished compost or they wind up being screened out prior to sale. What composters need are products that are explicitly "compostable."

Several standards now exist for determining whether a biodegradable plastic product is acceptably compostable. Generally, to earn this label, the product must: (1) disintegrate rapidly under thermophilic composting conditions; (2) biodegrade completely during composting or shortly after the finished product is used at a rate that is consistent with other materials composted; (3) generate no toxic compounds that can destroy the utility of the compost; and (4) not contain high levels of regulated metals. In Europe, EN 13432 is the standard for packaging products (EEA, 2020). In the United States, the corresponding standards are published by the American Society for Testing and Materials (ASTM). The standards are ASTM D6400 (specification for compostable plastics) and ASTM D6868

---

[4] The term "bioplastics" is intentionally excluded here. Bioplastics infers that the plastic resin has been derived from biological materials. It does not mean that the resulting product is degradable, biodegradable, or compostable (van den Oever, 2017).

FIGURE 4.10

Compostable products logo certified by the Biodegradable Products Institute.

Source: Biodegradable Products Institute, www.bpiworld.org.

(specification for biodegradable plastic used on paper and other compostable substrates). The ASTM specifications guarantee that there is no more than 10% of the product remaining on a 2 mm sieve after 12 weeks of composting (disintegration), that at least 90% of the plastic is converted to carbon dioxide in 180 days or less (biodegradation), the final product supports plant life when compared to comparable control compost (compost utility) and the final product does not have concentrations of regulated metals greater than 50% of federal mandates in the country where the products are sold. Products that meet compostable standards can be identified with a logo issued by a recognized industry organization or public authority. For example, in North America, the logo shown in Fig. 4.10 is certified by the Biodegradable Products Institute (BPI, 2020). Depending on the certifying organization, other restrictions may also apply. For instance, BPI prohibits the use of PFAS and restricts the use of carcinogens, reproductive toxins, and mutagens (CRMs).

8.12 Inorganic and organic extras

Some composters include miscellaneous and mostly inorganic ingredients to a composting mix. Examples include chemical fertilizers, lime, soil, peat moss, biochar, and recycled compost. In some cases, such materials are added to improve the composting process or product. A few additives, including biochar and zeolite, have been reported to reduce ammonia loss or greenhouse gas (GHG) emissions (see Chapter 11). In other cases, the motivation is to gain a fee for composting harmless by-products that need an outlet. Wood and coal ash and gypsum are examples in this case. Each of these materials has its own character and influence in the composting pile.

Fertilizer and urea, or other concentrated nitrogen sources, are sometimes added to lower the initial C:N ratio of a composting mix. While fertilizer does add nitrogen, and often speeds the process, the benefits are short-lived. Nitrogen from such sources tends to be available much more quickly than the carbon in the organic materials. Initially the available carbon and nitrogen are in balance; but as the easily available carbon is depleted, a surplus of nitrogen soon develops. Eventually the excess nitrogen is lost as ammonia.

Lime is added either to adjust pH upward or to control odors. Lime is rarely a necessary ingredient, as pH adjustment is rarely necessary, and can be detrimental. If lime is used for odor control, it can raise the pH enough to cause an excessive loss of ammonia. The odor-reducing effect of lime is temporary, and it requires a great deal of lime. The same effects should be expected for other concentrated sources of alkalinity, including cement kiln dust and wood ash.

Soil, especially clay soil, is a recommended ingredient for certain approaches to composting. The purpose of the soil is to provide additional microscopic surfaces for microorganisms to inhabit and metabolize the organic feedstocks. Clay also has the ability to hold onto positively-charged nutrients (i.e., cations). Also, advocates believe clay encourages humus accumulation. These purported benefits are not universally accepted. Still, adding soil has certain effects, both good and bad. On the positive side, soil helps to moderate moisture and temperature within the pile. It also adds weight to the finished compost, making the product feel more soil-like. Conversely, the soil decreases the organic matter level and raises the ash content. Abundant soil decreases the porosity and increases the bulk density of the composting mix. In short, adding soil is unnecessary but can help achieve specific aims.

Peat moss is an acidic fibrous material which has resulted from years of anaerobic decomposition beneath water or water-logged soils. It is low in nitrogen and highly absorbent of water, nutrients, and odors. It may hold over 10 times its weight in water. Except in regions where natural deposits exist, peat moss is expensive, partly because of its competing uses as an amendment for potted plants and other horticultural crops. Peat moss passes through the composting process virtually unchanged, producing a potentially high valued compost. Its odor- and water-absorbing qualities make it an excellent amendment, but cost limits its use.

Biochar is the dry charcoal-like residue obtained from "burning" biomass (usually wood) with insufficient oxygen so that combustion is incomplete and much of the carbon is retained. Biochar has abundant micropores and surfaces for adsorbing chemical compounds. It is normally promoted as a soil amendment, by itself and in combination with compost. However, it has also been proposed and used as an amendment for composting feedstocks. When used as a feedstock amendment, the main potential benefits of biochar are absorption of moisture, lower emissions of ammonia and GHGs and enhancement of the compost product (precharged with biochar). The carbon in biochar is largely unavailable to microbial decomposers. Research studies have generally shown that biochar can provide these benefits when the biochar comprises 2%−10% of the composting recipe, by volume (Camps and Tomlinson, 2015). However, there are several variables to consider including the relative amount of biochar used and the qualities of the biochar itself (e.g., how it was made and from what feedstocks). Overall, biochar appears to be a positive feedstock amendment, depending on the cost to purchase and handle it.

Wood ash is very dry with little or no carbon and nitrogen. It contains a fair amount of other nutrients, particularly potassium. The concentrations of trace metals may be a concern with some ash. In a composting mix, wood ash absorbs moisture. Like lime, it raises the pH of the mix and has also been used as an odor reducing agent. Handling is difficult as the ash is a fine powder which blows around and creates dust. Particles tend to cement together after they become wet. Wood ash can be used as a composting amendment for wet acidic mixes or for increasing the level of nutrients, especially potassium. It should not be used if the pH of the mix is already high.

Coal ash is what is left over after burning coal for fuel. Like wood ash, it is very dry and powdery but otherwise it differs from wood ash in content. Coal ash basically consists of oxides of silicon, aluminum, titanium, iron, calcium, magnesium, potassium, and sodium. It also contains molybdenum, sulfur, and a myriad of other elements that may be of concern. A 1994 study at Washington State University composted coal ash with manure and found the final product to show poor growth in greenhouse bioassays, but increased barley yield in field applications. Unless there is a real need for disposal of waste coal ash and a large tipping fee is involved, it is probably best to keep this out of the compost pile.

Zeolites are a group of minerals classified as aluminosilicates. Many types of synthetic zeolites are manufactured with different specific properties, while natural zeolites are mined, with properties that depend on the mineral deposits. They are used in a wide variety of applications from cat litter to water filtration.

Zeolites have very porous microstructures and high cation exchange capacities. They offer abundant internal surfaces that hold onto water, nutrients, and other compounds. Consequently, zeolite is of interest as an additive to limit the mobility of trace elements and reduce emissions of GHG and ammonia during composting, during storage and land application of raw manure, and as an agricultural soil amendment (Cataldo et al., 2021). In research studies for composting applications, zeolites have been applied both as a thin blanket (e.g., 2—4 cm or 1—2 in.) and as a component of the feedstock mix. Typically, zeolite comprises 5% to 10% of the mix on a dry weight basis (Soudejani et al., 2019). As Chapter 11 details, research results have generally reported substantial reductions in emissions.

The primary question regarding zeolite is the cost to obtain and handle it. On a per unit basis, the cost of zeolite products appears to be inexpensive. However, the large volumes required in composting applications can amount to a large expense.

Most of the *gypsum* available to composters is from wasted drywall, a layer of gypsum, 12—20 mm thick ($^1/_2$ to $^3/_4$ in.) sandwiched between thin sheets of paper. Also called, sheet rock or wallboard, it is generated in large quantities by construction and demolition activities. Therefore, composting drywall is considered a recycling option. Drywall needs to be broken up into small pieces before being composted. While the paper decomposes during composting, the gypsum merely disintegrates and adds its minerals to the compost. Gypsum is very dry with a moisture content around 5% and has almost no nitrogen. Gypsum is basically calcium sulfate, so the amounts of both calcium and sulfur increase in the finished compost. The additional sulfur can increase the emissions of sulfur-based odors. For the composter, the primary benefit of gypsum is the potential fee for taking waste drywall. It has little biochemical effect on the process, although gypsum has been reputed to reduce ammonia volatilization during manure and biosolids composting. Gypsum does not lower or raise pH, but it may change the consistency and feel of the compost.

Table 4.5 General and typical qualities of common composting feedstocks.

Feedstock	Moisture content	Nitrogen content	Bulk density	Structure	Degradability	Odor risk	Contamination concern	Notes
Yard trimmings (green waste)								
Mixed yard trimmings	Moderate	Moderate	Low-mod.	Good	Poor-fair	Low	Moderate	1,2
Deciduous leaves	Moderate	Moderate	Low	Good-fair	Good	Low	Moderate	2
Coniferous needles (e.g., pine)	Low-mod.	Moderate	Moderate	Fair	Good	Low	Low	3
Grass clippings	High	High	High	Poor	Good	High-mod.		4
Seaweed and aquatic plants	High	High	High	Poor	Good	Low-mod.	Low	2
Brush	Moderate	Low-mod.	Low	Good	Poor-fair	Low	Low	1
Wood and paper								
Wood and bark chips	Low-mod.	Very low	Low	Very good	Poor	Low	Low	5
Wood shavings and sawdust	Low-mod.	Very low	Low	Good	Poor	Low	Low	5
Waste paper	Very low	Very low	High-mod.	Poor	Good-fair	Very low	Low-mod.	6,7
Corrugated cardboard	Low	Very low	Moderate	Fair	Fair	Very low	Moderate	6
Paper mill sludges	High	Low-mod.	Low	Poor	Good-fair	Moderate	Low	

Continued

Table 4.5 General and typical qualities of common composting feedstocks.—*cont'd*

Feedstock	Moisture content	Nitrogen content	Bulk density	Structure	Degradability	Odor risk	Contamination concern	Notes
Municipal waste								
Biosolids	High	High	High	Poor	Good	High-mod.	High-mod.	8
Mixed solid waste	Low-mod.	Low-mod.	Moderate-low	Moderate	Moderate	High	Very high	9,10
Animal manure								
Caged poultry	High-mod.	Very high	High	Poor	Good	High	Low	11
Floor-raised poultry	Low-mod.	High	Moderate	Moderate	Good	High-mod.	Low	11,12
Dairy cattle	High-mod.	High-mod.	High-mod.	Moderate	Good	Moderate	Low-mod.	12,13,15
Beef cattle	High-mod.	High-mod.	High-mod.	Moderate	Good	Moderate	Low	12,13,14
Horses	Moderate	Moderate	Moderate	Moderate	Good	Low-mod.	Low	12
Goats, sheep, rabbit	High-mod.	Moderate	High-mod.	Moderate	Good	Low-mod.	Low	12,16
Swine	High	High	High	Poor	Good	High	Low-mod.	12,13,15
Fish manure (aquaculture)	High	High-mod.	High	Poor	Good	Moderate	Low	17

Agricultural crop and processing residuals

Hay	Low	Low-mod.	Low	Good-fair	Good-fair	Low	Low	11,18,19
Straw	Low	Low	Low	Good	Moderate	Very low	Low	18,19
Corn stover	Low	Low	Low	Good	Moderate	Very low	Low	18,19
Bagasse (e.g., sugarcane)	Low-mod.	Low	Low	Good	Moderate	Low	Low	11,18
Spoiled silage	High-mod.	Moderate	High-mod.	Moderate	Good	Moderate	Low	11,18
Rice hulls	Low	Low	Low	Good-fair	Poor	Very low	Low	
Nut shells	Low	Low	Low	Good	Moderate	Very low	Low	
Cotton gin trash	Low-mod.	Moderate	Moderate	Moderate	Good-fair	Low	Low-mod.	
Cranberry plant residues	Moderate	Low-mod.	Low-mod.	Moderate	Good-fair	Very low	Low	3
Culled fruit and vegetables	High	High-mod.	High	Poor-fair	Good	Low-mod.	Low	20
Potato culls	High	High-mod.	Low-mod.	Poor	Good	Low-mod.	Low	20
Apple, grape and cranberry filter cake	High-mod.	Moderate	Low-mod.	Poor	Good	Low-mod.	Low	21

Continued

Table 4.5 General and typical qualities of common composting feedstocks.—*cont'd*

Feedstock	Moisture content	Nitrogen content	Bulk density	Structure	Degradability	Odor risk	Contamination concern	Notes
Food waste								
Mixed food waste	Low-mod.	High-mod.	Low-mod.	Poor-fair	Good	High	High-mod.	22, 23
Food processing wastes	High-mod.	High-mod.	High-mod.	Poor-fair	Good	High-mod.	Low	22
Spent brewers' grains	High	High-mod.	High-mod.	Poor	Good	Moderate	Low	24
Waste beverages	Very high	Very low to none	Liquid	Liquid	High	Very low	Low	25
Fish and seafood								
Fish waste (e.g., frames, tails, dead fish)	High	Very high	Low	Poor	Good	High	Low	26,30
Fish gurry	Very high	High	Liquid	Liquid	Good	High-mod.	Low	27,30
Crustacean shells	Moderate	High	Moderate	Moderate	Good-fair	High	Low	28
Mollusk shells	Moderate	High-mod.	Low	Good	Poor	High-mod.	Low	29
Meat residuals								
Butcher and meat packing waste	High	Very high	Low	Poor-fair	Good-fair	High	Low	26,30
Blood	High-mod.	Very high	Low	Liquid	Good	High	Low	30
Paunch manure	High	Moderate	High-mod.	Moderate	Good	Moderate	Low	31

[1] Characteristics depend on the relative mix of woody and vegetative materials. The characteristics moderates with more vegetation.
[2] Physical contaminants like metal hardware, plastic bags, and other litter can occur with curbside collection.
[3] Can have a low pH (acidic).
[4] May carry residues of persistent pyridine herbicides.

5 Products derived from green wood have moderate moisture and low odor risk while those made from dried lumber (e.g., pallets) have a very low moisture content and odor.

6 Various types of wastepaper exist. Newsprint retains lignin and is therefore less degradable than other types of waste paper. Magazines have a high ash content due to the use of clay to impart a gloss.

7 Flat items, like sheets of paper, can stick together in layers, giving them a higher bulk density.

8 Some sources of biosolids may have high concentrations of trace metal.

9 Mixed solid waste is heavily contaminated with physical and some chemical contaminants and needs, which requires substantial pre- and/or postseparation operations.

10 Characteristics depend on the relative amount of paper included. Paper moderates the characteristics.

11 Depends on the amount of ambient drying occurs.

12 Depends on the amount of litter or bedding used. Litter/bedding decreases moisture and bulk density dilutes nutrients and increases structure and degradability.

13 Also, depends on methods of collection.

14 Manure from outdoor feedlots includes soil that is incidentally collected with the manure. Soil increases the bulk density and ash content.

15 Some manures may contain noticeable amounts of copper due to the use of copper sulfate in footbaths.

16 Typically occurs as a "bedded pack," manure mixed with bedding that accumulates for months.

17 Solids settled out from fish rearing ponds and raceways. May contain a fair amount of soil.

18 Moisture and nutrient content depend on exposure to weather.

19 Composting improves if chopped.

20 Structure disappears quickly, after plant cells begins to decompose and release moisture.

21 Residue remaining after juice is pressed out. May include pressing and filtering aides like diatomaceous earth. Grape pomace includes seeds that resist decomposition.

22 Highly variable, depending on the source and presence of other items, especially paper and cardboard.

23 May be highly contaminated with plastic if not separated at the source.

24 Highly variable depending on the product and processes.

25 Primary value is as a source of moisture.

26 May contain bones.

27 Fish waste ground into a liquid slurry.

28 E.g., lobster and crab shells decompose during composting.

29 E.g., clam, mussel, and scallop shells. Attached meat decomposes during composting but the shells remain intact.

30 May have other higher value uses compared to composting.

31 Stomach content of cattle and other ruminant animals at the time of slaughter. Silage-like in consistency.

References

Cited references

Amlinger, F., Line Diana Blytt, L.D., 2013. How to Comply with the EU Animal By-Products Regulations at Composing and Anaerobic Digestion Plants. European Compost Network. https://www.compostnetwork.info/download/good-practice-guide-comply-eu-animal-products-regulations-composting-anaerobic-digestion-plants/.

Arikan, O.A., Sikora, L.J., Mulbry, W., Khan, S.U., Foster, G.D., 2007. Composting rapidly reduces levels of extractable oxytetracycline in manure from therapeutically treated beef calves. Bioresour. Technol. 98, 169−176.

Arikan, O.A., Mulbry, W., Rice, C., 2009. Management of antibiotic residues from agricultural sources: use of composting to reduce chlortetracycline residues in beef manure from treated animals. J. Hazard Mater. 164 (2−3), 483−489.

Balch, A., Wilkinson, K., Schliebs, D., Hazell, L., 2019. Critical Evaluation of Composting Operations and Feedstock Suitability Critical Evaluation of Composting Operations and Feedstock Suitability, Phase 1 Report − Odour Issues. Arcadis Australia Pacific Pty Limited prepared for the Department of Environment and Science (DES Queensland). https://environment.des.qld.gov.au/__data/assets/pdf_file/0024/226293/phase-1-composting-study-report.pdf.

Beecher, N., Brown, S., 2018. PFAS and Organic Residuals Management, Parts I and II. Bio-Cycle. July and August 2018.

Brinton, W., Dietz, C., Bouyounan, A., Matsch, D., 2018. Microplastics in compost: the environmental hazards inherent in the composting of plastic-coated paper products. Woods End Lab. and Eco-cycle. http://ecocycle.org/files/pdfs/microplastics_in_compost_summary.pdf.

BPI (Biodegradable Plastics Institute), 2020. The Compostable Logo. https://bpiworld.org/BPI-Public/Program.html/.

Buyuksonmez, F., Rynk, R., Hess, T.F., Bechinski, E., 1999. Occurrence, degradation and fate of pesticides during composting: Part I: composting, pesticides, and pesticide degradation. Compost Sci. Util. 7 (4), 66−82.

Buyuksonmez, F., Rynk, R., Hess, T.F., Bechinski, E., 2000. Occurrence, degradation and fate of pesticides during composting: Part II: occurrence and fate of pesticides in compost and composting systems. Compost Sci. Util. 8 (1), 61−68.

Calisti, R., Regni, L., Proietti, P., 2020. Compost-recipe: a new calculation model and a novel software tool to make the composting mixture. J. Clean. Prod. 270 (10), 122427. https://doi.org/10.1016/j.jclepro.2020.122427. ISSN 0959-6526.

Camps, M., Tomlinson, T., 2015. The use of biochar in composting. Int. Biochar. Initiat. https://www.biochar-international.org/wp-content/uploads/2018/04/Compost_biochar_IBI_final.pdf.

Cataldo, E., Salvi, L., Paoli, F., Fucile, M., Masciandaro, G., Manzi, D., Masini, C.M., Mattii, G.B., 2021. Application of zeolites in agriculture and other potential uses: a review. Agronomy 11 (8), 1547. https://doi.org/10.3390/agronomy11081547.

Chandler, J.A., Jewell, W.J., Gossett, J.M., Van Soest, P.J., Robertson, J.B., 1980. Predicting methane fermentation biodegradability. In: Biotechnology and Bioengineering Symposium No. 10, pp. 93−107.

Chen, Z., Zhao, W., Xing, R., Xie, S., Yang, X., Cui, P., Lü, J., Liao, H., Yu, Z., Wang, S., Zhou, S., 2020. Enhanced in situ biodegradation of microplastics in sewage sludge using hyperthermophilic composting technology. J. Hazard Mater. 384 (2020), 121271. https://doi.org/10.1016/j.jhazmat.2019.121271. ISSN 0304-3894.

Coker, C., 2020a. Managing PFAS chemicals in composting and anaerobic digestion. Biocycle. January 21 and 28, 2020.

Dolliver, H., Gupta, S., Noll, S., 2008. Antibiotic degradation during manure composting. J. Environ. Qual. 37, 1245−1253.

Donoho, A.L., 1984. Biochemical studies on the fate of monensin in animals and the environment. J. Anim. Sci. 58 (6), 1528−1539.

EEA (European Environment Agency), 2020. Biodegradable and Compostable Plastics Challenges and Opportunities. https://doi.org/10.2800/552241. ISBN 978-92-9480-257-6 ISSN 2467-3196. https://www.eea.europa.eu/publications/biodegradable-and-compostable-plastics.

Ekinci, K., Keener, H.M., Elwell, D.L., 2000. Composting short paper fiber with broiler litter and additives: part i: effects of initial pH and carbon/nitrogen ratio on ammonia emission. Compost Sci. Util. 8 (2), 160−172. https://doi.org/10.1080/1065657X.2000.10701761.

EFSA J. 18 (8), 2020, 6226. https://www.efsa.europa.eu/en/efsajournal/pub/6226.

Frischmann, C., 2018. The climate change impact of food in the back of your fridge. Wash. Post. July 341, 2018.

Gouleke, C.G., 1972. Composting: A Study of the Process and its Principles. Rodale Press, Inc, Emmaus, PA.

Khan, S.J., Roser, D.J., Davies, C.M., Peters, G.M., Stuetz, R.M., Tucker, R., Ashbolt, N.J., 2008. Chemical contaminants in feedlot wastes: concentrations, effects and attenuation. Environ. Int. 34, 839−859.

Kayhanian, M., Tchobanoglous, G., 1992. Computation of C/N ratios for various organic fractions. Biocycle 33 (5), 58−60.

Lemmon, C.R., Pylypiw, H.M., 1992. Degradation of diazinon, chlorpyrifos, isofenphos and pendimethalin in grass and compost. Bull. Environ. Contam. Toxicol. 48 (3), 409−415.

McKellar, Q.A., 1997. Ecotoxicology and residues of anthelmintic compounds. Vet. Parasitol. 72, 413−435.

Oshins, C., 2019. Bulk Density and FAS Measurement. CCREF Compost Operators Training Workbook, Ithaca, NY. August 2019. https://www.compostfoundation.org/Education/COTC.

Paul, J., Geesing, D., 2015. Compost Facility Operator Manual. Transform Compost Systems, Abbotsford, BC. http://www.transformcompostsystems.com/learn-compost-operator-manual.php.

Pauwels, B., Wille, K., Noppe, H., De Brabander, H., Van de Wiele, T., Verstraete, W., Boon, N., 2008. 17α-ethinylestradiol cometabolism by bacteria degrading estrone, 17β-estradiol and estriol. Biodegradation 19, 683−693.

Puglisi, E., Cappa, F., Fragoulis, G., Trevisan, M., Del Re, A.A.M., 2007. Bioavailability and degradation of phenanthrene in compost amended soils. Chemosphere 67, 548−556.

Puyuelo, B., Ponsá, S., Gea, T., Sánchez, A., 2011. Determining C/N ratios for typical organic wastes using biodegradable fractions. Chemosphere 85 (4), 653−659. https://doi.org/10.1016/j.chemosphere.2011.07.014. ISSN 0045-6535.

Richard, T., 1996. The Effect of Lignin on Biodegradability. Cornell Waste Management Institute. http://compost.css.cornell.edu/calc/lignin.html.

Sanchez-Monedero, M.A., Cayuela, M.L., Roig, A., Jindo, K., Mondini, C., Bolan, N., 2018. Role of biochar as an additive in organic waste composting. Bioresour. Technol. 247, 1155−1164. https://doi.org/10.1016/j.biortech.2017.09.193. PMID: 29054556.

Soares, M.A., Quina, M.J., Quinta-Ferreira, R., 2013. Prediction of free air space in initial composting mixtures by a statistical design approach. J. Environ Manage. 128, 75−82. https://doi.org/10.1016/j.jenvman.2013.04.041. Epub 2013 May 27. PMID: 23722176.

Soudejania, H.T., Kazemian, H., Inglezakis, V.J., Zorpase, A.A., 2019. Application of zeolites in organic waste composting: A review. Biocatalysis and Agricultural Biotechnology 22, 101396.

Storteboom, H.N., Kim, S.-C., Doesken, K.C., Carlson, K.H., David, J.G., Pruden, A., 2007. Response of antibiotics and resistance genes to high-intensity and low-intensity manure management. J. Environ. Qual. 36, 1695−1703.

Strom, P.F., 2000. Pesticides in yard waste compost. Compost Sci. Util. 8 (1), 54−60.

Tchobanoglous, G., Theisen, H., Vigil, S.A., 1993. Integrated Solid Waste Management: Engineering Principles and Management Issues. McGraw-Hill, New York, NY.

USCC, 2020. PFAS in Compost. https://www.compostingcouncil.org/page/pfas.

Van den Oever, M., Molenveld, K., Van der Zee, M., Bos, H., 2017. Bio-based and biodegradable plastics - facts and figures. Wageningen Food Biobased Res. 172. ISBN-number 978-94-6343-121-7. https://doi.org/10.18174/408350.

Vandervoot, C., Zabik, M.J., Branham, B., Lickfeldt, D.W., 1997. Fate of selected pesticides applied to turfgrass: effect of composting on residues. Bull. Environ. Contam. Toxicol. 58, 38−45.

Wang, S.P., Zhong, X.Z., Wang, T.T., Sun, Z.-Y., Tang, Y.Q., Kida, K., 2017. Aerobic composting of distilled grain waste eluted from a Chinese spirit-making process: the effects of initial pH adjustment. Bioresour. Technol. 245 (Pt A), 778−785. https://doi.org/10.1016/j.biortech.2017.09.051.

Watteau, F., Dignac, M., Bouchard, A., Revallier, A., Houot, S., December 11, 2018. Microplastic detection in soil amended with municipal solid waste composts as revealed by transmission electronic microscopy and pyrolysis/GC/MS. Front. Sustain. Food Syst. https://doi.org/10.3389/fsufs.2018.00081.

Weithmann, N., Möller, J.N., Löder, M.G.J., Piehl, S., Laforsch, C., Freitag, R., 2018. Organic fertilizer as a vehicle for the entry of microplastic into the environment. Sci. Adv. (published online April 4, 2018) https://advances.sciencemag.org/content/4/4/eaap8060.

Consulted and suggested reading

Aikaitė-Stanaitienė, J., Grigiškis, S., Levišauskas, D., Čipinytė, V., Baškys, E., Kačkytė, V., 2010. Development of fatty waste composting technology using bacterial preparation with lipolytic activity. J. Environ. Eng. Landscape Manag. 18 (4), 296−305.

Allison, F.E., 1965. Decomposition of wood and bark sawdusts in soil, nitrogen requirements, and effects on plants. Agric. Res. Serv. USDA, Tech. Bull. No. 1332.

Awasthi, M.K., Wang, Q., Ren, X., Zhao, J., Hui Huan, H., Awasthi, S.K., Lahori, A.H., Li, R., Zhou, L., Zhang, Z., November 2016. Role of biochar amendment in mitigation of nitrogen loss and greenhouse gas emission during sewage sludge composting. Bioresour. Technol. 219, 270−280.

Barthod, J., Rumpel, C., Dignac, M., 2018. Composting with additives to improve organic amendments. A review. Agron. Sustain. Dev. 38. Article number:17.

Bayard, R., Benbelkacem, H., Gourdon, R., Buffière, P., June 15, 2018. Characterization of selected municipal solid waste components to estimate their biodegradability. J. Environ. Manag. 216, 4−12.

Bekier, J., Drozd, J., Jamroz, E., Jarosz, B., Kocowicz, A., Walenczak, K., Weber, J., 2013. Changes in selected hydrophobic components during composting of municipal solid wastes. J. Soils Sediments 2014 (14), 305−311. https://doi.org/10.1007/s11368-013-0696-0.

California Organics Recycling Council, 2011. Compostable Plastics 101. Available from: USCC Compostable Plastics Tollkit: https://www.cptoolkit.org/Portals/0/Documents/Compostable%20Plastics%20101%20Paper.pdf.

Chai, E.W., H'ng, P.S., Peng, S.H., Wan-Azha, W.M., Chin, K.L., Chow, M.J., Wong, W.Z., 2013. Compost feedstock characteristics and ratio modelling for organic waste materials co-composting in Malaysia. Environ. Technol. 34 (17−20), 2859−2866. https://doi.org/10.1080/09593330.2013.795988.

Cogger, C., Bary, A., Sullivan, D.M., 2002. Fertilizing with yard trimmings. WSU Bull.

Coker, C., January 21, 2020a. Managing PFAS chemicals in composting and anaerobic digestion. Biocycle. https://www.biocycle.net/managing-pfas-chemicals-composting-anaerobic-digestion/.

Coker, C., January 28, 2020b. Managing Treatment of PFAS-contaminated composting site runoff. Biocycle. https://www.biocycle.net/treatment-pfas-contaminated-composting-site-runoff/.

Esan, E.O., Abbey, L., Yurgel, S., March 25, 2019. Exploring the long-term effect of plastic on compost microbiome. PLoS One. https://doi.org/10.1371/journal.pone.0214376.

Evanylo, G.K., Sherony, C.A., May, J.H., Simpson, T.W., Christian, A.H., 2009. The Virginia Yard-Waste Management Manual, second ed. Virginia Cooperative Extension https://vtechworks.lib.vt.edu/bitstream/handle/10919/48325/452-055_pdf.pdf?sequence=1&isAllowed=y.

Food and Fertilizer Technology Center for the Asian and Pacific Region, 2005. Raw Materials Used for Composting. https://www.fftc.org.tw/en/publications/detail/1442.

Godley, A., Lewin, K., Graham, A., Barker, H., Smith, R., 2004. Biodegradability determination of municipal waste: an evaluation of methods. In: Proc. Waste 2004 Conf. Integrated Waste Management and Pollution Control: Policy and Practice, Research and Solutions. Stratford-upon-Avon, UK, 28-30 September 2004, pp. 40−49.

Godlewska, P., Schmidt, H.P., Ok, Y., Oleszczuk, P., 2017. Biochar for composting improvement and contaminants reduction: a review. Bioresour. Technol. 246, 193−202.

Guzman, A., Gnutek, N., Janik, H., 2011. Biodegradable polymers for food packaging - factors influencing their degradation and certification types a comprehensive review. Chem. Chem. Technol. 5, 115−122.

Hagemann, N., Subdiaga, N., Orsetti, S., de la Rosa, J.M., Knicker, H., Schmidt, H.P., Kappler, A., Behrens, S., February 1, 2018. Effect of biochar amendment on compost organic matter composition following aerobic composting of manure. Science Direct Sci. Total Environ. 613−614, 20−29.

Han, W., Clarke, W., Pratt, S., 2014. Composting of waste algae: a review. Waste Manag. 34 (7), 1148−1155.

Hills, K., 2019. The Devil Is in the Process: Co-composting Biochar Could Benefit Crop Growth and the Environment. CSANR.

Keener, H., Wicks, M., Michel, F., Ekici, K., 2014. Composting Broiler Litter. Cambridge Core 70 (4). Cambridge University Press.

Kissinger, W.F., Koelsch, R.K., Erickson, G.E., Klopfenstein, T.J., May 2007. Characteristics of manure harvested from beef cattle feedlots. Biol. Syst. Eng. Papers Pub 23 (3), 357–365. University of Nebraska.

Krishna, M.P., Mohan, M., 2017. Litter decomposition in forest ecosystems: a review. Energy Ecol. Environ. 2, 236–249. https://doi.org/10.1007/s40974-017-0064-9.

Krogmann, U., Westendorf, M.L., Rogers, B.F., 2006. Best Management Practices for Horse Manure Composting on Small Farms. Rutgers Cooperative Research & Extension, Bulletin, p. E307.

Lemus, G.R., Lau, A.K., 2002. Biodegradation of lipidic compounds in synthetic food wastes during composting. Can. Biosyst. Eng. 44, 2002.

Magri, A., Teira-Esmatges, M.R., 2015. Assessment of a composting process for the treatment of beef cattle manure. J. Environ.l Sci. Health Part B Pestic. Food Contam. Agric. Wastes 50 (6).

Manitoba Agriculture, Food and Rural Development, 2015. Properties of Manure. agrienv@gov.mb.ca.manitoba.ca/agriculture.

Massachusetts Department of Environmental Protection, 2003. Supermarket Composting Handbook. https://www.mass.gov/files/documents/2016/08/ql/smhandbk.pdf.

Mason, G., 2009. Predicting biodegradable volatile solids degradation profiles in the composting process. Waste Manag. 29 (2), 559–569. https://doi.org/10.1016/j.wasman.2008.05.001. Epub 2008 Jun 24.

Mercer, W.A., Rose, W.W., Champan, J.E., Katsuyama, A., 1962. Aerobic composting of vegetable and fruit wastes. Compost. Sci. 3-3, 9–19.

Mercer, W.A., Rose, W.W., 1968. Windrow composting of fruit waste solid. Compost. Sci. 9–3, 19–22.

Nakasaki, K., Nagasaki, K., Ariga, O., 2004. Degradation of fats during thermophilic composting of organic waste. Waste Manag. Res. 22 (4), 276–282. https://doi.org/10.1177/0734242X04045430.

Niladri, P., Utpal, G., Gourab, R., 2019. Composting. In: Larramendy, M., Soloneski, S. (Eds.), Organic Fertilizers - History, Production and Applications. IntechOpen. https://doi.org/10.5772/intechopen.88753. Available from: https://www.intechopen.com/books/organic-fertilizers-history-production-and-applications/compostingComposting.

Northup, B.K., Zitzer, S.F., Archer, S., McMurtry, C.R., Boutton, T.W., 2004. Above-ground biomass and carbon and nitrogen content of woody species in a subtropical thornscrub parkland. J. Arid Environ. 62 (2005), 23–43.

Pan, Y., Farmahini-Farahani, M., O'Hearn, P., Xiao, H., Ocampo, H., 2016. An overview of bio-based polymers for packaging materials. J. Bioresour. Bioprod. 1 (3), 106–113. https://doi.org/10.21967/jbb.v1i3.49.

Reid, I., 2011. Biodegradation of lignin. 2011·. Can. J. Bot. 73 (S1), 1011–1018. https://doi.org/10.1139/b95-351.

Ricci-Jürgensen, M., Gilbert, J., Ramola, A., 2020. Global Assessment of Municipal Organic Waste Production and Recycling. International Solid Waste Association. https://www.iswa.org/uploads/media/Report_1_Global_Assessment_of_Municipal_Organic_Waste_Compressed_v2.pdf.

Richard, C., 2013. Evaluation of Waste Gypsum Wallboard as a Compost Additive. Thesis. Dalhousie University, Halifax, Nova Scotia.

Rose, W.W., Katsuyama, A., Chapman, J.E., Porter, V., Roseid, S., Mercer, W.A., 1965. Composting fruit and vegetable refuse. Compost Sci 6−2, 13−25.

Scott, G.M., Smith, A., 1995. Sludge characteristics and disposal alternatives for the pulp and paper industry. In: Proceedings of the 1995 International Environmental Conference; Atlanta, GA. Atlanta, GA: Tappi Press: 269-279; 1995.

Taylor, J., Warnken, M., 2008. Wood Recovery and Recycling: A Source Book for Australia. Forest and Wood Products Australia, Melbourne. https://www.fwpa.com.au/images/marketaccess/PNA017-0708_Wood_Recycling_0.pdf.

Tuomela, M., Vikman, M., Hatakka, A., Itävaara, M., 2000. Biodegradation of lignin in a compost environment: a review. Sci. Direct Bioresour. Technol. 72 (2), 169−183.

Valzano, F., Western Sydney Waste Board, University of New South Wales., Recycled Organics Unit., 2000. A Literature Review on the Composting of Composite Wood Products Draft. University of New South Wales Resource Organics Unit, Sydney.

van der Linden, A., Reichel, A., 2020. Bio-waste in Europe — turning challenges into opportunities. Eur. Environ. Agency Rep. No 04/2020.

Van der Zee, M., Molenveld, K., June 2, 2020a. Fate of compostable products during industrial composting of SSO. Biocycle.

Van der Zee, M., Molenveld, K., June 15, 2020b. Compostable product disintegration rate. Biocycle.

Van Gestel, K., Mergaert, J., Swings, J., Coosemans, J., Ryckeboe, J., 2003. Bioremediation of diesel oil-contaminated soil by composting with biowaste. Environ. Pollut. 125 (3), 361−368.

Vandecasteele, B., Sinicco, T., D'Hose, D., Vanden Nest, T., Mondini, C., March 1, 2016. Biochar amendment before or after composting affects compost quality and N losses, but not P plant uptake. J. Environ. Manag. 168, 200−209.

Vidussi, F., Rynk, R., 2001. Meat by products as composting feedstocks, Parts 1 and 2. Biocycle. January and March 2001.

Walters, R., 2020. Composting basics: C:N ratio and recipe making. Technical Note 25. Dirt Hog's Compan.

Wang, X., Zhao, Y., Wang, H., Zhao, X., Cui, H., Wei, Z., 2017. Reducing nitrogen loss and phytotoxicity during beer vinasse composting with biochar addition. Waste Manag. 61, 150−156.

Passively aerated composting methods, including turned windrows

5

Authors: Frederick Michel[1], Tim O'Neill[2], Robert Rynk[3]

[1]*Ohio State University, Wooster, OH, United States;* [2]*Engineered Compost Systems, Seattle, WA, United States;* [3]*SUNY Cobleskill, Cobleskill, NY, United States*

Contributors: Jane Gilbert[4], Steven Wisbaum[5], Thomas Halbach[6]

[4]*Carbon Clarity, Rushden, Northamptonshire, United Kingdom;* [5]*CV Compost, Charlotte, VT, United States;* [6]*Extension Professor Emeritus, University of Minnesota, Minneapolis, MN, United States*

1. Introduction

The many types of feedstocks and applications of composting have given rise to many different composting methods. They cover a wide technological spectrum, from freestanding piles that are only occasionally manipulated to sophisticated composting bioreactors with automated discharge, temperature control, forced aeration, biofilters, and frequent agitation (Fig. 5.1). A composter's method of choice is largely determined by the feedstocks, scale of operation, financial constraints, regulations, and sensitivity of the neighbors. Management preferences and operator skills also play a role.

Many composting operations practice simple composting methods like passively aerated static piles and turned windrows. These methods prevail because they can be accomplished with little site alteration, engineering, and capital investment. For instance, a prospective composter, located on a farm or working within a municipality, may be able to start composting with almost no investment at all, using existing equipment and minimal site preparation. Aerated static pile (ASP) composting is another simple, widely practiced method with diverse applications. Some composting operations have adopted composting methods with advanced technology and automation in order to produce compost more quickly, save space, provide shelter from the weather, or reduce odors where sensitive neighbors are close. In some jurisdictions, regulations require additional controls and management techniques where food, animal by-products, or biosolids are composted. Many of these advanced methods are grouped into a category called in-vessel composting.

Composting methods can be labeled and categorized by several criteria including:

- the mode of aeration (passive vs. forced),

The Composting Handbook. https://doi.org/10.1016/B978-0-323-85602-7.00002-9

159

(A)

(B)

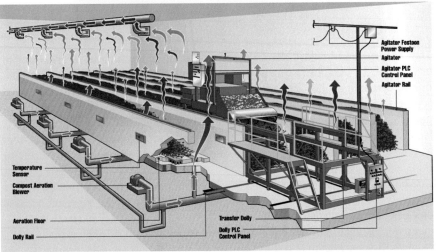

Agitator Festoon
Power Supply

Agitator

Agitator PLC
Control Panel

Agitator Rail

Temperature
Sensor

Compost Aeration
Blower

Aeration Floor

Dolly Rail

Transfer Dolly

Dolly PLC
Control Panel

FIGURE 5.1

The breadth of composting methods in scale and technology; from small on-farm windrows (A) to sophisticated indoor composting systems with automated aeration and turning (B). These aerated systems are presented in Chapters 6 and 7.

Sources: (A) R. Rynk. (B) BDP Industries.

- level of containment (open, covered, contained or "in-vessel"),
- degree of agitation (static vs. turned), and
- materials movement (batch vs. continuous).

Terms used to describe composting methods are presented in Table 5.1. A list of selected composting methods according to these categories is presented in Table 5.2. Most composting facilities use either passive piles, turned windrows, or ASPs.

Each of these composting methods are presented in this and the next two chapters. This chapter focusses on passively aerated methods, especially turned windrows. The ASP and similar forced aeration methods are described in Chapter 6. Chapter 7 covers contained composting methods, including in-vessel methods. Chapter 7 also includes a summary of all of the methods presented here.

Table 5.1 Terms used to classify or describe composting methods and systems.

Agitated or turned: Composting materials are mixed, agitated or "turned" regularly at intervals ranging from every day to every 2 months. The result is homogenizing materials by distributing moisture, moving materials from the oxygen poor interior to the oxygen-rich exterior, and mixing high temperature material with lower temperature material. Some turning methods can also reduce the particle size of compost and reform a windrow or pile. Turning generally refers to the process whereby composting feedstocks are lifted up into the air, mixed and allowed to drop back to the ground. A variety of equipment may be used, such as front-end loaders, augers, dedicated turning machines, etc.

Batch: Feedstocks are composted in separately identifiable lots, usually with little or no change in physical location through the process. After a batch is formed and starts to compost, little or no new material is added.

Bed or slab: This is a continuous layer of compost of near equal depth. Some of these systems are aerated from below, and they may or may not be regularly turned.

Bay: The space between, and defined by, physical barriers such as two parallel walls.

Continuous: Feedstocks physically move through the compost system in a nearly continuous fashion. Movement through the system corresponds with progressive stages of decomposition. Compost is removed at one point and new feedstock is added regularly and frequently at another point.

Forced aeration: One or more fans supply an air distribution network of ducts or pipes to deliver air to composting materials. These systems must be carefully designed and sized to allow even air flow distribution and optimize energy use. Aeration systems may be either "positive" (where air is blown into the composting mass), or "negative" (where air is sucked into the composting mass) pressure systems. Each has its own merits and drawbacks.

In-vessel or contained: Materials are composted within "reactors" or "vessels," which are usually totally enclosed. This enables the air above the composting materials and the resulting emissions to be controlled. Many methods employ forced aeration, some means of agitation, and often temperature control. Examples of reactors include aerated steel containers, vertically oriented reactors, and various enclosed bin configurations. Horizontal agitated bays are usually considered within this category.

Modular: Materials are composted in multiple, often relatively small, units or modules. Each module may represent an individual batch of material. Modules can be freestanding piles or windrows, bins, or enclosed reactors. As the number of modules increases, the system approaches continuous operation.

Open: Materials are composted in freestanding piles or windrows (i.e., long narrow piles) that are exposed to the atmosphere. In some systems, materials may be stacked in simple bins that are not fully enclosed. Systems may be enclosed within a building, but the composting environment is otherwise not controlled.

Passive or natural aeration: This includes systems that rely only on natural air movement as the means of aeration. Driving mechanisms include thermal convection, wind, and diffusion.

Static: Materials are composted without regular agitation or turning. Some turning may take place infrequently and irregularly when piles are moved or combined.

Table 5.2 Summary of selected commercial-scale composting methods.

Open passively aerated methods
Passively aerated static piles—Freestanding piles that are turned infrequently, if at all, and aerated passively without aeration aids.
Static piles and windrows with assisted passive aeration (e.g., passively aerated windrow system [PAWS] and naturally aerated static piles [NASP])—Static windrows and piles with passive aeration aids such as perforated pipes and aeration plenums.
Turned windrows—Long narrow piles that are regularly turned and aerate passively.
Open methods with forced aeration
Aerated static piles (ASPs)—Freestanding piles with forced aeration and little or no turning. Materials may be confined in bins or bunkers instead of freestanding piles.
ASPs under microporous covers—ASPs covered with a specially designed semiporous cover to protect it from the weather and help manage emissions and moisture.
ASPs within a plastic membrane—ASPs enveloped in a plastic tube with selected ventilation ports to isolate the pile from the surrounding environment. Some are made of specially designed semiporous material to protect it from the weather and help manage gas emissions and repel rainfall.
Methods that combine turning and forced aeration (e.g., aerated windrows, turned extended piles)—Freestanding piles or windrows, or simple bins with a forced aeration system. Materials are turned regularly.
Contained (in-vessel) methods
Horizontal agitated bays—Composting takes place between walls which form long, narrow channels referred to as bays. They combine forced-aeration, frequent to daily turning, continuous operation, and inherent materials handling.
Turned containers (e.g., Earth flow, hot rot)—Commercial systems in which composting takes place in vessels that are regularly turned and provide continuous movement of materials. These systems may or may not include forced aeration aerated with fans.
Aerated beds in buildings and halls—Composting takes place within a building in wide continuous beds that are aerated with fans and usually turned on a regular basis.
Silo or tower reactors—Vertically oriented forced aerated systems with top to bottom continuous movement of materials.
Rotating drums—Slowly rotating horizontal drums that constantly or intermittently tumble materials and move them through the system.
Aerated containers—Materials are enclosed within modular and moveable containers with forced aeration.
Tunnels—Fixed vessels with a long profile in which materials are contained and aerated. Many tunnels have an advanced aeration control system.

2. Passively aerated static piles

2.1 Basic principles

The passively aerated static pile method requires a patient approach to composting, relying on little more than convective aeration, natural decomposition, and an extended period of time to produce compost. Several feedstocks may be combined and mixed to adjust moisture, structure, porosity, bulk density, and/or the C:N ratio, but once a pile is formed, it is generally left undisturbed for months and then turned infrequently, if at all. This approach is best used for slowly decomposing feedstocks such as leaves, brush, bark, wood chips, and farm residues like horse stall bedding (Fig. 5.2). It is also the most common method used for composting animal mortalities (see Chapter 8).

In a passively aerated pile, aeration is driven primarily by thermal convection (Haug, 1993). The heat generated during composting increases the temperature of gases within the materials. The warm gases rise out of the composting mass, creating a lower pressure at the base or sides of the pile. Cool fresh air then enters from the sides. The aeration rate is determined by the temperature difference between the interior gas and the ambient air, the gas composition plus the resistance to air flow offered by the composting medium. Therefore, the keys to obtaining reliable passive air movement is creating the heat that drives thermal convection and maintaining a porous physical structure in the pile throughout the process (Veeken et al., 2003). For a well-aerated passive compost pile, nearly half of the pile volume should

FIGURE 5.2

Passively aerated static pile of deciduous leaves and horse manure.

consist of liquid-free airspace (FAS). For dense, wet materials like wet manure, food or biosolids, a large fraction of the pile must consist of a bulking agent to maintain this level of FAS.

Because static piles depend on passive air movement only, maintaining aerobic conditions is a challenge. The challenge increases as the piles get larger and dense feedstocks are included. Oxygen diffusion and air movements are restricted by the large mass of compacted materials (Shell, 1955; Fogiel, 1998). Therefore, oxygen concentrations within the core of large passive piles tend to be low (<1% oxygen by volume). Normally, aerobic conditions exist only within a meter or a few feet from the pile surface, and perhaps along air channels that form within the pile. Oxygen dispersion improves if the composting pile has good structure, ample FAS, and a low bulk density, and if it is uniform (e.g., well-mixed). As the feedstocks decompose and gradually demand less oxygen, aerobic conditions progressively penetrate deeper.

An important factor in composting by the passively aerated static pile method is *time*. Composting proceeds slowly in passively aerated static piles because interior conditions are characterized by low oxygen concentrations, and because the materials rarely receive the benefits of turning. Large passive piles turned only two or three times may require 6 months to a year to produce mature compost. With no turning, 1 year or more may be needed to harvest compost from large piles. Because of the long composting period, a second important factor is space. The site must have a large enough area to hold one or more years of partially composted feedstock, although the area requirement is reduced somewhat by the capacity of larger piles.

2.2 Pile configuration and size

Most static piles are simply freestanding piles. The size and configurations of these piles vary considerably, depending on the feedstocks and how operators choose to manage them. Piles tend to be moderate to relatively large, usually ranging from 2 to 5 m high (6–16 ft). Normally the width of a freestanding pile is slightly less than twice its height. However, wider extended piles are also used. A few composters work with extremely large passive piles that exceed even 10 m (30 ft) in height. These extremely large piles pose an elevated fire risk (Box 5.1).

The appropriate pile size for a given operation depends on the feedstocks, the desired composting period, and other management and site factors. The pile height is limited, in part, by the reach of equipment used to build the pile. A sensible approach is to limit the height such that the equipment does not have to drive onto the edge of the pile when placing feedstock or compost on top. Under this restriction, pile heights of roughly 2–2.5 m (6–9 ft) are possible using skid steer loaders and small tractors. Common front-end loaders can reach up to 4 m (13 ft) and up to 5 m (16 ft) with extended buckets. Still taller piles can be achieved with excavators or conveyors.

Static piles are sometimes enclosed in bins to more neatly contain the materials, segregate batches, or allow materials to be stacked higher in a narrow space (Fig. 5.3). A common configuration is a series of three-sided bins housed within open-sided buildings. Piles may be occasionally shifted between adjacent bins as a means of mixing.

FIGURE 5.3

Passively aerated bins.

FIGURE 5.4

A yard trimmings composting facility using passively aerated piles. Piles are periodically turned as they are moved. Note material flow through the facility.

Source: H. Hoitink.

Box 5.1 Massive passive composting

Some composters build massively large passively aerated piles exceeding 8 m (26 ft) in height and 30 m (100 ft) in width (Fig. 5.5). These extremely large piles have been referred to as composting "slabs," "blocks," or "tables." Many people, regulators and composters included, consider these large piles to be outside the definition of composting because they are hardly aerobic, and beyond the scale at which the composting process can be readily monitored and controlled. The critics

FIGURE 5.5

Large static pile of slowly composting land clearing debris.

Source: F. Michel.

claim that odors, fires, and poor-quality compost are the likely results. However, proponents counter that good compost can be produced within these large piles with less energy on a much smaller land area than smaller passive piles. Furthermore, the proponents argue that fewer odors and other emissions occur because piles are not disturbed, and the relative surface area of the piles is minimized by the large size. While some of the proponents' arguments have merit, the fact of the matter is that massive piles are a fire risk!

What about this "massive passive" approach to composting—can it legitimately be considered composting? Technically, the answer is yes. Legally, the answer depends on how a jurisdiction defines composting. Even if the oxygen concentrations within a pile are extremely low, decomposition continues steadily but slowly (assuming some moisture is present). As long as the product becomes aerobic in the end, such piles can yield compost. However, this says nothing about its quality. What about the odor and environmental impacts? Foul-smelling compounds and methane tend to accumulate within a large pile due to the anaerobic conditions. The methane and odorous compounds largely remain in the pile until they decompose or until the pile is disturbed. Otherwise, some portion continually escapes the surface of the static pile. Whether or not these passive emissions create odor or environmental problems depends on the specifics of the situation, including the amount of material present, the weather conditions, location and attitude of neighbors, other site conditions, feedstocks, and even the time of day. Hence, some operations with large piles are plagued by odors while some are not. Too much moisture, turning or otherwise disturbing the pile under the wrong conditions exacerbates odor problems.

Box 5.1 Massive passive composting—cont'd

Very large composting piles (e.g., more than 8 m high) are well insulated from their surroundings, including the human composters trying to manage them. With increasing pile size, moisture content, temperatures, and oxygen concentration can change beyond the composter's knowledge and control. It is this distinction that leads to the Achilles Heel of massive piles—*fire* (Fig. 5.6). The extreme size not only lengthens the composting period but also risks the development of fires from both spontaneous combustion and other forms of ignition. The risk is higher in part because the piles are large, static, in place for longer time periods, well insulated, and in part because it is difficult to know, and alter, what is happening inside the pile.

FIGURE 5.6

A spontaneous combustion fire in a large passive compost pile. The crews are beginning to extinguish the fire by removing material from the outside of the pile to reach the hot spot, with the fire department and their water hoses nearby.

Source: F. Michel.

2.3 Pile management

Once the composting process is well underway and the piles begin to shrink, they may be combined to save space and retain heat. Thereafter they may be turned occasionally with a bucket loader, backhoe, or excavator. Often, turning is performed only when piles are moved within the site (Fig. 5.4). Such turnings may occur monthly or only two or three times in a year, depending on the composter's situation. Each turning stimulates biological activity and reduces the time required to produce a uniform stable compost product. Although this composting method is classed as "static," it is still necessary to turn piles at least twice during the composting cycle, otherwise, the compost produced is inconsistent and may contain visible remnants of the feedstocks.

Given that the interior of the pile may be poorly aerated, there is a potential for odors with passively aerated static piles, especially when the pile is turned or otherwise disturbed. At the start, an exterior layer of finished compost, or slowly degrading amendment (e.g., wood chips), helps to contain odorous gases. This "biocover" is typically about 15 cm (6 inches) thick but may be as low as 10 cm (4 inches) or as thick as 30 cm (12 inches), depending on the stage of composting. After turnings, another new biocover can be applied if and as necessary. Gas permeable fabric covers can also reduce odor emissions somewhat, but their primary function is to exclude precipitation.

Large static piles are practical for coarse feedstocks like deciduous leaves, shredded brush, bark, and wood waste. Because of their slow rate of decomposition and high C:N ratios, these feedstocks are not normally malodorous, except when piles are disturbed. If necessary, when piles are established, the initial moisture content can be adjusted to the target level (e.g., 50%–65%). Occasional and timely turning of the piles, plus the addition of water when needed, avoids fires, and can result in high quality compost.

Large static piles are *generally* not appropriate for feedstocks that tend to decompose quickly, like grass clippings, biosolids, food processing residues. These materials rapidly create anaerobic conditions and produce odors. On farms, manure is sometimes allowed to compost in passive piles because odors are acceptable in some agricultural settings. However, this situation is changing. For many farms, manure cannot be composted in static piles without odor problems, and wider environmental awareness is driving the reduction of emissions of methane that may be released from such piles. Still, methane emissions from a passively aerated compost pile are low compared to an idle pile of raw manure or a lagoon of liquid manure.

Passive piles can however work for highly degradable types of feedstocks if they are mixed with generous quantities of amendments and a thick biofilter cover is used. Feedstocks that attract pests or have strong odors must remain beneath the pile surface until they are well decomposed. For example, passive piles are successfully used for composting animal mortalities and fish remains by keeping these materials completely surrounded by compost or coarse amendments that absorb and degrade odors (Chapter 8).

In summary, the static pile method can be an economical and successful approach to composting if time and space are available, and if odors can be controlled (Table 5.3). This passive method of composting entails minimal labor and equipment. The potential for odor problems is high when the piles are turned or moved. However, the success of composting animal mortalities in passively aerated static piles demonstrates that odors can be effectively managed. A long composting cycle is practical for many feedstocks, especially those that are generated almost entirely at a particular time of year, such as deciduous tree leaves. In this case, the material can remain composting on the site for a year and a half. Then, the compost can be sold or moved off-site in the spring and summer to make room for the incoming leaves in the fall.

Table 5.3 Summary of passive pile methods.

Key Features
• Incidental to infrequent turning.
• May be combined with a more-intensive first stage of composting.
• Low labor.
• Inexpensive equipment (loaders, excavators).
• Low-oxygen core.
• Slow—can take a year.
• Requires a large land base with long retention times.
• Little process control.
• Odors can be strong when piles are disturbed if materials have not completely decomposed.
• Wide variations in compost particle size.

Limited to:
• No hurry situations.
• Particular settings—isolated sites, farms with adequate space.

Works well for:
• Slowly degradable and/or porous feedstocks (generally those with a high C:N ratio and low bulk density).
• For mortalities, fish waste and wet food waste (because of encasement).
• Seasonal feedstocks that recur yearly (e.g., deciduous leaves).
• Facilities that can afford very long retention times (>6 months).

3. Techniques to improved passive aeration—passively aerated windrow system and natural aeration static pile

As explained in the previous section, passive aeration of static piles is imperfect. Convective air flow tends to concentrate beneath the envelope of the pile, leaving the core of the pile to obtain oxygen by the slower means of diffusion. To improve oxygen distribution, some composting techniques employ aeration aids such as pipes and air plenums to direct air flow deeper into the pile. These methods have been demonstrated to improve the rate of composting compared to conventional passively aerated static piles (De Silva and Yatawara, 2017).

Prominent among these techniques is an approach known as the passively aerated windrow system (PAWS), which was developed, and named, by researchers in Ontario Canada, circa 1990 (Mathur et al., 1988). The same research group later revised the PAWS approach and rebranded it the natural aeration static pile (NASP) (Mathur, 1995). PAWS was initially applied to composting of fish processing waste and then later used for composting manures (Mathur et al., 1990). Because of its local abundance, peat moss was initially used as an amendment and biocover.

Peat moss is very effective in both respects. However, for most composters, peat moss is impractical, because of cost and availability, and undesirable, because of environmental concerns associated with mining it. Therefore, other amendments are typically used, including straw, sawdust, wood shavings, and wood chips (Hayes et al., 1994).

The PAWS method, and its variations, involves relatively short windrows that are not turned but have some deliberate means of delivering air without fans. Four basic features common to most PAWS variations are:

- a homogenous and relatively low-density mixture of composting feedstocks,
- a delivery system for passive air flow, usually perforated pipes or a porous base,
- a base layer of stable absorbent material like straw, wood chips, or compost to absorb moisture and insulate the windrow (not always used), and
- an exterior layer of stable coarse material, at least 15 cm (6 in.) thick, that retains heat, moisture, odors, and ammonia (not always used).

The original PAWS method delivers air to the interior of the windrow through a series of perforated plastic pipes running parallel to one another *across* the width of the windrow (Fig. 5.7). The pipes rest on the absorbent base with the open ends of the pipes extending beyond the sides of the windrow. The premixed composting feedstocks are placed on top of the aeration pipes to the desired height, typically

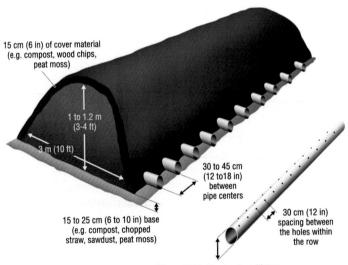

FIGURE 5.7

Basic features of the passively aerated windrow system (PAWS) as described by Mathur et al. (1990).

1–2 m (3–6 ft). The cover layer is then placed over the windrow. The windrow then composts in place for several weeks to several months without being turned. If and when the windrow is turned, the pipes are permanently removed. The base material is incorporated into the compost when the windrow is turned or broken down.

PAWS pipes are usually Schedule 40 PVC plastic (see Chapter 6). Each pipe is 10 cm (4 in.) in diameter and, along its length, it contains two rows of holes, 13 mm (0.5 in.) in diameter spaced about 75 mm (3 in.) apart. Recommendations suggest locating the holes near the top of the pipe at the two and ten o'clock positions. However, orienting the holes downward (e.g., five and seven o'clock) minimizes plugging and allows free water to drain from wetter feedstock mixes.

The NASP method eliminates the pipes and relies solely on the base to act as an air plenum. In this case, the base is a layer of highly porous material that facilitates air movement such as wood chips or bark. The thickness of the base layer is approximately 45 cm (18 in.), which raises the windrow height to roughly 2–3 m (6–9 ft).

Another variant of the original method replaces the pipes with a concrete platform which has a hollow core (Patni, 2000). The core serves as an air channel or plenum and eliminates the inconvenience of handling pipes when constructing or breaking down windrows. Other modifications include box-like containers, permeable covers, and varying windrow dimensions.

Like any passively aerated approach, PAWS and NASP rely on high temperatures to drive thermal convection. They have been shown to produce and maintain thermophilic temperatures with a wide variety of feedstocks including fish, several types of livestock manure, sawmill and paper pulp, food, and yard trimmings (Mathur, 1991). Both the lack of turning and the biocover help to contain odors.

4. Turned windrow composting

4.1 Basic principles

Windrow composting consists of placing a mixture of organic feedstocks in long narrow piles called "windrows" that are then agitated or "turned" on a regular basis (Fig. 5.8). The windrows are often built at a right angle to the prevailing wind direction and parallel to the slope of the site. Many sites include an improved surface consisting of concrete or asphalt that allows year-round manipulation of the windrows. Other sites consist of well-drained open land, but these may be inaccessible to heavy equipment during certain times of year.

The *active* composting stage for windrow composting typically lasts between 8 and 16 weeks, depending upon the feedstocks, turning intervals, and intended use for the compost. For instance, 8 weeks is a reasonable period for a manure composting operation in which the windrows are turned three times during the first week, and weekly or every other week thereafter. A typical windrow is turned about six to eight times before the windrow enters the curing stage. Through experience, the operator eventually gains a feel for the relationship between compost quality and the turning schedule and learns how to troubleshoot problems in the windrow.

(A)

(B)

FIGURE 5.8

Pictures of windrow composting: Aerial view of a windrow composting facility with a concrete pad, below grade aeration system, a covered feedstock storage area, and a wetlands treatment system for collection of pad run-off. An adjacent clay pad is used for leaf composting and long-term storage (A). Composting windrows with dairy manure on a natural soil pad (B).

Source: (A) F. Michel/T. Short. (B) R. Rynk.

Windrows aerate primarily by passive air movement (Buckner, 2002; Michel, 1999). The rate of air exchange depends on the porosity of the windrow and its temperature, which are related to pile size, feedstock characteristics, and ambient temperatures. A light fluffy windrow of leaves or tree bark (i.e., one that has a low bulk density) can be much larger than a wet, dense windrow containing manure. If a windrow is too large or wet, anaerobic zones develop and persist in the interior.

On the other hand, small windrows lose heat quickly and may not achieve suffi-ciently high temperatures necessary to evaporate moisture and kill pathogens and weed seeds.

Windrow heights typically range from 1 to 2 m (3–6 ft) but may be as high as 3–4 m (8–12 ft). The width varies greatly from as low as 3 m for short windrows to over to 8 m (25 ft) for large windrows turned with large machines (Fig. 5.9). As an approximation, freestanding windrows that are triangular or parabolic in shape tend to be *roughly* twice as wide as they are high. The width of trapezoidal windrows is determined by the turning equipment.

In principle, a windrow should be sized such that between turnings it establishes a gradient of oxygen concentration from the surface to the center that starts at ambient concentration (20.9%) and reaches 1%–2% only at the very center. If low oxygen concentrations are reached well before the center, then a large portion of the material is anaerobic and likely to release odors when turned. Maintaining ox-ygen concentrations in the center of the pile at 5%–6% will generally speed up the composting process.

In practice, the equipment used for turning largely determines the size, shape, and spacing of the windrows. However, the aeration characteristics of the feedstocks should strongly weigh on selection of that equipment. Bucket loaders and excavators with a long reach can build tall windrows. Windrow "turners," machines designed specifically for windrow turning, produce low, wide windrows. The relatively low profile of these windrows provides a large surface-to-volume ratio that favors pas-sive aeration.

The turned windrow method is a simple and flexible approach to composting that is appropriate for a wide range of feedstocks (Table 5.4). Windrow composting op-erations can be started with little investment if existing equipment is used and well-drained land is available. Primarily for these reasons, the turned windrow method is the most common method for composting yard trimmings, manures, crop residues, and many other feedstocks (see Box 5.3 at the end of this chapter).

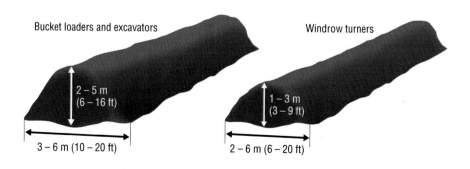

FIGURE 5.9

Typical windrow shapes and dimensions.

Table 5.4 Summary of turned windrow composting.

Key Features
• Simple, easily started, and managed.
• Little or no engineering.
• No utilities necessary.
• Turning strategy flexible and changeable—daily, weekly, monthly.
• Similar in nature to other materials handling and farming activities.
• Limited process control.
• Moderate composting time—months to year, depending on turning schedule.
• Large, well graded area required.
• Labor and fuel use can be substantial.
• Moderately expensive equipment.
• Equipment wear and maintenance required.
• Odors difficult to control, periodic intense emissions.
• Moisture management can be challenging.

Limited to:
• Sites that are not severely restricted by space or not very sensitive to odors or emissions.
• All but the most troublesome feedstocks.

Works well for:
• Large sites that are well buffered (e.g., farms).
• Start-ups and trials.
• Processing large amounts of material efficiently.
• Equipment-loving composters.

4.2 Turning principles

Turning—the inside-out/upside-down mixing of a pile—is the defining feature of windrow composting. It exposes all of the materials to a range of conditions within the windrow, which fosters more complete and uniform decomposition and other benefits. The conditions within a windrow vary from dry, high oxygen and low temperature conditions at the surface to high temperature, wet and low oxygen concentrations at the center, and many combinations in between. These conditions foster different rates and extents of decomposition.

Turning performs several useful functions that have the general effect of invigorating the composting process.

- Turning improves the uniformity of the feedstocks and the distribution of their moisture, nutrients, and microorganisms. For instance, turning can bring together isolated pools of nitrogen with isolated pools of carbon within the windrow.
- Turning breaks large clumps and particles into smaller pieces, which exposes more nutrients and surfaces to microorganisms.
- Turning exposes the interior of the pile to the surrounding environment and releases the trapped moisture and gaseous products of decomposition.

- Turning destroys the gradients and any inconsistent conditions that develop during composting (Fig. 5.10).
- Turning conveys the cooler material at the windrow's surface into the hot interior. With several turnings, eventually all the compost is exposed to the high pathogen-destroying temperatures inside the windrow. For this reason, the standards for sanitization in windrows call for a minimum of five turnings.
- Turning creates a more consistent product.

4.2.1 Turning and aeration

It is commonly, but erroneously, believed that turning is the primary way to aerate a windrow. Although turning does introduce some cool oxygen-rich fresh air, the effect is short-lived. When composting is active, microorganisms rapidly deplete the introduced oxygen, usually within hours (Fig. 5.11). Therefore, unless windrows are turned daily, they must rely on passive aeration, not turning, for aeration.

While the oxygen provided during turning is fleeting, turning can have longer lasting effects on aeration, for good or bad. First, the stimulating effects of turning help to maintain elevated temperatures within the windrow, which is the driving force for passive aeration. Second, in some situations, turning reshapes a windrow and, in doing so, restores the porosity and shape that has been lost due to settling. The increased porosity enhances passive air exchange, at least temporarily, until the windrow again settles. However, turning can also have the opposite effect. Turning decreases porosity and increases resistance to air flow if it substantially shreds feedstock particles to smaller sizes (Michel et al., 2002).

Whether or not turning increases or decreases porosity depends on the feedstocks and the turning machine used. Loose feedstocks with large particles, like leaves or straw, lose porosity after turning, while dense feedstocks that tend to compact, like

FIGURE 5.10

Gradients of moisture, oxygen, temperature and microorganisms are quickly established within active windrows.

Source: F. Michel.

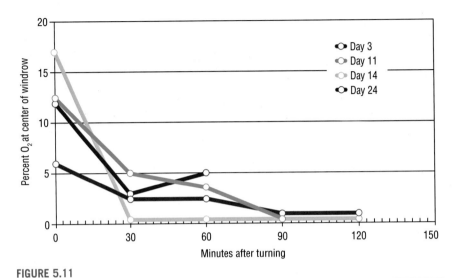

FIGURE 5.11

Oxygen concentrations at the center of a dairy manure/sawdust windrow just after turning and up to 120 minutes afterward at different stages of active composting.

Adapted from Tirado, S.M., Michel Jr., F.C., 2010.

dairy manure, may temporarily improve. The method of turning is also a factor. Some turning machines aggressively break apart particles while others are designed primarily to lift and repile the windrow contents with little cutting action. Bucket loaders and excavators fall into the latter category.

4.2.2 Turning and temperature

Turning stimulates biological activity and causes temperatures to rise to a level determined by the heat generated within, and the heat lost from, the windrow. When a windrow is turned, the temperatures at the core of the windrow do immediately decrease, but only for a short time. Afterward temperature rises rapidly to the preturned levels. Indeed, because of the process-stimulating effects of turning, temperatures often continue to rise above the preturned level.

The converse also can be shown. Leaving a windrow unturned does not necessarily lead to excessively high temperatures overall. As Fig. 5.12 shows, frequently turned and infrequently turned windrows have similar overall temperature profiles. Therefore, and in general, turning is not an effective way to lower temperatures when composting is at a vigorous stage. Reducing windrow size is the most dependable way to lower windrow temperatures.

There are important situations when turning can consequentially lower temperature. In a windrow that is nearing the curing phase, turning can permanently cool the windrow because there is not enough readily-available energy for microorganisms to harvest and release. Also, smaller windrows may permanently cool when turned during cold weather because the process cannot keep up with the heat loss

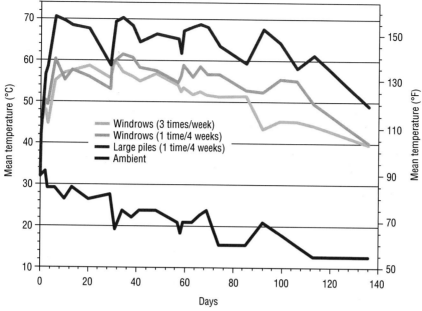

FIGURE 5.12

Average temperatures within yard trimmings windrows and piles (n = 20) made from a 4:1:1 mix of leaves, grass, and wood chips (by volume). Windrows were turned either three times per week or once every 4 weeks. Piles were also turned every four weeks but were much larger than the windrows. Ambient temperatures were taken at the time of measurement (Michel et al., 1996). Note: The abrupt increases in temperature on days 28, 56, and 84 shows the stimulating effect of turning in this composting system.

to the environment. Whether or not this permanent cooling takes place depends on the actual vigor of the process and the severity and persistence of the cold weather. In general, to retain the heat, it is better not to disturb maturing windrows during very cold weather.

Another consequence of turning concerns the uniformity of temperatures within the windrow. Turning is effective in eliminating isolated hot spots within a windrow or pile, especially a large pile. In doing so, turning interrupts a chain of reactions that could lead to very high temperatures and spontaneous combustion (see Chapter 14). Although the microorganisms are generally invigorated by the turning, the subsequent temperature climbs are more uniform. The hot spots do not readily reappear unless the turning procedure fails to disperse the materials surrounding the hot spot.

4.3 Windrow management

As with all composting methods, windrow composting involves managing moisture, porosity, feedstock mixing, bulk density, and temperatures. The job is complicated by the low profile and long length of windrows. The weather and surrounding environment are influential factors. Compared to large piles and contained systems, windrows tend to gain more moisture from precipitation, lose more moisture due to the effects of sun and wind, and lose more heat and gases from their larger exposed surface. In addition, windrow composters need to manage the timing and frequency of turning, which is discussed in the following section.

Windrows are commonly constructed from end-to-end, either in a single batch of feedstocks or in successive batches over days or a few weeks. Multiple feedstocks can be premixed or placed in the windrow in rough proportions and mixed in place with the turning mechanism. Bucket loaders, dump trucks, or manure spreaders place feedstock to the end of the windrow. Batch mixers, which are also common, can discharge mixed feedstocks onto the top or sides of a windrow.

After just the first few weeks of composting, windrow height can diminish appreciably. Usually, it is advantageous to combine or shorten windrows at this stage to maintain the size. Consolidation of windrows is also a good wintertime practice to retain the heat generated during composting. This flexibility is one of the advantages of windrow composting. It is a versatile system that can be adjusted to different conditions caused by seasonal, physical, and compositional changes.

4.3.1 Windrow temperature management

Feedstocks usually heat rapidly after being placed in a windrow. Temperatures below the windrow surface typically reach thermophilic levels within a few days. In all but the smallest windrows (e.g., less than a meter/yard high), temperatures should remain within thermophilic levels for weeks. Windrow temperatures should be monitored daily, and at several locations, at least through the first 2 weeks of composting. If the temperatures fall below the desired level, the windrow should be turned to stimulate composting and increase heat production. If a windrow struggles to maintain thermophilic temperatures, the feedstock mix may need to be adjusted and/or the windrow size may need to be increased.

Excessively high temperatures seldom occur in windrows, primarily because of their smaller size. If high temperatures do consistently occur, windrows should be made smaller and/or the feedstock mix should be adjusted to slow the rate of composting (e.g., increase the C:N ratio, add bulking agent). If exceedingly high temperatures occur only in isolated spots, the windrow should be turned and then monitored to see that the hot spots do not reoccur.

4.3.2 Windrow moisture management

It is very important to monitor and control the moisture content in the windrow during active composting. Feedstocks should be proportioned and combined to create a windrow that is moist enough to support active composting, ideally 55%−60%. Higher starting levels of moistures, up to 65% or even 70%, may be acceptable for some feedstocks, especially in arid locations. Through the active composting stage, the moisture content should be maintained in the 50%−60% range. As active composting wains, the moisture content can be allowed to decline toward 40% to improve compost screening and handling.

Moisture is affected by sun, wind, rain, snow, compost temperature, and rate of organic matter decomposition. Much of the water present initially in the feedstocks evaporates during active composting. Turning further increases evaporation. Modest precipitation (e.g., 25 mm, 1 inch per week) can replace much of the lost water, keeping the windrows moist enough to sustain composting while also yielding compost that can be easily handled or screened. Scant and heavy precipitation are both potential problems that require deliberate moisture management.

Under dry conditions, when precipitation cannot keep pace with moisture loss, it is necessary to add enough water to the windrow to reach a desired moisture content. There are several options for adding water, as discussed in Chapters 9 and 11. The best way to irrigate a windrow is to apply water during turning. Watering devices on windrow turners work well. Additionally, windrows are commonly watered from tank trucks or wagons, followed by turning.

If excess water is a problem, either from precipitation or wet feedstocks, windrows should be turned to stimulate heat generation and promote evaporation. Turning during warm, sunny and/or windy weather is especially effective. To exclude unwanted precipitation, some compost producers cover windrows with a semipermeable material that repels rainwater while still allowing air to penetrate the windrow (see Box 5.2). Covers are especially helpful after the active period of composting has passed because there is less internal energy remaining to evaporate added water from rain and snow.

Box 5.2 Introduction to compost covers

Author: Steven Wisbaum.

Macroporous covers, also known as compost "fleece," are breathable fabric blankets that shed liquid water. They are used *primarily* to shelter open windrows and piles from precipitation, although they also provide a barrier to other effects of the environment (Fig. 5.13). Fleece covers mostly find use with windrow composting because of the substantial exposure of windrows to the weather. However, they also can protect curing piles and storage piles of raw feedstocks, amendments, and finished compost from precipitation and wind-blown weed seeds.

FIGURE 5.13

Fleece covers used over windrows on a farm-based composting site in western Virginia.

Source: Eric Walter, Black Bear Composting.

Continued

Box 5.2 Introduction to compost covers—cont'd

Description, cost and life expectancy Macroporous fleece is a nonwoven fabric made from polypropylene fibers that is manufactured using either "spun" or "needle punch" technology. Fleece is different from microporous covers, which are "multilaminate" fabrics used primarily with aerated static piles (see Section 5.6.3). The macroporous fleece fabric is about 1.8 mm (1/16 in.) thick and weighs approximately 0.2 kg/m^2 (0.4 lb/yd^2). A piece that is 5.5 m wide and 30 m long (18 ft × 100 ft) weighs about 34 kg (74 lb). Some manufacturers offer a very limited selection of widths and lengths while others offer a much wider selection. Fleece covers typically cost \$3.20 to \$3.30 per square meter (\$2.65 to \$2.75/yd^2) plus freight (in 2020 USD).

The fabric is normally treated to resist UV-light degradation. In most cases, these covers are used only during the late fall, winter, and early spring and/or during periods of prolonged wet weather when UV-light intensity is lowest. Used in this way, the covers typically remain useable for 7 to 10+ years. However, in applications that require the covers to be used year-round, their life expectancy can be significantly reduced, especially in locations with long, dry summers, and high UV-light conditions. Therefore, fleece should be used only when conditions warrant, as opposed to being left in place continuously.

Functions and benefits The primary function of a fleece cover is to exclude precipitation from windrows and piles. The fleece sheds water through the combined action of capillary action, the cohesion, and surface tension of water molecules and gravity. When the fabric becomes saturated, the moisture "wicks" from fiber to fiber and migrates down from the pile's peak and sloped sides until this water reaches the bottom of the cover and drips onto the ground. For this reason, the water-shedding ability of the fabric is significantly reduced if used on piles with broad, flat tops.

As a protective surface, a fleece cover offers other potential benefits. They can reduce heat loss from windrows during very cold weather and thus increase internal windrow temperatures. Covers also reduce access to flies, scavenging birds, and mammals. They protect curing and finished compost from windblown weed seeds.

Emerging research and anecdotal evidence suggests that fleece covers can reduce odor emissions. The covers improve odors primarily by reducing the incidents of saturated/anaerobic conditions caused by excess rainfall. Odors can also improve from the filtering effect of the fleece. In this case, water vapor from the pile condenses on the polypropylene fiber matrix of the covers and traps some odor-related compounds. The physical barrier created by the covers provides a measure of containment that might prevent some odors from escaping. However, that barrier might also exacerbate odors released by turning by reducing the air flow through the windrow.

Handling and managing fleece covers Being relatively lightweight, fleece covers are typically deployed and removed manually. For compost operations using tractor-pulled turners, a device known as a "threading frame" can be attached to the turner which automatically raises and then lowers the cover as the turner is pulled through a pile. Hydraulic cover "winder" attachments can also be mounted on loaders and self-propelled turners to mechanically deploy and remove the covers. Large, self-propelled cover winder machines are also available at a cost of between \$75,000 and \$175,000.

Fleece covers are relatively easy to secure to piles because they do not build pockets of air pressure during windy conditions. The most common anchors are 7—9 kg (15—18 lb) truck tire sidewalls or sandbags placed at intervals of 3—9 m (10—30 ft) along the bottom edges. For large piles, multiple covers can be overlapped with anchors placed along the seam or by using grommets offered by some manufacturers to connect the covers with rope, cable, or plastic ties.

Since macroporous covers absorb water, they can be extremely heavy when wet, but they also dry out relatively quickly. Therefore, covers should ideally be allowed to dry out before handling to ensure they will be as lightweight as possible.

In winter, fleece covers can freeze into a solid sheet and/or stick to the underlying frozen compost and/or the ground. This occurs when ambient temperatures drop below freezing before the covers have had a chance to dry out, but it also depends on sun exposure and the amount of heat being released by a pile. For this reason, their use should be avoided on piles requiring access during extended periods of freezing temperatures.

Research studies concerning macroporous covers are summarized at https://www.cvcompost.com/mobile/researchandarticles.html. Also see Paré et al. (2000) and Keener et al. (2005).

4.3.3 Management of windrow odors and other emissions

Because of their expansive surface area, windrows are prone to liberate more gaseous compounds, including odors and ammonia, to the surrounding environment compared to large piles and vessels. Turning increases the emissions because it destroys the dry protective envelope that evolves at the outer layers of a windrow and because it exposes the more-active, wetter, and less-aerobic material within. The odor risks and ammonia losses are greatest just after turning and early in the composting process.

The first line of defense against unwanted emissions, at a given facility, is judicious selection and combination of feedstocks to optimize moisture content, particle size, bulk density, and C:N ratio. Afterward, management of process-related odors and other emissions is primarily a matter of managing turning and pile size. Turnings should be avoided when odors are likely to have an unfavorable impact on the neighborhood. For example, whenever possible, turnings should be precluded on weekends and holidays, late in the afternoon, when the wind is blowing toward sensitive neighbors, when other weather conditions prevent odors from diluting, and when the windrows are especially odorous. Very large piles where a large proportion of the compost is anerobic will release more odors than properly sized windrows that have smaller anerobic interior regions.

One viable strategy to control odors is to postpone, or at least minimize, turnings during the first weeks of composting. The opposite strategy is also viable—frequent turnings from the start. Very frequent turnings (e.g., daily) release ammonia and odorous gases but can greatly reduce the intensity of putrid smells associated with anaerobic conditions. The choice of whether to avoid or embrace turning depends on the feedstocks, the composter's ability, and desire to turn frequently and the odor situation of the facility (see following section).

If a windrow becomes noticeably odorous, either in-between or just after turnings, it should be covered with a layer of unscreened compost or dry stable bulking agent, like wood chips. Adding this "biocover" is an effective odor-control approach, but it is not easy. It consumes time, labor, equipment, and compost. Therefore, it is normally reserved for problem windrows.

4.4 Timing and frequency of turning

Choosing when and how often to turn windrows encompasses both the science and art of composting. The need for turning is influenced by the feedstocks, their rate of decomposition, the moisture content, particle size, bulk density of the windrow, ambient temperatures, desired product attributes, and the time available for compost production. At least a few turnings are necessary to assure that all of the materials in the windrow experience favorable conditions for decomposition—that dissimilar materials come into contact, that particle size is reduced and that all the material is subjected to high sanitizing temperatures.

When and how often turning takes place is usually dictated by the goals and preferences of the composter. In practice, the number of turnings and time between turnings varies greatly among composters, ranging from three to ten turnings over 4–12 months to forty turnings in a 2-month period. Some composters prefer to turn almost daily and may turn a given pile up to 40 times in the cycle. At the other extreme, relatively large windrows are turned only three or four times over a period of four months or more. At some operations, windrows are turned opportunistically—when operators have time, the weather is good, or the wind is blowing away from sensitive areas.

By monitoring the process conditions and the compost quality, operators learn the appropriate time intervals between turnings after gaining familiarity with the composting process and their feedstocks. At this point, turning often occurs at fixed time periods that accommodate the availability of labor and equipment. Weather and wind conditions at the site, and regulatory requirements, are also factors. The composting process affords a great deal of flexibility in this matter.

It should be noted here that although a lot of emphasis is placed on turning, windrow turning frequency appears to have less of an impact on the composting process than other variables such as feedstock composition (e.g., C:N ratio), moisture content, bulk density, and pile size (Tirado and Michel, 2010; Ogunwande et al., 2008).

4.4.1 The first turn

Determining when to first turn a windrow after formation can be a dilemma. It is a choice between managing the process at its most vigorous stage and avoiding odors, ammonia loss, and nuisances. Because the decomposition rate is greatest at the start of the process, and moisture and temperature gradients develop rapidly, the need for turning is generally greatest at the beginning (during the first 10–20 days) and decreases as the windrow ages. Easily degradable or high-nitrogen mixes can benefit from daily turnings at the start of the process so that moisture and nutrients are redistributed, carbon dioxide can be released, and all of the materials are exposed to oxygen. Wet mixes may require frequent turnings to promote drying.

However, turning early in the process has its disadvantages. Odorous compounds and ammonia are predominately formed in the first week or two of the process. Therefore, it may be preferable to delay turning for the first few weeks to allow the odors and ammonia to assimilate within the windrow. If the feedstocks include food, the food should be confined to the interior of the windrow by covering the windrow surface with dry amendment or compost. In such cases, turning should be postponed until the food is no longer attractive to animal and insect vectors. Alternatively, the freshly turned windrow can be again covered with a layer of compost or dry amendment so that no food is exposed, and odors are contained.

The choice of whether or not to turn early in the process depends on the feedstocks, the site conditions, the sensitivities to and consequences of odors, and the operator's knowledge of the general situation. In any case, observing smart odor mitigation and prevention practices tempers the dilemma. After the first one to 3 weeks of composting, the need for turning and the odor consequences diminish as the process slows. Thus, a turning routine can be established that is based on time intervals and/or process conditions.

4.4.2 Temperature-based turning triggers

A common and sensible strategy is to turn windrows according to temperature patterns in the windrow, because falling temperatures indicate that biological activity is also dropping off. For example, a windrow may be turned when the temperature falls to a prescribed level or if temperatures consistently decrease over a certain number of consecutive days. Examples of possible temperature triggers for turning include the following:

- Temperature pattern shows a steady drop in temperature over several days (e.g., an average of 1°C [2°F] per day over 7–10 days).
- The average temperature of the center of the windrow drops from a high temperature, about 65°C (150°F), for example, to below a desired level, such as 50°C (120°F).
- The temperatures measured at the same depth at several spots in the windrow are highly variable.
- The average temperature at the core of the windrow diverges from the average temperature 15 cm (6 in.) below the surface (e.g., the core becomes more than 10°C cooler).

Temperatures should not be considered alone in determining when a pile should be turned. Low or falling temperatures may be caused by conditions that turning cannot correct, such as an undersized pile, an overall lack of moisture, very low ambient temperatures, or that the process is nearing completion. Thus, the last two triggers in the preceding list may provide the best rationale for turning because they relate to disparities within the windrow, something turning does affect. High temperatures throughout the windrow are not a criterion for turning. As discussed earlier, turning has only a short-term effect on reducing high temperatures when composting is vigorous.

Developing a temperature-based strategy for turning requires frequent (e.g., daily) monitoring of windrow temperatures. Measurements should be taken, and recorded, at a sufficient number of locations within the windrow to characterize process behavior. A typical procedure is to measure temperatures at about 15 m (50 ft) intervals along the windrow length and at least two depths at each spot. After a while, particular spots in the windrow may prove to be representative of the overall process, at least for the purpose of triggering turning. It is worth noting here that "the temperature" of a composting pile or windrow is an elusive quantity. Temperatures range from ambient at the surface to 50–70 C at the center. Temperature measurements in a given windrow at similar depths can vary widely; by 5°C (10 F). Therefore, it is important to establish a consistent protocol for monitoring temperatures, using roughly the same set of locations and depths at roughly the same time of day. It helps to maintain a running graph of windrow temperatures to visualize the trends and identify anomalies (Fig. 5.14).

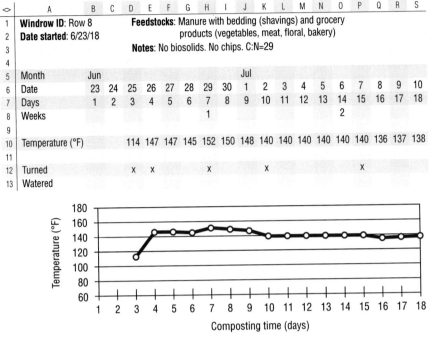

◇	A	B	C	D	E	F	G	H	I	J	K	L	M	N	O	P	Q	R	S
1	**Windrow ID**: Row 8			**Feedstocks**: Manure with bedding (shavings) and grocery															
2	**Date started**: 6/23/18						products (vegetables, meat, floral, bakery)												
3				**Notes**: No biosolids. No chips. C:N=29															
4																			
5	Month	Jun								Jul									
6	Date	23	24	25	26	27	28	29	30	1	2	3	4	5	6	7	8	9	10
7	Days	1	2	3	4	5	6	7	8	9	10	11	12	13	14	15	16	17	18
8	Weeks							1							2				
9																			
10	Temperature (°F)			114	147	147	145	152	150	148	140	140	140	140	140	140	136	137	138
11																			
12	Turned			x	x			x			x				x				
13	Watered																		

FIGURE 5.14

Example of a simple windrow data form for recording temperatures, turnings, and other information over time.

Source: K. Van Alstein.

4.4.3 Other turning triggers

Criteria other than temperature are also used to trigger turning. The most relevant criteria relate to inconsistencies in the windrow, which turning can correct. Examples include the following:

- The moisture content varies at different locations in the windrow,
- Feedstocks are noticeably poorly mixed,
- Wet spots on the windrow surface indicate hot spots or the presence of air channels, and
- Isolated cool or warm spots, indicating inconsistent conditions.

There are still other reasons to prompt turning of a windrow. For instance, during warmer seasons, when flies and other insects become problematic, active windrows should be turned at least once per week to break the flies' reproductive cycle, regardless of the windrow temperature. Since some species of flies develop into adults in as few as 5 days, windrows may require turnings every 4 days for fly control. Also, turning may be prompted by a rainstorm, as a convenient means to incorporate water into windrows and reduce runoff. Turning schedules may be dictated by regulations where composters are required to meet and document pathogen reduction requirements. For example, in the United States, a minimum of five turnings are required over a 15-day period during which the temperature remains above 55°C (131°F) in order to meet the requirements of the US-EPA "A Process To Further Reduce Pathogens."

4.5 Turning equipment

Windrow turning equipment both mixes the compost and reforms the windrow. Several different types of equipment can accomplish these tasks from conventional materials handling equipment to specially designed machines dedicated to turning.

4.5.1 Bucket loaders

For small- to moderate-scale operations, turning can be accomplished with the bucket on an excavator, wheel loader, skid steer loader, or tractor (Fig. 5.15). The bucket simply lifts the material from the windrow and lets it fall again, while mixing the materials and reforming a loose windrow. The loader can exchange material from the bottom of the windrow with material on the top by forming a new windrow next to the old one. Windrows turned with a bucket loader are often constructed in closely spaced pairs and then combined after the windrows shrink in size. Some operators prefer to work from the side of the windrow, "rolling" the working side of the windrow over the ridge of the opposite side (Fig. 5.16). Others work from the end of a windrow, placing the new windrow to the left or the right as they move down the length. This operation should be done without driving onto the windrow in order to avoid excessive compaction of the media.

When turning with a bucket loader, operators should allow the material in the bucket to pour out loosely rather than dumping the entire bucket at once. This approach allows the material to lose heat and gain oxygen more thoroughly and also leaves the pile at a lower-density. A short video program, produced by the former ROU organization in Australia, is currently available on the internet (e.g., You Tube) that demonstrates the technique (Campbell, 2012).

The time that it takes to turn windrows with a loader is roughly proportional to the size of the bucket. Typically, a loader can lift, maneuver, and empty a load of

(A)

(B)

FIGURE 5.15

Turning compost piles using a bucket loader and an excavator.

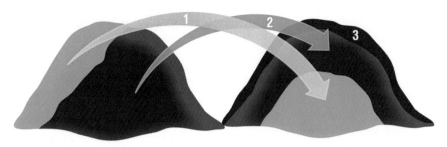

Existing pile   New pile

1. Working from the bottom   2. Flip the remaining pile on top   3. Add biocover for
edge upwards, flip about half   of the newly created pile. The   odor containment, if
the pile contents over the   existing pile into a new pile.   necessary.
existing pile into a new pile.

FIGURE 5.16

Procedure for turning windrows with a loader by rolling and flipping.

material in about one minute (Coker, 2020). General estimates of turning rates for tractor and skid loaders range from 15 to over 50 m³/h (20−65 yd³/h). Excavators can speed up this process as they can be positioned and then move materials by rotating left or right.

Turning with a bucket loader or excavator does not aggressively break up large particles and clumps or mix materials of different sizes. If additional mixing of the materials is desired, a loader can also be used in combination with a manure spreader. In this approach, materials from an existing windrow are loaded into the spreader. Spreader flails and augers provide a good mixing action for continued blending of raw materials. When the spreader is full, it unloads the material in a new windrow adjacent to the existing one. Although this approach provides better mixing than turning with a loader alone, it also involves additional equipment and more time.

Because turning with bucket loaders, excavators, and manure spreaders can be slow, these devices are usually limited to small-scale operations or where turning is infrequent (such as in passive composting). They remain good options for many on-farm and municipal composting situations because the equipment and sufficient land tends to be available. However, as turning consumes more time, dedicated windrow turners become more practical. A video program about turning windrows with a wheel loader.

4.5.2 Windrow turners

A large number and variety of specialized machines have been developed for turning windrows These windrow turners greatly reduce the time and labor involved, mix the materials thoroughly, and produce a more uniform compost than compost produced with a bucket loader. They come with various features like special power transmissions (e.g., hydraulic vs. mechanical), control systems, cab designs (including air conditioning and filtration), and other amenities. Windrow turners have also evolved to accomplish additional tasks, like adding water. A few machines also have the capability of loading trucks or wagons from the windrow. A self-driving GPS-guided turner cannot be too far into the future.

Windrow turners are distinguished by their mechanical action and their sources of power and travel. Mechanically, most models accomplish turning with a horizontal rotating shaft, 0.3—1.0 m (1—3 ft) in diameter. The shaft is referred to as a drum or is sometimes called a rotor. The drum has paddles, teeth, or auger flights that lift, move, and shred the compost to varying degrees and in varying directions (Fig. 5.17). The shroud or housing that surrounds the drum shapes the windrow and determines its dimensions. A few turner models principally rely on a wide backward-sloping conveyor to lift and reform the windrow contents. These models may include additional conveyors or augers to assist the main conveyor. Regarding power and travel, some machines are powered and/or moved by farm tractors or front-end loaders others are completely self-propelled.

Tractor-assisted turners can be either pulled or pushed by a tractor, skid-loader, or wheel loader. One type of turner uses augers attached to the front of a loader, similar in appearance to an auger-style snowplow (Fig. 5.18A and B). This machine turns over the windrows by shifting the material and the windrow to one side. Most

FIGURE 5.17

Close up of three windrow turner shafts with paddles and flights and the face of an elevating conveyor turner.

Sources: (A) R. Rynk. (B) Komptech Americas. (C) Scarab International LLLP.

FIGURE 5.18

(A—F) Variety of tractor-assisted windrow turners.

Sources: (A, B) Brown Bear Corp. (C, D, F) R. Rynk. (E) Vermeer.

models of tractor-assisted turners ride behind and to one side of the tractor, turning the windrow as the tractor travels in the aisle. Tractor-assisted turners are common and economical choice for farms and municipalities because an appropriate tractor is generally available and can be used for other in-house chores.

Most tractor-assisted windrow-turners are single-pass turners that straddle the windrow. In this case, aisle space for the tractor is required only between every other windrow. With straddle type turners, the windrow width and height must conform to the dimensions of the turner drum and housing. Typically, these machines handle windrows that range from 3 to 4 m (10—12 ft) in width and 1.2—1.8 m (4—6 ft) in height. Some tractor-driven machines turn only half the windrow in a single pass (Fig. 5.18F). Two passes are necessary for each windrow and an aisle is required on both sides of the windrow. Double-pass turners provide slightly more flexibility for the windrow width, allowing windrows up to 5 m wide (about 16 ft).

The least expensive turners rely on the tractor for both travel and power. Power is transferred through the power take-off (PTO) shaft of the tractor (Fig. 5.18C). The specifications vary among turner models, but generally the tractor must supply at least 60 kW (80 hp) to the PTO, although most turners require more than 75 kW (about 100 hp) of PTO power. At the same time, the tractor must travel very slowly while powering the turner. This combination of high power and slow speed requires a tractor with a creeper gear or hydrostatic drive. A less-convenient alternative is to use a second vehicle to tow the tractor/turner combination. If an appropriate tractor is not available or cannot be obtained economically, the next step is to purchase a self-powered turner that requires a tractor for travel only (Fig. 5.18A,E and F). These turners are powered by an on-board diesel engine that ranges from approximately 75 to 260 kW (about 100−350 hp). They otherwise operate in the same manner as the tractor-driven units.

Other compost turners are totally self-contained, providing their own power and travel. In addition to eliminating the need for a second piece of equipment, these machines also eliminate the large aisle required for tractor travel, allowing closer spacing of windrows. Most self-contained turners straddle the windrow and turn it with a rotating drum shaft (Fig. 5.19A−E through F). The elevating face conveyor type is also available as a self-driven unit (Fig. 5.19G).

As a group, self-powered turners are larger and accommodate larger windrows than their tractor-assisted counterparts, although several small self-contained turners also exist (Fig. 5.19E). Consequently, self-powered windrow turners generally offer more efficient use of land area and a greater turning rate (i.e., more cubic meters or cubic feet turned per hour). The largest straddle windrow turning machines turn windrows as large as 3 m (10 ft) high and over 8 m (25 ft) across. They can turn these windrows at a rate of more than 6000 m^3 (8000 yd^3) per hour (Ecoverse/Bachaus, 2021; Scarab, 2021). These advantages are made possible by more powerful engines and a substantially larger investment. Engine power is typically in the 75−190 kW (100−255 hp) range for the smaller models and up to 370 kW (500 hp) for the large machines. For applications with dense feedstocks, like soil remediation, a turner may have an engine that exceeds 450 kW (600 hp).

Large windrow turners are often custom built to the specifications and feedstocks used at a particular site. For instance, engine power and other features, are frequently adjusted for different feedstocks and operating conditions. On concrete or asphalt surfaces, two-wheel drive systems are generally used. However, turners with four-wheel-drive or track-drives are capable of turning compost in windrows at locations with poor surface conditions or uneven terrain.

Some windrow turners can fold allowing them to be towed on a trailer or pivot to become self-trailering. Increasingly, straddle style turners are being equipped with additional conveyors or augers to facilitate materials handling. For instance, a conveyor may allow a windrow to be shifted to one side, rather than turned in place. Auger style and elevating face turners inherently have this capability (Figs. 5.18B and 5.19F).

FIGURE 5.19

(A—F) Variety of self-propelled windrow turners.

Sources: (A) Komptech Americas. (B, E, F) R. Rynk. (C) Vermeer Corporation. (D) Scarab International LLLP.

Windrow turners can be fitted with spray nozzles and a plumbing system to introduce water during turning (Fig. 5.20). This feature can improve the moisture content and consistency of the windrow and also reduce dust and bioaerosols emissions. Tractor-towed turners usually use a tow-behind tank from which water (or stored leachate) is pumped to sprayers located under the turning deck. The tank must be filled periodically during turning, which slows the turning rate. Self-driven turners tend to tow a flexible hose that unwinds from a hose reel as the turner travels. The hose is connected to a water hydrant or pump at a run-off retention pond. The hose reel may be located on the ground or on the turner itself (see Chapter 9). In the latter case, the hose and water add substantial weight to the turner.

(A)

(B)

FIGURE 5.20

Tractor-towed windrow turner (A) and self-driven turner fitted with nozzles to add water while turning.

Source: (A) C. Oshins. (B) S. Gamble.

All windrow turners, regardless of their design, require regular maintenance. Routine maintenance is needed on the engine and hydraulic systems. Paddles and teeth also tend to break or wear and need periodic replacement. Broken or worn pieces can upset the balance of drum shafts and other rotating parts and lead to excessive vibration. Thus, many turner models have been designed for routine replacement and repair of these wear parts.

Box 5.3 Integrating windrow composting and animal housing

Authors: Frederick Michel and Harold Keener

Some animal housing designs incorporate composting directly into buildings that house swine or caged poultry. These systems usually feature slatted floors or wire cages in the animal living area and a composting pit or conveyor system below.

Manure is deposited in the lower level of these buildings from the animals residing above. A bed of amendment, like sawdust or corn stover, is supplied to balance the deposited manure. Small windrow turners have been adapted to operate within these pit type buildings (Fig. 5.21). The windrow turner is used to mix the deposited manure with the amendment periodically. A shield

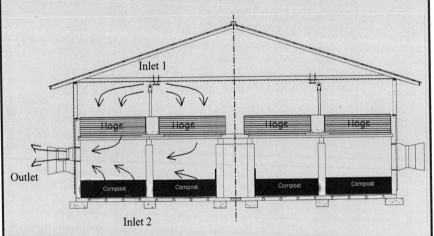

FIGURE 5.21

Pit-style animal barn with integrated composting system.

Source: H. Keener.

above the operator cabin is used to deflect falling manure. A forced aeration system in the floor of the buildings aerate and dry the manure to encourage composting and prevent anaerobic conditions and odor formation.

In another system, poultry manure is conveyed to a separate indoor composting facility where it is composted without amendment leading to a nitrogen rich final product that is custom applied as a crop fertilizer (Fig. 5.22). The air in the facility can be scrubbed to recover volatilized ammonia.

In deep bedded animal housing systems, animals are housed on top of a 75-cm bed of sawdust (about 2−3 ft). Manure is deposited on the surface and mixed into the bed usually daily or every other day using mechanical mixing equipment. The manure bedding mixture accumulates for months and is then removed, and land applied or sold as compost.

Another deep bed system uses the composting process as a biodryer to reduce the moisture content and mass of dairy manure and allow more land application options. Forced aeration and a roof is used to optimize this process. Forced air is used to compost and dry a layer of manure in 21 days. The heat generated by composting provides the energy to reduce the manure moisture content from 88% to 40% (w/w). Recycled compost, or a mix of compost and sawdust, or other

Box 5.3 Integrating windrow composting and animal housing—cont'd

FIGURE 5.22

Composting building with automated windrow turner integrated with manure management system for egg-laying hens.

Source: F. Michel.

amendment, removed from the bed at 40% moisture content is then spread in the cow alleyways about 8 cm (3 in) thick to absorb 1 day's production of 88% moisture content manure. The mixture is then scraped into the bio-drying shed, piled 2 m (6 ft) deep and aerated to produce 40% dry matter compost with a retention time of approximately 3 weeks.

References

Cited references

Buckner, S.C., 2002. Effects of turning frequency and mixture composition on process conditions and odor concentrations during grass composting. In: Proceeding of the 2002 International Symposium, Columbus Ohio, May 2002.

Campbell, A., 2012. Turning compost piles – with a wheel loader. CompostTV, episode 05a. Recycled Organics Unit, Vic., Australia. Available at: https://www.youtube.com/watch?v=VxWS4B1k0mc.

Coker, C., 2020. Composting business management: composting facility operating cost estimates. BioCycle.

De Silva, S., Yatawara, M., 2017. Assessment of aeration procedures on windrow composting process efficiency: a case on municipal solid waste in Sri Lanka. In: Environmental Nanotechnology Monitoring & Management. https://doi.org/10.1016/j.enmm.2017.07.008.

Ecoverse/Backhaus, 2021. https://www.ecoverse.net/a75/.

Fogiel, A.C., 1998. Experimental verification of the natural convective transfer of air through a composting media. MS Thesis. Michigan State University. https://d.lib.msu.edu/etd/27846.

Haug, R.T., 1993. The Practical Handbook of Compost Engineering. Lewis Publishers, CRC Press, Boca Raton, Florida.

Hayes, L.A., Richards, R., Mathur, S.P., 1994. Economic viability of commercial composting of fisheries waste by passive aeration. In: Compost Council of Canada Symposium, September 1994.

Keener, H.M., Wilkinson, T.F., Michel Jr., F.C., Brown, L.C., 2005. Evaluation of leaching from composting windrows using a rainfall simulator. In: 2005 Animal Waste Management Symposium, 10/5-7. Sheraton Imperial Hotel and Conference Center, Research Triangle Park, North Carolina. Published on CD Available at: http://www.cals.ncsu.edu/waste_mgt/natlcenter.

Mathur, S.P., 1995. Natural aeration static pile (based on Agriculture Canada's PAWS technology). In: Composting Technologies and Practices. A Guide for Decision Makers. The Composting Council of Canada, Ottawa, Ontario.

Mathur, S.P., 1991. Composting processes. In: Martin, A.M. (Ed.), Bioconversion of Waste Materials to Industrial Products. Elsevier, London, New York, pp. 147−186.

Mathur, S.P., Patni, N.K., Lévesque, M.P., 1990. Static pile, passive aeration composting of manure slurries using peat as a bulking agent. Biol. Waste. ISSN: 0269-7483 34 (4), 323−333. https://doi.org/10.1016/0269-7483(90)90033-O.

Mathur, S.P., Daigle, J.-Y., Brooks, J.L., Levesque, M., Arsenault, J., 1988. Composting seafood wastes - avoiding disposal problems. Biocycle 29, 44−49.

Michel, F.C., 2002. Effects of turning and feedstocks on yard trimming composting. BioCycle 43 (9), 46.

Michel, F.C., 1999. Managing compost piles to maximize natural aeration. BioCycle 40 (3), 56−58.

Michel Jr., F.C., Forney, L.J., Huang, A.J.-F., Drew, S., Czuprenski, M., Lindeberg, J.D., Reddy, C.A., 1996. Effects of turning frequency, leaves to grass mix ratio, and windrow vs. pile configurations on the composting of yard trimmings. Compost Sci. Util. 4 (1), 26−43.

Mitchell, R., 2012. Turning comost by temperature. Rodale Institute. https://rodaleinstitute.org/science/articles/turning-compost-by-temperature/.

Ogunwande, G., Ogunjimi, L., Fafiyebi, J., 2008. Effects of turning frequency on composting of chicken litter in turned windrow piles. Int. Agrophys. 22 (2), 159−165.

Paré, M., Paulitz, T.C., Stewart, K.A., 2000. Composting of Crucifer Wastes Using Geotextile Covers. Compost Science & Utilization 8, 36−45.

Patni, N., 2000. Videotape interview. In: The Future of Agricultural Composting, A Video Workshop. Compost Education and Resources for Western Agriculture. University of Idaho, Moscow, ID.

Scarab International, 2021. https://www.scarabmfg.com/products/.

Shell, B.J., 1955. The mechanism of oxygen transfer through a compost material. Doctotal thesis. Michigan Stae University Libraries. https://doi.org/10.25335/M56T0HB6S. https://d.lib.msu.edu/etd/45504.

Tirado, S.M., Michel Jr., F.C., 2010. Effects of turning frequency, windrow size and season on the production of dairy manure/sawdust composts. Compost Sci. Util. 18 (2), 70–80.

Veeken, A.H.M., Timmermans, J., Szanto, G.L., Hamelers, H.V.M., 2003. Design of passively aerated compost systems on basis of compaction-porosity-permeability data. In: Proceedings of 4th International Conference: Advances for a Sustainable Society, Perth, Australia. Published by: European Compost Network ECN e.V., ORBIT 2003.

Veeken, A., de Wilde, V., Hamelers, B., 2002. Passively aerated composting of straw-rich pig manure: effect of compost bed porosity. Compost Sci. Util. 10 (2), 114–128. https://doi.org/10.1080/1065657X.2002.10702072.

Consulted and suggested references

Bachert, B., Bidlingmaier, W., Wattanachira, S., 2008. Open Eindrow Composting, ORBIT ed. Bauhaus-Universität Weimar - Professur für Abfallwirtschaft.

Buckner, S.C., 2002b. Controlling odors during grass composting. BioCycle 43 (9), 42–47.

Chardoul, N., O'Brien, K., Clawson, B., Flechter, M. (Eds.), 2011. Compost Operator Guidebook. Michigan Recycling Coalition, Lansing, MI.

Cook, K.L., Ritchey, E.L., Loughrin, J.H., Haley, M., Sistani, K.R., Bolster, C.H., 2015. Effect of turning frequency and season on composting materials from swine high-rise facilities. Waste Manag. 39, 86–95. https://doi.org/10.1016/j.wasman.2015.02.019. PMID: 25752584.

David, A. (Project leader), S. Gamble (Project manager), 2013. Technical Document on Municipal Solid Waste Organics Processing, Canada Minister of the Environment, Ottawa.

Gamble, S., 2018. Compost Facility Operator Study Guide. Minister of Alberta Environment and Parks, ISBN 978-1-4601-4130-4 (PDF online).

Hyder Consulting, 2012. Understanding the processing options. In: Food and Garden Organics Best Practice Collection Manual. Department of Sustainability, Environment, Water, Population and Communities. https://www.environment.gov.au/protection/waste/publications/food-and-garden-organics-best-practice-collection-manual.

Keener, H., Elwell, D.L., Ekinci, K., Hoitink, H., 2001. Composting and value-added utilization of manure from a swine finishing facility. Compost Sci. Util. 9, 312–321.

Koenig, R.T., Miner, F.D., Miller, B.E., Palmer, M., 2005. In-House Composting in High-Rise, Caged Layer Facilities. USDA Sustainable Agriculture Research & Education. https://www.sare.org/resources/in-house-composting-in-high-rise-caged-layer-facilities/.

Lynch, N.J., Cherry, R.S., 1996. Design of passively aerated compost piles: vertical air velocities between the pipes. In: de Bertoldi, M., Sequi, P., Lemmes, B., Papi, T. (Eds.), The Science of Composting. Springer, Dordrecht. https://doi.org/10.1007/978-94-009-1569-5_93.

Mathur, S.P., Richards, R.W., 1999. Large-scale generation of hygenic, high-quality compost from beef feedlot manure and sawmill residues in natural aeration static piles (NASP) without pipes, plenums or envelope. In: Warman, P., Munro-Warman, T.R. (Eds.), Abstracts of the International Composting Symposium. September, 1999. Coastal Bio-Agresearch, Ltd., Truro, Nova Scotia, p. 36.

Michel Jr., F.C., Pecchia, J., Rigot, J., Keener, H.M., 2004. Mass and nutrient losses during composting of dairy manure with sawdust versus straw amendments. Compost Sci. Util. 2 (4), 323–334.

Michel Jr., F.C., Reddy, C.A., 1998. Effect of oxygenation level on yard trimmings composting rate, odor production and compost quality in bench scale reactors. Compost Sci. Util. 6 (4), 6–14.

Michel Jr., F.C., Forney, L.J., Reddy, C.A., 1993. Yard waste composting: studies involving different mixes of leaves and grass in a laboratory scale system. Compost Sci. Util. 1 (3), 85–99.

Misra, R.V., Roy, R.N., Hiraoka, H., 2003. On-Farm Composting Methods. Food and Agriculture Organization of the United Nations, Rome. http://www.fao.org/3/y5104e/y5104e00.htm.

Nelson, V., 2002. Assessment of Windrow Turners. Agritech Center. Alberta Agriculture, Food and Rural Development.

SKM Enviros and Frith Resource Management, 2013. Guidance for On-Site Treatment of Organic Waste From the Public and Hospitality Sectors. Waste and Resources Action Programme (WRAP), Banbury, UK.

Sun, H., Stowell, R.R., Keener, H.M., Michel Jr., F.C., 2002. Two-dimensional computational fluid dynamic (CFD) modeling of air velocity and ammonia distribution in a high-rise hog facility. Trans. ASAE 45 (5), 1559–1568. https://doi.org/10.1080/1065657X.2010.10736938.

The Composting Association (UK), 2008. Turners and mixers: the composters' guide to buying equipment and services compost. Compost. News 12 (1). Spring 2008.

The Compost Council of Canada, 2016. Best Practices for Operating an Aerated Windrow Composting Facility. The Compost Council of Canada for the Government of Manitoba. http://www.compost.org/wp-content/uploads/2019/11/Best_Practices_for_Operating_an_Aerated_Windrow_Composting_Facility.pdf.

Yu, S., Clark, O.G., Leonard, J.J., 2007. Estimation of vertical air flow in passively aerated compost in a cylindrical bioreactor. Can. Biosyst. Eng./Le Genie des biosystems au Canada 50, 6.29–6.35.

Zhou, J.-M., 2017. Effect of turning frequency on co-composting pig manure and fungus residue. J. Air Waste Manag. Assoc. 67 (3), 313–321. https://doi.org/10.1080/10962247.2016.1232666.

Forced aeration composting, aerated static pile, and similar methods

6

Authors: Frederick Michel[1], Tim O'Neill[2], Robert Rynk[3]

[1]Ohio State University, Wooster, OH, United States; [2]Engineered Compost Systems, Seattle, WA, United States; [3]SUNY Cobleskill, Cobleskill, NY, United States

Contributors: Jane Gilbert[4], Matthew Smith[5], John Aber[6], Harold Keener[7]

[4]Carbon Clarity, Rushden, Northamptonshire, United Kingdom; [5]USDA National Agroforestry Center, Lincoln, NE, United States; [6]Professor Emeritus, University of New Hampshire, Durham, NH, United States; [7]Professor Emeritus, Ohio State University, Wooster, OH, United States

1. Introduction

Forced aeration creates and controls airflow through a compost pile by means of a mechanical device—a fan, also referred to as a blower (Box 6.1). Compared to passive aeration, fan-induced airflow is a more certain way to supply oxygen and remove heat, moisture, and carbon dioxide. With fans supplying the aeration, the need for regular turning is substantially reduced, although not entirely replaced, because forced aeration is supplies oxygen to the core of the pile.

Fan-driven aerated composting first emerged in the 1950s in Europe, primarily as means of producing substrate for growing mushrooms. In North America, forced aeration took hold with the emergence of the aerated static pile, or "ASP" method of composting. The ASP method was first used in the 1970s as means to treat biosolids and evolved from research conducted at the USDA's Beltsville research station (Fig. 6.1). It became known as the Beltsville ASP method (Willson et al., 1980). Concurrent research at Rutgers University yielded another forced aeration approach known as the Rutgers strategy (Finstein et al., 1992). The Beltsville and Rutgers methods are similar but differ in their control strategies.

Together the Beltsville and Rutgers work stimulated an expansion of biosolids composting during the 1980s. Many elements of these original ASP practices persist today, especially at biosolids composting facilities. In the ensuing decades, the simplicity and advantages of ASP attracted the attention of composters of other feedstocks and began to complement and/or displace turned windrows. As ASP method expanded in use, it further evolved and diversified. ASP and similar methods now encompass various air delivery techniques, aeration control strategies, enclosures, covers, and combinations of feedstocks. Forced aeration continues to lure composters away from the turned windrow approach, largely because of its space-saving advantages.

The Composting Handbook. https://doi.org/10.1016/B978-0-323-85602-7.00007-8

Box 6.1 Fans versus blowers

In the composting realm, both "fan" and "blower" are used to describe the device that moves air through a pile or mass of feedstock. Composting practitioners tend to use the word "blower" while engineers prefer "fan." To be diplomatic, both words are correct. To be technically accurate, fan is the more correct and is used throughout this book. A blower is a fan, a certain type of fan (Fig. 6.2).

A fan is a mechanical device that moves gases, like air, by creating a pressure difference across itself. A fan does this by rotating blades that push

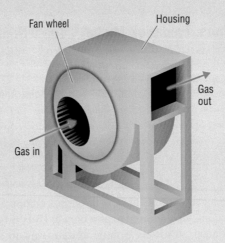

FIGURE 6.2

Typical centrifugal fan (aka blower) used for ASP composting.

air in one direction. The pressure that develops across the fan depends on the mechanics and power of the fan plus the resistance that develops along the path of air movement. The resistance is caused by friction, which is exacerbated by turbulence. It creates a back pressure, called "static" pressure, which pushes against the fan, making the fan work harder in order to move a given amount of air.

The fan versus blower argument rests on this pressure. Where there is little resistance and little back pressure, the familiar "axial" fans are common. Where substantial resistance to airflow occurs, such as in a network of airducts or through a compost pile, "centrifugal" fans prevail. Centrifugal fans rotate a series of tightly spaced blades on a wheel, pulling air in at the center of the wheel and discharging it sideways (e.g., 90 degrees) in the direction of the blades' rotation.

Generally speaking, blowers are centrifugal fans. Wherever moderate to high pressure differences occur, and centrifugal fans are used, people tend to call these fans "blowers." It is well-established jargon. Technically, however, blowers are centrifugal fans that develop very high pressures. The American Society of Mechanical Engineers (ASME) distinguishes fans and blowers by the ratio of the discharge (outlet) pressure and suction (inlet) pressure. According to the ASME, fans are "fans" if this ratio is *below* 1.11, which equates to a pressure difference of about 1136 mm w.c.[1] (45 in.). Fans are "blowers" if the ratio is between 1.11 and 1.2. Above 1.2 they are "compressors." Fan pressures in composting situations rarely exceed 400 mm w.c. (15 in.). Therefore, technically forced aeration composting uses fans, centrifugal fans.

[1] w.c. stands for "water column." The pressure expressed in millimeters or inches w.c. is equivalent to the pressure exerted at the base of a column of water so many millimeters or inches high.

This chapter describes the principles, practices, and components of ASP composting and methods that have similar attributes, including forced aeration systems under various types of covers. It also describes methods that combine forced aeration and regular turning, including aerated windrows and turned extended piles. The chapter does not cover fan-aerated methods that are "contained," such as agitated bays and in-vessel composting systems.

FIGURE 6.1

UDSA Beltsville Research Station composting site with individual and extended ASPs in the middle of the photo and turned windrows toward the bottom.

Source: US Department of Agriculture.

2. Aerated static pile

The ASP method is a versatile approach to composting. It is easily scalable in both size and technology (Table 6.1). It is suitable for small community-scale operations and very large ones. It is inherently simple but readily accommodates more sophisticated techniques when warranted.

ASP composting can be practiced at a relatively low level of technology, using rule-of-thumb design, off-the-shelf equipment, and little site modification. Following well-proven practices and guidelines, a small- to moderate-scale composter can implement an ASP system with common piping materials, an inexpensive fan, and an on/off timer (Fig. 6.3). This "basic" ASP approach has a long history of success. However, for its simplicity, a basic ASP approach sacrifices some control of the composting process. At times, temperatures can be either higher or lower than desired, and oxygen concentrations can fall below target levels.

To enhance the performance of ASPs, composting system designers have developed "advanced" practices that can increase process control in order to reduce air emissions and decrease processing times. Such systems are widely employed where the scale of operation, site conditions, and local constraints (e.g., odors, regulations, throughput) justify the associated initial expense and engineering effort. They are

Table 6.1 Summary of aerated static pile method.

Key features

- Maintains higher and more uniform oxygen levels.
- Can remove excess heat to moderate temperatures.
- Well-proven method.
- Long and varied body of experience.
- Can be technically simple.
- Requires less area (smaller footprint) compared to windrow composting.
- More space-efficient therefore can be less expensively enclosed within a bin or building.
- Easily scalable. Works for small or large systems.
- Requires attention to the type and amount of amendments used.
- Over-drying of piles is a common problem.
- It is difficult to add water when needed.
- Tends to suffer from compaction, short circuiting of air, and inconsistent decomposition within a batch if the initial mix isn't uniform and well bulked, and if piles are left static for too long.

Limited to

- Locations where power is available or where either a power generator or a solar photovoltaic system can be installed.
- Relatively homogenous mixes of feedstocks with good porosity and structure.

Works well for

- Biosolids, especially, because of a large body of experience.
- Feedstock mixes that require demonstrated pathogen and vector attraction reduction.
- Most feedstocks that carry risks of odors or vector attraction including food and manure.
- Space-constrained sites.
- Start-up, pilot, and research applications.
- Processing large amounts of material efficiently.
- Sites where surface water management considerations favor a small footprint.

most commonly used for large-scale composting of source-separated organics (SSOs), biosolids, anaerobic digestate, and food residuals. Also, specific regulations or air-quality agencies may require advanced features, even at a moderate scale. Advanced ASPs come with many variations, but almost always include a number of purpose-built aeration components, programmable controllers, and require engineering design (Fig. 6.4).

2.1 General principles

ASP uses fans to increase the airflow through the compost pile, compared to a passively aerated system, such as turned windrows. The increased airflow supports more efficient composting by limiting temperature rise, maintaining oxygen levels and by removing excess moisture, carbon dioxide, and ammonia. The standard practice with forced aeration is to adjust the airflow rate to match the rate of biological

FIGURE 6.3

ASP composting system using a single pipe-on-grade and laterals and a timer-controlled fan.

FIGURE 6.4

Advance ASP using extended pile, reversing aeration, and standalone biofilters.

Source: ECS.

heat generation, maintaining compost temperatures in the range of 60−70°C (140−160°F). More airflow is supplied early when the feedstock mix is rapidly consuming oxygen and generating high rates of heat. Later, as the biological activity slows, the rate of heat generation drops and need for aeration declines. Traditionally, ASPs have received little or no turning, except for what occurs when piles are moved. However, it is now common to turn piles at least once to remix and remoisten the piles (Bryant-Brown and Gage, 2015).

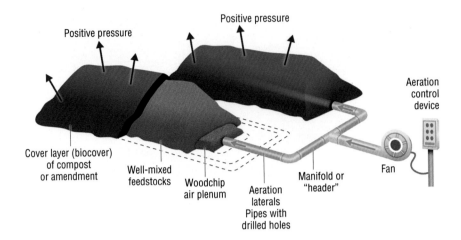

Positive pressure

Positive pressure

Aeration control device

Cover layer (biocover) of compost or amendment

Well-mixed feedstocks

Woodchip air plenum

Aeration laterals Pipes with drilled holes

Manifold or "header"

Fan

FIGURE 6.5

Aerated static pile primary components.

An ASP system includes the following primary components (Fig. 6.5). These components are described in more detail in the following sections.

- One or more fans,
- An aeration floor,[2] a set of lateral pipes or trenches that underlies and distributes air to the composting pile,
- An air distribution system that connects the fan(s) to the aeration floors,
- An optional "plenum," usually constructed of wood chips that covers the aeration laterals,
- An aeration control system that starts/stops or varies the rate, and possibly the direction, of airflow,
- Premixed composting feedstocks, and,
- An exterior layer or "biocover" made of unscreened compost, wood chips, screen overs, or alternative stable material.

During the early stage of composting, when the demand for heat removal is high, the fan operates continuously or intermittently. In the former case, fan speed is usually determined by a variable frequency drive (VFD), which is often controlled by temperature feedback, while in the latter case fan operation is determined by either an on/off timer or by a temperature feedback system.

---

[2] This book generally uses the term "aeration floor" to designate the network of channels and/or pipes that aerate a composting pile. Although not universally used within the composting industry, it is a concise and descriptive label.

The fan can either push or pull air through the pile. Pushing air, termed "positive aeration," is more prevalent and generally preferred because it is simpler and more energy-efficient. However, pulling air ("negative aeration") captures and concentrates the process exhaust air, allowing it to be directed to an odor treatment device, like an external biofilter. In some applications, both positive and negative (reversing) aeration is used.

The aeration floor is the part of the aeration system that either distributes air to the composting pile (positive aeration) or collects the compost air from the pile (negative aeration). The aeration floor contains one or more lines of pipe or channel, called laterals, which run underneath the pile. Two forms of aeration floors are common: "pipe-on-grade" (POG) and "below-grade." With a POG floor, the laterals are moveable and the perforated aeration pipes rest on the composting pad. The pipes are perforated only within the footprint of the pile. Because the pipes interfere with materials handling and are subject to damage, some facilities choose to invest in below-grade aeration floors. Below-grade laterals consist of trenches and/or pipes installed into and below the composting pad. Various techniques are used to deliver air to the pile from a below-grade aeration floor including perforated grates covering the channels and risers extending from the below-grade laterals to the composting pad surface. To achieve semiuniform air distribution from either type of floor, the size and number of perforations and dimensions and layout of the pipes and/or channels must meet certain design criteria. With both types of aeration floors, a layer of porous amendment, usually wood chips or screen overs, is often placed over the laterals as a plenum to help diffuse the air, filter particles, and protect pipes.

To achieve uniform airflow through the pile, ASP feedstocks need to be relatively porous (i.e., high free-air-space [FAS]) and homogeneous. To achieve homogeneity, the feedstocks are mixed before composting. This requirement might necessitate a feedstock mix with a high proportion of amendments or bulking agents with good structure and water content low enough to prevent filling of pore space in compost.

Freestanding ASPs are constructed either in individual piles, each holding a single feedstock batch, or in larger extended piles with multiple cells with one batch in each cell. ASPs are also placed in bins and bunkers.

As the pile is being built, a "biocover" is usually added to its surface. The biocover protects the surface of the pile from drying, insulates it from heat loss, discourages flies, and filters some of the ammonia, odors, and other volatile compounds generated within the pile. The biocover typically consists of 15–30 cm (6–12 in.) of unscreened compost, screened overs, or wood chips. This biocover is required if the materials need to complete a time/temperature pathogen reduction step. This cover can be used with both positive aeration and negative aeration.

Generally, the composting process can be divided into two distinct aeration phases, often termed primary composting and secondary composting. The primary phase is the most active and typically requires higher aeration rates. The primary phase typically lasts two to four weeks, depending on feedstocks. By the secondary phase, the rate of heat generation has decreased and a lower airflow can be used. The secondary phase typically goes on for another two to four weeks before transitioning to a curing period. The line between these two phases is not well defined and subject to the designers' and operators' judgments. Often the two phases involve moving a pile to a second location for the secondary phase, thereby providing an opportunity to rewet and mix the pile. The process generally benefits from a rewetting and remixing step after about two to four weeks.

2.2 Pile configurations and dimensions

Two forms of freestanding ASPs are common: individual (or standalone) piles and extended piles (or mass bed). In addition, feedstocks are sometimes piled in bunkers or bins, in which case they operate more like individual piles.

2.2.1 Individual piles

From an operational perspective, an individual pile holds a single batch of feedstocks that should generally consist of the same recipe and age (within three days, for example). Since the entire pile is aerated as a single unit, all the materials in the pile should have about the same demand for aeration. Individual piles are practical when feedstocks come available for composting at intervals rather than continuously—for example, if manure is cleaned from barns on a weekly basis or if short-term storage of feedstock is possible. Individual piles are also useful for separating batches of material for experimentation or to isolate specific feedstocks. The main drawback of individual piles is they require more space due to the sloping slides and extra aisle space. It also requires proportionally more material for a biocover layer compared to an extended pile.

Individual piles have a triangular to trapezoidal cross-sectional shape, depending on the width (Fig. 6.6). Side slopes vary with the feedstock characteristics but a slope of 45 degrees (1:1 rise:run) is a reasonable general assumption. Therefore, simple piles that have a triangular shape have a width roughly equal twice the pile height.

There is a *maximum* practical pile length that ranges from roughly 20—25 m (65—80 ft), depending on the aeration system. The maximum pile length is governed by the length of the aeration lateral, which is in turn limited by the need to maintain fairly even distribution of air along its length. Pile lengths can exceed this practical limit if the aeration system design is supported by engineering models. In general, the aeration lateral and the surrounding plenum should be somewhat shorter than the pile. To prevent air from short circuiting out the ends of the pile, the perforated sections of the lateral should begin and end a short distance inside of the pile as shown in Fig. 6.6. As a general rule, that short distance should be approximately equal to the pile height.

The initial height of piles generally should be in the range of 2—4 m (6—12 ft), not including the biocover. As piles become taller than 2.5 m (8 ft), the feedstock mix requires more structure and lower initial bulk density to minimize compaction and air channeling. Weather conditions are an additional factor. Extra height is advantageous in the winter to retain heat. In practice, the factor that usually limits pile height is the reach of the equipment used to build the pile. A pile is too high if a bucket loader must drive on any part of it to place the feedstocks or biocover.

Individual pile with multiple laterals

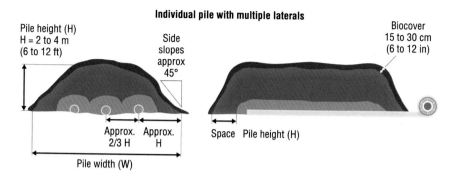

Individual pile with single lateral

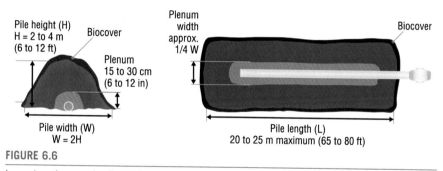

FIGURE 6.6

Layout and example dimensions of a Basic ASP with individual piles.

In most situations, the width of an individual pile is determined by the volume of feedstock to be composted in the batch, after the height and length of the pile are determined. Eq. (6.1) is the formula for calculating the width of a triangular or trapezoidal pile with 45° side slopes.

$$w = \frac{V}{(L \times h)} + h \qquad (6.1)$$

where: w is the pile width,
  h is the pile height,
  L is the pile length, and
  V is the required volume of the pile.

After the pile width is selected, the number of aeration laterals can be determined using the following guidelines.

- In a freestanding individual pile, the distance between the "toe" of the pile (i.e., outer edge) and the aeration lateral should be the approximately equal to the pile height. The underlying concept is that distance that the air travels to leave the pile is nearly the same whether it flows to the top or sides or bottom edge of the pile.
- The spacing between multiple laterals should be limited to about 2/3rds the pile height (0.67 × h). More general recommendations suggest that laterals should be spaced about 1.2—1.5 m apart (4—5 ft) (Moon, 2021; Gibson and Coker, 2013).

These rules indicate that one lateral should suffice for a pile with a width approximately twice its height. For a pile with a width about 2.67 times the height, two laterals are spot-on. Similarly, three laterals work perfectly for pile widths 3.33 time the height. For pile widths in between these points, the number of laterals becomes a judgment call. For example, the spacing between laterals can be set equal to the pile height in order to accommodate a given pile size. Example 6.1 illustrates the process of sizing a typical individual ASP by this approach.

An alternative approach is to settle on a pile width and lateral layout using the foregoing guidelines and then calculate the pile length required to achieve the desired volume. In this case, Eq. (6.1) can be rearranged to calculate the pile length (L).

$$L = \frac{V}{[h \times (w - h)]} \tag{6.2}$$

Example 6.1 Individual ASP sizing example

A large horse stable empties its manure storage bunker monthly and composts each batch of manure plus an equal volume of wood chips in individual static piles. With the wood chips, the total feedstock volume is 60 m^3 per batch (78.5 yd^3 or 2120 ft^3). The horse stable limits the piles to 2 m (6.6 ft) in height because the feedstock mix is moderately dense, even with the wood chips. The composting site limits piles to 10 m lengths (33 ft).

1. Determine pile width using Eq. (6.1).

$$w = \frac{V}{(L \times h)} + h = \frac{60 \text{ m}^3}{(10 \text{ m} \times 2 \text{ m})} + 2 \text{ m} = 3 \text{ m} + 2 \text{ m} = 5\text{m}$$

$$= \frac{2120}{(33 \text{ ft} \times 6.6 \text{ ft})} + 6.6 \text{ ft} = 9.7 \text{ ft} + 6.6 \text{ ft} = 16.3 \text{ ft}$$

2. Determine number of aeration laterals and their spacing.
   a. w = 5 m (16.7 ft), which is more than 2 × h, so assume two laterals.
   b. The spaces between the edges of the pile and the laterals accounts for 4 m (13.2 ft). That leaves 1 m or 3.1 ft of spacing between the two laterals, which is exactly ½ × h.
   c. Choices:
      i. Space laterals 1 m apart, even though this is less than the target of 2/3rds × h.
      ii. Space laterals 2/3 × h apart = 1.33 m (4.4. ft) and shorten the spacing between the laterals and edges of the pile accordingly.
      iii. Something between (i) and (ii).
      iv. Use a single lateral under the center of the pile.
   d. Judgment:
      i. For composting, these choice amount to "splitting hairs." It is difficult to place pipes so precisely regardless. All four choices are fine. The fourth choice (iv) is the cheapest and simplest. The third choice (iii) is a good compromise.

The foregoing guidelines are not hard-and-fast. With engineering analysis and/or experimentation, a particular ASP can be longer, wider, higher than these guidelines dictate or have aeration systems that defy the general rules.

2.2.2 Extended piles

An extended ASP consists of a series of cells, also called zones, which are stacked against one another to form a continuous mass bed (Fig. 6.7). This configuration also may have a push wall at the fan end. Each cell contains a single batch of material. The aeration floor can be POG or below-grade. Ideally, a cell should be filled in 1–3 days, but loading may occur over a period of a week or slightly more.

An extended ASP usually has a gap separating newly constructed cells from older cells nearing completion. The location of the empty cell shifts within the bed as compost is removed from a mature cell and feedstock is added to a previously empty cell. The space created by the empty cell(s) prevents sanitized-compost from mingling with fresh material that has yet to achieve sanitization (e.g., Process Further Reduce Pathogens). It can also serve to separate materials that are in different phases of composting.

The dimensions of an extended pile are determined by the same factors as an individual pile; cell lengths are generally limited to 20–25 m (65–80 ft) and cell heights fall in the range of 2–4 m (6–12 ft), without the biocover. Cell widths and/or lengths are determined by the volume of feedstocks intended for the cell.

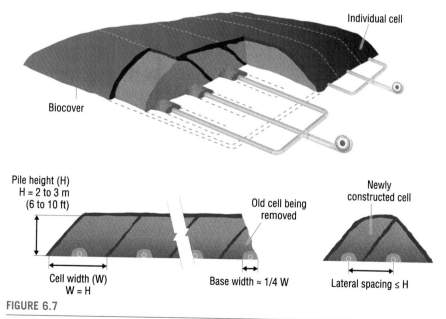

FIGURE 6.7

Layout and example dimensions of an extended ASP.

Commonly, cell widths are about equal to the cell height, although wider cells are also used to accommodate larger batches of feedstocks.

Each cell includes one or more aeration laterals. The number of aeration laterals is primarily determined by the width of the cell and the aeration rate required. Multiple laterals are used when a cell is wider than it is high (e.g., w = 1.5 × h) or if the airflow rate is high enough to justify reducing the air velocity with multiple laterals.

Although each cell can be aerated by its own fan and controlled by its own timer or temperature controller, it is more common and often economical to connect a series of cells to a single fan. The fan supplies air to a pipe manifold that connects to the aeration laterals.

Individual dampers modulate the airflow to each cell (Fig. 6.8). The dampers can be manually or automatically controlled. This practice requires a fan control strategy to vary its output to meet changing aeration demand across all cells.

2.3 Feedstock preparation, pile construction, and turning

The composition and uniformity of the initial feedstock mix are critical to ASP composting. Unlike a turned system, the feedstocks in an ASP becomes less homogeneous over time as the pile dries, degrades, and settles. Also, it is not possible to uniformly rewet the pile once it is formed without remixing the entire volume. Starting with a well-balanced and well-blended mix yields the best results and delays the eventual requirement to rewet and remix the pile.

Piles with good structure resist settling and compaction and can maintain porosity until the pile is turned and remixed. Such structure generally requires stiff

FIGURE 6.8

Positive-aeration ASP pipe manifold (upper) with damper-modulated aeration rates to the cell plenum below.

Source: ECS.

bulking agents. Wood chips are one of the most commonly used bulking agents due to their size, rigidity, and ability to create a porous medium. Because of their resistance to decay, large wood chips pass through the process only partially decomposed. They are usually screened from the finished compost and reused or discarded as bulking agents, disappearing slowly over subsequent cycles. Other possible bulking agents and amendments for ASP composting include sawdust, recycled compost, peat moss, tree bark, corn cobs, nut hulls, chopped straw, crop residues, bark, leaves, and shellfish shells. Large particles and nondecomposable material must eventually be screened from the compost and reused.

Since piles are turned infrequently, if at all, wet or otherwise dense primary feedstock (e.g., biosolids, manure, and food) must be thoroughly blended with a bulking agent before the pile is established (see Chapter 9). A bucket loader is the most common feedstock mixing device. It can do an acceptable job of mixing and building piles, especially with the aid of a push wall, and after the operator gains experience (Coker, 2018). However, a loader does little to reduce the size of large particles and they are ineffective at mixing materials that have very different densities, such as biosolids and sawdust. A bucket loader combined with a manure spreader can be used both to mix the feedstocks, break apart chunks of feedstock, and to form crude piles. If available, a windrow turner can also mix feedstocks prior to ASP composting.

Batch-type feed mixers (wagons or truck-mounted), pug mills, and other mixing devices work well for mixing. They do a better job of breaking up clumps and creating a uniform feedstock compared to bucket loaders. They can, however, reduce the porosity of a feedstock mix if material is overmixed. Some mobile mixers are equipped with conveyors that can discharge the feedstock mix directly into an ASP. However, it is more common to discharge the mixer into an intermediate stockpile, and then to build the pile in the ASP with a front-end loader.

If an aeration plenum layer is used, before the feedstock mix is placed in the pile, the aeration laterals should be covered with 15−30 cm (6−12 in.) of wood chips, screen overs, or other very porous material. The plenum is not essential, but it does provide some utility (Moon, 2021). It helps to disperse the airflow between the lateral and the pile. It also protects the lateral and filters small particles from the pile that might drop into and/or block the lateral orifices. Beltsvillen-era guidelines recommend the plenum width should be roughly ¼ of the pile width and that its length should start and end inside the pile by a distance roughly equal to the pile or cell height (see Section 1.2.1 and Fig. 6.5). More recent composting literature provides no other guidance for the plenum dimensions. Many composters simply use enough material to cover and surround the lateral, as shown in Fig. 6.9.

Feedstocks are formed into a pile with bucket loaders, dump trucks, manure spreaders, or conveyors. With POG aeration, equipment must be carefully maneuvered to avoid damaging the aeration pipes. Also, it is important to avoid compressing the pile itself by running the tires of the loader on the edge of the pile or by pushing the loader or blade into the pile without lifting at the same time.

Soon after the feedstock mix is in place, the biocover should be installed in sections as the pile is built. With an extended pile, only the horizontal surface of the newly formed cell needs to be covered, except for outside cells, which get fully

FIGURE 6.9

Plenum layers over aeration laterals, coarse overs placed over POG pipe (A) and wood chips covering below-grad aeration laterals (B). Note that in this picture, the plenum layer extends to the walls of a bunker. It is better to keep the plenum away from the wall to discourage airflow from following the wall surface.

Sources: (A) ECS, (B) C. Oshins.

covered. Since the biocover is typically placed with a bucket loader, it must be added in sections that are within the reach of the loader bucket to avoid driving on the pile. The biocover can be spread out with the bottom edge of the bucket, or manually raked on smaller piles. The thickness of the biocover should be 15 cm (6 in.) at a minimum. A thickness of 30 cm (12 in.) is preferable and commonly required by regulations for thermally reducing pathogens.

With any form of static composting, it greatly helps to turn the pile at least once during the thermophilic phase of composting. After composting in place for several weeks, a pile can become stratified and aeration channels develop. Turning provides an opportunity to add water, to homogenize, and recreate the initial porosity and moisture level of the mix. If necessary, additional amendments can be added.

When turning the pile, it can be shifted to an adjacent position, bunker or bin. Because turning can release odors, it should be delayed until the odor threat is reduced by the aeration system and temperatures have moderated to below 60°C. It is preferable to turn only after the pile has met pathogen and weed destruction conditions; otherwise, the biocover needs to be replaced. Typically, turning occurs 10–25 days after building the pile or during the transition to a second phase or aeration, but if or when turn takes place is often dictated by experience.

2.4 Aeration direction—positive versus negative

Aeration is said to be positive (i.e., positive pressure under the pile) if the fan pushes air through the pile. It is negative (i.e., suction or negative pressure under the pile) where the fan draws air through the pile. Positive aeration is simpler and less expensive to implement. It is generally preferred over negative aeration, unless the exhaust gases must be treated to reduce odors or other emissions or if heat recovery is desired. Positive and negative aeration can be combined by reversing the direction of airflow. Reversing aeration is complicated and often expensive to implement but has the advantage of minimizing thermal stratification in the pile to enhance process efficiency and to provide operational flexibility. Table 6.2 summarizes the advantages and disadvantages.

Positive aeration is the default approach for most ASP systems. It is more efficient at moving air compared to negative pressure because the fan moves air that is cooler, less humid and, consequently, more dense (higher density air takes less energy to move with a fan).

However, because the exhaust air leaves the compost pile over the entire pile surface, it generally cannot be collected for odor treatment or heat recovery. If better odor control is desired, the surface of the biocover layer can be made thicker and irrigated as needed. Otherwise, a positive pressure system can be contained in a building or other enclosure where the exhaust air can be captured and delivered to a biofilter. This approach requires an additional air-handling system and a larger biofilter. Positive aeration concentrates cool incoming air around the aeration laterals, which could lead to lower temperatures, at least temporarily, at the pile's base.

Negative aeration draws air into the pile from the outer surface and collects it in the aeration floor. Since the exhaust air is captured, it can be treated to reduce odors. The original Beltsville ASP procedure calls for the end of the discharge pipe to be inserted into a pile of finished compost. While this might work at small scale, modern negative aeration systems typically direct exhaust gas to a dedicated specially designed external biofilter (see Chapter 12).

With negative aeration, the system moves hot exhaust air and water vapor drawn in from the pile. The entire aeration system must be resistant to the high temperatures and corrosive effects of moisture, low pH, and ammonia. Furthermore, as the exhaust air cools, a large amount of condensate forms in the aeration pipes and channels and must be drained and collected. The majority of the condensate should be removed before it reaches the fan to reduce wear. An air-tight bulk storage container, like a typical industrial drum (e.g., 200-liter or 55-gallon), makes a simple inexpensive condensate trap for a small-scale ASP (Fig. 6.10). Larger systems require more elaborate condensate management.

Table 6.2 Aerated static pile aeration modes and their advantages and disadvantages.

Aeration mode	Advantages	Disadvantages
Positive	• Most energy efficient • Least expensive components • No fixed media biofilter	• Large vertical temperature gradient in pile • Least amount of odor control
Negative	• Nearly all process air is captured and can be scrubbed in a biofilter. • Better air quality inside a building • Opportunity to recover heat for other energy uses (e.g., space heating)	• Least energy efficient • Requires fixed media biofilter or other odor treatment. Aeration network must manage condensation • Higher temperature and corrosion-resistant components required throughout. • Greater chance of particles blocking orifices in aeration laterals
Reversing	• Least vertical temperature gradient to effect higher efficiency composting • Flexible operation: can keep aeration floor clear during loading (positive) then to minimize odor emissions early in the process (negative)	• Same as negative. • Most expensive option

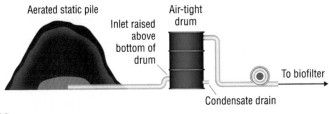

FIGURE 6.10

A 55-gallon drum condensate trap for a suction aeration system.

Another feature of negative aeration is the opportunity to capture heat contained in the hot moist air leaving the compost. There have been several schemes developed for capturing and using this heat (Box 6.2). Composters have found innovative ways to use the compost-derived energy including heating for anaerobic digesters, radiant floor heat for barns/shops/homes, drying for bagged compost, heating feedstocks in winter to jumpstart composting, winter heat for greenhouses/high tunnels, and a heating system for aquaponics (Smith and Aber, 2018).

Although the ability to contain exhaust gases for odor treatment is an important advantage of negative aeration, it pays a penalty for this in terms of fan power. Negative aeration requires 20%−50% more power than positive aeration in a comparable pile, partially due to the inefficiency of moving lower density air. Additional power is needed to move the exhaust air to and through a biofilter.

Systems that reverse the direction of air flow take advantage of the benefits of both modes. The negative mode prevails early in the process to capture gases for odor treatment Afterward, the aeration direction is periodically or automatically reversed to reduce moisture and temperature gradients that otherwise develop within single aeration direction static piles. Later in the process, the system can run consistently in the more efficient positive pressure mode with little concerns about odor emissions. In addition, the positive mode can be used when building and breaking down piles to minimize plugging of the aeration system. The advantages of reversing systems are countered by the additional expenses associated with the more complicated design, installation, and control (Fig. 6.11).

FIGURE 6.11

Reversing aeration system with dual manifolds, VFD-controlled fans, biofilter, and leachate collection tank.

Source: ECS.

Box 6.2 Heat recovery from composting—a brief review of history and current events

Authors: Matthew Smith and John Aber

The ability to recover energy from the composting process has a long history, dating back to hotbed systems used in ancient China 2000 years ago (Brown, 2014). In this type of compost heat recovery system (CHRS), a trench several feet deep is excavated and then filled with manure and crop residues, covered with topsoil and then planted with crops. From the 1600s through the 20th century, the hot trench concept was adapted and used to provide heat for greenhouses and growing season-extending high tunnels (Hong et al., 1997; Kostov et al., 1995).

In the 1970s, Jean Pain pioneered a type of CHRS that uses a long coil of water-filled tubing imbedded in a 49-ton mound of chipped brushwood (Pain and Pain, 1972). Pain's system was capable of warming water from 10 to 60°C (50−140°F) at a water flow rate of 3.8 L (1 gallon) per minute for six months, equivalent to roughly 14 kW (47,500 btu/h). While Pain's system upgraded the utility of the compost-derived heat, practitioners using this type of CHRS have reported problems, including the inhibition of microbial growth if the heat exchange is started too early in the composting process (Schuchardt, 1984; Seki, 1989).

Continued

Box 6.2 Heat recovery from composting—a brief review of history and current events—cont'd

Due to the limitations of within-pile heat exchangers, compost practitioners and researchers began exploring the compost vapor stream as a heat source. Thostrup (1985) was one of the first to create a CHRS using the vapor stream, by designing a pilot-scale composting plant to process manure from 200 pigs. The heat recovery unit was a tower design, with compost vapor being forced through a chamber containing an air-to-water heat exchanger. The system was capable of recovering roughly 5 kW (16,000 btu/h) with a system coefficient of performance of 6. However, the pilot system was shut down after six months due to high labor cost and mechanical issues. A second CHRS, developed by Fulford (1986), was an enclosed composting chamber running along the north end of a greenhouse. Inside the composting chamber were 10 composting bins with negative aeration. The blowers pushed the compost-heated air through perforated pipe below growing beds inside the greenhouse. Heat was captured from the condensing compost vapor, with the growing beds serving the dual purpose as a biofilter.

From the 1990s to the early 2000s, many compost heat recovery lab-scale studies were published (Smith et al., 2017). However, it wasn't until 2006 that a commercially viable CHRS was constructed. The composting system, designed by Agrilab Technologies Inc and installed at Diamond Hill Custom Heifers in Vermont, uses an ASP composting method to process 600–800 tons (wet weight) of agricultural wastes (Tucker, 2006). While the facility looks and operates like a typical ASP operation, its negative aeration system delivers exhaust air to a heat exchange with heat pipes that in turn transfer the energy to water in a storage tank (Fig. 6.12). This system is capable of recovering over 58 kW (200,000 btu/h). The heated water is used to warm milk formula and to provide radiant floor heating for a calf barn. This system was the inspiration for the construction of the first compost energy recovery research facility in 2012, which operates at the University of New Hampshire (UNH).

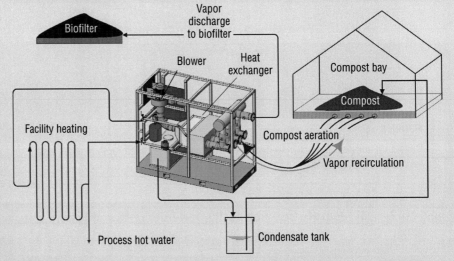

FIGURE 6.12

Schematic Agrilab Technologies heat recovery systems are built around a prefabricated mechanical unit. The Hot Box 8R, shown here, aerates eight compost piles (about 75 to 200 m³ each, or 100 to 250 yd³) in negative, recirculating and positive/drying modes. Heat captured in a hot water loop can be used for buildings, greenhouses or process heat, such as wash water. Odor can be managed by exhausting to a biofilter.

Reprinted by permission of Agrilab Technologies Inc.

The UNH facility was designed exclusively for research on heat production, recovery, and utilization. The facility composts manure from 50 dairy cows and uses a smaller version of the Diamond Hill's Agrilab Technologies heat recovery system (Fig. 6.13). Energy recovery rates peak around 10 kW (35,000 btu/h)

Box 6.2 Heat recovery from composting—a brief review of history and current events—cont'd

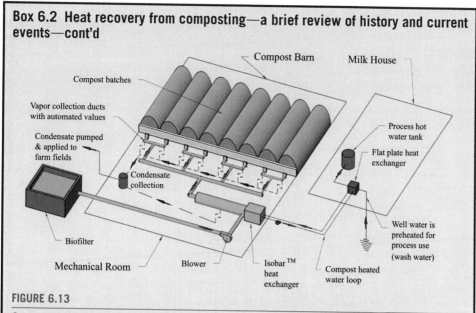

FIGURE 6.13

Conceptual diagram of the University of New Hampshire Energy Recovery Compost Facility (Smith and Aber, 2018).

for this pilot-scale system. As Fig. 6.14 shows, the energy recovery rates at the UNH facility primarily depend on the temperature and flow rate of the incoming compost vapor. To maximize heat recovery, feedstocks continually need to cycle through the system to take advantage of the highest heat composting phase.

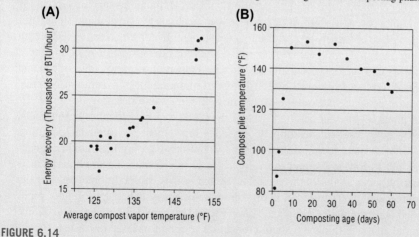

FIGURE 6.14

(A) Energy recovery rate by average compost vapor temperature for 17 trial runs and (B) typical temperature trends over time. In each trial, approximately 107 m³ (140 yd³) of feedstock was composted, consisting of a mix of cow manure (40%), bed horse manure (40%), and waste hay (20%). The aeration rate averaged 340 m³/h (200 cfm) (Smith and Aber, 2018).

Continued

> **Box 6.2 Heat recovery from composting—a brief review of history and current events—cont'd**
>
> Since the completion of UNH's research facility in 2012, the number of commercial composting systems with energy recovery has been increasing, with a majority using an Agrilab Technologies' heat recovery unit. These systems have proven to be successful, as heat recovery does not interfere with the composting process. Facility operators simply manage their systems for optimal compost quality or throughput and recover energy through the exhaust vapor. Furthermore, because a majority of the energy contained within composting feedstocks is in latent heat in the vapor stream, systems using this resource can have energy recovery rates as high as 60 kW (205,000 btu/h).
>
> A detailed description of the UNH composting facility, and its construction, can be found in Smith and Aber (2017). Brown (2014) present additional information about heat recovery from compost on a smaller scale.

2.5 Aeration system synopsis

The aeration system for an ASP includes the fan and an aeration floor, connected to one another by the air distribution system—a network of pipes, ductwork, and fittings. Aeration control devices, and possibly an external biofilter, complete the system. The aeration floor is the section of the system beneath the pile. A manifold or "header" distributes air from the fan to the individual "laterals" of the aeration floor by way of pipes and various fittings that divide the airflow, alter its direction and transition between pipes of different diameters. In smaller-scale ASPs, the manifold is often simply a tee connection at the fan to a single pipe that connects multiple laterals running under the pile. In larger systems, the manifold serves multiple piles and includes a damper at each branch (see Fig. 6.8).

As a benchmark, Table 6.3 lists aeration system specifications that are typical for many, if not most, ASP applications. These specifications are general and should not be used in place of facility-specific analysis. The main factor for selecting and sizing these components is the peak airflow rate. Other considerations for selecting these components can include corrosion resistance, temperature ratings, durability, and ease of use. Table 6.4 lists well-established rules of thumb for ASP aeration systems. Appendix F presents an example of a basic aeration system design.

The pipes in an ASP aeration system must perform in difficult environments, especially pipes used in aeration floors. They are exposed to high temperatures (80°C/176°F is not uncommon), corrosive conditions (e.g., high ammonia concentrations, pH as low as 5), and the weight of the pile above. Negative aeration is particularly harsh as the high temperatures and corrosive conditions apply to the manifold as well as the aeration floor. Additionally, many POG aeration floor pipes are pulled out from under a pile and across the pad by heavy equipment. The choice of material and wall thickness should reflect these conditions. Table 6.5 lists the dimensions and weight of pipe commonly used for ASP aeration systems.

Table 6.3 Typical aerated static pile aeration specifications.

Peak airflow rates[a]

First 10 days	6.5–22 m³/h per cubic meter of pile volume (3–10 ft³/min per cubic yard of pile volume)
10–20 days	3–6.5 m³/h per cubic meter of pile volume (1.5–3 ft³/min per cubic yard of pile volume)
After 20 days	1–3 m³/h per cubic meter of pile volume (0.5–1.5 ft³/min per cubic yard of pile volume)

Static pressure (without a biofilter), in centimeters and inches of water column (w.c.)

Lower peak aeration rates[b]	10–20 cm w.c. (4–8 in.)
Higher peak aeration rates[c]	15–30 cm w.c. (6–12 in.)

Typical fan power per individual pile or cell:

Lower peak aeration rates[b]	0.25–0.5 kW (0.33–0.5 hp)
Higher peak aeration rates[c]	0.75–3 kW (1–4 hp)

Typical aeration pipe diameter (lateral):

Lower peak aeration rates[b]	10 cm (4 in.)
Higher peak aeration rates[c]	15–20 cm (6–8 in.)

Notes:
[a] *Adapted from Coker and O'Neill, 2017b.*
[b] *Typical of basic ASPs with positive pressure and timer-based control.*
[c] *Typical of advanced ASPs with positive pressure and temperature-based control.*

Early ASP applications used inexpensive plastic pipe such as polyvinyl chloride (PVC) or low-density polyethylene (LDPE). These pipe materials are increasingly being supplanted by more durable alternatives, principally thick-walled high-density polyethylene (HDPE) pipe. Thin-walled corrugated LDPE pipe, made primarily for water drainage, is widely available and still used for very small ASP applications as a temporary POG. However, it is generally a poor choice. LPDE lacks strength. It is easily damaged by equipment and compressive force of the compost pile. It is often discarded after one or two composting cycles.

Table 6.4 ASP aeration system rules of thumb.

The following rules of thumb suffice for typical situations. They are not a substitute for detailed engineering design, and they can be over-ruled by engineering design.

- Pipes and channels should be sized to maintain the air velocity below 15 m/s (50 ft/s) and preferably below 10 m/s (33 ft/s).[a]
- To encourage uniform airflow down the lateral, the total area of the orifices in the lateral should be less than the cross-sectional area of the pipe or channel (assuming no taper). Preferably, the total area of the orifices should approximately equal half the cross-sectional area of the pipe or channel.
- The length of each aeration lateral, pipe or trench, should not exceed roughly 20 m (65 ft) unless the lateral is designed specifically to deliver uniform air distribution.
- Pipes should run straight for a minimum length equal to three pipe diameters before and after any change in pipe diameter or change in airflow direction (e.g., pipe tee or elbow).
- Fan outlet pipes should be at least six pipe diameters long with gradual tapers before encountering elbow or tee fittings.
- Elbow and tee fittings should have smooth transitions, rather than abrupt bends. "Street-type" fittings, which reduce obstructions at the pipe joint, should be used as needed.

[a] In the United States, air velocity is more often expressed in feet per minute or fpm. 33 ft/s and 50 ft/s are respectively 2000 fpm and 3000 fpm.

Table 6.5 Dimensions and weight of pipes commonly used for aerated static pile composting.

Type[a]	Nominal pipe size		Outside diameter		Inside diameter		Weight per unit length	
	DN[c] (mm)	NPS[d] (in.)	mm	in.	mm	in.	kg/m	lbs/ft
HDPE[a]	100	4	114	4.5	100	3.9	2.29	1.54
DR17[b]	150	6	168	6.6	147	5.8	4.96	3.33
	200	8	219	8.6	192	7.5	8.41	5.65
HPPE[a]	100	4	114	4.5	92	3.6	3.41	2.29
DR11[b]	150	6	168	6.6	136	5.3	7.39	4.96
	200	8	219	8.6	177	7.0	12.52	8.41
PVC/CPVC[e]	100	4	114	4.5	102	4.0	2.99	2.01
Schedule 40[f]	150	6	159	6.3	154	6.1	5.26	3.53
	200	8	219	8.6	203	8.0	9.52	6.39
	250	10	273	10.8	255	10.0	11.24	7.55
	300	12	324	12.8	303	11.9	14.91	10.01

Notes:
[a] HDPE is High Density Polyethylene.
[b] DR is the Dimension Ratio (outside diameter/wall thickness). The DR is an indication of the HDPE pipe strength. Pipe with lower DR is stronger because it has a thicker wall.
[c] DN stands for "diametre nominel." It refers to standard nominal pipe sizes used internationally.
[d] NPS stands for Nominal Pipe Size. It refers to standard pipe sizes given primarily in inches.
[e] PVC is Polyvinyl Chloride, CPVC is Chlorinated Polyvinyl Chloride. The weight given in the table is for PVC. CPVC is about 9% heavier.
[f] Schedule 40 is a standard rating of pipe strength. It relates to the thickness of the pipe wall. A higher schedule number is assigned to pipe with a thicker wall for a given outside diameter.

PVC pipe holds up better than LDPE. It is lightweight and easy to connect using glue or removeable rubber fittings. However, PVC is brittle, and breaks easily. Under positive aeration, PVC is a good choice for manifolds, which experiences less wear-and-tear. With careful handling, it also can be a low-cost option for POG material under positive aeration. PVC loses strength at the high temperatures typical of composting exhaust air (e.g., 60°C, 140°F). Chlorinated polyvinyl chloride (CPVC) is an alternative. It is more heat resistant, and slightly heavier, than conventional PVC. However, even CPVC pipe has limited use for systems that employ negative aeration. The strength of PVC and CPVC pipe is determined by the thickness of the pile wall, which is indicated by its "Schedule" number. Schedule 40 is the standard rating that is usually used under composting conditions.

HDPE pipe is generally the preferred choice for aeration floor laterals, especially POG laterals. Thick-walled HPDE is tough, heat resistant, and has the durability to be repeatedly pulled out from under a pile before needing repair or replacement. HDPE pipe is commonly rated by its dimension ratio (DR), which is the ratio of its outside diameter to its wall thickness. Stronger and heavier pipes have thicker walls and a lower DR number. Typically, DR17 HDPE pipes are used for ASPs. The stronger but heavier DR11 pipes are also used when needed.

Manifolds of some larger-scale ASP applications employ various forms of metal pipes, including mild steel, stainless steel, and aluminum pipe. However, these materials have found limited use in aeration floors. Mild steel pipe does not hold up well to the corrosive conditions of composting. Aluminum and 304-grade stainless steel pipes are more durable options but prohibitively expensive for most applications, especially aeration floors. All metal pipe is heavy and has high friction when pulled across a concrete or asphalt pad.

One of the primary rules for designing aeration pipes and trenches, both within the manifold and the aeration floor, is to make them large enough to maintain a low airflow velocity. The recommended maximum velocity is 10 m per second (m/s) or about 33 ft/s (2000 fpm). However, velocities slightly above the recommended maximum are tolerable if it means avoiding the jump to the next higher pipe size. Some designs allow for higher air velocities, up to 15 m/s (50 ft/s) to accommodate such situations (Bryant-Brown and Gage, 2015). With confidence backed by calculations and airflow models, design engineers occasionally exceed the recommended maximum velocity.

Eq. (6.3) is the formula for calculating the required cross-sectional area (A) of a channel based on volumetric flow rate (Q) and velocity (v). Eq. (6.4) calculates the corresponding the inside diameter (D_i) of a pipe, or other circular channel, assuming that A and D_i are in the same units (e.g., mm^2 and mm; $in.^2$ and in.).

$$A = CF \times Q/v \tag{6.3}$$

$$D_i = \sqrt{A \times 4\pi} \tag{6.4}$$

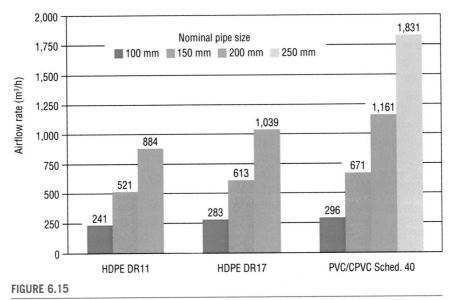

FIGURE 6.15

Airflow rates in m³/h corresponding to a velocity of 10 m/s in commonly used ASP aeration pipe.

CF is a conversion factor that varies with the units as follows:

- CF = 278 for A in mm², Q in m³/h, and v in m/s
- CF = 0.04 for A in in.², Q in ft³/h and v in ft/s
- CF = 144 for A in in.², Q in ft³/min. (cfm) and v in ft/min. (fpm).

Figs. 6.15 and 6.16 show the air flow rates (Q) that correspond to velocities of 10 m/s and 33 ft/s in commonly used aeration pipe. In the absence of an engineered design, these flow rates can be considered the maximum, or near maximum, for these pipes.

2.6 Aeration—how much?

In composting, the required rate of forced aeration is both wide-ranging and variable. The wide range comes from differences in the decomposition rates among feedstocks and the differences in process management goals among facilities. An open-air composting facility with a highly degradable food feedstock and nearby neighbors needs to optimize oxygen and temperature conditions during

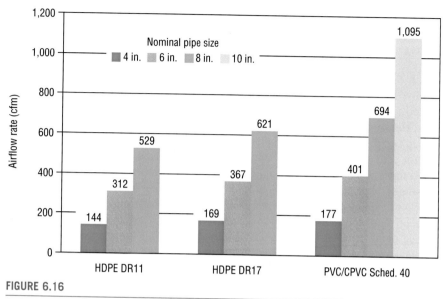

FIGURE 6.16

Airflow rates in ft^3/min (cfm) corresponding to a velocity of 2000 ft/min (fpm) in commonly used ASP aeration pipe.

the first three weeks of aeration. On the other hand, a rural facility handling a straw and manure mix can take a much more relaxed approach to aeration design and succeed.

The variability of the aeration rate is largely due to the pattern of decomposition. The rate of decomposition peaks early in the composting cycle then declines thereafter. It is common that over three times as much decomposition occurs in the first 10 days, as it does in the second 10 days. It continues to decrease as the pile approaches stability, although less dramatically. In addition to this longer-term trend, airflow demands vary sporadically over periods of hours as the complicated microbial activity rates ebb and flow (Fig. 6.30).

When composting is vigorous, a composting pile needs roughly 10 times greater airflow to remove the heat that accumulates than it needs to supply oxygen for microbial decomposition. If sufficient heat is not removed, temperatures rise.

Therefore, temperature control governs the airflow rate for an ASP (and most other forced-aerated composting methods). Providing aeration at a rate that achieves even modest temperature control supplies surplus oxygen for metabolic purposes.

2.6.1 "Peak" aeration

Ideally, the aeration system is designed to provide air at a rate that maintains compost temperature in a desired range when the composting process is at its peak. Hence, this rate is known as the "peak airflow rate." If an ASP system is operated in two distinct phases, each phase would have its own peak rate, with the peak rate for second phase would be lower than that of the first.

At times when the process requires less than the peak rate, aeration is curtailed. In some systems, the rate of air flow is reduced by varying the speed of the fan using a VFD or by adjusting dampers that restrict the air passage. In other cases, the fan turns off until the temperature subsequently rises to a prescribed level, or a prescribed time period has elapsed. The disadvantage of the latter approach is that it reduces oxygen to the microorganisms during the fan off-cycle. A preferable alternative is to provide some baseline level of airflow to supply oxygen continuously.

2.6.2 Estimating aeration rates

The amount of airflow required to stay under a given temperature threshold depends on the amount of heat generated in the composting pile. The amount of heat generated, in turn, depends on the pile volume, the process conditions (temperature, oxygen levels, moisture, etc.), and the potential decomposition rate of the feedstock mix. To stay under a given temperature threshold, a high energy mix of food waste and fresh ground wood can easily require five times the airflow compared to a mix of manure and composted screen overs. Process age also strongly determines the aeration rate requirements. Assuming an effective composting process, the peak aeration requirements for temperature control of the same mix should be reduced by roughly a factor of five or more between week one (primary) and week four (secondary).

Table 6.6 restates the ranges of airflow rates typical for ASP composting based on pile volume as suggested by Coker and O'Neill (2017b). These rates should provide a reasonable degree of temperature control for most composting situations. The selected aeration rate should move closer to high end of the ranges in Table 6.6, as

Table 6.6 Typical ASP design aeration rates.

First 10 days (peak)	6.5 to 22 m^3/h per cubic meter of pile volume (3 to 10 ft^3/min per cubic yard)
10 to 20 days	3 to 6.5 m^3/h per cubic meter of pile volume (1.5 to 3 ft^3/min per cubic yard)
After 20 days	1 to 3 m^3/h per cubic meter of pile volume (0.5 to 1.5 ft^3/min per cubic yard)

Adapted from Coker and O'Neill, 2017.

the feedstock mix becomes more energetic; that is, the feedstocks are more degradable or have low C:N ratios.

Ideally, when the pile is most active, the peak aeration rate would be able to prevent the pile temperature from ever going above 65°C (150°F). But for high-energy feedstocks, this level of temperature control might require airflows above even the top rate in Table 6.6. Most composting facilities don't need such a high level of process control. The fundamental question is, what level of process control (primarily the peak aeration rate) is both affordable and going to allow the facility to operate successfully?

In community-scale composting systems, McSweeney (2019) describes a procedure to determine aeration rates based on the concept of volumetric air exchange. It is the same concept used for building ventilation systems. The volume exchange rate is the frequency that the air volume of a pile (or cell or bin) is replaced with fresh ambient air, over a selected time period, usually 1 h. The air volume of a pile is estimated by multiplying the pile volume by its FAS, expressed as a fraction. For example, a pile with a total volume of 150 m³ (196 yd³) and an FAS of 36% (0.36) has an air volume of 54 m³. At an air exchange rate of 30 air exchanges per hour (AE/h), the aeration rate for this example pile is 1620 m³/h (953 cfm). McSweeney correlates high aeration rates with 30 to 40 air exchanges per hour (AE/h) and low aeration rates with 5 to 10 AE/h. High AEs could be used to estimate airflows during the first weeks of composting and give way to medium and then low AEs as biological activity declines (Box 6.3).

The volumetric air exchange method is a straightforward, uncomplicated way to estimate airflow rates, but it is not without flaws. At a given pile size, the method estimates higher air volumes and airflow rates for piles with higher FAS. However,

---

Box 6.3 How temperature affects aeration rate and vice-versa

Most of the heat generated during composting goes toward evaporating water in the pile. The ability of air to remove that heat is largely determined by how much water vapor the passing air can carry away. The amount of water vapor that air can hold increases sharply with increasing temperature. Since the air leaving an active composting pile is nearly always "saturated," which is to say it is at 100% relative humidity, modest increases in the exhaust air temperature represent large changes in the rate of heat removal.

As air moves through the pile, it is heated from ambient air temperature to the temperature near the exit point of the pile, acquiring water vapor along the way. The amount of water vapor that the air can carry increases geometrically with increasing temperature. Moisture-saturated air at 60°C (140°F) holds 10 times more water vapor than saturated air at 20°C (140°F), by either volume or weight. Saturated air at 70°C (160°F) carries almost twice the amount of energy and water as saturated air at 60°C. For this reason, maintaining a pile near 70°C requires a much lower air flow rate than maintaining a pile near 60°C (another reason is that biological activity and heat generation decrease from 60°C to 70°C). The temperature of the exiting air is therefore the dominant factor that determines both the drying rate and the rate of heat removal. The temperature and humidity of the *inbound* air are much less impactful.

piles with higher FAS likely have a lower amount of volatile solids, generate less heat, and need less airflow. In addition, the method does not account for the nature of the feedstocks. For example, a loose pile of chipped tree bark, with a high FAS, would generate relatively little heat yet earn a high aeration rate via this method. The art of using this approach rests on understanding the feedstocks and selecting the appropriate AE rate.

Ideally, aeration rates should be feedstock-specific and based on the dry weight of the primary feedstocks, or even better, their volatile solids or organic matter contents. Several research projects have investigated aeration rates for specific feedstocks. Table 6.7 summarizes the findings of some of these projects. It is important to note that these studies used laboratory scale composting reactors, which aerate more uniformly than a large pile. Also, the various studies evaluated the aeration rate based on different criteria, including the time to achieve a given stability level, nitrogen conservation, and emissions of greenhouses gases and other volatile compounds.

Table 6.7 Feedstock-specific recommended aeration rates based on selected research projects.

| Feedstock(s) | Dry Mass basis[a] | Aeration rate[b] (vol./time-dry mass basis) | | | References |
		L/min-kg	m³/h-kg	ft³/h-lb	
Vegetable/fruit waste	VS	0.62	0.037	0.596	Arslan et al. (2011)
Sewage sludge + maize straw	OM	0.56	0.034	0.538	Hu et al. (2012)
Grass clippings + vegetable waste	OM	0.41	0.025	0.394	Kulcu and Yaldiz (2004)
Pig manure + corn stalks	TS	0.48	0.029	0.462	Guo et al. (2012)
Dairy manure + rice straw	VS	0.25	0.015	0.240	Li et al. (2008)
Food waste + rice straw ash	TS	0.6	0.036	0.577	Zakarya et al. (2019)
Penicillin mycelial dreg	OM	0.5	0.030	0.481	Chen et al. (2015)
Sewage sludge + corn stalks	TS	0.2	0.012	0.192	Yuan et al. (2016)
Chicken manure + sawdust	OM	0.5	0.030	0.481	Gao et al. (2010)

Notes:
[a] OM, organic matter; TS, total solids (i.e., dry matter); VS, volatile solids.
[b] Research results were reported in Liters per minute per dry mass (VS, OM, or TS).

It is possible to mathematically determine the amount of air needed to evaporate water and thus remove the heat generated by a unit mass of a particular feedstock, given certain assumptions about properties of the pile. Such calculations are well presented by Haug (1993) and by Keener et al. (2007). The calculations usually provide a good estimate for the conditions assumed. However, because feedstocks vary in practice, most design airflow rates are derived from experiments and experience.

2.7 Fans and fan pressure

Forced aeration composting normally employs centrifugal fans, frequently referred to as blowers. Common industrial-grade centrifugal steel or aluminum fans are generally suitable for positive aeration. Small-scale ASPs have even used plastic fans normally intended for light-duty commercial applications. Negative aeration requires fans that are resistant to hot and corrosive air, even with a condensate trap in place. Some small- and moderate-scale facilities mount fans and their control devices on portable platforms for added flexibility (Fig. 6.17).

The required fan size and motor are determined primarily by the volume of feedstock to be aerated and the peak aeration, and by the aeration system's efficiency (airflow delivered ÷ motor power). The volume of feedstocks depends on the pile/cell size and the number of piles or cells the fan serves. For most ASPs, fans

FIGURE 6.17

Portable platform with the fan and its control devices.

FIGURE 6.18

A 1-kW (1.5 hp) fan feeding two laterals for a single 185 m^2 (2000 ft^2) aerated static pile of yard trimmings + food residuals.

generally range from 1/4 to 5 kW (1/3 to 5 hp). Small to moderate ASP systems typically require fans with less than $^3/_4$ kW (1 hp) when the fans serve individual piles (Fig. 6.18). High-level ASPs that use centralized fans to aerate a number of high-volume piles can have motors as large as 150 kW (200 hp) (Fig. 6.19).

In addition to necessitating multiple fans, there is an inherent disadvantage in aerating each pile with a dedicated fan. In this case, the peak aeration rate is needed for only a short period of time (one to two weeks). Afterward the fan runs less often or at a lower speed, squandering its capabilities. For this reason, it is common for a composter in this situation to select lower peak aeration rates (e.g., 5–7 m^3/h per m^3; or 2–3 cfm per yd^3) so that the fan is less oversized beyond the period of peak decomposition. When a fan serves multiple piles or cells, the designer can count on a range of pile ages and aeration demands, allowing the fan to operate more consistently. This approach affords higher peak airflow rates. However, it also requires a more complicated aeration system with flow-modulating dampers.

FIGURE 6.19

Large belt-driven fans supplying air to separate manifolds aerating multiple cells of an extended ASP for a mix of yard trimmings, food waste anaerobic digestate, and wine pomace (about 3000 m³ or 4000 yd³). Each fan is 45 kW (60 hp) working against a static pressure of approximately 39 cm w.c. (12 in.). Notice the gradual transition of pipe sizes and angles between the fans and the manifold to reduce friction loss.

Source: ECS.

In order to size or select a fan, it is necessary to know the required airflow rate and the air pressure loss, or static pressure "head," of the system (Box 6.4). It is also necessary to know the condition of the air entering the fan (temperature, humidity, and elevation). With this information, the appropriate fan can be selected from the manufacturers' fan performance curves. Since fan curves can be confusing, it is often easiest to contact a fan supply company with these operating conditions and ask them to select a fan to suite.

General estimates of the total static pressure for ASP composting range from 10 to 30 cm (4–12 in.) of water column (w.c.), depending on the design of the system and the pile density. A small-scale system can have a static pressure as low as 5 cm (2 in.). Air flow through the pipe holes or orifice usually accounts for about half of the total static pressure, assuming that feedstock mix is not overly dense. The pipe header, fittings, and lateral typically contribute 25–50 mm (1–2 in.) of pressure loss, if properly sized. These losses can double if the pipes are undersized or are overly tortuous. An external biofilter, if used, adds roughly 7.5–15 cm (3–6 in.) of pressure loss. Many advanced ASPs run between 25 and 50 cm (10–20 in.).

Static pressure losses through the composting pile tend to be lower than those of the air distribution network because the air disperses in the pile and travels at low

---

Box 6.4 Pressure points

A fan has to push air against a back pressure, or "static" pressure because friction occurs when the air moves. The friction occurs because:

(a) moving air molecules bounce against one another, especially if their flow is turbulent.

(b) air molecules move against stationary surfaces, like pipe walls and compost particles.

(c) moving air encounters stationary objects that change its direction, like pipe elbows; and

(d) airflow is constrained and changes direction as it exits the air stream, as it does at pipe holes and orifices.

"Pressure loss" is the term used to describe the pressure needed to overcome friction. It is considered a "loss" because this component of fan pressure does not directly move the air. It overcomes the energy lost to friction. Pressure loss is also called pressure "drop" and pressure "head," but these terms are essentially the same. They each refer to the difference in pressure between one point and another along the air's path.

Static pressure is the combined measure of pressure losses—the sum of the pressure losses along the air's pathway from the fan discharge to the point at which the air reaches its destination (e.g., the atmosphere). The pressure required to accelerate the air to the desired velocity is called "dynamic" or velocity pressure. Assuming that air compression is negligible (a reasonable assumption for composting applications), the pressure that the fan must produce is the dynamic pressure plus the static pressure.

Air velocity has a great effect on pressure loss because higher velocities increase the contact, collisions, and energy exchange among molecules. The magnitude of pressure loss is proportional to the velocity squared; hence a doubling of velocity quadruples the pressure loss. (The effect on fan power is even more dramatic as the power required is proportional to the velocity cubed.) Higher velocities also encourage turbulence. At low velocities, fluids, like air, flow in a uniform manner, termed "laminar" flow. With increasing velocities, the flow become more erratic, creating more friction. At a high enough velocity, the flow becomes predominately turbulent and pressure losses skyrocket. The velocity should be kept below this point, which is why guidelines recommend maximum velocities, such as 600 m/min (2000 ft/min).

At a given airflow rate (e.g., cubic meters or feet per minute), velocity is determined by the cross-sectional area of the conduits in which the air is traveling, typically pipes. Therefore, velocity, and pressure loss, can be reduced by increasing the area of flow by using larger pipes/channels or multiple pipes/channels. Increasing the diameter of a round pipe has a double benefit because the cross-sectional area increases with the diameter squared. This principle also applies to the open pore space within a composting pile. Air passing through a pile with more and larger pores has a lower velocity and loses less pressure loss compared to a dense pile.

In addition to velocity, pressure losses increase with anything that tends to create more friction including rougher pipe surfaces, longer pipe runs, and higher piles. Pressure losses also occur wherever the velocity changes (up or down) and where the direction of flow changes. Examples include changes in pipe diameters, any pipe fittings, and especially poorly designed fittings that produce a drastic change in diameter or direction of flow. Pipe holes and orifices are points of relatively large pressure loss because the passage narrows and the passing air changes direction. Elevated pressure losses at pipe holes and orifices are intentional. They are a means to achieve roughly equal airflow through each hole/orifice down the length of the pipe or channel.

---

velocities through a vast number of pore channels. Static pressure losses through the composting pile typically range from 0.6 to 2.5 cm (1/4 to 1 in) of water column. This loss varies with the bulk density or FAS of the feedstock mix, the pile height, and airflow rate. Several researchers have measured the pressure loss through various composting materials and developed mathematic models to predict static pressure loss based on pile characteristics and airflow (e.g., Sidelko et al., 2019; Barrington et al., 2002; Keener et al., 1993; Higgins et al, 1982). As an illustration of these relationships, Fig. 6.20 graphs *calculated* static pressure loss through composting piles of different heights at selected aeration levels for a relatively porous mix of biosolids and wood chips (McSweeney, 2019, after Haug, 1993).

Engineering calculations can fairly accurately predict the anticipated static pressure loss of air flowing through the mechanical components of an aeration system (ducts, fittings, channels, pipes, orifices). But the pressure losses that result from the every-changing mix porosity and the conditions at the aeration floor can only be estimated over a likely range. In general, it is safest to estimate high and adjust the fan output downwards if possible.

The flow rate produced by a given fan decreases as the static pressure it encounters increases. Every fan has a range where it operates most efficiently, and a cutoff pressure where it stops moving air. Therefore, to select a fan, one must estimate the static pressure of the system when the desired airflow rate is being produced. Engineers do this by calculating and/or modeling the pressure losses in the aeration system at a range of possible airflow rates. From these data, they produce a "system curve." The system curve is then plotted against the fan performance curve that is generated by the fan manufacturer, as shown in Fig. 6.21. The intersection of the

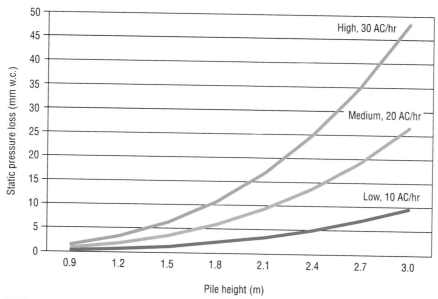

FIGURE 6.20

Calculated static pressure for various pile heights and aeration levels. Low, medium, and high aeration rates correspond to air exchange rates of 10, 20, and 30 air exchanges per hour (Section 1.6.2).

Adapted from McSweeney, 2019. after Haug, 1993.

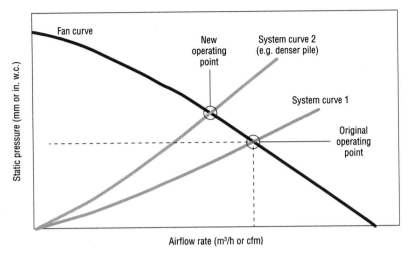

FIGURE 6.21

Graphic procedure for determining the operating point of a fan using the fan performance curve and the system curve.

two curves is the point at which the fan will operate in the system. This explanation is a simplification of the procedure that an engineer would do to select a fan. The fan efficiency, cost, air characteristics, operating conditions, and desired longevity and dependability also must be considered.

2.8 Aeration floor

The aeration floor is a key element in the success of an ASP. Assuming the feedstock mix is relatively homogeneous and adequately porous, the aeration floor is what determines the uniformity and reliability of air delivery to the pile. It is also the element that receives the most abuse and requires the most maintenance (Coker and O'Neill, 2017a).

The aeration floor consists of one or more pipes, and sometimes trenches, that transport air between the fan and the composting pile. In the case of positive aeration, the aeration floor delivers ambient air from the fan to the pile above. With negative aeration, the aeration floor draws ambient air down through the pile, collecting the warm moist exhaust air in the aeration system. Aeration floors have one more "laterals" that run underneath the pile. When there is more than a single lateral, a manifold connects the individual laterals.

Several types of aeration floors have been used for ASPs. They can be broadly divided into two categories: POG and below-grade. Regardless of the type, all aeration floors need to account for the fact that liquids accumulate at the base of piles and can inundate the pipes and/or channels.

2.8.1 Pipe-on-grade aeration floors

The original and most common aeration floor is comprised of one or more moveable pipes that rest on the composting pad (i.e., "on-grade"). These POG aeration floors are relatively inexpensive and simple to construct. The aeration floor can be assembled on compacted dirt, where regulations allow, or on an improved surface such as asphalt or concrete for both all-weather operation and to avoid mixing product with the base material (Figs. 6.22 and 6.23).

A POG lateral is a pipe, usually plastic and preferably HDPE, with a series of holes (orifices) along its length to distribute air to or from the pile. The pipe is capped on the far end and is connected to the fan outlet or header on the other end. The size, number, and spacing of the holes are designed to encourage uniform air distribution along the length of the lateral.

Pipe holes (orifices) should be in two rows facing downward, preferably at about 5 and 7 o'clock, as shown in Fig. 6.24. The 5/7 o'clock hole placement is approximate. It can be difficult to consistently drill or punch the holes, or rotate the pipe, in this exact alignment due to pipe twisting. Ideally the holes should be spaced no greater than 30 cm (12 in.) apart within a row though in long POG pipes it is often necessary to increase this spacing to limit the total number of holes. The number and size of the holes should provide a total hole area that is less than the cross-sectional area of the lateral pipe. To achieve an even more uniform air distribution down the lateral, the total hole area should equal half the cross-sectional area of the lateral pipe.

(A) **(B)**

FIGURE 6.22

POG aeration floor for an extended ASP with HDPE pipe (A). Damage that lateral pipe suffers tends to suffer in POG aeration floors (B).

Source: ECS.

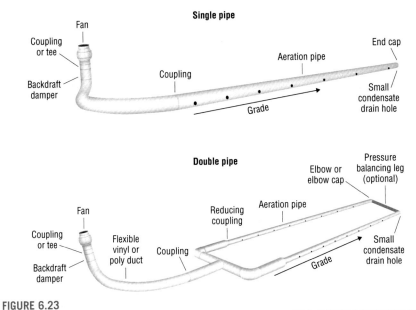

FIGURE 6.23

Diagrams of typical POG air distribution systems.

Adapted from McSweeney, 2019.

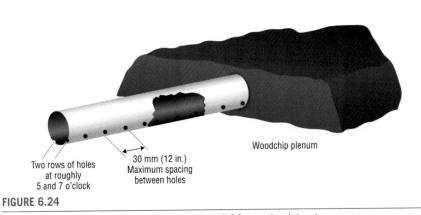

FIGURE 6.24

Recommended aeration pipe hole location for POG aeration laterals.

The purpose of restricting the hole size is to establish a static pressure loss across the holes (i.e., orifices) that is substantially larger than the pressure loss down the lateral pipe. In other words, the aim is to create more backpressure within the pipe lateral. If substantially more pressure is needed for air to squeeze through the holes, it does not preferentially exit the lateral through the holes nearest the fan; at least the difference in flow between the first hole and the last hole in the lateral is acceptably small. These "high-pressure" orifices effectively improve uniformity, but they also increase the static pressure and the required fan power for a given flow rate.

Table 6.8 gives the calculated hole diameters at various hole spacings and lengths of perforated pipe that would create a total hole area equal to the cross-sectional area of the associated pipe. To make the lateral pipe, a composter should use a drill bit or punch size nearest to the desired diameter, or the next size smaller. To create a total hole area equal to half the pipe's cross-sectional area, the diameters in Table 6.8 should be divided by four.

Another way to achieve uniform air distribution along the lateral pipe is to reduce the static pressure loss through the lateral by reducing the air velocity. This is most easily accomplished, by using large diameter lateral pipe, or by splitting the flow between two or more parallel laterals. Using pipe that yields a velocity well below the 10 m/s (33 ft/s) threshold is a good way to promote nearly equivalent flow through each hole along the pipe without sacrificing fan efficiency. Figs. 6.25 and 6.26 plot the airflow rates that correspond to velocities up to 10 m/s and 33 ft/s (2000 fpm) for pipes in the range of sizes used for POG laterals. The graphs are based on the inside diameters of HDPE DR17 pipe. They are suitable approximations for HDPE DR11 and Schedule 40 PVC/CPVC pipes. Both graphs are calculated from Eq. (6.5).

$$Q = A \times v \times CF = \left(\frac{1}{4} \times D_i^2 \times \pi \right) \times v \times CF \tag{6.5}$$

where Q is the airflow rate, A is the pipe cross-sectional area, v is the air speed, and D_i is the pipe inside diameter. Again, CF is a unit conversion factor that varies with the units as follows:

- CF = 0.0036, for Q in m^3/h, A in mm^2, D_i in mm, and v in m/s
- CF = 25, for Q in ft^3/h A in $in.^2$, D_i in inches, and v in ft/s
- CF = 0.0069, for Q in ft^3/min. (cfm) A in $in.^2$, D_i in inches, and v in ft/min. (fpm)

Dragging aeration pipe from composting piles requires a means to easily connect and disconnecting the pipe at the fan end. It requires a length of solid pipe to extend out of the pile to mate with the aeration system via a flexible connection. At the far end of the pile, the lateral pipe requires a means of attaching it to a wheel loader or tractor (Fig. 6.27). Often a "pulling eye" is mounted in a reinforced pulling head that is easily connected via a cable or chain to a tractor or wheel loader (Fig. 6.28). Some care must be exercised to make sure the pipe isn't too hot and the machine doesn't pull too hard. Even tough HDPE pipe can stretch and ultimately break.

There must be sufficient space in front of the ASP pile for pulling the pipe straight out and then turning. Unless the pipe is connected in sections, the space must be equivalent to the length of the pipe, plus the length of the pulling machine, plus some maneuvering room of the pulling machine. Pulling, replacing, reconnecting, and working around the POG pipes to build the piles is cumbersome and labor intensive, and can be somewhat dangerous.

Table 6.8 Required hole diameters for pipe-on-grade laterals for the total hole area to equal the cross-sectional area of the lateral pipe (based on HDPE DR17 pipe).

Pipe nom. Diameter DN (mm)	Inside Diameter (mm)	Cross-sec. Area (mm²)	Hole Spacing (cm)	Length of perforated section of pipe (m) Hole diameters in mm						
				3	6	9	12	15	18	21
100	100	7862	10	12.9	9.1	7.5	6.5	5.8	5.3	4.9
100	100	7862	20	18.3	12.9	10.5	9.1	8.2	7.5	6.9
100	100	7862	30	22.4	15.8	12.9	11.2	10.0	9.1	8.5
150	147	17,040	10	19.0	13.4	11.0	9.5	8.5	7.8	7.2
150	147	17,040	20	26.9	19.0	15.5	13.4	12.0	11.0	10.2
150	147	17,040	30	32.9	23.3	19.0	16.5	14.7	13.4	12.4
200	192	28,876	10	24.8	17.5	14.3	12.4	11.1	10.1	9.4
200	192	28,876	20	35.0	24.8	20.2	17.5	15.7	14.3	13.2
200	192	28,876	30	42.9	30.3	24.8	21.4	19.2	17.5	16.2

Pipe nom. Diameter NPS (in.)	Inside Diameter (in.)	Cross-sec. Area (in.²)	Hole Spacing (in.)	Length of perforated section of pipe (ft) Hole diameters in inches						
				10	20	30	40	50	60	70
4	3.9	12.19	6	0.62	0.44	0.36	0.31	0.28	0.25	0.24
4	3.9	12.19	9	0.76	0.54	0.44	0.38	0.34	0.31	0.29
4	3.9	12.19	12	0.88	0.62	0.51	0.44	0.39	0.36	0.33
6	5.8	26.41	6	0.92	0.65	0.53	0.46	0.41	0.37	0.35
6	5.8	26.41	9	1.12	0.79	0.65	0.56	0.50	0.46	0.42
6	5.8	26.41	12	1.30	0.92	0.75	0.65	0.58	0.53	0.49
8	7.5	44.76	6	1.19	0.84	0.69	0.60	0.53	0.49	0.45
8	7.5	44.76	9	1.46	1.03	0.84	0.73	0.65	0.60	0.55
8	7.5	44.76	12	1.69	1.19	0.97	0.84	0.75	0.69	0.64

Notes:

(1) The General formula is: $d = \sqrt{\left(Di^2 \times S / (2 \times L \times CF)\right)}$. Where d is the hole diameter, Di is the inside diameter of the pipe in mm or in., S is the spacing between holes within a row in cm or in., L is the length of the perforated section of the pipe in m or ft, and CF is a unit conversion factor. $CF = 100$ if S is in cm and L is in m. $CF = 12$, if S is in inches and L is in ft. (2) The calculations assume two roles of holes in the pipe. (3) Inside pipe diameter is for HDPE DR17.

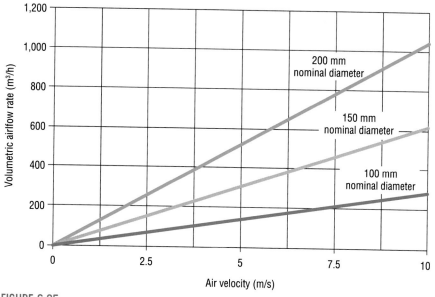

FIGURE 6.25

Airflow rates in m^3/h that correspond to velocities up to 10 m/s in HDPE pipe sizes commonly used for POG laterals (based on HDPE DR17).

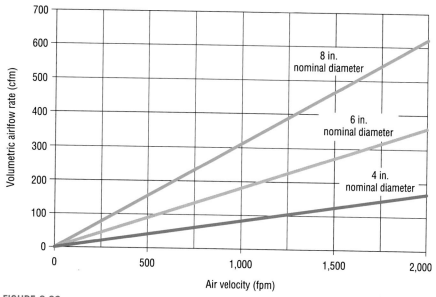

FIGURE 6.26

Airflow rates in ft^3/min that correspond to velocities up to 33 ft/s in HDPE pipe sizes commonly used for POG laterals (based on HDPE DR17).

FIGURE 6.27

Tractor pulling chain wrapped to the end of HDPE aeration pipe.

Source: O$_2$ Compost.

(A) **(B)**

FIGURE 6.28

Durable HDPE aeration pipe (A) capable of being towed out by the secure end cap
(B) before dismantling the ASP.

Source: ECS.

2.8.2 Below grade aeration floors

At many composting operations, both large and small, aeration floors are installed below the composting pad to reduce labor and avoid interference with the material-handling equipment. These "below-grade" aeration floors do not require the operators to repeatedly disconnect/move/reconnect pipes and they allow bucket loaders to run freely on the working surface of the ASP without having to avoid pipes. With any below-grade system, all of the aeration components should be installed slightly lower than the pad surface to avoid being damaged by loader buckets and other equipment. Air flow velocities within below-grade aeration floors can be kept low by using pipes and trenches with large cross-sections and/or by using a network of underground pipes to feed trenches at multiple points along their length.

Below-grade systems have lower operational costs, are capable of delivering airflow more uniformly and at higher rates, and can have other process advantages, such as improved drainage, potentially less plugging, and easier removal of solids. However, below-grade aeration floors generally have significantly higher initial costs than POG aeration floors. For large ASP applications, below-grade aeration floors tend to be custom engineered to fit the site layout and aeration rate requirements.

One approach to below-grade aeration is to form lateral trenches when the concrete pad is installed or cut them into an existing concrete composting pad (Figs. 6.29 and 6.30).[3] The trenches distribute air along their length and then through covers with orifices. The orifices are located and sized to provide the appropriate back pressure to achieve uniform air distribution through the lateral. The surface of the perforated covers can be blanketed with wood chips or screen overs to minimize plugging and better distribute airflow.

Another common method for building a below-grade aeration floor uses a set of lateral pipes, under the pad with a series of evenly spaced vertical risers, extending from the buried pipes up just below the pad surface (Fig. 6.31). A specifically sized hole, or orifice, at the end of each riser delivers or collects air from the pile above. The air flow among the risers is equalized by high flow resistance (i.e., pressure drop) at each orifice compared to the flow resistance to the air traveling down the pipe. Due to cost, this approach normally uses relatively fewer large orifices, spaced farther apart (2−4 ft is common), compared to trenches or POG pipes. This type of aeration floor is used with both positive and negative aeration. With negative aeration, the high flow rate into the orifices tend to pull solids quickly increase the resistance to airflow.

---

[3] Smith and Aber (2017) provide an excellent description, including numerous photographs, of the construction of a below-grade aeration floor in which the lateral pipe is formed into the concrete pad and capped with a perforated wooden cover-plate.

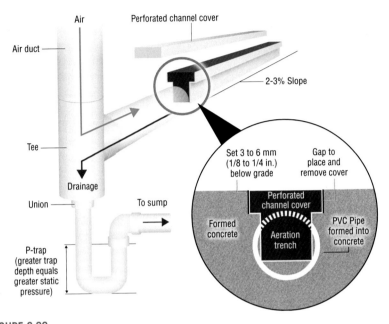

FIGURE 6.29

Diagram of a below-grade aeration trench with leachate drainage.

Adapted from McSweeney, 2019.

2.8.3 Aeration floor drainage

All aeration floors need to account for the fact that water collects at the base of piles and can inundate the aeration orifices and pipes. Some means of drainage is necessary to prevent liquids from interfering with aeration. Negative aeration is especially prone to accumulating water as the laterals are sucking water in and hot air is cooling and condensing.

POG aeration laterals can manage water by facing the holes downward. If necessary, the pipe laterals can be elevated slightly above the composting pad, by placing a few centimeters of the wood chip plenum underneath the pipes.

Aeration pipes and channels that lie below-grade collect some free water from the pile. Trench-style laterals collect much more water than pipes with risers. Both types need to drain these liquids without obstructing aeration. When air is moving, water lying in a lateral is pushed along by air flow down the lateral. Therefore, it

FIGURE 6.30

Below-grade aeration laterals in trenches. In the photo on the left (A), the laterals are in trenches formed by polymer concrete blocks for an ASP under a micro-porous covers (see Section 2.3). In the lateral on the right (B), a PVC pipe in the trench pulls air through holes drilled in the wooden cover planks.

Source: (A) W.L. Gore & Associates; (B) R. Rynk.

FIGURE 6.31

Below-grade aeration laterals. Lateral pipes with risers and before the concrete pad is poured (left). Risers and orifices formed into the pad at the same facility (right).

Source: ECS.

is best to keep the laterals flat and locate drains at the terminal end of the airflow since the moving air will push water up a modest slope. For a positively aerated system, the drain should be placed at the far end, away from the fan. With negative aeration, the drain should be located at the fan end, if possible. An alternative is to provide periodic 15—30 minute drain-down cycles when the fan is off and the lateral slopes toward a drain. Typical slopes range from 1 to 3%.

The drains in each lateral typically connect to a common drain line through a pressure trap. Fig. 6.29 shows a P-type pressure trap. Fig. 6.32 shows an S-trap. P-traps are generally preferred because S-traps are more likely to syphon (and may be prohibited by local plumbing codes). In the case of Fig. 6.32, a P-trap could not be used because of height restrictions. In both cases, the trap must hold back a depth of water that is greater than the pressure in the aeration pipe (in mm or in. w.c.).

The drain line is usually brought to a sump where it is either gravity drained or pumped into a holding tank or basin. The drained water is rich in nutrients and usually odorous. It must be recycled or discharged appropriately. It can be used to rewet composting piles that can still achieve temperature/time combinations for sanitization.

In addition to water, an aeration floor lateral and its drainage system collect solid particles and sediment. An aeration floor should include some means to regularly, and easily, flush or otherwise clear the solids that accumulate in pipes and trenches.

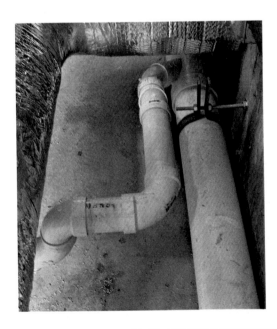

FIGURE 6.32

Drainage for a below-grade aeration floor using an S-trap. The larger pipe on the right is the aeration pipe. The smaller S-trap pipe drains water to the left to a condensate collection tank (see Fig. 6.13).

Adapted from Smith and Aber, 2017.

2.9 Aeration control

The compost process has been dubbed "the heat machine" by Sundberg (2005). But unlike a machine, the heat output is neither regular nor simply predictable over time. Fig. 6.33 shows the varying demand for cooling airflow (green line) which is synonymous with heat generation in the composting process. So, while the microbial activity is generally higher early, it quite suddenly and randomly ramps up and down during the composting cycle. This behavior is understood to be an artifact of sequential microbial communities becoming active/inactive. Therefore, while fixed amount of airflow to remove heat might keep temperatures below desired thresholds most of the time, it is also desirable to reduce airflow in response to a sudden decrease in heat generation so as to prevent over-cooling.

ASP aeration control strategies differ in how airflow rates are varied and what triggers changes to the aeration. Composters have a menu of options that determine whether aeration is continuous or intermittent, at a constant or variable rate, and whether the fan is prompted by time or temperature (Fig. 6.34). In addition, some strategies control the direction of airflow. The four most common aeration rate control strategies used with ASPs are as follows:

(1) Airflow is delivered intermittently, at a constant aeration rate when the fan is on. A timer turns the fan on and off according to a cycle set by the operator. The percentage of time that the fan is on over the total cycle time is called the "duty cycle."

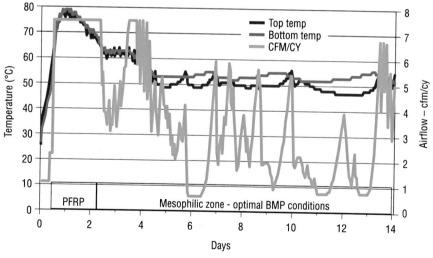

Data from ECS CV Composter in Omak WA
Control Set-points: #1=62C, #2=52C

FIGURE 6.33

Temperature pattern for a reversing aeration system with temperature feedback control of the airflow rate (green line) in response to demand for cooling to maintain set-point control temperatures. The maximum airflow for this system was 7.5 cfm/cy.

Source: ECS.

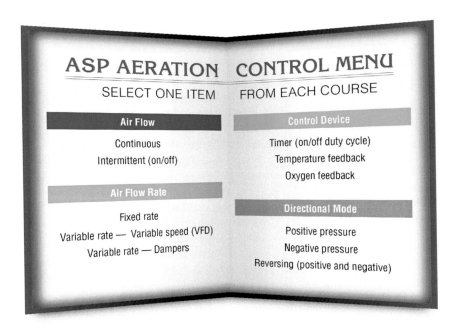

FIGURE 6.34

Menu of forced aeration control options.

(2) Airflow is delivered intermittently, at a constant aeration rate when the fan is on. A timer turns the fan on and off but with the addition of a thermostatic override of the "off" cycle when a temperature exceeds an operator-selected limit.

(3) Airflow is delivered intermittently, at a constant aeration rate when the fan is on. A thermostatic control system turns the fan on when the temperature surpasses a high set point and turns the fan off when it dips below the low temperature set point. If the fan is not triggered by temperature, the system engages the fan after a prescribed maximum off-time to ensure that oxygen is supplied.

(4) Airflow is delivered continuously at variable rates that are determined automatically using temperature feedback in an effort to maintain operator-selected temperature set points.

In the first case (1), a simple timer is commonly and successfully used for small- and moderate-scale operations that do not have strict process control requirements. Adding thermostatic control (2) to a timer-based airflow control approach provides additional cooling during peak heat generation to reduce peak temperatures. Actively controlling the aeration by temperature (3) has been practiced since the 1980s and can be quite effective. With the proliferation of relatively inexpensive VFDs that control the speed of an electric motor, it has become common to vary fan speed between low and 100% in response to temperatures (4). The low-speed setting is selected to continuously supply oxygen. The higher speeds increase airflow and therefore cooling. VFD-controlled fans also use less energy compared to turning motors on/off repeatedly to achieve the same average airflow.

With all intermittent airflow approaches, the oxygen level begins dropping during the off-cycle (Fig. 6.35). During vigorous composting, oxygen levels can drop to near zero within 15 min after the flow stops, even though temperatures remain somewhat stable due to the thermal mass and insulating nature of compost (Keener et al., 2005). Therefore, the length of time that aeration is turned off should be limited. Generally, the off-periods should ideally not exceed 10 minutes during the first few weeks of composting and 20 to 30 min later in the process (e.g., when decomposition and oxygen consumption rates are significantly lower). For this reason, some composting systems also include electronic oxygen sensors that activate or increase air flow when oxygen readings are low.

2.9.1 Timer-controlled on/off aeration cycles

Timer-controlled aeration is a simple way to switch fans on and off. The common control device is an inexpensive mechanical or electrical timer with an on-time input and an off-time input. Small mechanical on/off timers can directly switch smaller motors, typically less than 1.5 kW (2 horsepower). When larger motor loads are used, it is more common to use separate on and off timers located in the appropriate electrical enclosure (Figs. 6.36 and 6.37). Programmable logic controllers (PLCs) can also be used for time-based control for a group of motors.

Because timers do not directly respond to temperature changes in the pile, the usual goal is to provide moderate temperature control in addition to satisfying the process oxygen requirements. With timer-control, it is common to select a relatively low airflow rate, like those on the low end of the ranges in Table 6.6. In this case, the composter accepts high temperatures early in the cycle, but avoids over-cooling and drying later in the cycle.

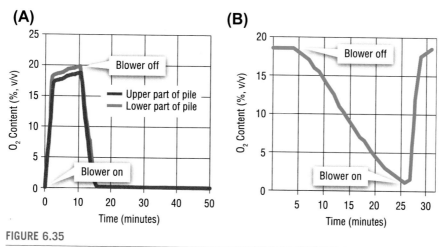

FIGURE 6.35

Two examples showing oxygen concentration in an ASP in response to fans turning on and off.

Adapted from: (A) Zheng et al., 2018; and (B) Murray and Thompson, 1986.

FIGURE 6.36

Single on/off mechanical timer with dual on-period and off-period inputs.

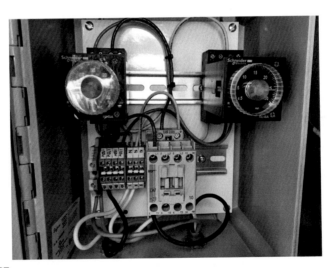

FIGURE 6.37

On and off timers + motor contactor architecture for controlling the fan duty cycle.

Source: ECS.

When a fan is timer-controlled, it runs at full speed whenever it turns on. It runs for a set interval of time and then remains off for another set interval. Each interval is decided by the operator. The duration of the "on" interval primarily determines the resulting temperatures while the "off" interval determines the minimum oxygen concentration.

The on and off intervals should be separately adjusted throughout the composting period based on manually measured pile temperatures, oxygen levels, or other conditions. Early in the process, the fan on-time is set longer to increase cooling and the off-time is shorter to limit the oxygen decline. Later, when temperatures indicate that the composting rate has fallen, the fan on-time can be shortened and the off-time lengthened. If temperatures remain below desired levels, the on-time should be decreased. If temperatures rise above desired levels, the on-time should be increased. If the composter is fortunate enough to have an oxygen probe, it can also be used to make sure the off-times are not so long as to let the oxygen levels hover below 10%.

There are only a few guidelines and general suggestions for determining the initial timer on and off intervals. First, to avoid oxygen depletion, the off-time in the cycle should not exceed 20 min early in the composting process (e.g., first two weeks) and 30 min thereafter.

Another constraint is fatigue on the motors and switches from too many on/off transitions. It is best to keep the total cycle time (on-time + off-time) above 10 min to avoid this fatigue. Recommendations found in composting publications suggest on/off intervals on the order of 5/15 to 10/20 on/off minutes and duty cycles in the range of 25%−33%. While these recommendations arise from a long history of practice, the relatively low duty cycles means that the available fan power is used inefficiently. Box 6.5 describes a site-specific procedure for setting on and off times based on experimentation.

Box 6.5 Finding your cycle

For a given application, the specific on/off cycle times usually are determined by on-site experiments and monitoring of the pile temperatures or oxygen concentrations. A good approach is to monitor the pile behavior using both an oxygen probe and a temperature probe. If an oxygen probe is not available, temperature alone can determine the on/off cycle times. The procedure is as follows:

1. Place the oxygen and temperature probes about one meter into the center of the pile. Monitor and record the oxygen and temperature at each step.
2. Turn on and run the fan until the oxygen probe shows that aerobic conditions are established (e.g., 15%−20% oxygen concentration).
3. Turn off the fan and allow the oxygen concentration to drop below the acceptable minimum (e.g., 5% or 10%). Record the time that it takes for this drop to occur. This time indicates the *maximum* amount of time the fan can remain *off*.
4. If no oxygen probe is available, monitor the temperature and record the time that it takes for the temperature to either rise or fall 5°C (10°F). It can go in either direction, depending on the conditions. In the absence of oxygen measurements, this time indicates the *maximum* amount of time the fan can remain *off*. If the temperature does not change, assume an off-time of 10 to 20 minutes as a starting point and monitor the pile for odors. If putrid odors develop, reduce the off-time. The off-time should not exceed 30 minutes in any phase of composting.
5. Turn on the fan again to determine the time required for the pile to return to aerobic conditions. This second time is the *minimum* length of time the fan should remain *on*.
6. Determine the *actual* on-time by continually monitoring the temperature behavior of the pile. The on-time should be set at the point that the temperature falls to the desired level (e.g., 57°C or 135°F)

This process should be repeated periodically (weekly) during the course of the composting cycle since oxygen demand and heat generation change, and generally decline, as composting progresses.

Another approach is to calculate a duty cycle based on the desired "average" airflow rate in relation to the airflow rate produced by of the available or selected fan. After selecting the desired average air flow rate, the next step is to identify a fan that has a larger output than desired average. Next, the required duty cycle can be calculated.

The average flow rate in a timer-controlled system is the airflow rate of the fan (when it is on) times the duty cycle, as shown in Eqs. (6.6) and (6.7). For example, if a fan supplies 200 m³/h (118 cfm), and it is on for 8 min and off for 12 min (a 40% duty cycle), then the average flow to the pile is 80 m³/h (47 cfm). Therefore, the duty cycle is the desired average airflow rate divided by the fan's airflow rate (Eq. 6.8). Example 6.2 demonstrates the procedure.

$$\text{Average airflow rate} = \frac{\text{fan airflow} \times \text{fan on minutes}}{\text{fan on minutes} + \text{fan off minutes}} \qquad (6.6)$$

$$\text{Average airflow rate} = \text{fan airflow} \times \text{duty cycle (as a fraction)} \qquad (6.7)$$

The duty cycle is therefore:

$$\text{Duty cycle (as a fraction)} = \text{Average airflow rate} \div \text{fan airflow rate} \qquad (6.8)$$

Once the initial timer cycle is set and operations begin, temperatures should be regularly monitored. If they are excessive, then the duty cycle should be increased. If the temperatures remain excessive even at a 100% duty cycle, then either the size of the pile should be reduced or a fan with a larger capacity should be installed.

Example 6.2 Determining ASP timer ON/OFF cycle based on average airflow rate

A 150 m³ (196 yd³) ASP is to be aerated with a timer-controlled aeration system. Based on Table 5.6, a reasonable starting point for the first week of composting is a flow rate of about 6.5 m³/h per cubic meter of pile volume (roughly 3 cfm per cubic yard).

The desired average flow rate for the pile is:

6.5 m³/h per m³ × 150 m³ = 975 m³/h (3 cfm per yd³ × 196 yd³ = 588 cfm)

The available fan has been tested in the composting aeration system and it produces an airflow rate of 2200 m³/h (1295 cfm). Therefore, the required duty cycle is:

Duty cycle = 975 m³/h ÷ 2200 m³/h = 0.44 = 44%
(588 cfm ÷ 1295 cfm = 0.45)

Therefore, the on-time and off-time are, respectively, 44% and 56% of the total cycle time. For the first week of composting, the desired off-time is 18 minutes. The total cycle time and the on-time are:

Cycle time = 18 minutes ÷ 0.56 = 32 minutes

On-time = 32 minutes − 18 minutes = 14 minutes.

A variation of the timer-control approach adds a thermostatic override to the off-cycle that engages the fan if the temperature in one or more sections of the pile reaches a high-temperature set point (option two in this section's introduction). This thermostatic override discourages the pile from accelerating to temperatures high enough to potentially set-back the composting process (e.g., > 75°C, 165°F). This approach adds a level of insurance in case the selected on-time period is too brief, possibly because the decomposition rate of the feedstocks is faster than anticipated. In addition to the timer, the control system requires at least one temperature sensor plus a control device that reads the senor signal and can activate the fan (see following section).

2.9.2 Temperature-controlled on/off aeration cycles

From the standpoint of process management, temperature control is the better aeration strategy, since it seeks to protect the process from being inhibited by high temperatures.

Temperature feedback control has the potential to maintain lower temperatures. Doing so necessitates greater peak air flow rates. Even partial temperature control early in the composting cycle can require relatively high peak airflow rates. For instance, in Fig. 6.33, a relatively high peak aeration rate of 16.6 m³/h per cubic meter (7.5 cfm per cubic yard) still allowed the temperatures to initially surpass the temperature set point by almost 18°C (32°F). High aeration rates require a larger more expensive aeration system. Also, compared to the time-control approach, temperature control involves more complicated and expensive control and monitoring devices.

Temperature sensors, such as thermocouples or thermistors, placed at selected locations in the pile provide a constant signal to a control device, such as a dedicated computer or PLC. A pile or cell needs only one sensor, but some systems use several sensors placed at various locations and depths in the pile (Fig. 6.38). The control device interprets the sensor signals, records them and, when needed, activates relay

FIGURE 6.38

Orange flagged temperature probes inserted in an extended ASP.

Source: ECS.

switches that turn the fan(s) on and off when the pile temperature cross selected temperature limits. In most cases, wires carry the signal from the sensor to the control device, which is normally at a central location. Because the wires can be awkward to work around, wireless sensors are sometimes used in large facilities.

The high temperature set point for temperature control systems typically falls in the range of 45–65°C (113–150°F), depending on the process goals. The system shuts the fan off when the piles cool below a low set point. The low set point is typically about 5°C (10°F) below the high set point. The temperature set points can be adjusted according to process conditions. For example, the low temperature set point can be set at 57°C during the three-day period that high-temperature sanitation conditions are being met. Afterward, the set point can be lowered.

If aeration is not triggered by high temperature within a certain time frame (e.g., 10–30 min), the fan control reverts back to a simple timer that operates in parallel to the thermostatic controls. The timer ensures that some oxygen is regularly supplied. This tends to occur during start up and late in the process or whenever the pile temperature is running below the low set point. While this control strategy can be successful, it will result some degree of process inhibition due to low oxygen levels and/or high temperatures.

When a single temperature sensor is used to activate the fan operation, it must be carefully placed to measure temperature representative of the mass of material in the pile. Finding a representative location can be a challenge because temperatures within most compost piles vary considerably. The typical location for a sensor is at least 45 cm (18 in.) below the pile surface and between 1/3 and 2/3 down the length of the pile (Fig. 6.39). Manual temperature measurements using a long-stem thermometer can help find a representative sensor location.

Periodic mapping of pile temperatures with a thermometer is recommended to verify that the feedstock mix and the aeration distribution are both relatively uniform and that the control system is providing the desired level of temperature control.

2.9.3 Continuous variable-rate aeration based on temperature feedback

Continuous variable-rate aeration is used in more advanced ASP designs that seek some combination of improved process control, lower power use, and better process visibility for system management. The control settings in these systems typically include a minimum baseline aeration rate that meets the oxygen demand and provides minimal cooling. The airflow rate is automatically modulated up and down, in response to temperature feedback, in an effort to achieve operator-selected set point temperatures (Fig. 6.33). A smart device, such as a PLC or a computer, monitors the temperature from one or more sensors and implements the control algorithms that control fans, and possibly dampers.

Continuous variable-rate aeration is produced using a VFD to control fan speed and/or a motorized damper. VFDs, in particular, have become quite affordable. The turndown ratio for these devices, which is the maximum flow/minimum flow, is on the order of 50/1 to 100/1. This is a wider dynamic range, or turn down, than can typically be achieved with intermittent aeration at a constant fan speed. VFD-controlled fans can provide high peak flow rates to manage the most energetic periods of composting, yet still not overly cool during periods of low heat production.

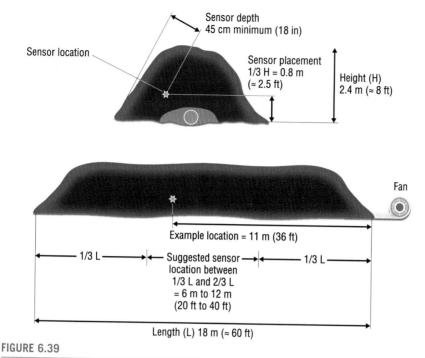

FIGURE 6.39

Suggested temperature sensor location for an aerated static pile using the example of a pile that is 2.4 m tall and 18 m long.

A VFD is device that electronically changes the frequency of the electricity supplied to an AC electric motor. A VFD makes it possible to operate a motor at numerous speeds (revolutions per minute or rpms) rather than the few fixed speeds that are feasible with unaltered electricity (Edison Tech Center, 2014) VFDs have made it practical to incrementally adjust the speed of composting fans to closely match a fan's output (e.g., m^3/h or cfm) to a pile's need for aeration (Box 6.6).

2.9.4 Other aeration control options

It is possible to operate a simple ASP at a continuous fixed aeration rate that supplies the required oxygen and some degree of cooling. However, this approach is generally ill-advised.

First, at a rate sufficient to cool the pile, the airflow is likely to overly dry the pile. Second, the fixed rate is unlikely to provide effective temperature control. Third, continuous fan operation leads to uneven pile temperatures. Unless a reversing airflow system is used, the areas near the air inlets remain cooler than the areas where the air

---

Box 6.6 Variable frequency drives (VFDs) for fans

A VFD is device that electronically changes the frequency of the electricity supplied to an AC electric motor (Fig. 6.40). A VFD makes it possible to operate a motor at numerous speeds (revolutions per minute or rpms) rather than the few fixed speeds that are feasible with unaltered electricity (Edison Tech Center, 2014). VFDs have made it practical to incrementally adjust the speed of composting fans to closely match a fan's output (e.g., m^3/h or cfm) to a pile's need for aeration.

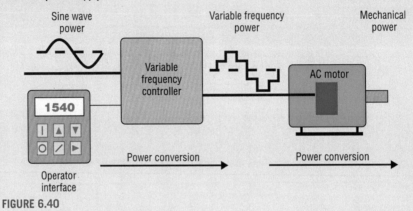

FIGURE 6.40

Schematic of a VFD motor control system.

Ordinarily, a fan rotates at a constant number of revolutions per minute (rpms), which is informally called its speed. The fan speed is determined by the rotational speed of its motor and its gear ratios, if any. The fan's rotation is constant because the motor rotates at a constant rpm that depends on the frequency of the electrical supply and the number of "poles" that the motor has. Because the frequency of the electrical supply is fixed, a given motor normally has a fixed speed determined by the number of poles. The device that a motor powers, like a fan, operate at the same rpm unless mechanisms like gears and pulleys are used to alter the rotational speed. Even with these mechanisms, the speed change takes place in rather large increments.

VFDs use electronic components like diodes to almost instantly transform the frequency of the electricity supplied to a motor in large and infinitesimal increments. Thus, a fan motor, and the fan itself, can be made to operate at half its full-frequency speed, or 10% of it, or 78.5% of it, or nearly any percentage of it.

---

leaves. With intermittent operation, temperatures in different sections of the pile tend to equalize after the airflow stops. A better option is to supply air continuously with one fan at a rate low enough to satisfy the oxygen demand and provide a second on/off fan to cool the pile when temperatures become undesirably high.

FIGURE 6.41

VFD fan supplied plenum supplying multiple aeration control dampers.

Source: ECS.

To achieve variable air delivery without a VFD, some ASP facilities control the fan output to a supply plenum with a pressure sensor/feedback and modulate flow to individual piles by adjusting the openings of louvered dampers on the lateral inlets (Figs. 6.41 and 6.42). The electric or pneumatic actuators that adjust the dampers are controlled by a temperature feedback from sensors in the pile. In this case, a single fan can supply a common plenum with multiple independent dampers modulating air to laterals serving individual cells or piles.

3. Variations of aerated static piles

The stripped-down ASP model is about piles—self-spreading freestanding piles, sitting in the open, exposed to the cruel outdoors. However, ASP-style forced aeration is routinely employed with a few frills. The four primary frills are: (1) enclosed buildings; (2) partial enclosures such as bins and bunkers; (3) blanket-like fabric covers; and (4) bags that envelope the pile. In these applications, the ASP method and its fundamentals remain the same. The practices previously described for ASP composting are generally followed. However, these applications come with some site- and method-specific practices.

FIGURE 6.42

Installation of a pneumatically adjusted gate valve (damper) that controls the airflow from the manifold to the laterals of a composting bay.

Adapted from Smith and Aber, 2017.

3.1 Bins and bunkers

Composting in bins and bunkers with forced aeration is like ASP composting, except the feedstocks are partially contained by walls (Figs. 6.43 and 6.44). The containment provides several attractive advantages. First, it allows vertical stacking of feedstocks at the perimeter to better use floor space than freestanding piles. Second, the exposed surface is smaller and flatter, which makes it easier to cover with a biofiltering layer of compost or a roof. During pile formation, the walls help keep feedstocks from escaping the fringes of the pile. Third, bins and bunkers are especially advantageous for composting food scraps, which notoriously escape the confines of the pile. Fourth, bins and bunkers define a discrete batch that can be managed and tracked through the process.

Overall, composting in bins and bunkers, rather than piles, makes the facility neater and easier to manage. On the negative side, bins and bunkers cost money and reduce the flexibility in using the site compared to piles.

Bins are commonly made of poured concrete, heavy concrete blocks, durable plastic, wood, or temporary materials like hay bales. Bunkers, which are simply

FIGURE 6.43

Examples of forced-aerated concrete bunkers.

bins in a more massive style, are usually concrete. A bin or bunker can have either four walls with a door, three walls with an open side, or even two walls. Units with two parallel walls can be loaded on one open end and unloaded on the other. Although some bins have a roof to exclude precipitation, they are otherwise exposed to outside ambient conditions. Most operations include multiple units, often adjacent to and sharing a common wall.

Together, bins and bunkers cover a wide range of sizes, from small one cubic meter bins (about one cubic yard) to large bunkers measuring 12 m wide, 24 m long, and 3 m high (40 × 80 × 10 ft). The number and size of bins are determined by the availability of feedstocks—how much and how often. Because each bin unit represents a batch, the size tends to accommodate the batch volume, which in turn is influenced by how often feedstocks are received or generated. With certain types of bins, the time span of a batch can be more flexible, compared to piles, because aeration can be started before the batch is complete. Some bins are loaded bottom

FIGURE 6.44

Examples of small-scale forced-aerated composting bins.

Source: O₂Compost.

to top in complete layers which makes it possible to aerate partially filled bins without short circuiting.

Bin dimensions must first provide the necessary volume and then accommodate the operation's feedstock, equipment, and space constraints. The minimum width of the bin is the width of the equipment used to load and unload it (Fig. 6.45). Normally that equipment is a front-end loader, wheel loader, tractor, or skid steer loader. The possibility of obtaining larger equipment in the future is a point that should be considered. The height of a bin is limited by the same factors that limit pile height—compaction, porosity, and the reach of equipment. Therefore, bins are typically high enough to allow materials to stack 1.5—2.5 m high (5—10 ft), including the compost cap. The walls should extend, roughly, an extra 30 cm higher (about 1 ft).

(A)   **(B)**

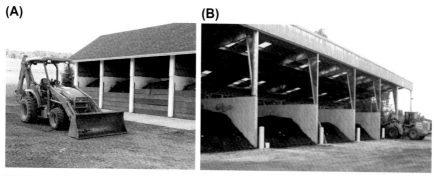

FIGURE 6.45

The width of bins/bunkers should match the width of materials handling implements.

Source: (A) R. Rynk and (B) F. Michel.

Most bins have positive aeration systems, although suction and reversing systems are used as well. Like piles, bins can be aerated with moveable pipe lying on the composting floor. However, most have permanent aeration system embedded in the floor. Aeration trenches are common (Fig. 6.46). Some smaller bin designs distribute air through wooden planks spaced roughly 12mm apart (about ½ in.). With a covering of wood chips, the boards form the floor surface. Beneath is an air plenum filled with gravel to disperse the air and drain water. Perforated plastic pipe, usually PVC, within the gravel distributes air from the fan. This approach delivers air to the feedstocks at relatively low pressures.

FIGURE 6.46

Aeration floor in aerated bins using below-grade basin covered with wooden planks.

Source: C. Oshins.

With batch-sized units, in-floor aeration and solid walls to push against, bins are suited to regular turning. Indeed, some bin systems are designed for turning on a weekly to biweekly schedule. To begin the process, a batch of mixed feedstocks is placed in the first bin. It is aerated and managed in the same manner as a conventional ASP. After one to two weeks, a bucket loader moves the batch from the first bin into the adjacent bin. The batch aerates in the second bin for several days to several weeks until it is shifted again to the next bin in the series. This is repeated two to five times until the material is harvested as compost from the last bin.

After the compost is harvested, in succession, bins are emptied and refilled with a batch from the adjacent bin. When a batch is shifted between bins, it can be mixed and adjusted for moisture if necessary. Two or more batches can be combined to compensate for volume loss. The series of bins also allows for finer design and control of the aeration system. Each bin represents a different aeration zone that can be sized and managed based on the stage of decomposition taking place in the bin. Frequent turning also can be accomplished in nearly the same manner by unloading materials from a bin, turning, and then reloading them into the same bin.

3.2 Buildings

Several composting facilities manage ASP composting systems inside a building. The building isolates the composting operation from the environment, protecting both the composting operation and its neighborhood. The piles can be either free-standing or confined in bins or bunkers.

Numerous small- and moderate-scale composting operations are housed in covered structures that are open on the ends or sides. There is little or no control of the interior conditions (Fig. 6.47). In these cases, the operation is merely conventional ASP composting under the convenience of a roof.

FIGURE 6.47

Aerated static pile composting in an open building.

Other composting facilities operate inside large fully enclosed and environmentally controlled buildings. Fully enclosed operations usually handle more odorous feedstocks such as biosolids, food waste, and municipal solid waste. One striking example of a fully enclosed ASP is the large Inland Empire Utilities Agency (IEUA) facility in California. This facility composts biosolids and wood chips in a former warehouse, renovated for composting (Fig. 6.48).

Fully enclosed composting applications differ only slightly from conventional ASP composting. Aeration floors are usually embedded channels or pipes with orifices in the floor. ASPs located inside building are mostly positive aeration systems. The hot moist process air exits the pile disperses in the headspace of the building. The building has to have a powerful and well-engineered ventilation system, which maintains a slight negative pressure in the building and exhausts to an outdoor biofilter or other odor scrubbing system. These ventilation systems often double the fan power required at a composting facility, greatly increase the size of the odor scrubbing system, and in cold environments, can add large heating loads to warm up the incoming air. Even with a well-designed building ventilation system, dust, odorous compounds, and high humidity tend to persist in the building. The high humidity commonly generates a thick fog, which is a concern for worker safety (see Fig. 12.11 in Chapter 12). The building structural components have to be very corrosion-resistant (e.g., high-build coatings, stainless steel).

FIGURE 6.48

The Inland Empire Regional Composting Facility. A fully enclosed aerated static pile system built inside of a renovated warehouse. The system's biofilter is outside the main composting building and runs parallel to it (lower part of the photo). The long fabric-covered building is for curing and product storage.

Source: IEUA.

3.3 Aerated static pile under microporous covers

One form of ASP composting features a semipermeable microporous fabric cover (Fig. 6.49) on top of positively aerated pile. The fabric cover adds some of the benefits of containment and protection without a building and without an external biofilter. As the pile's exhaust air hits the cover, moisture condenses on the underside. Water-soluble gaseous compounds are absorbed in the moisture and return to the composting pile as the water drips back. The fabric cover also provides a physical barrier that slows drying and shelters the pile from rain. It discourages birds and other pests that may be attracted to feedstocks like food scraps. Microporous fabrics that cover ASPs should not be confused with polyethylene pods (Box 6.7) or woven fabric covers used primarily for windrow composting (Box 5.2 in Chapter 5).

The covers are made of a microporous membrane that is laminated to the structural fabric. This membrane is permeable to gasses (oxygen, water vapor, and carbon dioxide) but relatively impermeable to liquid water since it occurs in large drops. Airflow through the cover is slowed by the tiny pore diameters (roughly 1 μm) and the water drops. The original cover material was an offshoot of the familiar Gore-Tex fabric used for outerwear. The subsequent composting system was also pioneered by W.L Gore and Associates. Microporous membranes are made of expanded polytetrafluoroethylene. This membrane is the core of a laminated system that includes two or more other membrane layers. The additional layers add tensile strength, durability, and ultraviolet resistance to the cover.

There are a number of microporous fabric producers around the world. In all cases, the initial investment in the covers is significant. Although relatively durable, the covers eventually need to be replaced as they are subject to damage and wear. Some cover suppliers offer integrated composting systems in addition to, or rather than, the covers alone. These systems may include the covers and tie-down apparatus, aeration and control systems, plus training and support services. Also,

FIGURE 6.49

Microporous fabric cover system over an aerated pile.

Source: W.L Gore & Associates.

FIGURE 6.50

Machine for handling fabric covers.

Source: ECS.

(A)

(B)

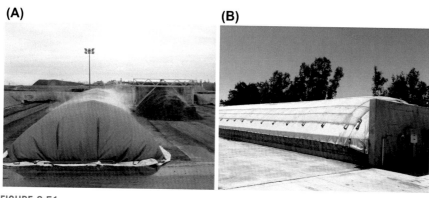

FIGURE 6.51

Microporous-fabric-covered piles in freestanding piles (A) and partially contained by concrete walls (B).

Source: W.L. Gore & Associates.

equipment is available to roll and unroll the covers, facilitating placement and removal (Fig. 6.50).

Operationally, fabric-covered ASP composting is much like the traditional ASP method. Both POG and below-grade aeration floors can be used. Like any ASP, each batch of feedstocks must be well-proportioned and well mixed with a good bulking agent and suitable moisture content (50%−60%). Piles can be freestanding or placed within walls (Fig. 6.51). They are usually built on a concrete pad. The covers are draped over the piles and held down with straps or weights.

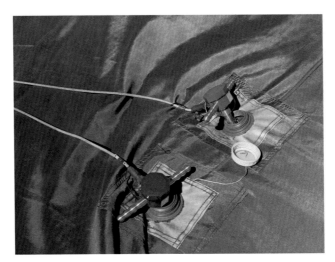

FIGURE 6.52

Placement of temperature and oxygen sensors.

Source: W.L Gore & Associates.

After a pile of feedstock is formed and covered, it composts undisturbed for four weeks or longer. Fabric-covered piles are often composted in two or more stages. In such cases, after the initial stage of two to four weeks, the pile is uncovered and then turned into a second stage for an additional two to four weeks. The covers are normally, but not universally, reapplied for the second composting stage. If additional turning takes place after the second stage, piles are typically left uncovered. Turning, especially after the initial stage, has the potential to release odors and should be scheduled to minimize impacts on the surrounding neighborhood.

Positive pressure aeration is used exclusively because the covers provide odor control. The tiny micropores resist the passage of air, especially once covered by water droplets that are many times larger in size. Pushing air through the micropores of the cover increases the system's static pressure, typically by 25−50 mm w.c. (1−2 in.), depending on the air flow rate. Due to the higher static pressure, fabric-covered ASPs maintain aeration rates at the lower end of the ranges specified in Table 6.6. Fans are turned on and off automatically based primarily on oxygen levels (Fig. 6.53) and manufacturer's proscribed duty cycles. The control system is generally trying to keep the oxygen levels above a minimum set points (e.g., 8%−10%), but is not intended to control temperatures by cooling the pile with a higher airflow.

Covers have counteracting effects on moisture control. On one hand, precipitation is excluded, which improves pile management when pile moisture is high, but worsens dry piles. On the other hand, the covers reduce pile moisture losses from surface evaporation and aeration. Therefore, water is a design and management issue for covered systems. Condensate should be prevented from running down the underside of the cover and collecting at the base of the pile.

Also, rain and snow melt must be drained from the outer surface of the covers. In cold weather, ice can freeze the covers to the composting pad. Some systems include hot water pipes to prevent this occurrence.

Box 6.7 ASPs in a pod

Another means to enclose ASP composting is to envelop the pile in a nonporous plastic bag, which is sometimes called a "Pod." The pods, and the machine used to load them, are an ironic adaptation of a method used to ensile livestock feed. The irony stems from the fact that making silage and making compost are opposite endeavors—eliminate air vs. aerate, dense-pack vs. loose-pile. Several companies have modified the bags and loading mechanisms to ingeniously convert them into an ASP within a plastic membrane (Fig. 6.53).

FIGURE 6.53

Modified ASP in a polyethylene pod.

Source: D. Inman, Ag-Bag Environmental.

The single-use pods are made of polyethylene film. The pod-filling machine pushes in the feedstock with limited force to maintain the feedstock's target bulk density. At the same time, it unrolls a length of flexible LDPE aeration pipe beneath the advancing feedstock. The loading machine has a hopper that holds a volume that corresponds to about one stroke of the hydraulic ram that pushes in the feedstock. The feedstock must be premixed before loading. The bag can accommodate several partial loadings on its way to being filled to capacity. After loading, the open end of the bag is tied off, the aeration pipe is cut and connected to the fan, and aeration is started. At the next loading, the existing pipe is joined to the next length of pipe.

The forced aeration system mimics that of traditional ASPs with airflow rates on the low side. Fans, usually timer-controlled, provide positive aeration through the perforated pipe that runs beneath the pile inside of the pod. The exhaust leaves the pod through a series of aeration outlets spaced along the top. These outlets also serve as ports for temperature probes and other monitoring devices. The feedstocks compost undisturbed in the bag for several weeks until containment is no longer needed. At this point, the bags are sliced open lengthwise to open the pile. The bags are either discarded or used for other purposes (e.g., agricultural mulch).

The appeal of this method is that it provides almost immediate containment and forced aeration without the need for an impervious pad or a building. However, pods rarely serve as a permanent composting system because of several significant shortcomings. The most consequential is the tendency to accumulate water at the floor of the bag. Water that condenses on the inside of the bag drains down the bag and tends to saturate the bottom of the pile. This situation worsens in cold climates if the water freezes inside the bag. Unless an effective means of drainage is added, odorous liquid accumulates inside the bag. Care must be taken to prevent it from draining out when the bag is sliced open. In addition, the plastic pods must be continually replaced. They cannot be reused for composting and create cumbersome plastic waste.

The plastic pod approach to ASP composting still has potential applications. It is an option where "enclosed" composting is required on a short-term basis for environmental compliance. Also, it can be a means to temporarily expand capacity of a facility in the case of an unexpected availability of feedstocks, for example, in the case of large amounts of food spoiled during a power outage.

4. Methods combining turning and forced aeration of windrows and piles

Several composting facilities employ methods that combine attributes of both turned windrows and forced aeration to overcome the deficiencies, and exploit the advantages, of each method. Two open-air methods that exemplify this strategy are:

- Forced-aeration windrows; and,
- Turned extended piles (also known as trapezoidal piles and mass-beds).

Several commercially developed composting systems also combine forced aeration and turning but are generally considered within the category of "in-vessel" composting. They are explained in Chapter 7.

4.1 Forced aeration turned windrows

Forced aeration turned windrows are typical windrows placed over an aeration floor built in the composting pad (Fig. 6.54). As with the standard windrow method, the windrows are turned periodically with a conventional windrow turner. Turning principles and frequency are usually the same as for nonaerated windrows. However, the reduced odor risk, due to forced aeration, affords the operator more flexibility in deciding when to turn. Fans push or pull air through the windrows via one or

(A)   **(B)**

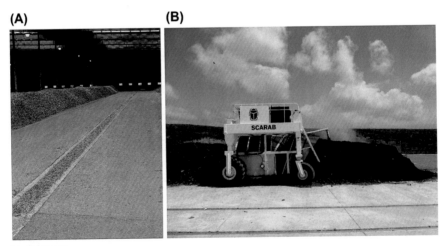

FIGURE 6.54

Two examples of forced-aerated windrows: biosolids + woodchips (A) and yard trimmings + food waste (B). Aeration trenches and a below-grade network of pipes provide positive aeration to each windrow.

Source: (A) R. Rynk and (B) ECS.

more aeration plenums. A windrow must be kept centered over the plenums to maintain even airflow on both sides. Otherwise, the same aeration principles that apply to ASPs also apply to aerated windrows.

Because the aeration system provides oxygen and temperature control, it reduces odors and decreases the time required to stabilize the compost compared to conventional turned windrow composting. Likewise, regular turning accelerates the process compared to ASPs by reestablishing porosity, redistributing moisture, and exposing new surfaces to bio-degradation.Forced-aeration windrows can also hasten the drying of very wet substrates like manures. However, this method also has the dual disadvantages of both windrows and ASPs. It requires the large pad area, a windrow turner, and the trappings of forced aeration. Therefore, the forced-aeration windrow approach is uncommon. It is primarily used at facilities that desire a short composting period and/or require a high level of process control (e.g., food waste, SSO). The Compost Council of Canada has a "best practices" guide specifically aimed at this method of composting (CCC, 2016).

4.2 Extended piles (aka. Trapezoidal, block, or table composting)

The turned extended aerated pile method resembles the previously described extended ASP in most respects, except that the pile is regularly turned and shifted to one side. The turning machine lifts a swath of the pile and drops it to the side (Fig. 6.55). The size of the swath and the distance it moves depend on the style and model of the turner. The machine makes repeated passes until the entire extended piles has move toward the more mature end and left an open space to receive fresh material.

Although forced aeration is the norm, a few operations rely only on passive convection and very frequent turning to minimize anaerobic conditions. Often referred to as trapezoidal, mass bed, table or block composting, the method supports large, relatively tall piles up to 3 m (9 ft) in height and 20 m (60 ft) wide. Facilities that employ this technique use either specially designed turners or an elevating-face turner fitted with side-discharging conveyors.

Once the pile is started, each batch of new feedstocks is added to one side of the existing pile, always at roughly the same location on the site. Each added batch is like a cell in an extended ASP. The feedstocks should be premixed to improve air distribution in the early stages. However, the subsequent turnings can correct poorly mixed batches. With each turning, the pile is shifted to the side in successive sections. All batches move together toward the discharge side of the pile. The compost is always removed from the pile on the opposite side from the fresh feedstocks. The number and frequency of turnings determine the composting period, at least in the extended pile. The compost normally continues to mature in a curing pile.

(A)

(B)

FIGURE 6.55

Two views of an extended pile with below-graded forced aeration and regular turning. In photo (B), notice the spray bar on the conveyor for adding moisture and reducing dust.

Source: (A) Jane Gilbert, (B) Dirt Hugger.

The turning procedure starts at the side containing the oldest material, or the compost. The turner moves along the edge of the pile and picks up the compost in a slice equivalent to the width of the turner. The side-discharge conveyor places the compost directly into a truck bed. Alternatively, a loader can be used to remove the end of the pile and move it to a curing or screening step. Next, the turner moves to the adjacent, newly exposed edge of the pile, picks up a slice, and deposits it over the strip of floor space previously occupied by the harvested compost. The turner progresses through the remaining slices, shifting each one over toward the harvested side of the pile. During, or prior to, turning, water can be added to sections of the pile that are too dry. Eventually, the entire pile is moved over by a distance equivalent to the width of the harvested compost. Finally, new feedstock is placed along the side of the pile, in the space left vacant when the last slice shifted over.

The forced aeration system differs slightly from that of the extended ASP. First, the air distribution components are necessarily recessed in the composting pad. Channels can be used but high-pressure orifices are more common to protect outlets from blockage. Second, the aeration system can be staged because the pile maintains fresh feedstocks on one side and less biologically active material to the other. Higher airflow rates can be delivered beneath the young cells while lower rates are applied to the older cells. Positive pressure fans are typically used. If odor is a critical issue, the operation is enclosed in a building.

The turned extended pile method provides the advantages of turning and forced aeration on a small footprint. Compared to extended ASPs, the composting time decreases, and so does the required area for the composting pad, per unit of feedstock and compost. Consequently, it is more efficiently covered with a roof or enclosed in a building. Regularly turning opens the large piles to inspection, measurement, and to the addition of water. Furthermore, this method adds a materials handling benefit as feedstocks are added, and composts are removed, from fixed locations. Conveyors and automated materials handling operations become more practical.

However, as with forced-aerated windrows, the turned extended pile method brings collective disadvantages. It requires investments in both turning and forced aeration equipment, labor, and facilities. Management of aeration is complicated by the ever-shifting pile of material above. While turning offers opportunities to add water, more water has to be replenished. The combination of frequent turning and regular aeration drives away more moisture. In some respects, the process is hostage to the turning schedule. One cell cannot be turned without also turning the other cells. Also, if turning does not occur, the pile does not move and create space for new feedstocks. There are, however, management options to deal with the restrictions imposed by turning. Overall, with the turned extended pile method, a bit of the simplicity of composting is forfeited.

As previously written, some operations use turned extended piles without providing forced aeration. In essence, these facilities practice turned windrow composting with one giant windrow. Without forced aeration, such large piles are not effectively aerated. Passive aeration cannot penetrate much beyond the exterior. To stave off anaerobic conditions, the pile has to be turned frequently, almost daily. This version of the method still offers the advantage of efficient use of space even if it takes longer, without forced aeration, to produce mature compost. (see "Massive Passive" sidebar earlier in this chapter).

References

Cited references

Alexander, R., 2019. BioCycle Trailblazers: USDA-EPA aerated static pile Team. BioCycle Connects. https://www.biocycle.net/biocycle-trailblazers-usda-epa-aerated-static-pile-team/.

Arslan, E., Ünlü, A., Topal, M., 2011. Determination of the effect of aeration rate on composting of Vegetable—fruit wastes. Clean-Soil Air Water 39, 1014—1021.

Barrington, S., Choinière, D., Trigui, M., Knight, W., 2002. SE—structures and environment: compost airflow resistance. Biosyst. Eng. 81, 433—441 (BioCycle, The JG Press. Emmaus, PA).

Brown, G., 2014. The Compost-Powered Water Heater. The Countryman Press. https://wwnorton.com/the-countryman-press.

Bryant-Brown, M., Gage, J., 2015. Lessons learned in aerated static pile (ASP) composting. In: Presentation at the USCC Annual Conference. Available from: https://www.slideserve.com/amal/lessons-learned-in-aerated-static-pile-asp-composting.

CCC, 2016. Best Practices for Operating an Aerated Windrow Composting Facility. The Compost Council of Canada (for the Government of Manitoba Conservation and Water Stewardship). http://www.compost.org/.

Chen, Z., Zhang, S., Wen, Q., Zheng, J., 2015. Effect of aeration rate on composting of penicillin mycelial dreg. J. Environ. Sci. ISSN: 1001-0742 37, 172—178. https://doi.org/10.1016/j.jes.2015.03.020.

Coker, C., 2018. Feedstock mixing systems. BioCycle. https://www.biocycle.net/feedstock-mixing-systems/.

Coker, C., O'Neill, T., 2017a. Aeration floor fundamentals. BioCycle. https://www.biocycle.net/aeration-floor-fundamentals/.

Coker, C., O'Neill, T., 2017b. Composting aeration floor functions and designs. BioCycle. https://www.biocycle.net/composting-aeration-floor-functions-designs/.

Edison Tech Center, 2014. The Electric Motor. https://edisontechcenter.org/electricmotors.html.

Finstein, M.S., Miller, F.C., MacGregor, S.T., Psarianos, K.M., 1992. The Rutgers strategy for composting: process design and control. Acta Hortic. 302, 75—86. https://doi.org/10.17660/ActaHortic.1992.302.7.

Fulford, B., 1986. The Composting Greenhouse at New Alchemy Institute: A Report on Two Years of Operation and Monitoring. Research Report 3. New Alchemy Institute, Cape Code, MA.

Gao, M., Li, B., Yu, A., Liang, F., Yang, L., Sun, Y., 2010. The effect of aeration rate on forced-aeration composting of chicken manure and sawdust. Bioresour. Technol. ISSN: 0960-8524 101 (6), 1899—1903. https://doi.org/10.1016/j.biortech.2009.10.027.

Gibson, T., Coker, C., 2013. Pipe and blower fan fundamentals in asp design. BioCycle 54 (2), 24. https://www.biocycle.net/pipe-and-blower-fan-fundamentals-in-asp-design/.

Guo, R., Li, G., Jiang, T., Schuchardt, F., Chen, T., Zhao, Y., Shen, Y., 2012. Effect of aeration rate, C/N ratio and moisture content on the stability and maturity of compost. Bioresour. Technol. 112, 171—178.

Haug, R.T., 1993. The Practical Handbook of Compost Engineering. Lewis Publishers, CRC Press, Boca Raton, Florida.

Higgins, A.J., Chen, S., Singley, M.E., 1982. Airflow resistance in sewage sludge composting aeration systems. Trans ASAE 25 (4), 1010—1014, 1018.

Hong, J.H., Park, K.J., Sohn, B.K., 1997. Effect of composting heat from intermittent aerated static pile on the elevation of underground temperature. Appl. Eng. Agric. 13, 679–683.

Hu, B., Wang, Z., Gao, M., She, Z.L., Zhao, C., 2012. Effect of aeration rate on forced-aeration composting of sewage sludge and maize straw. Appl. Mech. Mater. 178–181, 843–846.

Keener, H.M., Hansen, R.C., Elwell, D.L., 1993. Pressure drop through compost: implications for design. In: Paper Presented at the Summer Meeting of the American Society of Agricultural Engineers, Spokane WA, June 1993. Paper No. 93–4032.

Keener, H.M., Ekinci, K., Elwell, D.L., Michel Jr., F.C., 2005. Composting process optimization - using on/off control. Compost Sci. Util. 13 (4), 288–299.

Keener, H., Ekinci, K., Michel, F.C., 2007. Composting process optimization – using on/off controls. Compost Sci. Util. 13, 288–299.

Kostov, O., Tzvetkov, Y., Kaloianova, N., Cleemput, O.V., 1995. Cucumber cultivation on some wastes during their aerobic composting. Bioresour. Technol. 53, 237–242.

Kulcu, R., Yaldiz, O., 2004. Determination of aeration rate and kinetics of composting some agricultural wastes. Bioresour. Technol. 93 (1), 49–57.

Li, X., Zhang, R., Pang, Y., 2008. Characteristics of dairy manure composting with rice straw. Bioresour. Technol. ISSN: 0960-8524 99 (2), 359–367. https://doi.org/10.1016/j.biortech.2006.12.009.

Live Wire, 2019. Variable Frequency Drive Classroom Training. https://www.livewireindia.com/vfd_software_training.php.

McSweeney, J., 2019. Community-Scale Composting Systems. Chelsea Green Publishing, White River Junction, VT.

Moon, P., 2019. Aerated static pile composting, applications and opportunities. In: Presentation to the Maryland Department of Agriculture. Peter Moon, O_2 Compost. https://www.o2compost.com/.

Moon, P., 2021. Aerated static pile composting, introduction and overview. ASP Composting Webinar – Part 1. O_2 Compost. Available from: https://www.o2compost.com/asp-composting-webinar-pt1.aspx.

Murray, C., Thompson, J., 1986. Strategies for aerated pile systems. BioCycle 27 (7), 22–28.

Pain, I., Pain, J., 1972. The methods of jean pain: another kind of garden. Draguignan: Ancienne Imprimerie NEGRO.

Schuchardt, F., 1984. Heat loss during composting of sawtimber. Landbauforschung V€olkenrode 34, 189–195.

Seki, H., 1989. An investigation of practical process design and control of a soil warming system with heat generated in compost. J. Agric. Meteorol. 44, 259–267.

Sidełko, R., Janowska, B., Szymański, K., Mostowik, N., Głowacka, A., 2019. Advanced methods to calculation of pressure drop during aeration in composting process. Sci. Total Environ. 674, 19–25.

Smith, M.M., Aber, J.D., 2017. Heat Recovery from Composting: A Step-by-step Guide to Building an Aerated Static Pile Heat Recovery Composting Facility. University of New Hampshire Cooperative Extension; Research Report, Durham, NH, p. 64.

Smith, M.M., Aber, J.D., 2018. Energy recovery from commercial-scale composting as a novel waste management strategy. Appl. Energy 211, 194–199.

Smith, M.M., Aber, J.D., Rynk, R., 2017. Heat recovery from composting: a comprehensive review of system design, recovery rate, and utilization. Compost Sci. Util. 25 (S1), 11–22.

Sundberg, C., 2005. Improving Compost Process Efficiency by Controlling Aeration, Temperature and pH (Doctoral thesis). Swedish University of Agricultural Sciences, Uppsala Sweden.

Thostrup, P., 1985. Heat recovery from composting solid manure. In: Gasser, J.R.K. (Ed.), Composting of Agricultural and Other Wastes. Elsevier Applied Science Publishers, London, pp. 167–180.

Tucker, M.F., 2006. Extracting thermal energy from composting. BioCycle 47, 38.

Willson, G., Parr, J., Epstein, E., Marsh, P., Chaney, R., 1980. Manual for Composting Sewage Sludge by the Beltsville Aerated-Pile Method. U.S. Environmental Protection Agency, Washington, D.C. EPA/600/8-80/022.

Yuan, J., Chadwick, D., Zhang, D., Li, G., Chen, S., Luo, W., Du, L., He, S., Peng, S., 2016. Effects of aeration rate on maturity and gaseous emissions during sewage sludge composting. Waste Manag. 56, 403–410.

Zakarya, I.A., Khalib, S.N., Sandu, A., 2019. The study of different aeration rate effect during composting of rice straw ash and food waste in managing the abundance of rice straw at paddy field. Int. J. Conserv. Sci. 10 (2), 335–342, 8pp.

Zheng, G., Wang, Y., Wang, X., Yang, J., Chen, T., 2018. Oxygen monitoring equipment for sewage-sludge composting and its application to aeration optimization. Sensors 18 (11), 4017. https://doi.org/10.3390/s18114017.

Consulted and suggested references

Adams, Z., 2005. Understanding Biothermal Energy (Master's thesis). University of Vermont, Burlington.

Almeida, P., Silveira, A., 2008. Airflow pressure drop evaluation in forced aeration composting. In: compost and digestate: sustainability, benefits, impacts for the environment and for plant production. In: Fuchs, et al. (Eds.), Proceedings of the International Congress CODIS 2008. February 27-29, 2008.

Bhave, P.P., Kulkarni, B.N., 2019. Effect of active and passive aeration on composting of household biodegradable wastes: a decentralized approach. Int. J. Recycl. Org. Waste Agric. 8, 335–344. https://doi.org/10.1007/s40093-019-00306.

Cegarra, J., Alburquerque, J., Gonzálvez, J., Tortosa, G., Chaw, D., 2006. Effects of the forced ventilation on composting of a solid olive-mill by-product ("alperujo") managed by mechanical turning. Waste Manag. 26 12, 1377–1383.

Coker, C., Gibson, T., 2013. Design considerations in aerated static pile composting. BioCycle 54 (1), 30. https://www.biocycle.net/design-considerations-in-aerated-static-pile-composting/.

Das, K., Keener, H.M., 1996. Process control based on dynamic properties in composting: moisture and compaction considerations. In: de Bertoldi, M., Sequi, P., Lemmes, B., Papi, T. (Eds.), The Science of Composting. Springer, Dordrecht. https://doi.org/10.1007/978-94-009-1569-5_13.

Ekinci, K., Keener, H.M., Elwell, D.L., Michel Jr., F.C., 2004. Effects of four aeration strategies on the composting process. Part I - Experimental Studies. Trans ASAE 47 (5), 1697–1708.

Ekinci, K., Keener, H.M., Elwell, D.L., Michel Jr., F.C., 2005. Effects of aeration strategies on the composting process. Part II - Numerical modeling and simulation. Trans. ASAE 48 (3), 1203–1215.

Ekinci, K., Keener, H.M., Akbolat, D., 2006. Effects of feedstock, airflow rate, and recirculation ratio on performance of composting systems with air recirculation. Bioresour. Technol. ISSN: 0960-8524 97 (7), 922–932. https://doi.org/10.1016/j.biortech.2005.04.025.

Elwell, D.L., Hong, J.H., Keener, H.M., 2002. Composting hog manure/sawdust mixtures using intermittent and continuous aeration: ammonia emissions. Compost Sci. Util. 10 (2), 142–149. https://doi.org/10.1080/1065657X.2002.10702074.

Fulford, B., 1983. Biothermal energy: cogenerants of thermophilic composting and their integration within food producing and waste recycling systems. In: Leeds, E.S. (Ed.), Composting of Solid Wastes and Slurries. University of Leeds, England, pp. 1–19.

Gamble, S., 2018. Compost Facility Operator Study Guide. Minister of Alberta Environment and Parks, ISBN 978-1-4601-4130-4 (PDF online).

Hodakel, B., 2021. What Is Polytetrafluoroethylene (PTFE) Fabric: Properties, How It's Made and Where. Sewport. https://sewport.com/fabrics-directory/ptfe-eptfe-polytetrafluoroethylene-fabric.

Keener, H., Hansen, R., Elwell, D.L., 1997. Airflow through compost: design and cost implications. Appl. Eng. Agric. 13, 377–384.

Kulcu, R., Yaldiz, O., 2008. Effects of air flow directions on composting process temperature profile. Waste Manag. 28 10, 1766–1772.

McGuckin, R.L., Eiteman, M., Das, K., 1999. Pressure drop through raw food waste compost containing synthetic bulking agents. J. Agric. Eng. Res. 72, 375–384.

Michel Jr., F.C., Reddy, C.A., 1998. Effect of oxygenation level on yard trimmings composting rate, odor production and compost quality in bench scale reactors. Compost Sci. Util. 6 (4), 6–14.

Notton, D., 2005. Theoretical and Experimental Determination of Key Operating Parameters for Composting Systems (Doctoral thesis). Cardiff School of Engineering, Cardiff University.

Rasapoor, M., Nasrabadi, T., Kamali, M., Hoveidi, H., 2009. The effects of aeration rate on generated compost quality, using aerated static pile method. Waste Manag. 29 2, 570–573.

Shen, Y., Ren, L., Li, G., Chen, T., Guo, R., 2011. Influence of aeration on CH_4, N_2O and NH_3 emissions during aerobic composting of a chicken manure and high C/N waste mixture. Waste Manag. 31 1, 33–38.

Shimizu, N., 2017. Process optimization of composting systems. In: Robotics and Mechatronics for Agriculture. CRC Press, pp. 1–22. https://doi.org/10.1201/9781315203638-1.

Smith, M.M., Aber, J.D., 2014. Heat recovery from compost. BioCycle 55 (2), 27–29.

Sundberg, C., Jonsson, H., 2008. Higher pH and faster decomposition in biowaste composting by increased aeration. Waste Manag. 28 (3), 518–526.

Verougstraete, A., Nyns, E.J., Naveau, H.P., 1985. Heat recovery from composting and comparison with energy from anaerobic digestion. In: Gasser, J.R.K. (Ed.), Composting of Agricultural and Other Wastes. Elsevier Applied Science Publishers, London, pp. 135–146.

Viel, M., Sayag, D., Peyre, A., Andre, L., 1987. Optimization of in-vessel co- composting through heat recovery. Biol. Waste 20, 167–185.

Xiaoxi, S., Shuangshuang, M., Han, L., Ren-quan, L., Schlick, U., Chen, P., Huang, G., 2018. The effect of a semi-permeable membrane-covered composting system on greenhouse gas and ammonia emissions in the Tibetan Plateau. J. Clean. Prod. 204, 778–787.

Zeng, J., Yin, H., Shen, X., Liu, N., Ge, J., Han, L., Huang, G., 2018. Effect of aeration interval on oxygen consumption and GHG emission during pig manure composting. Bioresour. Technol. 250, 214–220.

Zhang, S., Wang, J., Chen, X., Gui, J., Sun, Y., i Wu, D., 2021. Industrial-scale food waste composting: effects of aeration frequencies on oxygen consumption, enzymatic activities and bacterial community succession. Bioresour. Technol. ISSN: 0960-8524 320 (Part A), 124357. https://doi.org/10.1016/j.biortech.2020.124357.

Contained and in-vessel composting methods and methods summary

7

Authors: Frederick Michel[1], Tim O'Neill[2], Robert Rynk[3]

[1]*Ohio State University, Wooster, OH, United States;* [2]*Engineered Compost Systems, Seattle, WA, United States;* [3]*SUNY Cobleskill, Cobleskill, NY, United States*

Contributors: Michael Bryant-Brown[3], Van Calvez[3], Ji Li[4], John Paul[5]

[3]*Green Mountain Technologies, NE Bainbridge Island, WA, United States;* [4]*Department of Ecology and Ecological Engineering, China Agricultural University, Beijing, China;* [5]*Transform Compost Systems, Abbotsford, BC, Canada*

1. Introduction

"Contained" composting methods encompass a diverse group of methods that confine the composting materials in whole or in part within a building, container, or vessel (Table 7.1). Such methods typically include agitated bays and other system that provide frequent turning; rotating drums; silos; enclosed aerated beds/bays; tunnels and containers with forced aeration; and mechanisms for on-site composting of food residuals.

Many of the methods included in this chapter traditionally have been grouped under the label of "in-vessel" composting (IVC). However, fully enclosed "vessels" are not involved in all cases. A key feature that distinguishes true in-vessel systems from other contained approaches is that the headspace in the vessel, above the composting materials, closely follows the composting conditions (e.g., temperature, humidity, oxygen concentration). It is generally assumed that temperatures within a vessel are uniformly hot in regard to sanitization requirements. This assumption does not hold for all contained methods. In some jurisdictions state/provincial or national regulations define what constitutes and "in-vessel" method.

The word "contained" is not an entirely satisfactory label either, as it can apply to an aerated static pile (ASP) in a bunker or beneath a cover. Nevertheless, containment is a common thread among this somewhat arbitrary grouping of methods. These methods rarely expose the composting process to the outdoors, and largely separate the composting materials (and their air emissions) from the human composters overseeing the process.

With the increasing diversion of food waste from disposal to composting, contained systems are attracting renewed interest because of the isolation that they inherently provide. In many cases, a proposed composting facility could not

Table 7.1 Common contained and in-vessel composting methods.

Method	Features
Agitated bays/beds	Long channels formed between walls with rails to support the turning machine. Regularly turned automatically without a human operator. Often includes forced aeration. Inherent materials handling advantages.
Turned vessels	Fully enclosed or covered vessels with some form of automatic agitation (e.g., augers). Often includes forced aeration. Inherent materials handling is possible.
Aerated bays in halls	Large bays inside a building with forced aeration floors and automated materials handling. Turning usually occurs as materials are relocated. Often used for biological treatment of waste prior to disposal.
Vertical silos	Vertically oriented vessels or reactors in which the feedstock mix is loaded at the top. Materials move downward as compost is removed from the base. Most silos systems have forced aeration where air flows from the base upward.
Rotating drums	Horizontally oriented cylinders of widely varying diameters and lengths. The drums rotate slowly; mixing, turning, and moving the materials within. Feedstocks are added at one end, compost is removed at the opposite end. Forced aeration is common but not always used for small drums.
Aerated tunnels/boxes	Enclosed containers or concrete tunnels that are longer than they are wide. Feedstocks are loaded and composted in distinct batches. A forced aeration floor pushes air upward through the pile, and exhaust air is captured in the headspace above. Air recycling is possible. No turning or agitation occurs.
Moveable/modular aerated containers	Fully enclosed boxes or containers usually made of metal that contain a forced aeration system. Containers are generally smaller than tunnels and handle smaller batches. Most containers are moveable on site. Some are small enough to be easily transportable. No turning or agitation occurs.
On-site food waste composting systems	A variety of commercial-developed reactors designed for composting food waste from kitchens, restaurants, institutions, etc. They are fully enclosed and isolate the decomposing feedstocks from the surrounding environment. Most systems automatically mix, aerate, and agitate. Some also include dewatering and supplemental heat.

otherwise go forward. For instance, in Europe, containment is mandatory for food residuals and certain types of animal by-products (see Chapter 4).

Although contained systems are often associated with large-scale facilities that handle, they are not limited to such situations. Many systems are designed for on-site composting applications that are small in scale. In China, contained technologies are used for accelerated composting of animal manures and biosolids as part of the country's efforts to promote circular agriculture (Box 7.1).

Over the past 50 years, many IVC and otherwise contained systems have beckoned the composting industry (The Composting Association, 2004; JG Press, 1986). Some have proposed new and innovative techniques; others promised improvements to existing practices. Many are no longer marketed. Service, simplicity, and adherence to fundamental principles (composting and engineering) appear to be the primary factors that separate the successful systems from those that disappear.

Box 7.1 Agricultural applications of contained composting in China

Author: Li Ji

China has a large population to feed, and with that, a large need for plant nutrients to grow the crops that feed the Chinese people. Therefore, the country strongly advocates the return of nutrients and organic matter to cropland. China Agricultural University, which is central to this effort, has established an Organic Recycling Research Institute in Suzhou.

Composting is a large part of the organics recycling strategy because it converts difficult to handle waste materials into an organic fertilizer that is easily applied. While windrow and static pile methods are used, contained composting systems have flourished in the agricultural section because they can quickly convert organics wastes into compost that is suitable for application to cropland and gardens.

In China, the "continuous dynamic bed" (CDB) composting system has been developed for large-scale composting for manure and sewage sludge (Fig. 7.1). It is a modified version of the "agitated bay" method. It consists of mixing, feeding, turning, aeration, discharging, and odor treatment, which can be operated automatically through the central controlling system. The high-temperature period of fermentation ranges between 10 and 15 days. Two CBD demonstration plants, in Wuhan and Tongliao city, can treat 300 tons of sludge per day. As of 2018, 66 CDB plants have been established over China.

(A)  **(B)**

FIGURE 7.1

The continuous dynamic bed composting system and turning machine.

A silo-style composting reactor, developed by a Chinese company, is increasingly used for on-site treatment of livestock and poultry manure in small and medium-scale farms in China (Fig. 7.2). The composting period generally ranges between 5 and 7 days, and the temperature can reach as high as 75°C. The volume ranges between 60 and 90 m^3, which can handle 5−8 tons of manure every day. After composting, the moisture content of the compost can be reduced to less than 35%.

Continued

Box 7.1 Agricultural applications of contained composting in China—cont'd

FIGURE 7.2

The Silo composting reactor developed in China for on-farm composting of livestock and poultry manure.

The following sections profile prevalent contained composting systems 2021. Some are known by brand names, or the names of the companies that developed them. Out of necessity, some brand names have been preserved here because the systems are unique to their brands. Detailed information about all of these systems can be found on their respective websites. BioCycle (2021), Composting News (2021), and other industry periodicals provide insights about past, present, and future techniques. Trade shows, organized by composting industry organizations, are another good way to learn about current composting systems.

The final section of this chapter is a general summary of composting methods discussed here and also in Chapters 5 and 6. It is intended to help the reader sort through the primary distinctions, advantages, disadvantages, and applications among the major methods of composting.

2. Basic principles

There are a variety of contained composting systems with different combinations of vessels, aeration devices, and turning mechanisms. Although they are diverse, these methods share many of the following characteristics:

- Forced aeration and/or frequent agitation—Many, but not all, contained methods employ both forced aeration and mechanical turning to control and speed the composting process. A few methods in this category lack automated agitation. Other methods do not use fans but turn the materials frequently.

- Investment—Fans, pipe, turning devices, containers, electronics, lots of concrete, buildings, architects, and/or engineers; the elements of contained composting require a heavy investment compared to windrows and piles. However, several low investment options are also available, primarily for smaller-scale applications and on-farm facilities.
- Process isolation—Isolating the process from the ambient environment reduces the impacts of weather, emissions of odors and other volatile compounds, moisture loss, and heat loss. At least, such isolation offers the opportunities to better manage these factors. Therefore, contained systems can effectively process "challenging" feedstocks, such as those with a high odor-generating potential or high pathogen loads. Many in-vessel methods isolate the composting environment without the need for a building. Other methods only partially enclose the process and a separate building completes the containment.
- Process separation—The flip side of process isolation is the lack of direct contact between the human composter and the composting process. The physical containment blocks some or all access to the process. Depending upon the specific methods, it can be difficult for the composter to discover disparities in the materials, determine whether and where moisture is needed, or have a practical means to add water.
- Process control—In general, process control is more rigorous, often including monitoring of oxygen or carbon dioxide and moisture as well as temperature.
- Automated monitoring and data management—Many contained systems are automated to measure, record, compile, and display process variables including temperature and oxygen concentrations. Data can be readily retrieved and viewed from various locations within the boxes, tunnels, or bays displayed on computer screens, giving an operator a quick snapshot of how the whole facility is performing. The automated data can allow a facility to easily verify process requirements are being met.
- Suitable electric power—Many systems run motors that require three-phase electrical power, often at higher voltage than standard electrical service.
- Short retention time—Contained methods tend to be used for only the early phase of composting. After being discharged from the system, the partially composted material is typically finished in a windrow, ASP, or protracted curing stage. A short retention time reduces the capacity required for the in-vessel components, which have higher capital costs. Retention times in contained systems range from 3 days for rotating drums to 3 weeks for some agitated bays.
- Operator expertise—Composting methods with more technology require operators with more skill and better understanding of the technologies being used. Automation and automated process controls do not make composting automatic for the operators. At best, they only reduce routine labor. Management tasks increase, and so does the need for diligence and expertise.
- Commercial dealers—Nearly all contained composting systems are supplied and supported by a commercial vendor. Most vendors are turnkey services providers. To varying degrees, they supply the system components, its design, supervision or consulting for construction, training, troubleshooting, etc.

3. Agitated bays

While it can be argued that agitated bays hardly isolate the composting process from its surroundings, the method does provide some measure of containment. Furthermore, most agitated bay facilities are housed under the cover of a building and separate facility operators from the composting materials during the height of the process.

With agitated bays, composting takes place between walls that form long, narrow channels referred to as bays, lanes, or beds. A compost-turning machine rides on the walls and reaches down into the bed of material below. The turner not only mixes but also moves the material in the bay to a fixed distance toward the discharge point. Forced aeration is supplied from beneath the bay. To protect equipment and improve composting conditions, the bays are housed in a building or a greenhouse or, in warm climates, just covered by a roof. Although uncommon, agitated bays can be operated in the open.

The agitated bay concept began as a method for composting municipal solid waste (MSW) with the Metro-Waste system in the 1960s (Haug, 1993). It later gained traction as means to compost livestock and poultry manure. Early agitated beds were developed in the 1970s for farm applications, one at the University of Guelph (Bell and Pos, 1971) and another at the University of Maryland (Hummel, 1973) (Fig. 7.3). Although the players have periodically changed, several companies have since refined the concepts and offered commercial systems (Naylor and Kuter, 2005).

Each agitated bay company provides its own approaches to machine design, aeration, bay construction, and facility size. Most systems combine frequent turning with controlled aeration. Others also include automated temperature monitoring and control, moisture addition, and alternating positive and negative aeration. The method is versatile in its scale of application. Operations with a single bay, as small as 3 m wide by 12 m long (about 4 ft × 40 ft), are commonly used for composting poultry manure. In contrast, this method is also used for composting MSW and biosolids in buildings housing over a dozen 30 m long bays (100 ft).

(A)   **(B)**

FIGURE 7.3

Early agitated bay systems developed for agricultural applications: University of Maryland in 1977 used for cattle manure (A), and the University of Guelph (1978) used for poultry manure (B).

Source: Robert Light (A); Jack Pos (B).

Nearly all agitated bay applications operate in a continuous mode (Fig. 7.4). The feedstocks are placed at the front end of the bay by a loader or conveyor. The turning machine starts at the back end of the bay, the compost end, and moves toward the fresh feedstock end. As it advances, the machine mixes the materials and discharges the mix behind itself. With each turning, the machine moves the compost a set distance toward the compost end of the bay.

Several varieties of turners are in use, which vary with the system vendors. They differ in size and mechanical details but all of them work much like windrow turners, using rotating paddles or flails to agitate the materials and break up clumps of particles (Fig. 7.5). A rail or groove on top of each wall supports and guides a compost-turning machine. Most machines work automatically without an operator. Their travel is restricted by mechanical or electronic limit switches. Also, a turner can be manually operated, within a climate-controlled cab.

Most commercial systems include an aeration plenum recessed in the floor of the bay, which is covered with grates and/or gravel or woodchips. Between turnings, aeration is supplied by fans to aerate and cool the composting materials. Most applications employ positive pressure aeration, although negative can also be used. Since the materials along the length of the bay are at different stages of composting, the bay can be divided into different aeration zones along its length. Several fans are

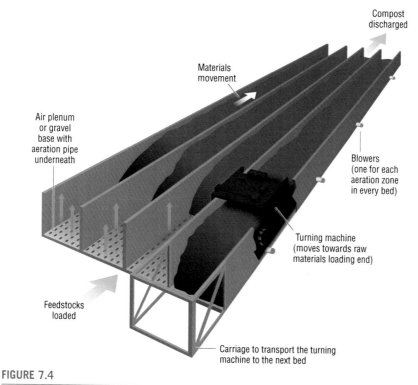

Compost
discharged

Materials
movement

Air plenum
or gravel
base with
aeration pipe
underneath

Blowers
(one for each
aeration zone
in every bed)

Turning machine
(moves towards raw
materials loading end)

Feedstocks
loaded

Carriage to transport the turning
machine to the next bed

FIGURE 7.4

General operational features of an agitated bay system.

FIGURE 7.5

An agitated bay with its turning device and aeration floor.

Source: BDP Industries.

used for one or more bays. Each fan supplies air to one zone of the bay(s). It is individually controlled by a temperature sensor or time clock. Thus, it is possible to supply negative aeration in the front portion of the bays and positive to the middle and end. Some agitated bay facilities forego forced aeration altogether. These facilities tend to have shallow bays and turn the bays more frequently (Fig. 7.6).

FIGURE 7.6

Shallow agitated bay systems without forced aeration composting various feedstocks in Hubei Province, China.

Considerable moisture is lost from the bays because of frequent turning, forced aeration, and absence of precipitation. However, turning allows water to be effectively added at several stages of the process. Some facilities have irrigation devices suspended above the bays for this purpose.

The capacity of the system depends on the number of bays and their dimensions. Bays are typically 20–30 m long (60–100 ft), although bays up to 90 m are currently in use (Nicolleti, 2020). Bay depths vary between 1 and 3 m (3 and 10 ft). The width of the bays must conform to the size of the turning machine, or vice-versa. Commercially available systems typically range in width from 1.2 to 3 m (4–8 ft), although bays, and turners, as wide as 6 m (20 ft) are available. Many agitated bay facilities have several adjacent bays. One turning machine can handle several bays, which is common in larger-scale applications. In this case, a cross track or transfer carriage shifts the turner from one bay to another (Fig. 7.7). Because the machine rides on top of the walls over such a long distance,

(A)

(B)

FIGURE 7.7

Large-scale agitated bay application with carriage to transfer the turner among the bays.

Source: BDP Industries.

the walls of any bay must be especially straight. Little deviation in the distance between the walls can be accepted from end to end.

The length of a bay, the frequency of turning, and the distance the turner moves the material determines the composting period. If the machine moves the materials 3 m (10 ft) at each turning and the bay is 30 m long, the composting period is 10 days with daily turning. It increases to 20 days if turning occurs every other day. Composting periods for commercial agitated bay systems are typically about 2 weeks, although some composters use them for up to four 4 weeks. Agitated bays are often followed by either a second stage of composting in separate piles or windrows or a long curing period.

Agitated bay systems have obvious appeal (Table 7.2). The combination of regular turning and forced aeration shortens composting time and produce compost with consistent quality. Also, this combination is advantageous when handling wet or dense feedstocks, such as where appropriate amounts of amendments are not available. With multiple bays, different feedstock mixes can be placed in separate dedicated bays, allowing them to be managed differently. A given bay can be set aside for experimental mixes. The required footprint of the system, per unit of production, is relatively small, due to the shorter composting period, the near elimination of aisle space, and the depth and rectangular profile of the composting units.

There are good reasons for placing agitated bays in a building. Protection of the process conditions and expensive equipment is obvious. In addition, turning frequently disturbs the bed of materials in the bays. Odors, ammonia, moisture, and dust are released into the space above the bays. Ammonia is especially prevalent with manure or biosolids as feedstocks. A building keeps those emissions from escaping uncontrollably. The conditions inside a room of an agitated bay system are harsh. A good ventilation system is necessary, along with corrosion-resistant construction materials. Even still, operators should not remain inside for more than brief periods (Chapter 13). Fortunately, and necessarily, the turners in enclosed applications operate automatically.

The capital expense associated with an agitated bay system, and its building, is substantial but also related to its scale. Smaller-scale systems cost roughly a few hundred thousand US dollars. Therefore, it remains a viable option for farms and other smaller-scale applications that gain value from faster and more controlled composting. Larger applications benefit from the economies of scale of big systems, but the capital costs still grow beyond the reach of most composters. Large-scale facilities invest many millions of dollars. Therefore, large-scale agitated bed applications are normally limited to municipal and commercial facilities that handle biosolids, food, and MSW.

Table 7.2 Summary of agitated bays.

Key features
• Good process control with forced aeration and moisture addition,
• Built-in materials handling and plug flow design,[a]
• Accelerates overall composting time,
• Overcomes stagnation common in static systems due to settling, air channeling, moisture variability, and inconsistent mixing,
• Produces little leachate,
• Can produce compost with a consistent texture without screening,
• Higher capital cost,
• Requires a vessel or building with corrosion-resistant materials and strong ventilation, and
• Requires engineering, site development and utility connections.

Limited to:
• Projects with a generous budget, except for small-scale applications.

Works well for:
• For most feedstocks, especially where odor control or an enclosure is advantageous,
• Where process and odor control are advantageous, and,
• As the first stage in a multistage system.

[a] *"Plug flow" designs refer to a process where materials physically travel through a reactor with little back mixing, like a "plug" in a pipe. In the case of composting, feedstocks are loaded at one end of the reactor and compost is taken out at the opposite end.*

4. Turned/agitated vessels

In a manner similar to agitated bays, several in-vessel systems convey materials through the composting vessel in a horizontal direction using an auger, paddles, or similar mechanisms. In most cases, turning moves the material through the vessel from loading to discharge points. Forced aeration may or may not be included, depending on the design. Although other details differ among these systems, they all generally provide continuous operation, frequent agitation, horizontal materials flow, and closed containment (Table 7.3).

Turned vessels share some operational features with rotating drums, including frequent agitation and horizontal conveyance of materials through the process (see Section 6). However, there also are distinct differences. Turned vessels are closer to other modes of composting in their turning mechanisms, flexibility, feedstock shredding action, and operator accessibility to the process inside the vessel. Turned vessel systems also have a longer residence time compared to most rotating drums—several weeks for the former, rather than several days for the latter. They produce compost that is closer to a finished product. Depending on the nature of the turning mechanisms, heavy or contaminated feedstocks can cause problems within turned vessels. For example, wet feedstocks tend to stick to auger flights and diminish their effectiveness. As it is with all turning devices, ropes, long fibers, and film plastics tend to wrap around shafts, and high proportions of sand and soil cause accelerated wear.

Table 7.3 Summary of turned vessel composting methods.

Key features
• Good process control with moisture addition,
• Built-in materials handling and plug flow design,
• Accelerates composting time,
• Overcomes stagnation common in static systems,
• Produces little leachate,
• Can produce compost with a consistent texture,
• Higher initial cost compared to open windrows and static piles,
• Requires corrosion-resistant materials,
• Requires some site development and utility connections, and
• Turning devices subject to wear and clogging.
Limited to:
• Relatively low-volume applications (e.g., 10 tons per day),
• Feedstocks with moderate moisture and bulk density (less than 560 kg/m^3 or about 950 lbs/yd^3).
Works well for:
• On-site composting,
• For most feedstocks, especially where odor control or an enclosure is advantageous,
• As the first stage in a multistage system, and,
• Where site appearance is very important.

Two current systems typify this approach to IVC—Earth Flow and Hot Rot. Other commercial-scale systems have comparable characteristics but tend to target on-site composting of food waste (Section 10).

4.1 Earth Flow

The Earth Flow composting system, developed by Green Mountain Technologies, is a composting vessel that uses a vertically oriented stainless steel auger to mix and move the feedstocks within and through a vessel (Fig. 7.8). This traveling auger design is the defining feature of the Earth Flow system. The auger is mounted to a gantry or carriage and extends vertically down into the bay, which is about 1.2−1.5 m deep (4.5−5 ft), depending on the system. The auger moves side-to-side along carriage rails that span the width of the vessel. The carriage also travels the length of the bay, riding on the walls. This system allows the auger to move both laterally and lengthwise within the bay. Also, the upright orientation of the auger mixes feedstock vertically, blending typically wetter material near the floor and core of the vessel with drier material near the top.

FIGURE 7.8

Earth Flow auger-turned vessel.

Source: Green Mountain technologies.

The auger's operation and movement is guided by a programmable control system that automatically positions the carriage lengthwise within the vessel and the auger laterally along the carriage. The angle of the auger is also adjustable and controls the rate of material flow from the front to the back of the vessel. The auger cycle can run multiple times per day.

The system can operate in either a batch or continuous (plug flow) mode. Feedstocks can be added to the container manually, with a tote dumping device, conveyor, or with a bucket loader, which can be driven into the vessel. The angled auger pushes composted material toward the discharge end of the container. In continuous mode, the auger gradually moves the material from the feedstock loading end toward the discharge end and eventually out the discharge door. The control system can direct the auger to move and lift material variably along the length of the vessel, adjusting for the loss of volume between the loading and discard ends of the vessel. The bay length, turning frequency, and auger angle determine the composting time, which commonly falls within the range of 14—21 days. In batch mode, the vessel is filled with fresh feedstocks all at once. The feedstocks compost in place for the desired time and the compost is then unloaded as a batch. Multiple vessels can be used in batch mode to smooth materials handling, and in continuous mode to add capacity.

The compost that leaves the vessel has typically completed a thermophilic process, but it is not fully mature. It can be further composted in a second system or cured in a separate pile or used directly if maturity is not critical.

The Earth Flow technology is offered in three configurations, as a greenhouse-covered steel vessel, a retrofitted shipping container, and as a local/site-built vessel (Fig. 7.9). Each configuration relies on the same mixing and aeration technologies. The Earth Flow auger technology also can be installed as free-standing unit within a new or existing building, in the same manner as an agitated bay.

Both the greenhouse and shipping container configuration are elongated rectangular steel containers lined with stainless steel. These vessels are available in various sizes, depending on the model. They are all roughly 2.5 m (8 ft) wide and range in length from 6 to 12 m (20−40 ft). The smallest unit holds about 15 cubic meters (20 yd^3), while the largest unit holds 34 cubic meters (45 yd^3). The daily throughput depends on the loading rate and composting time. The vessels do not include forced aeration through the composting mass. However, the headspace above the bed is ventilated, and the air can be directed to a biofilter if needed. The greenhouse and shipping container configuration are somewhat moveable.

Compared to a prebuilt vessel, the site-built configuration is more flexible in its dimensions and structure because it doesn't have to be transported over the road. Bays can be up to 6 m (20 ft) wide and 21 m (70 ft) long. Within these limits, the turning system can accommodate nearly any rectangular size. The site-built vessel walls can be constructed of concrete or wood, lined with stainless steel panels and can be installed inside of an existing building. There is also the option to include a forced aeration system in the site-built vessel.

FIGURE 7.9

Site-Built Earth Flow with concrete walls and greenhouse cover.

Source: Green Mountain Technologies.

4.2 Hot Rot

The Hot Rot composting system is horizontal, continuous, agitated, fan-aerated, enclosed, and modular (Fig. 7.10). It was originally developed in New Zealand to compost sludge from scouring wool. Because this sludge is highly degradable, process containment was a priority from its start. The system has since been used for a variety of feedstocks from yard trimmings to brewery wastes (Sierra Nevada, 2014).

The enclosed composting vessel is a long trough with a rotating shaft that runs the length of the trough. A series of tines extend outward from the shaft. The shaft rotates and intermittently on a time schedule, causing the tines to agitate and shift the material in the vessel (Fig. 7.11). Feedstocks are added to the vessel in the desired proportions by conveyor from a metered feed hopper. The shaft rotation is reversible to enhance mixing. However, the net movement caused by the tines is in the forward direction, which rolls the material toward the discharge end. At the discharge end, the rotation of the tines pushes the compost through an outlet port where it is collected and removed by a screw conveyor. Access to the vessel is limited to a few hinged portals along the vessel. Water can be difficult to add unless the vessel contains irrigation devices.

The bed of material is aerated by blowing air into the bed at selected points along the bottom of the vessel. Additionally, a suction fan evacuates air from the headspace above the bed and helps draw in the fresh air. The exhaust air moves through a condensate trap and then a biofilter (see Fig. 12.12 in Chapter 12). The aeration system can be controlled by either temperature or time.

FIGURE 7.10

Indoor hot Rot composting system at the Sierra Nevada Brewing company.

Source: Global Composting Solutions.

FIGURE 7.11

A look inside the Hot Rot vessel.

The amount of time that the material spends in the vessel depends on the rate of feedstock loading and the shaft turning scheme (direction, duration, and frequency). Retention times range from 10 to 20 days, with 10—12 days being most typical. Such short composting periods make the Hot Rot a two-stage operation. The resulting material needs additional time outside the vessel to compost and cure.

Most Hot Rot vessels are made of insulated steel and lined with stainless steel on the inside. The largest vessels can also be constructed of precast concrete. The shaft and other components are either stainless steel or treated to resist corrosion. The standard tines are made of mild steel and have an expected life of 5 years, although longer-lasting stainless steel tines can be purchased.

The composting vessels are available currently in three sizes. In its overall dimensions, the flagship unit (model 18011) is about 2.2 m wide, 2.3 m high, and 12.8 m long (7 ft × 8 ft × 42 ft). Its estimated throughput is 1.7—2.5 tonnes per day (1.9—2.8 tons), depending on particular features, and based on a 10- to 12-day retention time. The smallest unit, targeted for on-site applications, is narrower and 7.2 m long (23 ft). The large municipal-scale model is about 5 m wide, 4.25 m high, and 22 m long (16 ft × 14 ft × 72 ft). In all cases, processing capacity can be increased by adding additional units.

The Hot Rot method offers containment, agitation, and forced aeration, along with the associated benefits. The first benefit is the potentially short composting time. The composting process can be controlled and manipulated with agitation and/or turning. Materials move automatically through the first stage of the process. The reversible agitation permits back-mixing of feedstock batches, which is biologically interesting, if not ultimately advantageous. A building is unnecessary. Odors are confined, even with frequent agitation. Weather is a minor factor. Leachate is contained, if not eliminated. The primary tradeoff for these benefits is the investment required, both in money and land. The system has a relatively large footprint for its processing capacity.

5. Aerated beds and bays in buildings and halls

Several large-scale commercial systems compost premixed feedstocks in relatively wide beds or bays inside a building, sometimes referred to as "halls." Depending on the specific system, batches of feedstocks may be segregated in channels within the hall, or batches may form a single mass bed (Chiumenti et al., 2005). The composting materials typically remain in the hall until the compost is moved to the curing stage or its end use. In some applications, these systems are used for mechanical-biological treatment of waste materials that reduce the environmental impacts of wastes without necessarily producing marketable compost (Donovan et al., 2010).

Hall composting systems employ some combination of automated materials handling, moveable turning devices, and forced aeration. Materials are placed and removed by hoppers and conveyors that travel on rails and/or mobile beams. Turning devices move in a similar fashion among the bays or bed. The various systems turn the beds/bays using conveyors, augers, and rotors. Beds and bays can also be turned using mobile turners, like those used for agitated bays or extended piles. The beds and bays are nearly always aerated by fans from below. Like any composting building, the environment within the hall is harsh and must be adequately ventilated and the exhaust air should be directed through a biofilter or other treatment system.

The Paygro composting facility, near Columbus, Ohio, is an example of large-scale enclosed composting system. This facility consists of a very large, aerated bed that is fed a mixture of wet cow manure, food scraps, and sawdust (Fig. 7.12). The system has the appearance of an agitated bay but it is not; the bed is primarily static. The material can be turned in place by the conveyor when desired. A metering hopper loaded with a front-end loader fills the bed which has a depth of 3 m (10 ft) and capacity of 250 tons per day. The metering system consists of a flat bottom hopper and inclined drag conveyor, both hydraulically driven with variable speed controls. An operator controls the feed rate from the metering system to a conveyor belt located between two large bins. The center conveyor, a tripper car, and an indexing conveyor transport the feed material and deposit it evenly in the bed. A rail system mounted on the walls of the reactors permits movement of the tripper car, indexing conveyor, and extract conveyor to any point along the reactors. Once filled, the material is composted for a period of 14–21 days during which more than an 80% wet weight loss occurs.

Air is supplied to the Paygro reactors by a series of 6 kW (7.5 hp) fans positioned along the reactor walls, supplying between 85 and 142 cubic meters per minute (3000–5000 ft^3/min). The fans discharge to a plenum and an overlying gravel bed located in the bottom of the bin. A hydraulically operated extractor/conveyor,

(A)

(B)

FIGURE 7.12

Paygro systems used in Ohio for cow manure and food waste (A) and for biosolids (B). Material is mixed then placed in the bay by a conveyor. The compost is aerated from below. The compost is mixed by emptying and redistributing the bay with an extractor-conveyor.

Source: F. Michel.

made up of a rotary breaker, chain, and flight table conveyor and trailing conveyor, is used to remove material from the reactor. Once placed back on the central conveyor belt, the material can either be relocated to provide a means of turning (although this is usually only carried out once or twice) or discharged to storage and curing. This system reduces the amount of turning required compared to continuous agitated bed systems and also permits deeper piles to be used.

6. Silos

Silo-like composting reactors were among the first composting systems proposed for large-scale composting in the mid-20th century, initially for treating MSW (Diaz and de Bertoldi, 2007; Tuney, 1954). They have evolved various features since but have never become established as a mainstream composting method.

Silo composting reactors are vertically oriented, resembling a bottom-unloading grain silo (see Fig. 7.2). Each day an auger removes compost from the bottom of the silo and then feedstocks are loaded to the top, filling the space created by the descending material in the silo. Because materials receive little mixing in the vessel, feedstocks must be premixed and appropriately sized when loaded into these systems. Some systems use fans for aeration, while others rely on passive aeration. When used, the fan blows air up from the base of the silo through the composting materials (Fig. 7.13). The exhaust air can be collected at the top of the silo for odor treatment. Some silos have a series of floors so that the composting materials move down in stages. This feature reduces compaction. Temperatures are monitored at various depths in the silo.

Composting times for these types of systems are 14–28 days, so 1/14th to 1/28th of the silo volume is typically removed and replaced daily if operated in a continuous mode. After leaving the silo, the compost is cured, often in ASPs or windrows. Some systems complete a second stage of composting in a second silo.

The advantages of silo composting include automated materials handling, containment, and a small footprint. Silos minimize the land area needed because the materials are stacked vertically. However, the stacking also presents challenges with compaction, temperature control, leachate, material movement, and air distribution. Lack of access to the composting materials inside the silo is another concern. If a complication, such as low moisture or a hot spot, develops inside the vessel, the entire contents must be removed. It is also difficult to evenly distribute added water.

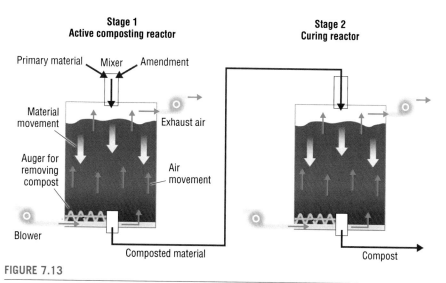

FIGURE 7.13

Diagram of a dual-stage silo composting system.

7. Rotating drums

Rotating drum reactors have long been used for composting municipal solid waste, starting with the Dano and Eweson drums in the mid-20th century (Haug, 1993). The initial drums were very long, a healthy fraction of a football-field long (Fig. 7.14). While long reactors continue to be used for composting MSW, rotating drum systems have generally downsized. They are now found in many situations for composting a variety of feedstocks including manure, animal mortalities, yard trimmings, source-separated organics (SSO), and food residues (Fig. 7.15).

Rotating drum systems use a slightly inclined horizontal steel cylinder to mix, aerate, and move compost through the system (Fig. 7.16). The drum is mounted on large bearings and turned through a large external gear. Feedstocks are loaded on one end via a conveyor and/or hopper. As the drum slowly rotates, the materials are upended and tumble toward the discharge end of the drum. The rotation can be continuous or intermittent. Some drums have blades or other projections inside for slicing open bags and breaking apart the materials as they tumble. In the drum, the composting process starts quickly because warm active compost is mixed with the fresh incoming feedstocks in the upper parts of the drum.

A fan usually supplies air through the discharge end of the drum. Oxygen is incorporated into the material as it flips over. As the air moves in the opposite direction to the material, the compost near the discharge is cooled by the fresh air. In the middle, it receives the warmed air, which encourages the process; and the newly loaded material receives the warmest air to initiate the process. Some short drums do not use forced aeration. Instead, air exchange occurs through the tumbling action and open ports at the ends of the drum.

FIGURE 7.14

Early rotating drum, Eweson Digester, in Big Sandy, TX.

(A)

(B)

(C)

FIGURE 7.15

Diverse applications of small-scale rotating drums: At a dairy farm for recycling manure into bedding (A); at a college for composting cafeteria wastes (B); at a soil products company for composting manures (C).

Source: (A, C) R. Rynk. (B) G. Evanylo.

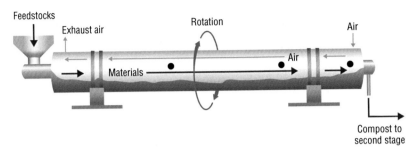

FIGURE 7.16

Flow through large-scale rotating drum.

Rotating drums can be either open or partitioned. An open drum moves the material through continuously in the same sequence as it entered. A partitioned drum, which may be divided into two or three chambers, offers more process-management options than an open drum. Each partition contains a transfer box equipped with an operable transfer door. At the end of each day, or other period of operation, the transfer door at the discharge end of the drum is opened and the compartment emptied. The other compartments are then opened and transferred in sequence, and finally a new batch is introduced into the first compartment. A threshold at each of the transfer doors retains about 15% of the previous charge to act as an *inoculum* for the succeeding batch.

Commercial-scale drums, used for MSW and/or biosolids, are huge, measuring 3—4 m (10—13 ft) in diameter and up to 40 m (about 130 ft) in length (Fig. 7.17). A drum of this size has a capacity of approximately 45 tons (50 tons) per day with a residence time of 3 days. Drums used for small-scale applications range in

FIGURE 7.17

Large rotating drum systems for composting municipal solid waste. Note the bearings in the foreground and the drive gear towards the back.

size from 1.5 to 3 m in diameter (4—10 ft) and 3—15 m in length (10—50 ft). To gain capacity and operational flexibility, two or three drums can be used in parallel.

With only a few exceptions, rotating drums are used as a first-stage composting. The residence time for rotating drums typically falls in the range of 3—5 days. Although the period in the drum is short, the high temperatures, continuous agitation, and containment of this system accelerate decomposition. Pathogen reduction requirements can be met quickly. In a few applications, the feedstocks pass through the drum in 1 day. When this occurs on a farm, the drum is used to kill pathogens and remove moisture from livestock manure. The manure is subsequently reused for animal bedding. In other cases, the drum primarily serves as a mixing device.

Besides its inherent containment, the primary advantage of a rotating drum system is rapid decomposition (Table 7.4). After 3 days in a well-working drum, the material discharged shows little evidence of vegetation, food, or paper. Woody, inert, and some large organic items remain intact. The material is drier, less "sticky," and can be more easily screened and sorted to remove physical contaminants. Further decomposition of the material is necessary to finish composting. The second stage of composting is usually accomplished in windrows or ASPs. Although the material leaving the drum requires additional composting, it has taken a big step toward stability. Therefore, rotating drums are especially advantageous for quickly converting feedstocks that attract pests into materials that can be further composted in the open.

Table 7.4 Summary of rotating drum composting systems.

Key features:
• Commercially purchased systems,
• Short retention time (3—5 days), usually followed by second stage of composting,
• Greatly accelerates early phase of decomposition,
• Second stage of composting needed,
• Almost constant agitation,
• Built-in materials handling,
• Mechanical challenges of rotating large heavy drum,
• Large capital investment, and,
• Requires engineering.

Limited to:
• Projects with a generous budget, except for small-scale applications.

Works well for:
• Most highly degradable feedstocks, especially where odor control or an enclosure is advantageous.
• Municipal solid waste, source-separated organics and small animal mortalities (e.g., poultry, fish).
• Livestock manure as a means to recover bedding.

8. Tunnels

Composting tunnels are totally enclosed vessels with a forced aeration system (Fig. 7.18). Tunnel composting systems have their origins in the mushroom industry where they are used primarily for the treatment of mushroom-growing substrate before seeding the mushrooms. Because of the containment and high degree of process control, composting tunnels are mainly used to compost highly putrescible feedstocks like SSO, food, biosolids, and MSW. Retention times are kept short (e.g., two weeks) to limit the number and size of tunnels needed. They are nearly always the first stage in a larger composting operation (Table 7.5).

A facility normally operates multiple tunnels. Each tunnel holds one batch of feedstock. The time span represented by a batch depends on the feedstock receiving/generation rate and the tunnel dimensions. Generally, heights and widths are on the order of 3—5 m (10—16 ft). Tunnels range in length from 5 to 15 m (16—50 ft). Usually, it is the tunnel length that is adjusted to accommodate the desired batch volume and time span.

Using a conveyor or bucket loader, premixed feedstocks are gradually loaded into the tunnel from back to front. The feedstocks are loosely piled within the tunnel, leaving space at the top. After the tunnel is fully loaded, the entrance door is closed and composting begins. The retention time in tunnels is typically about 2 weeks, although it can range from one to four weeks. After the desired composting period, the tunnel is emptied in the same manner as it is loaded. The partially composted material that emerges is well sanitized and no longer attractive to vectors because

FIGURE 7.18

Tunnel composting system for source-separated organic waste.

Source: ECS.

Table 7.5 Summary of composting tunnels and modular aerated containers.

Key features

- Total containment with controlled forced aeration,
- First stage of composting with moderate retention time (2–3 weeks),
- Second stage of composting needed,
- Wide range of scales (some modular),
- Usually, very good process control,
- Multiple control variables possible,
- Air and moisture recirculation possible,
- Abundant engineering,
- Requires electricity,
- Good odor and emissions control, and,
- Poor mixing and lack of agitation can lead to inconsistent composting.

Limited to:

- Projects (large and small) with a relatively generous budget.

Works well:

- As first stage of multistage system,
- For on-site composting in commercial, industrial, residential settings,
- In tight spots,
- In odor/emissions sensitive situations,
- For difficult feedstocks, and,
- Where amendments are scarce or expensive.

the tunnel easily attains the required temperatures for pathogen destruction. However, without agitation, decomposition is often uneven. The material can be remixed and reloaded into the same tunnel or a different tunnel. However, more often it is formed into a windrow or ASP to finish composting.

Positive pressure forced aeration is delivered to the feedstocks through an air plenum or set of channels built into the tunnel floor. The process exhaust air is collected from the headspace of the tunnel and channeled to a biofilter. Usually, aeration is controlled by temperature.

In many systems, a portion of the air removed from the headspace is recycled to the tunnel mixing it with fresh incoming air. Recycling the exhaust air adds a level of process control that most other composting approaches lack. It returns some of the heat and moisture emitted by the process. It also returns undecomposed gases back to the process for further composting and deodorizing. In general, recycling the air reduces the load on the biofilter. The air withdrawn from the headspace can be analyzed and monitored to gauge what proportion gets recycled and also for general process control.

Tunnel systems tend to be built on-site to the composters' specifications, so the materials and dimensions vary. Many tunnel systems are located within a building, others are constructed as a building. Some tunnels have flexible roofs. Reinforced

FIGURE 7.19

Self-constructed aerated tunnel for composing livestock manure on the farm. Water pipes built into the tunnel recover heat for warming an adjacent farm shop.

concrete, steel (protected for corrosion), and masonry are the most common construction materials, in that order. Wood has even been used for farm-based systems where occasional repair and replacement is not an issue (Fig. 7.19). Regardless of the construction, some means must be provided to drain and collect leachate from the tunnels. Often the leachate collected is sprayed back on to the pile inside the tunnel to replace lost moisture and nutrients.

9. Moveable and modular aerated containers

In essence, moveable containers are fan-aerated composting boxes, large enough to hold a batch of feedstock but small enough to be transported (Table 7.5). They have built-in forced aeration capabilities but no means of agitation. They are fully enclosed and operate in a batch and modular fashion. Different systems are transportable to different degrees. There are two main approaches. With one approach, containers are regularly moved only within the composting site. The second approach transports containers over the road between the point of feedstock generation and the composting facility.

9.1 Modular containers within a composting site

Containers that remain on site resemble detached composting tunnels (Fig. 7.20). They are constructed of insulated steel, with stainless steel or plastic lining the inside. The number of containers can be changed allowing flexibility and expandability. The containers are 2—4 m (about 6—12 ft) high and wide and 3—5 m long (10—16 ft). They hold 15—40 m^3 of material (about 20—50 yd^3). Forced air is distributed through an aeration plenum in the base of the container (Fig. 7.21). Aeration is centrally controlled and often provided alternately in forward and reverse

FIGURE 7.20

Modular containers lined up at a composting site.

Source: ECS.

FIGURE 7.21

Close-up of the aeration plenum of a modular container.

Source: ECS.

directions to minimize the development of moisture gradients. However, each container can be monitored and controlled using a single computerized control unit. The exhaust air is collected from the container's head space. The exhaust can be partially recycled and/or directed to a biofilter as needed. The biofilter can occupy another container, if desired.

FIGURE 7.22

Container tipping/unloading.

Source: ECS.

The feedstocks must be well mixed since no agitation occurs. Filling and emptying the containers can be time consuming. Some containers have hinged lids and can be loaded from the top with a bucket loader. Those that are opened only at the end are loaded with conveyors. After about 2 weeks of composting, containers are unloaded by tipping, using either special equipment or a standard roll-off collection vehicle (Fig. 7.22). The contents slide out, sometimes holding the shape of the container. At this point, the material is either taken to a second composting system or remixed and reloaded into the container.

These moveable modular container systems have similar advantages as tunnel composting systems—containment, close process control, the potential to recycle air, odor control, and batch operation. They also have similar disadvantages, including, lack of agitation and high capital and materials handling expense relative to feedstock volume. The smaller size and portability give them added flexibility. They can be used on temporary sites or temporarily while a given site is under construction. They can be used only for special feedstocks that require containment and odor control. They are particularly useful where space is limited and odor is a concern.

9.2 Transportable containers between sites

Aerated containers that are transported between sites are primarily intended to connect a central composting facility to one or more points of feedstock generation. Their primary function is to begin aeration and composting as feedstocks are collected, and control odors in the meantime. When the container is full, or on a prescribed collection date, it is picked up, loaded on a flat-bed truck, and delivered to the central facility to finish composting. At the same time, the feedstock generator is provided with another empty container to continue the cycle. The central facility empties the container and adds the contents to a windrow or ASP. The composting facility can host several feedstock generators, continuously providing one or two containers at each satellite location. This scheme works for many types of organic waste producers including farms, horse stables, grocery stores, markets, butchers, schools, and other food residual generators.

While some of the modular containers described above can be used, this approach is epitomized by a system developed in Australia called the BiobiN. This transportable vessel resembles a solid waste roll-off container. The container has aeration pipes in its base which are connected to a fan (Fig. 7.23). Positive pressure is used to aerate the contents. Exhaust air is drawn from the headspace and sent to a small external biofilter, which remains at the generation site until it needs changing. Aeration is controlled with an adjustable timer. At the generation point, the feedstocks and amendments are loaded daily into the container. The feedstocks are loaded from the top, either by hand, a mechanical bin tipper, or a bucket loader. Ideally, the feedstocks should be combined and blended with an appropriate amendment to the desired bulk density, moisture content, and C:N ratio, and then uniformly distributed in the bin. Between loadings, the lid is closed, and the contents are aerated until the container is picked up (Fig. 7.24).

Four BiobiN container sizes are currently available. While in operation at the feedstock generation site, the systems require the working area listed in (Table 7.6). Additional clearance is needed for operating equipment, like a bin tipper. The loaded-volume capacity of the bins is greater than the nominal volume because, at the point of generation, the feedstocks shrink in volume, by as much as 60% (and also lose moisture).

This approach occupies a potentially beneficial composting niche. Starting the composting process at the point of generation activates the composting early, and potentially improves the odor and vector situation from the onset, and onward. The staff of the composting facility manages the containers at the satellite locations. Ultimately, however, success depends on the diligence of the feedstock generators.

FIGURE 7.23

Transportable BiobiN with detachable aeration and biofilter unit stationed at an airport (Dubai).

Source: Peats Soils.

FIGURE 7.24

An empty BiobiN being loaded at the central composting site for transport to the waste generation site.

Table 7.6 BioBiN unit capacities and dimensions.

Nominal volume	Weight capacity	Dimensions length × width × height	Work area requirement
2 m³ (2.5 yd³)	1.3 tonnes (1.5 tons)	3.5 × 1.5 × 1.1 m (11 × 5 ft × 3.5 ft)	6 m² (65 ft²)
4.5 m³ (6 yd³)	3 tonnes (3.3 tons)	3.8 × 1.75 × 1.4 m (12.5 6 × 4.5 ft)	16 m² (172 ft²)
9 m³ (12 yd³)	6 tonnes (6.5 tons)	4.5 × 1.75 × 1.7 m (21.3 × 7.9 × 4.6 ft)	23 m² (248 ft²)
20 m³ (26 yd³)	14 tonnes (15.3 tons)	6.5 × 2.4 × 1.4 m (21.3 × 7.9 × 4.6 ft)	36 m² (388 ft²)

They and their staffs must fully accept the system, and learn something about composting. They must be willing to properly work the containers, regularly monitor them, remain alert to odors, add only appropriate well-balanced feedstocks, remove contaminants, and communicate with the composting facility. The feedstock generators and their staffs have to be trained and retrained. The system works best when one person at each satellite location is given the responsibility for the operation and takes ownership in it.

10. Methods for on-site composting of food waste

Driven by efforts to recycle food wastes, entrepreneurs have developed a family of small- to moderate-scale IVC machines designed for composting food scraps near the point of generation. Typical applications include restaurants, markets, multiresidential housing unit, schools, colleges, hospitals, and other institutions.

Most on-site composting machines aim to produce only partially stable compost in one to two weeks. A few strive for a four-week period. Often, only the initial stage of composting takes place on site. The process is either completed at a remote location or the compost is used without being fully mature. Legitimate on-site composting machines should not be confused with food waste recycling systems that merely macerate, cook, dewater, and/or dry the food waste. Any approach that promises biologically stable product in a few days is not composting.

On-site processing has great appeal for food waste composting. The feedstocks can be immediately processed with minimal handling, storage, and transportation. The on-site approach also has big risks, including safety, odors, flies, unsightly conditions, hygiene, space limitations, and lack of dedicated operators. On-site composting machines manage these risks, in part, with automation, enclosures, and/or biofilters.

The on-site food composting machines share certain characteristics (Fig. 7.25). They are generally intended to occupy a relatively small, protected area near the source of the food waste. They are enclosed units that integrate turning, aeration,

FIGURE 7.25

Four examples of vessels designed to compost food waste at the point of generation: (A) Big Hanna composter; (B) Earth Tub (no longer commercially available); (C) Rocket composter; and (D) Automated vessel developed in China.

Sources: (A) Berca Brand Srl./Cecilia Ek (A); Tidy Planet Ltd. (B) and R. Rynk (C, D).

and other operations within the envelope of the machine. Other operations might include mixing, shredding, loading, materials handling, and even screening. Nearly all of these machines operate in a continuous mode, relying on the turning mechanism to move the feedstock through the machine. That mechanism can be an auger, a rotating drum, or a rotating shaft with tines or paddles. Additional processing capacity is sometimes gained by adding heat and optional preprocessing operations like shredding and dewatering. Some manufacturers offer models that can be used on a sizable scale; for instance, servicing the needs of a food processing company. More details about composting systems on this scale can be found in McSweeney (2019).

11. Summary: comparing the composting methods

The selection of the appropriate method to use for a given composting application depends on a variety of factors, such as location, scale of operation, the consequences of occasional odors, access to capital, the nature of the feedstocks, environmental regulations, labor availability, and business objectives. As discussed in this and the previous two chapters, each composting method has its benefits and drawbacks, which need to be assessed on a case-by-case basis (Australian Government, 2012).

In terms of overall cost and ease of entry, the passively aerated pile, turned windrow, and ASP methods are on comparable level. With the possible exception of simple agitated bay systems, contained composting systems require much greater investment. However, they are usually more efficient in terms of labor and processing speed, and may be practical for larger operations, municipalities, and farms. For small to medium-sized operations, the choice of a composting method usually reduces to turned windrows, ASPs, and variants of ASPs (e.g., ASP in a bin).

Turned windrow composting is the most commonly used method among all composting operations, especially for yard trimmings, manure, and other agricultural feedstocks. Such feedstocks present less odor risk. In the rural setting, where windrows composting is common, land is not usually limiting, and odor risks are lower. In some cases, windrows can be built in agricultural fields where the compost may later be applied. Since no electricity is required, remote sites can be used. Windrow composting is similar in nature to other municipal and farming operations, and existing equipment can be used. Windrow composting also allows a greater choice of amendments. The turning process continues to mix and granulate the composting material, which produces a more uniform compost and reduces the need for secondary operations like screening and grinding.

A major disadvantage with windrow composting is the relatively large area required (i.e., footprint). The larger footprint means that there is more runoff to manage. Open windrows are at the mercy of the weather: rain, snow, and mud. The factors are more likely to cause problems with windrows than passive piles or ASPs. Asphalt or concrete surfaces and open-sided buildings have been added to some windrow facilities to better cope with adverse weather, but at a significant expense. Permeable windrow covers are also used to reduce weather-associated problems.

The ASP method is a more space-concentrated means of composting. It allows higher, broader piles and, therefore, requires less land area than either windrows or most passively aerated piles. The larger piles lessen weather impacts on open systems. The reduced area minimizes pad costs and makes it feasible to cover the system with a roof or enclose it within a building. Forced aeration accommodates automation, permits closer process control, and shortens the composting period. The biocover plus the larger pile size combine to reduce temperature variations, which improves conditions for destroying pathogens. The biocover and lack of turnings conserve nitrogen and limit the release of odors. With a suction aeration system, odors can be collected and treated. For all of these reasons, the ASP method is common among biosolids composting facilities. A disadvantage of ASP is the general lack of agitation. It can suffer from moisture gradients, short circuiting, and channeling of the airflow, which produces an unevenly composted product.

The passively aerated pile method shares features of both turned windrows and ASPs. Like the windrow method, it does not require electricity or engineering. Like ASPs, it has a relatively small area requirement and thus reduces the amount of runoff to be managed. Passive aeration can support aerobic conditions with well-balanced feedstocks (e.g., degradable, heat-generating, and low bulk density). Otherwise, the lack of forced aeration and regular turning can lead to varied and anoxic conditions within the piles. The primary consequences are slower decomposition and inconsistent compost. The larger passively aerated piles lose heat and moisture more slowly, which can be advantages or disadvantages depending on the circumstances. The greater potential for spontaneous combustion is the primary concern. The absence of turning, particularly early in composting process, tends to contain odors and shelter putrescent feedstocks. For this reason, passive piles are the usual choice for composting animal mortalities (see Chapter 8).

Contained and IVC methods are commonly commercial systems that are purchased, licensed for use, and/or specially designed by consultants. They usually carry high capital costs. Operation and maintenance also require greater expense and a higher level of knowledge and skill than the windrow and aerated pile methods. In exchange, contained systems offer several potential advantages including automated materials handling (in part), few weather problems, potentially better odor control, closer process control, faster composting, and consistent compost quality. However, a contained or IVC system does not, in itself, assure that these advantages will be realized. Many problems still occur, which often involve more expensive solutions. An increasing number of facilities handling food residues may find one of the lower-cost IVC systems worth the investment, especially if rapid decomposition is important. Agitated bays, turned vessels, rotating drums, and fan-aerated bays and tunnels, in particular, have found a place on small to moderate-scale operations including colleges, farms, municipalities, and food-processing facilities.

References

Cited references

Australian Government, 2012. Understanding the processing options. In: Food and Garden Organics Best Practice Collection Manual (2012). Department of Sustainability, Environment, Water, Population and Communities. https://www.environment.gov.au/system/files/resources/8b73aa44-aebc-4d68-b8c9-c848358958c6/files/collection-manual-fs5.pdf.

Bell, R.G., Pos, J., 1971. Winter high rate composting of broiler manure. Can. Agric. Eng. 13 (2), 60−64. https://library.csbe-scgab.ca/docs/journal/13/13_2_60_ocr.pdf.

BioCycle, 2021. The JG Press, Emmaus, PA. https://www.biocycle.net/.

Chiumenti, A., Chiumenti, R., Díaz, L., Savage, G., Eggerth, L.L., Goldstein, N., 2005. Modern Composting Technologies. JG Press, Emmaus, PA, ISBN 0-932424-29-5.

Composting News, 2021. McEntee Media. http://compostingnews.com/.

Diaz, L.F., de Bertoldi, M., 2007. History of composting. In: Compost Science and Technology. Elsevier Science.

Donovan, S.M., Bateson, T., Gronow, J.R., Voulvoulis, N., 2010. Characterization of compost-like outputs from mechanical biological treatment of municipal solid waste. J. Air Waste Manag. Assoc. 60 (6), 694−701. https://doi.org/10.3155/1047-3289.60.6.694.

Haug, R.T., 1993. The Practical Handbook of Compost Engineering. CRC Press, Boca Raton, FL.

Hummel, J., Equipment for composting of agricultural wastes, 1973. American Society of agricultural engineers. In: North Atlantic Regional Conference. August 1973. Orono, ME.

JG Press, Inc, 1986. The BioCycle Guide to In-Vessel Composting. Emmaus, PA.

McSweeney, J., 2019. Community-Scale Composting Systems. Chelsea Green Publishing, White River Junction, VT.

Naylor, L., Kuter, G., 2005. Agitated bed composting: past, present, and future. BioCycle 46 (7), 52.

Personal Communication Nicolleti, R., 2020. Compost Equipment Manager, BDP Industries. Sierra Nevada, 2014. Four years of super composting. https://www.bdpindustries.com/about-us/.

Sierra Nevada, 2014. Four Years of Super Composting. Sierra Nevada Brewing Co. https://sierranevada.com/blog/four-years-of-super-composting/.

The Composting Association, 2004. A Guide to In-Vessel Composting. The Composting Association, Wellingborough, England.

Tuney, W.G., 1954. A study concerning the establishment of a yardstick for garbage digestion by composting. Masters Thesis. Michigan State College. https://d.lib.msu.edu/etd/15928/datastream/OBJ/view.

Consulted and suggested references

Bonhotal, J., 2011. In-vessel composting for medium-scale food waste generators. BioCycle.

David, A., 2013. Technical Document on Municipal Solid Waste Organics Processing. Canada Minister of the Environment, Ottawa. https://www.ec.gc.ca'gdd-mwPDF.

Ekinci, K., Keener, H.M., Elwell, D.L., Michel Jr., F.C., 2005. Effects of aeration strategies on the composting process. Part II - Numerical modeling and simulation. Trans. ASAE 48 (3), 1203−1215.

Garrick, 2021. Paygro. https://www.garick.com/paygro.

Global Composting Solutions, 2021. Hot Rot Technology. https://www.globalcomposting.solutions/hotrot-technology.

Goldstein, N., 2021. Cannabis waste composting. BioCycle. https://www.biocycle.net/cannabis-waste-composting/.

Green Mountain Technologies, 2021. The Earth Flow Custom Vessel Composting System. https://compostingtechnology.com/in-vessel-composting-systems/earth-flow-custom-vessel/.

Hummel, J.W., Schwiesow, W.F., Willson, G.B., 1974. Trans. ASAE 17 (1), 0070−0073. https://doi.org/10.13031/2013.36790.

Kapnas, V., 2012. Springfield B2B: Paygro Turns Other People's Trash into a Profitable Commodity. Springfield News-Sun, May 6, 2012. https://www.springfieldnewssun.com/business/springfield-b2b-paygro-turns-other-people-trash-into-profitable-commodity/ARpU7AhvR8diPHI02pEFUI/.

Keener, H.M., Ekinci, K., Elwell, D.L., Michel Jr., F.C., 2005. Composting process optimization - using on/off control. Compost Sci. Util. 13 (4), 288−299.

Lindberg, C., 1996. Accelerated composting in tunnels. In: de Bertoldi, M., Sequi, P., Lemmes, B., Papi, T. (Eds.), The Science of Composting. Springer, Dordrecht. https://doi.org/10.1007/978-94-009-1569-5_141.

Noble, R., Gaze, R.H., 1998. Composting in aerated tunnels for mushroom cultivation: influences of process temperature and substrate formulation on compost bulk density and productivity. Acta Hortic. 469, 417−426. https://doi.org/10.17660/ActaHortic.1998.469.44.

SKM Enviros and Frith Resource Management, 2013. Guidance for On-Site Treatment of Organic Waste from the Public and Hospitality Sectors. Waste and Resources Action Programme (WRAP), Banbury, UK.

The Composting Association (UK), 2008. The buyers' guide to in-vessel and anaerobic digestion technologies. Winter 2008 12 (4). http://www.organics-recycling.org.uk/uploads/article1762/Buyers%20Guide%20to%20in-vessel%20and%20AD.pdf.

Traco Iberia, 2021. Vertical Composting Unit VCU. Traco Iberia S.L., Madrid. https://www.tracoiberia.com/english-version/vertical-composting-unit-vcu/.

USCC, 2021. Compost Equipment Guide for Small-Scale & Institutional Use. U.S. Composting Council. https://cdn.ymaws.com/www.compostingcouncil.org/resource/resmgr/images/USCC_Equipment_Guide_for_web.pdf.

Waste-C Control, 2021a. Composting 5: Composting in Closed Halls (Bays). http://www.epem.gr/waste-c-control/database/html/Composting-05.htm.

Waste-C Control, 2021b. Composting 4: Composting in Boxes. http://www.epem.gr/waste-c-control/database/html/Composting-04.htm.

Composting animal mortalities*

Authors: Jean Bonhotal[1], Mary Schwarz[1], Robert Rynk[2]

[1]*Cornell Waste Management Institute, Cornell University, Ithaca, NY, United States;* [2]*SUNY Cobleskill, Cobleskill, NY, United States*

Contributors: Johannes Biala[3], Jane Gilbert[4], Robert Michitsch[5]

[3]*Centre for Recycling of Organic Waste & Nutrients (CROWN), The University of Queensland, Gatton, QLD, Australia;* [4]*Carbon Clarity, Rushden, Northamptonshire, United Kingdom;* [5]*University of Wisconsin—Stevens Point, Stevens Point, WI, United States*

1. Introduction

Composting of animal mortalities is a specialized application of composting. The goal is to dispose of animal carcasses and animal parts in a secure, economical, and environmentally-benign manner. The composting system is dedicated to treating and managing the mortalities. The production of compost is incidental; except that the compost is a safe and beneficial medium for recycling animal remains.

Disposing of animal mortalities is a natural part of animal agriculture; however, it is not limited to on-farm applications. Meat processors and distributors, the fishing industry, public workers, and environmental mangers routinely need to dispose of carcasses or other animal by-products. As traditional methods of disposal—burial, incineration, and rendering—have lost favor, increased in cost, or are no longer a viable disposal option; farmers, butchers, and public works departments are finding it increasingly difficult to find biosecure, inexpensive, and environmentally safe disposal options. While renderers once paid farmers for animals, in many locations, the farms are now being charged for the service, or the service is unavailable. Many rendering operations have been curtailed by disease concerns, chemical and physical contamination, and the economy that has caused fluctuation in prices for hides, tallow, meat, bone meal, and the other useful commodities derived from animal carcasses. Some butchers process animal species (e.g., goats, sheep, ostrich, deer, and wild hog) that are not acceptable in the rendering stream so they are limited in their disposal options. The scarcity and cost of rendering is probably the biggest factor in the growing reliance on composting as an acceptable option for processing and disposing of animal carcasses. Hence, composting

---

* An early version of this chapter was issued by the Cornell Waste Management Institute under the title, "Composting Animal Mortalities" (Bonhotal et al., 2014).

The Composting Handbook. https://doi.org/10.1016/B978-0-323-85602-7.00004-2

provides an alternative to traditional carcass disposal as it is biosecure, self-sufficient, and can be less expensive.

Mortality composting practices are used for a number of purposes including handling of routine livestock and poultry mortalities, animals killed in catastrophic events (e.g., disease outbreaks, storms), road kill, beached sea mammals, meat from processers (e.g., butchers), fish from farms, docks and processing operations, and mass elimination of exotic pest animals (e.g., Australian National Carp Control Plan). The common factor among these applications is that they involve concentrated amounts of animal tissue, with high proportions of protein and moisture. Typically, organs, flesh, hides, feathers, and bones may be included. Handling and composting these materials demand care and special practices to accommodate their challenging properties and to control disease, odors, flies, scavengers, and leachate.

When considering the disposal of animal carcasses and by-products, one should remember that composting is not necessarily the most desirable method. All things equal, preference should be given to processes that recover nondiseased animal by-products for food or feed for animals and render them into commercial products. As an example, the raw and basted dog bone business is thriving in the United States and may offer outlets of local markets to farmers and butchers.[1]

This chapter provides an overview of composting animals and animal tissue in order to familiarize the reader with the associated applications, requirements, methods, and management practices. The chapter focusses on the most common application of mortality composting—treating animal carcasses, organs, and tissue that are generated on a regular and routine basis. Composting of animal carcasses generated in large numbers, from catastrophic events is a special case (Box 8.1). For readers who wish to pursue the composting option, numerous publications, video programs, and web sites are available that provide many details, such as state regulations, step-by-step procedures for various applications, and instructions for building composting windrows and structures A good starting point is the Cornell Waste Management Institute (CWMI, 2021), which guides readers to numerous resources including research articles, fact sheets, video programs and posters (see Consulted and Suggested References).

> ### Box 8.1 Composting animals from a mass casualty or emergency
>
> Death of large numbers of animals can occur for many reasons including fire, lightning, barn collapse, power interruptions, disease outbreaks, floods, fish kills, and marine mammal beaching. Composting can provide a timely solution for any number of animals if space and bulking materials are available.
>
> The procedures and protocols that are followed depend on the reason for the large loss of animals. In the worst case, large numbers of animals are intentionally euthanized to contain a contagious disease. In that situation, the composting procedures may be dictated, and possibly supervised, by

---

[1] By-products from animals that have been chemically euthanized CANNOT go into the pet food market.

Box 8.1 Composting animals from a mass casualty or emergency—cont'd

animal health authorities. In most other situations, the farmer or professional composter is responsible for composting procedures, although assistance may be available from public agencies, and regulations usually govern what must and must not be done.

In some situations, mass mortalities can be transported off-site to a local commercial, municipal, or farm composting facility. However, it is often more convenient, and certainly more biosecure, to compost moralities on site, especially on farms. Two approaches to composting mass mortalities are common—outside piles and "in-house" piles. In the latter case, composting piles are built in the building or barn in which the animals died (see Fig. 8.1). It is the most biosecure option where there is a concern about transmitting disease outside of the affected building.

FIGURE 8.1

In-house composting of poultry mortalities using a passively aerated pile.

Source: Eric Bendfeldt.

Whether animals are composted inside or outside, mass mortalities are constructed by the same pile building techniques and turning regimes described in this chapter for routine mortalities. The primary difference is the size of the batch. A large volume of dry low-density amendments are necessary.

Finding enough appropriate amendments can be tough in emergency situations, so a list of possible sources of amendments should be kept on site and up to date. In the event that conventional amendments are scarce, the Department of Agriculture in Maine recommends using immature biosolids compost as a supplemental amendment (Seekins, 2011). The heat of the biosolids compost jump-starts the mortality composting process.

Organization is the most important factor in responding to such an emergency. Plans should be developed with multiple disposal tools so that response time will be quick and effective. If there is a disease issue where typical thermophilic temperatures may not provide sufficient disease or pathogen control, additional measures can be taken.

Planning for potential catastrophic losses of animals should include gathering information from the governing agricultural and/or environmental government agencies as well as agricultural universities. A good deal of information for composting mass mortalities is available from national, state, and provincial publications like Ritz (2017), NC DAS (2016), and Tablante et al. (2004). At the time of this writing, an instructive series of video programs on mass mortality composting is accessible via YouTube, produced by Oklahoma State University Cooperative Extension Service (OkState DASNR, 2018).

2. Mortality composting—basic principles

To state it bluntly, mortality composting deals with animal flesh—whole dead animal carcasses and/or animal parts such as afterbirth, offal, organs, blood, fish wastes, bones, feathers, hides, and raw meat from butchers.[2] Composting this flesh waste differs from other categories of composting. While most of the fundamental principles are the same, specific practices evolved for composting mortalities for two reasons: first, the production of salable compost is usually not a primary goal and second, there is a higher concern for disease, hygiene, odors, flies, vermin, and scavengers.

Procedures for composting animal mortalities vary with the size of the carcass and the scale of the task; that is, the number of carcasses handled in each batch. The most common situation is composting of routine mortalities—the fraction of the flock, herd, or wildlife that periodically die and require ultimate disposal. Depending on the type of facility, and its size, each loading can range from a single large cattle carcass to several dozen fish or birds. Small carcasses can be loaded by hand, while a tractor or a loader is required to handle large animal carcasses or large quantities of small carcasses.

Farm mortalities and similar materials are composted by a number of different methods. Nearly all methods share distinct procedures that isolate the carcasses from the environment until they are well decomposed. In-vessel composting (IVC) in containers and rotating drums separates the carcasses through confinement in enclosed spaces. Passively aerated static piles, the most common approach, achieves that isolation by adding abundant amendment that facilitates air circulation. The carcasses are wholly contained within amendments (Fig. 8.2).

Most mortality composting methods use a two-stage process. With in-vessel methods, the first stage takes place in the vessels and it is relatively short. The second stage is more like conventional composting. With the passively aerated static pile method, the carcasses are layered or placed within amendments and then they are left to decompose undisturbed for weeks or months through stage one. Afterward, they may be turned, and composting continues through stage two for several months longer. The passively aerated static pile method is by far the most common approach and receives the most emphasis here (Box 8.2).

---

[2] For simplicity, within this chapter these materials are generally included in the terms carcass and mortalities, unless otherwise specified.

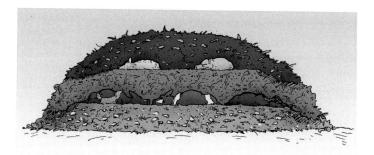

FIGURE 8.2

The most common method of composting animal mortalities and butcher waste is the passively aerated static pile method, which envelops the carcasses with dry porous amendments.

Source: Cornell Waste Management Institute. Artwork by Bill Davis.

Box 8.2 How to recycle your leviathan

In 1999, a Northern Right Whale in the North Atlantic became severely entangled in fishing equipment. About 6 months later, the whale was found dead off the coast of New Jersey. The US Coast Guard hauled the 15,000 kg (30,000-pound) whale to shore. A call went out to museums to see if there was interest to preserve this endangered whale species in some way. The Paleontological Research Institute (PRI) in Ithaca, NY, requested the whale carcass, primarily to recover the skeleton.

In New Jersey, some of the flesh and blubber was cut away from the carcass. The remaining carcass was hauled on a flat-bed truck to Ithaca. Behind PRI, the whale was laid in a large bed of horse manure and completely covered and left to compost in a large windrow. The pile was left for 6 months (October—April) and gently uncovered so the bones could be tagged. The pile was then turned by hand. The bones, bits of flesh and skin were again covered and left until October. With many volunteers, the bones were cleaned and weighed and ready to be assembled. In 1 year, the bones showed signs of pitting and degradation, suggesting that, for preservation purposes, the bones should have been removed from the pile sooner. The whale skeleton that was composted on their site is on display at the PRI.

Although composting a whale carcass still remains a subject of popular interest (Hubert, 2019; Reeves, 2017), the practice has lost a bit of its novelty. Composting is now a conventional option for recovery and/or disposal of whales and other marine mammal carcasses that are found along shorelines. For example, roughly five whale carcasses are composted annually in the state of Maine due to stranding, entanglement, disease, and other causes (King, 2021). The typical approach is to cut the carcass into pieces on or near the beach and then transport the pieces to a commercial or farm composting facility for composting (Fig. 8.3). Cutting the carcass on the beach is also necessary for conducting a necropsy.

With support from NOAA,[3] the University of New England's Marine Animal Rescue Center collaborated with Maine Department of Environmental Protection to study and evaluate

Continued

Box 8.2 How to recycle your leviathan—cont'd

composting as an option for marine animal carcasses. The effort resulted in the manual "A Guide to Composting Marine Animal Mortalities" (King et al., 2018).

FIGURE 8.3

(A, B) Whale carcass found in Trenton, Maine being cut during a necropsy (A) and for transport to a farm composting site where it was composted by encasing the carcass pieces in biologically active horse manure with wood shavings bedding (B).

Source: Mark King.

[3] National Oceanic and Atmospheric Administration (United States).

---

3. Pathogen elimination, risk management, and regulatory requirements

Warning: If there is suspicion that animals may have died from a reportable/notifiable or exotic disease, including prion diseases, the appropriate authorities should be contacted prior to making any decisions on disposal.

3.1 Pathogen elimination

Composting is known to control nearly all pathogens—viruses, bacteria, fungi, protozoa (including cysts), and helminth ova to acceptably low levels (Table 8.1). Exceptions to this are some spore-forming bacteria (e.g., *Bacillus anthracis*, "anthrax") and prions like bovine spongiform encephalopathy (BSE) and chronic wasting disease (CWD). Although *Bacillus* spp. spores and prions are highly resistant to both physical and chemical means of inactivating pathogens, there is some evidence that the microbial community in thermophilic compost has the ability to degrade recalcitrant proteins such as prions (Box 8.3). However, as stated in the warning above, the appropriate authorities should be contacted prior to making any disposal decisions when dealing with prion diseases.

Table 8.1 Inactivation or degradation of pathogens during composting.

Pathogen	Inactivation/survival during composting	References
Avian encephalomyelitis (AE)	Inactivation within 1 week	Glanville et al. (2006)
Avian influenza virus (AIV)	Not detected after 10 days Inactivation by end of composting infected birds Inactivation by 21 days 29 min to 6.4 h for 12-$\log_{10}$ reduction depending on temperature Infectivity loss between 15 min and two days depending on temperature	Senne et al. (1994) Flory and Peer (2009) Guan et al. (2009) Elving et al. (2012) Lu et al. (2003)
Bovine viral diarrhea (BVD) virus and classical swine fever (CSF) virus	Virus inactivation by the time the compost reached 35°C (95°F)	Guan et al. (2012)
Escherichia coli O157:H7	Not detected after seven days of composting	Xu et al. (2009)
Egg drop syndrome-76 adenovirus (EDS-76)	Not detected after 10 days	Senne et al. (1994)
Foot and mouth disease (FMD), picornavirus, and infectious bursal disease virus (IBDV)	No viral RNA detected in any specimens collected from compost at day 21	Guan et al. (2010)
Newcastle disease virus (NDV)	Rapid reduction of NDV during composting Inactivation by 21 days Inactivation within 1 week	Benson et al. (2008) Guan et al. (2009) Glanville et al. (2006)
Poliovirus and hepatitis A virus (HAV)	4 $\log_{10}$ inactivation in manure stored at 40°C (104°F) for 20–21 days.	Guardabassi et al. (2003)

Prions (PrP)
  PrP^{SC} (scrapie)
 	
  PrP^{TSE} (bovine spongiform encephalopathy)
  PrP^{CWD} (chronic wasting disease)
	Not detected after composting in the tissue remnants or surrounding sawdust
	Not detectable after 14 or 28 days of composting
	Reduced by at least 2-$\log_{10}$ after 28 days
	Reduced by 1-$\log_{10}$ after 28 days
	Reduced by 1-2-$\log_{10}$ after 28 days
		Hongsheng et al. (2007)
		Xu et al. (2013)
		
		Xu et al. (2014)
		Xu et al. (2014)
		
		Xu et al. (2014)

Pseudorabies virus (PRV)	Virus not detected in compost samples on days 7 and 14; pigs did not develop signs related to PRV; all tissues tested negative	Garcia-Sierra et al. (2001)
Salmonella enterica serotype Senftenberg	Not detected after 6–10 weeks	Collar et al. (2009)
Suid herpesvirus (SuHV-1)	Survival time of virus ranged from 34 to 44.5 h at 48°C (118°F)	Paluszak et al. (2012)

Box 8.3 Prion diseases

BSE or "mad cow disease," CWD in deer and elk, and Scrapie in sheep and goats are all diseases called transmissible spongiform encephalopathies (TSEs). TSEs are purported to be caused by prions. A prion is a protein chain that has become folded in an abnormal manner. It can infect and replace normal protein chains in the brain of an animal. Transmission among same and different species is not completely understood with prion diseases. Disabling this protein chain is difficult: it takes exposure to heat of 1000−1100°C (1800−2000°F) or alkaline digestion. Neither of these conditions occurs in a compost pile.

However, there is some evidence that the composting microbial community might have an effect on spore-formers and prions. Reuter et al. (2012) performed compost studies that investigate microbial communities linked to biodegradation. *Bacillus* spp. spores (related to anthrax outbreaks) carry exceptional resistance to heat, but spore survival times were magnitudes lower when exposed to wet-heat in compost as compared to dry-heat. Their data revealed that under composting conditions, a million-fold inactivation of Bacillus spores occurred and residual spores within compost biocontainment are unlikely to remain at an infectious concentration due to dilution. In addition, the use of molecular biology and microbiological assays revealed biodegradation of specified risk materials (SRMs) and a wide range of pathogens in combination with physiochemical compost conditions.

Hongsheng et al. (2007) investigated whether the abnormal prion protein (PrPSc) in tissues from sheep with Scrapie would be destroyed by composting. Before composting, PrPSc was detected in all the tissues by Western blotting (WB), but not detected in the first experiment after composting. It was detected in the second experiment but analysis showed there were more diverse microbes involved in experiment 1 than in experiment 2. It was suggested that the greater dominance of thermophilic microbes in experiment 1 may have value as a means for degrading PrPSc in carcasses and other wastes. In another experiment using PrPSc over 28 days in laboratory-scale composters, Xu et al. (2013) showed that prior to composting, PrPSc was detectable in manure with 1-2 log10 sensitivity, but was not observable after 14 or 28 days of composting. The authors state that this may have been due to either biological degradation of PrPSc or the formation of complexes with compost components that precluded its detection.

According to Xu et al. (2013), recent evidence has indicated that some bacterial proteinases exhibit the ability to degrade BSE prions, (PrPTSE). The bacterial species capable of this activity have been shown to be associated with compost. Also, their research group previously isolated a novel keratinolytic actinobacteria involved in the degradation of hoof keratin during composting. The microbial consortia in compost could carry out the biodegradation of recalcitrant proteins such as keratin or possibly PrPTSE, owing to the wide range of proteolytic enzymes produced by these complex microbial communities. Poultry feathers added to compost produced effective nonspecific proteolytic activity early in the composting process and promoted the growth of specialized keratinolytic fungi that degraded keratin in feathers during the latter stages of composting. Inclusion of feathers altered the composition of microbial community within the compost matrix, resulting in the establishment of communities that were more adept at degrading keratin and SRM. In 2014, Xu et al. further investigated degradation of prions associated with scrapie, CWD, and BSE in lab-scale composters and scrapie in field-scale compost piles. WB indicated that the prions for scrapie, CWD, and BSE were reduced by at least 2 log10, 1-2 log10, and 1 log10 after 28 days of lab-scale composting, respectively. Further analysis by protein misfolding cyclic amplification confirmed a reduction of 2 log10 in scrapie prions and 3 log10 in CWD. Addition of feather keratin (for proteolytic microorganisms) enhanced degradation of both scrapie and CWD prions. In field-scale composting, scrapie prions were removed periodically for bioassays in Syrian hamsters. After 230 days of composting, only one in five hamsters succumbed to TSE disease, suggesting at least a 4.8 log10 reduction in scrapie prion infectivity. Their research findings show that composting reduces TSE prion resulting in one 50% infectious dose remaining in every 5600 kg of final compost for land application. Microbial activity is likely part of the destruction of TSE prions (greater reduction in field scale which had longer periods of temperature above 55°C,

---

Box 8.3 Prion diseases—cont'd

than lab scale where it was only 1 or two days of >55°C—temp indicating greater microbial activity). Addition of chicken feathers (composed of β-keratin) enhanced protease activity in compost and promoted the growth of specialized keratinolytic fungi with the capacity to degrade feathers. As TSE prions share some structural similarities with feathers (both are rich in β-sheets), this may have helped in the destruction of prions.

Pathogens are killed during composting by multiple means, such as high temperatures, the direct and indirect effects of other microorganisms, and the presence of organic acids and ammonia in the compost process. Not only is temperature considered the most important factor in killing pathogens, it is also relatively easy to measure during composting. The heat required for the inactivation of pathogens is a function of both temperature and length (time) of exposure. Exposure to an average temperature during composting of 55−60°C (130−140°F) for a couple of days is usually sufficient to kill the vast majority of pathogens. To achieve efficient pathogen kill, all materials in a compost pile must be exposed to high temperatures for prolonged periods. In piles, there is great variation in the temperature profile from the cool outside layers to the hot central mass. The core of the composting pile is kept hot by a cap or cover of amendment (Fig. 8.4).

Diseases like avian influenza (AI) and foot and mouth disease (FMD) are caused by highly infectious viruses that are easily transmitted through the air. Extra precautions must be adopted that prevent the virus from spreading, such as composting in situ in barns and manipulating the carcass as little as possible.

In cases where composting may be less effective in eliminating vectors of diseases such as BSE or CWD, enclosed composting could still be used to reduce volume and moisture of the carcass before the resulting residuals, chips, and bones are

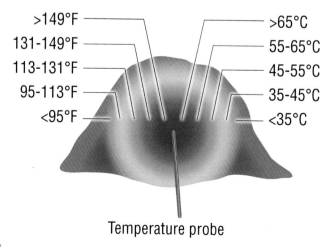

>149°F — >65°C
131-149°F — 55-65°C
113-131°F — 45-55°C
95-113°F — 35-45°C
<95°F — <35°C

Temperature probe

FIGURE 8.4

Temperature variation within a typical compost pile/windrow.

burned in a high temperature incinerator to destroy the prion. Burning whole animals at high temperatures is otherwise energy intensive, and it is difficult to achieve high temperatures throughout large carcasses.

3.2 Risk management

Proficient design and operation of mortality composting systems is very important to ensure that all material achieves adequate temperatures for long enough to kill pathogens. This is achieved for mortality composting piles provided that temperature is monitored. The current state of knowledge suggests that taking the following factors into consideration reduces the potential biosecurity risk associated with mortality composting:

- Attention to site design and layout to minimize scavenging and contamination of ground and surface water with pathogens,
- Using a minimum two-stage composting system followed by the use of a curing phase to properly complete the composting process,
- Fully enveloping mortalities in a "clean" amendment (i.e., coarse carbon source); use sufficient amendment to absorb liquids and odorous gasses produced during composting,
- Monitor and manage the composting process to maximize progress toward the full completion of composting (e.g., temperature, watering, and turning),
- Attention to basic standards of hygiene (e.g., minimize pooling of water at the site, regular sanitizing of equipment and keeping it separate from production facilities, and use of personal protective equipment by everyone involved in the composting operation).

Composting improves biosecurity. Biosecurity refers to possible introduction of potentially infectious organisms (pathogens) from outside sources, such as wild animals, people, and vehicles that have traveled from other farms and facilities. Composting is usually accomplished on site by in-house staff using their own machinery. It requires no or few outside personnel or vehicles that may be carrying pathogens from other farms or facilities. Dead animals can be isolated and incorporated into the compost pile promptly, rather than sitting on the ground waiting for pick up. Proper hygiene still needs to be observed as farm workers can carry disease organisms when they return home or otherwise travel off the farm.

3.3 Regulatory requirements

Regulations concerning carcass disposal vary greatly, internationally and within individual nations. In some jurisdictions, the practice is explicitly prohibited. In others, it is completely unregulated. Therefore, it is important to determine what restrictions and required practices are in place before pursuing a mortality composting program. Only a few examples are covered here.

Composting animal mortalities is rapidly spreading in the United States and Canada. Regulations concerning carcass disposal in the United States vary. Because of the instability in the United States rendering industry, many states are now addressing this in regulations.[4] Even in the United States, where mortality composting is generally well established, there are large regulatory differences among states. Generally, states adopt rules and practices developed or supported by their respective agricultural universities. Fortunately, these universities cooperate with one another and the recommended practices are fairly consistent across the United States. The primary differences among state regulations concern procedures for notification and permitting of mortality composting.

Due to the recent US outbreak of highly pathogenic avian influenza virus, the United States Department of Agriculture Animal and Plant Health Inspection Service (APHIS), in conjunction with composting subject matter experts, put together standard operating procedures for management of poultry and livestock (see APHIS references). For routine mortalities, USDA Natural Resources Conservation Services (NRCS) has an animal mortality facility practice code that was put together with agricultural university input and in many states, is the standard used for composting mortalities (see references).

In most European countries, unprocessed mortalities are treated as "animal by-products" and cannot be composted (see Chapter 4). Within the European Union, composting of mortalities is explicitly prohibited by the EU's Animal by-products regulation (Amlinger and Blytt, 2013). In the United Kingdom, dead animals can be composted if they are pressure-rendered and butcher/slaughter house waste can be composted without any pretreatment (GOV.UK, 2020). The argument is that the safety of mortality composting is unproven for prion diseases and some other animal diseases (see Box 8.3). The concern is that handling the carcasses and distributing the compost might contribute to the spread of diseases like CWD and BSE. Even where mortality composting is accepted, BSE and CWD remain a concern because of resistance to destruction of prions by heat.

Australia has not had a case of BSE and is free of FMD. Hence, mortality composting is not regulated but encouraged, and included as a disposal option for emergency animal disease incidents (Wilkinson, 2006). Yet, the Australian Ruminant Feed Ban was introduced in 1996 to minimize the risk of spreading BSE and dissemination of infections by placing restrictions on the storage and use of material that could contain "restricted animal materials" (RAMs) (Animal Health Australia, 2020). RAMs include any material taken from a vertebrate animal other than tallow, gelatin, milk products, or oils. Therefore, compost that contains dead animal carcasses, poultry litter, pig manure, or food wastes can contain RAM and

---

[4] 43 states have regulations in place listing composting as an acceptable method. California allows composting of nonmammals only. AK, CT, NV, NH, NJ, WV, and WI do not list composting as a method. AK, WV, and WI list every other method but composting. CT and NV list only burial and burning. NH lists burial only and NJ does not list any methods except rendering for hogs that die of cholera. For more on US regulations, visit http://compost.css.cornell.edu/mapsdisposal.html.

is covered by this regulation, which stipulates, among other aspects, that it is illegal to feed RAM to ruminant animals or to allow ruminants access to a stockpile of material (compost) containing RAM.

When products containing RAM are used on crop land, they must be incorporated into the soil and when used on pasture, they have to be spread evenly, and ruminants have to be prevented from having access to treated areas until a combination of rain or irrigation and pasture growth has minimized the risk of RAM ingestion when grazed by ruminants. A 21-day withholding period is considered adequate in good growing conditions for sufficient pasture growth to adequately minimize the risk of RAM ingestion, but longer ruminant exclusion periods are required when pasture growth or rainfall are low.

4. Feedstock characteristics and requirements

4.1 Carcass characteristics

The primary feedstocks in mortality composting are the animal carcasses or animal tissues separated from a carcass such as meat, offal, and organs. Carcasses also include bones, feathers, hide, and fur. The different components decompose at different rates and to a varying extent. Bones are the most resistant and large bones usually pass through the composting process intact, though they become brittle. Feathers, hides, wool, animal hooves, and poultry legs take longer to decompose than the soft tissue. It is essential to consider the possibility that a carcass may also contain residues of veterinary pharmaceuticals (Box 8.4).

While the various components of carcasses present varying characteristics, *in general,* animal carcasses can be considered sources of nitrogen and moisture. C: N ratios are in the 5:1 vicinity and moisture contents usually fall in the 60% −80% range. Most animal tissues, including feathers and fur, are a concentrated source of protein with relatively high nitrogen content. Cells of animal tissue hold

---

Box 8.4 Veterinary pharmaceuticals

In 2003, the US Food and Drug Administration added an environmental warning to animal euthanasia products stating that "euthanized animals must be properly disposed by deep burial, incineration, or other method in compliance with state and local laws to prevent consumption of carcass material by scavenging wildlife." Properly built and managed compost piles are an "other method." If the animal(s) being composted has been euthanized with a drug such as sodium pentobarbital, care should be taken to dispose of the remains as quickly as possible. They will contain potentially harmful residues. Wildlife and domestic animals may be attracted by the carcass and become intoxicated or die if allowed to feed on it. Properly built compost piles deter pets and wildlife from feeding on carcasses when the concentration of drug residue is at its highest. Sodium pentobarbital has been shown to degrade during the composting process so that by the time composting is finished (within 6 months), very low concentrations of the drug remain (Schwarz et al., 2013).

a great deal of water, which is released as the cells break down. Depending on the animal and source, carcasses can also include appreciable amounts of fats, which are a rich source of carbon. While carbon and nitrogen ratios are important, if the animal carcass is managed whole, the ratio is not in balance at the beginning of the process. As the animal carcass decomposes and/or turning occurs, the surrounding amendments mix together and the overall C:N ratio comes into balance.

An important characteristic is the high potential for nuisance issues, something that can and must be kept in check. Left exposed, carcasses attract scavenging animals and insect pests including rodents, skunks, racoons, birds, foxes, dogs, and flies. In addition, the wet and protein-rich composition makes animal tissue prone to putrid odors. The proteins break down to produce ammonia and odorous amines, especially at low C:N ratios and under anaerobic conditions. The odors produced by amines are exemplified by the names of two particularly descriptive amine compounds: putrescine and cadaverine. Both of these foul-smelling compounds occur in decaying animal tissue. Animal fats and oils contribute to the formation of volatile fatty acids, which are also odorous. Distinctive features of animal mortality composting, such as thick covers of amendment and restricted turning, have developed to control pests and odors.

Although any animal can be composted whole with minimal processing, preprocessing may speed up the composting process. It is a good practice to pierce the stomachs, especially the rumen, of cattle and other ruminants to prevent bloating and, possibly explosion, from accumulating gases.

Total size reduction of carcasses, like grinding or chipping, expedites the process but requires specialized equipment, as well as additional occupational health and safety requirements. This practice can be messy and unsightly. Size reduction can help solve problems with bones during processing and in the end-product. Even a well-minced and well-blended mixture of animal tissue and amendment attracts animals and liberates odors and spreads airborne disease. Piles must be covered with a minimum 60-cm (24-in.) layer of clean amendment or finished compost, or otherwise contained.

4.2 Amendments—carbon sources—bulking agents

Mortality composting relies on additional feedstocks, selected to make composting of the carcasses practical. Such feedstocks are commonly referred to as *amendments* because they are added to complement a primary feedstock, in this case, animal carcasses. In mortality composting circles, amendments are often called *carbon sources* because they balance the high nitrogen characteristic of the mortalities. This chapter uses the more general term, amendments.

The carcasses are commonly composted with a single amendment that serves all of the required purposes well, or well enough. The cap, base, and internal media can be the same material. In other cases, a combination of different amendments is used to improve the characteristics of the media, or because a single amendment is not available in sufficient quantity. Most often, the medium includes fresh amendments and/or some of the compost recycled from previous batches.

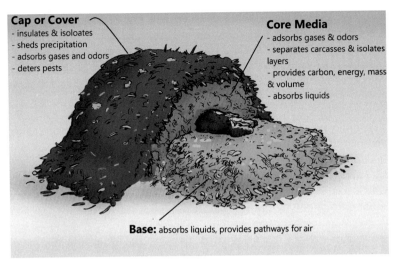

Cap or Cover
- insulates & isoloates
- sheds precipitation
- adsorbs gases and odors
- deters pests

Core Media
- adsorbs gases & odors
- separates carcasses & isolates layers
- provides carbon, energy, mass & volume
- absorbs liquids

Base: absorbs liquids, provides pathways for air

FIGURE 8.5

The functions of amendments in mortality composting.

Source: Cornell Waste Management Institute. Artwork by Bill Davis.

The amendments within the composting medium serve several functions (Fig. 8.5). They absorb moisture and trap and degrade volatile gases and odors from decomposing carcasses. Most importantly, the amendments provide carbon, structure, porosity for air exchange, and a barrier with the surrounding environment. Thus, good amendments have the following characteristics:

- Relatively dry (<50% moisture),
- Particles with some rigidity plus a large surface area,
- Low bulk density (porous),
- High C:N ratio (>40:1) with available forms of carbon,
- Not attractive to pests, and,
- Decompose well enough to produce useable compost.

Wood chips, wood shavings, chopped straw, immature or mature compost, and poultry litter are the amendments most often used as base materials or as the entire envelope of mortality composting pile. Table 8.2 indicates generic characteristics of potential amendments. Other amendments used as cover include horse stall bedding (with manure), feed refusal, old corn silage, old hay, dry manure, peanut hulls, corn cobs, corn stalks, shredded paper products, chipped or shredded yard trimmings, bark, feedlot manure, active compost, and mature compost (see Chapter 4 for detailed descriptions and characteristics). Chipping, chopping, or shredding improves, and may be necessary for, feedstocks with large particle size (e.g., straw, hay, corn stalks, cardboard, and vegetation residues). Amendments with particles in the range of 2–5 cm (1–2 in.) work well, even with some fines mixed in. Passive aeration is not effective or efficient if the particle sizes are too small. For example,

Table 8.2 Characteristics of typical mortality composting amendments.

Material	C:N ratio	Moisture (%)	Structure	Notes
Cattle manure	13–20:1	67–87	Poor	Nitrogen rich. High moisture content. Uniform, small particle size.
Poultry litter/ manure	3–16:1	22–75	Poor	Nitrogen rich. Moderately moist (depending on bedding).
Horse manure with shavings	25–30:1	55–75	Poor	High moisture content. Uniform, small particle size.
Wheat straw	100–150:1	10–14	Good	Best when chopped and mixed with denser materials. Needs moisture. Highly degradable source of carbon.
Corn silage	38–43:1	65–68	Good	Good particle size. Poor structure when wet. Strong odor.
Hay	15–32:1	8–10	Good	Best when chopped and mixed with denser materials. Needs moisture. Highly degradable source of carbon.
Wood shavings	100–750:1	19–65	Good Absorbent	Maintains structure when wet (with mixed particle size) but difficult to keep moist in dry climates. Low odor.
Wood chips	100–1000:1	–	Very good Promotes good air flow Most efficient	Adds structure. Excellent base under primary compost materials. Larger chips absorb less water and odor and are slow to degrade but can be reused.
Compost, preferably unscreened	20–40:1	–	Good	Low available nutrients. Good as absorbent base or biofilter cover.

wood shavings that are very fine, without larger pieces mixed in, absorb moisture, but then fill with water and no longer provide aeration. Base material must have a larger particle size to promote aeration of piles and bins and be thick enough to absorb the moisture generated above.

When considering amendments, it should be remembered that "not all amendments are created equal." For instance, particle size makes a difference: wood chips are larger and provide more porosity than shavings; and shavings provide more porosity than sawdust. In terms of degradability and the availability of carbon, sawdust would be the best of the three choices and woodchips the worst. However, woodchips are preferable because they give the pile the structure it needs to facilitate aeration. The condition of the amendment matters, perhaps more than the type of material. The moisture content, particle size, porosity, bulk density, C:N ratio, and age can vary for almost any type of amendment. Mature compost, aged manure, and very old silage may not provide enough carbon or energy to sustain efficient microbial activity while immature compost, fresh manure, and "fresh" silage would.

In most cases, the amendment of choice depends on the availability and cost of these materials. Because a relatively large volume of amendment is required, composters tend to use what is economically and locally available. However, the material must still be suitable for the purpose. Regardless of the materials selected, a ready supply of amendment should always be kept on site. It is also advisable to maintain a list of locations where additional amendment can be obtained for emergency situations.

As the size of the carcass increases, a greater volume of amendment is required per unit weight of carcass. The required volume of fresh amendment can be reduced by up to 50% by recycling the compost previously produced from mortality composting as long as particle size is adequate for aeration. Recycled compost with bones used for bases generally absorbs moisture well, provides good to fair porosity, and does not attract pests. It provides some carbon and energy (if it is not too degraded) and also inoculates the next pile with microorganisms (fresh woodchips can be used up to three times). It works particularly well for the base as the chips and partially decomposed bones add structure. If used as cover, this mixture may draw attention to the piles as bones are visible.

5. Methods and techniques

5.1 Passively aerated piles and bins

The passively aerated pile approach evolved from practices developed for composting poultry mortalities (Murphy, 1988; Murphy and Carr, 1991), but it is now the standard method for all types of mortalities and butcher waste. With this approach, a number of carcasses are placed in a layer on a base of dry amendment and capped with more amendment. Layers are built upward in succession until the pile or bin reaches its desired maximum height (see Fig. 8.2). Small carcasses (e.g., birds and fish), medium carcasses (e.g., pigs, small deer, and calves), and butcher wastes are generally layered. Whole large carcasses, such as mature sows, cattle, and horses are placed in a single layer (Fig. 8.6).

After the carcasses are layered or placed within amendments, they are left to decompose undisturbed for weeks or months, until the flesh is largely decomposed. This first stage is usually "static," that is, without agitation or turning. After stage one, the pile or material in the bin can be turned and composting continues for

FIGURE 8.6

Preparing a pile to compost a large carcass in a single layer.

Source: Jean Bonhotal.

several months longer during stage two. The second stage can include turnings to accelerate the process or combining piles and windrows as the piles reduce in height. The procedures are essentially the same or very similar whether freestanding piles or bins are used.

The dimensions of individual piles and bins vary with the animal species, mortality numbers, frequency, and equipment used. Freestanding windrows tend to be 3 to 4.5 m (10−15 ft) wide at the base, with the width increasing as the height increases. Heights typically range from 1.5 to 2 m (5−6 ft), although heights of up to 2.5 m (8 ft) are effective. Aeration becomes more difficult as the height increases. Maximum pile height is often determined by the equipment available to build piles.

Composting a 500 kg (1100 pound) cow under good conditions generally takes three to 6 months for the flesh to disappear. For small carcasses like birds, the flesh decomposes in two to 4 weeks. However, this time relates to just the first stage, as the compost itself does not reach adequate maturity for another 6 to 12 months. In regard to pathogen destruction, a composting period of 9 months appears to be a prudent target.

5.1.1 Passively aerated bins

Small to moderate carcasses are commonly composted in bins. Bins are especially popular for poultry and fish mortalities. Bins are usually fixed in size and shape. They are loaded until full, capped, and not touched until stage one of the composting process is complete. The time to load a single bin can be shorter than the stage one composting period so that multiple bins may be necessary for stage one. For stage two, the contents of two bins may be combined in a pile or windrow as volume is reducing during processing.

Composting mortalities in passively aerated bins follows the same process and nearly the same procedures as in passively aerated piles. The main difference is that bins offer the convenience of physical containment. They provide a measure of protection from scavengers, pests, the weather, and space is used more efficiently. A multi-bin system can be effective in many situations. It is an option for any size of operation with meat or seafood residuals, poultry, and small livestock. On the negative side, most bins are fixed in size and location, and the structural material for bins adds to the cost of composting and the operation.

There are many options regarding the configuration and construction of bins including:

- Temporary, portable, or fixed,
- Open, roofed, or inside a building,
- Four, three, or two-sided (open on each end),
- Square, wide, and shallow, or narrow and deep, and,
- Constructed of wood, poured concrete, concrete blocks, moveable barriers (e.g., Jersey barriers) or large bales (hay or straw, round or rectangular).

Bins that are fixed are usually made of wood or concrete and tend to have concrete floors. Temporary bins are often constructed from stacking hay bales or moveable fencing (Fig. 8.7). They are usually placed on native or amended soils. At least two and normally three or more bins are situated side-by-side. When turning occurs, the material is moved into an adjacent bin.

Several common styles of mortality composting bins are available. The choice depends on factors that include the type of operation, animal species, regularity and frequency of mortalities, weather, and available capital, including the availability of cost share funds. Covered wooden bins have been popular in the poultry industry (Fig. 8.8).

FIGURE 8.7

Example of a temporary mortality composting bin using moveable wire fencing.

Source: Josh Payne.

FIGURE 8.8

Example of wooden bins used for composting poultry mortalities.

The dimensions of individual bins vary with the animal species, mortality numbers, and frequency and equipment used. Some bins for poultry and fish are only 1 m (3 ft) high. Otherwise, maximum recommended heights typically range from 1 to 2 m (4–6 ft). Taller bins are often used for medium-sized carcasses, like swine. The width of the bin usually depends on the means of loading and unloading. In general, they range between 1 m (3 ft) for hand-loaded mini-bins and up to 5 m (16 ft) for large operations with temporary bins. The bin should be at least wide enough to fit the bucket of the loader that is used.

5.1.2 Mini composters

Mini composters are small portable bins made of pallets, timber, and/or plastic or wire mesh. They must be large enough to generate and retain adequate heat, with minimum dimensions of 1 m × 1 m × 1 m (3 ft × 3 ft × 3 ft). This size bin with a volume of 1 m^3 (35 ft^3) can hold a total of up to 300 kg (600 lbs.) of carcasses.

Mini composters represent a special application of passively aerated bins (Brodie and Carr, 1997). They were developed for poultry but can be used for composting small animals up to about 15 kg (30 lbs) including piglets, rabbits, and fish. The advantage of mini composters is that they can be located within the animal rearing areas, moved where needed, and they can be sized and multiplied according to need.

5.1.3 Procedures for passively aerated piles

Free-standing piles are used for composting all sizes of mortalities including whole carcasses, ground carcasses, and butcher waste. When dealing with routine and regularly occurring mortalities, piles are typically elongated into windrows by adding a new layered section on to the end of the previous one (Fig. 8.9). An entirely new pile or windrow is started when the desired cycle time elapses or the pile reaches the

FIGURE 8.9

Extended base of passively aerated static pile, forming a windrow with successive batches of mortalities.

Source: Mary Schwarz.

maximum length. Often, piles are constructed for roughly the same number of days that it takes to complete the first stage of composting. In this case, while one pile is being built, a second is progressing through the first stage of composting. If animals die sporadically (e.g., cattle, horses, road kill), multiple small piles are commonly constructed as the mortalities occur.

The *general* steps are as follows:

1. *Lay down a base layer* of amendment, approximately 30 to 60 cm (12–24 in.) deep, across the bottom of the pile. The thickness of the layer depends on the absorbency of the amendment and the size of the carcasses. Larger carcasses tend to contain a higher proportion of water and need a base of up to 60 cm (24 in.). Amendments with large particles, like wood chips, should be used for the base layer. If liquids leak from the base of piles, the thickness of the base layer should be increased in future piles ensuring that more absorbent chunky carbon is used.

2. *Place the carcasses* on the base layer. Small carcasses (birds, fish) can be spread over the layer evenly without being spaced apart. Animals of moderate size (pigs, sheep, deer, etc.) should be spaced close together but evenly over the base, separating carcasses by a few centimeters (inches) (Fig. 8.10). For large animals, like mature cattle and horses, only one carcass may fit the width of the pile. In all cases, carcasses should not be stacked on top of one another nor placed near the outer edges of the piles. A minimum buffer of 45 to 60 cm (18–24 in.) from the outer edges of piles should be maintained.

FIGURE 8.10

Poultry carcasses placed on the base layer of amendment.

Source: Mary Schwarz.

3. *Cover the carcasses with amendment*: How much cover and what material to use depends on whether additional layers will be added and how soon. Place 30 cm (12 in.) of cover over smaller animals and add another layer of carcasses (Fig. 8.11). If this is a continuous process with daily or weekly animal additions, it is best to stair step the animals into the pile. Build the first section at one end

FIGURE 8.11

Adding amendment to cover the most recent layer of mortalities, in this case poultry carcasses.

Source: Mary Schwarz.

of the windrow and extend a carbon bed out so that it is ready for additional animals (as in Fig. 8.9). In the case of large animals, the added amendment on top of the carcass is the final cover layer or biofilter and should be at least 60 cm (24 in.) thick. When animals are added to the pile, dates should be recorded, especially when the last animal was added to a pile. If it is a long windrow, built in sections, dated flags can be placed in the pile sections to keep track as one end of a long windrow will be finished before the other. Always make sure animals are well covered to avoid problems with odor, vermin, etc.

4. *Add water as necessary*: Although the carcasses supply considerable moisture, it may not be sufficient to support decomposition of the amendments. In addition, water released from the carcasses concentrates in small regions within the pile. Therefore, the amendments should also be moderately moist (40%−60%) at the time they are placed. The amendment should feel and look damp, not wet; this is especially important in arid climates.

5. *Cap the pile:* The final layer of carcasses should be covered with a thicker layer of amendment. For small to medium animals, a layer of 30 to 45 cm (12− 18 in.) is typical, and for large animals, a 45 to 60 cm (18 to 24 in.) layer is necessary (Fig. 8.11). The appearance of scavengers and flies means that the thickness is inadequate and should be increased. The same amendment used to cover intermediate layers can also be used for the final cap.

6. *Shape the pile:* Piles/windrows can be shaped to shed moisture or capture it. Taking precipitation into consideration, piles with peaked tops will shed moisture in high precipitation areas. Creating a flat top will allow moisture that falls on the pile to soak in. Creating a trough allows moisture to collect and soak in. When piles are working efficiently, it is hard to add moisture during stage one when piles are generating a lot of heat.

7. *Stage one composting:* The first stage of animal mortality composting can be described as above ground entombment within an organic biofilter. After the pile or bin is capped and shaped, it is left to compost undisturbed for roughly 4 weeks to 6 months, depending on the largest carcass size and management objectives. The carcasses are dense, wet, and have a low C:N ratio, while the surrounding amendments are relatively dry, porous, and typically have a high C:N ratio. At different times in the process, there are aerobic and anaerobic zones in the pile, producing liquids and gases that migrate into, and are absorbed within, the envelope of amendments. The liquids and gases continue to decompose along with the amendments. The surrounding media serves as a biofilter, adsorbing volatile organic compounds and trapping leachate, odors, and heat. The temperatures within the pile can reach thermophilic levels within a few days and remain high for prolonged periods of time. The pile should be monitored for temperature, odors, and signs of scavengers, flies, and leachate. Any problems that are detected should be corrected. At the end of the first stage, the partially composted material is largely free of soft tissue but inconsistent and littered with bones. While the interior is moist, dry sections persist throughout. At this point, the material cannot be considered compost, and

neither is it necessarily pasteurized. However, it can be land applied on the farm as long as there is no biosecurity risk, or it can be reused as the base for another composting pile.

8. *Turn and commence stage two composting:* The end of the first stage of composting may be indicated by falling temperatures or simply the end of the desired composting time period. Turning is typically accomplished with a bucket loader or windrow turner to mix the material, distribute moisture and nutrients, release trapped gases, and recharge the process. At this stage, composting continues more uniformly, like in conventional composting (Fig. 8.12). Moisture can be added at this stage if necessary. If the freshly turned pile is odorous or if any recognizable pieces of meat, organs, hide, feet, and bones remain exposed after turning, the pile should be covered with clean amendment. Temperatures usually return to thermophilic levels for several weeks longer. The material can be turned again or not depending on the management preferences. However, additional turnings during the second stage shorten the overall composting time. The composting time for stage two is typically less than that of stage one because decomposition is concentrated during stage one. However, the duration of stage two can equal or surpass stage one if mature compost is desired. In practice, it is best not to rush the composting process. Even after stage two has been completed and the compost is mature, bones persist, (unless the animals are immature or soft-boned). Clean bones can be recycled through the process. They add structure to the pile and become brittle with age and repeated composting. Feathers from poultry and wool from sheep take longer to decompose than meat. Large wood chips, feathers, and bones can be recycled as part of the media for future piles or bins.

FIGURE 8.12

Compost pile opened and turned after four months. Most of the soft tissue is gone, leaving hide, hair, and bones.

Source: Josh Payne.

6. Sizing guidelines for passively aerated piles and bins

The amount of overall space needed for a mortality composting operation depends on the animal species, mortality rate, number of animals managed, composting method, the type and amount of amendment, duration of composting, and other factors such as equipment used, additional feedstocks, and processing activities.

The required volume of an individual stage-one pile or bin is based on the size and number of mortalities generated over the loading period. The loading period is the number of days that an individual unit is loaded before the stage-one composting clock starts. With freestanding piles, the loading period is often roughly the same length of time as the first-stage composting period. In this case, only two first-stage piles are necessary, one for loading while the previous unit is composting. With bin systems, and sometimes with piles, the loading period is shorter than the composting time. Therefore, multiple first stage units are needed.

Most guidelines recommend that second-stage piles or bins should be allotted the same volume as the primary units. This recommendation is a safe and prudent one as some reduction in volume occurs during the first stage of decomposition. It is worth recognizing that some reduction in volume is likely between the two stages (from 20% to 50%). The volume loss depends on the proportion and nature of the amendments and the duration of the first stage of composting, with longer processing times and more degradable amendments resulting in more volume reduction.

Several approaches have developed for calculating the space requirements of composting sites that handle routine mortalities. The following steps outline one such procedure.

1. Determine the average daily mortality in mass or weight units using farm or business records or estimates of mortality rates from general data. Table 8.3 lists average weights and annual mortality rates for various animal species. The average daily amount of mortality to plan for routine mortality composting can be estimated with the following equation (Eq. 8.1).

Average Daily Mortality(kg/day or lbs/day) =

$$\frac{\text{Number of animals in herd or flock} \times \text{Carcass mass (kg or lbs)} \times \text{Annual mortality rate}}{365 \text{ days}}$$

(8.1)

2. Establish the time periods for loading the pile and the desired durations for the first and second stage composting. Also, consider storage time for amendments and finished compost.
3. Determine the volume required of the first stage unit based on the average daily mortality (Step 1) and loading time. One approach for calculating this volume is to assume a target animal tissue density (ATD, see below).

Table 8.3 Average on-farm mortality rates and carcass masses/weights.

Species	Annual mortality rate[f]	Average weight kg	Average weight lbs	Species	Annual mortality[f]	Average weight kg	Average weight lbs
Horses[a]				**Sheep**[b]			
Foals	0.049	45	100	Lambs	0.112	3.5	8.0
Adult	0.018	540–680	1200–1500	Adult	0.05	35–75	80–170
Beef cows[d]				**Dairy Cows**[c]			
Pre-wean	0.036	270	600	Still births	0.065	—	—
Feedlot	0.015	400	850	Pre-wean	0.078	40	90
				Heifers	0.018	270	600
				Cows	0.057	650	1400
Poultry[e]				**Swine**[e]			
Laying hen	0.14	1.8	4.0	Stillbirths	0.001–0.01	1.0–1.5	2.0–3.0
Broiler breeding hen	0.11	3.2	7.0	Farrow to wean	0.10–0.15	4.5	10
Broiler breeder pullet	0.05	2.0	4.3	Nursery	0.02–0.05	14	30
Layer pullet	0.05	1.3	2.8	Grow/Finish	0.02–0.08	75	160
Broiler	0.05	2.0	4.5	Breeding herd	0.02–0.06	160	350
Roaster	0.08	3.6	8.0				
Turkey hen	0.06	7.3	16				
Turkey tom	0.09	11	25				

[a] APHIS (2007): Trends in Equine Mortality (1998–2005).
[b] APHIS (2011): Sheep (2011).
[c] APHIS (2007): Dairy.
[d] APHIS (2008): Beef.
[e] Rozeboom et al. (2007).
[f] Average fraction of the herd or flock that die over a 365-day period.

4. Considering equipment, infrastructure, and other site conditions, establish the height and width of piles and bins. Calculate the cross-sectional area of the pile or bin (Chapter 7).
5. Based on required volume of the stage-one unit (Step 3), determine the required lengths of first-stage piles by dividing the required volume by the cross-sectional area. The same calculation gives the required depth of the first stage bins. However, this depth should be adjusted if multiple first-stage bins are used.
6. Add extra space to all dimensions as needed for equipment access, structures, and other components.
7. Add space for storage of amendment and compost.

A common way to estimate the required volume for a stage-one composting unit is to assume a target ATD. Successful animal tissue composting has been accomplished using a target ATD ranging from 8–240 kg/m^3 (0.5–15 lb/ft^3). That is, 1 m^3 (1 ft^3) of bulking material will be needed for every 8–240 kg (one-half to 15 lbs) of tissue. However, when ATD is greater than 160 kg/m^3 (10 lb/ft^3), intensive management of aeration and moisture is necessary. This is because the by-product is mostly fat trim (about 70% lipids), which will melt in the pile causing it to "slump" or lose its shape. Therefore, for by-product composting, an ATD of between 64 and 96 kg/m^3 (4 and 6 lb/ft^3) will probably work best. Higher ATDs generally apply to small and shredded carcasses; smaller ATDs to large carcasses.

The general formula for calculating the pile or bin volume using the ATD is given below by Eq. (8.2). It assumes the overall volume is approximately equal to the volume of amendment used, which is a reasonable approximation.

$$\text{Pile or Bin Volume } (m^3 \text{ or } ft^3) = \frac{\text{Average daily mortality } (kg/d \text{ or } lbs/d) \times \text{Loading time}(d)}{\text{ATD}(kg/m^3 \text{ or } lbs/ft^3)} \quad (8.2)$$

Several computer programs are available for calculating how much space is needed to compost animal mortalities and animal tissue (Schwarz, 2010). To use these programs, the user usually needs to know, or assume, at least the following: (1) the weight or volume of mortalities generated; (2) the weight or volume of amendment needed (or the target ATD); (3) the length of time needed for the primary composting phase (stage one); and (4) the type of composting system to be used.

7. Other mortality composting methods

7.1 Forced-aerated static piles and bins

Although passive aeration is the norm, forced aeration in both freestanding piles and bins is certainly feasible, and practiced (Fig. 8.13A). Methods for composting mortalities with forced air are similar to aerated static piles and bins for all feedstocks, with a few special considerations. Large and concentrated carcasses present an

FIGURE 8.13

Alternative methods of composting mortalities: Forced aerated bin for composting poultry mortalities mixed with manure and farm-generated amendments (A); Transportable aerated container (BiobiN, see Chapter 7) used for containing and composting poultry mortalities on site (B); Small-scale rotating drum composting reactor used for composting fish mortalities and greenhouse residues (C).

Source: (A, C) R. Rynk. (B) Peats Soils.

obstacle to evenly distributed air flow. Air likely finds a preferential pathway through the drier porous surrounding amendment. Concentrated odors from decomposing tissues within the pile can be carried out with the exiting air. For these reasons, forced aeration is more practical when carcasses are small or shredded and well mixed with amendment. Otherwise, forced aeration is better suited to stage 2 of mortality composting.

7.2 In-vessel composting

IVC systems offer the advantages of inherent containment and process control (see Chapter 7). A range of IVC technologies have been used for mortality composting including agitated beds, closed aerated containers, drums, tunnels, and vertical units. IVC systems vary in the level of sophistication and process control they offer. Most farm and commercial-scale IVC systems are aerated with fans. Some models have mixers on the front, built-in agitation, and size reduction, while others are loaded with a prepared mixture. Tunnel and silo composting systems can process mortality under 12.5 kg (25 lbs), size reduced mortality, or butcher waste. Because of the inherent containment, about 30% less amendment is required per unit weight of animal tissue compared to passively aerated piles. These systems can handle feedstock in batches or as a continuous feed depending on design. IVC systems have short retention times, ranging from a few days to a few weeks. However, the resulting compost needs additional composting, which usually takes place in a turned windrow, aerated static pile, or passively aerated pile.

The BiobiN system is unique in that it is a modular and transportable by design (Fig. 8.13B). The bin is first precharged with amendment as the base layer and subsequently gradually layered with amendment and poultry carcasses (or other small carcasses). While being gradually filled, the carcass/amendment mix is aerated by a fan that exhausts to a biofilter to control odor. When full, the entire bin is removed and taken to a centralized composting facility where compost production is finished in windrows. A new, precharged container is delivered to the site when the full one is collected. This system can also be used for on-farm composting at the point of mortality generation if two or more bins are used. While one bin is filling, the second can further compost before being transferred to stage two of the mortality composting operation.

7.3 Rotating drums

Rotating drums accelerate the first stage of composting carcasses. Tumbling of feedstock exposes the material to air and heat, thrashes the carcasses, and homogenizes the mix thereby creating conditions that promote rapid decomposition. Retention time for carcasses and amendments is typically 5 to 7 days, although longer periods, in some instances as long as 3 weeks, are used if the capacity exists. The drum itself also isolates the carcasses and odors from the surrounding area.

Poultry mortalities and other small carcasses (<15 kg or 35 lbs) are loaded into the drums intact and mixed with amendment. Carcasses over 30 kg (65 lbs) must be shredded to a small enough size to load into the drum, usually after mixing with amendment. The weight of amendment amounts to 20% to 40% of the total weight of the carcasses.

A number of drum models from several manufacturers have been used for mortality composting. These models are typically about 3 m (10 ft) in diameter and 5–15 m (16 ft to 50 ft) in length (Fig. 8.13C). The length is normally selected based on the number of mortalities to be composted, amendment ratio, and desired retention time. Most units used for composting carcasses have not been aerated with fans, although fans are used in many other applications. The drums are loaded with a screw conveyor through a port at one end. The compost mix is discharged through unloading doors or chutes at the opposite end. A well-blended mix emerges with nearly all of the soft tissue degraded beyond recognition. However, this partially composted material requires further composting, which can take place in passively aerated piles, bins, windrows, or mechanically aerated piles.

8. Managing mortality composting operations

By nature of the feedstocks, mortality composting deserves a heightened level of management. Management of mortality composting requires diligence, more so than labor. Success depends on keeping the process on track and keeping the animal tissue isolated from the surroundings. Site cleanliness is most important as it deters scavengers, helps control odor, and promotes good neighbor relations. Process monitoring and recordkeeping are also essential (see Chapter 11).

8.1 Process monitoring

A record of temperature, odor, vectors (any unwanted animals), leachate (liquid that comes out of the pile), spills, and other unexpected events should be kept as a record of the process. This allows the composter to see if sufficiently high temperatures are reached and adjust the process if there are any problems. Also, odor can be an issue and compost piles are an easy target for complaints. Maintaining records can help identify a problem and uphold or disprove a complaint.

Monitoring of the pile is done primarily by checking temperatures. Internal compost temperatures affect the rate of decomposition as well as the destruction of pathogenic bacteria, fungi, and weed seeds. The most efficient temperature range for composting is generally between 40 and 60°C (104 and 140°F); however, with butcher waste piles, it is not uncommon to reach temperatures of around 75°C (170°F). As these piles are high in moisture, spontaneous combustion does not seem to be a problem.

Compost temperatures depend on how much of the heat produced by the microorganisms is lost through aeration or surface cooling. During periods of extremely cold weather, piles may need to be larger than usual to reduce surface to volume ratio and surface cooling. As decomposition slows, temperatures gradually drop and remain within a few degrees of ambient air temperature. Thermometers with 1.0–2.0 m (3–6 ft) probes with or without data loggers are available for temperature monitoring, although care needs to be taken when inserting them into the pile, as they can bend or break when hitting rock, bone, or dense material.

Monitoring oxygen or carbon dioxide also indicate how well the process is progressing. With passively aerated static piles, the only way to keep oxygen levels high is by using bulky carbon sources. Ideally oxygen levels should be kept between 5% and 14%.

8.2 Pathogen control

Pathogen reduction (sanitization or pasteurization) during composting is primarily achieved through thermal disinfection by maintaining elevated temperatures, above 55°C or 60°C, or even 65°C, for a prolonged period. Sanitization is governed by an inverse timed temperature relationship. However, moisture content, ammonia concentration, and heat tolerance of pathogenic organisms are also key factors that affect pathogen inactivation. Turning of piles and windrows should be used in combination with time and temperature criteria. Turning facilitates adequate pathogen reduction as it mixes material from the outside into the hot center of the pile so that all material is exposed to thermophilic temperatures.

As an example, the USDA APHIS "Highly Pathogenic Avian Influenza" (HPAI) mortality composting protocol requires a 28-day composting process (APHIS, 2021). Composting operations that use a windrow or pile composting system must maintain a temperature of 55°C or higher for 10 days. The windrow can then be turned after the 10-day treatment period is achieved. To document compliance, temperatures must be monitored at depths of 45 cm and 90 cm (18 and 36 in.) at 3-m (10-ft) intervals.

Although thermal destruction dominates, pathogen reduction also occurs through nonthermal mechanisms including competitive interactions between microorganisms, nutrient depletion, by-product toxicity, and natural die-off. Microbial activity during composting contributes to the rapid killing of bacteria, viruses, and even to the inactivation of hardier pathogens. Greater microbial activity results in faster degradation. Proteolytic enzymes produced by bacteria and/or temperature and pH changes caused by microbial metabolism contribute to virus and other pathogen inactivation. High pH, low moisture, microbial activity, free ammonia, and high temperature are among the most unfavorable conditions for pathogen survival.

Mortality composting deactivates pathogens and limits the risk of groundwater contamination and air pollution. Composting on-site reduces the potential for farm to farm disease transmission, decreases transportation costs and tipping fees associated with off-site disposal. The composting process degrades carcasses into a soil amendment that has many productive uses.

References

Cited references

Amlinger, F., Blytt, L.D., 2013. How to Comply With the EU Animal By-Products Regulations at Composing and Anaerobic Digestion Plants. European Compost Network. https://www.compostnetwork.info/download/good-practice-guide-comply-eu-animal-products-regulations-composting-anaerobic-digestion-plants/.

Animal Health Australia, 2020. Australian Ruminant Feed Ban. https://www.animalhealthaustralia.com.au/what-we-do/disease-surveillance/tse-freedom-assurance-program/australian-ruminant-feed-ban/#:~:text=ruminant%20feed%20ban%3F-,Australia%20has%20an%20inclusive%20ban%20on%20the%20feeding%20to%20all,encephalopathy%20(BSE)%20in%20Australia.

APHIS, 2021. Highly Pathogenic Avian Influenza (HPAI). US Department of Agriculture. https://www.aphis.usda.gov/aphis/ourfocus/animalhealth/animal-disease-information/avian/avian-influenza/defend-the-flock-hpai.

APHIS Sheep 2011. Part II: Reference of Marketing and Death Loss on U.S. Sheep operations, U.S. Department of Agriculture. http://www.aphis.usda.gov/animal_health/nahms/sheep/downloads/sheep11/Sheep11_dr_PartII.pdf.

APHIS Beef 2008 Part IV: U.S. Department of Agriculture. http://www.aphis.usda.gov/animal_health/nahms/beefcowcalf/downloads/beef0708/Beef0708_dr_PartIV.pdf.

APHIS Dairy 2007 Part I. U.S. Department of Agriculture. http://www.aphis.usda.gov/animal_health/nahms/dairy/downloads/dairy07/Dairy07_dr_PartI.pdf.

APHIS Carcass Disposal — responsibilities and authorities. U.S. Department of Agriculture. https://www.aphis.usda.gov/emergency_response/downloads/hazard/Carcass%20management%20table.pdf.

APHIS, 2015. Carcass Management during a Mass Animal Health Emergency. U.S. Department of Agriculture. https://www.aphis.usda.gov/stakeholders/downloads/2015/eis_carcass_management.pdf.

APHIS, 1998—2005. Trends in Equine Mortality. U.S. Department of Agriculture. http://www.aphis.usda.gov/animal_health/nahms/equine/downloads/equine05/Equine05_is_Mortality.pdf.

Benson, E.R., Malone, G.W., Alphin, R.L., Johnson, K., Staicu, E., 2008. Application of in-house mortality composting on viral inactivity of Newcastle disease virus. Poultry Sci. 87, 627—635.

Bonhotal, J., Schwarz, M., Rynk, R., 2014. Composting Animal Mortalities. Cornell Waste Management Institute. https://hdl.handle.net/1813/37369.

Brodie, H.L., DW, Carr, L.E., 1997. Composting Animal Mortalities on the Farm. University of Maryland. Cooperative Extension, Fact Sheet 717.

Collar, C., Payne, M., Rossitto, P., Moeller, R., Crook, J., Niswander, T., Cullor, J., 2009. Pathogen reduction and environmental impacts associated with composting bovine mortalities. In: Proceedings: 3rd International Symposium: Management of Animal Carcasses, Tissue and Related Byproducts. Connecting Research, Regulations and Response. Davis, CA, July 21-23.

CWMI, 2021. Mortality Composting. Cornell Waste Management Institute. http://cwmi.css.cornell.edu/mortality.htm.

Elving, J., Emmoth, E., Albihn, A., Vinneras, B., Ottoson, J., 2012. Composting for avian Influenza virus elimination. Appl. Environ. Microbiol. 78 (9), 3280—3285.

Flory, G.A., Peer, R.W., 2009. Real world experience with composting confirms it as an effective carcass disposal method during outbreaks of Avian Influenza. In: Proceedings: 3rd International Symposium: Management of Animal Carcasses, Tissue and Related Byproducts. Connecting Research, Regulations and Response. Davis, CA, July 21-23.

Garcia-Sierra, J., Rozeboom, D.W., Straw, B.E., Thacker, B.J., Granger, L.M., Fedorka-Cray, P.J., Gray, J.T., 2001. Studies on survival of pseudorabies virus, *Actinobacillus pleuropneumoniae,* and *Salmonella* serovar Choleraesuis in composted swine carcasses. J. Swine Health Prod. 9 (5), 225—231.

Glanville, T.D., Richard, T.L., Harmon, J.D., Reynolds, D.L., Ahn, H.K., Akinc, S., 2006. Composting livestock mortalities. Biocycle 47 (11), 42—48.

GOV.UK, 2020. Using Animal By-Products at Compost and Biogas Sites. Department for Environment, Food and Rural Affairs (UK). https://www.gov.uk/guidance/using-animal-by-products-at-compost-and-biogas-sites.

Guan, J., Chan, A.M., Grenier, C., Wilkie, D.C., Brooks, B.W., Spencer, J.L., 2009. Survival of Avian influenza and Newcastle disease viruses in compost and at ambient temperatures based on virus isolation and real-time reverse transcriptase PCR. Avian Dis. 53 (1), 26—33.

Guan, J., Chan, M., Grenier, C., Brooks, B.W., Spencer, J.L., Dranendonk, C., Copps, J., Clavijo, A., 2010. Degradation of foot-and-mouth disease virus during composting of infected pig carcasses. Can. J. Vet. Res. 74, 40—44.

Guan, J., Chan, M., Brooks, B.W., Spencer, J.L., Algire, J., 2012. Comparing *Escherichia* coli O157:H7 phage and bovine viral diarrhea virus as models for destruction of classical swine fever virus in compost. Compost Sci. Util. 20 (1), 18—23.

Guardabassi, L., Dalsgaard, A., Sobsey, M., 2003. Occurrence and Survival of Viruses in Composted Human Feces. Sustainable Urban Renewal and Wastewater Treatment No. 32. Danish Environmental Protection Agency.

Hongsheng, H., Spencer, J.L., Soutyrine, A., Guan, J., Rendulich, J., Balachandran, A., 2007. Evidence for degradation of abnormal prion protein in tissues from sheep with scrapie during composting. Can. J. Vet. Res. 71, 34–40.

Hubert, M.-H., 2019. Whale Composting: Letting Bacteria Do the Hard Work of Specimen Defleshing. Canadian Museum of Nature Bolg.

King, M., Matassa, K., Garron, M., 2018. A Guide to Composting Marine Animal Mortalities. https://www.greateratlantic.fisheries.noaa.gov/policyseries/index.php/GARPS/article/view/7.

King, M., 2021. Personal Communication. Organics Management Specialist. Maine Department of Environmental Protection.

Lu, H., Castro, A.E., Pennick, K., Liu, J., Yang, Q., Dunn, P., Weinstock, D., Henzler, D., 2003. Survival of avian Influenza virus H7N2 in SPF chickens and their environments. Avian Dis. 47, 1015–1021.

Murphy, D.W., 1988. Composting as a dead bird disposal method. Poultry Sci. 67 (Suppl. 1), 124.

Murphy, D.W., Carr, L.E., 1991. Composting Dead Birds. University of Maryland. Cooperative Extension, Fact Sheet 537.

NC DACS, 2016. North Carolina Guidance for Composting of Mass Animal Mortality. N.C. Department of Agriculture and Consumer Services, N.C. Department of Environmental Quality.

OkState DASNR, 2018. Mass Mortality Composting. Video Series. Oklahoma State University, Disisiojn of Agricultural Sciences and Natural Resources. https://www.youtube.com/playlist?list=PLBX_xiiFGiPKN66fDmXwycr34GSFMkowx.

Paluszak, Z., Lipowski, A., Ligocka, A., 2012. Survival rate of Suid Herpesvirus (SuHV-1, Aujeszky's disease virus, ADV) in composted sewage sludge. Pol. J. Vet. Sci. 15 (1), 51–54.

Reeves, C., 2017. The time a 50-ton whale drove along the streets of Portland, Maine. Wash. Post.

Reuter, T., Gilroyed, B.H., Xu, S., McAllister, T.A., Stanford, K., 2012. Novel molecular and microbial insights into mortality composting. In: Proceedings: 4th International Symposium on Managing Animal Mortality, Products, By-Products, and Associated Health Risk: Connecting Research, Regulations and Response. Dearborn, MI, May 21-24, 2012.

Ritz, C., 2017. Composting Mass Poultry Mortalities, vol. 1282. University of Georgia Cooperative Extension Bulletin. https://secure.caes.uga.edu/extension/publications/files/pdf/B%201282_5.PDF.

Rozeboom, D.W., 2007. Suzanne Reamer and Jerrod Sanders. Michigan Animal Tissue Compost Operational Standard. Available at: https://www.msu.edu/~rozeboom/catrn.html.

Schwarz, M., Bonhotal, J., Bischoff, K., Ebel Jr., J.G., 2013. Fate of barbiturates and nonsteroidal antiinflammatory drugs during carcass composting. Vet. Anim. Sci. J. 4 (1), 1–12.

Schwarz, M., Bonhotal, J., Rozeboom, D., 2010. The Space it Takes - Footprint Calculator for Composting Butcher Waste. Cornell Waste Management Institute. https://hdl.handle.net/1813/22269.

Seekins, B., 2011. Best Management Practices for Animal Carcass Composting. Maine Department of Agriculture. http://composting.org/wp-content/uploads/2013/08/BEST-MANAGEMENT-PRACTICES-for-Carcass-Composting-2011-Complete.pdf.

Senne, D.A., Panigrahy, B., Morgan, R.L., 1994. Effect of composting poultry carcasses on survival of exotic avian viruses: highly Pathogenic Avian Influenza (HPAI) virus and adenovirus of egg drop syndrome-76. Avian Dis. 38 (4), 733−737.

Tablante, N.I., Carr, L.E., Malone, G., Patterson, P.H., Hegngi, F.N., Felton, G., Zimmermann, N., 2004. Guidelines for In-House Composting of Catastrophic Poultry Mortality. University of Maryland, Maryland Cooperative Extension.

Wilkinson, K., 2006. Mortality Composting: A Review of the Use of Composting. The State of Victoria, Department of Primary Industries.

Xu, S., Reuter, T., Gilroyed, B.H., Mitchell, G.B., Price, L.M., Dudas, S., Braithwaite, S.L., Graham, C., Czub, S., Leonard, J.J., Balachandran, A., Neumann, N.F., Belosevic, M., McAllister, T.A., 2014. Biodegradation of prions in compost. Environ. Sci. Technol. 48, 6909−6918.

Xu, S., Reuter, T., Gilroyed, B.H., Dudas, S., Graham, C., Neumann, N.F., Balachandran, A., Czub, S., Belosevic, M., Leonard, J.J., McAllister, T.A., 2013. Biodegradation of specified risk material and fate of scrapie prions in compost. J. Environ. Sci. Health A Toxic/Hazard. Subst. Environ. Eng. 48 (1), 26−36.

Xu, W., Reuter, G., Inglis, D., Larney, F.J., Alexander, T.W., Guan, J., Stanford, K., Xu, Y., McAllister, J., 2009. A biosecure composting system for disposal of cattle carcasses and manure following infectious disease outbreak. J. Environ. Qual. 38, 437−450.

Consulted and suggested references

Adams, D., Flegal, C., and Noll, S. Composting Poultry Carcasses. NCR-530, Purdue University, West Lafayette, IN. Available at: https://www.extension.purdue.edu/extmedia/NCR/NCR-530.html.

APHIS, 2021. Animal and Plant Health Inspection Service. U.S. Department of Agriculture. https://www.aphis.usda.gov.

Australian Government, 2021. National Carp Control Plan. Department of Agriculture, Water and the Environment. https://www.agriculture.gov.au/pests-diseases-weeds/pest-animals-and-weeds/national-carp-control-plan.

Bonhotal, J., 2002. Natural Rendering: Composting Livestock Mortality and Butcher Waste. Cornell Waste Management Institute, Ithaca, NY, 20-minute DVD. https://hdl.handle.net/1813/7870.

Bonhotal, J., Telega, S.L., Petzen, J.S., 2002. Natural Rendering: Composting Livestock Mortality and Butcher Waste. Cornell Waste Management Institute, 12-page fact sheet and 3 posters. Available at. https://hdl.handle.net/1813/2149.2.

Bonhotal, J., Harrison, E.Z., Schwarz, M., 2007. Composting Roadkill. Cornell Waste Management Institute, 12-page fact sheet. Available at. https://hdl.handle.net/1813/10866.

Bonhotal, J., Harrison, E.Z., Schwarz, M., 2007. Composting Roadkill. Cornell Waste Management Institute. DVD. Available at. https://hdl.handle.1813/11250.

Bonhotal, J., Schwarz, M., Brown, N., 2008. Natural Rendering: Composting Poultry Mortality. The Emergency Response to Disease Control. Cornell Waste Management Institute, 12-page Fact Sheet. Available at. https://hdl.handle.net/1813/11722.

Bonhotal, J., Schwarz, M., Brown, N., 2008. Natural Rendering: Composting Poultry Mortality. The Emergency Response to Disease Control. Cornell Waste Management Institute. DVD and poster. Available at. https://hdl.handle.net/1813/11663.

Bonhotal, J., Schwarz, M., Brown, N., 2008. Natural Rendering: Composting Poultry Mortality. The Emergency Response to Disease Control. Cornell Waste Management Institute. Literature review. Available at. https://hdl.handle.net/1813/44714.

Bonhotal, J., Schwarz, M., Williams, C., Swinker, A., 2012. Horse Mortality: Carcass Disposal Alternatives. Cornell Waste Management Institute, 8-page fact sheet. Available at. https://hdl.handle.net/1813/29008.

Bonhotal, J., Schwarz, M., 2012. Natural Rendering for Horses: Composting Horse Mortality. Cornell Waste Management Institute, 5:30 minute video. Available at. https://hdl.handle.net/1813/29538.

Collins Jr., E.R., 2009. Composting Dead Poultry. Virginia Cooperative Extension Publication 424-037. Available at: http://pubs.ext.vt.edu/442/442-037/442-037_pdf.pdf.

Co-Composter version 2a, 2001. Cornell University, Department of Biological and Environmental Engineering and Cornell Waste Management Institute, Ithaca, NY. Available at. https://hdl.handle.net/1813/44641.

Diamond, J., 2014. Composting Abbie: A Whale of a Story.

Higgins, S., Bruner, E., 2009. Composting: A Viable Alternative for Mortality Disposal. University of Kentucky College of Agriculture. http://equine.ca.uky.edu/news-story/composting-viable-alternative-mortality-disposal.

Kansas State University, Purdue University and Texas A&M Univeristy, 2004. Carcass Composting: A Comprehensive Review. Report Prepared by the National Agricultural Biosecurity Center Consortium Carcass Disposal Working Group for the USDA Animal and Plant Health Inspection Service. Available at: http://fss.k-state.edu/FeaturedContent/CarcassDisposal/CarcassDisposal.htm.

Keener, H., Elwell, D., Mescher, T., 1997. Composting Swine Mortality Principles and Operation. Available at: http://ohioline.osu.edu/aex-fact/0711.html.

Keener, H.M., Elwell, D.L., Monnin, M.J., 2000. Procedures and equations for sizing of structures and windrows for composting animal mortalities. Appl. Eng. Agric. 16 (6), 681−692.

Larson, J., 2006. Disposal of Dead Production Animals Bibliography. Available at: http://www.nal.usda.gov/awic/pubs/carcass.htm#2006.

Miller, L., et al., 2016. Mortality Composting Protocol for Avian Influenza Infected Flocks − FY2016 HPAI Response. USDA APHIS, pp. 1−31.

Nelson, V., 2011. Large Animal Mortality Composting. Alberta Agriculture and Rural Development. https://www1.agric.gov.ab.ca/$department/deptdocs.nsf/all/agdex13509/$file/400_29-4.pdf.

NRCS, 2021. Animal Mortality Facility. Practice Code: 316. For the Disposal of Dead Animals. Natural Resource Conservation Service, U.S. Department of Agriculture. https://www.nrcs.usda.gov/wps/portal/nrcs/detail/national/enespanol/?cid=nrcs144p2_027220.

Payne, J., Pugh, B., 2017. On-Farm Mortality Composting of Livestock Carcasses. Oklahoma State University Extension. https://extension.okstate.edu/fact-sheets/on-farm-mortality-composting-of-livestock-carcasses.html.

Ritz, C.W., Worley, J.W., 2012. Poultry Mortality Composting Management Guide. Available at: http://extension.uga.edu/publications/files/pdf/B%201266_3.PDF.

Schwarz, M., Bonhotal, J., Harrison, E., Brinton, W., Storms, P., 2010. Effectiveness of composting road-killed deer in New York state. Compost Sci. Util. 18 (4), 232−241.

Schwarz, M., Bonhotal, J., 2010. The Space It Takes. Available at: https://hdl.handle.net/1813/22269.

University of Maine Cooperative Extension. Safe Disposal of Backyard Poultry Flocks. http://extension.umaine.edu/publications/12e/.

Composting operations and equipment

Author: Scott Gamble[1]

[1]*Organic Waste Specialist, Professional Engineer, Edmonton, AB, Canada*

Contributors: Craig S. Coker[2], Frank Franciosi[3], Robert Rynk[4]

[2]*Coker Composting and Consulting, Troutville, VA, United States;* [3]*U.S. Composting Council, Raleigh, NC, United States;* [4]*SUNY Cobleskill, Cobleskill, NY, United States*

1. Introduction

The biological requirements of the composting process are critical and must not be neglected. However, the biological composting process is only one component of the larger system required to produce compost. Once those requirements are satisfied, producing compost becomes largely a matter of materials handling (Gamble, 2018). Most of the equipment and labor invested in a composting system involves moving, mixing, and otherwise manipulating the materials (Chiumenti et al., 2005). Therefore, the choice of equipment and procedures for handling materials can be as important as the choice of the composting method.

A compost production "system" can be viewed as a succession of unit operations, some of which may be repeated at intervals. As shown in Fig. 9.1, the typical sequence of operations at a composting facility is feedstock receiving and inspection, mixing feedstocks, composting, curing, and product storage. Several other operations are often necessary to condition the feedstocks for composting, to recover uncomposted materials from the finished compost, or to improve the compost's qualities for sale or use. Such operations include sorting, size reduction (e.g., grinding), product screening, drying, and bagging. Of these, size reduction and screening are most common, especially at municipal and commercial facilities that process wood, brush, and green waste. In all cases, materials handling permeates the system.

It is important to recognize that not all of the unit operations shown in Fig. 9.1 are necessary at a given facility. Indeed, variations in feedstocks and composting methods can render several unit processes unnecessary. For example, on-farm composting dairy cattle manure may involve no more than mixing of raw materials, forming and turning windrows, and application of the finished compost.

When multiple types of feedstock are received, or several compost products are produced for sale, there may be a need to augment receiving and inspection practices, record keeping, and product storage/shipping operations. Similarly, the

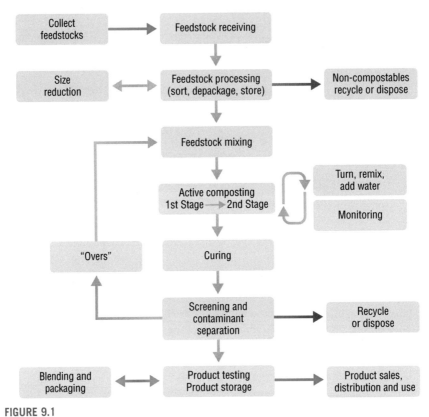

FIGURE 9.1

Flow chart of possible operations and steps in a compost manufacturing system.

Adapted from Oshins, C., 2020.

use of more specialized composting equipment may drive the need for increased sorting and preprocessing steps to prevent equipment damage and optimize the investment in composting technology.

2. Material handling equipment

The reality at most composting facilities is that, regardless of the composting method, heavy equipment is needed to handle and move materials, mix feedstocks, load grinders and screeners, and load final products into trucks. Since the volumes of material are often large, the overall productivity of the composting facility often depends on the reliability and efficiency of this equipment. There are many equipment options available, and the choice depends on material volumes, and the distance that materials need to be moved. Local availability of maintenance support and parts is also a consideration in equipment selection.

2.1 Front-end loaders

Front-end loaders, skid-steers, or farm tractors with bucket attachments are the workhorses at a composting facility and are used at most composting facilities on a full or part-time basis (Fig. 9.2). They are indispensable for transporting feedstocks, loading vessels or building windrows, turning and agitating piles, screening and loading finished product on trucks.

Front-end loaders, also called "wheel loaders" or "bucket loaders" are purpose built for the task of moving solid materials. They are generally more efficient and faster at moving large volumes of materials compared to skid-steers or tractors with bucket attachments. Front-end loaders are also available in a range of sizes and often the bucket can be removed from the unit and replaced with other attachments (e.g., pallet forks, grapples) to extend functionality.

FIGURE 9.2

Examples of "bucket" loaders: farm tractor (A); skid-steer (B); small wheel loader (C); and a large wheel loader (D).

Skid-steer loaders are built for moving solid materials, but with the added functionality of extreme maneuverability. Because they are smaller, skid-steer loaders can access much tighter spaces. Their steering system design also allows them to spin and turn around their central vertical axis. However, the size and maneuverability of these units is accompanied by a sacrifice in size and reach. Therefore, these units are best suited for smaller composting facilities, or supplementing front-end loaders at larger facilities.

Farm tractors fitted with bucket attachments can perform many operations at a composting facility including feedstock handling, pile building, turning, and site maintenance. Farm tractors can also be fitted with numerous other hydraulic or power-take off (PTO) attachments, which increase their versatility. However, tractors are generally much less maneuverable and nimble than front-end loaders and skid-steers. As a result, they are not as common at medium and large scale facilities.

Since loaders are available in such a wide variety of sizes, it is important to match the size of the unit to the facility. If the loader is too small, it may not have the ability to keep up with other processes, or the reach needed to load screens and trucks. Alternatively, if the unit is too large, it can be cumbersome to maneuver within the facility, or the operator may spend excessive time idling while waiting for other equipment.

Ideally, the layout of the composting facility should be developed with the size and turning capabilities of the front-end loader in mind. This minimizes space and safety constraints and avoids material handling bottlenecks. For example, receiving areas should have sufficient maneuvering room for the equipment even when feedstock stockpiles are present.

Careful consideration should also be given to the size and style of bucket used on the front-end loader. The productivity of front-end loaders is primarily determined by the capacity of the bucket that it carries. The "general purpose" buckets that most loaders are equipped with from the factory are sized for handling heavy materials such as sand or gravel. However, since compost is much lighter, loaders can often be equipped with "oversized" or "snow" buckets. These larger buckets can carry as much as double the volume of general purpose buckets, which substantially improves efficiency and saves money in the composting and curing operations (Fig. 9.3).

"Rollout buckets" are an option for larger front-end loaders and have become popular in the topsoil and composting industries. A rollout bucket is designed with support arms that attach to the loader arms, and a hinge connection near the front of the bucket where it attaches to the support arms. This allows the bucket to tilt up and forward from the front to unload, rather than pivoting downwards from the rear (Fig. 9.4). With an appropriately sized front-end loader, the higher effective dumping height of a rollout bucket allows for loading of screens, grinders, and dump trucks or trailers with side walls up to 4.5 m (15 ft) high without the use of a ramp. This greatly improves efficiency by reducing the loading time. To accommodate a rollout bucket, the loader needs to be equipped with an additional

FIGURE 9.3

Front-end loader with an "oversized" bucket for handling compost and light-weight feedstocks.

FIGURE 9.4

Roll-out bucket for extended reach.

Source: Scott Gamble.

hydraulic line to accommodate the cylinder needed to tilt the bucket. Adding rear counterweights, or filling rear tires with ballast (e.g., calcium chloride) may also be required to compensate for the additional leverage placed on the front of the loader.

(A) **(B)**

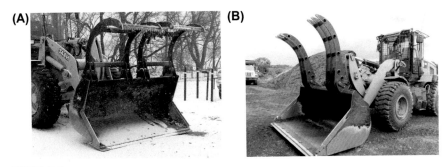

FIGURE 9.5

Examples of grapple buckets.

There are also other bucket types that can provide advantages in certain applications. For example, in areas where there is tight clearance and little room for loaders to maneuver, side-discharge buckets, ram-discharge buckets, and "four-in-one buckets" can improve performance. Side-discharge buckets scoop materials normally but discharge to the side by tipping the bucket to the left or right side rather than forward. Ram-discharge buckets are well suited in areas with low headroom, as the bucket does not need to tip to dump. Four-in-one buckets have a jaw assembly that allows the bucket to split open and dump from the rear of the bucket instead of having to tip forward.

Another attachment that can expand the capabilities and performance of a front-end loader at facilities that handle hay bales or brush is a grapple bucket (Fig. 9.5). This combines a typical bucket with a set of hydraulically operated "fingers" mounted to the top edge of the bucket. When closed, the fingers allow for larger amounts of loose materials to be held in the bucket without the risk of the material falling out.

2.2 Excavators

As a result of their long reach capabilities, dexterity, and rotational ability, excavators are well suited to many tasks at a composting facility (Fig. 9.6). They are most commonly used for loading brush and wood into grinding equipment. To allow the operator to "grab" brush and wood, the standard excavator bucket is often replaced with a set of grapples, or the bucket is equipped with a hydraulic "thumb" (Fig. 9.7).

Although it is not as efficient as a windrow turner, excavators can also be used to turn compost piles and windrows. For example, the excavator can be positioned on the ground alongside the pile or windrow, and the operator can pick up the material from the pile and swing/deposit it in a new pile beside the original one. The excavator can also drive onto the top of the old pile and work from there as it creates a new pile next to it. An excavator working on top of a pile can also mix and fluff the pile material in place, but the excavator must travel backwards in this case to ensure that any of the material it drives on top of is subsequently mixed and fluffed

FIGURE 9.6

(A, B): Examples of excavators used for loading diverse feedstocks including yard trimmings (A) and turning windrows (B).

Source: (A) Peterson Pacific. (B) Scott Gamble.

FIGURE 9.7

Examples of excavators with "thumb" attachments for handling course and large feedstocks, like yard trimmings (A) and logs (B).

Source: (A) R. Rynk. (B) Morbark, LLC.

so that process issues related to compaction and low free air space are avoided. Driving an excavator on top of a pile or windrow also requires that consideration be given to pile stability and operator safety.

2.3 Fork lifts

Fork lifts, also known as fork trucks, are not an obvious material handling device at composting operations but they can be indispensable. Fork trucks are normally used to move containerized materials and palletized materials (Fig. 9.8). At a composting facility, they can be put to use unloading feedstocks from flatbed trailers, van trailers or cube vans, carrying aeration pipes and other long items, moving pallets of bagged compost, and filling/moving cube bags of compost. They can also be used to aid in general maintenance tasks.

FIGURE 9.8

Fork truck unloading palletized feedstock (food processing residuals).

Fork trucks are especially maneuverable and nimble, but unless they are specifically configured for off-road use, they may not be appropriate for use at a compost facility. With some sacrifice in maneuverability, pallet forks can be attached to tractors, skid-steers, or front-end loaders. This may be more appropriate for the rough ground conditions that are typical at composting facilities. It also avoids the need for a dedicated fork lift.

2.4 Dump trucks, end dumps, and wagons

At some point, it becomes inefficient and cumbersome to move material from one point to another using a front-end loader. The distance beyond which it becomes inefficient is a factor of the size and speed of the loader, terrain, the size of the loader's bucket, and how much turning and maneuvering is required. Generally, it is more time- and energy-efficient to use one or more dump trucks or end dumps to move materials over long distances, or alternatively a farm tractor towing a self-dumping wagon (Fig. 9.9). Dump trucks and end dumps can often be hired locally, or it may be possible to rent a dump truck for short periods of time.

Often, a front-end loader can work effectively with two trucks, loading one while the other is driving and unloading. In this manner, the front-end loader works continuously and trucks do not have to wait to be loaded. With longer the haul distances, it may be possible to use three or even four trucks without any wait time for loading.

Dump trucks can carry anywhere from 4 to 12 m^3 (5 to 15 yd^3) of material, depending on the size of the truck. End dumps are able to carry even more material, but are generally not as maneuverable as dump trucks. End dumps can tip over if they are unloaded on uneven ground.

FIGURE 9.9

Farm truck transporting dairy manure to the composting site.

If large amounts of material need to be moved over very rough terrain, it may be cost-effective to rent or hire off-road trucks. These are smaller versions of the large dump trucks used at mining operations, and are designed with suspension systems that allow them to travel quickly over uneven roads and tracks, and are articulated to provide increased maneuverability.

A downside to using trucks or wagons to move materials is that when they unload, the height of the resulting pile of material is usually limited to about 2 m (6 ft). While this may not be problematic if the trucks are transferring material to a small composting pile or windrow, it may create a bottleneck if the truck is transferring materials to a distant screening system or to long-term storage piles. In this case, a front-end loader will probably be required to rehandle the material after it is discharged from the truck (i.e., to load it into the screening equipment or to restack the material into higher piles to save space). At larger facilities, it may be economical to use a drive-over truck unloading system to avoid the need for rehandling with a front-end loader.

2.5 Conveyors

Conveyors are efficient materials handling devices, requiring little energy and labor and relatively easy maintenance. While they are an essential component in many pieces of equipment used at composting facilities including grinders, screens, and even windrow turners, they are not commonly used for standalone material transfer applications. In contrast, conveyors are used extensively in sand and gravel pits, mining and cement production as they are capable of moving several thousand tons of material per hour over long distances.

FIGURE 9.10

Examples of materials handling belt conveyors: trough (A) and cleated (B).

Source: (A) Scott Gamble. (B) R. Rynk.

The primary applications for conveyor systems at compost facilities involve overland belt conveyors, stacking conveyors, and auger conveyors. Other conveyors that may find use in specialized applications include:

- Roller conveyors—for moving containerized materials like bagged compost,
- Vibrating conveyors—for moving dry bulk materials short distances when a moving belt is troublesome (e.g., because of dust or sanitation), and,
- Pneumatic conveyors—for moving dry particulate materials with the flow of air; such as blowing compost or mulch from a blower truck.

2.5.1 Belt conveyors

In its simplest form, a belt conveyor consists of a head or drive pulley, a take-up pulley, a continuous belt, and carrying and return idlers (Fig. 9.10). Various types of belts can be used depending on the material being conveyed and the application. For example, smooth belts are typically used in horizontal conveyors while inclined belts may have flights or cleats to discourage materials from sliding backwards. Conveyor systems are available in a range of widths, which in combination with the belt speed, determines the capacity of the system.

Belt conveyors are useful at larger composting facilities where operations are spread over many acres, or where processing equipment such as grinders or screeners are permanently located. Conveyors can be used instead of hauling materials from one location to another with front-end loaders, dump trucks, or wagons.

FIGURE 9.11

Examples of a stacking conveyors.

Source: (A) R. Rynk. (B) ECS.

2.5.2 Stacking conveyors

Stacking conveyors are belt conveyors specifically designed to convey and discharge materials to tall stockpiles. They are typically standalone pieces of equipment that are fed by front-end loaders, dump trucks, or another conveyor (Fig. 9.11). An increasing number of screens and grinders are being equipped with stacking conveyors to improve their versatility.

Most standalone stacking conveyors allow material to be heaped in stockpiles that are up to 15 m (50 ft) tall. Increasing the height of stockpiles is an effective way to store large volumes of finished product or amendments within a relatively small area.

The ratio of surface area to pile volume is also significantly smaller for stockpiles produced with a stacking conveyor than for flat-topped piles built with front-end loaders. This is an important consideration in areas of high precipitation as a smaller top-surface area means less moisture infiltration. Stacking conveyors are not limited to making discrete conical piles. By successively moving the stacking conveyor backwards away from the initial stockpile in 3 m (10 ft) increments, a very large pile can be created from an initial conical pile.

The true efficiency of this type of equipment becomes apparent with "radial" stacking conveyors. The midspan support wheels on a radial stacking conveyor can be rotated 90 degrees to the conveyor, which allows the conveyor to pivot in an arc around its lower end which is fixed in place. As shown in Fig. 9.12, the pivoting action allows for the construction of a large arc-shaped windrow without having to incrementally move the tail-end of the conveyor. Once an arc is completed, the radial stacking conveyor can then be moved backwards by about 10 ft, and a new arc can be created on the side of the previous one. By successively rotating and

FIGURE 9.12

Radial stacking conveyor (A) and the pile shape that it can produce (B). Notice the lateral wheels and engine to drive them.

Source: (A) EDGE innovate. (B) Scott Gamble.

moving the conveyor backwards, it is possible to create an 8 to 16 m high stockpile (15 to 25 ft) with a square or rectangular footprint instead of the 4 to 5 m high pile that could be built in the same location using a front-end loader.

2.5.3 Screw conveyors

Screw conveyors, also called auger or helix conveyors, use a helical shaft rotating within a trough or tube. As the screw rotates, the flights push materials through the tube (Fig. 9.13). Screw conveyors are commonly used in agriculture for handling grain, in food processing plants to move dry ingredients, and in wastewater treatment plans for handling grit and dewatered sludges.

While they are less common than belt conveys, screw conveyors do have potential uses at composting facilities, from handling of feedstocks and amendments to moving, elevating, and loading finished products. Screw conveyors are suitable for handling dry and semidry materials such as sawdust and fine wood chips, and

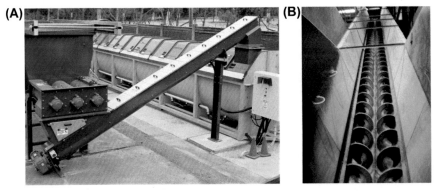

FIGURE 9.13

Screw conveyors loading feedstock into a composting vessel (A). Close-up of conveying auger (B).

Source: (A) Global Composting Solutions. (B) Scott Gamble.

excessively wet materials like biosolids and food residuals. One benefit of screw conveyors in these applications is that the material can be completely enclosed which reduces dust and/or spillage. Screw conveyors can also be set at any angle from horizontal to vertical (although their efficiency generally declines as the angle increases) which allows for more flexibility than conveyor belts when laying out and installing processing equipment.

3. Feedstock receiving and handling

The first step of the composting process is the receipt, inspection and handling of the various feedstocks that will be composted. In their unprocessed or "raw" form, many feedstocks are wet and/or odorous, and may already be undergoing anaerobic decomposition by the time they arrive at the compost facility. If not handled properly and with due care, these raw materials can be a significant source of odors, leachate, and other nuisance conditions. The manner and speed with which feedstocks are handled can greatly influence the neighbors' and the community's acceptance of the composting operation.

3.1 Feedstock inspection

Unless a feedstock supplier has a history of delivering clean material, or the risk of contamination is small, feedstocks need to be checked for unacceptable and prohibited materials before being processed. Ideally, materials obtained from a third-party source should be sorted by the supplier before delivery. However, this is not always possible, or it may require some negotiation of fees charged. Any composting facility that accepts feedstocks or amendments from third parties should establish policies regarding unacceptable and prohibited materials as well as the acceptable levels of contaminants. These policies should be communicated to the third-party suppliers. In some circumstances, it may be appropriate to incorporate these policies into procurement contracts and agreements.

The normal practice at most composting facilities is to inspect materials prior to preprocessing and composting. Preferably, this is done as the material is delivered and unloaded so that if there is a discrepancy with the material, it can be segregated before it contaminates other feedstocks, and the supplier can be notified.

At facilities which accept many loads during the course of the day, inspection of materials is often done by a "Spotter" who also directs traffic (Fig. 9.14). After directing vehicles into the unloading area, the Spotter watches materials as they are unloaded from delivery vehicles. When discovered, these "rejects" can be removed by hand or with the assistance of skid-steer or front-end loaders, and placed off to the side or in a waste bin for later disposal. Due to their nuisance and odor potential, rejects should not be allowed to accumulate at the facility. Rather, they should be removed and disposed of on a regular (i.e., at least weekly) basis.

FIGURE 9.14

Spotter checking bags for contamination while debagging material in noncompostable bags.

Source: Scott Gamble.

3.2 Receiving wet feedstocks

Wet feedstocks are particularly problematic during receiving because they may be odorous or liquids which seep from them can create a mess on the site. Examples of wet feedstocks include segregated food wastes and food processing residuals, biosolids, and fresh dairy manures. The first step in dealing with wet feedstocks is to keep on hand a comfortable amount of the dry bulky amendments (e.g., wood chips, shavings, chopped straw) or feedstocks (dry leaves) that are used in the composting recipe. It is a common practice to unload the wet feedstocks onto a bed of these dry materials (Fig. 9.15). Partially finished compost can also be used for this purpose. The depth of the bed is typically about 30 cm (1 ft) but may need to be thicker if more liquid needs to be absorbed. The wet feedstock and amendment should be mixed immediately and then either covered with more amendment or added to a pile or windrow. Ideally, this receiving step can serve as the entire feedstock mixing operation.

Some ingredients such as food waste, fish, or shellfish offal may be well suited to composting, but are also very odorous. These materials should be brought to the site on a scheduled basis so that operators are prepared to process them immediately, and can handle them in a manner that mitigates odor problems. In sensitive situations, it might necessitate a covered or enclosed receiving area (Fig. 9.16).

FIGURE 9.15

Unloading wet grocery produce onto a bed of wood chip amendment.

Source: Greg Gelewski.

FIGURE 9.16

Enclosed and partially enclosed receiving area.

Source: (A) Scott Gamble. (B) R. Rynk.

3.3 Handling liquid feedstocks

Examples of raw liquid wastes include manure slurries, fish processing wastes, dairy wastes, and wash waters from dairy barns and food processing plants. These materials might be a primary feedstock, a secondary material that the composting system is able to absorb, or they might be added for their nitrogen value. Liquid wastes can also be added to other feedstocks that lack adequate moisture.

Liquid feedstocks can pose several special handling problems. Some liquid wastes, such as spent milk and manure slurries, can cause significant odor problems if they are not processed quickly. Adding liquid wastes can also affect the free air space of the composting pile, by making other feedstocks and amendments soggy and less rigid and thereby allowing the pile to settle. For this reason, the composting mix must contain raw materials that are absorbent enough to hold the added liquid, and/or structured well enough that they do not lose their rigidity when wet. Usually this can be achieved by adding wood chip or oversized materials from compost screening operations.

Even when there is enough of a balance of absorbent and structure materials, care must be taken when adding liquids. Like water, if liquid wastes are added too rapidly or in too large a quantity, they can percolate through a composting pile and leach out the base adding to run-off and odors.

If the volume of the liquid ingredient is small, it can be added during the initial mixing. However, where the amount of liquid to be composted would make the initial mix overly wet, the liquid must be added regularly to an existing windrow, pile, or vessel as it loses moisture. This can be done with liquid-manure handling equipment or it can be sprayed out of vacuum trucks or water trucks (Fig. 9.17B). Turning is necessary soon after the liquid is added to blend it evenly into the windrow. To prevent liquid from running down the side, it may be helpful to create a furrow at the peak of the windrow and deposit liquid in the furrow (Fig. 9.17A).

When the liquid is odorous, it may be better to contain it within the windrow prior to turning. This has been successfully done with fish wastes by injecting it into the windrow with an apparatus mounted to the side of a tractor. In this case, a chisel plow creates a furrow in front of the hose which sprays in the liquid. A trailing disc then covers up the furrow. After the liquid is absorbed and begins to compost, the windrow is turned.

When adding liquid feedstocks to an existing windrow, due consideration must be given to pathogen reduction practices. After a liquid feedstock that does or could

(A)   **(B)**

FIGURE 9.17

Adding liquids to existing piles and windrows. Trough formed in a windrow to accept liquid feedstocks (A). Liquid brewery wastes pumped onto a parched windrow (B).

Source: (A) Scott Gamble; (B) Matt Cotton.

contain pathogens is added to a composting pile, that pile must go through the high-temperature pathogen reduction phase. If the pile has already gone through this phase, the liquid feedstock is essentially "reinoculating" the pile with pathogens and it must go through the high-temperature phase again. The further along the pile is in the active composting phase, the more difficult it may become to sustain the pile temperatures at or above 55°C.

3.4 Removing contaminants

Typical contaminants that can be found in composting feedstocks include plastic bags and containers, styrofoam food containers, wire, metal hardware and parts, glass bottles, rocks, concrete, and pallets. Some of these contaminants can impact the physical and chemical quality of the finished compost. Others can damage processing equipment and can pose a risk to the safety of those working at the facility. Mixers, windrow turners, and other processing equipment with rotating parts is especially prone to damage from long flexible items such as rope, wire, plastic strapping, and plastic shrink wrap.

There are a variety of materials separation operations available for removing these contaminants at either end of a composting system. Their feasibility depends on the qualities of the feedstock or compost, contaminant to be removed, and the subsequent operations (see Box 9.1).

3.4.1 Manual sorting

Manual sorting is common at all sizes of composting facilities. Facilities which accept large volumes of food waste from residential or commercial generators often use a combination of manual and mechanical sorting. At small facilities, manual sorting is often the only method of removing contaminants from incoming feedstocks.

In its simplest form, manual sorting involves having workers walk around the perimeter of piles of recently unloaded feedstocks, and remove visible contaminants by hand or using rakes. A front-end loader can be used to spread the material out for workers, making more contaminants visible and making the picking process easier.

Hand picking in this manner can be very effective and workers can be asked to target specific contaminants that are highly problematic (e.g., glass bottles, shrink wrap, concrete, and rocks). However, this method is also inefficient; a well-trained crew working with a dedicated front-end loader is still likely to miss many contaminants. As well as being unpleasant, this approach also introduces a number of safety hazards including having staff work in close proximity to mobile equipment, a heightened risk of slip/trip/falls or back injuries, and exposing workers to pathogens and sharps that might be present in the feedstocks.

Some of these safety issues can be mitigated through the use of sorting lines and picking stations, similar to those employed by the recycling industry (Fig. 9.18). Here, the material passes in front of one or more workers on a horizontal conveyor belt, and the workers can reach out and handpick contaminants. The sorting lines can be ergonomically designed to minimize the potential for injuries, and can be enclosed and heated or air conditioned to provide protection from the elements.

Box 9.1 Amicable separation options

The need to separate one material from another is universal to all industries. Scientists, engineers, and practitioners have developed numerous ways to do so that accommodate the materials at hand. In his June 2014 BioCycle article, *Contaminant Removal Strategies*, Craig Coker states, "Separation of materials requires identifying the appropriate characteristic by which separation can be done, and optimized. For example, color is the easiest characteristic by which to identify and separate a red bag of medical waste that was mistakenly mixed in with a stream of bagged organics. Magnetic attraction is the characteristic for optimizing separation of nails and screws from ground-up pallets. The key separation characteristics of all recyclables for processing fall into several categories: size, weight, density, hardness, magnetism, electrical conductivity and light refraction. Some materials are separated on the basis of more than one characteristic." Table 9.1, adapted from Mr. Coker's article, lists common composting contaminants and the characteristics and operations used to capture them.

Table 9.1 Composting contaminants and their separation characteristics and operations.

Contaminant	Substrate	Separation characteristic	Operation
Film plastic	Feedstock	Density and appearance	Flotation, air separation or hand picking
Stones, rock, concrete, etc.	Feedstock	Density or size	Flotation or screening (e.g., grizzly)
Soil, small stones, glass shards, etc.	Feedstock	Size	Screening (e.g., disk screen)
Large particles—branches, limbs, trash	Feedstock	Size	Screening (large screen mesh)
Large particles—wood chunks, chips, plastic	Compost	Size	Screening (small screen mesh)
Glass, stones, rock, concrete, etc.	Compost	Size or density	Screening or gravity separation
Light plastic	Compost	Density	Air separation

Adapter from Coker, C., 2014. Contaminant removal strategies. Biocycle 55 (5), 46.

Screening is by far the most heavily used separation operation in composting. Air separation and magnets are distant runners-up. Screening, magnets, and air separation are discussed in the body of this chapter. Flotation and gravity separation techniques are uncommon.

1. Screening: Screening separates materials by size. It is used on both the front-end of composting to remove unwanted contaminants and on the back-end to refine the compost. Screens used to remove contaminants from feedstocks typically have large openings in the mesh to capture the contaminates on the screen.
2. Air separation: This operation uses the force of air to entrain light particles into the air stream and capture them. It is most often used to remove plastic from screen overs, although some separation equipment uses air entrainment to remove packaging from feedstocks.
3. Magnets: Magnets attract and capture ferrous metals (e.g., iron and steel). Magnets are often integral parts of other operational machinery like grinders and screens. They are usually positioned above the feed conveyor. On the front-end, they remove metal objects that are potentially damaging to machinery. On the back end, they remove small ferrous objects, like nails from pallets, and from compost.
4. Gravity separation: Also called density separation, gravity separation relies on dense objects to fall into a collection container while lighter objects fly beyond it. Often there is a current or blast of air to push away the light objects. Some machines use the bouncing characteristics of the contaminant to achieve initial separation. In composting most are used to remove hard nonorganic objects from compost, including small stones and glass pieces. "Destoners and drum separators are two examples of gravity separating devices."
5. Flotation: Feedstocks placed in water-filled flotation separate by density. Contaminants with densities greater than water sink to the bottom while the lighter fractions float or partially float. Flotation tanks are practical for feedstocks that regularly include concrete and rock, like land clearing debris and construction and demolition waste. However, flotation can also be used to collect film plastic or wood as they float to the surface. The feedstock that leaves the tank is thoroughly wet, which can be an advantage if extra moisture is required for composting, a disadvantage otherwise.

FIGURE 9.18

Sorting line (A—front view, B—top view) at a composting facility to remove physical contaminants, such as plastic bags, from feedstocks prior to composting.

Source: Dirt Hugger.

At all facilities, regardless of size or the type of feedstock, it is a good practice for front-end loader operators to also inspect materials as they are being moved from the receiving area and loaded into preprocessing equipment. While it may not be possible to identify the source of the contaminant, a diligent equipment operator can help to prevent major equipment damage. Because of the lower likelihood of contaminants can be seen, relying on equipment operators as the sole means of identifying and removing contaminants is not recommended unless feedstocks are from a consistent group of generators and have a demonstrated history of being contaminant-free.

3.4.2 Magnetic separation

Nails, bottle caps, and wire are examples of ferrous metal contaminants that can find their way into composting feedstocks. If not removed, these items can clog or damage compost application equipment or their sharp edges can result in injury.

Magnets and electromagnets are commonly used to remove ferrous metals from feedstocks, amendments, and compost. Since the magnetic systems commonly used at composting sites require that the material is moved below or over the magnet on a conveyor belt, they are often installed as part of grinding, mixing, or screening equipment and systems.

The simplest ferrous metal removal system involves suspending a large magnet overtop a conveyor belt. As the material moves by on the conveyor, metals are pulled out and upwards by the magnet. The effectiveness of this arrangement depends on the strength of the magnet, the depth of the material passing by on the conveyor belt, and height of the magnet over the belt. The downside of this approach is the magnet must be periodically moved and cleaned off.

A more automated approach is to use a cross belt magnet (Fig. 9.19). This consists of a magnet and conveyor belt assembly that is suspended overtop and perpendicular to the material conveyor belt. Like the stationary magnet, metals are pulled up and out of the material as it passes below the magnet. The conveyor belt that covers the magnet surface then whisks the metal to the side and outside of the

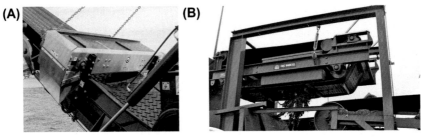

FIGURE 9.19

Cross-belt magnet on a grinder output conveyor (A). Magnetized belt capturing metal from a discharge conveyor (B).

Source: (A) Rotochopper, Inc. (B) Scott Gamble.

magnetic field, where it falls off into a collection chute. The performance of cross-belt systems is affected by the depth of material on the conveyor belt, and they work best if the material is of a consistent depth. Depending on their size, cross-belt systems may also generally require significant structural support to carry their weight.

As an alternative to a cross-belt system, a magnetic head pulley can be used at the top end of conveyor belts on screening and grinding equipment. The magnetic pulley is manufactured to the same size and specification as the standard pulley, but includes magnetic sources embedded on the pulley surface. Rather than being thrown off the end of the conveyor belt, metals in the material adhere to it and travel around the head pulley until they are clear of the magnetic field. Then they fall off the conveyor belt onto an included tray suspended underneath the conveyor, and slide down the tray into a hopper or bunker.

3.4.3 Depackaging

The rising interest in recycling food waste and source-separated organics has created a demand for machines that remove food products from their packaging (Coker, 2019). Both composting and anaerobic digestion (AD) facilities have a need for such equipment. As of 2020, depackaging equipment is available from about a dozen different companies worldwide. Many, if not most, are European in origin.

While the mechanics and details differ among manufacturers, all devices work by first opening the packages in some manner, extricating the food from the package, and then separating the food from the packaging materials (Fig. 9.20). The package-opening step is accomplished with either hammermills, shearing augers, or fast-rotating shafts with paddles. Food materials are freed from the packages either by the same mechanism that breaks apart the package or by additional mixing and agitation. The packaging materials are removed by screen with relatively small holes. Some machines use a screw press to force the food through the screen openings. Streams of water can be added to move the materials, improve separation, and/or clean the rejected packaging material. In composting applications, the amount of water added is minimized, if added at all. Without the added water, the recovered food waste emerges as a paste or pulp.

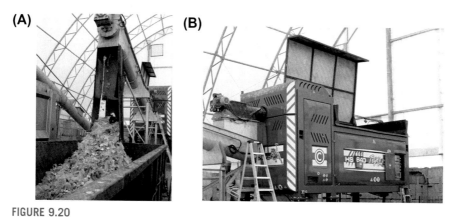

FIGURE 9.20

Packaging material separated from food waste by the depackaging machine pictured on the right (B). The auger conveyor on the left (A) moves the resulting food to the bed of a dump truck that will add the food to composting windrows.

To date, depackaging machines have been successful in capturing packaged food with relatively little packaging remaining. They have opened opportunities for more types of food waste to be composted.

3.5 Temporary storage of feedstocks

Once the feedstocks have been received and unacceptable and prohibited items are removed, they are ready for preprocessing (e.g., size reduction, moistening, mixing) and composting. In some cases, material flows continuously from the inspection stage through to preprocessing (e.g., size reduction and mixing). In other cases, the material is "staged" in short-term storage areas awaiting further processing (Fig. 9.21). In the latter case, it is recommended that feedstocks should not be stored for longer than one day if possible, as a means of avoiding nuisance conditions. It has been demonstrated that the sooner materials are processed and composting begins, the less likely it is that odors will develop. Exceptions to this advice include dry amendments that are stored on hand, waiting to be mixed with wet feedstocks. Wood such as dimensional lumber, tree trunks, and large branches awaiting grinding can also be stored for weeks with low risk of odors. Nevertheless, the general rule is: process all feedstocks as promptly as possible.

Ideally, feedstock (particularly wet feedstocks) should be stored in areas with concrete or asphalt working surfaces. Not only does this help to prevent contamination of soil and groundwater but also facilitates cleaning of the storage area. Periodic cleaning of storage areas by sweeping or pressure washing is recommended to prevent small amounts of materials from accumulating and leading to odors problems. It is common for storage areas that handle liquid manures and sludges to become coated with a hard layer of compacted and dried material called "hardpan" that will generate odors when rewetted.

FIGURE 9.21

Feedstock staging area—receiving and short-term storage of yard trimmings.

The base/floor of storage areas should be sloped to prevent liquids from draining and spreading into other locations beyond the storage area. If possible, the slope should be away from the point where equipment enters the storage area to minimize the amount of liquid that is tracked through the rest of the facility by equipment wheels. Adequate and controlled drainage devices should be provided to keep the liquid from saturating the feedstock, generating odors, and contaminating the surroundings.

Another means of controlling odors is to manage storage areas on a "first-in, first out" basis. That is, the oldest material should be removed from the storage area and processed first. If operators are not diligent in this practice, older materials can be buried under newer materials and can sit in storage for several days.

4. Amendment handling and storage

Amendments used in the composting process are often characterized by low moisture and high carbon content. Common examples include wood chips, straw, sawdust, and horse bedding.

Dry or high carbon amendments respond much more slowly to microbial activity. As a result, they can often be stored for an appreciable length of time before they begin to degrade. If they become wet, they may begin to compost but at a slow rate due to the lack of nitrogen. Handling and storage requirements for these types of materials are therefore less rigorous than for wet or high-nitrogen materials. Often, the amendments can be stockpiled outdoors with no resulting nuisance or environmental impacts. However, prior to making this decision, consideration should be given to the amendment's dust generating potential, to fire protection, and to prevention of spontaneous combustion. The impact of changes to the amendment's moisture content (MC) from exposure to rain and snow, and how this affects their use in a compost recipe, should also be considered.

Outdoor storage of fine amendments such as sawdust in windy locations can lead to the generation of dust. Not only is this a nuisance but also leads to loss of product. In very windy locations, larger particles can also blow onto neighboring properties, which may lead to complaints.

Litter may also be a problem with some amendments such as paper or cardboard. Such materials need to be stored and handled in a way that keeps them contained (e.g., fenced storage area, inside a building). Shredded paper and cardboard should be baled and/or stored inside if not composted immediately.

From a fire protection perspective, stockpiles of amendments should not be overly large, and should be limited in height. There should also be suitable aisles between and around stockpiles to allow equipment access in the event of a fire, and a nearby open area for spreading the material out if there is a fire. A reasonable practice is to limit the height of stockpiles to 5 or 6 m (15 or 18 ft), and to a maximum volume of 750 to 1500 cubic meters (about 1000 to 2000 cubic yards). Fire codes may also contain requirements or provide further guidance on the maximum size of stockpiles.

Composting facilities should always take care to manage materials to prevent conditions from developing that could lead to spontaneous combustion. For instance, operators should avoid driving equipment on the edges or tops of piles as internal heat does not dissipate from the piles if they have been compressed (see Chapter 14).

Amendments are often added to feedstocks as a means of reducing the overall MC of the composting mixture, such as when straw or woodchips are added to manures. If the amendment being used is too wet (e.g., because it was stored outside in the rain), a larger volume of material will be required to adjust the moisture and this may have implications on other aspects of the mixture (e.g., C:N ratio, bulk density [BD]). It may be necessary or beneficial to store amendments in a building to prevent precipitation from increasing their MC. The type of building needed for storing amendments need not be expensive. Simple pole buildings are suitable as are fabric-style buildings. The key is to ensure that the interior clearance within the building is high enough so that site equipment (e.g., front-end loaders, dump trucks) have enough room to work.

If storage in a building is not an option, amendment stockpiles can be shaped to limit the amount of precipitation that is absorbed; stockpiles with large flat top surfaces will absorb more moisture. Alternatively, stockpiles can be covered with tarps that are tied down or weighed down with tires, in a manner similar to hay bales or silage piles.

5. Feedstock preprocessing

As outlined previously, it is sometimes necessary to prepare feedstocks in some manner to help optimize the composting process. For example, dairy manure is too wet and sloppy to compost without mixing in some type of dry structural amendment. Similarly, tree trimmings often have a good balance of carbon and nitrogen, but cannot be composted effectively without being ground up. These two examples highlight the most common preprocessing steps: size reduction and mixing.

5.1 Size reduction

Particle size affects the composting process in two ways, as described in Chapter 3. Smaller particles offer greater overall surface area where biological action can occur. The size of individual particles, as well as their shape, also affects the size and continuity of the pore spaces between the particles, which in turn regulates the movement of air and moisture in the composting mass. The impact of particle size on the composting process should not be overlooked. Often, it provides a means of optimizing facility operations and gaining capacity without a significant investment of capital or increase in operating costs.

The optimal particle size depends on the feedstock and the amendments used, but particles in the range of 4 to 50 mm (1/8 to 2 in.) are generally acceptable. The type of composting method or technology that is used, and the resulting amount of control over aeration rates, also affects the choice of particle size.

In many cases, it is not necessary to size-reduce feedstocks prior to composting. For example, grass, leaves, and manures can usually be introduced into the composting process without grinding or shredding. Feedstocks that benefit from size reduction include newspaper, corrugated cardboard, straw, corn stalks, and other crop residues. Tree stumps, limbs, branches, brush, dimensional wood waste, and other large objects cannot be composted without size reduction. Reducing the particle size of feedstocks and amendments is normally done before the materials are placed in the composting pile.

Size reduction equipment and activities tend to be very loud. Hearing protection must be worn by operators working around this type of equipment. With some types of equipment, there is also a risk of flying debris. Furthermore, when grinding drier materials, such as wood waste, dust is often created which can pose both nuisance and health hazards.

5.2 Size reduction equipment

Size reduction equipment is available in a wide range of sizes and significantly different capabilities (Coker, 2017). Big, heavy-duty grinders are suitable for grinding large amounts of dry wood and brush. Portable units are available with diesel or gasoline engines ranging from about 200 to 1500 kW (about 300 to 2000 hp). Stationary units use diesel or electric engines. Most grinders and shredders have self-contained hoppers for feeding and conveyors for discharging the particles. They usually are loaded using a front-end loader, grapple, or excavator, but sometimes belt conveyor systems are used.

When selecting the appropriate unit, cost is an obvious factor. Since grinding and shredding equipment can be expensive to purchase and maintain, many composting

FIGURE 9.22

Textures of wood particles produced by the hammers of a horizontal grinder (A) and the knives of a chipper (B).

facilities outsource this aspect of their operations to grinding contractors. Grinding equipment can also be rented, although rentals for grinders are generally less available than other equipment (e.g., front-end loaders, screens). Due to the potential safety hazards created by grinding equipment, rental should be considered only if experienced and appropriately trained equipment operators are available.

The most common size reduction equipment used in the composting industry are tub grinders and horizontal grinders, hammermills, shear shredders, and chippers. Other equipment that can be used in certain situations includes paper shredders, large garden shredders, mowers, and forage choppers. When handling "soft" materials such as food waste or leaves, some mixing equipment and windrow turners can also provide the necessary degree of particle size reduction. Many size-reducing mechanisms can be matched with accessory equipment, such as balers, dust separators, conveyors, and screens.

One of the main distinctions between the different types of equipment is the method through which they reduce particle size, and the characteristics of the resulting particles (Fig. 9.22). Grinders (i.e., hammermills) reduce material size through impact, and as a result the particles tend to have ragged edges and are more irregular in shape. Shredders tear materials apart and produce thin elongated particles. Chippers use a cutting action to reduce particle size resulting in cleaner edges and flatter particles that are uniform in shape.

5.2.1 Hammermills

Hammermills use free-swinging "hammers" that are attached to a drum that rotates at a very high speed within an enclosed chamber. Material is introduced into the top of the chamber, and is broken down by the impact of the hammers until it is small enough to pass through openings in the discharge screen or "grates" at the bottom of the chamber (Figs. 9.23 and 9.24). The size of the particles produced is controlled by the size of the grate openings. Particle size can be increased or decreased by replacing the bottom grates with ones having larger or smaller openings. Some hammermills can be quite large and, as such, they are often stationary. Tub grinders and horizontal grinders are types of mobile hammermills. Hammermills also tend to be noisier than other size reduction equipment as a result of their pounding action.

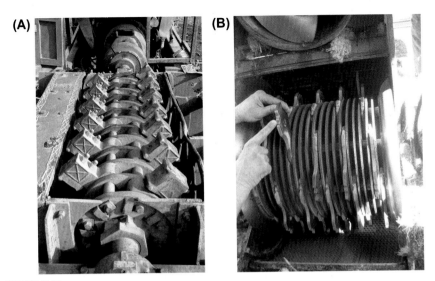

FIGURE 9.23

A large hammermill used for construction and landscape debris (A) and smaller hammermill for agricultural products and wood chips (B). Notice the fixed screens that control the size of the ground products.

Source: (A) Scott Gamble. (B) R. Rynk.

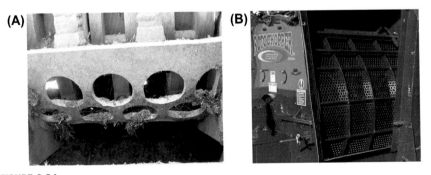

FIGURE 9.24

Examples of screens for hammermill chambers. The screen on the left (A) is for a coarse grind and the right is a fine grind (B).

Source: (A) R. Rynk and (B) Rotochopper, Inc.

5.2.2 Tub grinders

A tub grinder is a variation of a hammer mill that has a large rotating tub, typically 3 to 5 m in diameter (10 to 15 ft), which sits above the hammer chamber (Fig. 9.25). The rotation of the tub feeds material into the hammermill below. As material is ground, it falls through grates below the hammermill, and onto a conveyor belt which transfers it into a stockpile.

The large opening makes tub grinders easy to load, and the tub's rotation reduces clogging and bridging of materials across the face of the hammermill. However, a drawback of this design is the potential for material to be ejected or "thrown" from the tub during the grinding process. To reduce the chance of projectiles flying from the grinder, operators should, as much as possible, keep the tub filled with feedstock. Some manufacturers and operators also manage this safety concern by installing stationary baffles or hydraulically operated lids overtop the tub opening.

5.2.3 Horizontal grinders

Horizontal grinders are another variation of a hammermill. In this case, a metal chain conveyor built into the bottom of a low-sided feed hopper is used to move material laterally toward a cleated feed drum (Fig. 9.26). The feed drum in turn grabs the material and forces it against the hammermill. The hammermill shreds the materials into smaller particles that are small enough to fall through the grates below and behind the mill, onto a conveyor belt. The conveyor transfers and discharges the shredded material into a pile behind the grinder.

Horizontal grinders are available for a range of scales and applications (Fig. 9.27). Because horizontal grinders have a narrow throw zone, they have a lower chance of ejecting objects from the grinding chamber. This is one reason they are becoming more popular in the composting industry.

FIGURE 9.25

Rotating hopper (tub) of the tub grinder feeds material to hammer chamber below.

Source: Morbark, LLC.

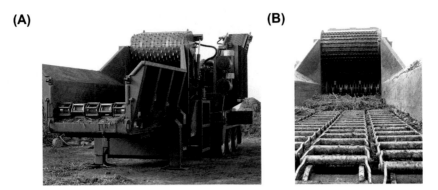

FIGURE 9.26

Horizontal grinder (A). Chain conveyor in the hopper bed carries material to the feed drum, which is currently raised (B). The drum delivers the material into the hammer chamber.

FIGURE 9.27

Examples of horizontal hammermill grinders at various scales and applications. The grinder picture in frame (D) is powered by an electric motor.

Source: (A) R. Rynk. (B) Morbark, LLC. (C) Rotochopper Inc. (D) Peterson Pacific.

5.2.4 Shear shredders

A typical shear shredder (also called a slow-speed shredder) consists of one or two horizontally mounted shafts on which fixed cutting discs or knives are mounted (Fig. 9.28). Materials are loaded into the top of the hopper, and the high-torque rotation of the shafts and gravity combine to pull the materials down and through the machine cutting and shearing material in the process.

With a dual shaft shredder, the two shafts rotate in opposite directions; this causes material to be pulled down through the center of the hopper between the two sets of discs. In a single-shaft system, the single row of discs work in combination with a stationary "comb" mounted on one side of the hopper to provide the needed shearing action. In both cases, discs are often modified with teeth to help pull material through. The size of the particles created by a shear shredder is determined by the size and spacing between adjacent discs on the same shaft, and the spacing between the discs in a two-shaft system, or between the discs and the comb in a single-shaft system.

Shear shredders can process a wide variety of material, including metals and tree trunks, depending on their power and design. Also, they can handle mixed feedstock streams, like municipal solid waste. Shear shredders are used both in stationary and mobile applications (Fig. 9.29).

5.2.5 Chippers

Chippers reduce particle size by slicing materials into small particles. Chippers are commonly used by tree service companies, public works departments, and forestry companies to reduce tree branches and limbs and tree debris into chips. The chips tend to be flatter, smoother, and more uniform in size and shape compared to particles from hammermills and shear shredders.

(A)   **(B)**

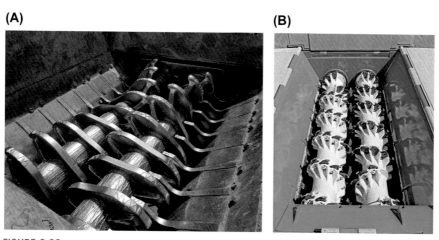

FIGURE 9.28

Examples of high-torque shredding shafts: double-shaft with shredding teeth (A); double-shafts with auger-like cutting flights (B).

Source: (A) EDGE innovate. (B) Komptech Americas.

FIGURE 9.29

Examples of mobile shear shredders.

Source: (A) C. Oshins. (B) Komptech Americas.

Chippers generally fall into two types: drum chippers and disc chippers. Drum-style chippers are the more familiar, often seen towed behind a truck on urban and suburban streets. These chippers have a cutter drum that is mounted horizontally in the feed chute (Fig. 9.30A). The drum serves as both the feed mechanism and cutter. It ejects the chips through an output chute that directs the chips into a container, truck bed, or pile. Drum chippers usually have a horizontal infeed chute that is a few feet off the ground, which is manually fed.

Disc style chippers are a more recent innovation and have cutting blades mounted on a disc that is oriented perpendicular to the material being feed through the infeed chute (Fig. 9.30B). As the disc rotates, it cuts the wood and drum into smaller chips which are ejected out of the rear of the unit. Disc chippers can be small or very large. Some manufacturers make larger units that are mounted on tracks and can be driven around a site. These large units can quickly reduce whole trees to a pile of chips.

5.2.6 Forage harvesters

Forage harvesters have been tried for shredding paper and cardboard with limited success. The harvester shreds paper well but corrugated cardboard tends to jam

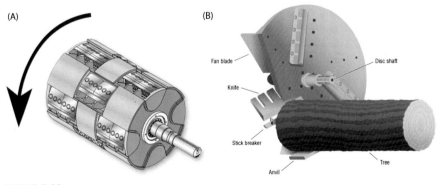

FIGURE 9.30

Drum (A) and disc (B) chipping mechanisms.

Source: (A) Peterson Pacific. (B) Doug Pinkerton.

the chopper. There is a good deal of wear-and-tear on the machinery, and trash from blowing paper can be a problem. Safety is probably the forage harvester's biggest drawback since there are no safety provisions protecting the operator feeding the chopper. For this reason alone, a forage harvester is not a good shredding device and they are not commonly used at composting facilities.

5.3 Mixing

One of the keys to success in composting is to ensure that the various feedstocks and amendments are thoroughly mixed together at the outset of the active composting stage. A homogeneous mixture is essential to ensuring that nutrients and moisture are evenly distributed, and that particles are distributed such that the required free air space is established. Proper mixing helps to prevent pockets of varying biological activity within the material during the composting process. A well-blended mix is most important for aerated static piles, aerated containers, and other composting systems that are not turned or otherwise agitated. Dedicated mixers are more likely to be used for these types of systems.

Blending of feedstocks and any amendments can be achieved in several different ways. Front-end loaders, manure spreaders, windrow turners, and horizontal or vertical mixers are the most common methods.

5.3.1 Front-end loaders

Front-end loaders are commonly used to concurrently move and mix feedstocks. At smaller sites, skid-steer loaders or farm tractors with bucket attachments are also commonly used.

When mixing materials with a loader, a common practice is to lay a bed of drier feedstock or amendment material out on the ground, and then pile the wetter feedstocks on top of this layer. The materials are then mixed by repeatedly "folding and back blading" them with the loader (Fig. 9.31). Push walls combined with a concrete or asphalt pad make mixing with a front-end loader significantly easier.

(A)

(B)

FIGURE 9.31

Food and shredded yard trimmings being mixed with a bucket loader (A) and the final mix (B) for an aerated static pile.

Source: OCRRA.

If sufficient maneuvering space is available, windrows and unaerated static piles can be mixed and formed in a single step on the active composting pad. First, the raw materials are deposited on the composting site in layers, forming a crude pile. Then, the loader is used to mix the materials together and form them into the desired pile shape. This method does not work for aerated static pile systems since the underlying porous base layer and aeration piping prevent mixing in place.

Front-end loaders are capable of producing a good mix, depending on the skill and experience of the operator. This is particularly true of dryer materials such as ground yard waste and wood. Care must be taken with wetter feedstocks such as manures and biosolids as over mixing can lead to the creation of "balls" several inches in diameter that are difficult to break up. Wet feedstocks can also become smeared onto the surface of the mixing pad, which can reduce traction and lead to odors if not cleaned up regularly.

5.3.2 Manure spreaders

Mixing using manure spreaders with horizontal or vertical beaters (as opposed to spinner beaters) is more common at agricultural composting operations since many farms already own this type of equipment. The cutting and slinging action of the beaters provides a good blend of feedstocks and amendments, and also provides a reasonable degree of particle size reduction for softer materials such as clumped manure and food residuals.

In practice, feedstocks and amendments are loaded into the manure spreader in the appropriate proportions. To improve the mixing process, the feedstocks and amendments should be loaded in the spreader in layers or alternating loads (for example, one bucket of feedstock followed by two buckets of amendment, one bucket of feedstock, and so on). Locating the amendment stockpile near the feedstock storage area reduces the time involved in loading the spreader in this manner.

Once loaded, the manure spreader can be driven to the location where the windrow or pile is being formed. The spreader is then slowly pulled forward as it is unloaded so the appropriate pile height is achieved (Fig. 9.32). With lighter materials, it may be possible to increase the height of the spreader's side walls (thereby increasing its capacity) and reduce the number of trips needed to move materials from the receiving area to the windrow.

It should be noted that the manure spreader's size can limit its efficiency; the decks of some rear discharge spreaders are too low to build a windrow high enough for efficient composting. In this case, a front-end loader is needed to restack the pile to the proper height, or to combine two smaller side-by-side piles created by the spreader into one windrow. Truck-mounted spreaders (which have a higher deck/discharge height) work well for building taller windrows, but require a larger front-end loader or farm tractor to load.

An alternative method of using the manure spreader is to operate it in a stationary location. In this case, the unit is parked in one spot and loaded with feedstocks and

FIGURE 9.32

Mixing feedstocks with manure spreader while building a windrow.

amendments, and the mixed materials are discharged into a bunker or pile behind the unit. The mixture is subsequently moved to the windrow with a front-end loader. Depending on the size of the spreader, the loader, and the terrain, this may prove to be a more efficient means of moving material and forming windrows.

One possible downside to using a manure spreader is that it can be inefficient if the tractor that powers the spreader is also used to load it; this necessitates coupling and decoupling the spreader from the tractor. Some materials also pose problems for the typical spreader mechanism. For instance, long straw is more difficult to mix than chopped straw.

5.3.3 Windrow turner

The agitation that results when a windrow turner passes through a pile is very effective at mixing materials together. The preferred method when mixing with a windrow turner is to construct a windrow with alternating layers of different feedstocks, beginning with a layer of amendment at the bottom. Once the windrow has been built to its full height in this manner, the windrow turner is used to mix and blend materials together (Fig. 9.33). The windrow should be turned a minimum of two times to ensure a good degree of homogenization.

Depending upon the types of feedstocks and amendments being handled, it may be possible to combine size reduction and mixing into one step. In this case, the appropriate amounts of unground feedstocks and amendments are layered in the windrow as described previously. The high speed of the windrow turner's flails would then be used to physically break down and homogenize the materials.

FIGURE 9.33

Windrow turner preparing to mix feedstocks placed and proportioned in a windrow.

Source: Johannes Biala.

Typically, two to three passes with the turner is needed to obtain the necessary results. This approach is generally only appropriate for materials such as newspaper, cardboard, and brush. Most windrow turners have sufficient power to go break up larger materials (e.g., wooden pallets), but the resulting particle size is usually too coarse for composting purposes, and this practice can lead to equipment damage and result in flying debris.

5.3.4 Horizontal and vertical auger mixers

Auger mixers, both horizontal and vertical types, migrated to the composting industry from agriculture, where they were used to mix animal feed rations. These mixers are efficient and well suited for use as mixers in composting operations that handle manures, biosolids, and food wastes. However, their ability to handle unground green waste is limited. Additionally, since these machines do not have many secondary uses, they are more common at larger composting sites.

Horizontal mixers have been used at composting facilities for decades. These mixers are loaded from the top, and rely on rotating mixing mechanisms. Some machines have two or three rows of horizontally mounted augers with diameters of roughly 12 inches to mix the materials (Fig. 9.34A and B). Other devices combine augers and rotating mixing bars (Fig. 9.34C and D). In addition to mixing, the rotation of the mixing mechanisms also moves material toward and out the unit's discharge door.

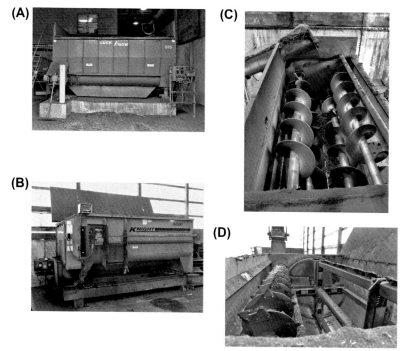

FIGURE 9.34

Stationary horizontal batch mixers for blending biosolids and wood chips. One machine uses multiple rows of augers (A, B), the second uses augers and mixing bars mounted on a rotating reel (C, D).

Source: (A, B) Scott Gamble. (C, D) R. Rynk.

Vertical auger mixers have long been used in the agricultural industry for mixing and distributing livestock rations. However, the adaptation of these mixers by the composting industry is a more recent development. Vertical auger mixers consist of one or two large augers (i.e., up to 4 feet in diameter) installed in the base of a large open-topped mixing tub (Fig. 9.35). The augers are often fitted with "knives" to improve their ability to cut materials and reduce their particle size.

Both types of auger mixers can be mounted on a heavy-duty trailer that can be towed by a farm tractor. In this case, the mixer is usually powered by the tractor's PTO shaft. Alternatively, the mixer can be mounted on a heavy truck chassis for greater mobility and powered hydraulically. Trailer and truck-mounted mixers eliminate the need for separate dump trucks or wagons to transport materials from receiving/mixing areas to composting areas. Some truck-mounted mixers are even capable of mixing materials while in transit, which can significantly increase productivity. Stationary auger mixers are generally electrically powered.

FIGURE 9.35

Mobile, PTO-powered vertical batch mixer (A) with dual augers (B) used for mixing bedded manure, cafeteria waste and yard trimmings.

Loading of these units is normally done with a skid steer or front-end loader, but stationary mixers can also be loaded with conveyor belts. Load cells are a common feature on vertical mixers and allow for more precise proportioning of feedstocks and amendments.

Horizontal auger mixers can be operated in continuous or batch mode. When operated continuously, feedstocks and amendments must be metered into the unit in an alternating fashion (e.g., one bucket of feedstock followed by two buckets of amendment, and so on). Even when operated in batch mode, a more homogenous mixture is achieved when feedstocks and amendments are loaded incrementally.

The design of vertical auger mixers and their ability to effectively mix materials added in layers eliminates the need for incremental loading; all of the amendments can be added at once, followed by all of the feedstocks afterward. However, the mixing process is generally improved if lighter materials are loaded first, and then heavier materials are added on top.

Vertical auger mixers are usually operated in a batch mode: feedstocks and amendments are loaded into the top of the mixing tub and allowed to mix for a short period of time before the unit is completely emptied. Material is discharged from the unit by opening a door in the tub's side wall; the continued rotation of the augers during unloading throws material out the discharge door in a matter of minutes.

When using vertical auger mixers, care must be taken to not over mix the materials, particularly when dealing with biosolids or manure slurries. Mixing for more than 5–10 min can result in the formation of "balls" of materials ranging in size from tennis balls to basketballs. In extreme cases, over mixed feedstocks can become lodged inside the mixing unit and have to be manually removed.

With most mixers, it is necessary to start the augers before or during the loading process as they do not have enough power to start the augers from a standstill when fully loaded. When the mixing process is completed, the materials are discharged directly onto the ground, or onto a conveyor belt which in turn discharges onto a larger stockpile. When equipped with side-discharge conveyors, trailer and truck-mounted mixing units can be used to create composting windrows.

5.3.5 Other types of mixers

A few other machines and techniques to mix composting feedstocks have been used and tested, including pug mills, screener-crusher attachments for front-end loaders and excavators, and rotating drums.

Pug mills have been used for mixing wood chips and biosolids prior to composting in aerated static piles. Pug mills use paddles attached to a rotating horizontal axle to mix materials. They consistently produce a good mix and are able to work on a continuous basis. These work faster than the batch-operated mixers, primarily because the materials are fed into them, and are discharged, continuously. However, the ingredients must be conveyed to the mixer in the proper proportions during its operation. This type of mixer is usually stationary and therefore lacks the mobility provided by trailer or truck-mounted mixers.

Mixing also can be accomplished with "screener-crusher" bucket attachments, which replace the standard bucket on a front-end loader or excavator (Fig. 9.36). The buckets contain a series of interlaced rotating cutting discs and fixed combs on their bottom sides, similar in concept to a slow-speed shredder. They allow an operator to scoop material as if they were a normal bucket, and then hydraulically engage the cutting discs to shred the material until it is small enough to fall between the combs and on to the ground. Screen crusher buckets reduce the size of some compost feedstocks such as bark, food waste, and cardboard, but they are more effective at mixing materials in advance of the composting process. The attachments can also be used to mix and break up clumps when turning windrows and piles, and to mix finished compost with topsoil or other soil products.

FIGURE 9.36

Screen crusher bucket attachment for an excavator (A) and detail of bucket cutting discs (B).

Source: (A) Scott Gamble. (B) R. Rynk.

Rotary drum mixers have been used with varying success for mixing organic feedstocks and amendments prior to composting. A drawback to this type of system is that material can tumble and roll down the inside the drum (instead of being lifted and dropped) which can lead to the formation of clumps and balls as large as basketballs. Wet feedstocks and slurries can also stick to the inside walls of the drum, creating the need for periodic cleaning.

6. Composting operations

The heart of any composting system is inherently the composting process, during which raw feedstocks are biologically converted into compost. The rate at which the process occurs depends on the composting method followed. The composting method, in turn, entails a number of supporting operations including pile or windrow formation, forced aeration and/or turning, and several maintenance operations such as combining piles or windrows and keeping the materials moist by occasionally adding water. Forced aeration and turning are discussed in Chapters 5 and 6.

6.1 General pile and windrow formation

Once the feedstocks and amendments have been prepared and mixed together, they must be transported to the active composting area and formed into piles or windrows. Windrow/pile formation can be accomplished in several ways, depending on the composting method used, available equipment, available labor, and how feedstocks are delivered and received.

Mixing and windrow/pile formation generally demand more labor than other composting operations. To reduce the labor involved or improve the performance, it may be advantageous to obtain new equipment or alter existing equipment—for example, upgrade to a larger manure spreader or purchase a larger bucket for the loader.

Having feedstocks delivered unloaded directly onto composting piles can be an efficient way of reducing labor and equipment requirements (Fig. 9.37) and should be considered where possible. However, there are reasons why this practice is not always practical. It may make inspecting and removing feedstocks more difficult, and delivery trucks may not always be able to access the composting areas during inclement weather. Generally, this practice is limited to feedstocks from sources that are well known to be free of contaminants.

While forming windrows and aerated static piles with a front-end loader is straight forward, some best practices should be followed to preserve the free air space of the composting material. For example, the operator should ensure the front tires of the loader do not drive on any part of the pile, and that the bucket doesn't

FIGURE 9.37

"Live-bottom" truck delivering horse manure directly to windrow.

push down on the top of the pile. Both of these practices can compress the freshly mixed materials and collapse pore spaces. If necessary, the loader can be refit with "high dump" or "rollout" buckets to increase its reach and avoid driving on the edge of the pile. Similarly, materials should not be dropped out of the bucket from an excessive height, or dropped as a single lump. Instead, material should be slowly cascaded from the loader bucket positioned low over the pile. The material should also be unloaded at the top of the pile's side-slope, so that the material tumbles down the side.

Manure spreaders and batch mixers can also be used for forming windrows and piles, and for adding additional feedstock to them (Fig. 9.38). Depending upon the skill of the operator, some cleanup of the windrow or pile may be necessary with a front-end loader. Skid-steer loaders are particularly useful for this task because of their maneuverability.

When the compost site is remote from the mixing area, it may be more efficient to use dump trucks or wagons to transport the materials from the mixing area to the

FIGURE 9.38

Batch mixer unloading mixed feedstocks, forming the beginning of a windrow. A longer conveyor can be fixed to the batch mixer to form higher windrows.

active compost area, and for forming windrows or piles. The materials are unloaded by backing up to the end of the existing windrow and tilting the bed of the truck or wagon while slowly moving the vehicle forward (see Fig. 9.9). The speed at which the vehicle moves forward, and truck or wagon width, determine the windrow/pile heights. If necessary, a loader can reshape or enlarge the pile/windrow formed.

Regardless of how the windrow is formed, it is important that care be taken to keep them straight and parallel with other windrows. A series of crooked or misaligned windrows can end up using significantly more space than a series of straight parallel windrows, and this can affect the capacity and throughput of the composting operation. Permanent or temporary sign posts, flagging tied to fences, pavement markings, painted lines, or some other type of marker can be used by operators to align piles as they are constructed.

Once constructed, it is a good practice to label each individual windrow or pile with a unique tracking number (Fig. 9.39). This allows each pile to be quickly located by all site personnel, eliminates confusion, and eases management and record keeping. The downside to labeling each windrow is that the markers or signs need to be removed and replaced every time the windrow or pile is turned or moved. If used, pile markers should be made from a material that does not contaminate the compost if they are not removed prior to turning and are mistakenly incorporated into the composting pile. A simple and inexpensive solution is to use a wooden survey lathes that are available at most hardware stores. The windrow or pile number can be written on the lathe with a marker.

FIGURE 9.39

Simple and noncontaminating pile label.

6.2 Windrow turning

As outlined in Chapter 5, turning windrows and composting piles releases heat, helps redistribute moisture, and fluffs up the materials to reestablish free air space in the pile. The frequency that composting piles need to be turned depends in part on the type of feedstocks as well as the size of the pile, but generally less frequent turning is required as a pile ages. Initially, it may be necessary to turn piles two or three times per week to meet the pathogen reduction requirements that are often contained in governing regulations. Following the pathogen reduction period, it is generally advised that piles be turned based on the results of temperature and moisture monitoring. It may also be advantageous to turn piles following major rainfalls to take advantage of and evenly distribute the moisture within the pile.

When turning the piles and windrows, equipment operators should take care to avoid driving overtop any part of the windrows as this will result in compaction and a reduction in free air space. After piles are turned, it is a good practice to walk along them and remove any visible contaminants that may have surfaced.

6.3 Combining piles and windrows

As the active composting progresses, and materials are biochemically converted, the volume of material being composted is reduced. The reduction is greater and more noticeable early in the process, and more gradual during curing. To make more space available within the composting facility for fresh materials, it is a common practice to combine two or more smaller windrows or piles together into a single larger pile. Combining piles and windrows may also be advantageous to reduce the effects of the weather. In the winter, combining piles can ensure piles are large enough to be self-insulating and conserve heat within themselves. In dry climates, combining piles (and thus decreasing the amount of surface area per unit volume of material) can reduce moisture loss.

From a site management perspective, it is important that the piles combined are roughly the same age (i.e., formed within a few weeks of each other) as the youngest pile generally dictates the length of time required to stabilize the newly created windrow/pile. Combining a fresh pile with a pile that is partially cured will result in the latter material having to go through the active composting process again, and ultimately having to remain on site for a longer period than is necessary.

Combining piles involves moving and turning both piles. If the facility uses a windrow turner, piling one windrow on top of the other, and then turning them once or twice is appropriate. Without a turner, this method is not advisable as the materials will not be sufficiently mixed and process management problems may arise. Operators should therefore follow the same procedures used when turning single windrows: the outside of old windrows is used to form the inner core of the new windrow, and the inside of the old windrows becomes the outer layer of the new windrow. The size of combined windrow or pile should not exceed the maximum height, width, and length of newly constructed windrows or piles.

For record keeping purposes, the newly formed windrow or pile should be assigned a new unique tracking number, and the records should reflect the tracking numbers of the piles that it was formed from.

6.4 Adding water to windrows and piles

During active composting, moisture is lost from the materials through leaching and evaporative cooling. Some moisture is added back by precipitation but usually there is a net loss. If the net water loss is significant, it is necessary to add water back to the pile to maintain optimal composting conditions. Adding water back to the pile is difficult for several reasons. First, a very large volume of water may be necessary. Second, the water needs to be well-distributed throughout the composting material, which can be a challenge because of the heterogeneity of materials (i.e., some materials absorb more water than others) and the configuration and size of composting piles.

6.4.1 Adding water—how?

In most in-vessel composting systems, water addition systems are incorporated into the equipment design. Operators often need only monitor the MC of the materials (through sampling or indirect measures) and activate and monitor the water addition system.

Adding water into windrows or static composting piles is normally a manual process and can be time consuming. Therefore, it often helps to take advantage of natural precipitation by turning windrows and piles during or after a rainstorm. Although it may be cumbersome to do with a front-end loader, creating windrows and piles with a flat or concave top can be used to increase the amount of rainfall that is trapped and absorbed into the pile (see Chapter 14). Operations with windrow turners can mount a plow on the backside of the turner to simplify this task.

FIGURE 9.40

Spray bar on a windrow turner supplied with water from a tank wagon trailing behind the turner.

Source: Cary Oshins.

Because precipitation alone is seldom sufficient and reliable, composters must add water directly to piles and windrows. Common approaches include water nozzles on turners, water tanks, manually directed hoses, stationary irrigation sprinklers, traveling sprinklers, and low-flow irrigation.

Turners—One of the most effective ways of watering windrows is to add water while the row is being turned. This method immediately distributes the water evenly through the windrow and reduces the potential that it will pool at the bottom of the windrow or on the ground next to the windrow. This method requires permanently installing a spray bar on the turner. Water is supplied to the spray bar either from a trailer-mounted water tank that is towed behind the turner (Fig. 9.40) or by an irrigation hose reel parked at one end of the windrow (Fig. 9.41). Due to the volume of water normally applied, it is not practical to mount a large enough tank directly onto the windrow turner. It can take more time to turn a windrow when supplying water in this manner if water tanks need to be repeatedly refilled.

Mobile water tank—Another common method of applying water to windrows directly from a water truck or a tractor with a water trailer. The truck or tractor drives alongside the windrow. An on-board pump sprays water through the hose or pipe onto the top and one side of the row. To be thorough, water may need to be applied to both sides of the windrow. Typically, a short flexible hose is attached to a boom mounted onto the back or side of the water tank and positioned to spray water sideways (Fig. 9.42A). With a large enough sprinkler, this approach can also be used to

FIGURE 9.41

Water supplied to a windrow turner via a mobile hard-hose reel.

Source: Scott Gamble.

FIGURE 9.42

Delivering water to windrows with a sprinkler (A) or hose (B) from a water truck or wagon.

Source: (A) Johannes Biala. (B) R. Rynk.

add water to the surface of large piles. Alternatively, water can be delivered to the spine of a windrow via a metal pipe with support brackets affixed to the wagon (Fig. 9.42B).

Watering a windrow directly from a water tank allows large volumes of water to be added to in a short time and in a controlled manner. However, this method has its drawbacks. Despite the high application rate, it often takes multiple passes to apply the amount of water needed to a windrow. In addition to being labor intensive, there is a risk that the rate at which water is added will exceed the windrow's ability to absorb the water. Also, some portion of the water sprayed onto the side of the pile also invariably splashes backwards and creates puddles in the aisle. Using a fan-shaped or "duck bill" nozzle helps reduce the splashing. Still, it is common to see water pooled along the bottom edge of the windrow. In addition, water applied in this manner does not penetrate far or evenly into the windrow. For these reasons, it is best to turn the windrow shortly after watering.

Hand-held hoses—At small facilities with only a few piles, water can be manually added to composting piles using a 75-mm or 100-mm hose (3- or 4-in.) connected to a fire hydrant, water truck, or other pressurized water source. In this case, the water could be added in advance of turning, or at the same time that the pile is being turned. The latter approach is more labor intensive since it requires one person to spray the material while another person mixes it with a loader.

Sprinklers—A more efficient method of watering windrows and static composting piles is to spray water over the pile surface using sprinklers. This approach works particularly well if the goal is to moisten a biocover layer that has been placed overtop the static pile so that it is more effective in capturing and decomposing volatile compounds.

Ideally, stationary sprinkler(s) selected for watering windrows and piles should have a radius of throw that minimizes watering the aisles. Sprinkler placement should allow the sprinkler patterns to nearly overlap from one sprinkler head to the next, but no more. Too much or too little overlap leads to uneven watering (Fig. 9.43).

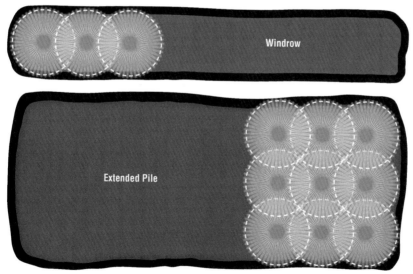

FIGURE 9.43

Recommended sprinkler placement for conventional sprinklers.

FIGURE 9.44

High-volume rotary sprinkler mounted on a water tank.

Source: Scott Gamble.

High-volume rotary sprinklers are common because they allow for relatively large amounts of water to be applied in a matter of a few hours with a relatively low amount of effort (Fig. 9.44). These sprinklers can usually be adjusted to provide a circular or semicircular pattern. However, care must be taken with this method to balance the rate of water application to the rate at which the water is absorbed by materials in the pile. If more water is applied than can be absorbed, water may percolate down through the middle of the pile and saturate its base. A downside to using high-volume rotary sprinklers is that a large amount of water often lands on the ground beside the pile as well as on the pile itself. Not only is this an inefficient use of water, but it can also create puddles (which can subsequently lead to odors) and can make the aisles between adjacent piles muddy enough that they are inaccessible to equipment.

To increase water penetration and improve irrigation efficiency, it has been reported that some composters have temporarily leveled windrows during the time sprinklers are applying water (Rynk, 2001a). In this case, the upper sections of the windrows are pushed into the aisle space separating the windrows. After the water is applied, the windrows are reformed.

Even when water is uniformly distributed over the surface of a composting pile with a sprinkler system, it may not penetrate fully into the pile or it may find channels as it migrates down through the pile. It is therefore best to mix or turn the pile shortly after watering to help distribute the water evenly through the pile.

Traveling sprinklers—Another high-rate alternative to using water trucks is to use auto-rewinding irrigation hose reels and sprinkler gun carts. With this approach, the hose reel is positioned at one end of the windrow and the hose is unspooled so it

FIGURE 9.45

Traveling sprinkler gun at a compost site.

Source: Scott Gamble.

sits parallel to and runs the full length of the windrow. When water is supplied to the reel, its auto-rewind function engages and the hose slowly pulls the sprinkler gun cart along the length of the windrow (Fig. 9.45). Aisle spacing must also be adjusted accordingly to allow for access, and if traveling systems are used, the aisles must be clear and reasonably level.

It is possible to use lateral-move irrigation systems to water windrows, but only when the height and width of the windrows and the aisles between them are consistent throughout the year. Such a system could be fitted with standard sprinkler heads that broadcast water over both the windrows and the aisles, or it could be customized so that water is just applied to the windrows. However, this approach is costly relative to the other methods described, and requires that aisles are kept relatively level and clear. As a result, this approach is rarely used.

Low-flow irrigation—Applying water with a system of low-flow sprinklers or industrial drip tubes can be used to overcome some of the problems encountered with using high-volume rotary sprinklers (Fig. 9.46). These devices apply the water at a slower rate, which allows more of it to be absorbed by the material in the pile, and generally results in less water being sprayed on the ground next to the pile. However, the slower rate means that watering with these systems may have to be done over the course of two or more days. They also normally involve a larger network of hoses, which means there is more work involved with placing and relocating them.

FIGURE 9.46

Watering a static pile using low-flow irrigation.

Source: Cary Oshins.

6.4.2 Adding water—how much?

The amount of water needed to achieve a target MC can be calculated from the following equations:

$$W_w = DM/s_t - DM/s_i \tag{9.1}$$

$$DM = TM_i \times (1 - m_i) = TM_i \times s_i \tag{9.2}$$

where: W_w is the mass or weight of water needed

DM is the dry mass (weight) in the material or pile

TM_i is the initial total mass of the material (wet or "as is" mass or weight)

m_i is the initial or current moisture content of the material on a wet basis

s_i is the initial solids content as a decimal fraction ($s_i = 1 - m_i$)

s_t is the target solids content as a decimal fraction ($s_t = 1 - m_t$)

m_t is the target moisture content on a wet basis.

In Eq. (9.1), the first term (DM/s_t) is the final weight of the material after water is added. The second term (DM/s_i) is the initial weight of the material before water is added. The difference is the weight of water added to bring the material up to the target MC. The key is that the dry mass remains constant and can be used to calculate the total, or "wet," mass at both sides of the moisture addition. The equations are cleaner by using the solids contents instead of the MCs but either can be used ($s_i = 1 - m_i$, and $s_t = 1 - m_t$).

If it is more practical to handle water by volume (V_w), the weight (W_w) can be converted by dividing it by the density of water. Convenient values for the density of water are 1 kg per liter or 1000 kg per cubic meter or 8.34 lbs per gallon.

Table 9.2 summarizes the calculations for 1000 kg and 2000 lbs of material at selected MCs. Example 9.1 demonstrates the use of the equations. Example 9.1 demonstrates the use of the equations. Table 9.2 is provided as a quick reference tool that can be used to quickly approximate water requirements in lieu of completing the detailed calculations.

Example 9.1 How much water?

After four weeks of composting, the moisture content (MC) has fallen to 45% in a windrow with a mixture of cattle manure and corn stover. The composter would like to know how much water is needed to raise the MC to 60%.

The average dimensions of the windrow are 4.5 ft height × 12 ft width and 120 ft length. The windrow has a roughly trapezoidal shape. The bulk density (BD) was measured to be 800 lbs/yd^3.

Step 1: Calculate the total volume (V) of material in the windrow.

Note: For a trapezoidal shape, the cross-sectional area (A_x) is: $A_x = h \times (w-h)$

Volume is: $V = A_x \times L = [4.5 \text{ ft} \times (12 \text{ ft}-4.5 \text{ ft})] \times 120 \text{ ft} = 4050 \text{ ft}^3 = 150 \text{ yd}^3$

Step 2: Calculate the initial total mass (weight) of the windrow (TM_i).

$TM_i = V \times BD = 150 \text{ yd}^3 \times 800 \text{ lbs/yd}^3 = 120,000 \text{ lbs}$.

Step 3: Calculate the dry mass (DM) of the amount of material selected using Eq. (9.2).

$DM = TM_i \times s_i = 120,000 \text{ lbs} \times (1-0.45) = 66,000 \text{ lbs}$.

Step 4: Using Eq. (9.1), calculate the amount of water needed (W_w) to raise the material to the target moisture content.

$W_w = DM/s_t-DM/s_i = [66,000 \text{ lbs}/(1-0.60)]-[66,000 \text{ lbs}/(1-0.45)] = 45,000 \text{ lbs}$.

Step 5: Convert the weight of water (W_w) to volume using the density of water.

$V_w = W_w/D_w = 45,000 \text{ lbs}/8.34 \text{ lbs/gal} = 5396 \text{ gallons of water}$

Similar, but not equivalent, metric units for this example are:

Windrow dimensions = 1.5 m high × 4 m wide × 40 m long.

$BD = 475 \text{ kg/m}^3$	$A_x = 3.75 \text{ m}^2$	$V = 150 \text{ m}^3$
$TM_i = 71,250 \text{ kg}$	$DM = 39,188 \text{ kg}$	
$W_w = 26,719 \text{ kg}$	$V_w = 26,719 \text{ L} = 26.7 \text{ m}^3$	

---

7. Curing

Following active composting, compost requires a curing period of several weeks to finish the process and allow the compost to develop the desired characteristics for its intended use (Shaffer, 2010). Many facilities cure the compost in the same location and in the same windrows as the active composting step. This approach reduces material movement and handling. At facilities that use in-vessel or aerated static pile composting systems, curing is normally considered a separate step and materials are moved from the active composting area to a new location. Some windrow

Table 9.2 Volume of water required to increase the moisture content of composting materials per metric ton (1000 kg) and per US ton (2000 lbs) of material.

Liters (or kgs) of water to add per metric ton (1000 kgs) of pile material

Initial Moisture	Dry Mass DM (kg)	Target moisture content								
		30%	35%	40%	45%	50%	55%	60%	65%	70%
30%	700	0	77	167	273	400	556	750	1000	1333
35%	650		0	83	182	300	444	625	857	1167
40%	600			0	91	200	333	500	714	1000
45%	550				0	100	222	375	571	833
50%	500					0	111	250	429	667
55%	450						0	125	286	500
60%	400							0	143	333
65%	350								0	167
70%	300									0

Gallons of water to add per US ton (2000 lbs) of pile material

Initial Moisture	Dry Mass Weight (lbs)	Target moisture content								
		30%	35%	40%	45%	50%	55%	60%	65%	70%
30%	1400	0	18	40	65	96	133	180	240	320
35%	1300		0	20	44	72	107	150	206	280
40%	1200			0	22	48	80	120	171	240
45%	1100				0	24	53	90	137	200
50%	1000					0	27	60	103	160
55%	900						0	30	69	120
60%	800							0	34	80
65%	700								0	40
70%	600									0

facilities also relocate materials to a separate curing area where they are finished in larger windrows; doing so frees space on the composting pad for the active windrows and piles which are more intensively managed.

When a separate curing area is used, it should be situated up-slope from receiving and active composting areas, or otherwise out of their path of drainage. This precaution prevents the curing piles from being contaminated by run-off from these areas.

Forced aeration or frequent turning of curing piles is typically not necessary so long as active composting had been sufficient and the curing piles are small enough and porous enough to permit adequate natural air exchange. A maximum pile height of 3.6 m (12 ft) is suggested. Since the curing piles are not turned (or watered) as frequently, it may be possible to place them closer together to conserve space.

The compost curing process does not generate enough heat to evaporate any moisture that accumulates from heavy precipitation or run-off. As a result, excessive moisture or water can accumulate in the base of the pile, and anaerobic conditions can arise. To prevent this from occurring, surface runoff from other areas should be channeled away from the piles. The curing pad itself should also be well drained, and the length of the piles should run parallel with the slope of the pad surface. Curing piles can be covered with tarps to exclude precipitation, given their relative comparatively low oxygen demand (Fig. 9.47). Preferably, the tarps should be made from a material that allows some oxygen exchange.

Since curing piles are still undergoing decomposition, albeit at a much slower rate, it is still important to encourage aerobic conditions. Odors from anaerobic conditions in curing piles can be as offensive as anaerobic conditions elsewhere

FIGURE 9.47

Curing piles covered with tarps to shed rain, reduce dust, and exclude weed seeds.

in the process. Anaerobic conditions in curing piles can also result in generation of compounds that are toxic to plants. For this reason, it is prudent to allow compost taken from curing and storage piles to "air out" briefly in smaller uncovered piles before being used or sold.

8. Postprocessing

Depending upon the feedstocks and the application, it may be possible for the cured compost to be used or sold without any further processing. For example, compost made from leaves does not necessarily need screening. Similarly, compost made from manures amended with sawdust, wood shavings, or straw may be suitable for application to agricultural lands without screening. However, if larger amendment particles were used, or if the composts are to be sold or used in nonagricultural applications, some form of postprocessing is generally required. The most common postprocessing step is screening. Drying, magnetic separation, and blending are other, less common steps.

8.1 Screening

Screening separates materials based on size. It is done mainly to improve the quality of the compost product and/or to recover bulking agents. Screening is the most common method of adding value to the compost. It creates compost grades with particle size ranges suited to the intended use of product (e.g., soil conditioner, mulch, top dressing, ingredient for manufactured topsoil, and growing media). Also, it removes physical contaminants that may diminish the compost's value or use.

For many composting facilities, screening is only an occasional operation, required once or twice yearly for only days at a time. Therefore, it often makes sense to rent screening equipment or hire out the task to a screening contractor. At larger facilities, it is generally more cost effective for the facility to purchase or lease and operate the screening equipment.

Screening at most composting facilities is performed after the curing step. At some composting facilities, screening is sometimes used as a preprocessing step to remove unwanted objects such as rocks, metal, and plastic bags from waste-derived feedstocks like leaves, yard trimmings, and food. Screening equipment can also be modified to open bags and remove plastic from feedstocks, or it can be used to blend various final products together.

Screens filter material through a moving mesh or network of discs or fingers. Smaller particles pass through the openings in the mesh, or the space separating discs/fingers, and then fall onto a conveyor, pile, or container. Most screens separate materials into two fractions: oversized material or "overs," and undersized materials, the "unders" or "fines." Normally the undersized fraction is the desirable product. The size ranges of the fractions are determined by the size and shape (i.e., circular, square, or rectangular) of the openings in the screen surfaces and the motion or the screen. Some equipment uses multiple screening surfaces with different opening sizes to create three or more size fractions.

Common screen sizes used at composting facilities range from 6 to 25 mm (¼ to 1 in.), although sizes up to 50 mm (2 in.) are not uncommon when removing large contaminants. The choice of screen size depends upon the material to be separated out and the end use for the compost. For producing finished compost products, the most common screen sizes are 10, 12, and 16 mm (3/8, ½, and 5/8 in.). Wider screens, 50 mm and larger, are used for coarse mulch products.

Smaller screening openings provide better separation but, for a given screen, reduce the throughput capacity of the screen (e.g., cubic yards per hour) and increase the chances of "blinding" and "flooding." Blinding refers to the condition when the screen openings become blocked with material. Most screens include some method to reduce blinding, like brushes, vibration, or bouncing balls. Flooding occurs when the capacity of the screening equipment is exceeded, and also when considerable blinding occurs. When screens are flooded, much of the material runs through the equipment without touching the screen and ends up in the oversized fraction. Flooding can be reduced by slowing the feed rate.

Screen performance is judged primarily by two measures: throughput capacity and screen effectiveness. Throughput is how fast the screen processes a material, and usually varies from material to material (e.g., peat vs. sand and gravel vs. compost) for the same piece of equipment. It also depends on the MC of the material, feed rate, screen movement, angle of incline, and the screen size (width or diameter and length). In general, the feed rate determines the throughput. In turn, the feed rate can be increased with more vigorous screen movement, larger openings, and/or greater screen surface area (e.g., length, diameter, or width). In practice, the throughput rate is also determined by how fast and consistently the screen is loaded. Because so many variable factors affect the throughput, manufacturers are often the best source of throughput estimates. Where possible, manufacturer estimates should be checked against the experiences of other composting facilities that handle similar materials under similar conditions.

Screen effectiveness relates to the success of separating the particles into the desired fractions: Effectiveness decreases when particles smaller than desired are retained in the oversized and when particles larger than desired pass through the screen into the undersized fraction. For a given material (e.g., particle shape and size distribution), effectiveness is impacted by the screen opening size, the thickness of the screen wire or discs, blinding, throughput, and screen surface area. For example, as the throughput rate increases, more small particles may end up in the overs pile.

Generally, a balance must be struck between throughput and effectiveness when selecting equipment parameters like screen opening size, screen diameter, and length. When processing compost, keeping the material in the screen longer, either by slowing throughput or using a longer screen is preferred since this captures more compost. However, it does allow more time for oversized particles to "spear" through the screen openings and end up in the product.

Screens generally perform better with dryer material, which is one reason why compost is usually screened after the curing stage. If the compost is too moist, material can clump together and inadvertently end up as part of the oversized fraction. Moist compost can also stick to the screens, slowly reducing the effective size of the openings and affecting screen throughput and performance. The MC should generally be less than 50% and preferably less than 45%. However, if the MC of the compost is too dry (<35%), the screening process can generate significant amounts of dust which can lead to nuisance and other problems.

8.2 Screening equipment

There are many different types of screens, with a variety of add-on options and features, available from a large number of vendors. Common features include portability, feed hoppers, conveyors, and devices to reduce blinding. The types of screens most commonly used to separate compost and other soil-like materials are trommel screens and star screens. Other types of screens used by composters include disc, shaker, vibrating and orbital screens.

Most screens used in composting facilities are portable. They are intended to be set up close to the material being screened, and loaded with a front-end loader. Larger facilities may have stationary screening equipment installed inside full or partial enclosures; material is delivered to the stationary screens with trucks, front-end loaders and/or conveyor systems. Stationary systems usually have a higher capacity, and the enclosure allows them to operate more effectively in the rain or during the winter or during windy conditions.

Feed hoppers are an integral feature of almost all portable screens. The feed hopper balances the flow of material to the screens and prevents flooding. They generally range in capacity from 1 to 10 cubic meters or yards, roughly increasing with the screen size. A given screen may be available with different feed hopper capacities. The capacity of the feed hopper is particularly important when the equipment is filled in surges, as occurs with a front-end loader. For efficient operations, the hopper should have, at a minimum, the same capacity as the bucket on the loader. The width of the hopper should also be the same or greater than the width of the loader bucket so the equipment operator can unload without spilling material. The height of the hopper's top lip above ground is another key consideration: if the hopper is too high for the loader to reach, then it is necessary to use a ramp to access the hopper. Requirements for safe operation of loaders on ramps generally slows the loading process.

Nearly all screens have at least one conveyor for discharging the screened product. Many have a second conveyor for oversized material. Screens that produce multiple product sizes have multiple conveyors. Some manufacturers offer useful options for discharge conveyors like covers, to minimize dust and odors, and water nozzles, to reduce dust and add moisture.

8.2.1 Trommel screens

A trommel screen is a rotating horizontal screening drum (Fig. 9.48). The screening drum normally has a slight decline so the material flows along its length as the drum rotates. Unscreened material is fed into the interior of the drum at one end, and as it travels the length of the drum, small particles fall through the screen openings in the drum's surface. Material that is too large to pass through the screen openings is discharged from the other end of the drum.

The drum on a trommel screen normally consists of a circular steel frame to which screening panels are secured (Fig. 9.49). Stationary trommel screens normally use steel panels with holes punched in them. A few manufacturers of mobile screens also use these "punched-plate" screen panels, but it is more common to use wire mesh screen panels which wrap around the circumference of the drum. Some trommel screen manufacturers also install internal flights inside the drum to help move the materials through. The flights also help to stiffen and reinforce the drum. Trommel screens typically have brushes to clear the mesh of particles.

Although a drum can be fitted with screen panels with two slightly different opening sizes, it is more common to use the same sized openings along the entire length of the drum. If it is necessary to change the screen size, the entire drum is lifted out of the screen bed with a loader or excavator and replaced with a different drum with a different screen size. Lifting and replacing a drum typically takes less than one hour. However, removing and replacing screen panels from a drum can take many hours or even days, depending on crew size. Since replacing the screen panels on a drum can result in lost productivity, it may be worth investing in a second drum if screen changeouts are necessary on a regular basis.

The throughput capacity of the trommel is dictated by the length and diameter of the drum as well as the screen opening size. Trommel screens range from 1.5 to 2.5 m (5 to 8 ft) in diameter. Lengths range from 2 m (6 ft), for small-scale screens

FIGURE 9.48

Trommel screen; screening in operation.

Source: Komptech Americas.

FIGURE 9.49

Trommel screen input and output conveyors.

Source: USCC.

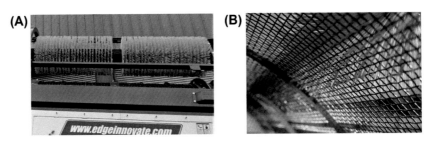

FIGURE 9.50

Trommel brushes for clearing screen openings on trommel screen drum.

Source: (A) Edge innovate. (B) Vermeer.

to over 18 m (60 ft), for large stationary systems. A typical trommel screen for compost processing would measure about 2 m (6 ft) in diameter and roughly 4.8 to 8.5 m (16 to 28 ft) in length. The top of the trommel screen drum is normally exposed and a rotary brush is fixed to the outside surface to clear the screen openings and overcome blinding (Fig. 9.50).

One advantage of trommel screens is that they are simple machines. They can be made to suit a wide range of scales and applications, from a small farm to a large manufacturer. A few ambitious and mechanically skilled composters have successfully designed and constructed their own trommel screens.

8.2.2 Star screens

Star screens were introduced to composting in the late 1990's and have since become as common as trommel screens. These screens rely on a bed of interlaced hard-rubber star-shaped discs mounted on a series of parallel horizontal shafts (Fig. 9.51). The shafts rotate in a common direction causing the stars to also spin. The material is introduced at one end of the screening deck, and the coarser material stays on top of the stars and is thrown toward the other end and off the screening deck. Smaller pieces fall between the stars as they rotate, and onto a conveyor belt located below the screening deck. The size separation depends on the separation distance between the stars and their speed of rotation. Some models of star screens have two decks of stars to produce two product sizes along with the overs.

Some screens use discs with simpler shapes in place of the rubber stars. Although these screens are sometimes referred to as "disc" screens, they are like star screens in their operation and functions. They differ from the disc screens discussed in the next section.

The throughput capacity of a star screen is determined by the rotational speed of the star shafts, the size and spacing of the stars, and the width and length of the screening deck. The rotational speed can be adjusted, which allows for fine-tuning of the screens.

Star screens have several advantages, and several disadvantages, over other types of screens. They tend to be more forgiving of higher MC materials than trommel and deck screens. Thus, they are well suited for screening moist compost. Also, the size separation is more flexible. Instead of changing screen mesh, the desired separation

(A) 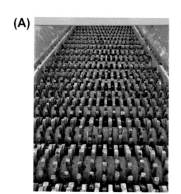   **(B)**

FIGURE 9.51

Star screen in operation (A) and its 247 mm star deck (B).

Source: Neustarr.

can be adjusted in minutes by changing the screen's operational settings. The centrifugal force on the spinning stars works to keep material from building up on the stars. However, it is not completely effective; stickiness happens. Thus, some star screens have separate "fingers" to clean the stars. On the negative side, star screens are more complex machines. There are more components that move, more components that wear, particularly the stars, and more maintenance to be performed. Multiple rotating shafts provide multiple opportunities for stringy objects, like plastic film, to wrap around and plug the equipment.

8.2.3 Disc screens

Disc screens are very similar in appearance to star screens. However, instead of hard-rubber stars, disc screens use steel discs and rotate more slowly (Fig. 9.52). The steel discs are more resistant to damage and are much more suitable for handing larger heavy objects. As with star screens, material is loaded on the screen deck and moves down the deck by the rotation of the discs. Undersized material falls between the disks and the shafts due to the separation distance between the shafts. On disc screens, the rotation is primarily for conveyance.

The primary use of disc screens is for course separation. Disc screens are generally used at the preprocessing stage to remove contaminants from feedstocks or prepare materials for grinding, rather than for finishing compost products. For example, disc screens can be used to exclude soil, rocks, and/or leaves from yard trimmings prior to grinding, thus reducing wear on grinding equipment parts. They can also be useful in sorting construction and demolition debris. Disc screens also are prone to being clogged by plastic film and other stringy items wrapping around shafts.

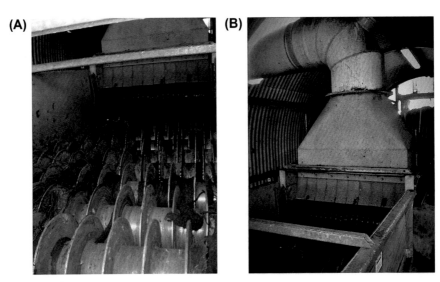

FIGURE 9.52

Disc screen surface (A) with a self-built vacuum system (B) for pulling off light plastics.

8.2.4 Vibrating and shaker deck screens

Deck screens are often used to separate fine materials in industrial processes, but some models have been adapted specifically for handling compost, peat, and other soil-like materials. Deck screens use a flat screening deck that is set on a slight incline. The screens are constructed of wire mesh, perforated panels, or "piano wire." The screen deck moves rapidly in an oscillating (shaking) or orbital (vibrating) motion to enhance the separation. The slope and/or the motion of the screen moves the oversized particles along the screen deck while small particles fall through onto a conveyor (Fig. 9.53).

The throughput of deck screens depends on the nature of the material being screened, the screen's mesh size, and the deck area. For a given material and screen mesh, a larger deck area allows for a faster feed rate and faster throughput.

Deck screens are prone to blinding with moist materials like compost. Some deck screens have cleaning balls that bounce against the underside of the screen to help dislodge material blinding the screen openings.

Shaker screens are deck screens that oscillate horizontally at a moderate frequency. Many shaker screens have two or three decks stacked one overtop the other. This allows the undersized fraction from the first screen to be further separated into two sizes. The decks are sloped to discharge the screened fractions onto conveyors, although the overs may simply fall into a pile on the ground. Shaker screens are common for separating and sizing aggregate materials like stone, compost, and mulch.

Vibrating screens are deck screens that undulate at a high frequency. The orbital movement of the screen conveys the material horizontally along the surface of the screen. The motion also lifts the material vertically which helps separate intertwined particles from each other and increase screening efficiency. Vibrating screens are normally used for separation of dry fine particles like sand and soil. They are less common at composting facilities.

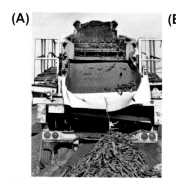 

FIGURE 9.53

Deck screen shaker deck (A) and in operation (B).

8.2.5 Orbital screens

Also called a dish or satellite screen, an orbital screen consists of screen panels that are mounted on a rotating bowl-like frame or dish (Fig. 9.54). During operation, the dish is set on an incline, and material is fed from a hopper into the unit by a conveyor belt. Fine materials fall through the screen and are conveyed away. Coarse materials fall over the lower lip of the screen onto a separate conveyor. The rotation moves material over the screen surface and separates fine and coarse materials. In the most common units, the dish rotation and the conveyor are powered by an onboard diesel engine. However, some units are also driven by electric motors and tractor PTO drives.

Like trommels, orbital screens are flexible and mechanically simple, with few moving parts. The screen panel can be lifted out with a loader and replaced with a different-sized mesh. Orbital screens are typically portable with the hopper and feed conveyor integrated into the unit. These screens are suited to a wide range of materials. However, they have lower throughput capacity and tend to be used at smaller facilities, like on-farm facilities.

FIGURE 9.54

Orbital screen.

8.3 Film plastic removal

Film plastic contamination in feedstocks and finished products is one of the most vexing problems faced by the composting industry. Small bits of film plastic, particularly white or light-colored plastics, significantly affect the aesthetics of a finished product and reduce its market value.

The most effective way to deal with this contamination is to keep plastic out of feedstocks and out of the composting facility. This can be achieved through education of feedstock generators and through the design of collection programs (i.e., using paper bags or collection container instead of plastic bags). Inspecting loads of feedstocks as they arrive at the composting facility and rejecting those that contain plastic is another practice that can be implemented.

Once they have been mixed with feedstocks and gone through the composting process, there is no easy way to remove film plastics. While larger pieces of partially shredded plastic can be removed through hand picking, it is time consuming. Hand sorting to remove the coin-sized pieces of film plastic that result from grinding and turning is not practical. Instead, most facilities rely on running compost through 6 to 12 mm (¼ to ½ in.) screens to remove film plastic from finished product. While some smaller pieces of film plastic invariably pass through the screens, it usually results in acceptable levels of film plastic contamination in the product. The downside to this approach is that all of the film plastic is concentrated in the oversized material from the screening step. It reduces the ability to reuse the oversized material as a bulking agent in new composting piles.

In recent years, several air-separation systems have become available, with the primary purpose of cleaning plastic from screen overs and coarse feedstocks like ground wood. Using vacuum or "wind sifting" technologies, these systems have shown good success in removing a large portion of the film plastics (Goldstein, 2005). Plastic recovery rates fall in the range of 80%—95%. The success of these units in a particular application depends on the size and density of the plastic particles; smaller and lighter particles being easier to capture. The MC of the overs stream also affects performance; air separation is less effective when the MC increases greater than 45%.

Vacuum systems use suction hoses or nozzles to lift and remove film plastics and other lightweight materials from heavier particles as they pass by on a conveyor belt underneath the hose or nozzle (see Fig. 9.52B). The light materials are sucked through the vacuum fan and discharged into a nearby container or a mesh bag. Alternatively, the hose or nozzle can be situated to capture film plastic as the material cascades off the end of the conveyor belt.

Vacuum systems are usually mounted directly on the overs belt of the screening equipment and in some cases can be powered by the screen itself. The suction device is usually mounted on an adjustable support so it can be moved closer or further away from the conveyor belt. This, in combination with adjusting the speed of the vacuum fan, is used to fine tune the extraction process. These systems work best if the material on the belt is spread out in a thin layer so that light material is "visible" to the vacuum instead of being buried underneath heavier material particles.

In contrast, wind sifters are typically standalone pieces of equipment. Wind sifters are more complex than vacuum systems, but this added complexity allows them to remove more film plastic. A wind sifter is typically fully enclosed and employs two fans (Fig. 9.55). One fan either blows air across the material as it bounces along a shaker deck or a conveyor belt, or blows air through the material as it cascades off a conveyor belt. This extracts the light material from the heavier materials and makes it airborne. The second fan provides a strong vacuum to remove the

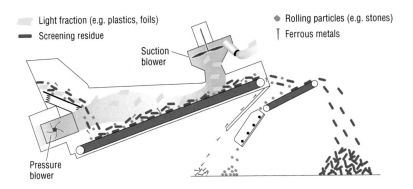

Light fraction (e.g. plastics, foils)

Screening residue

Rolling particles (e.g. stones)

Ferrous metals

Suction blower

Pressure blower

FIGURE 9.55

Air separation schematic.

Adapted from Rynk, R., 2001b with permission of Komptech Americas.

airborne material from the enclosure. In addition to separating film plastic and other lightweight materials, wind sifters can be equipped with drum separators or incline conveyors to also remove very heavy items such as rocks and concrete from the screening overs.

Since these units often have a feed hopper, they can be fed with material from a stockpile using a front-end loader. However, it is more efficient to use a wind sifter in combination with the screen, placing it such that the overs conveyor from the screen discharges into the wind sifter's feed hopper (Fig. 9.56). This approach eliminates the effort required to stockpile and subsequently rehandle/reprocess the screening overs.

FIGURE 9.56

Air separation receiving screen overs directly from the screen conveyor and discharging the separated fraction to a waste container.

Source: Komptech Americas.

8.4 Drying

A reasonable goal is to produce compost with a MC between 35% and 45%. It is more efficient to screen compost when it is dry, but when the MC is below 35%, handling and screening the material results in excessive dust. When the MC starts to approach 50%, screening efficiency will be noticeably reduced. Mature compost does not generate enough heat to drive off any excess moisture from rainwater or overwatering, so in order to meet these specifications, it may be necessary to dry the material through other means.

In aerated static pile composting systems, drying can typically be achieved by increasing the aeration rate toward the end of the process, or extending the length of time that materials remain in the aerated piles. In windrow and static pile systems, drying can be achieved with the help of the sun and wind by turning more frequently during the later stages of composting. Alternatively, during warm, dry weather, the windrows and piles can be broken down into smaller piles and allowed to dry naturally.

If the compost produced is consistently wetter than desired, this may be a symptom of other problems. For example, wet compost can result from the MC of the initial mixture being too high; a lack of energy in the raw materials; infrequent turning/aeration; drainage problems at the composting site; or cold, wet weather.

Compost that is intended for bagged products often needs to be drier than compost sold in bulk. Some compost producers have used forced hot air driers to remove excess moisture. While this approach involves additional handling and the cost of fuel, it may be the only option when the weather is consistently uncooperative.

9. Finished compost storage

In most areas, the use and distribution of compost is seasonal, with peak periods in the spring and fall and little to no sales activity in the winter. The mismatch of seasonal usage and production can easily create the need to store between three and six months worth of production. To avoid operational problems, this space requirement needs be understood and included in the initial site layout and design of the facility.

Finished compost that has been properly composted and cured is still subject to microbial decomposition, albeit at a very low rate. Since the potential for development of anaerobic conditions and odors still remains, storage conditions need to be managed. This primarily means that storage pile heights need to be reasonable, and that free air space of the material be maintained by not driving on or otherwise compacted it.

At facilities without stacking conveyors, storage pile heights are generally determined by the reach of front-end loaders. Heights greater than this can be an indicator that equipment operators are driving on the piles and compacting the compost. Using stacking conveyors, sites can achieve product stockpile heights up to 15 m (50 ft), which is an efficient use of space and can help reduce the amount of precipitation that infiltrates into the stored material. However, as the pile height increases, the risk of the material in the base of the pile becoming compacted from the weight of the material on top, and thus the development of anaerobic conditions, increases.

If products are stored in very large piles, it may be necessary to break down and restack the compost into smaller piles a few weeks prior to use, sale, or shipping. This allows the stored compost to aerate naturally and dissipate any phytotoxic compounds that may have been generated in anaerobic pockets within the storage pile.

An important but sometimes overlooked practice is to prevent finished compost from becoming contaminated while it is in storage. Poor housekeeping, poor litter control, and improper segregation/containment of feedstocks or residuals can lead to contamination of product stockpiles. This in turn could mean that the product's value is reduced or it has to be rescreened prior to shipping.

Similarly, pathogens reintroduced into the product can render it unusable. Pathogens can be reintroduced by handling the product with dirty equipment or coming in contact with contaminated surface water run-off. Like curing piles, storage piles should therefore be located in an area where they cannot be contaminated by run-off from feedstock receiving and active composting areas. As a general practice, the buckets and tires of loaders should be cleaned prior to handling the product.

Weeds are another common problem at compost facilities. Storage piles provide a very good environment for weed propagation and can quickly be overrun. Storage piles should be regularly inspected, and any weeds that are found should be manually removed. Storage piles can be covered with tarps to exclude weeds and precipitation. Ideally, the tarps should allow some oxygen exchange.

If the compost produced is to be applied to cropland, the curing and/or storage piles can be located near the appropriate fields, similar to a manure stack. Again, poorly drained sites and steep slopes should be avoided to minimize anaerobic conditions and the loss of compost and nutrients from surface runoff. There also may be local or state regulations that disallow this practice.

10. Blending compost products

One of the more common and accommodating markets for finished compost is in soil blends. Soil blenders are active in most urban areas, working with land developers, landscapers, and landscape depots to supply various products for commercial and residential uses. For these markets, compost is commonly blended with topsoil, sand, and/or peat to create "manufactured topsoils." Compost is also blended with marginal topsoils or native soils to improve their organic matter content and other basic properties. Blending compost with biochar and mineral additives, such as gypsum or sulfur, is also becoming more popular. Such blends are commonly targeted for agricultural applications. In these cases, the minerals used and the ratio of minerals to compost are often specified by an agrologist based on the crop type and soil conditions at the receiving farm.

Compost can be blended with other soil products and minerals in a number of ways. The simplest method is to mix the materials together in small batches using a front-end loader. The same batch mixers used to blend feedstocks can also be used to blend soil products, although the units must be thoroughly cleaned prior to being used to prevent contamination of the finished products. Larger amounts of material can be blended using windrow turning equipment in the same manner that initial feedstocks and amendments are mixed; the windrow is constructed by placing the materials in alternating layers and then turning them two to three times. Again, the turner should be cleaned before using it to mix finished products. Another method of blending larger quantities of material is to use screening equipment, loading the hopper in the appropriate ratio of materials (e.g., one bucket of compost, two buckets of topsoil, and so on).

Before compost is used in the production of blends, all of the required compost testing and quality assurance steps should be completed. Not doing so places the entire blended product at risk of needing to be reprocessed or disposed of if there is a problem with the compost's quality.

11. Bagging

Packaging compost in 20 to 50-liter bags[1] makes it more convenient for customers who require only small volumes, or to sell compost through retail outlets. Bagged compost can bring a higher price than compost that is sold in bulk, and is normally practiced when the sales volume justifies the increased effort and equipment requirements (see Chapter 18).

Customers generally expect that compost sold in bags will be of a higher quality than compost sold in bulk, and will have no objectionable odors. Therefore, it is critical that compost be fully matured before being packaged. Bagged compost should have a reasonably low MC (<40%) and the bags should have small perforations to allow airflow; compost may become anaerobic as it continues to decompose in an airtight bag.

For very small volumes of bagged sales, special equipment is not necessary. Hand bagging with a shovel, though laborious, works well on a small scale (Box 9.2). For larger quantities, a simple hopper with a hand or foot operated valve can be used to increase production (e.g., 100 to 200 bags per hour) and make the work of filling bags much easier. Bag filling equipment can be purchased, or can be fabricated from salvaged or obsolete equipment.

---

[1] In the United States and Canada, bags are often sold by weight, commonly in sizes that range from 10 to 20 kg or 25 to 40 lbs.

Box 9.2 Easy manual bagging

Manual bagging is made significantly easier by using an old 5-gallon or 20-liter plastic bucket. After cleaning the bucket, use a saw to remove the bottom. File the cut edge if necessary to make sure there are no sharp edges that could catch and rip a bag.

Place the pail inside the empty compost bag with the bucket handle up, and then use a shovel to fill the pail with compost. When the pail is full, grab the handle and lift it straight up and out of the bag, leaving the desired volume of compost behind (Fig. 9.57).

FIGURE 9.57

Easy metering device for manual bagging.

For high-volume operations, automated bagging systems that include large product storage and feed hoppers, scales, baggers, bag sealers, and transfer conveyors with integrated scales can be purchased. Since many buyers require bags to be palletized and wrapped, a palletizer may be necessary. The cost of a complete automated bagging system can easily cost in excess of $100,000. Bagging systems are often housed in a building, although outdoor mobile bagging systems are also used (Fig. 9.58).

FIGURE 9.58

Mobile semiautomated bagging system.

Source: Rotochopper, Inc.

As an alternative to investing in bagging equipment, it may be possible to outsource this step to an existing bagging operation in the area, thereby allowing smaller quantities of material to be packaged and palletized for a reasonable price.

References

Cited references

Chiumenti, A., Chiumenti, R., Díaz, L., Savage, G., Eggerth, L.L., Goldstein, N., 2005. Modern Composting Technologies. JG Press, Emmaus, PA, ISBN 0-932424-29-5.

Coker, C., 2019. Food waste depackaging systems. BioCycle. July 10, 2019. https://www.biocycle.net/food-waste-depackaging-systems/.

Coker, C., 2017. Particle size reduction strategies. BioCycle. March 8, 2017. https://www.biocycle.net/particle-size-reduction-strategies/.

Coker, C., 2014. Contaminant removal strategies. BioCycle 55 (5), 46. https://www.biocycle.net/contaminant-removal-strategies/.

Gamble, S., 2018. Compost Facility Operator Study Guide. Minister of Alberta Environment and Parks, ISBN 978-1-4601-4130-4 (PDF online).

Goldstein, J., 2005. Plastics separation in compost. BioCycle 46 (9), 58.

Oshins, C., 2020. Compost manufacturing system (presentation slide). In: CCREF Compost Manufacturing: Principles and Practices. Elizabeth, NJ, September 15-16, 2020.

Rynk, R., 2001a. Getting moisture into the compost pile. BioCycle 42 (6), 51−56.

Rynk, R., 2001b. Air separation strategies tackle plastics contamination. BioCycle 42 (9), 52–56.

Shaffer, B., 2010. Curing compost: an antidote for thermal processing. Acres USA 40 (11).

Consulted and suggested references

Anderson, K., Slaughter, S., 2020. Incorporating depackaging into composting operation. BioCycle. May 19, 2020. https://www.biocycle.net/incorporating-depackaging-composting-operation/.

Composting News (UK), 2009. The Buyers' guide to shredders and grinders. Compost. News 13 (1).

Composting News (UK), 2007. The Buyers' guide to screens and baggers. Composting News 11 (3).

Coker, C., 2010. Controlling costs in compost manufacturing. BioCycle 51 (9), 26.

ConBelt, 2021. Basic-Components of a Conveyor Belt. https://www.conbelt.com/how-does-a-conveyor-belt-work/basic-components-of-a-conveyor-belt/.

Elfatih, A., Arif, E.M., Atef, A.E., 2010. Evaluate the modified chopper for rice straw composting. J. Appl. Sci. Res. 6 (8), 1125–1131.

Elm, G., 2011. Screening insights for compost and mulch producer. BioCycle 52 (4), 30.

Fenner Dunlop, 2009. Conveyor Handbook. Fenner Dunlop, Australia.

Gibson, S., 2017. The Difficulties of Screening Compost. Waste Advantage Magazine. https://wasteadvantagemag.com/the-difficulties-of-screening-compost/.

Goldstein, N., 2005. Screening strategies to optimize compost quality. BioCycle 47 (6), 44.

MHI, 2021. The Ten Principles of Material Handling. The Materials Handling Institute. https://www.mhi.org/downloads/learning/cicmhe/guidelines/10_principles.pdf.

Organic Recycling, 2009. Screening and Bagging Equipment. Organics Recycling Group. Renewable Energy Association. Summer 2009. ORG homepage (organics-recycling.org.uk).

Perks, D., 2010. The importance of safe, reliable, materials handling equipment for use in the compost industry. Organics Recycling. Spring 2010. http://www.organics-recycling.org.uk/uploads/article1762/Materials%20Handling%20Equipment%20Guide.pdf.

StarsPlus, 2021. Horizontal Screw Conveyor. http://www.starsplas.com/horizontal-screw-conveyor/.

The Cat Rental Store, 2021. Different Types of Loaders. https://www.catrentalstore.com/en_US/blog/different-types-of-loaders.html.

Thomas Publishing Company, 2019. Understanding Conveyor Systems. https://www.thomasnet.com/articles/materials-handling/understanding-conveyor-systems/.

WRAP, 2009. Report: Review of Food Waste Depackaging Equipment. The Waste and Resources Action Programme. https://wrap.org.uk/content/food-waste-depackaging-equipment.

Yepson, R., 2009. Historical perspective: screens & air classifiers. BioCycle 50 (3), 20.

Site planning, development, and environmental protection

10

Authors: Matthew Cotton[1], Robert Rynk[2]

[1]*Integrated Waste Management Consulting, Richmond, CA, United States;* [2]*SUNY Cobleskill, Cobleskill, NY, United States*

Contributors: Lorrie Loder-Rossiter[3], Andrew Carpenter[4]

[3]*Revinu, Inc., Fleming Island, FL, United States;* [4]*Northern Tilth, LLC., Belfast, ME, United States*

1. Introduction

Choosing the location of a composting facility is perhaps the most consequential decision one can make regarding the sustainability of the composting efforts. The decision is important, whether it is locating the site on an existing piece of property or finding a new location for a dedicated composting facility. Choosing a site that has all of the required components for a successful operation can be a lengthy and time-consuming process. Unfortunately, siting is often hastily considered and underanalyzed, resulting in sites which do not survive the myriad challenges of changing land use. Composting is often an adjunct to another business (like landscaping, farming, or solid waste management) and in many cases a composting site can be located adjacent to the existing business. While this offers convenience and efficiency, it can also present some challenges due to limitations of the existing location. In practice, the ideal site rarely exists, so the goal is to choose a site that checks as many of the boxes as possible and then make the best of the situation at hand.

A composting site must provide the required area (now and in the future), efficient materials movement, truck and equipment access, and conditions for all-weather composting while simultaneously protecting the surrounding environment (Fig. 10.1). The latter requirement includes protecting surface water and groundwater and limiting the impacts of the facility on its neighbors. Those impacts include noise, dust, traffic, and especially odors. In most cases, operating a composting facility requires approval from local, state, provincial, and/or national planning and environmental agencies. Larger and more complex facilities typically may require oversight from numerous regulatory agencies.

Because protecting the environment is imperative, most composters receive "guidance" from governmental agencies in the form of environmental regulations. Composters must comply with, and the site must conform to, the provisions of

FIGURE 10.1

There are many elements to a composting site, in addition to the actual composting pad. Chief among those elements are provisions to insulate the composting operation from its surroundings, and vice-versa.

Source: Dustin Montey, Shakopee Mdewakanton Sioux Community.

these regulations, which vary considerably from one jurisdiction to the next. Compliance typically involves obtaining one or more permits, or another form of approval from one or more government agencies.

During the site planning stage, there may be potential support provided by governmental agencies for developing a composting site. For example, in the United States, the USDA Natural Resources Conservation Service (NRCS) offers assistance to agricultural composters with site planning, design and cost recovery, including soils information, drainage control, pad construction, and other conservation structures. Other agencies may similarly provide site development assistance. Financial incentives may be available because of the environmental and waste management benefits of composting.

In addition to the site regulatory requirements, it is important to be aware that starting or expanding a composting operation will likely raise concerns among neighbors and local public officials. Experience has shown that composting facilities are best served by working with neighbors and community groups *from the start*. Although it can be a frustrating and painful undertaking, a critical part of getting started smoothly is openly sharing information with these groups about the composting facility, its benefits to the community, and its realistic impacts.

2. Compost-site regulations

At its core, composting uses *wastes* or by-products to create soil amendments and recycle nutrients and organic matter back into the soil. Because of this, it can be a highly regulated activity in some places, though this varies widely. Even within some countries, different states and provinces have very different regulations.

Regulatory requirements usually differ with the type and source of feedstocks, nature of the composting enterprise, scale of the operation, and whether or not the compost is sold. Making compost, from one's own self-generated feedstocks, for one's own use, is typically exempt from regulation. On-farm composting facilities are often exempt, or the requirements are greatly relaxed, especially if the primary feedstocks come from the farm. Importing feedstocks from outside sources can trigger regulations that were previously excused. Developing a commercial composting operation on a farm (one that charges for feedstock and sells compost) may change the land use status of the farm. Publicly owned facilities may have obligations that complicate the site-approval process (Box 10.1).

Box 10.1 Siting a publicly owned facility

Private compost facilities must do a lot of homework to find a long-term sustainable site for their composting business. While a site-evaluation is still needed, a publicly owned facility often has a dedicated site in mind, a site located adjacent to infrastructure that the jurisdiction already owns (e.g., the county landfill or the regional wastewater treatment plant). However, if a public agency wants to site a stand-alone composting facility, the process may be more involved. In order to justify why a public agency is purchasing land and developing what is generally perceived as an unpopular land use, the jurisdiction maybe obligated to conduct a formal, written siting study. Depending on the scale of the facility, this process may involve public hearings and/or other processes to gain citizen input. For large regional sites, the jurisdiction may develop its own set of siting criteria. Typically, these criteria are weighted and an analysis is done to rank one site over another. Examples of siting criteria for public facilities include:
1. Transportation impacts—transportation distance, traffic, and air quality.
2. Neighborhood impacts—air quality and noise.
3. Environmental impacts—biological resources, cultural resources, hydrology, and water quality.
4. Site costs—acquisition, population and housing, on-site and off-site development costs, utilities and service systems.
5. Land use designation.
6. Visual impacts.
7. Multiuse potential.

Composting regulations govern more than just the site but the features of the site figure prominently in regulatory schemes. Primarily through the mechanism of permitting, regulations can dictate the location of a site (e.g., industrial or agricultural zoning), the size of a site (or its maximum volume), separation distances to other land uses (e.g., residences, streams), pad construction, road construction, pile dimensions, layout, runoff management, and site components like buildings, fencing, entrances, and signs.

The primary tool for regulating compost sites is the permit, which may also be called an "approval," a "consent," or a registration. A permit is like a contract between the facility and the particular regulatory authority issuing it. It is not uncommon for a facility to need multiple permits from multiple public agencies (waste, water, air quality) and perhaps at multiple levels (local, state, national). A permit might be specific to a particular facility, or it might be general, applying to all facilities in a common situation (e.g., moderate-scale facilities, composting only yard trimmings). The latter situation is called a "Permit by Rule" in some places, or a "general" permit. Permit by Rule is a way of lessening the permitting burden by having clear rules that certain types of facilities must follow in order to remain in the Permit by Rule tier.

Composting regulations are more or less burdensome depending on the facility's situation (e.g., size, feedstocks). Typical thresholds that jurisdictions use to regulate the design and operation of a compost facility include:

- **Feedstocks:** In general, regulatory oversight increases based on the perceived potential to cause environmental impact. Many jurisdictions have few regulations when only yard trimmings are composted, but impose more stringent requirements if the facility processes manure or food. Often jurisdictions will have a "tiered" structure for permitting, recognizing that in addition to the feedstock characteristics, the *scale* of the operation also plays a role in potential impacts. In jurisdictions with significant agricultural operations, composting of some manures may be exempt (especially, for example, if they are composted within the footprint of the dairy on which they were generated). Similarly, biosolids composted at the treatment plant may have some reduced permitting requirements. Smaller amounts of low-threat feedstocks (like leaves and brush) may be exempt under certain conditions. However, in most jurisdictions, once the facility starts accepting off-site feedstocks, especially feedstocks like food scraps, mixed solid waste, or biosolids, it may fall into a permitting scheme.
- **Volume processed/volume on site:** Some jurisdictions limit the daily incoming feedstocks and may also limit the total volume of in-process compost a facility can have on-site. Tiered regulatory approaches may also be based on volume thresholds (which are usually arbitrarily selected volumes). Regulatory schemes for defining the volume at which a permit is required vary considerably; and they do not seem to be based upon much analysis or evidence. In addition, in many jurisdictions, the agency that permits the design and operation of the compost facility may be different from the agency that governs the sale of compost. Typically, regulatory oversight increases as volumes accepted and/or processed increases. Some US states exempt from permitting facilities with less than 2300 m^3 on-site (3000 yd^3). Others have a threshold of 7600 m^3 (10,000 yd^3) on-site for a lower tier permit. These volume thresholds are generally arbitrary and vary considerably from jurisdiction to jurisdiction.

- **Location:** The location of the site (relative to other land use types) usually determines whether or not local land use or zoning approvals may be required. Some states prohibit composting in certain land use types.
- **Selling (or giving away):** Many jurisdictions make a distinction for sites that make compost for their own use. A facility is more likely to be regulated once it starts selling (or even giving away) compost. In addition, almost all jurisdictions have specific regulations where certain marketing or performance claims are made; especially nutrient claims and especially on a packaged product (Chapter 18). However, these regulations have little to do with the site features.

Sources of specific information about regulations in particular countries are available from public agencies and composting industry associations. For example, the US Composting Council maintains a web page dedicated to regulations in individual states (USCC, 2021).

3. Environmental and community considerations

The features of the composting site determine the risk associated with its potential impacts to the environment and surrounding community—odors, noise, dust, and runoff. For example, a site located near neighbors is more vulnerable to odor complaints, all other things being equal. Also, a site that is at the base of a long slope has to divert more rainwater away from the working areas of the site. The feedstocks, composting method, and management also impact these environmental concerns.

Odor from the composting process is minimized through good management only if the composting system is properly sited, designed, operated, and laid out (Chapter 12). In siting the facility, consider the factors that influence odor migration including the direction of prevailing winds during warm weather, "air drainage" patterns, topography, vegetation, and site visibility.

Consideration must be given to the noise and dust resulting from the composting operations and from transport vehicles traveling to and from the site. This can be addressed somewhat by selective scheduling of activities during the day and by road use selection. Grinding is a particularly noisy operation and should be performed when noise will have the least impact. Noise from site operations extend for longer periods as the size of the operation increases. Depending upon the material being composted or the type of compost enterprise, noise may be only a seasonal factor. It is of greater concern during mild weather conditions when windows are open, and neighbors are outside.

Site visibility and appearance influence human perceptions. Fewer neighborhood complaints may be received if the composting site is less visible. To shield the composting site from public view, take advantage of natural landscape features such as trees and shrubs; otherwise establish new plantings. If the site is visible to the public, it must be kept neat. Sloppy sites are perceived to have greater problems. A composter can make use of some of the compost produced to landscape the site and make it attractive. A community-savvy composter keeps grass around the site mowed, controls weeds, and maintains plantings in good condition.

Management of on-site water is a very important site consideration and is foremost on the minds of environmental regulators. Rain and snow melt percolate through the materials and into the ground and/or create runoff, which can carry away contaminants that can impair the environment or well water. Possible contaminants from composting runoff include nitrate-nitrogen, ammonia, phosphorus, pathogens, pesticides, and organic and inorganic oxygen-consuming compounds. Most of these contaminants are associated with sediments carried with runoff. Fortunately, there are effective management practices to collect, contain, reuse, and treat site runoff. These practices are discussed in detail in Section 10.

At a minimum, the following community- and environmental protection measures should be observed. The remainder of the chapter provides more detailed information about these practices.

1. Select a site that poses little risk of potential impacts to the community and the environment. Observe the recommended separation distances to surface and ground water.
2. Develop site-specific measures to further reduce the risks of impacts (i.e., berms to manage run-on and runoff and retention ponds to manage on-site surface water).
3. In general, follow good composting practices. Establish a balanced mix of feedstocks, including a carbon to nitrogen ratio above 25:1 to limit the loss of nitrogen. Maintain windrows and piles below the maximum recommended moisture content (i.e., 65%) to minimize anaerobic conditions and leaching (especially during the "wet" season—if there is one).
4. Divert water entering the site (run-on) from upgradient areas away from the composting pad and away from storage areas.
5. Do not allow runoff from the composting pad and storage areas to empty directly into surface water. Runoff can be collected in a storage pond and later used to moisten dry feedstock, windrows, and piles. Depending on regulations, collected runoff also may be channeled to cropland or a vegetated infiltration area or used to irrigate crops. Many of the potential contaminants that pose problems for streams, ponds, and lakes can be effectively treated by the soil.
6. Use sedimentation or filtering devices (e.g., compost socks and berms) to separate solids from runoff prior to a pond or infiltration area.
7. Store feedstocks and finished compost away from surface water and drainage paths. Wet feedstocks that are prone to leaching are best stored under-cover or on an impervious surface with a method to collect and safely dispose of process water.

4. Site selection/evaluation

Site selection is usually a matter of adaptation—making the best of the limited and imperfect sites that are available. It is rare that a prospective composter can search the local terrain for the best site, checklist in hand. This situation is particularly true for on-farm composting because the site choices are usually within the limits of the farm area itself. Still, other options may exist, such as purchasing additional property or renting land on a neighboring farm. Plus, some site features can be altered including surface soils and drainage patterns. Despite the limitations, a checklist is not a bad tool to use when considering a potential composting site. Table 10.1 is a checklist of sorts, itemizing some of the important considerations in locating a suitable composting site.

The criteria for evaluating and selecting a good site for composting include location, proximity to critical land uses (e.g., streams, schools), available land area (i.e., space), visibility, access (roads and utilities), soil type, groundwater level, land slope, surrounding topography, and prevailing wind direction.

Table 10.1 Criteria for a good composting site. Check as many of the boxes as possible.

- ☐ Secluded land: Remote from residential and commercial neighbors, now and for the foreseeable future.
- ☐ Vacant, relatively flat land: Preferably 0%–5% slope.
- ☐ Open field: Open lots are generally less expensive to develop but forested or wooded areas around the perimeter are an advantage (to minimize the neighbors' ire, cut down as few trees as possible).
- ☐ Alternatively, previously developed industrial or agricultural property with appropriate building and/or paved surfaces.
- ☐ Sufficient size: No one ever wishes for a smaller site (i.e., volumes handled on successful composting sites tend to increase over time).
- ☐ Proper zoning or land use designation: Most compost sites are located in either agricultural or industrially zoned land, but others can work. Some jurisdictions allow composting in more land use designations than others.
- ☐ Minimal environmental impacts: Avoid wetlands, floodplains, high groundwater table, endangered species, etc.
- ☐ Minimal cultural impacts: Avoid historic sites and sites of archeological importance, sites important to indigenous cultures, etc.
- ☐ Distant from sensitive populations: Avoid site locations near schools, hospitals, facilities housing elderly populations and communities that are already overburdened by [pollution (i.e., environmental justice)].
- ☐ Sufficient distance from airports: Check locally applicable regulations.
- ☐ Good truck access: Preferably close to a freeway or other commercial routes.
- ☐ In close proximity to feedstock sources and/or compost users.
- ☐ Visual buffer or potential for one.
- ☐ Availability of water (and other utilities) on site.
- ☐ No drainage problems.

4.1 Location

When locating a stand-alone composting site, a key decision that needs to be made is whether it is advantageous for the site to be closer to sources of feedstocks (particularly those generating tip or gate fees) or closer to the target markets for compost. Generally, these advantages exist in populated areas. On the other hand, populated areas have the disadvantages of greater impacts from odors and traffic, higher land cost, and increasingly resistant neighbors. For these reasons, it is appealing to find a remote location, isolated from houses and development. Thus, the *ideal* composting location is usually elusive. In most cases, the selection involves compromises that balance the contrary factors of siting near population centers and yet far from homes and sensitive businesses. All other things being equal, a location that is isolated from neighbors tends to be the better choice. In populated and growing areas, encroaching, or changing land use may also be a pitfall for siting a facility. So, it is important to understand the planned growth of a community to the extent possible.

For most on-farm composters, the question of location is less complicated, reduced to what location within the farm area is best for a composting site. Proximity to neighbors remains an important consideration but less so because farms are typically well-buffered from neighbors. If neighbors are in close proximity, they tend to be accustomed to the atmosphere and operations of farming, which are similar to those of composting. Usually, the most convenient composting site on the farm is near the barn or manure storage—the point where manure is collected. However, the convenience of a particular site must be weighed against other factors such as those discussed below and in Table 10.1.

Potential composting locations near ground and surface water resources deserve extra scrutiny. Composting facilities are usually allowed where ground water, wetlands, streams, and lakes exist but at a separation distance to avoid negative impacts. Additional mitigation measures may also be required. Land that is interspersed with wetlands and streams may either be impossible to develop or too expensive to adequately protect the water resource.

Other locational features to avoid include the following:

- *Floodplains:* Most regulations prohibit or restrict composting facilities from being located within floodplains, with a 100-year floodplain being the most common stipulation (land that statistically is expected to flood once in 100 years). In some cases, composting operations with lower-risk feedstocks, like yard trimmings, are allowed in floodplains (Ohio EPA, 2017). In other cases, sites may be allowed in floodplains with approved mitigation measures. Because many farming practices are allowed in floodplains, on-farm composting operations may also be allowed where composting is deemed a farming activity.

- *Airports*: In selecting a potential site, composters should be aware of the proximity of nearby airports. Both food-based feedstocks and open water attract birds (and other wildlife), which can be a hazard to aircraft. In the United States, Federal Aviation Administration (FAA) guidance limits outdoor facilities handling food waste from close proximity to jet-served airports (FAA, 2007). This guidance suggests that composting facilities that handle food waste cannot be located within 3 km (10,000 ft) of a jet-serving airport and 1.5 km (5000 ft) of an airport service piston-engine aircraft. These separation distances apply only to composting facilities that accept food waste (FAA, 2007; Transport Canada, 2014), although regulators may choose to apply the same rule to all composting facilities. In any case, local and regional regulations should always be consulted. Significant open water (like from a retention basin) can also be a bird attractant.
- *Sensitive populations:* All composting facilities should avoid locations near sensitive populations such as schools, hospitals, nursing homes, and assisted living communities for elderly populations. Where environmental justice is a relevant concern, it is conscientious to steer clear of communities that already suffer an environmental burden; unless the composting operation represents an environmental improvement, as in the case of community composting (Gaia, 2021). In California, CalRecycle provides a web page that generates an environmental justice map for siting a composting facility (CalRecycle, 2019).

Given this myriad of limitations, what types of locations are potentially good for a composting site? The answer is primarily locations that are undeveloped or already developed for uses with similar impacts as composting. Farms are prime candidates.[1] Many sites locate or relocate to rural areas served by a highway or land on the perimeter of towns and cities (Fig. 10.2). Other examples of candidates include landfills and former landfills, wastewater treatment facilities, vacant or underused industrial sites, quarries, vacated military facilities (beware of existing site contamination), and suitable land owned by a feedstock supplier (e.g., food processing factory). None of this guidance precludes composting from taking place in urban and suburban locations. Indeed, many successful composting operations operate among the populace. A composting facility might even offer the potential to revitalize blighted urban properties. However, it often requires contained composting methods (e.g., enclosed or in-vessel), environmental protections (e.g., roofs, biofilters), and community good will.

---

[1] In some locations, there are restrictions on land considered "prime" farmland or otherwise recognized as exclusively for farms.

FIGURE 10.2

A site near major highways is generally good for business but highways also bring residential and commercial development.

4.2 Proximity to sensitive land uses

As discussed above, separating a compost operation or facility from neighbors is a key to developing a sustainable site. Similarly, making sure there is some distance between other sensitive areas is also key to locating a long-term site. The separation distance, or buffer zone, between the composting operation and adjacent land features is one means of protecting water sources from the potential impacts of the composting operation and minimizing nuisances such as odor and equipment noise. Usually, separation distances are measured from the edge of the activity of concern (e.g., composting pad, grinding pad, holding pond), but sometimes the baselines are set at the facility boundaries. Local, state, and/or provincial regulations may recommend minimum separation distances, or they may prescribe a procedure for setting them with local officials. In some cases, required separation distances depend on the feedstocks being composted and may also depend on the scale of the operation. The appropriate environmental agencies should be consulted for requirements and advice.

For surface water protection, the minimum horizontal separation distance is the distance between a compost facility and the boundary of a surface water body or wetland. For groundwater protection, it is the vertical distance from the compost pad surface to the seasonal high water table (Fig. 10.3). In the absence of mandated separation distances, a soil or hydrology professional (i.e., a civil engineer) can help determine a safe distance to minimize stormwater or groundwater impacts.

Table 10.2 lists setbacks commonly recommended for small- to moderate-scale composting sites in the United States. Outside of the United States, the required setbacks between composting facilities and neighbors tend to be substantially greater, typically in the range of 500–1000 m (1640–3280 ft). The values listed in Table 10.2 are guidelines in the absence of regulations. Local requirements and conditions should always prevail.

Although mandated setbacks are often arbitrary judgments or "best guesses," they generally provide a decent rule-of-thumb for locating a composting site in relation to sensitive areas. Meeting required separation distance does not ensure that the composting facility will not negatively impact water and neighbors. In fact, no reasonable setback can guarantee that neighbors will never detect odors. Good operation and site maintenance, plus sensitivity to the surrounding environment, remain essential. Developing structural stormwater controls (like ditches or lined ponds) can be expensive, so careful siting to avoid areas that may require costly mitigation can save on site development costs.

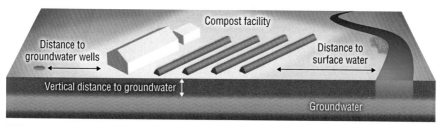

FIGURE 10.3

Separation distances for water protection (CA State Water Resources Control Board).

Table 10.2 Common minimum setbacks in the United States for composting activities.

Sensitive land use	Minimum separation distance	
	(m)	(ft)
Property line	15–30	50–100
Residence or place of business	60–150	200–500
Private well or other potable water source	60–150	200–500
Wetlands or surface waters (streams, ponds, lakes)	30–60	100–200
Subsurface drainage pipe or drainage ditch discharging to a natural water course	8–10	25–30
Water table; (seasonal high)	1–2	2–6
Bedrock	1–2	2–6

Based on information compiled from USCC, 2021. Setbacks in other countries tend to be larger than in the U.S. In general, required setbacks can vary considerably among localities, states, provinces and countries. The setback requirements for any specific location must be determined on a case-by-case basis.

4.3 Available land

Space is the first frontier of site design and a major factor in site selection. A potential site must provide sufficient area to compost the volume of feedstocks for the expected time period (retention time) and also to accommodate the other requirements that come with operating a composting facility (Fig. 10.4). Few operators ever wish for a smaller site. To some extent, the required space can be decreased by reducing the retention time of the compost, but that has other impacts to the process.

First, a composting site needs to be large enough to fit the combined footprint of the windrows, piles, or composting vessels. This amount of space differs with the composting method and equipment. Composting sites also need *additional* space for all of the activities associated with operating a compost facility. Additional space can be needed to receive and process feedstocks, for access roads, for curing and storage of compost, for supplemental operations like grinding and screening, office

FIGURE 10.4

Allocation of activity space on the Dirt Hugger composting site in Dallesport, WA.

Source: Reprinted with the permission of Dirt Hugger.

and employee facilities, equipment storage and maintenance, and for handling of stormwater runoff. Procedures for determining the area required for these features and activities are discussed later in this chapter. Many commercial facilities tend to grow as the business grows and as more customers visit the facility. Therefore, future growth is another important factor in finding a suitable site.

4.4 Electricity

Most composting operations use diesel-powered equipment for the majority of their processing. Depending on the compost method, and air quality regulations, electricity also may be necessary. Aerated static pile (ASP) and in-vessel composting systems require electricity while passive pile and windrow composting do not. However, electricity may be needed for supplemental purposes including powering conveyors, office functions, computers, lighting, alarm systems, and/or electric motors. When electricity is required, three-phase service is preferred, if not necessary. Thus, access to appropriate electrical service can be an important site selection feature. It can be expensive to bring or upgrade utility electricity to a distant site. Solar power is an increasingly viable option for operating many electrical components of a composting facility, including fans for an ASP system (Spencer, 2009; SJV APCD, 2013). Solar energy is perhaps the best option for remote locations. A diesel- or gas-powered generator is another alternative. Solar energy can work well for smaller loads, but larger loads like grinders or other similar high-horsepower engines are not practical for on-site solar power.

4.5 Water

Like electricity, an on-site water supply may not be necessary for composting. However, it is an advantage to have a sufficient supply available. In some jurisdictions, access to a minimum amount of water may be required for fire suppression.

The need for water at a composting site depends on the climate, feedstocks, composting methods, scale of operation, and other factors such as how quickly the compost must be produced. In arid climates, during the dry season, water must be added routinely to sustain the pace of composting (Chapter 9, Section 6.3). In humid climates, enough precipitation may fall regularly to replace the water lost from piles and windrows. Water may only be needed during periods of low rainfall. The moisture condition of raw feedstocks is another primary determinant. In some cases, feedstocks may contain enough moisture to sustain the composting process to its conclusion (with help from precipitation). For dry feedstocks, water is needed initially and again during dry, warm weather. If employees are based at the site, then potable water will be needed.

Depending on the foregoing factors, it may not be a necessity to have a preexisting source of water on the site. However, if a dedicated supply of water is needed, it will have to be developed from water resources (e.g., groundwater, surface water) on or near the site. Developing a water source often includes obtaining water rights and/

or water use permits. Municipal or community water service is also an option, but it can be expensive as a source of water to support the composting process. All things considered, it is desirable (if not necessary) to have a significant source of water on the site—at the very least because having to haul water from off-site can be costly and time-consuming.

4.6 Access and infrastructure

The ideal site should be reasonably accessible for vehicles delivering feedstocks and supplies (e.g., fuel) and for vehicles removing compost products. A location close to the barn or manure storage that is or can be connected by a road is generally a good choice on livestock farms. Close proximity to major highways is an advantage if feedstocks or compost products are carried long distances by large trucks (e.g., greater than 50 km or 30 miles).

A site that has existing infrastructure—buildings, roads, paved surfaces, and utilities—can also be a good candidate for a composting site. The cover of buildings is appreciated for storage of compost, amendments, and/or equipment. On a farm, such a site could include obsolete barns, unused silage bunkers, and paved barnyards. An abandoned commercial site that is becoming an eyesore and a liability for the community can be developed for composting, perhaps with community support. Gravel mines, abandoned lumber yards, old military bases, and even defunct airstrips have been used successfully as composting sites (Fig. 10.5).

FIGURE 10.5

This composting facility is located on a former gravel site.

Source: Tracy Material Recovery Facility and Solid Waste Transfer.

4.7 Mobile phone and internet services

Just a few years ago, the advice given here would have been to seek a site location with landline telephone service. However, mobile phone service has made that advice nearly obsolete (there remain valid reasons to continue useful landline communication). While mobile phones have expanded the communications map, service is not yet available everywhere. It is advantageous, if not necessary, to have such service at a composting site. In addition, sufficiently fast and reliable internet service is important if the composting site includes an office, or is a base of operation for some employees. Increasingly some site monitoring services require a strong WIFI signal (for remote sensing of compost temperatures for example).

4.8 Site surface

The ability of the site surface to support composting operations year-round is an important characteristic. Many composting sites are established on native soil, so the soils beneath the composting pad must be able to support heavy equipment in nearly all-weather conditions and seasons (Fig. 10.6). Furthermore, the soils must provide a safe barrier to groundwater, allowing any water that does infiltrate to be absorbed and sufficiently cleaned of pollutants.

FIGURE 10.6

Many composting facilities are able to work on native soil surfaces, but they are at the mercy of the weather. This site in Upstate NY is workable most of the time.

Moderately well-drained soils are preferable for all sites. Soils that drain quickly (e.g., gravelly) are workable but increase the risk for groundwater contamination. Poorly drained soils (e.g., clay) form ruts and puddles and lead to muddy conditions. A muddy composting pad is perhaps the most common site-related complaint of composting operators. Muddy site conditions limit access by equipment and can interrupt the composting operation for several weeks. Also, the site should have few rocks, which can get mixed into the composting materials and damage machinery. If the soils are not suitable to composting, the composting pad and other work areas can be altered or paved. Many successful composting sites have been developed on abandoned pavement from previous activities (e.g., former airport, factory storage yard).

Whether or not a facility can operate on native ground depends on the nature of the existing soils and may also be defined by local, state, or provincial regulations. Some jurisdictions may insist on a paved or engineered surface for composting facilities of a minimum size, or if specific feedstocks are composted (for example, biosolids or food wastes).

4.9 Depth to groundwater

The layer of soil between the surface and groundwater filters and treats nutrients and other contaminants infiltrating from above. As the depth to groundwater decreases, less filtration and decomposition takes place, and more contaminants find their way to the groundwater. Thus, sites with shallow groundwater, or a "high water table," should be avoided. Once again, local, state, or provincial regulations may dictate a minimum depth to groundwater. This required depth to groundwater can vary considerably among jurisdictions. A water table near the ground surface suggests the proximity of wetlands, which raises additional concerns. Choosing a site with shallow groundwater can increase site development costs substantially.

To determine both the groundwater level and the soil conditions, a site soil investigation should be conducted by a soil science professional. Deep-hole checks should accompany a site soil investigation. A backhoe is normally used for this purpose, though drilling rigs are also common. The hole excavations are made at the compost-processing site location to determine the presence of bedrock or groundwater. If groundwater is not detected, then the soil profile (e.g., mottled color) is used to evaluate a fluctuating water table. Hole depths are commonly 2−5 m (7−13 ft).

4.10 Drainage, land slope, and topography

An ideal site is located on flat land with positive drainage. Although less than flat sites can work, it is harder on equipment and makes for a more challenging layout. Good drainage at a composting site is a must. Poor site drainage leads to ponding of water, saturated composting materials, muddy site conditions, and excessive runoff from the site. Drainage is determined by the soils, the slope of the land on the site, and also the surrounding topography.

Topography refers to the features of the terrain on and around the site. In addition to stormwater drainage, land features affect the site's visibility, noise level, its microclimate, wind patterns, and air movements generally. All of these effects influence how, when, and how much neighbors are impacted by odors, noise, and the aesthetics of the site. Relevant land features include land slope, valleys, hills, elevation, canyons, large water bodies (e.g., lakes, ocean), forests, and other types of vegetation. With the exception of maintaining and establishing trees and vegetative borders, there is little that a composter can do to change the lay of the land.

4.11 Site visibility

Visibility, or rather the lack of visibility, can be an important site feature. A facility that is visible to neighbors and others who pass by is more noticeable generally. Out-of-sight, out-of-mind can be a benefit to a compost site despite the fact it could raise suspicions among some neighbors. A site that is hidden by trees and other vegetation is a good candidate. The trees and vegetation provide an attractive and functional barrier that makes the site less objectionable. Coniferous vegetation is better than deciduous as it obscures a site through the winter. Other natural land features, such as hills and canyons, also provide good cover. Artificial visibility barriers, such as mounded soil and solid fences, are better than nothing but are not as effective as natural features. Sites with little or no vegetation can be planted to create barriers. However, shrubs, and especially trees, take a long time to grow into a barrier (Fig. 10.7). In the meantime, the site will be noticeable and may create a negative impression that lasts even after the planted vegetation reaches its desired height.

FIGURE 10.7

A row of fast-growing poplar trees and more slowly growing evergreen trees planted along the exposed perimeter of a farm composting site.

There are several reasons why visibility influences odors and other nuisances; some are psychological and sociological phenomena; others are real physical consequences. A site with trees around the border creates more air turbulence, which dilutes odorous air more effectively (Chastain and Wolak, 2000). In general, a composting site does not provide an attractive landscape to gaze upon. Few neighbors want to see it outside their windows. Thus, opposition to the composting activities may begin simply because the site is there and may escalate if odors become the dominant complaint. Similarly, natural features like trees give neighbors and the community a positive attribute to associate with the site.

If a site has no visual barriers and is open to neighbors or nearby roads, it is critical to maintain good housekeeping and keep the site looking like an intentional manufacturing facility as opposed to a junkyard. Being mindful of the site's appearance can go a long way to neighbor's acceptance. Feedstock should be contained; litter should be frequently collected; and broken or surplus equipment should be stored out of sight.

5. Site development

Most composting sites need some *development*—modifications that make them more suitable for composting. Typically, the soil has to be amended to support heavy use and the terrain may need slight to moderate *grading* (i.e., alterations to the slope). Some sites require substantial engineering in these respects. Site development is usually concentrated on the composting pad, the area occupied by the windrows and piles that are actively worked. Other heavily worked areas, like a grinding or screening area, may also be substantially altered.

Depending on the nature and scale of the operation and site conditions, additional development tasks can include establishing needed facilities and utilities (Table 10.3). Typical municipal and farm-based sites that use passive piles or windrow composting systems require little more than sources of water and fuel, and these items are commonly available within the municipality or farm already. In comparison, large-scale operations that charge fees to accept feedstocks and sell compost may need fencing and gates, truck scales, security lights, and a building for office functions and equipment maintenance.

5.1 Access and security

The site must be accessed from surface streets and must provide for safe queuing of vehicles, if queuing might occur. Delivery vehicles and customers must be able to find the site entrance and navigate the site with a minimum of communication. In addition to establishing a safe road turn-off, the access point(s) typically require a gate and possibly other access controls (e.g., signs, lights, security cameras).

Table 10.3 Summary of common composting site development items.

Item	Purpose
Access and security	
Roads	To access the site and move around within it
Fencing and gates	To prevent unauthorized access, safety
Signage	To direct traffic, inform employees, instruct visitors, educate users,
Lighting, security cameras	Site safety and security
Drainage and runoff controls	
Diversion structures (e.g., berms, swales, drains)	To divert clean runoff and seepage away from the composting pad and other working areas
Roof gutters	To keep clean precipitation off the composting pad and other working areas
Runoff collection (e.g., berms, swales, drains).	To collect and control the flow of contaminated site runoff
Storage ponds	To store runoff and provide a source of water for process and site management
Infiltration area	To discharge and treat both clean and site-contaminated runoff
Sedimentation and filtration devices	To remove solids from runoff before storage ponds and infiltration areas
Utilities	
Water source	Dust control, compost process moisture, fire prevention, water for facilities and employees
Electricity	For fans, lights, some processing equipment, communications, computers
Internet/mobile phone service	Communication, data collection and storage, on-site research
Fuel (diesel, gasoline)	Diesel for most processing equipment. Gasoline for small trucks and service vehicles
Trash	Some feedstocks come with contaminants that must be removed regularly and disposed, especially for sites handling municipal or commercial food wastes
Facilities	
Operator comforts	To accommodate workers, safety, hygiene
Office	A place to greet customers and suppliers, to store important records and to conduct business
Scales or load scanning	To accurately track incoming and outgoing loads
Signage	To direct traffic, educate users
Shop	To facilitate repairs and maintenance, storage of parts and consumables

5.1.1 Roads

Like an all-weather surface for the operating pad, a compost site needs an all-weather access road to assure ongoing access for deliveries of feedstock and outgoing deliveries of compost products. Interior access roads aren't always paved, though paved areas are easier to work on. Another consideration is to make sure that interior access roads are adequate enough to prevent vehicles accessing the site from forming a line that would back cars out onto the public road.

5.1.2 Signs

Printed signs tend to be undervalued and underused at composting facilities, farms, and many other places of work and commerce. Signs foster safe and efficient operation of the site. Good signs clearly and explicitly communicate important information to workers, customers, and visitors (Figs. 10.8 and 10.9). Signs intended for workers communicate rules, rights, procedures, and safety information. Signs tell customers about business policies, operating conditions, and may also provide instruction. Signs for visitors tend to give basic information about the business, including contact information, but also direct visitors on how to safely access the site. Some types of signs are mandated by regulations, particularly regulations that apply to worker health and safety.

5.1.3 Security—fencing, gates, etc

Larger sites usually need to erect fences and gates if they are not already present. Fencing and locking gates prevents unauthorized access and limits the potential for trespassing. Depending on the site and its operations, other security features to consider include lighting and security cameras.

5.2 Drainage and stormwater control

To avoid standing pools of water, the compost site should have positive drainage. A slope between 2% and 4% is generally recommended but topography, site conditions, and scale of operation may dictate a slope outside of this range. Sometimes slopes as slight as 1% are used to limit the elevation difference that must be created for greater slopes. Slopes above 5% can be hard on equipment and provide stormwater management challenges. Except for arid locations, windrows and piles should run parallel to the slope to prevent runoff from collecting on the uphill side of windrows/piles, creating saturated conditions at the base (Fig. 10.10). In arid locations or in dry seasons, orienting windrows/piles across the slope can help capture scarce water (which should be mixed into the entire windrow/pile shortly after it rains).

The site should be graded for handling stormwater runoff without creating erosion. Depending on local, state, or provincial restrictions, runoff from the composting site can be directed to an infiltration area or collected and stored in a holding pond for

FIGURE 10.8

Signs for the public: Visitors and customers need a lot of information. The signs displayed here just "crack the surface" of the body of information that needs to be communicated to the public.

FIGURE 10.9

Facility signs: This collection of signs is a sampling of the messages that need to be communicated to and among facility staff. Signs in dual languages are often needed.

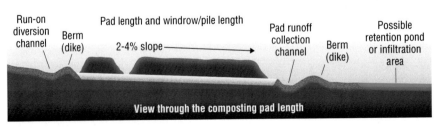

FIGURE 10.10

Composting pad drainage principles.

later use. On some farm sites, an adjacent pasture or cropland can receive the site runoff assuming there is adequate erosion control and protection of surface water. However, the receiving area for site runoff may be governed by local, state, or provincial regulations. More detail on handling stormwater runoff is contained later in this section.

Runoff or seepage from surrounding land that drains onto the site should be diverted away from the composting pad and storage areas. Such diversion can be accomplished by using diversion ditches, interceptor drains, or dikes. Buildings should have roof drainage gutters or perimeter drains if the roof runoff would otherwise empty onto the site. The land area that receives this clean diverted water must also be protected from erosion.

In many locations, public agencies offer technical services, and sometimes financial assistance, for surface water protection, stormwater, and erosion control projects. In the United States, such assistance is available from local soil and water conservation organizations and, nationally, through the USDA NRCS. In any locale, public resources should not be overlooked when pursuing soil and water conservation projects.

5.3 Utilities and facilities

The utilities and facilities that must be developed vary greatly among composting facilities. Typical farm-based sites that use passive piles or windrow composting systems require little more than sources of water and fuel, and these items are commonly available within the farm already. In comparison, large-scale operations that charge fees to accept feedstocks and sell compost may need fencing and gates, truck scales, security lights, and a building for office functions and equipment maintenance. The requirements depend on the scale of operation, the nature of the enterprise and composting methods, and operations employed.

The utilities and facilities that a given composting site *might* require are listed in Table 10.3. Some of the more common items are discussed below.

5.3.1 Water source

Possible sources of water for composting sites include:

- Ponds collecting runoff water from the composting site,
- Runoff collection ponds from other areas of a farm, municipality, or enterprise,
- Liquid manure storage ponds and tanks,
- Surface water bodies (e.g., streams, ponds),
- Ground water wells,
- Liquid feedstocks (e.g., dairy wastewater, food processing waste),
- Municipal/community water supplies,
- Local off-site sources.

A pond that collects runoff from the composting site should be among the first choices for process water. A runoff storage pond must be drawn down before the cold or wet season, in any case, unless the evaporation rate remains high. Furthermore, the pond water contains nutrients and organic matter that return to the composting system when the water is applied. A disadvantage of relying on runoff collection ponds is that they contain the least amount of water during dry spells when the composting piles need water the most. Another challenge of using stored stormwater runoff is that it likely contains pathogens and shouldn't be added to windrows and piles after completing the pathogen reduction process.

When the pond cannot supply enough water, a secondary source on-site may be used, such as manure storage, surface water, or a well. Alternatively, water may be available from off-site sources, including a neighboring farm, nearby food processing businesses, or a vendor that supplies water in a tank truck for a fee. The last option is an expensive one. Sources of nonhazardous liquid wastes or wastewater should be considered with the caution that accepting wastewater may violate the facility's operating permit and negate pathogen reduction efforts. In arid parts of the United States, when available, some composters can take advantage of reclaimed water projects.

For composting operations on livestock farms, another good water source is a liquid or slurry manure storage on the farm. Manure tends to be available regardless of the weather. Liquid manure harbors a fair amount of organic matter, nutrients, and microorganisms. It can invigorate or reinvigorate the composting process. If added in the latter stages of composting, the composting time may need to be extended (but with any addition potentially carrying pathogens, be mindful of where the material is in the pathogen reduction process).

It can be expensive to establish systems to use natural sources of water—groundwater wells and surface water. More importantly, government permits might be necessary to tap these sources. Permit requirements tend to be more common for surface water than groundwater. The need to obtain a permit delays site development. Permitting requirements can also open the composting facility to additional scrutiny by public agencies and the local community.

5.3.2 Electricity

Electrical power is necessary to operate blowers for aerated piles and to run certain materials handling equipment like augers or conveyors. Electrical motors approaching 7 or 8 kW (10 hp) should run on three-phase electrical service. During certain times of year, portable lights may be desirable to safely conduct some operations after daylight hours.

In order to reduce noise and/or fuel emissions, some composting facilities opt for electrically powered equipment. Screens, mixers, conveyors, and even grinders are available with electric power. In these cases, the electrical power requirements can be substantial. It is essential to provide three-phase power at a higher voltage service (e.g., 240/V, 480 V). Electric demand charges also become significant when large motors are used.

Developing a power source is usually a matter of working with the local electrical utility to bring the appropriate electric power lines (i.e., cables) to the site from the nearest point of access. The cost to do so is roughly proportional to the distance that the lines must be extended.

5.3.3 Fuel storage

A reliable and ample source of fuel is necessary at any site that employs large equipment—tractors, wheel loaders, windrow turners, grinders, etc. Diesel fuel is used for such equipment. Gasoline may be needed for small trucks, maintenance vehicles, and other small equipment. Depending on how much maintenance is done on-site, supplies of engine oil and hydraulic fluid may also be stored on-site. In some locations, environmental and safety agencies require permits, collision protection, and secondary containment for storage of fuel and other potentially hazardous liquids (Fig. 10.11).

FIGURE 10.11

Sturdy and well-protected fuel storage tanks with secondary containment to contain potential leaks. The features of this particular storage building were dictated by the regulatory authority.

5.3.4 Trash

No feedstocks are completely free of debris, and some contain substantial amounts of noncompostable material which must be removed, if not prior to composting, then before the compost is used. A composting facility should expect to generate a fair amount of trash that will be separated from the feedstocks or compost product. Most facilities need at least a "dumpster-sized" container and a contracted trash removal service, if not self-serviced or shared with the farm generally. Consider establishing a "waste-by-pass" area in which the trash is contained and stored apart from the rest of the site. Having an area dedicated to trash reduces litter and the possibility of recontamination of cleaned feedstocks and compost. It demonstrates that the facility is serious about high quality compost. Also, it shows feedstock haulers and generators that their feedstocks handling practices can be improved.

5.3.5 Load measurement—scales and such

Even the smallest composting operation needs to assess the volume of material arriving at and being composted. On small sites this can be done by eyeballing trucks and piles using simple estimation techniques. As operations increase in size, they may consider adding a vehicle scale, also called a weighbridge (Fig. 10.12). A scale makes tracking incoming and outgoing weights more manageable (and considerably more accurate). In some cases, product marketing regulations require certified scales for the sale of bulk products. An alternative to scales is load scanning technology that uses lasers to measure the volume of bulk material in a truck or trailer bed (Chapter 14).

FIGURE 10.12

Truck scales and scale house at a commercial/farm composting facility that handles a variety of feedstocks, primarily yard trimmings.

5.3.6 Employee comforts

A composting facility must accommodate the needs of its workers. At a minimum, such needs include toilets, facilities for washing, drinking water, access to shade in the summer and heat in the winter, a means of communication, and safety items (e.g., first aid kits, fire extinguishers). Sites that are close to the home office or farm or that are worked for only short periods may require little or nothing in this regard. Workers can use the facilities at the other base of operation. Stand-alone facilities with one or more full time employees require little to many accommodations, depending on the scale and nature of the operation. At many composting sites, workers are commonly served by temporary and mobile facilities such as portable toilets, cell phones or walkie-talkies, and vehicles that have air conditioning, heat, and carry water, safety kits and fire extinguishers. Moderately large sites that serve customers often include an office trailer that also serves worker needs while large-scale facilities tend to have dedicated buildings housing office and maintenance buildings. Labor and workplace safety rules can dictate what "comforts" and accommodations employers have to supply to employees.

5.3.7 Office

All facilities need to keep good records of incoming and outgoing materials, tasks to be completed, compost process data (temperature, turnings, etc.). Larger facilities need to develop a professional office for a multitude of purposes—as a place to greet customers, receive payments, to store and manage administrative and process records, and to provide sanitary facilities. Smaller scale facilities can often meet these administrative needs with a simple shed, fixed or transportable. If the facility has a truck scale, the scale house often serves as an office. Otherwise, mobile or portable structures, such as an office trailer, are common at both large and small-scale composting facilities (Fig. 10.13).

FIGURE 10.13

Two examples of composting facility office buildings, (A) a trailer style office, common to many composting facilities; and (B) a well-styled office constructed from steel shipping container.

5.3.8 Maintenance shop

It is at least advantageous, if not necessary, for a composting facility to have dedicated space for maintenance of equipment, and/or storage of the associated tools. Where a shop is needed on site, it is commonly housed in a barn, three-sided pole barn or metal shed, a dedicated garage or part of a larger multifunctional building (Fig. 10.14). In addition to tools, the shop should accommodate storage of essential supplies including oil and other mechanical fluids and lubricants, filters, wear parts, etc.

FIGURE 10.14

Multipurpose utility building serves as a maintenance shed for equipment.

There are several scenarios for which a facility can function without a maintenance shop including any of the following:

- Equipment is serviced by a dealership or other service company
- Equipment is serviced at a shop or garage located off-site on another part of the enterprise (e.g., farm or municipal shop)
- Equipment can be safely serviced outside, possibly with the support of a tool shed or a tool/service truck or van.

6. Site layout

The design of a composting site starts with its layout. The layout establishes the locations and areas of the various workstations and land features relative to one another. A good site layout accomplishes the following objectives:

- Provides sufficient area to perform composting and associated activities safely and efficiently,
- Minimizes materials handling,
- Prevents materials handling bottlenecks,
- Allows for safe movement of people and equipment,
- Minimizes stormwater run-on and runoff,
- Separates clean runoff from dirty (compost particle containing) runoff,
- Facilitates housekeeping and maintenance,
- Prevents cross-contamination of feedstock with curing and finished compost,
- Takes advantage of natural features,
- Controls access, and,
- Protects the surrounding environment.

A few of the guiding rules for laying out a composting site include the following:

- Establish the largest practical separation distances from critical land features (e.g., streams, wetlands, residences, property lines, neighbors),
- Take advantage of natural vegetation and barriers to shelter the most intrusive work areas (e.g., odorous, noisy, unsightly), if not the entire site,
- Segregate feedstocks from curing piles and finished compost products,
- Locate curing piles and finished compost products up-slope from the feedstock storage area and composting pad (or otherwise prevent the drainage from these areas from running toward curing and finished compost),
- Locate operational areas such that materials move in a straight line or circular pattern. Materials should move through the operation without crossing paths or backtracking,
- Leave sufficient space and aisle widths for truck drivers and loader operators to clearly see other vehicles and people on the site. Also follow spacing requirements that may be required by fire departments.
- Generally, the length of windrows and piles should run parallel with the slope of the pad to shed water. In dry climates, windrows and piles can run across the slope to retain water,
- Stormwater run-on draining toward the site should be diverted around it,
- Water draining off working areas of the site should be contained, stored, and/or treated,
- Locate water storage ponds and runoff treatment areas at the downhill side of the site,
- Place fences and gates at strategic points to control access,
- Use signs to direct traffic, and,
- Locate receiving areas, office facilities, scales, etc., near the site entrance.

The site layout process starts with a preliminary sketch of the proposed compost site and facility showing all key areas. Using online tools, like satellite images (e.g., Google Earth) can be very useful in this preliminary stage. The sketch should show the prevailing wind direction, traffic flow patterns, the land slope, runoff patterns, surrounding land uses, floodplain and wetland boundaries, and other pertinent environmental factors. A circle diagram is a simple technique to begin the layout. The diagram can be superimposed over a satellite image, as shown in Fig. 10.15.

The next step is a more detailed drawing that is drawn to scale, such as Fig. 10.16. This drawing should include the same features as a circle diagram but with more detail, dimensions, and the relative areas of the sections of the site. The volume of feedstocks and the composting method have the biggest influence on site layout.

7. Composting pad construction

The composting pad is the surface occupied by windrows and piles during the active composting period. A firm all-weather surface is necessary, although it does not need

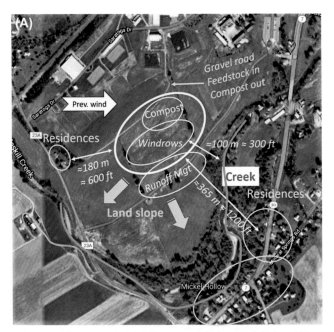

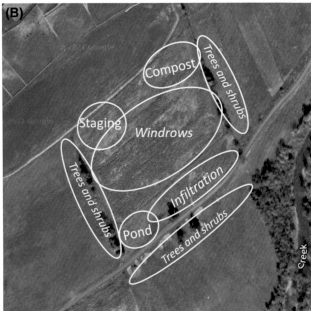

FIGURE 10.15

Circle diagrams of a proposed on-farm composting site drawn over satellite images for quickly envisioning the general layout of a facility. The sizes of the circles are relatively and roughly proportional to the areas of their activity.

FIGURE 10.16

Conceptual site layout drawn approximately to scale, developed on the site shown in Fig. 10.15.

to be paved, at least from an operational standpoint. Moderate- to well-drained native soils are satisfactory for many composting situations, and especially for on-farm composting. Typically, the existing ground surface requires some modification for drainage and efficient equipment handling. However, the nature of the composting surface is sometimes dictated by regulations, even on farms. An impervious pad might be required when particular feedstocks are composted, or within environmentally sensitive areas.

Paved composting pads are the preferred preference by compost operators (e.g., asphalt, concrete). A paved surface reduces problems related to mud, equipment operation, and pad maintenance (Box 10.2). A paved pad allows operations under wet conditions when an unpaved pad might be unworkable (Fig. 10.17). With paved surfaces, the finished compost acquires less soil, rocks, and/or stone. These materials are scraped up from unpaved surfaces when the feedstocks and compost are turned and moved. Unfortunately, the cost of paving is often prohibitive, and more pad runoff must be managed. Fortunately, there also is middle ground between native soil and conventional paving. Where the existing soils are unacceptable and paving

FIGURE 10.17

Asphalt pavement allows these windrows to be turned even in the pouring rain.

Source: Cary Oshins.

is impractical, alternative pad surfaces include amended soils, engineered surfaces, and surfaces with soil hardened with lime or cement. The common composting pad surfaces are summarized in Table 10.4.

Good information about pad surfaces can be found in aggregate and paving construction publications and web pages. Several are included in the References section.

Particularly on farms, there may be opportunities to adapt composting pads from under-used facilities (e.g., barnyards, fallow fields). Another cost-saving alternative is to limit concrete or asphalt surfaces to activity areas that especially benefit from a solid surface. These include areas where feedstocks are moved with a bucket loader; areas that receive feedstocks, and surfaces that store wet feedstocks. These areas are smaller than the composting pad, so the cost of installing concrete or asphalt may be acceptable.

Compost pads are subjected to heavy equipment, rough tires gripping with high torque, scraping blades and buckets, soil, sediment, erosion, and weather changes, sometimes extreme. In short, a lot of wear and tear should be expected. Therefore, regardless of the construction, all composting pads require regular maintenance to put the pad back into shape. Typical maintenance includes regular grading, filling of ruts and puddles, and replacement of the surface material (e.g., gravel). In general, less costly pad-construction options require more maintenance. Soil-covered pads need frequent regrading. Engineered and paved sites need to be periodically repaired, replenished, and/or repaved. All pad surfaces build up a layer of compost over time. This layer occasionally needs to be scraped off when it begins to interfere with the site operations (e.g., becomes slippery to tires).

Box 10.2 Preparation and anatomy of a composting pad

Regardless of the eventual surface material, nearly all pads (and surrounding operational areas) must be "prepared." Preparation involves grading the site to the desired contour and slope (e.g., 0%–5%), compacting the remaining soil and installing runoff diversion, collection and conveyances as needed. The topsoil and even the subsoil may also need to be excavated, depending on its qualities and the type of pad to be installed. In general, a more "engineered" pad demands more site preparation.

Engineered or paved pads are built up in layers on top of the existing soil, subsoil, or fill material. At its most basic configuration, the first layer is a compacted subbase that supports the surface layer above (Fig. 10.18). The subbase is the foundation of the construction. It spreads out the loads on the pad to the ground below. It is often the thickest layer in the construction, typically ranging from 150 to 250 mm (6–10 in.) thick for a composting pad. In addition to spreading out the loads from above, the subbase stabilizes and preserves surface aggregate while allowing water to drain. It also hinders the surface layer from sliding laterally under pressure.

A popular subbase is compacted soil or aggregate placed over geotextile fabric. Many types of geotextile products are available, serving different functions and meeting different performance specifications. A suitable product for a composting pad subbase is flexible fabric, either woven or nonwoven, depending on drainage and strength requirements. Product suppliers or local agricultural advisors can recommend appropriate products. Soil-based options include compacted native soil (if appropriate), crusher run, gravel fines, and manufactured soil. Suitable material must compact well and have particles that are generally smaller than the aggregate above.

Above the subbase, the details and the number of layers depends on the surface material. For example, aggregate surfaces are generally laid directly on the subbase while asphalt and concrete usually have multiple layers that include a "base course" above the subbase.

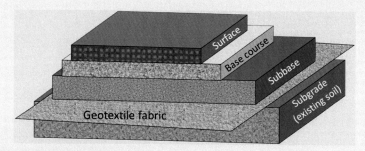

FIGURE 10.18

Anatomy of a simple composting pad. Note that the base course typically is not used with some construction.

Table 10.4 Summary of alternative composting pad surfaces, generally from least to most expensive (Chapter 19).

Surface type	Advantages and disadvantages
Existing soil	Depends on suitability of native soil. May become unworkable when wet. May develop ruts and depressions. Needs regrading.
Amended soils	Extra cost for aggregate amendment but improves upon existing soils. Usually better under wet conditions. Requires maintenance and replacement of aggregate surface.
Engineered soil/aggregate	Pad surface remains workable under a wide range of conditions. Drainage is better and more certain. Extra costs for engineering design and construction. Compromise between cost and performance.
Lime-stabilized soil Soil cement Roller-compacted concrete	Firm workable surface under all weather conditions for less cost than asphalt. Requires suitable soil to start with. Installation requires some expertise and/or experience. Little soil or rocks added to the compost.
Asphalt	Impervious and flexible all-weather surface. Protects underlying soil from leachate. Improves pad drainage but produces more runoff. Little to no soil added to the compost.
Concrete	Most durable and most expensive. Well-suited to particular areas that receive the most wear and tear (e.g., mixing, grinding). Little to no soil added to the compost

7.1 Existing soils

In many situations, existing surfaces provide an acceptable surface for composting (Fig. 10.19). Existing soils can work well as long as leachate and runoff do not impair surrounding waters (including groundwater), the soils do not erode, and equipment can operate when necessary.

Composting pads are best located on well-drained to moderately well-drained soils; for instance, soils rated in Group B of the NRCS hydrologic classification (Table 10.5). Often little more than minor grading, removal of topsoil and possibly precompaction of the existing ground is necessary to create a composting pad.

(A) **(B)**

FIGURE 10.19

Two examples of composting sites on native soils. (A) is a farm in Ohio and (B) is a small commercial composting site in upstate NY.

Table 10.5 NRCS hydrologic soil classification (NRCS, 2007a).

Classification	Description
Group A	Soils having a high infiltration rate (low runoff potential) when thoroughly wet. These consist mainly of deep, well drained to excessively drained sands or gravelly sands. These soils have a high rate of water transmission.
Group B	Soils having a moderate infiltration rate when thoroughly wet. These consist chiefly of moderately deep or deep, moderately well drained, or well drained soils that have moderately fine texture to moderately coarse texture. These soils have a moderate rate of water transmission.
Group C	Soils having a slow infiltration rate when thoroughly wet. These consist chiefly of soils having a layer that impedes the downward movement of water or soils of moderately fine texture or fine texture. These soils have a slow rate of water transmission.
Group D	Soils having a very slow infiltration rate (high runoff potential) when thoroughly wet. These consist chiefly of clays that have a high shrink-swell potential, soils that have a permanent high water table, soils that have a claypan or clay layer at or near the surface, and soils that are shallow over nearly impervious material. These soils have a very slow rate of water transmission.

The soil becomes further compacted as the pad is repeatedly worked and driven on by equipment. If the underlying soil is primarily firm but soft or slippery at the surface, a thin layer of aggregate material can be spread on the surface layer of soil and then compacted with a roller or other heavy wheeled vehicle.

Natural soils that are excessively well-drained (e.g., gravel) are best avoided because they can convey nutrient-laden water to ground water or adjacent surface waters with little filtering. Natural soils that are poorly drained (e.g., clay) make poor composting pads because they become and remain soft and muddy. Equipment is difficult to operate and may even get stuck in the mud. Soft soils form ruts and depressions that hold water and may become sources of odor and fly or mosquito nurseries. Sites with heavy soils may be usable if the site is worked relatively infrequently. An example is a passive pile that is only sporadically turned. Such soils may also be acceptable as temporary or seasonal pads, used when precipitation is sparse.

On-farm composting sites can be located in fields that are actively being cultivated, if the environment is protected. One example is to place compost windrows in under-productive areas, such as the corners of land irrigated by center-pivot irrigation circles. Another interesting approach is to establish a moveable composting site, set it in fallow fields in between crops (Fig. 10.20). The composting site rotates with the crops to the next fallow ground. The ensuing crop uses any nutrients left behind by the composting windrows and piles. The approach works well because the migration of nutrients through soils tends to be shallow, within the root zone of a crop. For this system to be successful, the ground on all fields must be firm enough to support the composting equipment. Also, the fields must be sufficiently

FIGURE 10.20

Farm composting operation in which windrows are placed on fields that are fallowed. The site rotates with the crop's year-by-year.

distanced from neighbors and surface water, and sufficiently accessible for trucks delivering the feedstocks and removing the compost. If the compost is used within the farm, it can be applied directly to the field on which composting took place.

7.2 Amended soils

Existing soil can be improved for composting by reworking the surface and amending it with aggregate materials. The existing soil is not removed, at least not permanently; instead, aggregate material is incorporated into the existing soil. Typically, the existing soil is graded to the desired slope and the aggregate is added as a shallow layer on top of the soil (100–200 cm or 4–8 in.). Afterward, the pad is compacted with a roller to press the aggregate into the soil (Fig. 10.21). If the soil is very poor, the top layer can be excavated, remixed, mixed with the aggregate, and then reapplied and compacted. A common soil to aggregate ratio is 2 parts soil to 1 part aggregate, by volume.

Several materials that make suitable aggregate include gravel, stone, slag, recycled asphalt, and recycled concrete. Wood chips also have been used as an amendment for working surfaces. However, wood chips are a temporary solution. They gradually decompose and lose their utility unless they are regularly replaced. Also, wood chips support biological growth and become slippery.

Since the aggregate is mixed with the soil, the particle size distribution is not important, although very large pieces are undesirable. Usually, the least costly material is whatever aggregate is either locally abundant or in close proximity. Often road construction and demolition contractors are looking for places near the job site to deposit ground asphalt or concrete. This situation presents an opportunity for free or cheap pad material.

Amended soils can wear away or become completely covered with compost. It is not unusual to apply additional aggregate to the surface on an annual or biennial

FIGURE 10.21

Amended soil pad with recycled asphalt added as the surface layer and compacted.

basis. Amending existing soils is not nearly as reliable or durable as an engineered pad, but it is substantially less expensive in the short run.

7.3 Engineered pad surfaces

The intent with an engineered surface is to replace the existing soil, not improve it. The existing soil is covered with the new pad material, usually after excavating the topsoil. The constructed pad is typically 30—45 cm deep (12—18 in.), deeper if a new subbase is installed beneath the surface material.

Engineered pads have a designed aggregate surface installed over a prepared subbase. The choices for the finished surface aggregate material are similar to those for amended soils—sand, various types and grades of gravel, stone and recycles, slag, glass, asphalt, and recycled concrete. In engineered applications, the aggregate size and size distribution are important. Designers specify a certain particle size and size distribution from the material supplier, depending on the aggregate functions. Relatively large particles that are uniform in size are desired where good drainage is a function (Fig. 10.22). For the surface of a composting pad, the aggregate should contain wide-ranging particle sizes (i.e., "well-graded"). The smaller particles serve to bind the larger particles keeping them from shifting and form a strong compact surface. Bank gravel, crusher run,[2] and unscreened recycled asphalt and concrete are commonly used (Fig. 10.23).

---

[2] Crusher run is a mix of coarse aggregate and fine stone particles that compacts well and forms a solid firm surface. It goes by several names including dense grade aggregate, road stone graded aggregate base (GAB), and aggregate base course.

FIGURE 10.22

This gravel pad slopes to the right and is stabilized by geotextile fabric below (CWMI, 2005). Runoff drains above and within the gravel to the collection pond, located out of the picture to the right (A). The inset shows the relatively large gravel used for the pad's surface (B).

FIGURE 10.23

Engineered compost pad with mixed and smaller aggregate than what is shown in Fig. 10.22. The tighter surface allows runoff to drain off the pad surface.

Most engineered soil pads are permeable surfaces. Water drains through the surface material, geotextile fabric, and soil subbases. However, an impermeable soil/aggregate pad can be created by underlying the pad with an impermeable membrane, such as liners used in pond or landfill construction. Layers of sand sandwich the plastic liner to protect it from puncture. The upper layer of sand drains water and conveys it to a collection point. If the percolating water is stored in a pond, the liner can be a continuous membrane that underlies the pad and forms the bed of the pond (See section 10.4.1).

7.4 Lime-stabilized soil

Several composting facilities have pads constructed of lime-stabilized soil, most notably the composting facility on the USDA's research farm in Beltsville, Maryland (Sikora and Francis, 2000a). Lime-stabilized soil is a hard and nearly impervious pad material that forms from a chemical reaction between the existing soil, lime, and water. At a pH higher than 11.5, the minerals in the soil (silicates and aluminates) and the lime chemically react and cement the soil particles together. Over several days, the pH drops and the soil hardens. The permeability of the surface is similar to tight clay soils.

The steps for creating a lime-stabilized soil pad include grading the existing soil as necessary, and then the soil is tilled and mixed in-place a predetermined amount lime and water. The mixture is compacted and allowed to cure for several days. If performed correctly, the process produces a durable all-weather surface, strong enough to support the wear and tear of composting equipment. The depth of the mixture usually falls in the range of 6—12 in. (15—30 cm), depending on the underlying soil, climate, and anticipated loads. A lime-stabilized soil pad should be roughly 20% thicker than a conventional concrete pad for the application at hand.

The soil type is a factor in the success of achieving a hard stable surface. Generally, clay soils are preferred (10% minimum clay content). In a sense, it turns a poor soil situation into a good one. Soils with less than 10% clay might still be appropriate if fly ash or other amendment is added to increase the level of silicates and aluminates. The process also needs to be modified if the existing soil contains higher levels of organic matter (>1%), sulfates, and sodium and potassium hydroxides. In these situations, the amount of lime or water is typically increased, or the procedures are more deliberate.

To raise the pH of the soil above 11.5, a highly reactive type of lime is required. Quick lime (calcium oxide) or hydrated lime (calcium hydroxide) is suitable for soil stabilization while normal agricultural lime (calcium carbonate) is not. The amount of lime necessary is determined in advance via simple laboratory test (e.g., ASTM C977-18). The basis of the test is to determine the minimum amount of lime needed to raise the soil pH to 12.4 when the lime is added to a representative sample of the soil.

Lime-stabilized soil composting pads have generally performed well over many years, providing the same benefits as a paved surface at a lower cost. At some facilities, the hard surface failed to set up, or it cracked soon after it was put into service. The failures can be explained by the wrong type of lime, poorly suited soil or excessive drying before the chemical reaction was complete. The January 2000 issue of BioCycle contains a detailed article on the lime-stabilization procedure and the experience at the USDA Beltsville farm (Sikora and Francis, 2000b).

7.5 Soil cement

Soil cement is similar to lime-stabilized soil in that existing soil is typically mixed in place with a chemical binding agent in this case, Portland cement. It forms a durable impervious concrete-like surface at a lower cost than concrete. It is widely used in road construction, primarily as a base course for the roadway surface.

Because cement requires only water for its chemical reaction, almost any type of soil is suitable. Granular soils, like sand and gravel, are preferred because they tend to be more uniform and require less cement. Laboratory tests determine the cement to soil ratio (on the order of 90% soil by weight).

The procedure for constructing a soil cement pad is much the same as lime-stabilized soil—(a) the site is graded; (b) the soil is loosened with tillage equipment; (c) a predetermined amount of cement is spread over the soil and mixed in with a mixing machine; (d) the surface is shaped and compacted; and (e) the surface is cured by adding water and sealing it with a tarp or bituminous compound to prevent drying during the curing period. The soil cement pad will continue to strengthen with age. An alternative method is to mix the soil and cement at a central plant and then transport it to the site. After placing the mixture on a prepared surface, the surface is shaped, compacted, and cured. In this case, the soil used is typically granular and not native to the composting site.

7.6 Roller-compacted concrete

Roller-compacted concrete (RCC) is another paving alternative that occupies a niche somewhere between soil cement and conventional concrete. It is stronger than soil cement, less expensive than conventional concrete but with fewer applications. It is commonly used for roads, dams, and paved areas running heavy equipment, including composting facilities.

RCC is a relatively dry stiff mixture of Portland cement and aggregate (e.g., gravel). It contains less water than conventional concrete, so it develops higher strength but is more difficult to place and spread. It does not flow so concrete forms and reinforcing steel are not used. Instead RCC is installed by pavers and vibratory roller compactors, the type of machinery typically used for asphalt paving. The cement and aggregate are usually mixed at or near the site in pug mills and transported by dump trucks to the paver. The mix is placed in lifts ranging from 20 to 30 cm thick (8−12 in.).

7.7 Asphalt (Macadam)

Asphalt is a common and popular working surface for composting because it is both largely impermeable and forgiving (Fig. 10.24). While it is generally not as long lasting and strong as concrete, it is generally less expensive, for a given use, and quicker to install. Asphalt is more resistant than concrete to certain chemicals, such as salts, but less resistant to others, like solvents and organic acids. The ability of asphalt to flex accommodates small variations of the underlying soil due to frost or uneven loads. However, it may collapse if large voids develop (e.g., potholes) whereas concrete may not. Asphalt is relatively easy to patch, repair, and resurface. Still, for smaller scale composting facilities, asphalt is a relatively costly alternative. Installation requires specialized equipment and contractors. It is not a do-it-yourself option.

The term, "asphalt," refers to the tar-like binder (i.e., bitumen) that holds together aggregates, sand, and/or other materials to form a solid composite surface. Usually, the term suggests the most common type of asphalt pavement, "hot mix asphalt" (HMA), or hot rolled asphalt. HMA is a mixture of small aggregates, sand, miscellaneous additives, and the liquid binder. The mix is heated and applied in a nearly continuous layer by special machinery. The combination of materials used in the mix vary with the intended application and expected loads. Paved asphalt surfaces may also be referred to as "macadam," which originally was a method of forming aggregate into a tight firm road surface.

FIGURE 10.24

Asphalt is a popular pad surface among composters. It is clean, firm, and drains well.

Asphalt pavement is applied in at least two layers—a base course and a wearing course at the surface. In most cases, the base course is applied on top of a 10–20 cm (4–8 in.) compacted subbase. The thickness of the layers, and their composition, depends on the soil conditions, with poorly drained soils requiring thicker layers. Geotextile liners may also be included below the subbase. The expected vehicle load on the pavement also affects the design thickness. For composting pads, a 5–10 cm (2–4 in.) asphalt base course and 5 cm wearing course usually suffice.

Under normal wear and tear, an asphalt pad should last 10–20 years, depending on its design, quality of construction, and maintenance. Typical maintenance includes filling of cracks and potholes plus regular sealcoating. However, such maintenance can be difficult to carry out at a busy composting facility. Therefore, conservatively assume that an asphalt pad, or sections of it, will need to be replaced every 10 years and repaired occasionally in between.

7.8 Concrete

Concrete makes an excellent composting pad. It is strong, durable, impermeable, easily cleaned, and relatively low maintenance. The concrete formula can be adjusted to change its character or to suit varying conditions during its installation and use. For instance, it can be formulated to be more or less strong, abrasion resistant, slip resistant and porous. It is usually the most expensive pad surface. With the right experience, a concrete pad can be self-constructed. In this case, the right experience includes building forms, laying steel reinforcing and spreading, and finishing concrete.

Concrete is a combination of Portland cement, aggregate, sand, water and air. The Portland cement and water chemically react to bond the aggregate together. The proportion and size of the aggregate and sand plus the ratio of cement and water determine the concrete's properties and strength. Water is a decisive element. A higher proportion of water weakens the concrete but allows the concrete to flow and makes it easier to place. Do-it-yourselfers need to resist the temptation to add more than the recommended proportion of water.

Several opportunistic composters have been able to establish a paved composting site with OPC ("Other People's Concrete"). OPC is gained by establishing the composting site on the unused concrete surfaces at lumber mills, airport runways, military bases, factories and similar abandoned properties. OPC providers a long-wearing surface at a very low cost.

8. How much space? Estimating the area for composting

The land area needed for composting depends first on the scale of the operation—the volume of feedstocks to be composted. After that, the required area is influenced by the feedstock characteristics, feedstock delivery and storage patterns, the composting method and equipment selected; vehicle traffic patterns; compost curing and storage practices; areas need for ancillary operations (e.g., screening, grinding), and buffer areas for odor, noise, and environmental protection.

There are rules-of-thumb for estimating the required size of a compost pad. For example, Michigan's *Compost Operator Guidebook* (Chardoul et al., 2015) suggests that, for most facilities, at least 1 acre is needed to effectively process 5000 yd^3 per year, although "intensively managed" facilities may be able to process 8000 yd^3 per year on an acre. These estimates translate to approximately 9500 m^3/year per hectare, for most facilities, and 15,000 m^3/year per hectare, for intensively managed sites. These rough estimates are good for imagining the magnitude of a given facility, but they are not good enough for determining its actual size. For that task, even more math is required.

The composting pad dominates the land area at most composting facilities. Therefore, it usually serves as the basis for estimating the required area of the entire site. The areas needed for the additional activities and buffers are added onto the required footprint of the pad. However, the storage area for compost products can occupy the largest share of the site where demand for compost products is sporadic or highly seasonal, or where outside products are acquired and then stored on site.

8.1 Estimating the composting pad area

The following information applies primarily to turned windrows, free-standing piles (e.g., ASPs), and rectangular bins and bunkers. The space requirements for in-vessel composting systems are best determined by working with the system supplier. However, it is important to recognize that in-vessel composting is usually followed by a second stage of composting in windrows or ASPs.

The area required for the composting pad depends on the following factors:

- the volume of feedstocks composted,
- the time that the materials remain on the pad,
- shape and length of the windrow, pile or bin/bunker,
- the feedstocks' volume shrinkage,
- whether windrows, piles or bins/bunkers are combined due to the shrinkage; and,
- the space needed to maneuver equipment.

The windrow/pile shape is determined by the composting method and equipment used to build and turn windrows/piles. Fig. 10.25 depicts typical shapes, dimensions, and spacings for common composting methods and equipment. The width of turned windrows is determined by the windrow turner. Otherwise, recommended widths and heights can be found in Chapters 5 and 6. Table 10.6 provides the formulas for calculating the cross-sectional area of the corresponding shapes. With this information, the required composting pad area can be calculated by a number of

Free-standing piles turned with bucket loaders

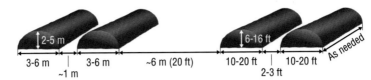

Tractor-assisted straddle-style windrow turners

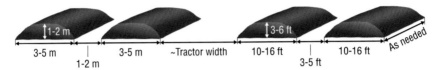

Self-driven straddle-style windrow turners

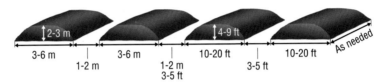

Individual aerated static piles

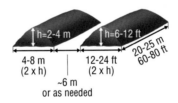

Extended aerated static piles

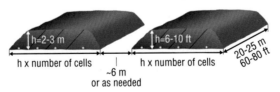

FIGURE 10.25

Dimensions and spacing of typical piles and windrows.

approaches that take different steps but use the same principles. Box 10.3 presents one commonly followed procedure.

Example 10.1 follows the procedure outlined in Box 10.3 for a hypothetical windrow composting facility. Example 10.2 estimates the pad and curing areas for an extended ASP operation using a slightly different set of steps.

Table 10.6 Formulas for calculating the cross-sectional area of piles and windrows of various shapes.

Method and equipment used	Approximate shape	Cross-sectional area
Windrows/piles turned with a bucket loader	h = 6-12 feet, b = 10-20 feet	$A = 2/3 \times b \times h$
Small tractor-drawn windrow turners or any turners with wet materials	h = 3-4 feet, b = 9-18 feet	$A = 2/3 \times b \times h$
Self-propelled and tractor-drawn windrow turners	t, h = 4-9 feet, b = 10-20 feet	$A = h \times (b - h)$ or[a] $A = h \times \dfrac{(b + t)}{2}$
Individual aerated static piles and other piles with little or no turning	h = 5-8 feet, b = 2 x h	$A = 1/2 \times b \times h$
Extended aerated static piles	h = 5-8 feet, cell width b = h (approximately)	Cell area: $A = b \times h$ Pile area: $A = n \times a = n \times b \times h$ where n is the number of cells

[a] This formula is an approximate and is valid only when the width is greater than or equal to twice the height.

Box 10.3 A procedure for estimating compost pad area for windrows, piles, bins, and bunkers

1. *Estimate the volume of material to be composted.*

   For moist primary feedstocks, like manure and biosolids, the volume of amendments required is often three to four times the volume of the primary feedstock. For manure composting, the amount of manure generated on a daily basis can be estimated from one of the many good publications about livestock and poultry waste management (e.g., ASABE, 2005).

2. *Account for volume loss when feedstocks are mixed together.*

   Usually when composting materials are mixed together, the volume of the mixture is less than the combined volume of the individual ingredients Therefore, the volume of material in newly formed piles/windrows can be estimated by adding together the volumes of the individual ingredients and multiplying this sum by a factor that accounts for this loss in volume. A factor of 0.80 (80%) is reasonable starting point, which corresponds to a volume reduction of 20%. Preferably, this factor

Continued

Box 10.3 A procedure for estimating compost pad area for windrows, piles, bins, and bunkers—cont'd

should be determined by experimentation, or using existing data with similar feedstocks. For a conservative estimate, the individual feedstock volumes can be added together without applying this factor.

3. *Calculate the total volume of material on the pad.*

   Multiply the daily volume of feedstock by the number of days the material will remain on the composting pad (Chapter 3). This result volume is the volume of material that the composting pad must hold, assuming no shrinkage or combining of windrow/piles or bins/bunkers.

4. *Account for volume loss during composting.*

   Because the feedstocks lose volume during composting, windrows/piles and bins/bunkers are often consolidated after a few weeks. In this case, the volume obtained from step 2 can be multiplied by a shrinkage factor if desired. As a general approximation, a shrinkage factor of 0.75 can be used if specific data are lacking. The actual shrinkage depends on the raw materials, so use a more specific value if known (Chapter 3).

5. *Estimate the probable dimensions of the windrows, piles, bins or bunkers.*

   Based on the proposed equipment and composting method, determine the pile shape and dimensions. Determine the available length at the site for windrows or piles. Account for space at ends for vehicle access (approximately 3 m or 10 ft) and separation distances from property lines, wetlands, streams, and so on. Also account for space between separate piles/windrows lined up end-to-end.

6. *Determine the volume of a single windrow or pile.*

   Calculate the cross-sectional area of a windrow/pile from the formulas in Table 10.6. Multiply this area by the estimated windrow/pile length to determine the windrow/pile volume.

7. *Calculate the number of individual windrows, piles, bins or bunkers.*

   The number of windrows, piles, or cells required equals the total volume (from step 2) divided by the volume per windrow/pile/cell (step 5). Round off to a reasonable whole number.

8. *Account for the spacing between windrows and piles.*

   Refer to Fig. 10.25 for spacing of windrows/piles. The width plus spacing times the number of windrows/piles gives the approximate pad width.

9. Add the perimeter aisle/road space to the length and width to determine the overall pad dimensions.

10. Sketch the layout of the composting pad to see the full picture.

8.2 Curing area

In many cases, compost is cured on the active composting pad, not in a separate curing area. This situation is common among facilities that compost by turned windrows or passive piles. ASP and in-vessel facilities are more likely to use separate curing space. The advantage of doing so is that the curing piles can be somewhat taller (up to 4 m or 12 ft) and wider with less aisle required (Fig. 10.26). Where a separate curing area is used, it is typically less than 20% of the composting pad area. The volume of compost produced is generally less than half the original feedstock volume, although it can be, as low as 25% for loose, degradable feedstocks like leaves. Also, curing is a relatively short step, lasting one to two months.

FIGURE 10.26

Curing piles can be wider and/or taller/spaced closer than active composting piles. Smaller curing piles are necessary if active composting is rushed.

The space requirement for curing can be calculated in the same manner as the compost pad, based upon the amount of compost, the length of time the compost is cured, and the shapes, heights, lengths, and spacings of the piles. Most curing piles are formed by loaders and tend to have a parabolic or trapezoidal shape, depending on the height and width (Table 10.6). Alternatively, the required area for curing can be roughly estimated by dividing the estimated compost volume, in m^3 of ft^3, by the *average* pile height in m or ft.

Example 10.1 Composting pad area for a turned windrow composting facility

The municipality has proposed a new yard trimmings composting facility to recycle yard trimmings that are currently transported to a private facility 200 km away (125 miles). The average weekly generation of yard trimmings varies month-by-month as shown in Fig. 10.27. The turned windrow composting method will be used, with windrows remaining on the composting pad for 12 weeks (about three months). There will be no separate curing area. Mature compost will cure in the windrows. After 5 or six weeks, adjacent windrows will be combined due to volume shrinkage. A straddle-type windrow turner will be used that creates windrows that are 5 m wide and 2 m high with a spacing of 1.5 m between windrows (16.5 ft w × 6.5 ft h × 5 ft spacing).

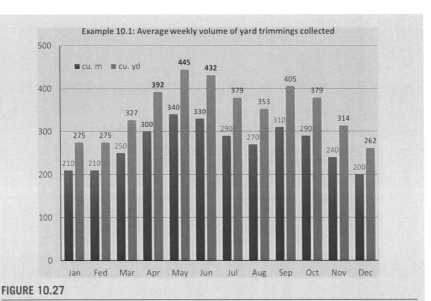

Example 10.1: Average weekly volume of yard trimmings collected

FIGURE 10.27

Average weekly generation of yard trimmings collected. This data will be used to calculate the composting pad area for the facility in this example.

1. *Estimate the volume of material to be composted.*
   Fig. 10.27 shows that the average weekly collection of yard trimmings varies monthly.
   The pad needs to be sized for the peak deliveries of feedstock. Because the the windrows of yard trimmings will remain on the pad for 3 months, the area should be calculated based on the three consecutive months with the highest volume. In this case, those months are April, May and June. The average weekly collection for these three months is approximately 325 m³ or 425 yd³.

2. *Account for volume loss when feedstocks are mixed together.*
   Not applicable. No mixing is taking place.

3. *Calculate the total volume of material on the pad.*
   Multiply the weekly volume of feedstock by the number of weeks the material will remain in on the composting pad:
   12 weeks × 325 m³/week = 3900 m³ 12 weeks × 425 yd³/week = 5100 yd³.

4. *Account for volume loss during composting.*
   Assume a shrinkage factor of 0.75: 3900 m³ × 0.75 = 2925 m³
   5100 yd³ × 0.75 = 3825 yd³ × 27ft³/yd³ = 103,275 ft³

5. *Estimate the probable dimensions of the windrows.*
   The windrows will be 5 m (15 ft) wide × 2 m (6 ft) high. The site suggests an appropriate windrow length is 40 m (about 130 ft).

6. *Determine the volume of a single windrow.*
   Calculate the cross-sectional area of a windrow, then multiply by the length. The turner creates windrows with a trapezoidal shape. Assume that the angle of the side slopes is roughly 45 degrees.

Example 10.1 Composting pad area for a turned windrow composting facility—cont'd

$$\text{Area of a trapezoid} = h \times (w - h) = 2 \text{ m} \times (5 \text{ m} - 2 \text{ m}) = 6 \text{ m}^2 \times 40 \text{ m} = 240 \text{ m}^3$$
$$= 6.5 \text{ ft} \times (16.5 \text{ ft} - 6.5 \text{ ft}) = 65 \text{ ft}^2 \times 130 \text{ ft} = 8450 \text{ ft}^3.$$

7. *Calculate the number of windrows.*
   The number of windrows required equals the total volume divided by the volume per windrow. Round off to a reasonable whole number.

$$\text{No. of windrows} = 3825 \text{ m}^3 \div 240 \text{ m}^3 \text{ per windrow} = 12.2 \text{ windrows} \rightarrow \text{assume } 13$$
$$= 103,275 \text{ ft}^3 \div 8450 \text{ ft}^3 \text{ per windrow} = 12.2 \text{ windrows} \rightarrow \text{assume } 13$$

8. *Account for the spacing between windrows.*
   The windrow width plus spacing times the number of windrows gives the pad width.

$$\text{Pad width} = (5 \text{ m} + 1.5 \text{ m}) \times 13 \text{ windrows} = 84.5 \text{ m} \approx 85 \text{ m}$$
$$= (16.5 \text{ ft} + 5 \text{ ft}) \times 13 \text{ windrows} = 279.5 \text{ ft} \approx 280 \text{ ft}$$

9. *Add the perimeter aisle/road space to the length and width to determine the overall pad dimensions.*
   Assume a 5 m space on all sides of the composting pad for access (about 15−16 ft)

10. *Sketch the layout of the composting pad to see the full picture (Fig. 10.28).*

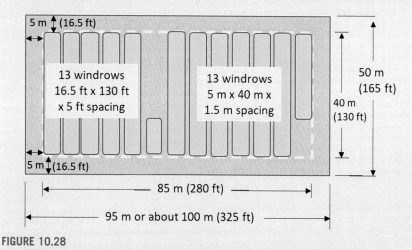

FIGURE 10.28

Sketch of composting pad area for the example turned windrow composting facility.

Example 10.2 Composting pad area for an extended aerated static pile (ASP) containing yard trimmings and source separated organics

The municipality (from Example 10.1) is also considering an eventual expansion of their compost operation. There will be an opportunity to compost source separated organics (SSO) from households and food service businesses along with additional yard trimmings. The SSO, which is collected by a small for-profit business, is primarily food waste with trivial amounts of paper products. During peak periods, 8 metric tons (tonnes) per day are collected (8.8 U.S. tons). The SSO has an average bulk density of 420 kg/m^3 (708 lbs/yd^3). The volume of yard trimmings is expected to increase as neighboring towns divert materials to the city's proposed composting facility. The plan is to compost the feedstocks in a ratio of 4 volumes of yard trimmings to 1 volume of SSO. Mobile mixing equipment will be used to blend the feedstocks.

The municipality will use the extended ASP composting method in two phases. In Phase 1, each cell of the ASP will hold one day's volume of feedstocks. After approximately two weeks in Phase 1, cells will be removed and added as new cells to the Phase 2 pile, where the material will compost for a second two weeks before being transferred to curing piles. Due to material volume shrinkage, which is anticipated to be 50% in Phase 1, two cells from Phase 1 will be combined into a single cell in Phase 2. Shrinkage during Phase 2 is anticipated to be 20% of the volume delivered to Phase 2. Approximate expected dimensions of the piles are listed below.

Pile	Retention Time (days)	Number of cells	Cell/pile Width	Cell/pile Height
Phase 1	14	14	2 m (6.5 ft)	2 m (6.5 ft)
Phase 2	14	7	2 m (6.5 ft)	2 m (6.5 ft)
Curing	28	—	6 m (19.7 ft)	2.5 m (8.2 ft)

1. *Estimate the daily volume of materials to be composted.*

$$\text{SSO: } (8 \text{ t/d} \times 1000 \text{ kg/t}) \div 420 \text{ kg/m}^3 = 19 \text{ m}^3/\text{day}$$
$$(8.8 \text{ tons/day} \times 2000 \text{ lbs/ton}) \div 708 \text{ lbs/yd}^3 = 24.9 \text{ yd}^3/\text{day} \times 27 \text{ ft}^3/\text{yd}^3$$
$$= 671 \text{ ft}^3/\text{day}.$$

$$\text{Yard trimmings: } 4 \times 19 \text{ m}^3 = 76 \text{ m}^3/\text{day} \qquad 4 \times 671 \text{ ft}^3 = 2684 \text{ ft}^3/\text{day}.$$

2. *Account for volume loss when feedstocks are mixed together.*

In this case, applying a mixing loss factor does not work well. The yard trimmings dominate the mix and the SSO will nearly disappear within it. Therefore, a reasonable and conservative assumption is that half of the SSO adds to the overall volume.

$$(\tfrac{1}{2} \times 19 \text{ m}^3) + 76 \text{ m}^3 = 86 \text{ m}^3/\text{day} \qquad (\tfrac{1}{2} \times 671 \text{ ft}^3) + 2684 \text{ ft}^3 = 3020 \text{ ft}^3/\text{day}$$

Example 10.2 Composting pad area for an extended aerated static pile (ASP) containing yard trimmings and source separated organics—cont'd

3. *Calculate the required cell length.*

It is time to make a logical diversion from the procedure of Box 10.3. From here forward the steps differ. The required length of the cell can be calculated because the cell dimensions are already known; and the volume of each Phase 1 cell is also known. It is 1 day's volume of feedstock.

PHASE 1:

a. Determine the cross-sectional area of a cell.

    From Table 10.6, the area of a cell is width × height.

$$A = h \times w = 2 \text{ m} \times 2 \text{ m} = 4 \text{ m}^2 \qquad 6.5 \text{ ft} \times 6.5 \text{ ft} = 42.25 \text{ ft}^2$$

b. Determine the required cell length by dividing the cell volume (daily feedstock volume) by the cross-sectional area.

$$L = 86 \text{ m}^3 \div 4 \text{ m}^2 = 21.5 \text{ m} \qquad 3020 \text{ ft}^3 \div 42.25 \text{ ft}^2 = 71.5 \text{ ft}$$

Note that this cell length is less than 25 m (80 ft), the maximum aeration length recommended in Chapter 6.

PHASE 2: Phase 2 cells are the same length as Phase 1 cells because:
- Phase 2 cells are the same width and height as Phase 1,
- The volume shrinkage during Phase 1 is assumed to be 50%; and,
- Each Phase 2 cell is holds two Phase 1 cells, therefore:
- Phase 2 cells hold the same initial volume as Phase 1 cells.

4. *Reduce volume to account for volume shrinkage. Assume a shrinkage factor of 0.75.*

Not applicable. Shrinkage is included in the Phase 2 volume calculation.

5. *Estimate the probable dimensions of the cells.*

    Each cell is 2 m wide × 21.5 m long (6.5 ft wide × 71.5 ft long)

6. *Determine the overall width of each extended ASP.*

$$\text{Phase 1: Pile width} = \# \text{ cells} \times \text{cell width} = 14 \text{ cells} \times 2 \text{ m/cell} = 28 \text{ m}$$
$$= 14 \text{ cells} \times 6.5 \text{ ft/cells} = 91 \text{ ft}$$

$$\text{Phase 2: Pile width} = \# \text{ cells} \times \text{cell width} = 7 \text{ cells} \times 2 \text{ m/cell} = 14 \text{ m}$$
$$= 7 \text{ cells} \times 6.5 \text{ ft/cells} = 45.5 \text{ ft}$$

7. *Determine the length and number of curing piles.*

a. Determine the daily volume to curing pile

On average, one half of a Phase 2 cell is sent to curing each day. Each Phase 2 cell initially holds 86 m³ (3020 ft³) and shrinks 20% through Phase 2 composting (80% remains).

$$\text{Curing daily volume} = 1/2 \times 86 \text{ m}^3 \times 0.80 = 34.4 \text{ m}^3/\text{day}$$
$$= 1/2 \times 3020 \text{ ft}^3 \times 0.8 = 1208 \text{ft}^3/\text{day}$$

b. Determine the total volume on the curing pad over 28 days

$$34.4 \text{ m}^3/\text{day} \times 28 \text{ days} = 963.2 \text{ m}^3 \quad 1208 \text{ ft}^3/\text{day} \times 28 \text{ days}$$
$$= 33,824 \text{ ft}^3 \ (1253 \text{ yd}^3)$$

c. Determine the cross-section area of the curing piles, assume a trapezoidal shape.

$$A = h \times (w-h) = 2.5 \text{ m} \times (6 \text{ m} - 2.5 \text{ m}) = 8.75 \text{ m}^2 \quad 8.2 \text{ ft} \times (19.7 \text{ ft} - 8.2 \text{ ft})$$
$$= 94.3 \text{ ft}^2$$

d. Determine the total length of curing piles needed. Divide the total volume by the pile cross-sectional area.

$$963.2 \text{ m}^3 \div 8.75 \text{ m}^2 = 110 \text{ m} \quad 33,824 \text{ ft}^3 \div 94.3 \text{ ft}^2 = 359 \text{ ft}$$

e. Layout the curing area. Select the number of curing piles and their respective lengths. This step is best done with the layout of the entire site in mind. However, for the sake of this example, assume that the compost will cure in eight separate piles. Therefore, the average length of the curing piles is:

$$110 \text{ m} \div 8 \text{ piles} = 13.75 \approx 14 \text{ m per pile} \quad 359 \text{ ft} \div 8 \text{ piles} = 44.9 \approx 45 \text{ ft}$$

8. *Add the desired spacing between piles and around the perimeter to determine the overall pad dimensions.*
   Assume:
   - 5 m of space on all sides of the composting pad for access (about 15−16 ft),
   - 5 m (minimum) separating the two asp extended piles,
   - 5−10 m between the extended ASPs and the curing area, and,
   - 1.5 m (5 ft) between adjacent curing piles.

9. *Sketch the layout of the composting pad to see the full picture* (Fig. 10.29).

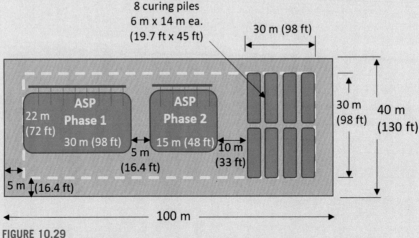

FIGURE 10.29

Composting pad area for the example extended aerated static pile (ASP) containing yard trimmings and source separated organics.

8.3 Compost storage area

The area required for storing finished compost products varies with the facility. It is typically 10%–25% of the composting pad but can range from nearly zero to twice the area of the pad. A large area is needed when: (a) several distinct compost products are created; (b) products are stored for long periods due to seasonal patterns of use or sales; and/or (c) compost is blended with additives in the storage space. The composting storage area can serve several functions, each needing a piece of the space (Table 10.7).

Table 10.7 Possible elements of a compost product storage area.

Relatively common elements	Less common site-specific elements
Bulk storage of one or more batches (lots) of basic compost	Sales yard
Bulk storage of separate compost products (e.g., course-screened, fine-screened, topsoil blend, premium potting mix)	Storage of additives for blended products (e.g., soil, sand, biochar)
Space for vehicles to maneuver	Area for screening and plastic removal (e.g., windsifter)
Bulk compost loading area	Compost product pick-up and loading area for public use
Area occupied by conveyors	Bagging and pallet loading area Storage of bags and pallets

In addition to the regular volume of compost produced, the area required depends on the storage period and the number of separate piles and size of piles. The storage period can last for more than 6 months, but piles can be made large, up to 15 m (50 ft) tall, by building them with stacking conveyors. If built with conveyors, storage piles are often conical in shape. However, the cones can be amassed to form a longer pile with a somewhat triangular or trapezoidal cross section (Chapter 9, Section 6). If built with a loader, piles heights are parabolic or trapezoidal in shape with heights limited by the reach of the loader (unless they are driven upon, which is discouraged).

The compost storage area can be estimated in same manner as the composting pad and curing areas, with the primary difference being the conical shape of piles in the total area, allowance for movement and loading of vehicles must be included along with space for blending products and amendments if that occurs at all.

8.4 Areas for feedstock receiving, mixing, and storage

Because of the diversity of feedstocks, and the patterns of feedstock generation and delivery, it is not possible to make general estimates of the area need for feedstock receiving and storage. For example, bulky feedstocks that store well, like yard trimmings and wood debris, often accumulate on site, sometimes for long periods waiting for a custom grinding service. In contrast, feedstocks that contain food or biosolids are typically processed within a day of delivery. Farm manures can be transported directly from the barn to a windrow without a receiving or storage step. Nevertheless, a few general recommendations can be made.

The feedstock receiving area should accommodate the receiving practices of the facility. Depending on those practices, space may be needed for a combination of the elements listed in Table 10.8.

Table 10.8 Possible components of a feedstock receiving and storage area.

Relatively common elements	Less common site-specific elements
Short-term feedstock storage for feedstocks needing prompt processing (e.g., biosolids).	Covered, enclosed, or partially enclosed structure for receiving wet or putrescible feedstocks (e.g., food waste).
Long-term storage for amendments and stable feedstocks (e.g., yard trimmings).	Feedstock drop-off area for public use.
Space for vehicles to maneuver.	Scale (weighbridge).
Short-term staging area for inspecting and sorting.	Scale house or spotter's station.
Trash containers.	Depackaging equipment.
Space to temporarily store sorted contaminants or rejected loads.	Area for cleaning and storing collection containers.
Feedstock mixing pad, push walls, and/or equipment.	Tanks for holding liquid feedstocks.

For any element within the receiving area, the volumes used to estimate area requirements should be based on expected *peak* volume, not expected average volume. Feedstock loads can be variable and seasonal. The receiving and storage components must be able to handle the peak conditions that occur when feedstocks are generated or delivered in their greatest amount; examples include deciduous leaves in the autumn, grass clippings in early summer, and food waste during holiday seasons.

For feedstocks that are normally processed daily, the short-term storage area should be able to hold more than a day's volume, at least to 2, and preferably 3 days, of volume. The extra area affords flexibility in case normal operations are interrupted (Gamble, 2018).

Dry stable low-nitrogen amendments should be stored on-site, so they are available for mixing soon after wet primary feedstocks arrive. The amount of amendment to keep on hand is site specific. Descriptions of biosolids composting facilities report amendment storage capacities ranging from 1 week to 1 month (USEPA, 1989; LAWPCA, 2020). With moist amendments, like green wood chips, longer storage periods (e.g., >1 month) raises the risk of fire.

If feedstocks are premixed before composting, it is commonly done within or near the area that they are received and stored. Mixing does not *necessarily* require additional space. For instance, wheel loaders can mix feedstocks on the receiving pad. Also, mobile mixing equipment can simply move in and out of the receiving area as needed. When a dedicated mixing station is used, the space required depends on the mixing method. When mixing is done by a wheel loader, the core area required is roughly the combined volume of the batch of feedstocks mixed divided by an estimated average depth of the materials being mixed (e.g., about 1 m or 3 ft). When stationary, mixing equipment is used, the size of the mixer is the basis for the core area. It is reasonable to double the core area to estimate the total area to account for equipment movement, push (buck) walls, bins, and temporary piles.

8.5 Areas for specific operations

Usually only a small percentage (e.g., 10%—20%) of the active composting area is occupied by operations like size reduction (grinding, shredding), screening, and plastics separation (e.g., windsifter). The required space for each of these operations depends on the volume of material processed, but the required area is not proportional to that volume. As shown in comparing Fig. 10.30A with B, much depends on the method of screening and how fast, the materials are handled. The processing equipment anchors the space, with the conveyors, vehicles, and piles situated around it. Therefore, a reasonable way to estimate the total required space is to multiply the footprint of the anticipated equipment when it is in operation, (including conveyors) by a factor of 4—8, depending primarily on the space needed for piles. If piles are removed promptly, 4 is a reasonable factor. Equipment dealers and manufacturers publish the dimensions of their equipment.

One important consideration regarding space for hammermill grinders, especially tub grinders, is the buffer distance required because of the possibility of

FIGURE 10.30

Ample space allotted for this star-screening station, is primarily for piles (A). The trommel screen conveys compost directly to a truck and the overs are removed promptly requiring relatively little space (B).

Source: (A) NeuStarr; (B) R. Rynk.

projectiles. This concern might dictate a large space for grinding, perhaps an acre or more. Much depends on the equipment, grinding frequency and schedule, and the site layout.

9. Building—roofs and enclosures

Whether it affords full enclosure or just a roof, a building can deliver attractive advantages to composters. It can keep materials, people, and equipment dry; provide all weather operations in cold and/or rainy climates; and hide some or all of the operation. Having a roof, alone, reduces the amount of stormwater runoff to store and treat. Full enclosure allows for containment and treatment of odors. A building, especially a fully enclosed building can improve the facility aesthetically. A building may allow a facility to operate in urban and suburban locations or just fit well into the landscape (Fig. 10.31).

Buildings are used increasingly to house parts of the composting operation, if not the entire facility. In particular, facilities that handle food waste commonly have fully enclosed, ventilated, and air-filtered buildings for the receipt and mixing of food with other feedstocks. In some jurisdictions, regulations require roofs or enclosures for composting facilities that handle difficult feedstocks, like food, or operate in sensitive neighborhoods. Outside of these situations, buildings continue to be used sparsely and for primarily particular purposes like, storing compost and/or dry amendments, bagging compost products, and for maintaining equipment.

New buildings are a large expense. Existing buildings are an opportunity. Underutilized buildings on a farm or a municipal or industrial complex can be useful in providing an all-weather option for some purposes, like maintaining equipment, storing amendments, or compost. However, even these functions can be harsh on building components, so existing buildings need to be appropriately constructed, and/or renovated.

FIGURE 10.31

Composting is happening here, but there only a few signs of it because of the concealment made possible by the buildings.

Several large-scale composting facilities operate almost totally within buildings, but these are sophisticated endeavors developed by architectural and engineering companies. *Design of Composting Plants*, de Bertoldi-Schnappinger, 2007 uniquely presents the requirements for composting structures including architectural and engineering design topics that are rarely discussed in the composting literature.

9.1 The environment within

The environment within a building can be harsh. Composting buildings are susceptible to corrosion due to moisture condensation from feedstocks, and especially composting. In addition, ammonia, emitted from nitrogenous feedstocks, can be especially corrosive when the ammonia-laden vapor condenses on metal supports, hardware, and fasteners. Even galvanized steel and aluminum are prone to corrosion in the presence of ammonia.

Buildings should be ventilated and designed to withstand the corrosion. Roof-covered open-sided buildings are inherently ventilated and therefore less susceptible to the harsh conditions than enclosed structures. Nevertheless, even open buildings cannot escape the corrosion.

Once a working space is enclosed, human exposure to that air requires that certain air quality standards be maintained. In addition to decomposing organic materials, trucks and loaders exhaust carbon monoxide and other potentially harmful gasses. It is critical to provide adequate ventilation for employees working inside of the building (Fig. 10.32). Most building codes and worker protection standards mandate a minimum number of air changes per hour. The combination of high air exchange rates and large building volumes results in a big air handling and air

FIGURE 10.32

The round air duct above the conveyor to the left (A) is the air intake for ventilating a building that houses an agitated bay composting system. The blowers on the right (B) draw air out of the building and deliver the air to a biofilter. Even with this power ventilation system, the air inside the building is pungent.

scrubbing system, and the consumption of a significant amount of electrical power. Some facilities avoid the air handling requirement by constructing roofs supported by pillars but without sidewalls. Roof-covered structures provide significant weather protection without requiring significant air handling equipment. However, open-sided buildings may require additional design features to make sure they are structurally sound under local wind conditions.

9.2 Building options

There are many types of roofs and buildings used for various purposes at composting sites (Fig. 10.33). They range from small farm-style wooden pole buildings to industrial-scale metal structures. In between these extremes are greenhouse structures and clear span fabric-covered buildings.

Typical farm structures, like open-sided pole buildings and greenhouses, work well for composting functions. Many farm buildings are designed for high-moisture environments. Fabric-covered buildings have become popular at composting sites. They can span a fairly large width, making it easy to create large piles and move equipment. When open at the ends, these buildings provide good ventilation but are not temperature controlled. Their primary purpose is protection from wind, sun, and precipitation. Conventional industrial metal buildings offer excellent service but are the most expensive and must be corrosion-resistant or limited to storage of equipment and dry materials.

Selection of building materials is governed by their corrosion resistance. Fabric- or other plastic-covered structures can be used with less fear of corrosion failure. However, *all* components of such structures must be corrosion resistant. For instance, an off-the-shelf greenhouse with an aluminum skeleton is not acceptable. Reinforced concrete is generally good for most purposes including, floors, columns, bunkers, and sidewalls. Wood (for poles, beams, and trusses) and plywood (for siding and bins) are acceptable in many applications, if pressure treated with preservatives or suitably protected with coatings. Unprotected wood in frequent contact with moist materials or condensate eventually decomposes.

For metal buildings, any steel used in buildings for compost facilities must be of a quality that prevents the corrosion inherent in an indoor compost facility—stainless steel (which is expensive) or steel with a sufficiently corrosion-resistant coating. A number of large indoor composting facilities have had roof failures due to corrosion in steel construction. This problem is of particular concern when retrofitting an existing building (designed for another purpose) as a composting building. Some facilities have had to improve the support features and provide an anticorrosive coating in order to use existing buildings. Galvanized steel or aluminum do not provide adequate protection from corrosion. In particular, these materials are prone to corrosion by ammonia. In all cases, the choice of the fasteners and connectors is equally important. For instance, galvanized screws and aluminum truss plates cannot withstand harsh composting conditions.

FIGURE 10.33

Examples of the types of buildings, building materials and building uses on composting facilities: (A) clear-span fabric building protecting feedstock receiving and mixing area in Upstate New York; (B) partially enclosed ASP bunkers on a dairy farm in Northern Virginia; (C) greenhouse-like building over agitated bays in Hebei Province, China; (D) wooden pole building containing passively aerated composting bins in North Carolina; (E) multi-bay roof supported by steel beams covering an aerated windrow system in Colorado; and (F) open rigid-frame steel building for storing and bagging finished compost in Maryland.

10. Handling run-on/runoff

Properly managing water that emerges from the compost site is likely the most significant thing an operator can do to prevent potential environmental impacts. Runoff and leachate from composting facilities can contain many substances that can impair the quality of surface waters and drinking water if they get that far. It is important to know the nature of feedstocks brought on to the composting site because the feedstocks greatly influence the character of the water. However, feedstocks are not the only factor.

Runoff sources at a compost facility are distinguished by the degree of contact that the water has with the feedstocks and in-process composting materials, and therefore the risk of carrying contaminants. All of these sources must be managed so as not to create a nuisance or impair surface and ground water. Although the terms are not universal, the following descriptions encapsulate the various definitions used by different jurisdictions and publications (ODEQ, 2004; Coker, 2008a; Gamble, 2018).

- *Runoff*—Runoff encompasses all the sources of water that drain from a composting site due to rain and snowmelt, regardless of its path. It usually excludes water that remains confined and not a result of precipitation. For instance, water collected in aeration condensate traps, under-pile drains, and wash water sumps is not runoff unless it is released.
- *Stormwater*—Also called "nonprocess" stormwater or "clean" runoff, stormwater is the precipitation that falls within the boundaries of the composting site but does not touch the feedstocks, in-process composting materials, or other activity areas. Stormwater that tends to drain onto the active site is called *run-on*. Run-on should be intercepted and diverted away from the working site. Drainage from roofs is considered stormwater if it is captured and redirected away from the site. Also, runoff from finished compost storage piles may be considered stormwater, depending on the opinion of the regulating authority (Gamble, 2018). Stormwater can be handled as natural stormwater. However, if stormwater mingles with process water, it becomes process water.
- *Process water*—Process water, or process stormwater, comes into contact with the workings of the composting site and the material within it—feedstocks, windrows, piles, composting pad, grinding, and screening areas, curing area, drainage and aeration pipes, and equipment. It is the precipitation, applied water, and runoff that has been lightly to heavily contaminated as it passes through the site. Process water carries sediment, nutrients, pathogens, and other constituents of possible concern. It includes runoff from aisles and unoccupied pad areas, aeration system condensate, and equipment wash water. Many authorities count leachate as process water; others consider leachate outside the definition of process water.

- *Leachate* is the water that percolates through piles/windrows of feedstocks and actively composting materials. It usually is the most contaminated runoff water source from a composting site. In open piles and windrows without under-pile drains, leachate flows from the windrow/pile into the aisle and combines with pad runoff. In this case, leachate is integrated with process water. If the composting pad is pervious, some leachate can also percolate directly into the soil. Under- or in-floor drains are commonly used with ASPs, bunkers, vessels, and covered composting systems (e.g., roofs, microporous covers). In these cases, leachate can be handled separately from the more-dilute process water.

In addition to this section, Coker (2008a,b,c) presents insights and guidance about managing stormwater in a series of articles in *BioCycle* magazine. Additional useful references on the topic have been published by the Ministry of Alberta Environment and Parks (Gamble, 2018), Oregon Department of Environmental Quality (ODEQ, 2004), and the Clean Washington Center (CWC, 1997).

10.1 Stormwater pollution prevention planning

Managing water at a compost site is very site specific. For larger sites, regulations typically require the facility to prepare and maintain a written plan that guides the facility in its management of stormwater, process water, and runoff. The plan may have to be a stand-alone document, or it might be included as a component of the facility's standard operating procedures manual. In the United States, any facility that discharges stormwater to a water body must obtain a National Pollutant Discharge Elimination System (NDPES) permit (Box 10.4). The NDPES permit requires a written Stormwater Pollution Prevention Plan (SWPPP or SPPP).

The SWPPP is a written plan which provides site-specific design and operating procedures to minimize discharges to surface water and requires regular sampling of stormwater and regular monitoring of stormwater controls. The critical part of an SWPPP is a thorough site description. A site description would include the following:

- A site map or drawing showing process and stormwater drainage,
- The locations of all receiving waters,
- The direction that stormwater flows from the facility,
- Locations of all ditches, pipes, swales, drains, inlets, outfalls, or other stormwater conveyances,
- Locations of secondary containment structures,
- Locations and quantities of potential stormwater pollutants,
- Stormwater monitoring points,
- Locations of fueling stations, offloading areas, outdoor maintenance and equipment storage, tanks, and transfer areas,

- Sources of run-on from adjacent properties that could cause pollution,
- Reuse and treatment and handling of stormwater, process water, and leachate, and,
- Design of structures such as a retention pond.

The SWPPP typically contains a list of management practices that the facility uses to minimize impacts to surface or groundwater. There are scores of site-specific management practices a compost facility can use to manage stormwater. The US EPA maintains an NPDES website with up-to-date information and links to Management Practices, Stormwater Pollution Prevention information and state contacts. https://www.epa.gov/npdes.

Most on-farm composting operations are exempt from stormwater planning requirements. However, in some jurisdictions, accepting off-farm feedstocks or selling farm-produced compost removes that exemption. Local permitting authorities should be consulted. Whether it is required or not, preparing a stormwater management plan is a sensible practice and an enlightening exercise.

The US Environmental Protection Agency has a useful publication for developing SWPPPs for construction sites (USEPA, 2007). It is relevant to composting facilities as well.

Box 10.4 NPDES FYI

The NDPES is a part of the US Federal Clean Water Act which seeks to manage discharges of pollutants that could degrade water quality. If stormwater is discharged to a water body, it falls under the NDPES umbrella. Management of the NPDES is the responsibility of the US EPA. However, nearly all state governments have accepted US EPA's invitation to issue permits and oversee enforcement (EPA, 2020). Each state follows the basic guidelines and specific requirements of the Federal rules but augments them to suit the individual situations and goals of the state.

An NDPES permit specifies pollutant limits for the discharge waters. Usually, the limits govern both the concentrations of pollutants and/or the accumulated amounts over a specified time period. At a minimum, the permit also includes requirements for monitoring and reporting the discharges. For stormwater discharges, a given permit is usually a "General Permit," containing general requirements that are common across a particular industry (Coker, 2008c). In unique cases, "Individual Permits" are required that have site-specific limits.

If a composting facility discharges runoff to surface water, as many do, it needs an NPDES permit. Obtaining a general permit requires the facility to submit a Notice of Intent and/or an application to the appropriate state environmental regulatory office. The process requires the facility to complete and implement a SWPPP.

US EPA maintains an NPDES website with up-to-date information and links to Management Practices, Stormwater Pollution Prevention information and state contacts. https://www.epa.gov/npdes.

10.2 Runoff characteristics and constituents

There is little published data characterizing the composition of process water and leachate from composting facilities. It varies substantially with the scale of the operation, the feedstocks, composting technology, stage of composting and other factors (Brewer et al., 2013; Larney et al., 2014). As an example, ranges of analytical testing results from four composting facilities are shown in Table 10.9. Krogmann and Woyczechowski (2000) report similar, and similarly variable, measurements in runoff from composting facilities in Germany. Kannepalli et al. (2016) also report wide ranges of contaminants in runoff from wood mulching facilities. These numbers demonstrate that process water and leachate can be as polluting as municipal wastewater (Coker, 2008a). These findings affirm that runoff that contains process water cannot be allowed to drain into natural water bodies, at least not without pretreatment.

Table 10.9 Range of constituents in stormwater from four facilities (CWC, 1997).

Parameter	Range (mg/L)
Biochemical oxygen demand (mg/L)	20–3200
Total solids (mg/L)	1100–19,600
Volatile solids (mg/L)	430–9220
Color (color units)	1000–70,000
Fecal coliform (MPN/100 mL)	200–24,000,000
Copper (ppb)	33–821
Zinc (ppb)	107–1490
Total Kjeldahl N (mg/L)	14–3000
Ammonia N (mg/L)	32–1600
Nitrate + nitrite N (mg/L)	0–8
Total phosphorus (mg/L)	4–170
Ortho phosphate (mg/L)	0–90
Potassium (mg/L)	167–4640
Chloride (mg/L)	52–2100
pH (standard units)	6.7–9.5
Conductivity	887–16,500

Table 10.10 summarizes the contaminants of greatest concern in stormwater runoff. Most contaminants impact surface water bodies rather than groundwater. Sediments or suspended solids can change the habitat conditions of receiving waters and contribute to turbidity. In addition to these direct impacts, sediments also carry many of the other contaminants described in Table 10.10. Organic substances in runoff consume oxygen in surface waters and contribute to turbidity. Surface water quality also are impacted by nutrients including various forms of nitrogen and phosphorus.

Table 10.10 Possible contaminants of concern in process water and leachate from composting sites.

Contaminant	Consequence	Composting comments
Sediments, measured as suspended solids	Sediments alter aquatic habitat and contribute to turbidity of water. They also carry nutrients and oxygen-consuming substances.	Sediments are the most consequential pollutant in composting runoff.
Pathogens (human and animal)	The presence of pathogens is usually determined by enumerating indicator organisms like *Escherichia coli* and total fecal coliforms. Pathogens can cause illness. Thus, their presence disrupts surface water uses, leading to drinking water alerts (e.g., boil water orders), swimming restriction, beach, and shellfish bed closure, etc. Pathogens are not a groundwater threat, but they can contaminate flawed shallow wells.	Pathogens are present in all composting feedstocks, some more than others. Although composting effectively destroys pathogens, they can be carried away in sediments within runoff from feedstocks and composting piles that have not yet been sanitized. It may be useful to separately handle process water from the early stages of composting.
Oxygen-consuming substances	Deplete dissolved oxygen in water, harming or killing fish and other aquatic life.	Assessed by measuring BOD and COD.
Biochemical oxygen demand (BOD)	General measure of the strength of a substance to consume oxygen. Oxygen is consumed in a sample by bacteria over a prescribed time period, usually five days (i.e., BOD_5).	Caused by organic matter in the runoff and also some inorganic chemicals, like sulfides and ammonia.
Chemical oxygen demand (COD)	Another measure of oxygen-consuming potential. Strong chemical oxidant is used to oxidize organic and inorganic substances. Because the chemical oxidant reacts faster and more completely than bacteria, a COD test consumes more oxygen and is usually higher than BOD (e.g., lignin is oxidized).	COD is less often used than BOD for natural process. Comparing BOD to COD for a particular sample reveals information about degradability. A BOD_5/COD_5 ratio below 0.1 indicates resistance to biological decomposition (Coker, 2008a)
Nutrients	When a given nutrient is naturally limited in surface water, nutrient enrichment leads to growth algae and other organisms. Excessive algae growth increases turbidity and oxygen depletion.	Primary nutrients in composting process water are ammonia-N, nitrate-N, and phosphate.

Table 10.10 Possible contaminants of concern in process water and leachate from composting sites.—*cont'd*

Contaminant	Consequence	Composting comments
Nitrate-N (NO_3)	Nitrate is water soluble and readily moves with runoff and leachate. In addition to enriching surface water, nitrate is a regulated contaminant is drinking water.	Runoff from active composting piles contains relatively low concentrations of nitrate. Nitrate increases as compost matures. Piles of curing or stored compost are usually greater potential sources of nitrate.
Ammonia-N (NH_3 and NH_4)	Both forms of ammonia-N are soluble. Both forms contribute to nutrient enrichment, and may be converted to NO_3 in soils, storage ponds, and surface water. In addition, elevated concentrations of NH_3 are toxic to fish and other aquatic life.	Ammonia is most abundant during the early stages of composting and gradually decreases to low levels in mature compost. Ammonia is prevalent with high-N feedstocks. A high C:N ratio feedstock mix releases little ammonia.
Phosphate (PO_4)	Phosphorus is often the limiting nutrient in surface waters. Thus, when it arrives with runoff, it stimulates the growth of algae and aquatic vegetation.	Most, but not all, forms of phosphate in feedstocks and composting piles are not water soluble. Phosphate movement is more closely tied to sediment than leaching.
Turbidity—lack of clarity or cloudiness of water	Some turbidity in water is natural and healthy. However, high turbidity clouds the water for aquatic life and absorbs sunlight, blocking it from aquatic vegetation and causing the water to warm. High turbidity is an indication of other problems including sediments and excessive algae growth.	Composting runoff can potentially impact the turbidity of surface water due to sediments, BOD, and colorful chemicals like tannins.
Tannins	Tannins are not toxic. They contribute to turbidity, due to their dark color, and BOD, due to their organic nature. Tannins are not a health threat to surface or ground water. They can be an aesthetic issue for drinking water from shallow wells (taste, stains).	Tannins are organic by-products of the natural decomposition of vegetative and woody materials. Composting process water is often copper or brown in color due to tannins, especially at yard trimmings composting facilities.

Continued

Table 10.10 Possible contaminants of concern in process water and leachate from composting sites.—*cont'd*

Contaminant	Consequence	Composting comments
Trace metals (heavy metals)	Trace metals include copper, iron, cadmium, zinc, mercury, lead, arsenic, and others. In low concentrations, most trace metals are essential nutrients, but in high-enough concentrations, they are toxic to aquatic life and a threat to human health. Some forms of some metals are water soluble and can seep into groundwater. Most trace metals are included in drinking water regulations. Some metals, like mercury, bioaccumulate in the food chain.	The presence of trace metals in composting process water depends on their presence in the feedstocks (Chapter 4). In composting piles, most trace metals compounds are bound to organic matter and not soluble in water. Therefore, sediments are likely the main source of trace metals in process water.
Pesticides	Innumerable pesticides exist with varying properties and effects on the natural, environment. Many are toxic to aquatic life and/or persist in the environment. Many are water soluble to varying degrees and can potentially contaminate wells and drinking water.	Pesticide residues are carried by yard trimmings and agricultural products. Most decompose during composting but in the meantime soluble pesticide compounds can travel with leachate.

The primary groundwater contaminant of concern is nitrate because it is a health risk in drinking water. It is possible that soluble compounds, like trace metals and pesticides, can impact groundwater. However, this potential problem has not arisen in practice, and related research is scarce. Richard and Chadsey (1989) found no elevated metal concentrations in soil water collected by lysimeters beneath leaf composting piles.

In general, the risks of groundwater contamination are low compared to the risks to surface waters for several reasons. First, sediments are a nonissue. Second, composting windrow/piles tend to retain a large amount of water from precipitation, releasing relatively little water to the ground below (Richard and Chadsey, 1989; Zbinden et al., 2015).[3] Third, the surface of unpaved composting sites become well compacted over time. These arguments do not suggest that soluble contaminants don't leach into the ground. Nitrate, in particular, has been found to migrate into the soil profile from piles with low C:N ratios (Nienaber and Ferguson, 1992; Butler and Connolly, 1994). Confesor et al. (2007), found that soil nitrate levels

---

[3] Phase I part of the research only.

increased under a gravel composting pad even though it was well compacted. However, that migration of nitrate below composting sites appears to be slow, and possibly small (Nienaber and Ferguson, 1992).

10.3 Runoff system design considerations

The design of how water flows on your composting site depends on several site-specific factors. Usually, a system consists of a series of separate devices that perform different functions (Bartlett, 2006). Common components include berms and channels to divert run-on, collection devices to collect process water, a storage pond, and a means to safely infiltrate excess water into the soil. Each component has to be sized and designed to handle the flow of runoff from a "worst case" storm.

More complex sites and larger facilities typically engage professional engineers or hydrologists to design the site and calculate runoff volumes, design ponds and other stormwater features. Some states require licensed professionals to plan and design stormwater features. Core design principles are described below.

10.3.1 Reduced runoff/process water volume

Runoff volumes increase as the managed land area increases. Therefore, one means to reduce the amount of runoff to manage is to minimize the area that creates contaminated runoff (i.e., process water). A generously sized composting pad has its advantages, but it also increases the amount of process water that must be managed. This rationale is an argument in favor of ASPs over turned windrows (and extended ASPs over individual ASPs). Not only do ASPs require a smaller footprint but they fill the composting pad with a greater proportion of precipitation-absorbing material, typically retention times in ASPs are also shorter than in windrows thus allowing more material to be managed in the same or less space.

Another effective way to reduce process water is to exclude precipitation from the composting area, or parts of it. Rain and snow can be excluded by composting in a vessel or under a roof. These are expensive options but effective. Less expensive, and easily overlooked, accumulated snow should be plowed and relocated outside the working areas of the site before it begins to melt and turns into process water.

10.3.2 Prevent run-on

Run-on is water that flows onto the working area site from upslope areas. Run-on can be intercepted using berms, channels, and/or drains to eliminate clean stormwater from the site's process water burden. Run-on can be diverted to natural stormwater drainage pathways. It should be done in a way that avoids erosion. Drainage channels should be vegetated or otherwise protected. The channel's slope along its length should be low enough, and its cross-sectional area large enough, to accommodate the expected peak runoff flow rate (volume/time) while keeping the peak velocity below erosive levels. The peak flow rate is determined by the drainage area and its characteristics during a presumed severe storm. In this case, a severe storm is presumed to be a 24-hr storm with a recurrence interval of 10, 25, or 50 years, depending on the application. The allowable peak velocity is determined based on the erodibility of the soil and the conditions of the channel (NRCS, 2009a). It general falls in the range of 0.3—1.5 m/s (1—5 ft/s).

As a very general guide, the longitudinal slope of the channel should be between 1% and 2%, if workable, but not more than 6% (ODEQ, 2003). The side slopes of the channel cross-section are commonly 1:3 (rise:run), as depicted in Fig. 10.34, or 1:5 if the side slopes are to be mowed. Guidance for designing and constructing diversion channels can be found in the *Natural Resource Conservation Service*[4] *Engineering Field Handbook*, Chapter 9 (NRCS, 2009b).

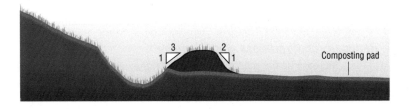

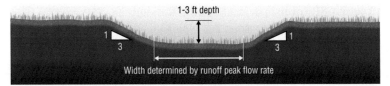

FIGURE 10.34

General dimensions for stormwater diversion berms (dikes) and channels.

10.3.3 Separate "clean" stormwater and "tainted" process water

Runoff that passes through, or otherwise comes in contact with composting materials, is process water. It picks up contaminants that can decrease the runoff quality. Although many site-specific factors are at play, keeping clean stormwater separate from process water is a key tool to reducing the quantity of process water that needs to be captured, stored, and/or treated. Regulations for what is or isn't "process water" can vary among states, provinces, and countries. Isolation of clean stormwater is achieved by diverting run-on, grading the site to drain unworked surfaces away from

---

[4] The Natural Resource Conservation Service, or NRCS, is part if the US Department of Agriculture.

the process water collection devices, collect and divert roof drainage, minimizing the area holding piles and equipment, and keeping the nonworking areas as clean as possible.

10.3.4 Maintain positive drainage

Several composting publications, including this one, generally recommend that composting pads have a slope between 2% and 4% (CWMI, 2005; Chardoul et al., 2015). If the pad is paved, a 1% slope is acceptable. The intent of this guidance is to assure that the site has positive drainage (Fig. 10.35). However, depending on the site, positive drainage may be achieved at less than a 1% slope on a paved surface. On large composting pads, a 1% slope creates a significant amount of rise. The main point is to make sure that stormwater hitting the site runs off and doesn't create long lasting puddles, rills, or other erosion.

FIGURE 10.35

A gentle slope to the pad allows process water to effectively drain away from windrows and piles.

10.3.5 Settle the sediment

Many of the constituents of concern in stormwater runoff include the composting materials themselves. Small organic particles are easily swept up in runoff and contribute oxygen demand, nutrients, pathogens, or other constituents to turnoff. Removing the sediment removes much, if not most, of the contamination. Solids entrained in runoff can be removed relatively easily, either by filtering or settling.

Filtering simply presents a physical barrier preventing some or most solids from passing through to the collection device or pond. Composters have several good filtering options at their immediate disposal including berms made of compost or wood chips, compost-filled filter socks, and bales of straw or other amendments (Fig. 10.36). After a time, the filtering media and the trapped solids can be removed and added to the composting piles, replacing the spent media with new media. Filtering runoff in this manner is an appealing first choice. With apologies for the pun, it is "dirt cheap" and effective at capturing large- and moderate-sized particles. However, these filters impede the flow of runoff from heavy precipitation, and they are not fully effective at removing small particles of sediment.

FIGURE 10.36

Runoff contamination can be substantially reduced by filtering sediment at the edge of the pad. Runoff from the pad shown in this photograph is filtered through a series of compost socks before it enters the drains. Sediment barriers can also be berms, composed of unscreened compost, screen overs, wood chips or hay bales.

Source: St. Louis Composting.

Settling, or sedimentation, removes solids by slowing the velocity of water, allowing the suspended solids to settle by gravity. Settling devices can be concrete or earthen basins, ponds, or damming devices, like compost filter berms. They are commonly used on livestock farms to remove solids from farmyard runoff and liquid manure. Settling devices are designed to handle the peak runoff flow rates based on a design storm (e.g., 10-year return period).

FIGURE 10.37

Process water at this facility runs into the settling basin pictured above (A). As the sediment falls out, cleaner runoff either overtops the spillway into the adjacent pond or flows through the stone in the gabion barrier (B).

Settling devices must have a means for the water to pass while holding back the solids. After the flow of runoff stops, the solids that are left behind need to either be pumped out or dewatered and removed with a tractor, wheel loader, or excavator. For that purpose, many settling basins have a ramp with a firm base for equipment to enter the solids collection basin (Fig. 10.37). The recovered solids can be added back to the composting piles.

The runoff retention pond at a facility can serve the purpose of a sedimentation device. However, it is better to remove sediments before the pond. In this way, the pond has more storage capacity, a much lower BOD load, and is less likely to develop odors or require aeration.

10.3.6 Reuse the water

Reusing collected runoff is a good practice. It supplies water for drying piles and windrows while also reducing the amount of process water that must be treated and/or discharged. Unfortunately, those facilities that have the most runoff to use, tend to need less of it, and vice-versa. Therefore, water reuse usually requires a storage pond to capture runoff from this month's wet weather for use during next month's dry spell. In addition to using the water to remoisten windrows and piles, lightly contaminated runoff can be used to water the site for dust control or even wash equipment.

There are several cautions to observe when reusing runoff, especially process water. Runoff storage ponds can harbor pathogens. It should be reused only for windrows, piles, and vessels that will still remain hot enough for a long enough period to meet sanitization requirements. Only stormwater that does not come into contact with the working site should be used in the later stages of composting (i.e., after sanitization). Highly contaminated water, such as leachate, should not be reused for any purpose unless it is diluted or treated.

10.3.7 Avoid discharge to surface water, if possible

The restrictions and requirements grow substantially when runoff, even treated runoff, is discharged to any surface water body. In most locations, regulations require a facility to get a permit, or other form of approval in order to discharge off site. The permit is not only cumbersome and/or costly to obtain, but it also carries treatment, monitoring, and reporting requirements. In addition to reusing runoff, there are several options available to composters to avoid discharging. For instance, many composting facilities use infiltration approaches including swales, filter strips, and basins. Facilities located on farms can usually irrigate surplus water onto crops if nutrient management and hydraulic loading limits are observed.

10.3.8 Export, if necessary

If on-site storage or infiltration is not feasible, it is possible to send collected stormwater off site for treatment. With permissions and specifications from the local wastewater authority, contaminated runoff can be discharged into the sewer system, presuming the facility has access to it. A common alternative is to pump collected stormwater into mobile storage tanks and haul it to a wastewater treatment plant, though this can be an expensive option. Some composters are able to transport collected runoff to off-site agricultural facilities.

10.3.9 Various control and treatment options

There is an entire industry dedicated to managing stormwater that has developed a myriad of stormwater management practices. A menu of practices is listed in Table 10.11. Fig. 10.38 shows selected practices. In 2004, Oregon Department of Environmental Quality published a list of runoff best management practices ranked according to their cost-effectiveness (ODEQ, 2004). An abundance of guidance is also available from many other public and industry organizations. Several good publications are included in the reference section.

Table 10.11 Management practices reducing the amount and/or the impacts of stormwater.

Practice	Brief description
Green roof	Roof planted with selected vegetation to absorb and store precipitation, using it to irrigate plants. It reduces the amount of runoff and also cools the surrounding environment.
Cistern	Large tank, usually below ground, that collects and stores stormwater from roofs and drainage areas, often paved. It reduces the amount of runoff and stores water for positive uses (e.g., irrigation, toilet flushing).

Table 10.11 Management practices reducing the amount and/or the impacts of stormwater.—*cont'd*

Practice	Brief description
Permeable pavement	Water-permeable surface that replaces asphalt or concrete to allow stormwater to infiltrate through into the soil. Where a solid surface is required, permeable surfaces are usually constructed of pavers made of brick, concrete, or stone.
Bioretention area	Generally small, shallow, and vegetated depression that collects and retains runoff for a short period. The water infiltrates into the soil and is used by the vegetation. Bioretention areas are used for clean stormwater, not process water. Small bioretention areas are called rain gardens.
Swale or bioswale	Drainage channel that conveys and retains stormwater runoff while encouraging the water to infiltrate into the soil bed of the swale. Dry swales are grasses or otherwise vegetated and do not contain water most of the time. Wet swales are often filled with water. Swales can be designed as treatment devices for process water, especially after sediment is removed.
Vegetated filter strip (infiltration strip)	An area of vegetated land through which runoff is directed. The filter strip is nearly level in one direction and gently sloped in the other to spread out the flow. They over include a means to introduce the water in a level fashion (e.g., level lip spreader). Filter strips can be treatment devices for process water, especially after sediment is removed.
Infiltration basin	A natural or constructed impoundment that captures, temporarily stores, and infiltrates runoff. The basins are typically 1–2 m deep (2–6 ft). They are created by excavating depressions, building banks around the infiltration area, or both. Infiltration basins usually have little or no water in them. The floor of the basin, which is often vegetated, is sufficiently permeable to infiltrate the design runoff volume within the desired drawdown time.
Sand and organic filters	A bed or vessel full of sand or other media through which runoff flows to filter sediment and other contaminants. Instead of sand, organic media, including compost and peat moss, are sometimes used to enhance removal of dissolved contaminants (e.g., trace metals). Sand/organic filters are a treatment device for process water after sediment is removed.
Constructed wetland	Capture and carry stormwater and pretreated process water through a designed and created wetland system. Constructed wetlands mimic the substantial contaminant-cleansing effects of a natural wetland while also creating diverse wildlife habitat.
Riparian buffer	Dedicated space between a water body and the site. The purpose is to simply distance the surface water from potential contamination. Regulations usually compel composting facilities to maintain a buffer with a minimum distance. A larger buffer is more protective than a smaller buffer.

Adapted from USEPA, 2021. NPDES Permit Basics, U.S. Environmental Protection Agency. https://www.epa.gov/npdes/npdes-permit-basics, see actual text for more detail.

FIGURE 10.38

Examples of runoff management and treatment options: (A) vegetated filter strip accepting runoff from a lime-stabilized soil composting pad in Maryland; (B) a vegetative filter strip behind a level-lip spreader in Maine; (C) bioswale at the base of composting pad in Pennsylvania; (D) artificial wetland in Ohio.

Source: (A, C, D) R. Rynk and (B) Andrew Carpenter.

10.4 Runoff storage (ponds)

Many composting sites store collected runoff on-site, usually in a pond. In fact, many facilities have multiple ponds. Multiple ponds might be used to separately store different runoff sources (e.g., clean storm water and process water). Otherwise, a facility might have multiple ponds because it is easier to drain different parts of the site in different directions. On large sites, a single pond can result in an impractically long drainage path. It leads to a large elevation difference between the head of the slope and the pond, which complicates pad construction.

Although they look similar, there are two types of ponds used at composting facilities, a *retention* pond, and a *detention* pond. A *retention* pond is the more common. It holds water indefinitely until it is reused or evaporates (Fig. 10.39). A *detention* pond is designed to hold water for a short period of time, on route to its next destination. The detention pond settles sediment, buffers surges in runoff flow rates, and equalizes concentrations of contaminants ahead of additional

FIGURE 10.39

Typical soil-based retention pond at a composting facility.

treatment or discharge (e.g., swale). It typically has a pipe or spillway to discharge water at a measured rate.

Local stormwater and/or composting facility regulations often provide specific guidance about the design of the pond and whether or not it has to be a detention pond or a retention pond. Regulations might specify whether or not the pond has to be lined, the maximum infiltration permeability of the liner, how much freeboard is required, and the length of time the stormwater must be contained can be state-specific. At a minimum, regulations prescribe the amount of rainfall that a pond must accommodate, such as a 25-year/24-hour storm event or simply 100 mm or 4 in.

Of the two types of ponds, a detention pond is the easier to size because it is intended to hold only the runoff from the design storm. It is largely empty between storm events, which is why it is also called a "dry" pond. However, detention ponds can also be designed to continually retain a minimal level of water.

In addition to capturing the runoff from the design storm, a retention pond stores normal runoff over an extended time period (minus the amount of water removed for reuse) plus accumulated sediment. Usually, another 30 cm (1 ft) of depth is added as "freeboard," which is an extra measure of defense against water overtopping the pond. Fig. 10.40 illustrates the basic components and layers of a simple retention pond that is commonly found at composting facilities. Fig. 10.41 shows the components of an engineered pond that is likely to be found on larger facilities subject to more stringent stormwater requirements.

10.4.1 Additional pond designs

- *Shape:* Removing sediment is an important function of detention ponds so they should be longer than they are wide, to slow water flowing through the pond and allow for more solids to settle out. The minimum recommended length to width ratio is 2:1, but higher ratios are preferred. The presumption is that the inlets and

FIGURE 10.40

Layers and general specifications for a retention pond.

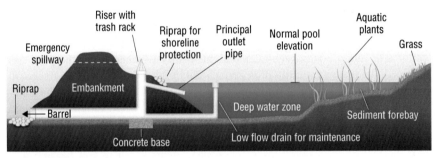

FIGURE 10.41

Features of an engineered retention pond.

Adapted from Clemson Cooperative Extension, 2021.

outlets are on opposite ends of the pond. Shape is not an important factor for retention ponds.

- *Soil and liners:* Unless a pond is designed for infiltration (e.g., infiltration basins), it needs to hold water. Native soils, in hydrologic groups C and D, for example, are often suitable for ponds, especially after self-sealing (ME DEP, 2016). If the native soils are not suitable, the pond can be lined with either clay or a synthetic liner. Clay bases are compacted and installed to a thickness of about 50–70 cm (18–24 in.). Alternatively, and frequently, ponds are lined with synthetic membranes, such as 30 to 60 mil vinyl or polyethylene sheets (Figs. 10.42 and 10.43). It is easier to cover sloping banks with a membrane than placing and compacting clay. The different liner materials have various

FIGURE 10.42

Plastic membrane liner for a retention pond.

Source: Cary Oshins.

FIGURE 10.43

Lined retention pond. This facility illustrates one continuous membrane underlying the composting pad and the pond. Water seeping through the pad flows to the pond.

Source: Matt Cotton.

advantages in regard to durability and methods of sealing adjacent sheets. Sand or rock-free soil is placed on top of the membrane to a depth of 30–50 cm (12–18 in.) to protect the liner from physical damage. Regulations may spell out the thickness and/or permeability of liner materials and construction procedure.

- *Inlets and outlets*: A number of different devices are used to let water in and out of ponds including pipes, inverted pipes, weirs, level lips, and spillways. These inlet and outlet structures are important and fragile pond components. They must be constructed for stability, otherwise leaks easily occur. The locations where inlets and outlets release water should be protected against the erosive action of streaming water. For this reason, riprap (gravel or stone) should be placed around these points of discharge.
- *Embankment slopes:* Except in arid locations, the banks of a pond should be vegetated, and the vegetation needs to be maintained (e.g., mowed). For this reason, and slope stability, outside side slopes should be modest, no more than 18 degrees (1:3 rise:run). Inside slopes can be somewhat steeper.
- *Prior removal of sediment*: Even though sedimentation is one of the functions of a pond, it is still wise to remove as much sediment as possible from the runoff before it enters a pond. Doing so reduces the amount of sediment that must eventually be cleaned from the pond, or increases the time between cleanings. Limiting sediment also lowers the nutrient and BOD load, improving the ponds performance in treating runoff and reducing the risk of odors.
- *Post removal of sediment:* Depending on the character of the runoff, and prior sedimentation, ponds eventually accumulate sediment. At some point, the sediment needs to be removed. Options include: (a) mixing the pond with a circulation pump to resuspend the solids and then pumping out the solids-ladened pond water; (b) using an excavator to reach in and dig out the solids while carefully avoiding the pond base or liner; and (c) driving a loader into the pond base by way of a ramp built into the pond design (Fig. 10.44). In all cases, the pond should be designed with sediment removal in mind. For example, the

FIGURE 10.44

This pond has a concrete ramp that allows a wheel loader to access the base of the pond for cleaning accumulated sediment. This particular pond has heavy sediment load so the base is concrete and the embankments are lined with a synthetic membrane.

pond design might include a flat base to situate the excavator or a concrete pad for the circulation pump. The sediment and/or water removed from the pond can be added to feedstock or used to moisten windrows or piles. Alternatively, it can be pumped into a tank wagon or truck and land applied or irrigated onto fields, if farm fields are available and accessible, and if permits/regulations allow. The last resort is to discharge it into a wastewater treatment system.

- *Aeration:* Aeration is necessary if a pond becomes overloaded with organic matter, turns anaerobic, and generates odors. Pond aeration is also used as an intentional treatment step for high-strength runoff. Aeration of ponds is most often accomplished by surface aerators which churn the surface or the water mixing it with the air above. Floating pumps, paddle wheels, and fountains are examples. Bubblers and spargers that inject air into the water below the surface are also used. Often, a pond needs just a little help with aeration so off-the-shelf aerators can be used. However, for deliberate treatment, and for permanent odor control, the aeration device should be selected based on the pond BOD level. The services of an engineer or environmental consultant would be beneficial.

10.5 How much runoff?

In designing stormwater conveyances and containment structures, it is critical to estimate the amount of runoff water to expect. The amount of runoff is determined by the amount of rainfall (precipitation), the surface features of the site (including materials management), and the total area of the site from which runoff is collected.

10.5.1 Rainfall

The amount of rain, and when it arrives, varies significantly from one geographic region to the next and even within the same geographic region. To manage this variability, a given rainstorm can be characterized by its "return period," also referred to as a "recurrence interval" or "frequency." A return period is a statistical tool that attempts to predict, for a particular geographic location, the amount of rain that will occur over a given time period under a "worst-case" scenario. It is not truly the worst case but a scenario that is accepted as the worst case for planning purposes. For example, a "25-year storm" is the biggest rain producer that is likely to occur once every 25 years. In any given year, a 25-year storm theoretically has a 4% chance of occurring ($1 \div 25$). The "duration" of the return storm also matters. The duration is the time period over which the rain occurs for the predetermined worst-case scenario. Table 10.12 shows the amount of rain expected from a given return storm and a given duration in Riverside, California. The table indicates that a 25-year, 24-hour storm can be predicted to produce 108 mm (4.24 in.) of rain. If it was more relevant to worry about the amount of rain that might fall every 25 years over a 12-hour period, then one would design for 83 mm (3.26 in).

    This example is for a site in Riverside County, California (NOAA, 2021). Similar tables for any location are available from various public agencies that are charged with monitoring the weather and/or its environmental impacts. For example, in the United States, this information is developed and distributed by the National Oceanic and Atmospheric Administration (NOAA) and its regional offices.

Table 10.12 Rain event volume prediction for riverside, CA, in mm (in.). The highlighted cells correspond to the preceding text.

Return period	Design rainfall depth in mm (and in.) Duration					
Years	1 h	2 h	3 h	6 h	12 h	24 h
1	10 (0.39)	15 (0.57)	18 (0.72)	26 (1.04)	35 (1.36)	42 (1.68)
2	13 (0.50)	18 (0.72)	23 (0.90)	34 (1.32)	44 (1.74)	55 (2.18)
5	17 (0.66)	24 (0.93)	30 (1.16)	43 (1.70)	58 (2.29)	74 (2.91)
10	21 (0.81)	28 (1.11)	35 (1.38)	51 (2.01)	69 (2.70)	88 (3.47)
25	26 (1.02)	35 (1.37)	43 (1.69)	62 (2.44)	83 (3.26)	108 (4.24)
50	31 (1.20)	40 (1.59)	50 (1.95)	71 (2.78)	94 (3.70)	123 (4.83)
100	36 (1.41)	47 (1.83)	56 (2.22)	80 (3.14)	106 (4.16)	138 (5.44)

Data gathered from predecessor to NOAA, 2021. NOAA Atlas 14 Point Precipitation Frequency Estimates: CA, US Department of Commerce, National Oceanic and Atmospheric Administration, National Weather Service. https://hdsc.nws.noaa.gov/hdsc/pfds/pfds_map_cont.html?bkmrk=ca.NOAA. 2021. NOAA ATLAS 14 POINT PRECIPITATION FREQUENCY ESTIMATES: CA. US Department of Commerce, National Oceanic and Atmospheric Administration, National Weather Service. https://hdsc.nws.noaa.gov/hdsc/pfds/pfds_map_cont.html?bkmrk=ca.

10.5.2 Runoff volume and flow rate

In order to design a runoff management system, one needs to know the expected volume and the volumetric flow rate of the runoff. For instance, the flow rate is necessary to design water conveyance components, like diversion channels and infiltration swales. The total volume is required for sizing components that hold water for a longer term, like detention ponds. The amount and flow rate of runoff depends on hydrologic and meteorological conditions, topography (e.g., slope), pad construction and coverage (how much is absorbed vs. runs-off), and materials and material coverings (if any). These factors are taken into consideration by hydrologic models that estimate the fate of precipitation that falls from a design storm.

There are numerous models available to predict likely stormwater runoff. A classic and uncomplicated approach is called the rational method, which was first proposed in 1839 (Kalaba et al., 2007). The rational method uses a basic calculation:

$$Q_t = C \ i_t \ A \times K \tag{10.1}$$

where:

Q_t = peak runoff discharge rate (m^3 or ft^3 per second),
C = Rational runoff coefficient,
i_t = Rainfall intensity (mm or in. per hour) for the design storm,
A = Drainage area (hectares, acres, m^2 or ft^2), and
K = conversion factor to correlate the units of measurement.

Example 10.3 demonstrates the use of the Rational Method for estimating runoff volume and flow from the hypothetical composting operation described in Examples 10.1 and 10.2. This purpose of this example is to demonstrate how runoff estimates are determined, and the factors that influence the results. In practice, a good deal of hydrologic and engineering thought enters the design.

The runoff coefficient (C) is a key factor. It reflects the proportion of rain that is expected to become runoff, rather than infiltrate into the soil, evaporate, or be absorbed by the material that it hits (e.g., compost). For example, 70% of the rainfall theoretically becomes runoff on land surface with a runoff coefficient of 0.70. The C number depends on the permeability of the material but also other factors like condition and land slope. Mostly impermeable surfaces like asphalt have a high C number, in the range of 0.85–0.99, depending on their condition and coverage. The C number for composting windrows and piles depends on the materials in the pile, moisture content, and stage of composting among other things. Although composting piles have not been widely studied, various references have suggested values between 0.5, for a dry windrow and 0.68, for a saturated windrow (Kalaba et al., 2007; Wilson et al., 2004). For drainage areas with varying surfaces, the C number is calculated as a weighted average for the surfaces.

The rainfall intensity (i) is based on the return period of the design storm with a duration usually equal to the "time of concentration" (t_c) of the drainage area. The t_c is the time it takes for water to drain from the furthest point in the drainage area (hydrologically) to the point of discharge (e.g., detention pond). At this point in time, runoff from the entire area is contributing to the flow and the flow is at its maximum. Finding the t_c requires knowledge about the hydrology of the drainage area. For a composting pad, Kalaba et al. suggest the following empirical equation proposed by US Federal Aviation Agency for paved airport runways:

$$t_c = 1.8(1.1 - C)L^{0.5}/S^{0.333} \qquad (10.2)$$

where: t_c = the time of concentration (min); C = rational method runoff coefficient; L = overland flow length (ft); and S = surface slope (percent). Given that this equation is empirical, these particular units must be used.

To estimate the volume of runoff generated by a given storm requires an average flow rate, not the peak flow, plus a time period for the storm (i.e., duration). This estimate can be made using the Rational Method equation by using the rainfall depth from the design storm (e.g., 25-year/24-hour) in place of precipitation intensity.

Example 10.3 Estimation of runoff volume and peak flow using the rational method

The proposed composting facility from Examples 10.1 and 10.2 is near Riverside, California. The facility manager needs to estimate runoff for sizing two retention ponds. The facility will compost yard trimmings in turned windrows and is considering future expansion in which it will compost yard trimmings plus food using extended ASPs. In the larger scheme, the windows will occupy one half of the asphalt composting pad (Pad A) and the ASPs and its curing piles will be on the other half (Pad B). Each half of the pad (Pad A and Pad B) will have a separate retention pond. The layout and dimensions of the system are depicted in Fig. 10.45. Regulations require the retention ponds to be designed to handle runoff from a 25-year/24-hour storm.

Example 10.3 Estimation of runoff volume and peak flow using the rational method—cont'd

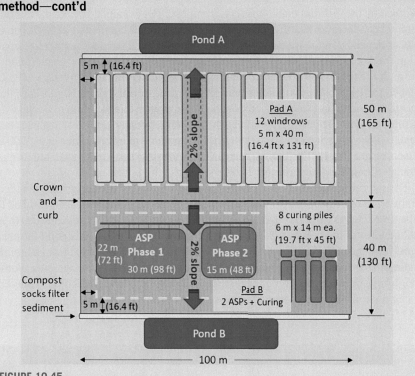

FIGURE 10.45

Layout and anticipated pile dimensions for the proposed composting facility, including possible future expansion.

Step 1: Select the curve numbers for the surfaces. Based on the foregoing information, the following C n values will be used.

Asphalt C = 0.90 Compost in windrows and ASPs C = 0.60 Curing piles C = 0.5.

Step 2: Calculate the area of each surface.
Pad A

Total area of Pad B: 50 m × 100 m = 5000 m² (53,820 ft²).

Windrow area: 12 windrows × 5 m × 40 m = 2400 m² (25,833 ft²).

Exposed asphalt: 5000 m²−2400 m² = 2600 m² (27,986 ft²).

Pad B

Total area of Pad B: 40 m × 100 m = 4000 m² (43,056 ft²).

ASP area : Phase 1 pile : 22 m × 30 m = 660 m² (7104 ft²).

Phase 2 pile : 22 m × 14 m = 330 m² (3552 ft²).

Area of curing piles: 8 piles $\times$ 4 m $\times$ 15 m = 448 m^2 (ft^2).

Exposed asphalt: 5000 m^2−2800 m^2−300 m^2 = 1900 m^2 (20,451 ft^2).

Step 3: Determine the composite C number for each pad by finding the area weighted.

Pad—surface	C number	Area	C × area	Composite C number
Pad B Windrows	0.6	2400	1440	3780/5000 = 0.756 ≈ 0.76
Exposed asphalt	0.9	2600	2340	
Totals		5000	3780	
Pad A ASPs	0.6	990	594	3124/4000 = 0.781 ≈ 0.78
Curing piles	0.5	448	224	
Exposed asphalt	0.9	2562	2306	
Totals		4000	3124	

Step 3: Calculate the volume of runoff, using the Rational Method formula, substituting the rainfall amount for the design storm in place of the intensity. In this case, the formula calculates volume rather than peak flow rate. To make this distinction, the subscript v is added to the equation.

$$Q_v = C\, i_v\, A \times K$$

The rainfall for a design storm can be found from tables and charts. For this example, the rainfall amount (i_v) for a 25-year/24-hour storm in Riverside, CA, is 108 mm (4.24 in.), conveniently found in Table 10.12. All of the information needed for the equation is now in hand. A suitable conversion factor (K) is used to calculate Q_v in the desired units (m^3 or ft^3). In this example, the only difference between pads A and B is the composite curve number.

For Pad/Pond A:

$$Q_v = C\, i_v\, A \times K = 0.76 \times 108 \text{ mm} \times 5000 \text{ m}^2 \times 1\,\text{m}/1000 \text{ mm}$$
$$= 410 \text{ m}^3 = 14{,}479 \text{ ft}^3.$$

For Pad/Pond B:

$$Q_v = C\, i_v\, A \times K = 0.78 \times 108 \text{ mm} \times 4000 \text{ m}^2 \times 1\,\text{m}/1000 \text{ mm}$$
$$= 337 \text{ m}^3 = 11{,}901 \text{ ft}^3.$$

Step 4: Determine the time of flow concentration (t_c). This step is an intermediate step that is required before calculating the peak flow rate (Q). The time of concentration is a characteristic of the drainage system. For this case, on a composting pad with a uniform slope, the equation below can be used.

$$t_c = 1.8(1.1 - C)L^{0.5}/S^{0.333}$$

In this example, the slope (S) and the flow length (L) are the same for both pads. S is 2%. L depends on where the flow concentrates to reaches its peak, and that depends on how the runoff is conveyed from the edge of the pads to the pond. For this example, it is expedient to assume that peak flow occurs at downslope edge of the pad, which is 50 m or 164 ft. (Pad A) and 40 m or 131 ft (Pad B) from the furthest point in the drainage area. This reasoning further assumes that the pad is fairly level across the main slope. So, for:

$$\text{Pad A}: t_c = 1.8(1.1 - C)L^{0.5}/S^{0.333} = 1.8(1.1 - 0.71)164^{0.5}/2^{0.33} = 7.2 \text{ min, and}$$

$$\text{Pad B}: t_c = 1.8(1.1 - C)L^{0.5}/S^{0.333} = 1.8(1.1 - 0.78)131^{0.5}/2^{0.33} = 5.2 \text{ min.}$$

Step 5: Determine the storm rainfall intensity (i). What happens next is that t_c is assumed to be the storm duration and along with the desired return period, in this case, 25-years, the intensity can be determined from tables and charts for a particular location or region. Such charts are not presented here but they are available from public agencies, hydrology-minded organizations, and books.

Based on an Intensity-Frequency-Duration chart in Gribbin (2007), the intensity from a 25-year/7-minute storm (Pad A) is about 2.8 in./h or 71 mm/h for the region in this example (Southern California). The intensity for a 25-year/5-minute storm (Pad B) is slightly higher, 3.2 in./h or 81 mm/h.

Step 6: Calculate the peak flow rate using the Rational Method formula. For Pad/Pond A:

$$Q = C \, i \, A \times K = 0.71 \times 71 \text{ mm/h} \times 5000 \text{ m}^2 \times 1 \text{ m}/1000 \text{ mm} = 252 \text{ m}^3/\text{h}$$
$$= 8900 \text{ ft}^3/\text{h.}$$

For Pad/Pond B:

$$Q = C \, i \, A \times K = 0.78 \times 81 \text{ mm/h} \times 4000 \text{ m}^2 \times 1 \text{ m}/1000 \text{ mm}$$
$$= 253 \text{ m}^3 = 8935 \text{ ft}^3/\text{h.}$$

The peak flow rates calculated here could be used to design a channel that takes water from the edge of the pads to the inlet of the pond. However, additional design steps and methods are needed design that channel. The peak flow rate at the end of the pad is just a start. This is one reason why engineers or hydrologists usually get involved. Also, note that the diagram depicts compost socks at the bottom of each pad to filter sediment. The compost socks confound the hydrology at this point and beyond.

One advantage of the Rational Method is that it is easy to understand its equation and underlying rational (hence the name). However, as Kabala et al. (2007) describe, the Rational Method has a number of limitations, including the fact that it is intended for a single rainfall event, not season of storms. It may be acceptable for estimating the size of a drainage channel or culvert but not a detention pond. Therefore, more detailed and complicated models are often used to predict runoff flow rates and volumes. In addition to computer simulation models, there are computational techniques that generate a synthetic hydrograph, which is a calculated graph of the flow of runoff (or any water course) over time at a given point in the drainage system (Fig. 10.46). The hydrograph is generated from the characteristics of the drainage area based on the rainfall of the design storm.

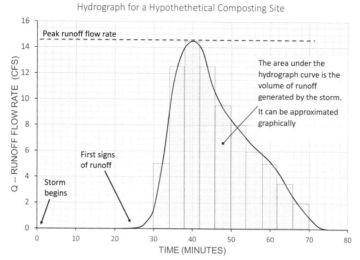

FIGURE 10.46

Runoff hydrograph for a 5-hectare (13.25-acre) gravel composting pad in Virginia. The hydrograph was modeled for a 10-year, 1-h storm (56 mm/h or 2.2 in./h). Notice that runoff did not occur until almost 30 min after the rain began.

The hydrograph directly shows the peak flow from a given rainfall event. The volume of runoff generated at any point in time is the area under the hydrograph curve from the start to the time in question. Using a hydrograph technique, Kabala et al. (2007). were able to get good agreement between predicted runoff and actual runoff for 17 storm events.

The NRCS method is a commonly used procedure for generating a synthetic hydrograph for small watershed. The method is similar to the Rational Method in that it uses a coefficient, "curve number" (CN), to represent the proportion of rainfall that becomes runoff. The CN has been determined from data for various land uses, surface covers, and soil types, specifically hydrologic soil type (Table 10.5). CNs can be selected from published tables (NRCS, 2004b). The CN for an area with mixed characteristics is the area-weighted average for the entire drainage area. The NRCS method equation determines runoff volume, rather than flow rate, based on the precipitation of design storm. After determining the CN, it is relatively easy to determine the runoff volume from a storm. In fact, graphs are available that skip the math altogether. However, generating a hydrograph is complex as it requires detailed information about the drainage area. An eager person with a fair knowledge of algebra can accomplish the task with guidance from a hydrology textbook or the hydrology section of the *NRCS Engineering Handbook* (NRCS, 2021a), especially Chapter 10 (NRCS, 2004).

References

Cited references

ASTM C977-18, 2018. Standard Specification for Quicklime and Hydrated Lime for Soil Stabilization. ASTM International, West Conshohocken, PA. https://www.astm.org/Standards/C977.htm.

ASABE, 2005. Standard D384.2: Manure Production and Characteristics. American Society for Agricultural and Biological Engineers. https://efotg.sc.egov.usda.gov/references/public/NE/ASABE_Standard_D384.2.pdf.

Bartlett, J., 2006. Storm water treatment options at composting facilities. BioCycle 47. https://www.biocycle.net/storm-water-treatment-options-at-composting-facilities/.

Brewer, L., Andrews, N., Sullivan, D., Gehr, W., 2013. Agricultural Composting and Water Quality. Oregon State University Extension Service. https://catalog.extension.oregonstate.edu/em9053.

Butler, R.M., Connolly, W.M., 1994. Windrow Composting of Municipal Leaves and Dairy Manure. Department of Agricultural Engineering Technology, University of Wisconsin-River Falls.

CalRecycle, 2019. Environmental Justice Compost Facility Map. California Department of Resources Recycling and Recovery. https://www2.calrecycle.ca.gov/SolidWaste/environmentaljustice.

Chardoul, N., O'Brien, K., Clawson, B., Flechter, M. (Eds.), 2015. Compost Operator Guidebook, Best Management Practices for Commercial Scale Composting. Michigan Recycling Coalition. https://www.michigan.gov/documents/deq/deq-oea-compostoperatorguidebook_488399_7.pdf.

Chastain, J.P., Wolak, F.J., 2000. Application of a Gaussian Plume model of odor dispersion to select a site for livestock facilities. In: Proceedings of the Odors and VOC Emissions 2000 Conference, Sponsored by the Water Environment Federation, April 16-19, Cincinnati, OH.

Clemson Cooperative Extension, 2021. Stormwater Pond Design, Construction and Sedimentation. Clemson University. https://www.clemson.edu/extension/water/stormwater-ponds/problem-solving/construct-repair-dredge/index.html.

Coker, C., 2008a. Managing storm water. BioCycle 49 (2), 28. https://www.biocycle.net/managing-storm-water/.

Coker, C., 2008b. Operator insights: storm water treatment. BioCycle 49 (4), 29. https://www.biocycle.net/operator-insights-storm-water-treatment/.

Coker, C., 2008c. Storm water management regulations. BioCycle.

Confesor, R., Hamlett, J., Shannon, R., Graves, R., 2007. Movement of nitrogen and phosphorus downslope and beneath a manure and organic waste composting site. Compost Sci. Util. 15, 119−126.

CWC, 1997. Evaluation of Compost Facility Runoff for Beneficial Reuse. Prepared by E & A Environmental Consultants. Published by the Clean Washington Center.

CWMI, 2005. Selecting, Siting, Sizing and Constructing Compost Pads. Cornell Waste Management Institute, Ithaca, NY. https://ecommons.cornell.edu/bitstream/handle/1813/2540/compostfs6.pdf?sequence=1&isAllowed=y.

de Bertoldi-Schnappinger, U., 2007. Chapter 6 Design of composting plants. In: Diaz, L.F., de Bertoldi, M., Bidlingmaier, W., Stentiford, E. (Eds.), Compost Science and Technology, Waste Management Series, vol. 8. Elsevier, ISBN 9780080439600, pp. 89−117. https://doi.org/10.1016/S1478-7482(07)80009-1. ISSN 1478-7482. https://www.sciencedirect.com/science/article/pii/S1478748207800091.

FAA, 2007. Hazardous Wildlife Attractants on or Near Airports. Federal Aviation Administration, U.S. Department of Transportation. https://www.faa.gov/documentLibrary/media/Advisory_Circular/AC_150_5200-33B.pdf.

Gaia, 2021. Cultivating Climate Justice through Compost: The Story of Hernani. https://www.no-burn.org/cultivating-climate-justice-through-compost-the-story-of-hernani/.

Gamble, S., 2018. Compost Facility Operator Study Guide. Minister of Alberta Environment and Parks, ISBN 978-1-4601-4130-4.

Gribbin, J.E., 2007. Introduction to Hydraulics and Hydrology with Applications for Stormwater Management. Thomson Delmar Learning, Clifton Park, NY.

Kalaba, L., Wilson, B.G., Haralampides, K., 2007. A storm water runoff model for open windrow composting sites. Compost Sci. Util. 15 (3), 142–150.

Kannepalli, S., Strom, P., Krogmann, U., Subroy, V., Giménez, D., Miskewitz, R., 2016. Characterization of wood mulch and leachate/runoff from three wood recycling facilities. J. Environ. Manag. 182, 421–428.

Krogmann, U., Woyczechowski, H., 2000. Selected characteristics of leachate, condensate and runoff released during composting of biogenic waste. Waste Manag. Res. 18 (3), 235–248.

Larney, F., Olson, A., Miller, J.J., Tovell, B., 2014. Nitrogen and phosphorus in runoff from cattle manure compost windrows of different maturities. J. Environ. Qual. 43 2, 671–680.

LAWPCA, 2020. Lewiston-Auburn Water Pollution Control Authority Sludge Composting Facility Operation and Maintenance Manual. Penley Corner Road, Auburn, ME.

ME DEP, 2016. Maine Stormwater Management Design Manual. Maine Department of Environmental Protection, Augusta, ME. https://www.maine.gov/dep/land/stormwater/stormwaterbmps/index.html.

Nienaber, J.A., Ferguson, R.B., 1992. Nitrate movement beneath a beef cattle manure composting site. In: 1992 International Winter Meeting. December 15-18, Nashville, TN. Paper No. 922619. The American Society of Agricultural Engineers.

NOAA, 2021. NOAA Atlas 14 Point Precipitation Frequency Estimates: CA. US Department of Commerce, National Oceanic and Atmospheric Administration, National Weather Service. https://hdsc.nws.noaa.gov/hdsc/pfds/pfds_map_cont.html?bkmrk=ca.

NRCS, 2004. Estimation of Direct Runoff from Storm Rainfall, Chapter 10; Part 630 Hydrology National Engineering Handbook. Natural Resource Conservation Service. U.S. Department of Agriculture. https://directives.sc.egov.usda.gov/OpenNonWebContent.aspx?content=17752.wba.

NRCS, 2007a. Hydrologic Soil Groups, Chapter 7, National Engineering Handbook, Part 630. Natural Resource Conservation Service. U.S. Department of Agriculture. https://directives.sc.egov.usda.gov/OpenNonWebContent.aspx?content=17757.wba.

NRCS, 2009a. Conservation Practice Standard: Vegetated Treatment Area. Code 635. Natural Resource Conservation Service. U.S. Department of Agriculture.

NRCS, 2009b. Diversions, Chapter 9, Engineering Field Handbook. Natural Resource Conservation Service. U.S. Department of Agriculture. https://directives.sc.egov.usda.gov/OpenNonWebContent.aspx?content=25756.wba.

NRCS, 2021a. National Engineering Handbook, Part 630, Hydrology Chapters. Natural Resource Conservation Service. U.S. Department of Agriculture. https://www.nrcs.usda.gov/wps/portal/nrcs/detailfull/national/water/manage/hydrology/?cid=stelprdb1043063.

ODEQ, 2003. Biofilters for Stormwater Discharge Pollution Removal. Oregon Department of Environmental Quality. https://www.oregon.gov/deq/FilterPermitsDocs/biofiltersV2.pdf.

ODEQ, 2004. Commercial Composting Water Quality Permit Development. Prepared by CH2MHill for the Oregon Department of Environmental Quality.

Ohio EPA, 2017. Siting Criteria in the Solid Waste Program. Ohio Environmental Protection Agency. https://www.epa.ohio.gov/portals/34/document/guidance/gd_693.pdf.

Richard, T.L., Chadsey, M., 1989. Croton Point Compost Site Environmental Monitoring Program. Final Report. Westchester County Dept. of Public Works, p. 15.

Sikora, L.J., Francis, H., 2000a. Lime-Stabilized Soil for Use as a Compost Pad. U.S. Department of Agriculture-Agricultural Research Service Beltsville. U.S. Department of Agriculture, MD.

Sikora, L., Francis, H., 2000b. Building a pad from lime stabilized soil. Biocycle 41, 45–47.

SJV APCD (San Joaquin Valley Air Pollution Control District), 2013. Greenwaste compost Site Emissions Reductions From Solar-Powered Aeration and Biofilter Layer. San Joaquin Valley Technology Advancement Program, p. 132. https://www.o2compost.com/Userfiles/PDF/VOC-Emissions-Report.pdf.

Spencer, R., 2009. Solar power at composting facilities. BioCycle 50 (9), 32. https://www.biocycle.net/solar-power-at-composting-facilities/.

Transport Canada, 2014. Aviation — Land Use in the Vicinity of Aerodromes. Transport Canada. https://tc.canada.ca/en/aviation/publications/aviation-land-use-vicinity-aerodromes-tp-1247.

USCC, 2021. State Regulations. U.S. Composting Council. https://www.compostingcouncil.org/general/custom.asp?page=StateRegulations.

USEPA, 1989. Summary Report: In-Vessel Composting of Municipal Wastewater Sludge, Technology Transfer Document EPA/625/8-89/016. U.S. Environmental Protection Agency, Cincinnati. https://cfpub.epa.gov/si/si_public_record_report.cfm?Lab=NRMRL&dirEntryId=50677.

USEPA, 2007. Developing Your Stormwater Pollution Prevention Plan: A Guide for Construction Sites. U.S. Environmental Protection Agency, Washington, DC. https://www3.epa.gov/npdes/pubs/sw_swppp_guide.pdf.

USEPA, 2021. NPDES Permit Basics. U.S. Environmental Protection Agency. https://www.epa.gov/npdes/npdes-permit-basics.

Wilson, B.G., Haralampides, K., Levesque, S., 2004. Storm Water runoff from open windrow composting facilities. J. Environ. Eng. Sci. 3, 537–540.

Zbinden, M., Kish, K., Halbach, T., Ludvik, A., Black, B.G., 2015. Scientific Evaluation of Potential Environmental Impacts of Contact Water Generated from Composting Source Separated Organic Material. Minnesota Pollution Control Agency, St. Paul, MN.

Consulted and suggested references

Black, G., Zbinden, M., 2014a. Adding SSO to yard trimmings composting operations. BioCycle. https://www.biocycle.net/adding-sso-to-yard-trimmings-composting-operations/.

Black, G., Zbinden, M., 2014b. Understanding the chemistry of compost contact water. BioCycle. https://www.biocycle.net/understanding-the-chemistry-of-compost-contact-water/.

Bradford, A., Sundby, J., Truelove, A., Andre, A., 2019. Composting in America, A Path to Eliminate Waste, Revitalize Soil and Tackle Global Warming. U.S. PIRG Education Fund. https://uspirg.org/reports/usp/composting-america.

Chatterjee, N., Flury, M., Hinman, C., Cogger, C.G., 2013. Chemical and Physical Characteristics of Compost Leachates — A Review. Washington State University, Puyallup, WA. Report prepared for the Washington State Department of Transportation. https://www.wsdot.wa.gov/research/reports/fullreports/819.1.pdf.

Cleary, E.C., Dolbeer, R.A., 2005. Wildlife Hazard Management at Airports: A Manual for Airport Personnel. University of Nebraska- Lincoln. https://digitalcommons.unl.edu/cgi/viewcontent.cgi?article=1127&context=icwdm_usdanwrc.

Coker, C., 2014. Organics recycling facility siting. BioCycle. https://www.biocycle.net/organics-recycling-facility-siting/.

Cole, M.A., 1994. Assessing the impact of composting yard trimmings. BioCycle 35 (4), 92.

Confesor, R.B., Hamlett, J.M., Shannon, R.D., Graves, R.E., 2009. Potential pollutants from farm, food and yard waste composts at differing ages: leaching potential of nutrients under column experiments. Part II. Compost Sci. Util. 17, 6—17. https://doi.org/10.1080/1065657X.2009.10702394.

Dirx, T., 2021. Efficient layout at A composting facility. BioCycle. https://www.biocycle.net/efficient-layout-at-a-composting-facility/.

Dow, R., 2009. On-farm Fuel Storage. Michigan State University Extension. https://maeap.org/wp-content/uploads/2019/03/On-FarmFuelStorageWQ-59.pdf.

du Plessis, R., 2010. Establishment of Composting Facilities on Landfill Sites (Master's thesis). University of South Africa. http://uir.unisa.ac.za/bitstream/handle/10500/4904/dissertation_du_plessis_r.pdf?sequence=5.

Emerson, D., 2005. Innovations in compost facility structures. BioCycle 46 (1), 23. https://www.biocycle.net/innovations-in-compost-facility-structures/.

Environmental Protection Authority Victoria, 2017. Designing, Constructing and Operating Composting Facilities. Victoria State Government, Carlton, VIC. https://www.epa.vic.gov.au/about-epa/publications/1588-1.

Evanylo, G., Sherony, C., May, J., Simpson, T.W., Christian, A.H., 2009. The Virginia Yard Waste Management Manual. Virginia Polytechnic Institute and State University. http://vtechworks.lib.vt.edu/bitstream/handle/10919/48325/452-055_pdf.pdf?sequence=1.

Faucette, L.B., Jordan, C., Risse, L.M., Cabrera, M., Coleman, D., West, L., 2005. Evaluation of stormwater from compost and conventional erosion control practices in construction activities. J. Soil Water Conserv. 60, 288—297.

Faucette, L.B., et al., 2009. Storm water removal performance of filter socks. J. Environ. Qual. 38 (3), 1233—1239.

Faucette, B., et al., 2013. Performance of compost filtration practice for green infrastructure stormwater applications. Water Environ. Res. 85 (9), 806—814.

FHWA, 2015. Gravel Roads Construction & Maintenance Guide. Federal Highway Administration, U.S. Department of Transportation. https://www.fhwa.dot.gov/construction/pubs/ots15002.pdf.

FHWA, 2017. Bases and Subbases for Concrete Pavements. Federal Highway Administration, U.S. Department of Transportation. https://www.fhwa.dot.gov/pavement/concrete/pubs/hif16005.pdf.

Goldstein, N., 2012. Building solutions for composting, anaerobic digestion. BioCycle 53 (6), 30. https://www.biocycle.net/building-solutions-for-composting-anaerobic-digestion/.

Goldstein, N., 2016. Critical operations go under cover. BioCycle. https://www.biocycle.net/critical-operations-go-cover/.

Johnston, M., 2014. Compost specific storm water management guidance. BioCycle. https://www.biocycle.net/compost-specific-storm-water-management-guidance/.

Keener, H.M., et al., 2007. Flow-through rates and evaluation of solids separation of compost filter socks versus silt fence in sediment control applications. J. Environ. Qual. 36, 742−752.

Kennedy/Jenks Consulting, 2007. Compost Leachate Research. Report prepared for the Oregon Department of Environmental Quality, Land Quality Division.

King, M.A., MacDonald, G.M., 2016. Guide to Recovering and Composting Organics in Maine. Maine Department of Environmental Protection. https://www.maine.gov/dep/sustainability/compost/compost_guide2016.pdf.

Last, S., 2006. Information for buyers of site construction related products. Compost. News 10 (3). http://www.organics-recycling.org.uk/dmdocuments/Buyers'_Guide_to_Site_Constructions_Related_Products.pdf.

Last, S., 2007. The buyers' guide to composting pads and drainage systems. Compost. News 11 (4). http://organics-recycling.org.uk/dmdocuments/Buyers'_Guide_to_Composting_Pads_&_Drainage_Systems.pdf.

Longstroth, M., 2012. Analyzing and Improving Your Farm's Air Drainage. Michigan State University Extension, Van Buren County. https://www.canr.msu.edu/news/analyzing_and_improving_your_farms_air_drainage.

Martin, D., Dewes, T., 1992. Loss of nitrogenous compounds during composting of animal wastes. Bioresour. Technol. 42, 103−111.

Martinson, S., van de Kamp, M., Tso, S., 2017. Guide to Agricultural Composting. Massachusetts Department of Agricultural Resources. https://www.mass.gov/doc/guide-to-agricultural-composting/download.

McCay, T., 2019. The anatomy of a gravel road. The NorthStar Monthly. http://www.northstarmonthly.com/features/the-anatomy-of-a-gravel-road/article_3b0c6b48-e37c-11e9-8846-5f95f4be9e8a.html.

Michelle, A.M.H., Adella, M.K., Jason, R.V., Glenn, O.B., 2021. Pollutant removal in stormwater by woodchips. Int. J. Environ. Sci. Nat. Res. 26 (5), 556200. https://doi.org/10.19080/IJESNR.2021.26.556200.

Millner, P.D., 1995. Bioaerosols and composting. Biocycle 36 (1), 48−54.

MPCA, 2021. Minnesota Stormwater Manual. Minnesota Pollution Control Agency. https://stormwater.pca.state.mn.us/index.php?title=Main_Page.

Nienaber, J.A., Ferguson, R.B., 1994. Nitrate concentration in the soil profile beneath compost areas. In: Proceedings Great Plains Animal Waste Conference on Confined Animal Production and Water Quality. October 19-21, Denver, CO. Great Plains Agricultural Council Publication Number 151.

NRCS, 2000. Agricultural Waste Management System Component Design, Chapter 10, Agricultural Waste Management Field Handbook. Natural Resource Conservation Service, U.S. Department of Agriculture. https://directives.sc.egov.usda.gov/OpenNonWebContent.aspx?content=25756.wba.

NRCS, 2005. Conservation Practice Standard: Composting Facility Code 317. Natural Resource Conservation Service, U.S. Department of Agriculture.

NRCS, 2007b. Guide for Use of Geotextile. Natural Resource Conservation Service, U.S. Department of Agriculture. https://www.in.gov/idem/stormwater/files/stormwater_manual_apndx_c.pdf.

NRCS, 2011. Conservation Practice Standard: Pond. Code 378. Natural Resource Conservation Service, U.S. Department of Agriculture.

NRCS, 2021b. Agricultural Waste Management Field Handbook. Natural Resource Conservation Service, U.S. Department of Agriculture. https://www.nrcs.usda.gov/wps/portal/nrcs/detailfull/national/water/?&cid=stelprdb1045935.

NJDEP Science Advisory Board (SAB), 2020. Outdoor Food Waste Composting. New Jersey Department of Environmental Protection. https://www.nj.gov/dep/sab/sab_food_composting.pdf.

Oregon Department of Revenue, 2009. Cost Factors for Farm Buildings. Oregon Department of Revenue, Salem, OR. https://www.oregon.gov/DOR/forms/FormsPubs/303-417.pdf.

Peigné, J., Girardin, P., 2004. Environmental impacts of farm-scale composting practices. Water, Air, Soil Pollut. 153, 45–68. https://doi.org/10.1023/B:WATE.0000019932.04020.b6.

Platt, B., Goldstein, N., Cooker, C., 2014. State of Composting in the US: what, Why, where & How. Institute for Local Self-Reliance (ISLR). Available at: http://ilsr.org/wp-content/uploads/2014/07/state-of-composting-in-us.pdf.

ReTap, 1998. Evaluation of Compost Facility Runoff for Beneficial Reuse. Phase 1 & 2. Technology Brief. CM-98-1. CM-97 -4. NIST MEP Environmental Program, Seattle, WA.

Richard, T., Chadsey, M., 1990. Environmental impact of yard waste composting. BioCycle 31 (4), 42–46.

Rymshaw, E., Walter, M.F., Richard, T.L., 1992. Agricultural Composting Environmental Monitoring and Management Practices. New York State Agriculture and Markets, Albany, NY.

SA EPA, 2019. Compost Guideline. South Australia Environment Protection Authority, Adelaide. https://www.epa.sa.gov.au/files/7687_guide_compost.pdf.

Savage, G., 2008. Calculating capacity at composting sites. BioCycle 49 (3), 38. https://www.biocycle.net/calculating-capacity-at-composting-sites/.

SCAPA, 2020. Asphalt Pavement Design Guide. South Carolina Asphalt Pavement Association. https://www.scasphalt.org/uploads/5/6/3/8/56382339/scapa_design_guide_3rd_edition__6-30-20_.pdf.

Shipitalo, M.J., et al., 2012. Sorbent-amended compost filter socks in grassed waterways reduce nutrient losses in surface runoff from corn fields. J. Soil Water Conserv. 67 (5), 433–441 (Impact of Grassed Waterways and Compost Filter Socks on the Quality of Surface Runoff from Corn Fields).

Sullivan, C., 2013. Siting and Operating Composting Facilities in Washington State. Washington Department of Ecology. https://apps.ecology.wa.gov/publications/documents/1107005.pdf.

Tao, W., Hall, K., Masbough, A., Frankowski, K., Duff, S., 2005. Characterization of leachate from a woodwaste pile. Water Qual. Res. J. Can. 40, 476–483.

Tollner, E.W., Das, K.C., 2004. Predicting runoff from a yard waste windrow composting pad. Trans ASAE 47 (6), 1953–1961.

USCC, 2013. Model Compost Rule Template (Currently Being Updated). U.S. Composting Council. https://cdn.ymaws.com/www.compostingcouncil.org/resource/resmgr/images/advocacy/Model_Compost_Rule.pdf.

Virginia Department of Conservation and Recreation, 1999. Virginia Storm Water Handbook. Sec. 4, Hydrologic Methods.

WAPA, 2020. Asphalt Pavement Design Guide. Wisconsin Asphalt Pavement Association. https://www.wispave.org/wp-content/uploads/dlm_uploads/WAPA-Design-Guide-2020-Final.pdf.

WasteMINZ, 2009. Consent Guide for Composting Operations in New Zealand. Waste Management Institute of New Zealand. https://www.wasteminz.org.nz/pubs/consent-guide-for-composting-operations-in-new-zealand-2009/.

Webber, D., Mickelson, S.K., Richard, T.L., Ahn, H.K., 2016. Vegetative buffer and fly ash pad surface material system Application for reducing runoff, sediment and nutrient losses from livestock manure windrow composting facilities. In: Daniels, J.A. (Ed.), Advances in Environmental Research, vol. 47. Nova Science Publishers, New York, pp. 39–60 (Chapter 3).

WI DNR, 2012. Storm Water Pollution Prevention Plan (SWPPP) Worksheet for Licensed Compost Sites that Submit Plans of Operation to the DNR. Wisconsin Department of Natural Resources. https://dnr.wi.gov/files/PDF/pubs/wa/wa1587.pdf.

Williams, D.J., 1991. Environmental Implications of Composting. Cooperative Extension Service University of Illinois at Urbana-Champaign.

Willson, G.B., Hummel, J.W., 1975. Conservation of nitrogen in dairy manure during composting. In: Managing Livestock Wastes, Proceedings of the 3rd Int. Symp. On Livestock Wastes. April 22-24; Urbana-Champaign, IL. ASAE St. Joseph, MI.

Yesiller, N., Vigil, S.A., Hanson, J.L., 2011. In: 26th International Conference in Solid Waste Technology and Management Proceedings: Philadelphia, PA, March 27, 2011, pp. 1281–1291. https://digitalcommons.calpoly.edu/cenv_fac/259/.

Process management

Authors: Robert Rynk[1], Jeff Ziegenbein[2]

[1]*SUNY Cobleskill, Cobleskill, NY, United States;* [2]*Regional Compost Operations, Inland Empire Utilities Agency, Chino, CA, United States*

Contributors: Cary Oshins[3], Nanci Koerting[4], James Hardin[5], Jeff Gage[6]

[3]*U.S. Composting Council, Raleigh, NC, United States;* [4]*Environmental Compliance, Grant County Mulch, Boonsboro, MD, United States;* [5]*SUNY Cobleskill, Cobleskill, NY, United States;* [6]*Green Mountain Technologies, Bainbridge Island, WA, United States*

1. Introduction

The foundation of any composting operation is the composting process—the biochemical action that transforms the raw feedstocks into compost products. The biochemistry happens by itself, and, if left undisturbed, it proceeds on its own path. The pathways it follows on its own are not necessarily consistent with the goals of the composting operation. For instance, if the feedstocks are too moist, anaerobic conditions set in. The process slows and can generate obnoxious odors. High temperature can also retard the process and possibly lead to a fire. Hence, facility managers need to exert some measure of process control to produce the desired compost product in the desired time frame while avoiding nuisances and fires. Managers do so by monitoring key process indicators and adjusting as necessary, based on what those indicators indicate (Fig. 11.1). Such adjustments include adding water, turning, mixing in bulky feedstocks, or altering forced aeration.

Process monitoring requires keen observation, a critical sense of smell, plus a feel for moisture. Temperature is the chief *measured* indicator. Therefore, a good thermometer is essential. Moisture is the next priority. Additional information can be gained from monitoring oxygen or carbon dioxide (CO_2) concentrations, bulk density, and pH. In some instances, oxygen monitoring is mandated by process protocols or by regulations.

Most composters monitor the process manually, using their own senses, a dial thermometer, and the hand-squeeze test for moisture. However, digital electronic technology is increasingly used, either replacing manual tasks or making them easier (Coker, 2007). Many larger facilities have systems that monitor process factors automatically and then automatically make adjustments, like changing fan speed. Such systems are known as Supervisory Control and Data Acquisition (SCADA). However, even with the most sophisticated SCADA system, manual process

The Composting Handbook. https://doi.org/10.1016/B978-0-323-85602-7.00011-X

501

monitoring should continue to be a daily routine, at least to some degree. Manually monitoring temperatures every day has the advantage of forcing the operator to walk the site and make observations. A lot can be noticed in the time that it takes a thermometer to stabilize.

This chapter provides guidance for adjusting the composting process in response to observed process indicators and site conditions. Table 11.1 summarizes this guidance. Appendix G is a more detailed troubleshooting guide for process management.

Table 11.1 Process monitoring guide.

Process indicator	Possible causes
Temperature	
Low temperatures at the start of composting	• Drop in pH inhibiting microbial activity • Poorly degradable feedstocks • Feedstocks too dense or too wet
Consistently lower than normal temperatures throughout	• Materials too dry • Materials too wet or dense • Insufficient aeration • Windrow/pile is too small
Consistently higher than normal temperatures throughout	• Highly degradable feedstocks • Low C:N ratio • Moderately low moisture • Insufficient forced aeration • Large and well insulated pile (e.g., wood chips)
Highly variable temperatures readings	• Feedstocks poorly mixed • Development of aeration channels • Forced aeration poorly distributed
Large temperature difference between outer layer and inner core	• Windrow/pile compacted, needs turning • Low rate of forced aeration • Possibly a mechanical problem with forced aeration; a strong cooling wind; or sudden drying of the windrow/pile
Very high temperature readings (>70°C or 160°F)	• Insufficient cooling due to moderately low moisture, insufficient forced aeration rate, intermittent aeration off-time too long; and/or an overactive process due to highly degradable feedstocks
One or more temperature readings above 75°C (170°F)	• The process of spontaneous combustion might have started in one or more sections of the pile. • Pile may be too large.
Temperature fails to increase after turning	• Moisture content is very low or very high • Active composting phase is nearing completion

Table 11.1 Process monitoring guide.—*cont'd*

Process indicator	Possible causes
Moisture	
Moisture content >65%—70% throughout	• Windrow/pile is too wet. It needs to be dried by turning and/or mixed with dry amendments
Moisture content <45%—50% throughout	• Windrow/pile is too dry. Water or wet feedstocks need to be added.
Variable moisture readings	• Feedstocks are poorly mixed. Turning needed.
Oxygen	
O_2 measurements <5%—10% throughout	• Anaerobic or anoxic conditions due to: • Highly degradable feedstocks • Insufficient forced aeration • Pile too large
O_2 measurements <5%—10% near center but higher near surface	• Windrow/pile compacted, needs turning • Low rate of forced aeration
Highly variable O_2 measurements	• Feedstocks poorly mixed • Development of aeration channels • Forced aeration poorly distributed
Consistently high O_2 measurement (>16%)	• Over-aggressive forced aeration • Microbial activity slowed by low moisture, high C:N ratio, lack of secondary nutrient or some toxic factor
Sharp drop or rise O_2	• Possibly a mechanical problem with forced aeration; a strong cooling wind; or sudden drying of the windrow/pile
pH and EC	
pH < 5.5 during the first few days of composting	• Accumulation of organic acid due to highly degradable feedstocks and/or insufficient aeration
pH > 8.5 to 9 near the end of active composting	• Abundance of high pH and/or high N feedstocks
Higher than desired EC measurements in finished compost	• Feedstocks are rich in nutrients (e.g., NH_4, NO_3, PO_4, K) and/or have high concentrations of chemical elements and compounds that form salts (e.g., Na, Ca, Cl, SO_4) • Frequent irrigation of piles and windrows without drenching precipitation

FIGURE 11.1

Operators and managers need to be keen observers of the process and site conditions.

2. Odor

Odor is a key process control parameter. The presence of odors and their character indicate how well the process is being managed. Indeed, a primary process-related goal should be the minimization of offensive odors.

The character of an odor can be an important process-control indicator. Some odors are feedstock-related, some are process-related. Odors described with a putrid, rotting, or septic character are indicative of anaerobic conditions. Some compounds, like hydrogen sulfide (rotten egg smell), are products of anaerobic decomposition. In general, the odor from anaerobic compounds are more offensive. Their presence not only threatens the facility but also suggests the process needs attention. Under certain weather conditions, some of these odors can travel great distances creating even more risk of offending neighbors.

There is no better odor monitoring tool available to composters than their own sense of smell, even when it grows dull. Every composter should have their noses work hard while they are on and near the facility. In fact, composters should enlist the noses of neighbors in an effort to optimize the process and the facility. A good odor-measuring instrument is not yet available for composting facilities. However, there are some devices available to help the nose perform better and zero in on order sources. Such devices attempt to expose the human nose to odors from a particular source and/or filter out background odors. The topic of odor detection and mitigation is discussed in more detail in Chapter 12.

3. Temperature

Because the heat produced during composting is directly related to the microbial activity, temperature is the primary benchmark for assessing the status and condition of the composting process. It should be monitored in each pile/windrow regularly, if not daily. Temperature measurements should be recorded, evaluated, and preferably graphed.

For a given set of feedstocks and conditions, composting produces typical temperature levels and patterns. Abnormally low or high or inconsistent temperatures at any given stage suggest that there is a glitch in the process, and some corrective action is needed. Action may also be needed if temperatures trend upward or downward more than expected.

As discussed in Chapter 3, a healthy temperature for thermophilic composting hovers near 60°C (140°F). A good goal during the thermophilic stage is maintaining temperatures between 55 and 65°C (131 and 150°F). In addition, there are a few temperature thresholds that merit attention (Table 11.2). If any single temperature measurement violates one of these thresholds, some type of process correction should be taken. Such thresholds include the level required for meeting sanitization requirements (e.g., Pasteurization, PFRP), high temperatures that impair the overall microbial population and very high temperatures that risk the evolution of spontaneous combustion. The following section, "Temperature Signals," discusses what actions should be taken when temperature measurements are out of bounds.

Temperature is the best guide for managing the composting process, but it is not without its quirky patterns and peculiarities, as described in Box 11.1. An understanding of these patterns and peculiarities helps in interpreting the temperature signals coming from a pile.

3.1 Temperature monitoring procedures

The temperature of each pile should be monitored frequently and on a regular schedule. Freshly created piles and windrows merit daily monitoring until temperature trends become familiar. Inexperienced operators and technicians should also

Table 11.2 Composting temperature guidelines.

Transition from mesophilic to thermophilic microbial populations	40 to 45°C 104 to 113°F
Healthy range for thermophilic composting (considering pathogen and weed seed destruction)	55 to 65°C 131 to 150°F
Minimum temperature for sanitization or pasteurization (depending on the jurisdiction and feedstock)	55 or 60°C 131 to 140°F 65°C (149°F) in certain cases
Minimum temperatures recommended for destroying most weed seeds at 24-hour exposure. (Dahlquist et al., 2007)	50°C or 122°F
Level at which microbial populations suffer the effects of high temperatures and the process slows	Roughly 70°C or 160°F

Box 11.1 Temperamental temperatures—patterns and peculiarities of pile temperatures

The first thing to understand about temperature is that it depends on the amount of heat generated and the amount of heat lost from a pile to its surroundings. Temperature increases when the heat generation exceeds the heat loss, which is the usual situation during active composting. Temperature decreases when heat loss surpasses heat generation. For instance, a small pile that is actively composting can have a lower temperature than a large pile that is decomposing at a slower pace because the large pile retains more heat. Second, temperatures and spatial temperature patterns continually change. They change with process conditions, stage of composting and slightly with the weather. Third, the temperatures at specific locations within the pile can vary considerably, depending on the consistency of the feedstocks, the composting method, and the stage of compost. It is not unusual to measure a difference of 10°C (18°F) between two points that are only 30 cm (12 in.) apart. Still, there are a few common and general patterns to note when monitoring pile temperatures (Rynk, 2016; Irvine, 2002; Lukyanova, 2012; Stegenta et al., 2019; Yeh et al., 2020).

- Temperatures increase from the exterior surface inward. Pile temperatures are near ambient temperature at surface. Temperatures rise abruptly within a short distance below the surface layer (e.g., 20–30 cm or 8–12 in.). The pile is generally hotter further inward, but the temperature rise is neither consistent nor continuous toward the pile center. The region of highest temperature is often near the center of the pile, but not always.
- Sometimes the highest temperatures are found nearer to the surface (30–60 cm or 12–24 in. deep) and then get progressively cooler toward the pile center. This situation occurs when oxygen is scarce near the center; for example, if the core of the pile is wet and/or dense.
- Temperatures are generally lowest near the pile base and highest just below the ridge or peak of a pile. In addition to the cooling effect of the ground, limited oxygen and compaction near the base, the air flow warms as it travels to its exit at the top of the pile. This situation is typical for passively aerated piles and windrows.
- For forced aeration, temperatures are routinely highest near the area where the air flow exits the pile and coolest where the fresh air enters. With positive forced aeration, the pile is warmest just below the surface. A small cool pocket can form around the air inlets with positive aeration. This pocket can persist if aeration is continuous but tends to dissipate when aeration is intermittent. With negative forced aeration, the warmest temperatures are at the air outlets near the base of the pile.
- Air flow patterns affect temperatures, especially when air channels form within the composting mass. In most cases, locations surrounding air channels show higher temperatures due to the stimulating effects of oxygen. In some cases, locations near air channels show lower temperatures due to the cooling and drying effects of the air flow.
- Wind has a cooling effect on pile and windrow temperatures, at least near the windrow/pile surface (approximately as deep as 1 m). Temperature measurements can differ from the windward side of a pile to the leeward side. In cold weather, the windward side is cooler. However, deeper within the pile, temperatures can spike during windy conditions. A high wind event can transform an unnoticed smoldering fire into a flaming fire.
- Temperature measurements can differ from the sunny side of a pile to the shaded side. The sunny side tends to be higher, but the difference depends on pile size and stage of composting.
- Cold spots occur where dense or moist clumps of feedstocks exist due to poor initial mixing.
- Hot spots can develop due to diverse moisture and nutrient distribution. The situation occurs when feedstocks are not well blended (e.g., pockets of grass clippings among yard trimmings). These hot spots can later become the nucleus for spontaneous combustion fires if not eventually broken apart by mixing or turning.
- Moisture mediates temperature changes and differences. As the moisture content increases, temperatures rise and fall more slowly and are generally more uniform.
- Dry feedstocks and compost are excellent heat insulators. Pockets or layers of dry materials slow heat loss and can cause temperatures to increase. However, large sections of dry materials may remain cold due to the lack of biological activity.
- Turning and other forms of mixing homogenizes materials and temperatures. Just after turning temperatures are uniform. Soon after turning, temperatures begin to diverge according to the patterns noted above.

monitor temperatures more frequently until they acquire a strong feel for the process. Also, if pathogen destruction must be documented, then daily temperature records are necessary, at least for the duration of the pathogen destruction period. Operating permits and regulations may dictate a frequency or schedule for monitoring temperatures.

Temperature measurements should be recorded and stored until the data are no longer relevant (e.g., one year after the associated compost batch has been used). When temperatures are monitored manually, standard temperature recording forms ("logs") reduce errors and ease the task for the operators and manager. They can also be used for recording other observations including moisture and odor. McSweeney (2019) provides examples of pile/windrow monitoring logs.[1] The field logs should align with computer spreadsheet programs that summarize and store the data. Even with manual monitoring, measurements can be easily recorded digitally by entering the measurements directly into hand-held computers, like a smart phone or tablet. Many electronic devices offer the advantage of transferring temperatures directly into a data recording instrument (i.e., data logger), from which the data can be downloaded directly into a central computer. Some devices have the ability to transmit data wirelessly to a computer using radio frequency signals.

The most common approach for monitoring temperature is to use a portable thermometer to measure temperatures in each pile at numerous selected locations. A standard temperature monitoring procedure should be established for operators to follow. The procedure should consider the following recommendations:

- For the best indication of trends and changes within the pile, measure temperatures at roughly the same locations and depths each time temperatures are recorded.
- Mark the measurement locations with clear and easily visible markers, such as colored flags. If it is inconvenient to put markers in the piles, then create a legible diagram for operators to follow. Label each location with an appropriate name or number (e.g., 1A, 1B, 2A …).
- Unless conditions prove to be consistent, monitor temperatures at a minimum of two representative locations along the length. For long windrows and piles, measure temperatures at fixed intervals between 15 and 45 m (50 and 150 ft). Ideally those locations should correspond to different batches or ages of feedstocks within the pile. If pathogen destruction must be documented, regulations may dictate the maximum distance between measurements, typically 45 m (150 ft) or at intervals that encompass a pile volume of about 200 cubic meters (or 200 cubic yards).
- Insert the thermometer at a height of about two-thirds to three-quarters the height of the pile and angle it toward the midpoint of the pile's core (see Chapter 6, Fig. 6.39).

---

[1] The On-Farm Composting Handbook also has examples of monitoring logs.

- At each location, measure the temperature at two depths, such as 0.5 and 1 m (1.5 and 3 ft). The desired depths can be marked on the probe stem with durable paint or tape, although both will gradually wear away. The temperatures at one depth relative to those at the other depth can indicate the state of the process.
- Most thermometers record the temperature at the tip of their stems. When reading temperatures at multiple depths, first measure the shallowest depth (closest to the surface) and then push the stem to the deeper spots, in turn.

The proceeding recommendations apply to passive piles, windrows, and individual aerated static piles (ASPs). Portable thermometers are cumbersome for monitoring extended piles because they cannot reach the interior cells. Instead, electronic instruments are typically used for extended ASPs, and also many individual ASPs. The instruments remain in place, recording temperatures continually or at regular time intervals and sending the data directly, and often wirelessly, to an electronic display and/or a computer (Fig. 11.2; also see Fig. 6.38 in Chapter 6). Regardless of the device, each batch or cell in the pile should be monitored, preferably in at least two different locations and at two different depths.

Temperature monitoring instruments are integral to most in-vessel composting systems and are usually electronic and permanently placed. In some cases, removeable thermometers are used either as back-up devices or the main measuring tools. The vessel usually has ports for placing temperature monitoring devices at predetermined locations.

FIGURE 11.2

Electronic temperature (red) and oxygen (blue) measuring devices used for an ASP under a microporous cover.

3.2 Temperature measuring devices

A mechanical (bimetal) dial thermometer is the iconic device for monitoring temperatures (Fig. 11.3A). It is inexpensive, accurate, and ubiquitous within the composting industry. A dial thermometer is durable, even in a composting environment, although the stem can break if it is abused or pushed too hard through dense material. The main drawback of a dial thermometer is that it is slow to reach a stable measurement (e.g., 2 min.).

A dial thermometer used for composting measurements has a few specific requirements. It should measure temperatures in the range of 0−100°C or 0−200°F. The dial face is typically 75 mm (3 in.) wide. A convenient feature is the dial face that displays both the Celsius and Fahrenheit scales. A detachable handle can be purchased that protects the dial from damage and generally makes the thermometer easier to use (Figs. 11.3 and 11.4).

The stem of the thermometer should be at least 1 m long (3-ft). Thermometers with longer stems, in the range of 1.2−2 m (4−6 ft), are available and often advantageous. These longer stems should be used where pile heights exceed 2.5 m (8 ft). A thick stem, 8 mm (5/16 in.) or greater, with a pointed tip is needed to push the thermometer through dense clumps of material and lower the chance of breaking the stem. Longer probes should have thicker stems. The stems can be narrowed at the tip, where the temperature is sensed, to improve the response time. A metal sheath is available that further protects the stem during use and in storage. Otherwise, the thermometer should be stored in the plastic tube that thermometer is shipped in.

(A)　　　　　　　　　　　　　　**(B)**

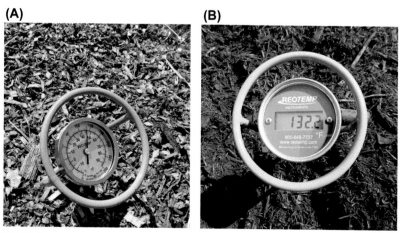

FIGURE 11.3

Hand-held composting temperature probes: (A) the ubiquitous bimetallic thermometer with a dial readout with its orange handle and (B) its digital cousin, which measure temperature via electrical resistance.

Source: (A) R. Rynk. (B) ReoTemp Instruments.

(A)   **(B)**

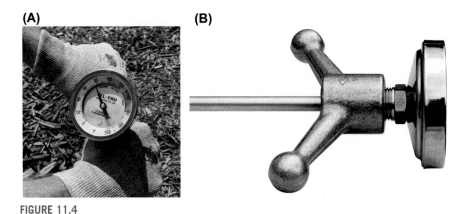

FIGURE 11.4

Dial thermometer with detachable handle that is used to push the probe into the pile, avoiding damage the to the dial.

Source: Tel-Tru Manufacturing Co.

Portable electronic temperature probes with digital readout are also commonly used in the composting industry. They are convenient to use and slightly faster than dial thermometers in reaching a stable temperature reading. Some devices can record and store data for a short time. In turn, they are more expensive, especially since the probe/sensor is separate from the meter. An electronic temperature probe is otherwise used in the same manner as a dial thermometer. Temperature is measured via a sensor that sends an electrical signal to a hand-held meter. A slight drawback is the difficulty reading the electronic display in the outdoor glare. Electronic probes offer the option of transferring temperatures directly into a data recording instrument (i.e., data logger), from which the data can be downloaded directly into a computer spreadsheet program.

3.3 Temperature signals and responses

A thorough set of measurements on any given day includes temperature readings at several different locations along the length of each pile and temperatures at two depths at any location plus whatever spot checks are made. Which one should be used to indicate the performance of the pile? The answer is all of them. The various measurements offer different clues about the status of the composting process.

1. Look at each temperature measurement individually to discover potential abnormalities, like a hot or cold pocket of material.
2. To understand the variations from one end of the pile to the other, compare temperatures at the different locations along the length. They can be compared at one selected depth, at both depths and/or an average of the two depths.
3. To understand the effectiveness of aeration, the moisture condition or the need for turning, compare the temperatures of the two depths at each location or at one selected location that is indicative of the pile's behavior. Turning or mixing can be triggered when the temperature gap between the two depths grows to an unacceptable level, such as 5−10°C or 10−20°F.

4. To assess the health, behavior, and stage of the composting process, establish a characteristic temperature benchmark based on experience with particular feedstocks. Use this benchmark to graph the temperature trends of the pile. The temperature benchmark can be an average all of the temperature measurements, the average of the temperatures at a given depth for all locations, or the temperature at one location that is generally indicative of the pile's behavior (e.g., where the stationary thermocouple should be located) (Fig. 11.4).

Tracking the characteristic temperature on a regular basis reveals a trend in the composting process as the pile ages. This trend suggests how well composting is progressing, when turning is needed and how aeration should be managed. Over time, normal patterns of temperature emerge that represent successful composting. Deviations from the normal patterns, such as a sharp drop or rise in temperatures, indicate conditions that might need correcting. Figs. 11.5–11.7 show three example temperature patterns and the conditions that prompt process corrections.

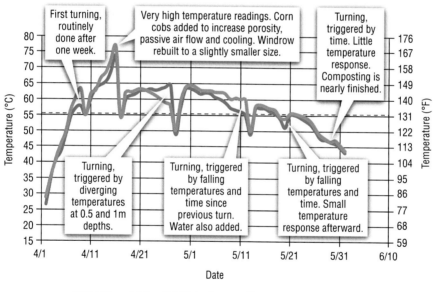

— Temperature at 0.5 m (1.5 ft) depth — Temperature at 1 m (3 ft) depth

FIGURE 11.5

Example temperature trends and management responses to temperature signals for a hypothetical turned windrows system handling cattle manure.

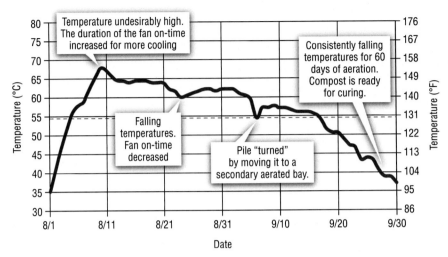

FIGURE 11.6

Example temperature trends and management responses to temperature signals for a hypothetical *aerated static pile composting biosolids and wood chips. Aeration is intermittent* and controlled by a clock timer.

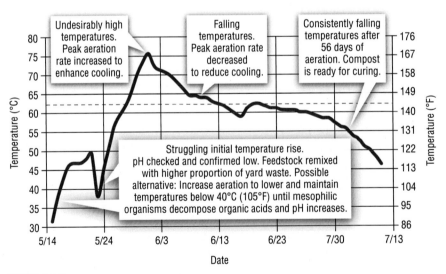

FIGURE 11.7

Example temperature trends and management responses to temperature signals for a hypothetical extended aerated static pile composting a mixture of yard trimmings and food waste.

Temperatures greater than 70°C (about 160°F) are considered detrimental to the overall microbial population and signal a need for cooling. Individual temperature readings in the 70—75°C range are not uncommon, nor a source for alarm. The microbial population and the composting process can readily recover. Nevertheless, high temperatures signal the need for greater cooling. Depending on the composting method and the stage of the process, a pile may be cooled by increasing forced aeration, making the pile smaller, lowering bulk density, or in some cases turning. Turning, however, is only effective at reducing temperature during the latter stages of composting or during extremely cold weather. It otherwise stimulates microbial action and raises temperatures. For effective cooling during active composting, a passively aerated pile must be rebuilt to increase heat loss, either by reducing its size or by adding bulky feedstocks to increase its porosity and air flow. With forced aeration, temperatures can be reduced by increasing the air flow rate or by lengthening the fan on time when aeration is controlled by a timer.

Any single temperature measurement greater than 75°C (or 170°F) should raise attention regarding the potential for a fire. At such temperatures, a smoldering fire may soon develop, if it has not already established itself. The prudent course of action is to deconstruct the pile to investigate and eliminate a potential hot spot with a source of water readily available. Operators should use great caution when deconstructing a pile at temperatures above 75°C. Flames might erupt when the core of the pile is suddenly exposed to oxygen. Loader buckets should be removed slowly, and ample water should be made available nearby to douse burning material. At a minimum, a very high temperature measurement calls for heightened awareness. Temperatures should be monitored closely and in more pile locations. If temperatures continue to rise, the pile must be taken apart. Chapter 14 discusses fires and fire prevention in more detail.

Temperatures are considered low if they are well below the expected level, especially if they linger in the mesophilic range. Most sanitization requirements (e.g., pasteurization or PFRP) call for composting temperatures to meet or surpass 55 or 60°C (131 or 140°F) for a prescribed amount of time. If such requirements are mandatory and the pile is not reaching 55°C at any measured point, then the process needs to be adjusted to generate or conserve more heat. Depending on the underlying cause, possible options include adding higher nitrogen feedstocks, adding porosity, adding moisture, making the pile larger, adding an insulating biocover of compost to the surface, and adjusting aeration.

Abnormally low or falling temperatures signal reduced aerobic microbial activity or excessive cooling. They can result from any one of several possible deficiencies including compaction, the lack of some nutrient, excessive moisture, or excessive drying. In the absence of strong odors, low temperatures often suggest that the pile is dry. The remedy is to add water or a moist feedstock. If there is sufficient moisture present, and the pile is aerated with fans and the problem may be overly aggressive aeration that is removing heat as fast as it develops. This situation often leads to dry piles. Low temperatures accompanied by odors usually point specifically to a lack of oxygen, which can mean that the materials are too wet or poorly mixed. Increasing the frequency or rate of forced aeration, turning and/or adding porosity are possible remedies.

Low temperatures are often due to a local deficiency, such as depleted nitrogen or moisture or poorly distributed air flow. It is not unusual to find substantially lower temperatures at some locations of the pile while other locations are well-heated. Uneven mixing and/or airflow channeling are common causes of such differences. Therefore, isolated or scattered low temperatures often call for turning or mixing to replenish locally deficient nitrogen or moisture and break up air channels. If the pile temperature does not recover after turning, either the process is nearing completion, or a more general problem exists. For instance, turning does little to speed the process and raise temperatures in a pile that is short of nitrogen throughout. Also, temperatures can vary along the pile length if the differing sections have been composting for different lengths of time (i.e., different batches).

With rapidly degrading feedstocks, like food waste, temperatures can be slow to increase early in the process. In this case, the problem may be caused by a rapid drop in pH. A sour odor further indicates this condition. If you suspect this to be the case, measure the pH. More information about pH measurement and correction are provided later in this chapter.

4. Monitoring moisture content

The primary reasons to measure moisture content are to:

- discern whether a feedstock is too wet or too dry for composting,
- balance the moisture content of feedstock mixes,
- determine the moisture status of composting piles and windrows,
- determine the moisture content of finished products,
- calculate nutrient concentrations in feedstocks (expressed on a dry-weight basis) to balance the C:N ratio, and
- calculate nutrient concentrations in finished products for customers or land application.

Of these, only the last item requires enough accuracy to merit analysis in a commercial laboratory. The other reasons in the list require only moderate accuracy. In such cases, moisture content can be measured in-house with simple procedures and inexpensive drying and weighing equipment.

Moisture is perhaps the most overlooked factor in composting. Both too little and too much moisture adversely affects the process and the facility. Composting can come to a near stop if operators neglect to monitor piles when they are dry, or if they are unable to add enough water to keep the process moving along. A lack of moisture especially plagues facilities located in arid climates and all facilities during a long stretch of dry weather. Overaggressive forced aeration can also lead to excessive drying. The problem of too much moisture occurs if feedstocks are not properly balanced and also during extended periods of prolonged precipitation. It is typically much easier for an operator to know when the pile is too wet. Still, diligent monitoring takes away the guesswork, especially when excess moisture is due to the feedstocks, not the atmosphere.

A composting pile can be highly variable in its moisture content. In a well-blended pile, the base and core are typically the wettest regions. From the surface inward and the top downward, the moisture content increases. The outer surface is typically dry, except after a rain, in which case it can be wetter than the interior. Light to moderate rain rarely penetrates more than a few centimeters or inches below the surface. The pile moisture can also vary due to uneven air flow. The pile material is drier near preferential air flow channels and wetter in the areas that air flow bypasses. This condition develops under both passive and forced aeration, but the differences tend to be greater with forced aeration, especially if the feedstocks are not well blended or the aeration system is unevenly laid out. A pile can also vary in moisture along its length.

Because of the possible moisture variations, getting a representative sample is important when monitoring moisture. If the objective is to determine the moisture content of the entire pile, or a large section of it, then the best practice is to obtain and evaluate a composite sample from multiple locations and depths in the pile, altogether avoiding the top layer (e.g., top 15 cm or 6 in.). To determine the moisture content at a particular spot in the pile, a grab sample should be taken at that spot. Taking multiple grab samples and evaluating each one individually can show inconsistencies within the pile.

The options for evaluating the moisture of composting materials during composting include the look and feel approach, the squeeze test, weighing and drying samples, and electronic sensors.

4.1 Evaluating moisture content by hand and experience

One way to gauge whether a composting pile has the correct amount of water is simply to visually inspect the pile and handle the material in it. An experienced operator can see and feel when a pile is excessively dry or excessively wet. Dust is a sure indicator of a dry pile. Freely draining liquid, odors, and flies suggest that a pile is too wet. Estimating the wetness or dryness of materials in the midrange is more difficult. The backyard composting adage states that materials at the right moisture content should have the feel of a wrung-out sponge. The look-and-feel approach depends on the feedstocks. Loose, fibrous materials feel drier than dense materials with small particles at the same moisture content. Also, because the pile changes texture and consistency as composting progresses, the look and feel of a moist pile also changes. Over time, a composter can acquire the visual and tactile sensitivity to know when a pile needs water or needs to dry.

The advantage of the look-and-feel approach is that it gives an immediate assessment. Also, the method is based on experience with the specific feedstocks and conditions at the facility. This approach gives only the general sense that a pile is too wet, too dry, or just about right. It is a qualitative rather than a quantitative approach. Moreover, it can be deceptive. Lacking experience with a given set of feedstocks, most people consistently underestimate the amount moisture in a pile, except at the extremes.

The squeeze test is a step beyond the look-and-feel approach; it is a *formalized* look-and-feel. The procedure is simple—grab a handful of the material from the pile, squeeze it firmly in your hand, open your hand, and observe material (Fig. 11.8). If the material remains clumped together, your hand is moist but not wet, then the material has about the right amount moisture for composting. If squeezing causes water to stream out, the material is too wet. If the material cracks or breaks apart in your hand, and your hand is not moist, then the material is too dry. This is the general interpretation for the squeeze test. Some practitioners have applied other interpretations including the following. The material is too wet if water drips from your hand without squeezing; or, squeezing causes the material to greatly deform or press through your fingers. The material is too dry if no water drips from the material when squeezed; or the material simply does not feel moist (e.g., like a wrung-out sponge). The material is at a good moisture content if only a trickle of water emerges from your hand when squeezed, or if squeezing leaves a wet sheen of water on your hand (your hand glistens). Box 11.2 describes a procedure for conducting the squeeze test and quantifies the results of the test against broad ranges of moisture contents.

If the squeeze test feels uncomfortably vague, an evaluator can "calibrate" her/his hand by practicing the squeeze test on samples of feedstocks and compost. Then, the squeeze test experiences can be compared with the moisture content of the same materials determined by the weighing and drying technique described in the following section.

(A)   **(B)**

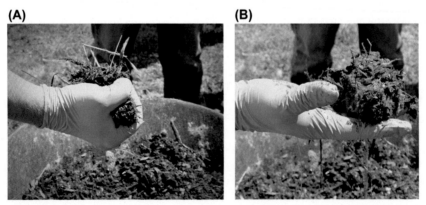

FIGURE 11.8

Examples of the hand squeeze test. This particular example holds together in a ball after releasing the squeeze indicating that it has adequate moisture, approximately 50%—55% (Table 11.2).

Box 11.2 Procedures and interpretation for the hand-squeeze test for moisture content

1. Reach into the pile or composite sample bucket and grab a handful of composting material,
2. Squeeze the material hard with your hand, check for drips,
3. Release your grip and allow the material to stay in your hand, smear some between your finger and thumb,
4. Inspect the material and your hand, and,
5. Use the "Rules of Thumb" in Table 11.3 to estimate percent moisture content.

Table 11.3 Rules of Thumb for estimating moisture content by the hand-squeeze test.

Observation—description	Estimated moisture content
• Water flows freely out of your hand	Greater than 65%
• A few drops of water are visible between your fingers	60%–65%
• You don't see any water between your fingers. When you open your hand, a sheen of moisture is clearly visible.	55%–60%
• No sheen of water is visible and a ball of compost remains in your hand. If you tap the ball gently, it remains intact.	50%–55%
• A ball of compost forms but break apart during tapping.	45%–50%
• After squeezing, the compost does not remain in a ball when opening your hand	40%–50%
• No ball forms and a dry talcum-like feel remains on your hand after discarding the material.	Less than 40%

Source: *Oshins, C., 2021. COTC Compost Operator Training Notes. Compost Research and Education Foundation, Raleigh, VA.*

4.2 Determining moisture by weight and drying

Weighing a sample of composting material before and after drying is the most certain way to accurately determine its moisture content. It provides an objective quantitative value. This method is formally referred to as the "gravimetric" method. The weighing and drying procedures are simple, and little can go wrong to harm the measurement's accuracy, at least for the purpose of monitoring pile moisture. The problem with weighing and drying samples is that it requires more effort and time than on-the-spot evaluations, especially if multiple samples must be tested. Depending on the drying method used, results are not available for hours or even days.

The sample size for the weighing-and-drying approach is not especially important as long as the sample can dry in a timely manner. It typically ranges from 100 cubic centimeters to 1 L (about 4 fl. oz to 1 quart). The larger size is common when testing many feedstocks, which tend to be inconsistently moist. Small samples are appropriate for finished compost. In general, a larger sample is more representative but takes longer to dry.

The basic procedure is to weigh the sample as it is, which is the "wet" weight; after subtracting the weight of its empty container. It is also called the "as is" weight. Next, the sample is dried until no moisture is left and then it is weighed again, giving the "dry" weight; again, after accounting for the container weight. The difference between the wet weight and dry weight is the weight of water that the sample contained. The weight of water divided by the wet weight of the sample equals the moisture content, expressed as a decimal fraction. Technically, this value is the moisture content on a "wet basis," which is the criterion used for composting. The "dry basis" moisture content is the weight of water divided by the dry weight.

$$\text{Weight of water} = \text{Wet weight} - \text{Dried weight} \qquad (11.1)$$

$$\text{Moisture content (as a fraction)} = \text{Weight of water} \div \text{Wet weight} \quad (11.2)$$

$$\text{Dry matter, or "Solids," content (as a fraction)} = \text{Dry weight} \div \text{Wet weight} \quad (11.3)$$

To determine when the sample is dry enough, dry and weigh it in stages. The sample is dry when its weight remains constant between two consecutive drying stages. For composting purposes, the sample can be considered dry if its weight changes by less than 1% of the original wet weight (for example, 1 g for a 100-g sample).

The goal in drying a sample is to remove the water while minimizing the loss of volatile dry matter; compounds such as ammonia and volatile organic acids. Ideally, samples should be dried at relatively low temperatures over a long time period because high temperatures increase the dry matter loss, especially if a sample burns. However, there is a trade-off between accuracy and speed. Lower temperatures and larger samples generally improve accuracy but increase drying time. After a number of experiments, typical drying times can be established. General guidelines, such as those given below, provide starting points, but experimentation is still helpful to establish routine procedures for specific sample characteristics and drying methods.

Four common methods of drying composting samples are used. These are air drying, forced hot air, conventional oven drying, and microwave oven drying.

Air drying is a no-frills method of determining the moisture of a sample. After determining the weights of the wet sample and its container, the sample material is spread in a thin layer (1–2.5 cm, or ½ to 2 in.) on a paper, plastic sheet, or aluminum foil surface in a warm room (weigh the plate or foil beforehand). A fan is used to improve air circulation and decrease drying time. The sample should be allowed to dry for 24 to 48 h, stirring occasionally to obtain uniform drying of all particles. The dried material is poured back into the sample container and weighed again. It may be necessary to repeat the above steps, weighing every several hours, until the weight loss is negligible. Air-drying removes most but not all of the water contained in the sample material and, therefore, tends to underestimate the actual moisture content. However, for most composting situations, air-drying works well enough.

Several *forced hot air* drying devices, primarily used to test the moisture of forages and grains, are well suited for composting materials as well (Koster, 2021). Such devices use a fan and an electric heater to move heated air through the sample

(A)  **(B)**

FIGURE 11.9

Forced hot air moisture tester with scale and timer.

Source: Koster Moisture Tester/Sprucewood Farms.

placed in a fitted basket (Fig. 11.9). Samples containing long particles (e.g., sticks, straw) need to be chopped in order to fit in the basket. The tester has a scale that reads the loss of water after drying. With the prescribed sample size of approximately 100 g, the drying times ranges from 30 to 60 min, but the time may be 50% shorter or longer depending on the initial moisture content. For a given material, a specific drying time can be established by alternately heating and weighing the sample, until the weight loss becomes negligible. These types of testers are small and portable enough to be used in the field. However, a source of electricity is needed to power the fan and heater. Hot-air moisture testers are generally accurate if operators are diligent in drying the sample completely.

A *laboratory oven* is the conventional drying device. Samples can be more thoroughly dried at temperatures between 60 and 110°C (140 and 220°F). An oven temperature of 100°C (212°F) is the normal temperature; a good compromise between speed and accuracy for composting materials. At temperatures near 100°C, samples are typically left in the oven to dry overnight or for a full 24 h. Drying is expedited by spreading the sample in a thin layer. Although oven drying takes a day or so to provide results, it offers several advantages. Most ovens are spacious enough to accommodate several samples and samples of different sizes, including large samples with long particles (Fig. 11.10). Also, oven drying can take place without the continual oversight of an operator.

A *microwave oven* considerably shortens the drying time, compared to a hot air oven (Fig. 11.11). Samples must be placed in microwave safe containers (e.g., not metal!). A paper plate is a good container because it is light-weight, and the sample can be spread out. Initially, it may take a few trials to determine the drying time for a

(A)

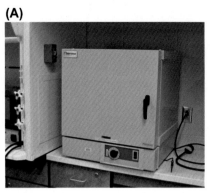

(B)

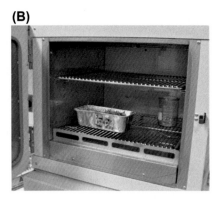

FIGURE 11.10

Laboratory drying oven provides a stable and even drying temperature for multiple samples.

(A)

(B)

(C)

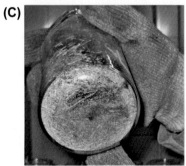

FIGURE 11.11

Microwave oven dries samples quickly but unevenly. Samples are susceptible to charring.

given microwave oven and sample size and type. The following procedure is a reasonable starting point.

1. Use a 100-gram sample of moist material and heat it for 6 min at 30% power in a microwave oven with 1000 W of power. For relatively dry materials, such as finished compost, reduce the heating period to four minutes. For a less powerful microwave oven, adjust the power setting proportionally higher (e.g., 40% power in an 800-watt oven).
2. After this initial heating, remove the sample from the oven and weigh it.
3. Reheat the sample for another 2 min, rotating it 90 degrees from its original position when replacing it in the oven.
4. After reheating, weigh the sample again.
5. Repeat the cycle of heating and weighing at one-minute intervals until the weight change is negligible.
6. If the sample is burned or charred, start a new trial using less power and/or shorter heating times.
7. After determining the total required drying time for a particular microwave oven, sample size, and material, a continuous drying period can be used.

Microwave drying is a convenient and reasonably accurate method of determining moisture content. However, microwaves heat the material unevenly. Care must be taken to avoid overheating and spot burning of the sample. After a charred sample is removed from the oven, it may gradually start to smoke and eventually catch fire. The resulting smoke and odor are unpleasant, if not unhealthy. Drying samples in a microwave oven should be done outside or under a fume hood. For all of these reasons, someone needs to stand by when using a microwave for drying.

4.3 Electronic moisture probes

When the weighing and drying method is too slow or cumbersome, and the squeeze test is too subjective, instruments with electronic moisture sensors can be used. These devices correlate moisture with some quickly measurable property of the material that changes with the material's moisture content. They provide an instantaneous reading that indicates the proportion of water contained in a given material.

The most common type of composting moisture sensor has a long probe with two pointed electrodes on the end. As a probe is inserted into the pile, the material in the pile is pushed into the space between the electrodes. The meter measures the electrical resistance of the material between the electrodes. The meter has a battery to generate electrical current. This device is essentially the same as the moisture probes used to test the moisture content of hay, but with a wider moisture measurement range.

Readings from electrical resistance sensors should be considered rough estimates of "wetness," rather than measurements of moisture content. Indeed, some meters display the moisture content on a general scale (e.g., 1−10), not as a percentage. The reason for the limited accuracy is that moisture is only one of several factors that affect the electrical resistance of the material being measured. Other factors include: the nature of the feedstocks; concentration of salts; bulk density; and the degree of contact between the material and the probe's electrodes. Therefore, to get reasonably accurate results, an electrical resistance sensor must be calibrated using the specific material it is intended to measure. Each manufacturer provides instructions for calibrating a particular device. Generally, the calibration procedure involves measuring a water-saturated sample of the material and setting the meter at the highest level and then measuring a dry sample the material and setting the meter at the lowest level. A particular calibration procedure may call for the calibration to be made with samples of known moisture content, predetermined by weighing and drying the samples.

Several commercially available electrical moisture sensors, based on different electrical properties, are used for applications similar to composting but are not widely used at composting facilities. Such instruments measure the material's electrical capacitance or dielectric constant. In fact, sensors that measure a material's dielectric constant hold promise because the dielectric constant is predominantly determined by the amount of water present. Nevertheless, the sensors also need to be calibrated with the specific material that they are intended to measure.

In general, electrical moisture-measuring instruments trade accuracy for ease of use and quick results. They offer the ability to measure moisture at many locations, in numerous piles, much faster than sampling, weighing, and drying. Because it takes experience to interpret the results in relation to this the specific materials and conditions at hand, using an electrical moisture sensor is comparable to conducting the squeeze test. It comes down to whether you are more confident with results that come from the feel of your hand or reading a meter.

4.4 Moisture signals

If moisture measurements show that the moisture content is low (<50%) throughout the pile, then water or a wet feedstock should be added to raise the moisture content to at least 60% (see below). However, there is an exception. If the compost is nearing completion, the threshold for adding moisture is lower because drier compost is easier to screen and generally preferred by users. The threshold for adding water to a nearly finished pile is about 40%. The moisture content should be raised to 45%–50% to keep the process active until it is complete, and also to keep the compost from getting dusty. Chapter 9 covers the method of adding water to windrows and piles, and the amounts of water needed to raise moisture content to the desired levels. Chapter 14 discusses options for minimizing the weather's impact on windrow/pile moisture levels.

If the moisture measurements show that the moisture is high (>65%) throughout the pile, the moisture content of the pile should be reduced somehow. One option is to dry the pile by turning it frequently during favorable weather (e.g., sunny, windy). A second option is to mix in dry feedstocks. If new feedstocks are mixed in an actively composting pile, the time and temperature criteria for sanitization must be restarted. With forced aeration, excess moisture can be driven off by increasing aeration, especially if temperatures remain high.

If the pile moisture is highly variable, the remedy is to turn or remix the pile. Water can be added during turning and mixing if the overall moisture content is low. Similarly, dry feedstock can be added if the overall moisture content is high.

5. Oxygen and carbon dioxide monitoring

Because aerobic decomposition is a defining feature of composting, it would seem logical that oxygen (O_2) is vigilantly monitored at most composting facilities; but it is not. Only a minority of composting facilities measure O_2 concentrations, or its counterpart—CO_2. Given that O_2 is critical to the composting process, why isn't O_2 or CO_2 monitoring more widely practiced? One reason is that temperature monitoring is generally good enough. The additional information provided by O_2 or CO_2 measurements, while potentially helpful, is otherwise unnecessary. In most cases, an operator can make sound decisions based on temperature and moisture alone. A second reason is that obtaining an accurate O_2 concentration is more difficult and expensive than measuring temperature (Paul, 2013a). A third challenge is interpreting the measurements. For example, it is common for measurements near the core of the pile to show almost no O_2 while temperatures are healthy and ample O_2 is present at moderate depths. How should the operator respond to this condition? Sometimes more information confuses matters. Many composters are unconcerned if the interior of the pile is largely anaerobic as long as the exterior is aerobic, and odors are contained. These practitioners are not likely to monitor for O_2.

Still, having information about the aerobic status of the pile could be potentially useful in managing the composting process. Some composting approaches rely primarily on O_2 measurements for process management (Edem et al., 2012; Gore, 2012). Knowing the O_2 concentrations can give an operator additional evidence about the condition of a pile and how to respond—when to turn, how to adjust the forced aeration system and if the pile is too dense, too wet or poorly mixed. Perhaps O_2 monitoring is most valuable for troubleshooting problems and irregularities, rather than routine process management. In addition, O_2 monitoring is also a way to manage the process to minimize greenhouse gas (GHG) emissions. Low O_2 measurements signal that the pile is decomposing anaerobically, which encourages the formation of methane (CH_4), a significant GHG (Section 9.2).

5.1 O_2 and CO_2 monitoring devices

There are two options for measuring O_2 concentrations in the compost pile, directly by measuring O_2 itself, and indirectly by measuring CO_2. During composting, O_2 and CO_2 have a direct and inverse relationship. The consumption of one O_2 molecule produces one CO_2 molecule. So as O_2 disappears, CO_2 appears. As the O_2 concentration decreases from about 21% (by volume, in fresh ambient air) to near 0%, the CO_2 concentration increases in step from near 0% toward 21%. Thus, by subtracting the measured CO_2 concentration from 21%, one gets a good estimate of O_2 concentration.

The basic combustion gas analyzer is an instrument for measuring CO_2 concentrations that is popular among composters. For using in composting situations, it requires minor modifications to the sampling tube, which can be found in Appendix H. The device contains a selected fluid that absorbs CO_2 and subsequently expands after mixing (Bacharach, 2021). The air sample is pumped into the fluid by means of a hand-operated aspirator that is connected to a long pipe, which is inserted into the

(A) **(B)**

FIGURE 11.12

Combustion gas analyzer for measuring CO_2 concentration (Fyrite Classic Gas Analyzer, Bacharach, 2021). The sample is drawn out through a tube inserted into the composting mass by pumping and aspirator (A). The assembly of PVC fittings is a homemade handle (see Appendix H). Notice the moisture filter on the collection hose after the handle. This filter is important for accurate readings and also prevents condensation in the instrument. The gas is absorbed by the red fluid in the body of the analyzer (B). This particular sample appears to be aerobic (4% CO_2, 17% O_2).

pile (Fig. 11.12A). The instrument's operating manual provides specific instructions regarding how many times the aspirator should be squeezed and how to mix the gas before the CO_2 measurement is read from the scale on the fluid chamber (Fig. 11.12B). The entire procedure takes only a few minutes. The indicator fluid is good for approximately 300 measurements, more or less, depending on the concentrations of CO_2 absorbed over time. In the meantime, the fluid can and should be checked regularly to determine if it is still good. Some combustion gas analyzers can also measure O_2 concentrations directly if a different absorption fluid is used. However, the O_2 absorption fluid is affected by ammonia and it does not last as long as the CO_2 fluid.

The just-described device is primarily intended for analyzing combustion efficiency.[2] It is a manual device, simple to operate but measurements cannot be automatically recorded. There is wide selection of CO_2 (and O_2) measurement instruments for combustion analysis that can be adapted to composting; most are electronic, handle data digitally, and can automatically record and transmit data (Grainger, 2021).

---

[2] This type of device is also used to measure CO_2 inside controlled atmosphere cold storage buildings, although the CO_2 range is much narrower in this application (0%−8%). It is another potential source of instruments not yet marketed directly to the composting industry.

Composters have a variety of direct O_2-sensing probes at their disposal to use for process monitoring. Most O_2 sensors are electrical cells that generate a voltage that is related to the O_2 concentration. Gas samples are drawn across the sensor where the O_2 molecules penetrate a selective membrane or other semiporous barrier of the cell. The membrane is susceptible to fouling so these instruments require occasional maintenance and regular calibration. Some O_2-measuring instruments are manually operated hand-held devices for making spot measurements, usually with an embedded temperature sensor. Other sensors are stationary and send readings to a computer for displaying, recording and process control. Still others measure O_2 concentrations in the exhaust from a tunnel, aerated container, or negatively aerated ASP.

Most manually operated O_2 monitoring instruments resemble compost thermometers—long metal probes with the sensor at the stem tip and meters connected at the top (Fig. 11.13). Many instruments are multifunctional, measuring O_2 and temperature, plus moisture in some cases. Gas from the pile is drawn out and across the sensor using an aspirator or hand pump. The reading is interpreted and displayed by the meter connected at the top of the device. Some instruments can download readings to a hand-held computer.

Stationary sensors may be inserted into a pile on the end of the probe or permanently installed at a key point in the composting system, such as the exhaust vent of a forced aeration system or the headspace of a vessel. Like the stationary temperature sensors, signals are sent to the computer through wires or wirelessly. Stationary O_2 sensors contribute to process control by triggering fan operation and/or fan speed by either turning fans on and off, controlling a fan's variable frequency drive, or adjusting valves in the aeration system.

(A)   **(B)**

FIGURE 11.13

Portable digital O_2 and temperature probe.

Contained/in-vessel composting systems, and negatively aerated ASPs, can make use of "in-line" CO_2 and/or O_2 monitoring devices that sense the concentration of gases in the forced aeration exhaust. However, the resulting measurements differ from measurements drawn from the pore spaces of piles and windrows. The interpretations need to be system specific.

5.2 Oxygen signals

As noted in Chapter 3, the process is considered anaerobic when the O_2 concentration is less than 5%, although many composters believe that O_2 levels should be maintained a good deal higher, in the 8%–16% range (O'Neill, 2021). Still, it can go as low as 1% at the core of the pile during active composting with no adverse effects on the process or product (Wang et al., 2013). Therefore, an operator's reaction to a given O_2 reading depends on the process management philosophy of the facility. Regardless of the minimum O_2 level desired, low O_2 readings suggest that aeration is insufficient. The response for correcting the problem fundamentally differs for facilities that rely on forced aeration (e.g., ASPs) versus passive aeration (e.g., turned windrows). Possible responses to low O_2 measurements with forced and passive aeration are discussed below.

5.2.1 In passively aerated piles

In passively aerated piles, O_2 levels are typically low at the center the pile and gradually increase outward toward the pile surface. They jump to over 20% somewhere within roughly 15 cm (6 in.) of the exterior. Such a pattern should be expected when taking O_2 measurements. However, if O_2 readings are sporadic at different places in a pile, especially at a given depth, it suggests poorly mixed piles and/or the natural flow of air into and out of the pile is occurring in channels. It is not necessarily a bad condition in the short run. Nevertheless, turning homogenizes the materials and evens the O_2 pattern, although sporadic O_2 measurements may return in between turnings.

O_2 readings may be consistently near zero in the interior of passively aerated piles. The hard truth is that it is difficult to maintain O_2 at a concentration above 5% with passive aeration. To do so requires either small piles, ample free air space (FAS), or very frequent turning (e.g., every other day). This condition does not necessarily mean that the pile is perpetually anaerobic. Instead, O_2 is being consumed as fast as it is migrating into the pile. The aerobic microorganisms retain a foothold and contribute to the decomposition, as do the anaerobes. The situation, and the O_2 concentrations, improve as the process matures and the rate of decomposition slows. In the meantime, there are options for improving passive aeration.

With turned windrow composting, the apparent response to low O_2 levels is to turn the pile. Turning briefly refreshes the pile with O_2-rich air and reestablishes aerobic conditions. However, the aeration effects of turning are temporary, lasting only hours to a few days, depending on the feedstocks and state of composting. If O_2 concentrations are to be maintained above 5%, frequent turning is necessary. Composters who subscribe to this management philosophy may turn a pile 30 to 40 times. Their goal is usually to produce compost with particular characteristics and quality.

There are other longer-lasting responses to low O_2 conditions in passively aerated piles. One option is to make smaller piles, thereby allowing more O_2 penetrate into the core. A second option is to add additional bulky amendments to the pile in order to increase its FAS. With greater FAS, the rate of airflow and O_2 delivery also increases. Low O_2 may also be a consequence of a wet pile. In this case, turning can have a lasting effect by accelerating the evaporation of the excess moisture. The dry pile, on the other hand, is characterized by unusually high O_2 readings.

5.2.2 With forced aeration

O_2 measurements can be used to fine tune a forced aeration system so that the system works efficiently and the desired O_2 levels are maintained. In the usual situation with ASPs, and similar composting methods, fans run intermittently and are turned on and off by a timer. Low O_2 measurements in this case should first suggest that the fans should be engaged more frequently (i.e., shorter off-time). If O_2 readings remain low even when fans are operating, the rate of airflow should be increased. Low O_2 readings can also signal a malfunction within the mechanics of the aeration system—a bad relay switch, blown circuit breaker, damaged fan, or plugged or broken aeration pipe. An increasing number of composting systems runs fans continuously at a minimum rate, and then the rate increases based on signals from temperature or O_2 sensors. In this case, consistently low O_2 measurements point to a mechanical problem or the aeration rate schedule is miscalculated.

It is also a problem when O_2 levels in a forced aeration system are higher than necessary because overaerating suppresses temperatures, dries out the materials, and increases emissions of ammonia (NH_3). With fans controlled by a timer, again, the first response to high O_2 measurements is increasing the time that the fans remain off. When O_2 measurements are consistently high in any system, the aeration rate should be checked, and if necessary, adjusted downward. It is also possible to observe higher than expected O_2 readings if the composting process has slowed. The usual culprit is low moisture, but it can also happen when the process is nearing completion.

Forced aeration systems also have a characteristic distribution of O_2 concentration within the pile or vessel. The highest O_2 concentration is at the point where the fresh air enters—at the inlet pipe where channel in a positive aeration network and near the surface where negative aeration is used. O_2 measurements are usually taken at some mid-depth that is indicative of the general O_2 status of the pile or vessel. If O_2 measurements are taken in several places and the measurements are uneven, then either the feedstocks are poorly mixed, short-circuiting air channels have developed, or there is a mechanical problem with the aeration system. For example, a broken or blocked aeration pipe may cause the air flow to be directed unevenly.

With forced aeration, consistently low O_2 readings also might indicate a poor compost recipe. For example, if the feedstock mix lacks porosity or is too dense, air may not be able to efficiently penetrate the pile, resulting in low O_2. To confirm this possibility the FAS and/or bulk density should be check (see following section).

6. Bulk density and free air space

As described in previous chapters, bulk density and FAS are important factors related to aeration during the composting process. Either one can be used as an indicator of how well materials aerate, primarily via passive aeration. These parameters are most useful at the start of composting when combining feedstocks. Nevertheless, if other parameters such as temperature and O_2 suggest that something needs to be modified, it may be enlightening to also check bulk density and/or FAS even after composting is well-underway. If the bulk density proves to be too high, or the FAS too low, then the remedy is to add dry bulky amendments to the existing mix.

Both bulk density and FAS are easily tested in the field using the procedures described in Chapter 4. The key is finding a way to reproduce the consistency and compaction of the material in the pile or container in question. In this procedure, repeatedly dropping the bucket from a fixed height is an attempt to mimic the compaction of the material in a pile. The material in the bucket may not actually achieve the same level of compaction. Nevertheless, the procedure provides a consistent, repeatable measure to compare feedstocks and composting materials and evaluate their performance during composting.

7. Monitoring pH

pH is rarely monitored on a routine basis during composting. When it is monitored, it is usually during the start of the process and with quickly decomposing feedstocks like fruits, starchy vegetables, and food wastes generally. The decomposition of such feedstocks can potentially depress the pH enough to inhibit microbial activity, especially during the transition from the mesophilic to thermophilic stage (Sundberg and Jönsson, 2008). The pH tends to be lower in the core of the pile because more organic acids accumulate where oxygen is scarce. If the process is starting slower than expected, checking the pH is a logical move. Later in the process, the pH rises as organic acids are oxidized and ammonia forms and volatilizes. If excessively high pH is a concern, it may be useful to check the pH after thermophilic temperatures have set in.

7.1 pH measurement methods

Accurately measuring pH involves taking a representative sample, adding a measured amount of water to create an extract and testing the pH of this extract by one of several methods that vary in their accuracy and precision (Box 11.3). Water is added because the pH detection methods require a saturated consistency to disperse the acidic and basic ions. pH testing methods include color-indicating chemicals (e.g., soil pH test kits and paper strips), hand-held pH probes, and laboratory-style pH meters (Fig. 11.14).

Box 11.3 Making an extract for measuring pH

When making an extract for pH measurement, distilled water, or otherwise deionized water, should be used. Also, the slurry container must be clean, preferably rinsed with distilled water.

TMECC (USCC, 2012) recommends making an extract that is 5 parts water to 1 part dry matter (by weight). Doing so requires either accounting for the water in the sample or drying the sample until it is bone dry. If the sample is completely dry, the amount of water to add is five times the weight of the dried sample. If an "as is" or "wet" sample is to be used, the amount of water to be added (W_{add}) is five times the dry weight (W_{dry}) of the wet sample minus the weight of water already present in the sample (W_{h2o}), as shown by Eq. (11.4). Once the moisture content (mc, expressed as a fraction) of the sample is known, Eq. (11.5) can be used to calculate using the wet sample weight (W_{wet}). If the sample is greater than 80% moisture, no water needs to be added.

$$W_{add} = (5 \times W_{dry}) - W_{h2o} \qquad (11.4)$$

$$W_{add} = W_{wet} \times (5 - (6 \times mc)) \qquad (11.5)$$

As an example, suppose that a sample of composting material weighs 200 g (about 7 ounces) and its moisture content has been determined to be 45% (or 0.45 as a fraction), then:

$$mc = 0.45 \; W_{wet} = 200 \text{ g } W_{dry} = 200 \text{ g} \times (1 - 0.45) = 110 \text{ g } W_{h2o} = 200 - 110 = 90 \text{ g}$$

If the sample is dried beforehand, then the amount of distilled water to add is five times the dry weight (5×90 g) or 450 g, which equals 450 mL of water (conveniently 1 g of water = 1 mL). If the wet sample is used, the amount of distilled water can be calculated by either Eqs. (11.4) or (11.5).

$$W_{add} = (5 \times W_{dry}) - W_{h2o} = (5 \times 110 \text{ g}) - 90 \text{ g} = 550 - 90 = 460 \text{ g or 460 mL}$$

$$W_{add} = W_{wet} \times (5 - (6 \times mc)) = 200 \text{ g} \times (5 - (6 \times 0.45)) = 200 \times 2.3 = 460 \text{ g or 460 mL}$$

(A)   **(B)**

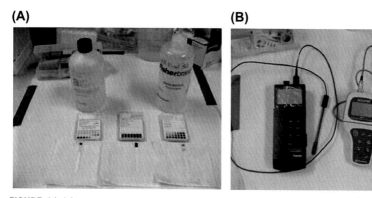

FIGURE 11.14

Options for measuring pH: (A) pH test strips and (B) pH meters (Bruckner, 2021). The instruments can also measure electrical conductivity (EC).

Source: Monica Bruckner.

Color-indicating soil test kits and strips are widely available for testing soil pH in gardens. Both are inexpensive and easy to use. Soil pH kits includes a clear container and a chemical that is combined with water and a small sample of soil or composting material (i.e., the extract). The resulting color of the solution correspond to the pH level. Similarly, pH-testing strips are impregnated with indicator chemicals that also change color with pH. The end of a strip is dipped into an extract made with the composting sample and allowed to briefly dry while the strip changes color. The results of these color-indicating methods are imprecise and subject to the user's interpretation of color, but they are indicative of the general pH level (Bruckner, 2021).

Inexpensive electronic probes, sold for measuring soil pH, also can be used for composting materials. These devices measure the electrical characteristics of the sample and relate that measurement to pH. Although the probes can be inserted directly into a pile or sample of material, the resulting measurement is not necessarily accurate. They work better when inserted into an extract made from the sample.

The most accurate and precise way to test pH is with a calibrated laboratory-style pH meter. This device measures the electrical voltage difference between a pair of electrodes that are immersed in the sample extract. The voltage difference is well correlated with pH and the meter displays pH units directly. The meter must be calibrated periodically using reference liquids ("buffers") of known pH. The prices of benchtop laboratory meters range from $500 to $1000 (US). Portable handheld meters of the same type are also available for about the same price. The portable devices must be calibrated and handled gently, like the benchtop devices.

7.2 pH signals

pH measurements preferably should fall in the range of 5.5–9; the closer to neutral (pH $= 7$), the better. If the pH falls outside of this range, it is more likely to be on the low side, especially at the start of the process. In this case, there are several options, the first of which is to do nothing and trust that the pH will rise naturally. This option is a reasonable course of action, but the potential downside of doing nothing is that the process may continue to lag, and the accumulation of organic acids may lead to unpleasant odors.

Low pH is principally a feedstock problem. It suggests that the feedstock mix is overburdened with quickly degradable compounds. The usual culprit in this case is food waste or other foods, like raw fruit and vegetable by-products (e.g., potato culls, fruit peels). If low pH is a frequent occurrence, options include adjusting the feedstock recipe, reducing the proportion of food (or other fast-degrading ingredient), and increasing the proportion of more slowly degrading feedstocks, preferably those materials with more structure and bulk. For instance, food scraps can be mixed with additional mixed yard trimmings, chopped straw or dry leaves to temper the pH drop. Another option is to add lime, wood ash, or finished compost to better balance and/or buffer the pH. Lime or wood ash should be used sparingly as they could elevate the pH too much and lead to ammonia nitrogen loss.

As noted in Chapter 3, low pH can be corrected, and possibly avoided, by increasing aeration. The enhanced aeration has multiple effects; it provides more oxygen for oxidizing the organic acids and it keeps the temperatures lower, extending the duration of conditions suitable for mesophilic organisms to decompose the organic acids. With forced aeration, the remedy is to increase the fan aeration rate or extend the fan on time, if an intermittent cycle is used. The greater air flow drives off more moisture so more water may have to be added back. Passively aerated piles/windrows can be reduced in size or density to lower their resistance to natural convection.

A high pH measurement is a concern if it is detrimental to the quality of the compost, which depends on its intended use (Chapter 15). A second problem related of elevated pH is ammonia volatilization, which increases dramatically above pH 8 to 8.5 (Fig. 11.16). High pH also is mainly a feedstock issue. If the pH is higher than desired, then the feedstock mix can be adjusted to include fewer alkaline ingredients, such as animal manures, in favor of more acidic or neutral ingredients. A high C:N ratio with readily available carbon tends to limit pH rise while an abundance or nitrogen-rich feedstocks tend to increase it. An immediate correction is adding sulfur-based products that are used to lower soil pH. However, a large quantity of sulfur is required to appreciably affect the pH of composting materials.

8. Monitoring soluble salts (electrical conductivity)

In-house monitoring of soluble salts is uncommon at composting facilities. If done at all, soluble salts are checked for product quality control rather than process management. Finished compost is measured, and occasionally raw feedstocks, but soluble salts are rarely measured during the process.

The concentration of soluble salts is indicated by measuring the electrical conductivity (EC) of a compost after diluting it with deionized water to form a paste or diluted solution. It is the same procedure used to measure pH (Box 11.3). In fact, most laboratory and many handheld pH instruments usually have the capability of also measuring EC. In the latter case, a voltage is created between the electrodes and the meter reads how well the compost paste conducts electricity. The measurement is indicative of the concentration of all soluble salts, including those salts that are also plant nutrients. Laboratory analysis is necessary to determine the concentration of any particular element of salt (e.g., Sodium, Na).

Chapter 15 describes the relevance of EC values in regard to compost quality. The soluble salts in a given feedstock generally become more concentrated after composting unless the feedstock is diluted by mixing with other less-salty feedstocks or ample rain or irrigation water.

9. Conservation of nitrogen and organic matter

Decomposing organic matter emits chemical compounds either as volatile gases or with leachate. Some of these compounds carry away carbon, nitrogen, and other nutrients that are otherwise valuable to the compost while others can be worrisome environmentally (e.g., GHGs). In general, conserving nutrients and organic matter is a matter of establishing and maintaining balance in the composting process, minimizing anaerobic conditions, and observing other best practices. Highly adsorbent additives, like zeolite and biochar, have been shown to help retain some volatile and mobile compounds in the composting pile. Some volatile compounds can be captured and retained in the layers of a biocover or beneath fabric covers.

9.1 Conserving nitrogen

During composting some amount of N is lost to the environment. Depending on the circumstances, barely any loss of N may occur, or the loss could be substantial, even exceeding 50% of the initial N in the feedstocks (Szántó, 2009). More typically, the N loss falls in the range of 10%−30%. N loss during composting is primarily due to ammonia (NH_3) volatilization (Box 11.4). Other contributors include denitrification and subsequent volatilization of nitrous oxide (N_2O) and elemental nitrogen (N_2). Additional N can be lost if precipitation washes away both soluble N compounds (ammonium, NH_4 and nitrate, NO_3) and particulates.

Box 11.4 Nitrogen dynamics during composting

Nitrogen (N) is a pivotal element in biochemical processes, and it has an active chemistry. Nitrogenous compounds are continually changing and changing back due mostly to the efforts of microorganisms. Even highly stable nitrogen gas in the atmosphere (N_2) is brought back into the nitrogen cycle by N-fixing organisms.

A newly formed pile of feedstock contains primarily organic nitrogen and smaller amounts of nitrate (NO_3) and ammonium (NH_4). The latter two forms of N are considered "available N" because they are soluble in water and immediately usable by plants and microorganisms. Before composting even begins, the organic N in the feedstocks has already started to biologically decompose. In a process called mineralization or ammonification, microorganisms convert the organic N into NH_4. Some of the NH_4 further evolves into ammonia gas (NH_3). These two "species" of ammonia readily shift back and forth, with NH_3 occupying the gaseous pore spaces and the NH_4 residing in the liquid portions of the pile. Their respective concentrations settle into an equilibrium that is determined by the pH and other surrounding conditions (Fig. 11.14). If the pH is below 7, NH_4 dominates, comprising more than 90% of the total ammonia. If the pH is above 10, the reverse is true; NH_3 dominates. The two are at equal concentrations at roughly pH 8.5 to 9.5, depending on the temperature (pH 9.2 at 25°C; pH 8.3 at 60°C). Higher temperatures favor more NH_3 at the expense of NH_4. As NH_4 is used or converted by microorganisms, some NH_3 converts to NH_4 to maintain the equilibrium. Likewise, as NH_3 volatilizes, some NH_4 replaces it. As the pH and temperatures change, the equilibrium shifts accordingly.

The NH_4—NH_3 equilibrium is at the center of the N dynamics during composting, as depicted in Fig. 11.15. Disturbing a pile by turning or other means causes much of the trapped NH_3 to evaporate. Also, the accumulated NH_3 is readily carried away by passive and forced air currents. As the NH_3 travels through the pores of the pile, highly adsorbent particles capture some of the molecules. Some of molecules condense on cooler surfaces. A portion of the molecules that remain in the pile changes back to NH_4. Still others remain as NH_3 and continue on their outward trek.

Box 11.4 Nitrogen dynamics during composting—cont'd

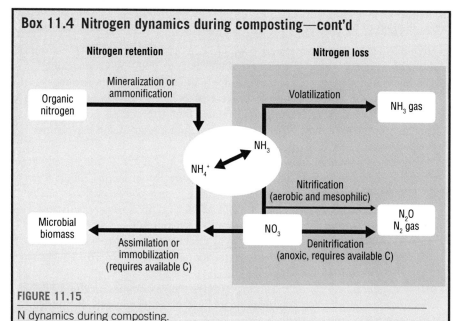

FIGURE 11.15

N dynamics during composting.

Adapted from Veeken, A.H.M., Hanajima, D., Hamelers, H.V.M., 2000.

In the meantime, some molecules of NH_4 take other paths, including assimilation and nitrification. NH_4 molecules are assimilated by microorganisms for cellular protein. Generally, the assimilation of inorganic compounds, like NH_4, by microorganisms is termed immobilization. Nitrification is a two-step process in which microorganisms convert NH_4 first to nitrite (NO_2) and then almost immediately convert the NO_2 to NO_3. Some N_2O also can form in this process. Nitrification is performed by specific genera of aerobic mesophilic bacteria. Thus, mineralization primarily occurs toward the end of composting and in the curing stage, after temperatures fall below 40°C (105°F). However, even during thermophilic periods, NH_4 molecules may still mineralize in low-temperature areas of the pile (e.g., near the surface).

NO₃ is water soluble. Once formed, it is readily assimilated by microorganisms within the pile (or plants outside the pile). If there is little oxygen present, microorganisms use the NO_3 as an oxygen-substitute in a process termed denitrification. The process successively strips away an oxygen molecule, forming NO_2, then NO, N_2O, and eventually N_2. Denitrification is the source of most of the N_2O formed during composting. Denitrifying microorganisms need a source of available carbon. If available carbon is in short supply, denitrification can stall at the N_2O step.

9.1.1 Minimizing ammonia loss

The way to conserve ammonia is to encourage microorganisms to assimilate it soon after it appears. This is accomplished using a two-prong strategy which includes: (a) supporting the organisms with carbon and (b) allowing the ammonia to remain in place long enough for the organisms to metabolize it. The C:N ratio of the feedstock mix is the crucial factor in determining the degree of NH_3 loss, but it is not the only factor. Carbon degradability, pH, aeration rate, frequency and timing of turning, moisture, and temperature are also influential. The following factors and conditions reduce the loss of NH_3 and conserve N in the compost (Wang and Zeng, 2018).

- *High C:N ratio*: The most effective way to conserve N is to create a feedstock mix with a high C:N ratio (Tiquia et al., 2002; Szántó, 2009; Whatcom County). The microorganisms need a sufficient supply of carbon to metabolize the ammonia. As Fig. 3.13 in Chapter 3 suggests, C:N ratio of 30:1 or higher results in minimal loss of N. However, establishing a high C:N ratio is not always enough. Sufficient C must be *available* to the microorganisms to balance the available N.

- *Degradable C*: A high C:N ratio minimizes N loss if the C-compounds are available to the microorganisms (Li et al., 2013). Feedstocks with C in the form of cellulose and hemicellulose provide available C. In woody feedstocks, much of the C is locked up with lignin; the *effective* C:N ratio is substantially lower than the ratio of total C to N. If a recipe relies on wood-based amendments, a "higher-than-normal" total C:N ratio can be established to compensate for the C that does not lend itself to microbial degradation (Chapter 3 [Section 3.5.1] and Chapter 4 [Section 4]).

- *Moderate and timely turning*: Turning exposes the gas-rich interior of a pile or windrow, accelerating the volatilization and escape of its contents, including NH_3 (and N_2O). More turning leads to more exposure and more losses (Parkinson et al., 2004; Cook et al., 2015; Yang et al., 2019). Turning during the first two weeks of composting results in the greatest N losses because concentrations of NH_3 are highest then. However, turning has positive effects on composting; effects that can help conserve N, like stimulating biological activity and assimilation of NH_4 and NH_3 (Szántó et al., 2007). Again, the point is not to stop turning but to turn only when necessary and, if possible, when NH_3 concentrations are low.

- *Moderate forced aeration rate*: All other things being equal, higher rates of forced aeration generally increase N loss (Osada et al., 2000; Shen et al., 2011). As the airflow increases, more NH_3 is flushed out. High aeration rates can even undermine the positive effect of a high C:N ratio (Szántó, 2009). There is balance to be achieved, however, because reducing aeration might also result in higher temperatures or anaerobic conditions. The former tends to increase NH_3 volatilization, the latter leads to more N_2O. Like many things in composting, the challenge is to find the right balance and to avoid aerating the pile more than necessary.

- *pH below 8*: The equilibrium between NH_4 and NH_3 is highly pH dependent (Fig. 11.16). Low pH keeps more NH_4 dissolved in the liquid fractions of the composting materials. Higher pH pushes it toward more gaseous, and volatile, NH_3. At low pH levels, up to approximately 7.0, nearly all of the NH_4–NH_3 complex is in the NH_4 form. A transition occurs between pH 7 and pH 8, at which the proportion of gaseous NH_3 rises quickly. For this reason, the addition of lime or wood ash to composting feedstocks is discouraged. To conserve N, the pH should be below pH 8 if possible. The best option may be to start with a feedstock mix that is inherently acidic and has a high C:N ratio (see Section 3.9

in Chapter 3). Fortunately, during the first few weeks of composting, when the ammonia levels are highest, the pH tends to be lowest.

- *Sufficient moisture*: As moisture becomes scarce within a composting pile, the concentration of NH_4 increases within the remaining liquid. This drives the equilibrium toward forming more NH_3. Thus, more N is conserved if moisture is maintained in the range generally recommended for composting (e.g., 50%–60%).
- *Lower temperatures*: Temperatures that are higher than necessary are generally adverse for composting, but they also increase NH_3 volatilization (Eklind et al., 2007). Higher temperatures shift the NH_4—NH_3 equilibrium toward more NH_3. Higher temperatures, in effect, reduce the pH levels at which more NH_3 forms at the expense of NH_4.

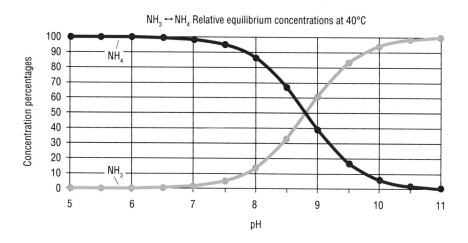

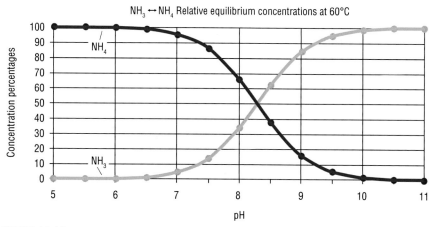

FIGURE 11.16

Effect of pH on relative equilibrium concentrations of NH_4 and NH_3 at 40 and 60°C.

9.1.2 Minimizing nitrous oxide losses

N_2O loss usually accounts for only a small share of the N lost during composting, although it can approach 10% of the initial N content (Szántó, 2009). N_2O arises from the nitrification of NH_3 to NO_3 and the subsequent denitrification of NO_3 (Box 11.4). Some N_2O can also form during the nitrification step, but this pathway seems to contribute little N_2O compared to denitrification (Maeda et al., 2010). Therefore, a first step in limiting N_2O formation is to limit NH_3 accumulation by establishing a sufficiently high C:N ratio with available forms of carbon and maintaining adequate moisture.

Because denitrification occurs when oxygen is scarce, a second step is to maintain aerobic conditions as much as possible. Smaller piles forced aeration and/or feedstock mixes with good FAS *generally* lead to higher oxygen levels and should also discourage denitrification (Paul, 2013b). As explained in Chapters 3 and 5, turning only increases O_2 concentrations for a short time. Unless turning also improves FAS, more frequent turnings should not be expected to reduce N_2O formation (Szántó et al., 2007; Parkinson et al., 2004). Otherwise turning facilitates volatilization of the N_2O that accumulates. A similar effect can occur with forced aeration; greater airflow might increase volatilization of the N_2O that has already formed.

In summary, minimizing N_2O emissions is largely a matter of following good composting practices—establish a good C:N ratio, provide available C, avoid higher-than-necessary temperatures, maximize aerobic conditions with small piles, good FAS, and well-distributed forced aeration.

9.1.3 Minimizing N loss via runoff

While not depicted in Fig. 11.15, runoff from heavy precipitation is another possible pathway for N loss. Runoff can carry away soluble N dissolved in the water (NO_3 and NH_4) and N-rich particles entrained in the flow. The solution is to thwart the flow of precipitation through and around the pile with proper moisture management and appropriate site design (Chapter 10). Various types of covers are effective in shedding water from pile surfaces, including "fleece" covers, and various process-containment methods from microporous fabrics to composting vessels. A biocover of compost or dry porous amendment also helps to shed precipitation.

9.2 Conserving carbon and organic matter

The conversion of organic matter to CO_2 is inevitable during composting. Indeed, it is necessary and desirable. It is part of the equation that defines composting. This conversion creates heat and enables self-pasteurization. However, it should not go further than necessary to create compost that is safe and suitably stable for its intended use. Because organic matter contributes value to the compost, it should be conserved. The goal should be to produce a stable compost product without phytotoxic elements while otherwise minimizing the conversion of organic matter to CO_2. The following practices conserve organic matter:

- *Suitably short composting period*: Composting beyond the required level of stability only drives off additional organic matter without improving the compost quality. The active compost phase should be curtailed when the compost is stable enough for the intended use.

- *Lower temperatures*: Higher temperatures hasten decomposition during any stage of composting. Lower temperatures tend to conserve organic matter without sacrificing compost maturity for a given set of feedstocks (Adler and Sikora, 2005; Eklind et al., 2007). After sanitization requirements are met, lower process temperatures may help to retain more organic matter.
- *Lignin-rich amendments*: Lignin is resistant to biological decomposition and feedstocks that have a high lignin content do not substantially decompose during composting. In general, the degradability of a feedstock is proportional to its lignin content (Chapter 4). The resulting compost retains a high proportion of organic matter and organic C. Lignin-rich feedstocks are primarily used as amendments and bulking agents. The most prominent are made of wood, including wood chips, wood shavings, and even sawdust. Other examples include lightly processed paper (e.g., newsprint, cardboard), fruit pits, and nut shells. Smaller particles decompose to a greater extent but much of the lignin-bound organic matter still persists in the compost. There are possible negative consequences to relying on lignin-rich amendments. They contribute less C to the composting process and therefore create a deceptively high C:N ratio. The resulting compost tends have woody or fibrous texture and may have a higher than desired C:N ratio for general compost use. These characteristics restrict its uses and markets. Large particles can be removed by screening and reused to serve as amendments for another batch of compost.
- *Aerobic conditions*: When conditions are anaerobic, some organic compounds are converted to other volatile carbon compounds, in addition to CO_2. CH_4 is the prime example. In addition to contributing to nonproductive organic matter loss, CH_4 is a GHG. Under aerobic conditions, CH_4 does not form (Brown et al., 2009). In general, maintaining aerobic conditions requires adequate forced aeration, moderate turning, high FAS, and/or moderate moisture (Ermolaev et al., 2019). When CH_4 does form, it can be assimilated by some microorganism and converted back into organic carbon, if O_2 and moisture are present. The same is true for most volatile C compounds. Therefore, some organic matter can be conserved by hindering the migration of CH_4 out of the pile with a sufficiently moist biocover.

9.3 Feedstock additives that conserve nitrogen and greenhouse gasses

A substantial body of research suggests that a variety of chemical and microbial additives can effectively reduce losses of N and GHGs during composting. Most of this research focused on adsorbent additives like zeolite or biochar, but other additives have also been studied, including lignite, peat, gypsum, clay, chemical salts, and also specific microbial inoculants. Overall, the researchers have reported positive results. A meta-analysis of the research through February 2018 by Cao et al. (2019) found that, on average, the additives reduced NH_3, N_2O, and CH_4 by 44.5%, 44.6%, and 68.5%, respectively. A reading of the individual research articles further suggests that the additives do not negatively affect, and may positively affect, the composting process or the compost quality.

The N- and GHG-conserving additives are summarized below. Several of these additives are described in more detail in Chapter 4.

- *Biochar* particles have enormous microsurfaces that become active locations for microbial processes. It is also dry and absorbs moisture. These qualities generally improve the environment for composting and specifically attract and hold many chemical compounds, like NH_4 (Godlewska et al., 2017; Guo et al., 2019). Numerous studies, involving various composting and biochar feedstocks, have consistently found that including biochar as a feedstock amendment reduces losses of total N, NH_3, N_2O, and CH_4 (Sanchez-Monedero et al., 2018). Reported reductions in total N and CH_4 losses range widely from roughly 5% to over 80%. The meta-analysis by Cao et al. suggests that the use of biochar typically reduces the losses of N and GHG by 50%−60%. Under a given set of conditions (e.g., feedstocks, composting method), these losses were generally proportional to the amount of biochar used. Most of the research projects studied biochar doses between 2% and 15% of the feedstocks' dry weight. References from the biochar industry suggest applications between 5% and 20%, depending on the feedstock C:N ratio (Pacific Biochar, 2021). Biochar derived from different feedstocks appear to perform differently, although the research does not favor one biochar feedstock over another. It is important to recognize that the C in biochar is not available-C. It should be ignored when estimating or testing feedstocks for C:N ratio.
- *Zeolite* is natural mineral (aluminosilicate) that, like biochar, has very porous microstructure and a high cation exchange capacity (CEC). Due to these qualities, zeolite has a high capacity to capture and store chemical compounds before they volatilize (Soudejani et al., 2019). The research suggests that zeolite produces similar results as biochar in regard to conserving N and GHG. Cao et al.'s (2019) analysis shows that zeolite is slightly better at conserving N_2O but less effective at reducing NH_3 loss. As with biochar, the zeolite dose and composting feedstocks influence the amount of N and GHG conserved. To conserve N and GHG, Awasthi et al. (2016) concluded that a zeolite dose amounting to 30% of the feedstocks dry weight should be used when composting biosolids.[3]
- *Peat moss (peat) and lignite* have the potential to reduce losses of N and GHG when used as a composting amendment. Both materials have a high CEC, are slightly acidic, and provide little or no available-C. While peat is fibrous, lignite

---

[3] Alwasthi et al. supplemented the zeolite with a slight dose of lime, but the benefit of the lime is not apparent from the research results.

is dense and soil-like in texture. Several researchers have studied the effects of lignite or peat moss (peat), on N and GHG conservation. For example, Impraim et al. (2020) found the addition of lignite to unamended cattle feedlot manure reduced losses of NH_3 by 35%−54%; N_2O by 58%−72%, and CH_4 by 52%−59%. Two different sources of lignite were used, each comprised 20% of the feedstock mix by weight. He et al. (2020) reported a reduction in NH_3 loss of up to 24% when peat comprised 15% of the dry mass of a feedstock mix that included food waste and sawdust with a C:N = 25. The positive effect of the peat appears to have been due more to its ability to adsorb ammonia than reduce pH.

- Many other *inorganic chemicals* have been proposed and studied as additives to conserve N and GHG. Such chemicals include gypsum, superphosphate, clay, acids (e.g., citric acid), and various salts including phosphate, magnesium, calcium, and aluminum salts. These chemicals are intended to either interfere with the conversion of NH_4 to NH_3 or lower the pH to reduce NH_3 formation. The meta-analysis by Cao et al. (2019) reported that many chemicals successfully conserved N and GHG but the different chemicals favored different N and GHG components. For example, phosphate (PO_4^{3-}) and magnesium (Mg^{3+}) salts were most effective of all tested additives, including biochar and zeolite, at conserving N, due to reductions in NH_3 losses, but less effective at conserving GHG. Gypsum scored well at conserving CH_4 but increased N_2O losses. Because many of the effective chemical amendments are salts, their impact on compost quality and marketability is a concern.

- *Microbial inoculants* are another class of additives that can potentially conserve N and GHG during composting. In this case, specific organisms or enzymes are added that are adept at either promoting the conversion of N_2O or assimilating CH_4 or NH_3. Often these microbial inoculants are proposed along with other additives like biochar. According to Cao et al. (2019), research studies have shown that microbial additives have generally positive effects in conserving N and GHG but not as strong as biochar and zeolite. However, in some individual research reports, microbial additives appear to be highly effective (Duan et al., 2019; Tu et al., 2019).

The research concerning the N and GHG benefits of the foregoing additives is positive. However, the practical implications are less clear. These additives cost money to purchase and handle while the resulting N and GHG savings do not normally have monetary value to compost producers; not unless carbon-trading credits are available to composters. Perhaps, a given compost product can earn a higher price because of the higher N content, but most compost buyers do not care. The use of biochar, and perhaps zeolite, is an exception as the soil-improving benefits of biochar are becoming widely recognized. Therefore, biochar might be the *external* additive of choice to conserve N and GHG. Internally, research has demonstrated that amending feedstocks with finished compost also tends to reduce losses of N, GHG, and other volatile gases (Büyüksönmes and Evans, 2007; Hwang et al., 2020).

It is relevant to note that many of the research studies examined composting of unamended or lightly amended animal manures and food waste. For these feedstocks, the starting C:N ratios are low, making the composting systems prone to N loses and challenging to keep aerobic. In these cases, the additives only improve a bad situation. It is likely that adding amendments with moderately available-C and good structure can bring the same N- and GHG-conserving benefits as the additives described above (Chowdhury et al., 2014; Chang et al., 2020). In addition, the overall climate change impact of these additives in composting has yet to be determined. It may be that the mining, manufacture, and transportation of the additives adds more C to the atmosphere than it saves during composting.

9.4 Capture and containment

Another way to reduce the loss of volatile gases is to delay their escape from the composting pile. In doing so, those gases remain longer within the composting environment, where they can be assimilated by microorganisms. Various types of permeable covers, including biocovers and semipermeable fabrics, impede and then capture some portion of the outward-bound NH_3, N_2O, and CH_4 molecules. Condensation and adsorption are prominent among the physical/chemical mechanisms that capture the volatile molecules. Research studies generally support the use of covers as a means to reduce losses of NH_3, N_2O, CH_4, and other volatile compounds.

Biocovers are organic blankets, typically 15−30 cm (6−12 in.) placed on the surface of windrows and piles. They are made of unscreened finished compost, screen overs with entrained compost particles or wood chips with fines. Biocovers have been shown to be effective at capturing many volatile organic compounds (Büyüksönmez and Evans, 2007). The research regarding NH_3, N_2O, and CH_4 is less clear. In the laboratory, Büyüksönmez et al. (2012) found that a biocover lowered NH_3 emission but in field studies N_2O or CH_4 emissions increased when biocovers were used on passively aerated windrows (Büyüksönmez, 2011). The latter case, the biocovers likely reduces passive airflow and O_2 concentrations. Luo et al. (2014) studied the effect of both a continuous biocover and a biocover that was mixed into the feedstocks after six days. In both cases, the biocover decreased N_2O and CH_4 emissions, and also reduced NH_3 emissions during the first week of composting. However, the biocover that remained in place lead to higher cumulative NH_3 emissions in the end, apparently due to the accumulation and eventual release of NH_3.

The larger picture suggests that effectively managed biocovers can be effective at capturing gaseous products including NH_3. They appear to be most advantageous during the first week or two of composting. Afterward, the biocover material should be turned into a pile or windrow, which has additional emission-reducing benefits. The biocover should be moderately coarse in texture to minimize interference with passive airflow. Compost that is rich in N is a poor choice because its N compounds may mineralize and/or denitrify. Also, a biocover should be kept moist so that the resident microorganisms can assimilate the captured gases.

Fabric covers also appear to have a positive effect on reducing emissions (Fig. 11.17). Both microporous (ePTFE) covers and macroporous covers (fleece) interfere with the movement of water vapor (Chapter 5). Much of the water vapor

(A)

(B)

FIGURE 11.17

Research has shown that microporous (A) and macroporous (B) fabric covers can conserve nitrogen and/or reduce emissions of VOCs and greenhouse gas.

Source: (A) W.L. Gore & Assoc. (B) CV Compost.

condenses on the underside of the covers and drips back into the pile or windrow. The returning water carries some NH_3, N_2O, and CH_4, which are each water-soluble. Both types of covers exclude precipitation, prevent runoff, and greatly reduce the associated nutrient losses.

Paré et al. (2000) studied the effects of macroporous covers on composting mixtures of cauliflower, broccoli, sawdust, and wheat straw. In two trials, the N-content of compost produced in covered piles was approximately 40% higher than that of the compost in uncovered piles. The authors attributed the difference to rainfall and leaching of nutrients in the uncovered piles.

Microporous covers also serve as a barrier to diffusion and convection of specific volatile compounds. For example, Sun et al. (2018) found that the microporous covers greatly reduced the migration of NH_3, N_2O, and CH_4 across the membrane. Other research projects have consistently reported that microporous covers reduced emissions of NH_3 and GHGs (Ma et al. 2018, 2020; Al-Alawi et al., 2019).

References

Cited references

Adler, P., Sikora, L., 2005. Mesophilic composting of Arctic char manure. Compost Sci. Util. 13, 34−42.

Al-Alawi, M., Szegi, T., Simon, B., Gulyás, M., 2019. Investigate biotransformation of green waste during composting by aerated static windrow with GORE(R) cover membrane technology. Jordan. J. Eng. Chem. Ind. 2, 32−40. https://doi.org/10.48103/jjeci252019.

Awasthi, M.K., Wang, Q., Huang, H., Ren, X., Lahori, A.H., Mahar, A., Ali, A., Shen, F., Li, R., Zhang, Z., 2016. Influence of zeolite and lime as additives on greenhouse gas emissions and maturity evolution during sewage sludge composting. Bioresour. Technol. 216, 172−181.

Bacharach, 2021. Combustion Analysis. Bacharach Inc. https://www.mybacharach.com/product-category/combustion-analysis/.

Bruckner, M.Z., 2021. Water and Soil Characterization - pH and Electrical Conductivity. Carlton College Science Education Resource Center (SERC). https://serc.carleton.edu/microbelife/research_methods/environ_sampling/pH_EC.html.

Brown, S., Cotton, M., Messner, S., Berry, F., Norem, D., 2009. Methane Avoidance From Composting. Climate Action Reserve. http://faculty.washington.edu/slb/docs/CCAR_Composting_issue_paper.pdf.

Büyüksönmez, F., Evans, J., 2007. Biogenic emissions from green waste and comparison to the emissions resulting from composting Part II: volatile organic compounds (VOCs). Compost Sci. Util. 15, 191−199.

Büyüksönmez, F., 2011. Comparison of Mitigation Measures for Reduction of Emissions Resulting from Greenwaste Composting. San Diego State University. https://sdsu.academia.edu/Departments/Department_of_Civil_Construction_and_Environmental_Engineering/Documents.

Büyüksönmez, F., Rynk, R., Yucel, A., Cotton, M., 2012. Mitigation of odor causing emissions—bench-scale investigation. J. Air Waste Manag. Assoc. 62 (12), 1423−1430. https://doi.org/10.1080/10962247.2012.716808.

Cao, Y., Wang, X., Bai, Z., Chadwick, D., Misselbrook, T., Sommer, S., Qin, W., Ma, L., 2019. Mitigation of ammonia, nitrous oxide and methane emissions during solid waste composting with different additives: a meta-analysis. J. Clean. Prod. https://doi.org/10.1016/j.jclepro.2019.06.288.

Chang, R., Li, Y., Chen, Q., Gong, X., Qi, Z., 2020. Effects of Carbon-Based Additive and Ventilation Rate on Nitrogen Loss and Microbial Community during Chicken Manure Composting. BioRxiv.

Chowdhury, M.A., de Neergaard, A., Jensen, L.S., 2014. Composting of solids separated from anaerobically digested animal manure: effect of different bulking agents and mixing ratios on emissions of greenhouse gases and ammonia. Biosyst. Eng. 124, 63–77. https://doi.org/10.1016/j.biosystemseng.2014.06.003.

Coker, C., 2007. Compost data tracking and analysis. BioCycle 48 (7), 26.

Cook, K.L., Ritchey, E.L., Loughrin, J.H., Haley, M., Sistani, K.R., Bolster, C.H., 2015. Effect of Turning Frequency and Season on Composting Materials from Swine High-Rise Facilities. Publications from USDA-ARS/UNL Faculty, p. 1498. https://digitalcommons.unl.edu/usdaarsfacpub/1498.

Dahlquist, R., Prather, T., Stapleton, J.J., 2007. Time and temperature requirements for weed seed thermal death. Weed Sci. 55 (6), 619–625.

Duan, Y., Awasthi, S., Liu, T., Zhang, Z., Awasthi, M., 2019. Evaluation of integrated biochar with bacterial consortium on gaseous emissions mitigation and nutrients sequestration during pig manure composting. Bioresour. Technol. 291, 121880. https://doi.org/10.1016/j.biortech.2019.121880.

Edem, K., Louis, J., Baba, G., Ludington, G., Moursalou, K., Tchegueni, S., Matejka, G., 2012. Urban waste management: composting control by oxygen content measurement. Int. J. Emerg. Trends Eng. Dev. 2, 102–113.

Eklind, Y., Sundberg, C., Smårs, S., Steger, K., Sundh, I., Kirchmann, H., Jönsson, H., 2007. Carbon turnover and ammonia emissions during composting of biowaste at different temperatures. J. Environ. Qual. 36 (5), 1512–1520. https://doi.org/10.2134/jeq2006.0253. PMID: 17766831.

Ermolaev, E., Sundberg, C., Pell, M., Smårs, S., Jönsson, H., 2019. Effects of moisture on emissions of methane, nitrous oxide and carbon dioxide from food and garden waste composting. J. Clean. Prod. 240, 118165. https://doi.org/10.1016/j.jclepro.2019.118165.

Godlewska, P., Schmidt, H.P., Ok, Y., Oleszczuk, P., 2017. Biochar for composting improvement and contaminants reduction. A review. Bioresour. Technol. 246, 193–202.

Gore, 2012. The Principle of Organic Waste Treatment with GORE® Cover. W. L. Gore & Associates GmbH.

Grainger, 2021. Grainger Catalog. W.W. Grainger, Inc. https://www.grainger.com/content/general-catalog.

Guo, X., Liu, H., Zhang, J., 2019. The role of biochar in organic waste composting and soil improvement: a review. Waste Manag. 102, 884–899.

He, Z., Li, Q., Zeng, X., Tian, K., Kong, X., Tian, X., 2020. Impacts of peat on nitrogen conservation and fungal community composition dynamics during food waste composting. Appl. Biol. Chem. 63, 1–12.

Hwang, H.Y., Kim, S.H., Shim, J., Park, S., 2020. Composting process and gas emissions during food waste composting under the effect of different additives. Sustainability 12, 7811.

Impraim, P., Weatherley, A., Coates, T., Chen, D., Suter, H., 2020. Lignite improved the quality of composted manure and mitigated emissions of ammonia and greenhouse gases during forced aeration composting. Sustainability 12, 10528. https://doi.org/10.3390/su122410528.

Irvine, R., 2002. Temperature and Physical Modelling Studies of Open Windrow Composting (Doctoral thesis). University of Abertay University, Dundee.

Koster, 2021. Our Story. Koster Moisture Tester, Inc. https://buykoster.com/our-story.

Li, Y., Li, W., Wu, C., Wang, K., 2013. New insights into the interactions between carbon dioxide and ammonia emissions during sewage sludge composting. Bioresour. Technol. 136, 385–393.

Lukyanova, A., 2012. Spatial Modeling of the Composting Process (Master's thesis). University of Alberta.

Luo, W.H., Yuan, J., Luo, Y.M., Li, G.X., Nghiem, L.D., Price, W.E., 2014. Effects of mixing and covering with mature compost on gaseous emissions during composting. Chemosphere 117, 14–19. https://doi.org/10.1016/j.chemosphere.2014.05.043. Epub 2014 Jun 7. PMID: 25433989.

Ma, S., Sun, X., Fang, C., He, X., Han, L., Huang, G., 2018. Exploring the mechanisms of decreased methane during pig manure and wheat straw aerobic composting covered with a semi-permeable membrane. Waste Manag. 78, 393–400. https://doi.org/10.1016/j.wasman.2018.06.005. Epub 2018 Jun 23. PMID: 32559926.

Ma, S., Xiong, J., Cui, R., Sun, X., Han, L., Xu, Y., Kan, Z., Gong, X., Huang, G., 2020. Effects of intermittent aeration on greenhouse gas emissions and bacterial community succession during large-scale membrane-covered aerobic composting. J. Clean. Prod. 266, 121551. https://doi.org/10.1016/j.jclepro.2020.121551.

Maeda, K., Toyoda, S., Shimojima, R., Osada, T., Hanajima, D., Morioka, R., Yoshida, N., 2010. Source of nitrous oxide emissions during the cow manure composting process as revealed by isotopomer analysis of and amoA abundance in betaproteobacterial ammonia-oxidizing bacteria. Appl. Environ. Microbiol. 76 (5), 1555–1562. https://doi.org/10.1128/AEM.01394-09.

McSweeney, J., 2019. Community-Scale Composting Systems. Chelsea Green Publishing, White River Junction, VT.

O'Neill, T., 2021. Best Practices in Managing Compost Facility Design and Operations Are the Keys to Success. Waste Advantage Magazine. https://wasteadvantagemag.com/best-practices-in-managing-compost-facility-design-and-operations-are-the-keys-to-success/.

Osada, T., Kuroda, K., Yonaga, M., 2000. Determination of nitrous oxide, methane, and ammonia emissions from a swine waste composting process. J. Mater. Cycles Waste Manag. 2, 51–56.

Oshins, C., 2021. COTC Compost Operator Training Notes. Compost Research and Education Foundation, Raleigh, VA.

Pacific Biochar, 2021. Biochar in Composting. Pacific Biochar Benefit Corporation. https://pacificbiochar.com/how-to-work-with-biochar/in-composting/.

Paré, M., Paulitz, T., Stewart, K.A., 2000. Composting of crucifer wastes using geotextile covers. Compost Sci. Util. 8, 36–45.

Parkinson, R., Gibbs, P., Burchett, S., Misselbrook, T., 2004. Effect of turning regime and seasonal weather conditions on nitrogen and phosphorus losses during aerobic composting of cattle manure. Bioresour. Technol. 91 (2), 171–178.

Paul, J., 2013a. Controlling aeration of a compost pile by oxygen or by temperature. Transform Compost Syst. http://www.transformcompostsystems.com/blog/2012/12/05/controlling-aeration-of-a-compost-pile-by-oxygen-or-by-temperature/.

Paul, J., 2013b. Nitrous oxide emission during composting. Transform Compost Syst. http://www.transformcompostsystems.com/blog/2013/06/14/nitrous-oxide-emission-during-composting/.

Rynk, R., 2016. Unpublished pile monitoring data.

Sanchez-Monedero, M.A., Cayuela, M.L., Roig, A., Jindo, K., Mondini, C., Bolan, N., 2018. Role of biochar as an additive in organic waste composting. Bioresour. Technol. 247, 1155−1164. https://doi.org/10.1016/j.biortech.2017.09.193. Epub 2017 Sep 30. PMID: 29054556.

Shen, Y., Ren, L., Li, G., Chen, T., Guo, R., 2011. Influence of aeration on CH_4, N_2O and NH_3 emissions during aerobic composting of a chicken manure and high C/N waste mixture. Waste Manag. 31 (1), 33−38.

Soudejani, H.T., Kazemian, H., Inglezakis, V.J., Zorpas, A., 2019. Application of zeolites in organic waste composting: a review. Biocatal. Agric. Biotechnol. 22, 101396.

Stegenta, S., Sobieraj, K., Pilarski, G., Koziel, J., Bialowiec, A., 2019. Analysis of the spatial and temporal distribution of process gases within municipal biowaste compost. Sustainability 11, 2340.

Sun, X., Ma, S., Han, L., Li, R., Schlick, U., Chen, P., Huang, G., 2018. The effect of a semipermeable membrane-covered composting system on greenhouse gas and ammonia emissions in the Tibetan Plateau. J. Clean. Prod. 204. https://doi.org/10.1016/j.jclepro.2018.09.061.

Sundberg, C., Jönsson, H., 2008. Higher pH and faster decomposition in biowaste composting by increased aeration. Waste Manag. ISSN: 0956-053X 28 (3), 518−526. https://doi.org/10.1016/j.wasman.2007.01.011.

Szántó, G.L., Hamelers, H.V., Rulkens, W., Veeken, A.H., 2007. NH_3, N_2O and CH_4 emissions during passively aerated composting of straw-rich pig manure. Bioresour. Technol. 98 (14), 2659−2670.

Szántó, G., 2009. NH_3 Dynamics in Composting − Assessment of the Integration of Composting in Manure Management Chains (Ph.D. thesis). Wageningen University, Wageningen, the Netherlands.

Tiquia, S., Richard, T., Honeyman, M., 2002. Carbon, nutrient, and mass loss during composting. Nutrient Cycl. Agroecosyst. 62, 15−24. https://doi.org/10.1023/A:1015137922816.

Tu, Z., Ren, X., Zhao, J., Awasthi, S.K., Wang, Q., Awasthi, M.K., Zhang, Z., Li, R., et al., 2019. Synergistic effects of biochar/microbial inoculation on the enhancement of pig manure composting. Biochar 1, 127−137. https://doi.org/10.1007/s42773-019-00003-8.

USCC, 2012. Test Methods for the Examination of Compost and Composting (TMECC). U.S. Composting Council. https://www.compostingcouncil.org/page/tmecc.

Veeken, A.H.M., Hanajima, D., Hamelers, H.V.M., 2000. Rates of ammonia assimilation and ammonification during composting: development and methodology. In: Hartmans, S., Lens, P. (Eds.), Proceedings of the 4th International Symposium on Environmental Biotechnology, Noordwijkhout, The Netherlands.

Wang, W., Wang, X., Liu, J., Ishii, M., Igarashi, Y., Cui, Z., 2013. Effect of oxygen concentration on the composting process and maturity. Compost Sci. Util. 15, 184−190. https://doi.org/10.1080/1065657X.2007.10702331.

Wang, S., Zeng, Y., 2018. Ammonia emission mitigation in food waste composting: a review. Bioresour. Technol. 248 Pt A, 13−19.

Whatcom County. (Date unidentified). Composting fundamentals − Reclamation of nitrogen and other nutrients, Whatcom County Extension, Washington State University. http://whatcom.wsu.edu/ag/compost/fundamentals/consideration_reclamation.htm.

Yang, X., Liu, E., Zhu, X., Wang, H., Liu, H., Liu, X., Dong, W., 2019. Impact of composting methods on nitrogen retention and losses during dairy manure composting. Int. J. Environ. Res. Publ. Health 16.

Yeh, C.K., Lin, C., Shen, H.C., Cheruiyot, N., Camarillo, M.E., Wang, C.L., 2020. Optimizing food waste composting parameters and evaluating heat generation. Appl. Sci. 10, 2284.

Consulted and suggested references

Al-Jabi, L.F., Halalsheh, M.M., Badarneh, D.M., 2008. Conservation of ammonia during food waste composting. Environ. Technol. 29 (10), 1067–1073. https://doi.org/10.1080/09593330802175872. PMID: 18942574.

Awasthi, M.K., Duan, Y., Awasthi, S.K., Liu, T., Zhang, Z., 2020. Influence of bamboo biochar on mitigating greenhouse gas emissions and nitrogen loss during poultry manure composting. Bioresour. Technol. 303, 122952. https://doi.org/10.1016/j.biortech.2020.122952.

Bakx, W, Date Not Identified. USCC Composter Training Program, U.S. Composting Council.

Brown, S., Kruger, C., Subler, S., 2008. Greenhouse gas balance for composting operations. J. Environ. Qual. 37 (4), 1396–1410.

Brown, S., Beecher, N., Carpenter, A., 2010. Environ. Sci. Technol. 44 (24), 9509–9515. https://doi.org/10.1021/es101210k.

Brown, S., 2016. Greenhouse gas accounting for landfill diversion of food scraps and yard waste. Compost Sci. Util. 24 (1), 11–19. https://doi.org/10.1080/1065657X.2015.1026005.

Buggeln, R., Rynk, R., 2002. Self-heating in yard trimmings: conditions leading to spontaneous combustion. Compost Sci. Util. 10 (2), 162–182. https://doi.org/10.1080/1065657X.2002.10702076.

Büyüksönmez, F., 2012. Full-scale VOC emissions from green and food waste windrow composting. Compost Sci. Util. 20, 57–62.

CCME, 2005. Guidelines for Compost Quality. Canadian Council of Ministers of the Environment, Winnipeg, Manitoba.

Chan, M.T., Selvam, A., Wong, J.W., 2016. Reducing nitrogen loss and salinity during "struvite" food waste composting by zeolite amendment. Bioresour. Technol. 200, 838–844. https://doi.org/10.1016/j.biortech.2015.10.093.

Chou, C.-H., Büyüksönmez, F., 2006. Biogenic emissions from green waste and comparison to the emissions resulting from composting (Part 1: ammonia). Compost Sci. Util. 14 (1), 16–22. https://doi.org/10.1080/1065657X.2006.10702258.

Fraser, B.S., Lau, A.K., 2000. The effects of process control strategies on composting rate and odor emission. Compost Sci. Util. 8 (4), 274–292. https://doi.org/10.1080/1065657X.2000.10702001.

Fukumoto, Y., Osada, T., Hanajima, D., Haga, K., 2003. Patterns and quantities of NH_3, N_2O and CH_4 emissions during swine manure composting without forced aeration——effect of compost pile scale. Bioresour. Technol. 89 (2), 109–114. https://doi.org/10.1016/S0960-8524(03)00060-9.

Fukumoto, Y., Suzuki, K., Waki, M., Yasuda, T., 2015. Mitigation option of greenhouse gas emissions from livestock manure composting. Jpn. Agric. Res. Q. 49, 307–312.

Hemm, G., 2007. The composting conundrum: just how aerobic Is"Aerobic"? WasteMINZ conference 2007. WasteMINZ. https://www.wasteminz.org.nz/pubs/the-composting-conundrum-just-how-aerobic-is-aerobic/.

Hao, X., Chang, C., Larney, F., Travis, G., 2001. Greenhouse gas emissions during cattle feedlot manure composting. J. Environ. Qual. 30 2, 376–386.

Highfields Center for Composting, 2015. Compost Site Management Monitoring Piles: Why and How. Vermont Agency of Natural Resources, Department of Environmental Conservation, Solid Waste Program.

Janczak, D., Malinska, K., Czekała, W., Cáceres, R., Lewicki, A., Dach, J., 2017. Biochar to reduce ammonia emissions in gaseous and liquid phase during composting of poultry manure with wheat straw. Waste Manag. 66, 36–45.

Jianfei, Z., Shen, X., Sun, X., Liu, N., Han, L., Huang, G., 2018. Spatial and temporal distribution of pore gas concentrations during mainstream large-scale trough composting in China. Waste Manag. ISSN: 0956-053X 75, 297–304. https://doi.org/10.1016/j.wasman.2018.01.044.

Khalil, A.I., Hassouna, M.S., Shaheen, M.M., Abou Bakr, M.A., 2013. Evaluation of the composting process through the changes in physical, chemical, microbial and enzymatic parameters. Asian J. Microbiol. Biotechnol. Environ. Sci. 15 (1), 25–42.

Kristanto, G., Raissa, S., Novita, E., 2015. Effects of compost thickness and compaction on methane emissions in simulated landfills. Proc. Eng. 125, 173–178. https://doi.org/10.1016/j.proeng.2015.11.025.

Linzer, R., Mostbauer, P., 2005. Composting and its impact on climate change with regard to process engineering and compost application-A case study in Vienna. In: Sardinia 2005 Tenth International Waste Management and Landfill Symposium, pp. 59–60.

López-Cano, I., Roig, A., Cayuela, M.L., Alburquerque, J.A., Sánchez-Monedero, M.A., 2016. Biochar improves N cycling during composting of olive mill wastes and sheep manure. Waste Manag. 49, 553–559.

Maeda, K., Morioka, R., Osada, T., 2009. Effect of covering composting piles with mature compost on ammonia emission and microbial community structure of composting process. J. Environ. Qual. 38 (2), 598–606. https://doi.org/10.2134/jeq2008.0083. PMID: 19202030.

Malinowski, M., Wolny-Koładka, K., Vaverková, M., 2019. Effect of biochar addition on the OFMSW composting process under real conditions. Waste Manag. 84, 364–372.

McCartney, D., Eftoda, G., 2005. Windrow composting of municipal biosolids in a cold climate. J. Environ. Eng. Sci. 4 (5), 341–352.

NRCS, 2004. Use of Reaction (pH) in Soil taxonomy. United States Department of Agriculture, Natural Resources Conservation Service. https://www.nrcs.usda.gov/wps/portal/nrcs/detail/soils/ref/?cid=nrcs142p2_053575.

Paul, J., Geesing, D., Date not identified. Aerated Windrow Composting (Un-) Covered. Presentation Slides. Transform Compost Systems Ltd.

Phong, N.J., Cuhls, C., 2016. The effect of turning frequency on methane generation during composting of anaerobic digestion material. J. Viet. Env. 8 (1), 50–55. https://doi.org/10.13141/jve.vol8.no1.pp50-55.

Puyuelo, B., Gea, T., Sánchez, A., 2014. GHG emissions during the high-rate production of compost using standard and advanced aeration strategies. Chemosphere 109, 64–70. https://doi.org/10.1016/j.chemosphere.2014.02.060.

Siebert, S., Auweele, W.V., 2018. European Quality Assurance Scheme for Compost and Digestate. European Compost Network. https://www.compostnetwork.info/ecn-qas/ecn-qas-manual/.

Sullivan, D., Bary, A., Miller, R., Brewer, L., 2018. Interpreting Compost Analyses. Oregon State University Extension Service. https://catalog.extension.oregonstate.edu/em9217.

Sunar, N.M., Stentiford, E.I., Stewart, D.I., Fletcher, L.A., 2009. The process and pathogen behaviour in composting: a review. In: Proceeding UMT-MSD 2009 Post Graduate Seminar 2009. Universiti Malaysia Terengganu, Malaysian Student Department UK & Institute for Transport Studies University of Leeds, ISBN 978-967-5366-04-8, pp. 78–87.

USCC, 2021a. Seal of Testing Assurance (STA). U.S. Composting Council. https://www.compostingcouncil.org/general/custom.asp?page=SealofTestingAssuranceSTA.

USCC, 2021b. Combatting Climate Change. U.S. Composting Council. https://www.compostingcouncil.org/general/custom.asp?page=ClimateChangeBenefits.

Vergara, S., Silver, W., 2019. Greenhouse gas emissions from windrow composting of organic wastes: patterns and emissions factors. Environ. Res. Lett. 14, 124027.

Vitinaqailevu, R., Rajashekhar Rao, B.K., 2019. The role of chemical amendments on modulating ammonia loss and quality parameters of co-composts from waste cocoa pods. Int. J. Recycl. Org. Waste Agric. 8, 153–160. https://doi.org/10.1007/s40093-019-0285-3.

Wang, M., Awasthi, M.K., Wang, Q., Chen, H., Ren, X., Zhao, J., Li, R., Zhang, Z., 2017. Comparison of additives amendment for mitigation of greenhouse gases and ammonia emission during sewage sludge co-composting based on correlation analysis. Bioresour. Technol. 243, 520–527. https://doi.org/10.1016/j.biortech.2017.06.158. Epub 2017 Jun 30. PMID: 28697454.

Wisconsin Department of Natural Resources, 2012. Temperature Monitoring at Licensed Compost Facilities. https://dnr.wi.gov/files/PDF/pubs/wa/WA1585.pdf.

Woodford, C., 2021. How Do pH Meters Work? ExplainThatStuff. https://www.explainthatstuff.com/how-ph-meters-work.html.

Zambra, C., Moraga, N., Escudey, M., 2011. Heat and mass transfer in unsaturated porous media: moisture effects in compost piles self-heating. Int. J. Heat Mass Tran. 54, 2801–2810.

Zheng, G., Wang, Y., Wang, X., Yang, J., Chen, T., 2018. Oxygen monitoring equipment for sewage-sludge composting and its application to aeration optimization. Sensors 18, 4017. https://doi.org/10.3390/s18114017.

Odor management and community relations

12

Authors: Tim O'Neill[1], Robert Rynk[2]

[1]*Engineered Compost Systems, Seattle, WA, United States;* [2]*SUNY Cobleskill, Cobleskill, NY,
United States*

Contributors: Ginny Black[3], Anna F. Bokowa[4]

[3]*Compost Research and Education Foundation (CREF), Raleigh, NC, United States;*
[4]*Environmental Odour Consulting Corporation, Oakville, ON, Canada*

1. Introduction

Odor is the "soft underbelly" of the composting industry.[1] The perception of bothersome odors to the surrounding community is the principal reason composting facilities prematurely close or are required to undergo costly modifications (Coker, 2012a). It is why neighbors with metaphorical pitch forks and flaming torches pack into public hearings to oppose even the slightest change to a facility's operational permit. More than any other factor, the perception that composting facilities are inevitable sources annoying odors stands in the way of their acceptability and a broader realization of composting's environmental benefits.

Fortunately, there is a well-established body of science-based knowledge that provides a basis for preventing nuisance odor events, and simultaneously limiting the emissions of all volatile organic compounds (VOCs), as explained in Box 12.1. Unfortunately, this knowledge has not been uniformly understood nor adopted. It is the goal of this chapter to foster a better understanding of how to manage composting odors and so increase public acceptance.

Successful odor management requires both technical competency and smart public relations. Even well designed and operated composting facilities will likely have the occasional odor event. As Tim Haug has written within his sagacious theorems about composting odors: "You can stop all of the odor some of the time, but you can't stop all of the odor all of the time." (Haug, 1993, 2020). It is important to remain on good terms with the people who might be affected when an odor situation occurs—neighbors, community officials, and regulators.

This chapter covers the successive elements that might cause a composting odor to become an odor nuisance, shown diagrammatically in Fig. 12.1 and methods to avoid this outcome. These elements and methods include:

---

[1] With respect and apologies to Winston Churchill.

The Composting Handbook. https://doi.org/10.1016/B978-0-323-85602-7.00019-4

- How odor is generated in the composting process and the role process conditions play in determining both their strength and their nature?
- How to evaluate odor risk and select the process sequence and best management practices (BMPs) to appropriately limit the generation and the emission of odors?
- How odor emissions can be reduced through capture and control practices?
- The transmission and dispersion of emitted odors across a landscape from the point of generation to the receptor.
- The odor receptor—human factors of odor perception, permitting, and politics.

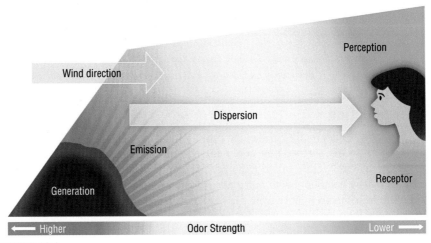

FIGURE 12.1

Sequence from odor generation to perception by a receptor, possibly leading to an odor complaint.

The final section of the chapter is a synopsis of odor characterization and measurement, important background information for understanding how odors are viewed by scientists, engineers, regulators, and the public. Odor science is, however, a large topic with breadth and depth that surpasses this chapter. A great deal of knowledge can be found in professional and scientific publications including Balch et al. (2019); Pinasseau et al. (2018); Ma et al. (2013); Nordic Council of Ministers publication (2009); CalRecycle (2007), and a series of articles in BioCycle by Coker (2012a—e, 2013a—c). Good information can also be gleaned from publications and web pages that address odor management for livestock farms, wastewater treatment plants (WWTPs), and odor assessment generally (see References section).

Box 12.1 The connection between VOCs and odor

VOCs are carbon-based chemicals that easily evaporate into the air at room temperature. Along with carbon, they contain elements such as hydrogen, oxygen, fluorine, chlorine, bromine, sulfur, or nitrogen. Research has demonstrated a strong relationship between odor and VOC generation in the composting process (Krzymien et al., 1999). This makes sense given that most of the odorous chemicals emitted in composting exhaust air are VOCs. While there have been many studies of the chemical composition of odors, the complexities of the human olfactory and the mix of odorous chemicals have prevented the development of a simple quantitative relationship between VOC concentrations and measured odor concentrations. The research has shown that increasing/decreasing odor concentrations correlates to increasing/decreasing VOC concentrations; and that changes in the rates of VOC generation and emission correlate fairly closely to changes in the rates of odor generation and emission (Sundberg and Jönsson, 2008). Roughly speaking, if a process improvement produces a 50% reduction in VOCs, it also reduces odors proportionally.

VOC's are becoming an increasing point of concern for air quality regulators who permit larger composting facilities. In the United States, regulators are required to implement the Clean Air Act which regulates the emissions of VOCs with the goal of protecting human health. Despite the fact that total VOC concentrations in compost exhaust air are very low, typically in the low parts per million range, the annual total of VOCs emitted from a large facility can trigger a long and complicated permitting process and add recurrent fees and require expensive source testing.

While these permit conditions may seem burdensome, the good news is that the same process conditions that minimize the generation of VOCs also minimize odor generation. In other words, preferred feedstock mixes, temperatures, and oxygen levels both help avoid odor complaints and produce very low VOC emissions.

VOC emissions are expressed in mass of VOCs per wet mass of composting feedstocks. Typical units include kg of VOC per metric ton (t) of feedstocks or, in the United States, lbs of VOCs per short ton (ton) of feedstocks. This measure is called an emission factor (EF). Multiplying the permitted annual tons of feedstock by the assumed, or measured, EF yields the annual mass of VOCs presumed to be emitted. It forms the basis of the air quality permit. Commercial source testing at full scale composting facilities have measured EFs that range from 0.04 to over 15 kg/t (0.08 to over 30 lbs/ton). This extreme range represents the extreme range of compost process conditions that exist and highlights the importance of managing these conditions.

To develop a default EF for regulatory purposes, the California Air Resources Board (ARB) chose a data set of VOC EFs measured at passively aerated windrow composting facilities. These data range from 0.43 to 5.0 kg/t (0.85 to 10.03 lb/ton). The ARB chose to average all the data, including several very high outliers, to develop a default value of 1.79 kg/t or 3.58 lb/ton (ARB, 2015). Source tests of "well-designed" composting processes have measured EFs in the range of 0.05–2.5 kg/t (0.1–0.5 lbs/ton).[2] These facilities very likely emit very little odor.

[2] In this case, the "well-designed" process are characterized by forced aeration systems that maintain oxygen saturation in the film layer above 3 ppm, avoid sustained temperatures above 65°C (150°F), and start with feedstock mixes with high FAS, low bulk density, and balanced C:N rations.

2. Odor regulations—"All Over the Map"

Internationally, regulations that govern nuisance odors are as diverse as the social and economic conditions that create the need for odor regulations. In largely rural societies that rely heavily on basic industries (e.g., agriculture, fishing), often there are no national regulations at all, although some cities may impose their own rules. In some industrial countries, detailed regulations precisely define what constitute an odor nuisance and how to measure it. In between, many political jurisdictions have regulations that specify little more than minimum separation distances.

Most odor regulations are not necessarily specific to composting. They govern a wide array of potential odor sources including manufacturing, food processing, food service, fisheries, and agricultural. However, even in many industrial countries, selected sectors of the economy are exempt from nuisance regulations, most notably agriculture. Nevertheless, regulations specifically for composting facilities do exist, established by municipalities, states, provinces, and/or countries. State and provincial governments usually are primarily responsible for odor regulations, although local or regional agents often have the task of managing compliance and complaints. In the case of Europe, regulations have a multinational influence through the European Union (EU) and its "Industrial Emissions Directive" (IED).

The basic approaches to nuisance odor regulations are summarized below. This summary is based on a recent review of regulations by Bokowa et al. (2021). This review article is an excellent source for more details about how odors are regulated in particular countries. Many of the odor science terms introduced below are explained in Section 13.

1. *Defined nuisance criteria:* The prevailing approach in much of Europe, Australia, New Zealand, Canada, Japan, and a few other countries is to establish "odor impact criteria" (OIC). OICs are specific criteria that define a nuisance odor (Box 12.2). OICs are most often expressed as maximum odor concentration, measured or predicted either at the facility boundaries or at one or more receptor locations. OIC odor concentrations are usually expressed in European odor units per cubic meter (ou_e/m^3). At a receptor location, a typical OIC ranges from 1 to 10 ou_e/m^3, depending on the situation. In some cases, an OIC is defined as the collective amount or percentage of time that an odor is detected or recognized. In a given regulation, the OIC typically varies with conditions, such as the nature of the receptors (e.g., population density) or the odor source and its typical hedonic tone or intensity. Once an OIC is established, the remaining challenge is how to measure compliance, either for routine investigation or in response to an odor complaint. Various countries use various techniques including sampling and laboratory olfactometry, field olfactometry, field investigation, and atmospheric dispersion models, alone or in combination (see Section 13).

2. *Specific odorant concentrations:* A few countries base odor performance on concentrations of specific odorants, like hydrogen sulfide or acetaldehyde, measured at either the source or the facility property line. This approach has mostly fallen out of favor because odorants can be malodors at concentrations below conventional detection limits. Also, this approach does not consider the character of the odor, and the complexity of odors when several odorants coexist.

3. *Permits/Approvals and setbacks:* In many places, including most states or municipalities in the United States, composting odors are addressed through permitting, which is termed "approval" or "registration" in some cases. When a

facility is granted a permit, certain obligatory conditions come with it. A permit may be specific to a particular facility, or it may be general, applying to all facilities in a common situation (e.g., moderate scale, composting only yard trimmings). "Setbacks" are almost always among these conditions. Setbacks are separation distances between the facility and surrounding land features (Chapter 10). In addition, permit conditions typically dictate facility capacity, either the maximum amount of feedstock a facility can accept or the maximum amount of material it can keep on-site. In some cases, permits specify performance criteria, like a maximum odor concentration at some location on or near the property. Permit conditions often require facilities to submit an odor prevention, management, and/or odor mitigation plan, which spells out how the facility expects to minimize odor emissions and respond to odor complaints if they occur. Monitoring for compliance is usually a matter of periodic (e.g., annual) inspections by local or state compliance agents. If odor complaints are received, the local agents tend to make informal investigations through neighbor visits and on-site inspections, attempting to identify and resolve the problem. In some jurisdictions, a complaint automatically triggers a mandatory inspection and odor measurements by the inspector. Some situations call for more thorough field investigation techniques, including field olfactometry and neighborhood surveys. If the local agency deems that nuisance or other violation has occurred, the composting facility is urged, or required, to change its practices. Offended neighbors can also seek a remedy via lawsuits.

4. *No regulations/complaint response:* In some countries and states, there are no regulations specifically addressing composting odors. In these situations, if an odor complaint arises, local, state, or national authorities respond on a case-by-case basis. They follow similar procedures described in the previous paragraph for investigating and resolving odor problems.

In all cases, governments expect the potential odor generator to operate responsibly. Increasingly operators are required to develop odor prevention and/or management and mitigation plan.

Most operating permits/approvals require odor generators to follow accepted codes of practice, BMPs, or adopt best available technologies (BATs). Accepted practices are often codified in an official document recognized by the state, province, country, or region. The EU's IED is accompanied by a set of BATs for various odor-emitting industries, termed BREFs in the EU. Composting is covered under the BAT for waste treatment (Pinasseau et al., 2018). With BATs as a guiding influence, the tendency, and current trend, is for regulatory bodies to encourage composting facilities to transition to greater levels of technology and containment. Passively aerated windrows are converted to aerated static piles (ASPs) or agitated bays. Outdoor ASPs acquire fabric covers or move inside a building. In-vessel systems are sometimes required from the start when food is composted.

Box 12.2 Quantifying odor—dilutions-to-threshold and odor units

Without a specific odorant to measure, determining the strength or concentration of an odor presents a conundrum. How can the concentration of an odor be quantified if there is no way to assign it a parts per million or grams per liter? The answer that has been established is to rely on the human nose, or more accurately a panel of human noses. The general approach is to dilute a sample of odorous air with fresh, often filtered, air, and ask each member of a panel of human detectors whether or not he/she can detect it or recognize it (Section 13.2.1). A critical point is reached when half the panel can no longer detect it (or recognize it). This point is what has been used to define an odor concentration.

One measurement commonly used to quantify the strength of a general odor is called the *dilutions-to-threshold* (D/T), also called effective dilutions or EDs. The D/T concentration is the number of dilutions required for some proportion of the panel members to no longer detect the odor. Usually, the concentration is set at dilution level at which 50% the panel no longer detect the odor. It is labeled as D/T50 or ED50. A strong odor requires more dilutions to become undetectable; hence strong odors have high D/T50 concentrations.

A more widely used odor concentration parameter is the *odor unit* (ou or ou_e). It is an important parameter that is used in most odor dispersion models and also forms the basis of many odor regulations. Basically, one odor unit is the number of dilutions required for 50% of an odor panel to no longer detect the odorant in question. It is essentially the D/T50 concentration plus 1. An odorant with a concentration of 20 odor units has to be diluted 20 times to escape detection by half the odor panel.

Understanding of odor units is somewhat confused by the fact that there are two separate designations—traditional odor units (ou) and European odor units (ou_e). Although they have the same values, they are distinguished by nuances in their definitions. The definition in the previous paragraph suffices for the traditional ou, which is more likely to be used in the United States. The European ou_e is more precisely defined, and more widely and commonly used. An informal definition[3] is "the number of dilutions with neutral air that are necessary to bring the concentration of the sample to its odor detection threshold (OT), i.e., the threshold at which the odor is perceived by 50% of the examiners" (Bax et al., 2020). In practice, the odor concentration is usually expressed on a per unit volume basis such as ou_e/m^3 or ou/ft^3. By multiplying the concentration (e.g., ou_e/m^3) by airflow rate leaving an odor source (e.g., m^3/s), one can calculate an odor emissions rate (OER) or "odor flux" (ou_e/s).

---

[3] A more technical definition of a European odor unit is "1 ou_e/m^3 is defined as the amount of odorant that, when evaporated into 1 m^3 of gas air at standard conditions, causes a physiological response from a panel (detection threshold) equivalent to that of n-butanol (reference gas) evaporated into 1 m^3 of neutral gas" (Bratolli et al., 2011; after CEN, 2003).

---

3. The nature of composting odors

An odor, or smell, is the sensation generated by the brain in response to certain chemicals breathed in through the nose (Dalton, 2003). Airborne chemicals, "odorants," react with olfactory receptors in the nose and sends signals to the brain, which produces odor sensations, both good and bad.[4]

---

[4] Any excellent graphic of the human olfactory system accompanies an article in BioCycle titled, Odor Monitoring and Detection Tools, https://www.biocycle.net/odor-monitoring-and-detection-tools/ (Coker, 2013c).

Odors occur at composting sites because organic materials decompose, and composting sites accrue large quantities of organic material within a relatively small area. More organic material means more odors. Organic compounds—carbohydrates, proteins, and lipids (fats and oils)—inherently decompose into various intermediate compounds as microorganisms harvest energy and nutrients from the parent materials. Some of these compounds are *odorants*—they are volatile, and they have an odor that at least some people dislike.

Volatility is necessary for a chemical to be odorous. A compound's volatility—its conversion to a vapor and subsequent migration into the air—is what allows it to be sensed by human noses. The tendency of a chemical to volatilize is indicated by its vapor pressure. Chemicals with high vapor pressures volatilize easily. As temperature increases, vapor pressure also increases; hence liquids (and solids) vaporize faster at higher temperatures.

Table 12.1 summarizes the chemical compounds that are most often responsible for odors at composting facilities. This group of "usual suspects" includes ammonia (NH_3), hydrogen sulfide (H_2S), organic sulfides, amines, volatile fatty acids (VFAs), and terpenes." Appendix I describes each of these categories in more detail. In composting situations, no odorant occurs alone. Rather, combinations of odorants emerge that produce a diverse menu of aromas. Each odorant lends its fragrance

Table 12.1 Common odorants released during handling and decomposition of organic materials.

Compound name	Odor descriptors	Chemical formula	Vapor pressure at 20 or 25°C (mmHg)
Volatile fatty acids (VFAs)			
Formic acid	Biting, pungent	$HCOOH$	42.8
Acetic acid	Vinegar-like, pungent	CH_3COOH	15.7
Propionic acid	Rancid, pungent	C_2H_5COOH	3.53
Butyric acid	Rancid butter, body odor	C_3H_7COOH	1.65
Valeric acid	Unpleasant, sweat	C_4H_9COOH	0.1
Isovaleric acid	Rancid cheeses	$(CH_3)_2C_2H_3COOH$	0.554
Caproic acid (hexanoic acid)	Pungent	$C_5H_{11}COOH$	0.2
Volatile nitrogen compounds: ammonia, amines, indoles			
Ammonia	Pungent, sharp, irritating	NH_3	7510
Putrescine	Putrid, nauseating	$NH_2(CH_2)_4NH_2$	4.12
Cadaverine	Putrid, decaying flesh	$NH_2(CH_2)_5NH_2$	1.0

Continued

Table 12.1 Common odorants released during handling and decomposition of organic materials.—*cont'd*

Compound name	Odor descriptors	Chemical formula	Vapor pressure at 20 or 25°C (mmHg)
Methylamine	Putrid, fishy, rotten fish	CH_3NH_2	1810
Dimethylamine	Fishy, rotten fish	$(CH_3)_2NH$	1520
Trimethylamine	Fishy, pungent	$(CH_3)_3N$	2650
Ethylamine	Ammonia-like, irritating	$C_2H_5NH_2$	
Indole	Fecal, nauseating	$C_6H_4(CH_2)_3NH$	0.012
Skatole	Fecal, nauseating	C_9H_9N	0.015

Volatile sulfur compounds: hydrogen sulfide, mercaptans, organic sulfides

Hydrogen sulfide	Rotten egg	H_2S	15,600
Methyl mercaptan	Pungent, rotten cabbage, skunk, garlic	CH_3SH	1510
Ethyl mercaptan	Rotten cabbage, leek-like	C_2H_5SH	529
Carbon disulfide	Disagreeably sweet, rotten pumpkin	CS_2	360
Dimethyl sulfide	Sulfurous, rotten cabbage	$(CH_3)_2S$	502
Dimethyl disulfide	Putrid, sulfurous	$(CH_3)_2S_2$	28.7

Terpenes

α-Pinene	Sharp, turpentine	$C_{10}H_{16}$	4.75
Limonene	Sharp, lemony	$C_{10}H_{16}$	1.98

Ketones and aldehydes

Phenol	Medicinal	C_6H_5OH	0.38
Acetone	Pungent, solvent	CH_3COCH_3	225
Methyl ethyl ketone (butanone)	Sweet, solvent	$CH_3COCH_2 CH_3$	72
Formaldehyde	Acrid, medicinal	H_2CO	3883
Acetaldehyde	Green, sweet, fruity	CH_3CHO	900

Adapted from CalRecycle (2007) and compiled from various sources including Brant and Elliott (2004), Chiumenti et al. (2005), Willmink and Diener (2001), Epstein (1997), and Engineering Toolbox (2006).

to the resulting odor soup, although one particular odorant, or class of odorants, can decide the general character of the smell.

As an organic material decomposes, many volatile chemical compounds are formed, destroyed, and/or emitted due to the innumerable combinations of raw feedstocks and the diverse and ever-changing process conditions. As decomposition proceeds, and organic compounds stabilize, the mix of volatile compounds changes, and so does the characteristic odor. In general, odors are most prevalent in feedstocks that are highly degradable and during the first 10—14 days of active composting.

The odorants that tend to be most offensive develop and accumulate oxygen (O_2) is lacking (i.e., anoxic or anaerobic). Given time and sufficient oxygen, these compounds decompose further into innocuous products. Thus, one key to limiting odors is to promote aerobic conditions wherever organic materials are actively decomposing. However, some odorants can also develop via aerobic decomposition.

Time and oxygen are not always sufficient to decompose intermediate compounds. The temperature, moisture content, pH, and nutrient balance must also be conducive. For example, proteins decompose into nitrogen-carrying compounds known as "amines." Some amines have putrid or fishy odors. If O_2 is present, amines decompose into carbon dioxide (CO_2) and water (H_2O) while the nitrogen (N) is either assimilated by microorganisms or volatilized as gaseous ammonia (NH_3). If there is a sufficient supply of available carbon (C), the N is more likely to be assimilated (an amine molecule, in itself, does not supply enough C).

Generation of odors is not the same as causing an odor issue. First, most odorants are transient. After forming, or being liberated, they decompose, immobilize, change phase, and/or disperse, depending on the environmental conditions. Second, odor emissions are reduced by practices that capture and control the odorants. Even processes with high emission rates (OERs) don't become a nuisance unless tons of material are present to cause it to surpass a site-specific threshold (Box 12.3). It is wasteful to require optimal control measures beyond what is needed to avoid nuisance odors. The key is to understand the constraints of the facility, then design a plan to maintain the conditions that will prevent the perception of annoying odors off-site.[5]

Box 12.3 Measuring and quantifying the odor emissions rate (OER) or "odor flux"

OER, also called "odor flux," is the measurement of the amount of odor emitted per unit of time by an odor source. The odor flux is the key input for dispersion modeling, which is used to predict the odor impact on neighbors. Assessing odor flux requires measuring both odor concentration in, and the mass flow rate of, the air carrying the odor away from the process. It is usually reported in odor units per second (ou/s or ou_e/s). However, it can also be expressed as a concentration of a particular chemical odorant. The most common reason for estimating OERs is to supply data for atmospheric dispersion models. It is also measured in support of regulatory compliance and for conducting experiments (e.g., testing composting recipes for their odor impacts). The job of measuring OERs is usually hired out to an odor measurement professional.

Continued

---

[5] This paragraph also holds true for the emission of excessive VOCs, which have many similarities to odorants in their generation, emissions, containment, dispersal, and control.

Box 12.3 Measuring and quantifying the odor emissions rate (OER) or "odor flux"—cont'd

OER data are *relatively* easy to determine when an odor source is a "point source," that is, it has a distinct point of outflow, such as a building ventilation outlet or an air exhaust pipe of a composting vessel or a negative pressure ASP. A sample is collected at such a point and measured by an odor panel, or other appropriate method, to obtain the odor concentration (C_{od}). The concentration multiplied by the air flow rate (Q_{air}) at the point of sample collection equals the OER, as in Eq. (12.1) (Capelli et al., 2013).

$$OER = C_{od} \times Q_{air} = (ou_e / m^3) \times (m^3 / s) = ou_e/s \qquad (12.1)$$

Many, if not most, odor sources at open air composting facilities are "area sources" (e.g., windrows, curing piles, feedstocks). Odors escape over broad surfaces; there are no concentrated air flows to tap for a sample. With a positive-pressure ASPs, there is usually enough air pressure within the pile to collect samples by placing a hood-like cover over part of the pile. The sample is taken, and the air flow is measured, at the outlet of the hood. Because the hood covers only a portion of the pile, the odor emission calculation must be adjusted to account for the entire area of the pile (Capelli et al., 2013). Since odor flux varies widely across most compost pile surfaces, multiple sample locations are generally required to get representative data. The first step is to calculate the *specific odor emission rate* (SOER) from the odor concentration (C_{od}) and the air flow rate inside the hood (Q_{hood}) divided by the area of the hood (A_{hood}). The OER is the SOER multiplied times the total area of the pile (i.e., emitting surface, A_{em}).

$$SOER = \frac{C_{od} \times Q_{air}}{A_{hood}} = \frac{ou_e/m^3 \times m^3/s}{m^2} = ou_e/m^2 s \qquad (12.2)$$

$$OER = SOER \times A_{em} \qquad (12.3)$$

The situation is more difficult for passively aerated odor sources like windrows or static piles since there the flow rate of air is difficult to measure (Fig. 12.2). Common sampling devices include wind tunnels, hoods, and tent-like structures that use fans to induce an air flow across a portion of the pile surface (Capelli et al., 2013; Balch et al., 2019). The ou is sampled, and the air velocity is measured, at the inlet and outlet ducts. This allows the OER to be calculated from the SOER as described above. Another method for measuring odor flux on these surfaces is to employ a flux chamber that adds an odorless carrier gas at precisely controlled rate to the odor sample. When the odor sample is analyzed at the lab, the concentration of the carrier gas in the sample is measured in addition to the ou. Since the balance of the gas in the sample came from the pile, and both the flow rate of the carrier gas and the sampling time are known, the effective mass flow rate from the pile can be calculated.

(A)  **(B)**

FIGURE 12.2

Two methods of sampling air emissions, either VOCs or OER odor flux from an area source: flux chamber (A) and wind tunnel (B).

Source: ECS.

Box 12.3 Measuring and quantifying the odor emissions rate (OER) or "odor flux"—cont'd

No surface flux measurement devices are useful at measuring the air emissions when a pile is broken down or a windrow is turned. Attempts at estimating these episodic odor fluxes have included measuring average pore space odor units in the pile, estimating the total air volume in the pile, and assuming that all the air space is released upon turning.

A review of emissions measurement methods is presented by Carpenter et al. (2020).

4. The anatomy of an odor problem

A compost odor incident occurs when a source generates sufficiently strong odorants that are emitted, first from the point of generation, and then from the confines of the facility. Then the collective odorants must migrate to the environment beyond, with insufficient dispersal/dilution, to a person who can perceive the odor. The incident becomes a problem if the affected person finds the odor disagreeable and undesirably intense for a sufficiently long time or if the incident occurs repeatedly. Thus, the necessary steps in the anatomy of an odor problem are odor generation, emission, migration/dispersal, perception, and annoyance, as depicted in Fig. 12.1.

There is one more step in this train of events that begs to be added—*resolution*. To avoid major problems, the composter must make a good faith effort to resolve the issue by addressing the offended person's complaint, at the very least. As necessary, the composter should also investigate the cause of the odor problem and attempt to prevent a reoccurrence. Any of the steps in the "odorants" journey can be attacked to resolve the problem, but composters have better options when addressing the odor formation and emissions steps. How the composter reacts to an odor complaint greatly affects the outcome.

Fortunately, there is a long list of tools and proven practices available for composters to minimize the formation of odors, contain or capture the odorous compounds that do form, and control (treat) the captured odors. Over the past 30 years, the composting industry has grown more adept at odor management and control. Facilities can operate successfully in urban and suburban locations with appropriate facility design and best practices. Although rural composting facilities operate in a more forgiving environment, they also must operate sensibly.

5. The nature of the nuisance

When intense odors are detected and experienced either frequently or over a prolonged duration, they create a *nuisance*. Although the legal definition of a nuisance differs among jurisdictions, *generally*, a nuisance is considered a condition that unreasonably interferes with another's enjoyment of life, conduct of business, or use of property (Charles, 2015). Whether or not a particular odor constitutes a nuisance is determined by criteria defined by odor regulations, where such specific regulations exist. Where odor regulations do not define an odor nuisance, the decision is ultimately determined by the legal system. In these cases, which are more common in the United States and the United Kingdom, courts apply the concept of "reasonableness" (Brant and Elliott, 2004). For instance, courts may determine if a composting operation is reasonably well-managed and took reasonable steps to prevent odor impacts. Courts may also question whether the complaining neighbors are being reasonable in their demands. In many places, farms, and on-farm composting, are afforded more protection against nuisance complaints because the nature of farming and its historical connection to the land. However, farms must operate in a reasonable manner to retain that protection.

With the possible exception of enclosed spaces and vessels, odors from composting do not pose a health threat, at least not directly (CalRecycle, 2007; Ward and Wiens, 2018). The chemical substances responsible for odors are not present in high enough concentrations to cause physiological injury. People may legitimately feel symptoms, like headache and sore throat because of odors. However, such symptoms are brought on by psychological stressors, not the odors themselves, or possibly a coincidental factor like bioaerosols (Shiffman and Williams, 2005; Pelosi, 2003).

The chain of events that leads to a nuisance complaint depends on the attributes of both the odorant(s) and the discerning individual (receptor). First, an odorant (or mix of them) must be in a high enough concentration for an individual to detect the resulting odor. Second, to reach the complaint level, the odor also must be sufficiently annoying. How an individual interprets an odor is largely determined by its intensity and quality. Third, the odor must persist, either continually for an uncomfortable duration or repeatedly for short periods. If the odor dissipates relatively quickly and doesn't return, then the odor will likely have little effect. However, if it remains long enough to be bothersome, or reoccurs repeatedly, then odor complaints are likely, either with the present odor occurrence or the next one. The timing of the incident is another factor. A neighbor will have little tolerance for odors that occur during a summer barbeque.

The factors that influence a nuisance odor are often referred to as FIDO or *frequency, intensity, duration,* and *offensiveness.* Some authors add an L or R to the end of the acronym to bring an additional factor that accounts for the people impacted—*location* or *receptor* (Bokowa et al., 2021). How these factors interplay determines the nuisance.

Frequency: People are less tolerant of odors that come-and-go frequently. There are no data in the literature to suggest what frequency triggers an odor complaint or how the interval between incidents affects the situation. The grid method of field investigation (Section 13.2.4) has criteria that raises concern if an odor is detected 10% of the time that it is assessed, no matter how intense.

Intensity: Odor intensity relates to the strength of the odor. Intensity is strongly affected by the odor concentration, but the quality of the odor is also a factor. Intensity reflects an odor's staying power or pervasiveness. Some evidence suggests that odor complaints begin at odor intensities above 3.5 on the butanol intensity scale. Ratings from 4 to 6 correspond with a possible to probable nuisance and 6 to 8 would definitely be a nuisance (Haug, 1993). Das (2000) suggests that odor complaints tend to occur at concentrations that are five times the detection threshold. An odor concentration of 5 D/T at the location of a receptor is a reasonable target for minimizing off-site odor impacts. It is a target for managing, not a guarantee that odor complaints won't occur.

Duration: Generally, the longer an odor lingers, the less tolerant the affected person becomes. The duration required to push a person over the tolerance line depends on the odor's character, intensity, and the personal tolerance level of the individual. Odor science literature provides no numerical guidance regarding duration and odors. There are no criteria that indicates how long the average person will endure an odor of given strength and hedonic tonic tone before she/he takes action or lodges a complaint. Ironically, an odor of a long duration can increase tolerance. When exposed to a constant odor level, people tend to adapt to it. Their sensitivities and perceptions of the odor decrease (Dalton, 2003). This adaptation may explain why the last people to recognize an odor are often the composters themselves. Repeated odors appear to have the opposite effect. When an odor disappears and then returns, an individual recognizes the returning odor more readily than the initial odor of the same strength.

Offensiveness: Offensiveness is strongly related to the odor quality as perceived by the person experiencing it. However, odor concentration is also a factor. An otherwise pleasant smell can become unpleasant at high concentrations. There is a threshold concentration at which any odor becomes offensive. It is lower for odors that are commonly considered unpleasant, like waste-related odors. Hedonic tone, which is described later, is a key parameter in determining the offensiveness of an odor.

Location/receptor: Location or receptor refers to the character of the neighborhood and the sensitivity of the neighbors near the facility. The combined composition of the people who might experience an odor makes a difference. The impacts are potentially greater where there are more people nearby. The sensitivity, expectations, attitudes, and experiences of the odor-perceiving individual influence whether or not the individual complains, and how forcefully (Box 12.4). In fact, the expectations and attitudes of people can lead to odor complaints when the odors are fleeting, and even when odors are absent.

As an example, in response to an odor complaint, the Texas Commission on Environmental Quality (TCEQ) uses the FIDO model to confirm whether a nuisance odor exists when more direct evidence is lacking (e.g., evidence gathered by the inspector). Fig. 12.3 is the FIDO chart from the TCEQ website (TCEQ, 2016). The chart is true to its name as it uses a combination of odor frequency, intensity, duration, and offensiveness data to determine if a nuisance condition exits. The data for the chart are collected by a trained odor incident inspector. Any colored block in the FIDO chart is considered a nuisance. The website includes instructions and forms for making the assessments but no information about the basis for the criteria in the chart (e.g., why an unpleasant odor of moderate intensity is considered a nuisance when it occurs weekly at a four-hour duration).

> **Box 12.4 People's perception of odors**
>
> Odor perception is an individual matter. How an individual perceives odors is greatly influenced by personal experience, gender, psychology, and societal factors. A person's feelings and beliefs about an odor affect his/her response. Human psychology has evolved such that once a specific odor has conditioned an emotional response, it is generally very difficult to change this response. All future exposure to the same nuisance odor, even when short-lived or faint, will often cause a strong negative reaction.
>
> In an excellent article, Pamela Dalton, psychologist with the Monell Chemical Senses Center, writes "Research has shown that people's reaction to odor and their beliefs about the effects from odor are influenced by a diverse set of factors including personality traits, personal experience and information or social cues from the community and media. These factors can increase, or in some cases, decrease a person's sensitivity and awareness of environmental odors" (Dalton, 2003). In the same article, Dalton summarizes research that shows that expectations and social clues can determine a person's response to odors. People are more likely to detect an odor, or interpret it as bad, when they have negative expectations or have received disparaging information beforehand.
>
> When neighbors are preconditioned to expect malodors from a composting facility, they are more likely to notice odors and react negatively. An activist neighbor protesting a facility is conditioning other neighbors to perceive the situation in a negative manner. This fact should motivate a composter to take part in the neighborhood conversation.

6. Minimizing odors through site selection and management

Most of the odor management practices presented in the following sections of this chapter address specific activities that generate and emit odorants. However, there are a number of facility-wide measures that can improve a facility's position within the community, thereby making it more likely to withstand an odor controversy.

6.1 Site selection and layout

Strategic site-related factors to minimize odor complaints include:

- *Select a good site, distant from neighbors*: A facility location that is far from neighbors is perhaps the best defense against odor complaints. If remoteness is not possible, select a site that has other good characteristics, like predictable and strong winds in a favorable direction or a vegetated perimeter that promotes wind turbulence.
- *Minimize the visibility of the facility*: A site that is less visible is also less susceptible to complaints. A perimeter that includes trees is good; preferably evergreen trees (Fig. 12.4). If not trees, other vegetation and berms or fences can be helpful.
- *If the site must be visible, make it as visually attractive as possible*: Again, trees, vegetation, and gardens. Keep the site itself as neat and organized as possible. Control and collect trash and other debris regularly.
- *Strategically locate odorous processes*: Consider locating the most odorous processes, likely tipping raw feedstocks and active composting, at location as far as possible from neighbors. If feedstocks are likely to be highly odorous, consider locating tipping and preprocessing inside a building.
- *Install an impermeable concrete or asphalt pad where odorous feedstocks are to be unloaded*: Odorous liquids can soak into soil-based tipping areas and passively emit odors.
- *Receive and store liquid feedstocks in tanks or basins*: Odorous liquids must be well controlled and not allowed to run off and emit odors.

Odor Intensity Legend

VS = Very strong	M = Moderate	VL = Very Light
S = Strong	L = Light	NA = None

Offensiveness: **Highly Offensive**

Duration	Single Occurrence	Frequency Quarterly	Monthly	Weekly	Daily
1 minute	NA	NA	VS	S	M
10 minutes	NA	VS	S	M	L
1 hour	VS	S	M	L	VL
4 hours	S	M	L	VL	VL
12 hours +	M	L	VL	VL	VL

Offensiveness: **Offensive**

Duration	Single Occurrence	Frequency Quarterly	Monthly	Weekly	Daily
1 minute	NA	NA	NA	VS	S
10 minutes	NA	NA	VS	S	M
1 hour	NA	VS	S	M	L
4 hours	VS	S	M	L	VL
12 hours +	S	M	L	VL	VL

Offensiveness: **Unpleasant**

Duration	Single Occurrence	Frequency Quarterly	Monthly	Weekly	Daily
1 minute	NA	NA	NA	NA	VS
10 minutes	NA	NA	NA	VS	S
1 hour	NA	NA	VS	S	M
4 hours	NA	VS	S	M	L
12 hours +	VS	S	M	L	VL

Offensiveness: **Pleasant**

Duration	Single Occurrence	Frequency Quarterly	Monthly	Weekly	Daily
1 minute	NA	NA	NA	NA	NA
10 minutes	NA	NA	NA	NA	NA
1 hour	NA	NA	NA	NA	VS
4 hours	NA	NA	NA	VS	S
12 hours +	NA	NA	VS	S	M

FIGURE 12.3

TCEQ's FIDO Chart used to confirm nuisance odors. A combination of conditions that result in a colored block indicates a nuisance odor.

Source: TCEQ (2016).

FIGURE 12.4

Odor problems are less likely at sites that are isolated by either distance or the landscape.

6.2 Facility management

Facility-wide management factors that can help to minimize odor complaints include the following.

- *Know the limits of your chosen composting system*: Some feedstocks are inherently odorous and/or quickly degradable. Food waste, biosolids, poultry manure, grass, grease trap waste, septage, and other highly putrescible materials are not appropriate for all process methods or all sites. Have a policy that defines which feedstocks your facility will accept based on what it is designed to handle successfully. If necessary, avoid or discontinue accepting troublesome feedstocks.
- *Avoid overloading the site with materials*: In general, off-site odors increase with the amount of material on the site. If it means staying out of trouble with the community, stay small.
- *Increase collection frequency if feedstocks too often arrive in an odorous state*: Feedstocks which are collected every other week (or less frequently) in warm weather are more likely arrive at the facility with strong odors. If possible, ask your haulers to increase the collection frequency to reduce the amount of time the materials are allowed to degrade in an uncontrolled manner.
- *Maintain equipment and have contingency plans*: Bad things can happen when key pieces of equipment are unavailable. Unprocessed raw feedstocks, un-amended biosolids, unturned windrows, and huge stockpiles waiting for screening are likely sources of unmanaged odors. Establish contingency plans for each critical piece of equipment (e.g., grinder, loaders, mixers, turners) that might break down. Maintain equipment in good repair. Keep an inventory of critical spare parts. Develop preemptive agreements with equipment dealers, other composting facilities, or customer service contractors to obtain equipment or continue operations in the event of equipment failures.
- *Clean up at the end of the shift*: Remove organic material from aisles and roadways; clean and sweep paved surfaces of dust and debris; pick up trash

around the site perimeter; cover fresh feedstock piles and odor-emitting piles with a biocover as needed.

- *Control dust*: Use water trucks, surface irrigators, misting systems, and/or sweepers to control dust. Dust leaving the facility is never good for neighborhood relations and can absorb/desorb odors and transport them off-site.
- *Divert incoming feedstocks to alternative outlets when the facility is overwhelmed*: In advance, identify and establish protocols for limiting feedstocks over and above what the facility can successfully process. Examples of possibly diversion outlets include other composting or anaerobic digestion facilities, farms for direct land application, and/or landfills.
- *Actively maintain good relationships with neighbors and the community* (Section 12).

7. Odor generation during composting

Odors can arise from any location at a composting facility where organic materials are stored, processed, and treated; or were once stored, processed, and treated. As discussed in later sections, feedstock handling, curing, and organics-ladened process water can emit impactful odors. However, because the active composting process accounts for most of the decomposition, it deserves the most scrutiny and management.

The strength of odor emissions during composting are primarily determined by the physical, chemical, and thermal process conditions in the pile. These conditions include oxygen, temperature, pH, and feedstock properties such as moisture, bulk density, structure, and nutrient balance (e.g., C:N ratio). Arguably, the most impactful factor in odor generation is the character of the initial feedstock mix. The characteristics of the mix determine the process requirements to remove excess heat and moisture, deliver oxygen, and moderate pH.

7.1 Oxygen

Composting voraciously consumes oxygen. The oxygen level measured at the center of a freshly turned windrow can drop from near 21% (ambient) to below 1% within 20 min of turning. Research has shown that the availability of oxygen has a profound impact on odor generation (Michel and Reddy, 1998).

When available oxygen drops low enough, facultatively anerobic bacteria predominate.[6] These bacteria produce higher concentrations of odorous compounds during the degradation of organic matter (Sunberg et al., 2013). Low oxygen levels also slow the rate of composting. As a result, unstable compost may be prematurely moved to curing and product storage piles, where it continues to consume oxygen and generate odorants.

---

[6] Facultative bacteria can switch from aerobic to anaerobic conditions in response to the presence or absence of oxygen.

7.2 Temperature

The temperature of the compost influences odor emissions in many ways, both directly and indirectly. The following factors explain why a hot compost pile tends to emit more odors than a cool compost pile:

- *Vapor pressure:* All odorants have some degree of water solubility that keeps them dissolved in the liquid film surrounding solid particles (Chapter 3). As the temperature of the liquid increases, the kinetic energy of the molecules in the liquid phase increases, driving more molecules into the vapor phase (which is the same as increasing the vapor pressure). Thus, the concentration of odorants within the pile's free air space (FAS) increases.
- *Oxygen saturation:* As temperature increases, the solubility of oxygen in the liquid film decreases (because the oxygen's vapor pressure increases). There is less dissolved oxygen (DO) in the liquid film available to the decomposer organisms (Box 12.5).
- *Ammonia:* Higher temperatures generally lead to greater volatilization of ammonia. As temperature increases (at a given pH), the equilibrium between soluble ammonium (NH_4) and gaseous NH_3 shifts toward NH_3 (Chapters 3 and 11).
- *pH inhibition:* When quickly degradable organic compounds decompose, an abundance of organic acids can form that depress the pH levels of the composting process when combined with temperatures above 45°C (113°F) (Sunberg et al., 2013). The persistence of these organic acids leads to strong odors. High temperatures inhibit the decomposition of organic acids (Box 12.6).
- *Thermal inhibition:* Thermophilic microorganisms begin to be inhibited at temperatures approaching 65°C (150°F), and then are sharply inhibited at temperatures in excess of 70°C (160°F). The effectiveness of thermophilic bacteria to fully metabolize organic matter to CO_2 is diminished, producing more partially decomposed and odorous compounds.

Box 12.5 Combined effect of temperature and oxygen concentrations on dissolved oxygen

As discussed in Chapter 3, the composting activity takes place in the water film at the surface of a particle of organic matter. The aerobic bacteria active in that film layer rely on oxygen being dissolved into the water from the surrounding pore space air. When the dissolved oxygen (DO) is plentiful, aerobic activity can be uninhibited. When DO levels drop low, aerobic activity is inhibited, and facultatively anerobic bacteria predominate. It is standard practice for waste water treatment plant operators to focus on maintaining the DO levels to continuously match the "biological oxygen demand (BOD)" of the wastewater. When the DO levels fall, so does the rate at which organic matter is consumed and, conversely, the rate at which odors are produced tends to increase. Composting is affected by the same fundamental principles.

Box 12.5 Combined effect of temperature and oxygen concentrations on dissolved oxygen—cont'd

The level of DO in the water film layer is a function of both the temperature of the water and the concentration of oxygen in the pore space air at the surface of the water film. This relationship was first elaborated by the early 19th century chemist William Henry. As temperatures rise, gas solubility in liquids decreases (this why warm beer tends to be flat). Nick Sauer, a compost researcher for the UK Environment Agency was tasked with developing protocols to reduce odor emissions at composting facilities. One of the phenomena he studied was the relationship between the *available* oxygen and odor emission from composting. Using Henry's law, he calculated the DO level in the water film at various temperatures and oxygen levels in the pore space air (Sauer and Crouch, 2013). The pore space oxygen level is what is measured when an oxygen probe is inserted into a pile. The recommendation from this work is to maintain the DO concentrations above 3 ppm for odor sensitive facilities, and 2 ppm as a general guideline.

Table 12.2 is a recreation of Nick Sauer's work. DO levels below the recommended 2 and 3 ppm guidelines are highlighted in yellow and orange, respectively. The impact of temperature on available oxygen is evident from the Table. For example, if a pile has a pore space oxygen level of 10%, it has an acceptable DO concentration of 3.16 ppm at 40°C (104°F); but at 70°C (158°F), the DO concentration is only 1.97 ppm, below the 2 ppm recommended threshold.

Most composting facilities without forced aeration often operate in the "yellow zone"; and many can do so successfully. On the other hand, many facilities with odor issues, regardless of forced aeration or not, operate too much of the time in the "yellow zone," which is the hallmark of a process inhibited by a lack of available oxygen.

Table 12.2 Dissolved oxygen concentration in water as a function of temperature and oxygen concentration in pore space air based on calculations using Henry's law (Sauer, 2012).

Pore Space O2	Temperature												
	20°C 68°F	25°C 77°F	30°C 86°F	35°C 95°F	40°C 104°F	45°C 113°F	50°C 122°F	55°C 131°F	60°C 140°F	65°C 149°F	70°C 158°F	75°C 167°F	80°C 176°F
20%	9.17	8.32	7.57	6.91	6.33	5.81	5.35	4.94	4.57	4.24	3.94	3.67	3.42
19%	8.71	7.90	7.19	6.57	6.01	5.52	5.08	4.69	4.34	4.02	3.74	3.48	3.25
18%	8.25	7.49	6.82	6.22	5.70	5.23	4.82	4.44	4.11	3.81	3.54	3.30	3.08
17%	7.80	7.07	6.44	5.88	5.38	4.94	4.55	4.20	3.88	3.60	3.35	3.12	2.91
16%	7.34	6.66	6.06	5.53	5.06	4.65	4.28	3.95	3.65	3.39	3.15	2.93	2.74
15%	6.88	6.24	5.68	5.18	4.75	4.36	4.01	3.70	3.43	3.18	2.95	2.75	2.57
14%	6.42	5.82	5.30	4.84	4.43	4.07	3.75	3.46	3.20	2.96	2.76	2.57	2.39
13%	5.96	5.41	4.92	4.49	4.11	3.78	3.48	3.21	2.97	2.75	2.56	2.38	2.22
12%	5.50	4.99	4.54	4.15	3.80	3.49	3.21	2.96	2.74	2.54	2.36	2.20	2.05
11%	5.04	4.58	4.16	3.80	3.48	3.20	2.94	2.72	2.51	2.33	2.16	2.02	1.88
10%	4.59	4.16	3.79	3.46	3.16	2.91	2.68	2.47	2.28	2.12	1.97	1.83	1.71
9%	4.13	3.74	3.41	3.11	2.85	2.62	2.41	2.22	2.06	1.91	1.77	1.65	1.54
8%	3.67	3.33	3.03	2.77	2.53	2.32	2.14	1.98	1.83	1.69	1.57	1.47	1.37
7%	3.21	2.91	2.65	2.42	2.22	2.03	1.87	1.73	1.60	1.48	1.38	1.28	1.20
6%	2.75	2.50	2.27	2.07	1.90	1.74	1.61	1.48	1.37	1.27	1.18	1.10	1.03
5%	2.29	2.08	1.89	1.73	1.58	1.45	1.34	1.23	1.14	1.06	0.98	0.92	0.86
4%	1.83	1.66	1.51	1.38	1.27	1.16	1.07	0.99	0.91	0.85	0.79	0.73	0.68
3%	1.38	1.25	1.14	1.04	0.95	0.87	0.80	0.74	0.69	0.64	0.59	0.55	0.51
2%	0.92	0.83	0.76	0.69	0.63	0.58	0.54	0.49	0.46	0.42	0.39	0.37	0.34
1%	0.46	0.42	0.38	0.35	0.32	0.29	0.27	0.25	0.23	0.21	0.20	0.18	0.17
0%	0.00	0.00	0.00	0.00	0.00	0.00	0.00	0.00	0.00	0.00	0.00	0.00	0.00

> ## Box 12.6 pH effects—inhibition, odors, and mesophiles
>
> Sundberg et al. (2005, 2008, 2013) have conducted extensive research on the role of pH on generation of odors and VOCs in acidic feedstocks, primarily in food waste rich feedstocks that arrive with a low pH. Their work has led to a set of composting odor control guidelines that are broadly followed in Nordic countries (Nordic Counsel of Ministers, 2009). A key finding of this research is that composting process is inhibited by the combination of low pH and temperatures above about 45°C (113°F). This inhibited state both slows stabilization and generates much higher odor emissions than if the feedstock pH is near or above neutral (>6.5).
>
> In the cited research, air emissions from uninhibited composting treatments measured to be about 50 times lower than those from the pH-inhibited treatments. The rate of stabilization (measured by the generation of CO_2) showed an even larger increase in the uninhibited treatments. Better stabilized material has less odor-producing potential during the subsequent process steps of curing, screening, storage, and load out.
>
> The key to avoiding this inhibition is to achieve a rise in pH during the first few days of composting. The pH can be shifted toward neutral in two ways. The first way is with the addition of significant volumes of more alkaline materials (including recycled compost). The second way is by keeping temperatures below 45°C so mesophilic bacterial are predominate during the first process 2—3 days. To keep temperatures below this threshold, the recommendations are either to build small piles with abundant surface area per mass to promote ambient cooling, or to provide high aeration rates. The correct application of one or more of these methods allows the mesophilic bacteria to metabolize the organic acids and raise the pH toward neutral. Thereafter temperatures can be allowed rise into the thermophilic range without the deleterious effects of pH inhibition.

7.3 Feedstock mix properties

The properties of the mix of feedstocks have a profound influence on the generation of odors. Moisture content, FAS, and pile structure are important due to their impact on oxygen delivery. The DO levels in the water film layer around the composting particles are a key determinate of the rate of odor generation. The oxygen in the film layer is determined by the concentration of oxygen in the pore space air and the amount of water that it needs to diffuse through.

- *FAS:* The ability of air to move via either forced or natural convection through the pile is determined by the amount of air-filled pore space (i.e. FAS). When that pore space is occupied by excess water, or reduced by compaction, it is difficult to supply enough air to keep up with microbial oxygen consumption and oxygen levels drop. These conditions occur when the bulk density is high. Also, it is difficult to remove excess heat and water from a dense pile—further compounding the problem.
- *Moisture content:* When moisture levels become excessive, the oxygen must travel through thicker film layers. Oxygen diffusion through air is rapid, but

through water, it is thousands of times slower, effectively reducing the available oxygen levels to the microorganisms residing in the outer layers of the film.

- *Structure:* To minimize odors, the initial mix guidelines given in Chapter 4 should be followed. However, even starting with a good mix, the pile FAS decreases over time due to compression and degradation. The deeper the pile and the less structural the ingredients of the mix, the faster FAS is lost. A few ways to counteract this effect is to limit pile height, add more structural amendments (such as coarse ground wood), and/or remix the piles with a frequency to match the rates of densification.

- *Carbon to nitrogen ratio:* In regard to C/N ration, odors tend to arise only when the C/N ratio is low (e.g. <20). Feedstocks such as yard waste and some agricultural residuals have desirable C/N ratio without the addition of amendments. Feedstocks such as biosolids, food, digestates from an AD facility, and manures have inherently high nitrogen content and low C/N ratios that are well below the desirable range. An initial low C/N increases odor emission in several ways. First, the excess N provides fodder for the formation of ammonia (both NH_4 and NH_3) and odorous amines. Second, abundance of NH_3 can inhibit stabilization (Guo et al., 2012), leading to odor generation in subsequent process stages. Although NH_3 itself is seldom a source of off-site nuisance odors, low C:N ratio and high NH_3 emissions are associated with higher emissions of odorous sulfide compounds (Zhu et al., 2021). The later effect may be due to an insufficient pool of available C, which is need for microorganisms to assimilate both N and sulfur (S) compounds.

- *Carbon to sulfur ratio (C/S):* The C/S ratio is a concern only if sulfur-based chemicals are the dominant odors, which is most likely in composting biosolids and a few other residuals. Sulfur-base odorants (e.g., mercaptans, sulfides) are more likely to be immobilized, and less likely to volatilize, if a sufficient pool of C is available (e.g., C/S ratio >100:1). While there is little research supporting this practice, the principle is sound, and similar to the logic applied to C/N ratios (Miller, 1993). Stable organic matter has a C/S ratio of approximately 100:1 suggesting that materials with lower C/S ratios harbor excess sulfur that could potentially volatilize.

---

8. Strategies to reduce the generation of odors

It is almost always more efficient and cost-effective to minimize the generation and emission of odors than it is to rely primarily on capturing (enclosing) and controlling (scrubbing) them. Poor process conditions can generate 50 to 200 times more odor emissions per ton of feedstock compared to a well-designed process. By contrast, most capture and control methods, such as an enclosed composting system with a free-standing biofilter, offer only a 5 to 20 times reduction in air emissions, yet requires much more infrastructure. Thus, odor management at composting facilities should first focus on the root cause of odor generation, generally through process improvements, and then consider the capture and control options.

A plan for avoiding nuisance odors begins with identifying potential sources of significant odor emissions at composting facility. The most common sources are listed below.

1. The receiving (tipping) and preprocessing area where incoming loads of feedstock are already odorous when they arrive.
2. The active composting process where oxygen is rapidly consumed, and significant heat generation can cause excessively high temperatures.
3. In curing, storage, and postprocessing areas when significant amounts of semistabilized materials have accumulated and remain anoxic (very low oxygen).
4. In process water that contains high levels of nutrients and organic materials, leading to low DO levels.

Odor reduction, or mitigation, methods for each of these facility elements are presented in the following subsections.

8.1 Reducing odor generation from receiving and preprocessing

Who hasn't noticed the odor from a restaurant dumpster or a pile of decaying grass on a warm day? These are examples of odorous material that can arrive at a composting facility. Highly degradable feedstocks can become odorous within the receiving area if they are not processed promptly.

The following is a list of operational considerations intended to help reduce odor generation during the receiving and preprocessing of raw feedstocks. When these measures, together with the design and management steps are inadequate to manage odors, enclosing these operations may be necessary (Section 9).

- *Mix materials off-site before delivery to the composting site (if possible)*: Have feedstock generators add amendments at the point of collection. For animal manure, generous amounts of bedding can be added to poultry and livestock rearing areas.
- *Process feedstocks promptly*: As a rule, moist degradable feedstocks should be mixed with amendment to create a balanced mix and actively composted as soon as possible. Highly degradable feedstocks, like food waste, should be processed, or at least covered, on the same day they are received. Feedstocks

that arrive in an odorous state should be processed as soon as they arrive immediately.

- *Make sure to practice first in, first out handling*: To reduce the possibility of odors developing in storage, process the feedstocks that arrived yesterday before those that arrived today (unless today's load is particularly odorous).
- *Directly incorporate odorous loads into actively composting windrows, piles, or vessels*: If the incoming material has reasonable mix characteristics, then less handling and time outside of active composting is desirable to minimize odor emissions. If necessary, the receiving windrow or pile should be capped with a biocover.
- *Upon arrival, amend wet and odorous feedstocks with dry carbonaceous materials* (Chapter 9): Mixing wet feedstocks with an adequate amount of a dry bulky amendment generally checks odors for at least 24 h, even without further processing. Process freshly tipped feedstocks quickly employing one or more of the following methods (Fig. 12.5).
  - ○ Unload wet feedstocks on to a bed of dry amendment or composted material,
  - ○ Promptly blanket odorous feedstocks with a biocover of dry amendment or compost, or,
  - ○ Mix odorous feedstocks with woody absorbent amendments.
- *Consider adding lime, wood ash, or other alkaline materials to exceptionally sour-smelling, low pH materials*: Alkaline additives can raise the pH and/or temporarily slow decomposition (Lystad et al., 2002). Note it often takes a large amount of alkaline material to significantly raise the pH (figures on the order of 15%–30% by weight are common). Also, odors can eventually return when the pH moderates and the high pH may lead to increased ammonia loss.
- *Check to see that the receiving floor and preprocessing area is kept clean*: Make sure all incoming odorous feedstocks have been processed and moved to the active composting stage and that there is no standing water, and that the drainage system is clear.

8.2 Reducing odor generation from active composting

As previously discussed, the first line of defense against odors forming during composting is to follow good basic composting practice—adequate aeration, balanced carbon and nitrogen, high FAS, balanced moisture content, and moderate temperatures and non-acidic pH. With these basics fulfilled, the focus of odor management becomes limiting odor emissions and their potential impacts. The following subsections address differing practices among the various methods of composting.

8.2.1 Addressing odor from windrows and other passively aerated piles

Odor generation in passively aerated methods is especially influenced by the quality of the initial mix. Pile geometry. FAS and bulk density are particularly important

(A)

(B)

FIGURE 12.5

Incoming food waste tipped onto a bend of ground wood (A) and immediately mixed and covered (B).

Source: Greg Gelewski.

(Michel and Reddy, 1998; Buckner, 2002a). If passively aerated windrows and piles are generating nuisance odors, consider the following list of remediations for process conditions and operational practices:

- *Low oxygen and high temperature*: If oxygen measurement is possible, and the average oxygen levels in the pile are below 5% or if the temperature is above the target ranges given in Section 7.2, consider implementing one or more of the following remediations:

- Reduce bulk density/increase FAS to enhance natural convection and diffusion,
- Reduce pile depth and breadth (increase surface area to increase heat loss and reduce the amount of insulation), and/or,
- Add recycled compost to decrease rates of oxygen consumption.
- *Low pH*: If the pH remains below 6.5 for period of 1−2 weeks, implement one or more of the three options given above for reducing high temperature.
- *Bulk density*: If the bulk density measures above 550 kg/m^3 (930 lbs/yd^3), add a sufficiently coarse and structural bulking agent to increase the FAS.
- *C/N ratio*: If the C/N ratio is below 25, add carbonaceous materials such as ground wood or sawdust to increase it into the BMP range. Note: not all wood is a great source of carbon; very coarse wood, kiln dried wood and overs do not provide much bioavailable carbon.
- *C/S ratio:* If the C/S ratio is below 100, amend with more carbonaceous materials.
- *Build/turn windrows when they are first formed such that they have a homogenous mix:* Inadequate blending of odor-prone feedstocks (e.g., food, grass) leads to anaerobic pockets which can cause significant odors.
- *Irrigate windrows:* Spraying water onto a windrow while it is being turned helps to prevent the release of dust and to wet the newly formed surface layer. Irrigating windrow after turning has been shown to cool the surface and decrease air emissions (Section 10.1).
- *Avoid turning when the atmosphere is stable and still*: Emissions from windrows and piles disperse poorly when there is little atmospheric mixing, due to temperature inversions, overcast skies, and/or calm winds. The odors remain more concentrated for a longer distance. If possible, reduce or eliminate disturbing windrows/piles during these conditions.
- *Avoid turning activity when wind is in the direction of nearby receptors*: If possible, turning should take place when the wind is favorable—i.e., away from the most objecting receptors.
- *EITHER Turn early and frequently*: Turning windrows frequently during the first weeks of composting hastens the decomposition of highly degradable feedstocks while discouraging anaerobic conditions. With this strategy, turning should be frequent; every 1−2 days during the first week and 2−3 days during the second week (Gage, 2020). This strategy runs the risk of releasing strong odors regularly. Its effectiveness depends on the odor quality of the feedstocks and the situation in the neighborhood.
- *OR Postpone turning for the first 2−4 weeks after building windrows*: This strategy allows the feedstocks decompose within the confines of the windrow without turning for a few weeks. It presumes that many of the odorants formed also decompose within the windrow. After 2−4 weeks, the next turning is likely

to release the odorants remaining in the windrow. This turning can be scheduled during favorable wind conditions. Thereafter, the generation of odor from the windrow is lower and a more regular turning schedule (e.g., weekly) can be adopted. This strategy works best if the windrow has a relatively low bulk density and high FAS. A biocover can be applied to the windrow, first when it is built, and then after the initial turn.

- *Include finished compost in the feedstock recipe*: Recycling some finished compost back into the feedstock mix has been reported to reduce the emissions of odors (CalRecycle, 2007). This positive effect may be due to the added microorganisms or to a slowing of the overall rate of decomposition. Unscreened compost is the logical choice, although screen overs with a heavy load of fine compost particles is another option.
- *Consider additives intended to reduce odors*: Several additives have been reported to reduce emissions of VOCs, including odorants, during composting (Chapter 11). Such additives include biochar, zeolite, wood and fly ash, and specific microorganisms (inoculants). The research is mixed regarding the amounts and the effectiveness of such additives, but some positive results have been reported (Barthod et al., 2018; Zhu et al., 2021).
- *Consider adding forced aeration to the active composting phase*: If odors are not being acceptably managed with windrows or passively aerated piles, forced aeration may be the solution. A properly designed aeration system (Chapter 6) is a much more certain way of supplying oxygen, moderating temperatures, and avoiding the odor producing process conditions discussed in Section 12.7.

8.2.2 Addressing odors from forced aeration processes

A forced aeration processes doesn't guarantee low odor composting. The initial mix and pile geometry are still important, as is the functionality of the aeration system. If one of these processes are generating nuisance odors, consider the following list of remediations for process conditions and operational practices:

- *Low oxygen.* If oxygen measurement is possible, and the average oxygen levels in the pile are below 15% during active composting (Chapter 6 and Section 7.2), the aeration rate is significantly below optimum. Consider the following remediations:
  - Increase the airflow if possible.
  - For timer-controlled airflow, adjust the control cycle so the off time is always less than 10−15 min.
  - Reduce pile depth/volume over the aeration floor to increase the unit air flow rate (cfm/cubic yard).
  - Increase the capacity of the aeration system.

- *High temperature.* It takes 5 to 10 times as much airflow to control temperature as it does to supply oxygen; some excursion into higher than optimum temperatures are normal in most forced aerated systems. The goal is to avoid sustained elevated temperatures (Chapter 6). If temperatures remain above 65°C (150°F) for more than a few days, then consider the implementing one or more of the following remediations:
    - o Increase the airflow if possible
    - o Reduce pile depth/volume over the aeration floor to increase the unit air flow rate
    - o Increase the capacity of the aeration system
    - o Add recycled compost to decrease rates of oxygen consumption/heat generation
    - o Add a coarse plenum layer over the aeration floor to increase flow rates and the uniformity of air distribution.
- *Low pH*: If the pH remains consistently below 6.5 for the first three days implement one or more of the options given above for reducing high temperature. This can take the form of an initial three-day retention period in ½ height pile to keep temperatures in check, allowing the pH to rise prior to moving it to a full-size pile where temperature can be allowed to increase without causing significant odors.
- *Bulk density*: If the bulk density measures above 550 kg/m³ (930 lbs/yd³), add a sufficiently coarse and structural bulking agent to increase the FAS.
- *C/N ratio:* If the C/N ratio is below 25, add carbonaceous materials such as ground wood or sawdust to increase it above 25 (and preferably to 30 or above). Note: not all wood is a great source of carbon; very coarse wood, kiln dried wood and overs do not provide much bioavailable carbon.
- *C/S ratio:* If the C/S ratio is below 100, amend with more carbonaceous materials.
- *Uneven aeration*: If possible, measure either oxygen or CO_2 level around pile to see if the aeration floor is delivering relatively uniform air flow. The oxygen levels in a properly performing aeration floor with adequate flow should be approximately +3% of the pile average throughout. Poor air distribution or "dead spots" are generally the result of inadequate aeration floor design and should be corrected. One corrective action is to add a coarse plenum layer over the aeration floor to produce more uniform flow.
- *Irrigate pile surfaces*: The cooler moist outer layer of an ASP captures and decomposes odorants from the exiting air. If piles are already irrigated, consider increasing the irrigation frequency, especially during the 5–10 days of composting and/or when the weather is warm.
- *Apply a biocover*: See Section 10.2 and also Chapter 6.
- *Consider capture and control of air emissions*: When the above remediations do not sufficiently reduce odor generation, consider the capture, and control options discussed in Sections 9 and 10.

8.3 Reducing odor generation from curing

Curing is an additional phase of composting in which organic compounds continue to decompose, though at lower rates than in active composting (Fig. 12.6). Usually curing piles are not aerated. Often, they are stacked high.

If the active compost process hasn't achieved adequate stabilization (measured by a low enough respiratory rate), the process conditions in the curing phase, and beyond, are likely not adequate to maintain desirable process conditions (oxygen, temperature, etc.). As a result, curing and product storage can become a significant source of odors. These odors are emitted slowly and passively during the curing stage, and then suddenly when curing piles are opened or moved. It is not uncommon that compost facilities with consistent odor issues find that large curing and storage piles are a major source. This finding points to the need to improve the process conditions in one or both of the active and the curing phases to address the sources of process inhibition discussed earlier in this section.

- *Extend active composting time prior to curing*: Ideally curing should begin when temperatures in the active composting system fall to near ambient conditions. If the composting time is cut short, compost in the curing pile still demands oxygen and still generates heat. A large passively aerated curing pile may not provide enough air exchange to maintain aerobic conditions.
- *Monitor curing pile temperatures*: Curing piles that develop thermophilic temperatures (>40°C, 105°F) are likely holding unstable compost. If temperatures are consistently high, the piles should be broken apart into smaller piles. High temperatures during curing indicate the active composting process is not long enough or otherwise inefficient (e.g., poor aeration, too dry, too wet, poor mix, etc.).
- *Lightly irrigate curing pile surfaces*: Just as it does for composting piles, a moist and cool exterior surface helps absorb and degrade odorants before they are emitted from the curing pile.

FIGURE 12.6

Curing piles aerate passively in piles that are usually larger than active composting piles and windrows. If the compost is not sufficiently stable, curing piles can develop odors.

- *Check and correct for dry conditions during active composting*: Compost may be prematurely moved to curing because low moisture inhibits temperature, rather than compost stability. The low moisture may be in only part of the active pile. Simply moving this compost to the curing pile can invigorate the process as the dry partially composted material is mixed with moist material.
- *Screen compost after curing to maintain porosity*: Screening compost after curing retains larger particles in the curing piles, improving FAS and passive aeration.
- *Decrease curing pile size (height)*: Depending on the stability level, very large curing piles (over 3 m, 10 ft) may not aerate properly. Decreasing the pile size improves diffusion of oxygen into the pile interior.
- *Use forced aeration for curing piles*: If necessary, curing piles could be aerated with fans to provide needed oxygen. The aeration rate should be substantially lower than that required for active composting.

8.4 Reducing odor generation from stormwater

Open pools of water on composting sites include puddles from precipitation, drainage channels, and retention ponds, all of which can contain enough organic matter to create an odor. The greatest odor risk comes from "process water," water that has been in contact with the composting materials. The keys to managing the odor are to first eliminate the water that is unnecessary (e.g., puddles) and keep organic matter out of the water that is stored on-site. A third option is to treat the water, so it does not develop odors.

Most compost facilities are not permitted to discharge runoff water that has been in contact with composting feedstocks or active composting piles. Therefore, process water needs to be collected, carefully managed, and reused on-site at nearly every composting facility (Chapter 10). When designing a runoff collection system, a key focus should be removing the solids that are inherent in process water. The goal is to minimize the BOD load by delivering the fewest solids possible to the retention pond or storage tank.

- *Inspect the site after precipitation*: Inspecting the site regularly, especially after rain events, identifies areas where water may have pooled, leading to possible pockets of odor. Severe rain events can also saturate piles leading to odor. Liquid freely draining from windrows and piles is a sign of a possibly saturated pile, at least at its base. Turning after a rain distributes added water throughout the pile.
- *Remove puddles, depressions, and wheel ruts where water collects*: Regularly grading the site not only reduces these potentially odorous pools of water but also improves site appearance. Alternatively, wood chips, soil, or another absorbent material can be used to soak up pools of water and fill potholes and ruts.
- *Remove sediments from water draining into stormwater retention pond*: Most of the organic matter and nutrients in stormwater is associated with sediment. Sediment can be removed by with a filter berm, compost sock, settling basin, or

other mechanism that slows the flow of stormwater runoff into a retention basin. Collected sediment can be added to feedstocks and composted.

- *Clean the pond of sediment and vegetation during dry season*: A stormwater retention pond should be cleaned out prior to the next season's rainy period. This will prevent loss of retention capacity and also reduce the nutrient loading from the previous season's sediment.
- *Increase retention pond capacity*: If the stormwater retention pond is a constant source of odor due to organic matter, the pond may be undersized for the facility. Increasing the pond size dilutes the organic load and increases the surface area for oxygen diffusion into the pond.
- *Use retention pond water*: With some restrictions, water from the retention pond can be used for several positive purposes, including adding moisture to dry composting windrows and piles, crop irrigation, and cleaning equipment and site surfaces. Reusing water with a high BOD level (low DO) without creating odor can be challenging. Reuse generally means mixing the water into raw feedstocks that have extra water carrying capacity and/or injecting it into active composting piles. Applying odorous water with sprinklers on top an open pile would be a practice to avoid. The idea to is to minimize the amount atmospheric exposure that would allow the odorants in the water to evaporate.
- *Aerate the retention pond(s)*: Pond aerators have been used effectively in some compost facility stormwater retention basins to provide oxygen to the pond and encourage decomposition of the pond organic matter. A common approach is to keep the DO elevated. A typical minimum design threshold for DO is 4 mg O_2 per/1 kg of water. The water treatment industry offers a wide range of aeration devices to do this. An effective system matches influent BOD and flow to achieve the desirable DO and retention time (just like matching a composting aeration system to the feedstocks and the site conditions). Estimating BOD and flow from a composting site is challenging because it is so variable. But it is helpful to keep in mind that low flow generally means high BOD and vice versa (i.e., a big rain event result in high flow of low concentration).
- *Collect and store high BOD process water in an enclosed tank*: Water collected from vessels, feedstock-receiving areas, and under pile drains can be especially strong. Odorous water can be stored in a tank to minimize the escape of volatized odorants. Depending on the facility's situation, air in the headspace of the tank may need to be diluted or scrubbed before release to atmosphere.
- *Cover some or all of the composting site*: One of the better ways of minimizing process water is to cover working areas with a roof—not necessarily an enclosed building (Fig. 12.7). The roof must include gutters or another precipitation collection system to keep the roof water off the working site. Rainwater from the roof can also provide a great source of clean water for adding to dry windrows and piles after sanitization is complete.
- *Cover open piles and windrows with fabric covers*: Less effective than a roof, but also less expensive, fabric covers prevent precipitation from passing through open windrows and piles (Chapters 5 and 6). The water that these covers shed is still being considered process water, but it carries less organic matter (i.e., lower BOD).

FIGURE 12.7

This composting facility has a 2-ha (5-acre) roof supplying clean water to a 20,000 m³ pond (about 5-million gallons).

Source: ECS.

9. Capture and control of odors once generated

Odor strength is naturally diluted between the point generation and the receptor. However, if odor emissions are too high for the dilution effects to sufficiently reduce the odor intensity at the receptor, additional controls are required to avoid odor issues. Fig. 12.8 shows an odor emission capture device placed around the generation source. The captured air is then routed to a control mechanism (i.e., treatment device) to reduce the concentration of odorants prior to discharge.

Capture and control methods vary from simple to elaborate. At one extreme, the surface of a composting is simply irrigated. An example of the other extreme is constructing structural enclosures that contain the composting processes and ducting all of exhaust airflow through elaborate air scrubbing technologies then up a tall stack for increased dilution. The breadth of capital expenses associated with these control methods is correspondingly broad, ranging from less than a dollar per ton composted, up to thousands of dollars per ton composted. Prior to committing to expensive enclosures and air scrubbing technologies, it is worthwhile to first assess the costs/benefits of reducing odor generation by investing in process improvements instead. Because engineered capture and control methods come at both a capital and an operational cost, they should only be applied to areas with the potential to generate high odor emissions. These areas include feedstocks receiving and preprocessing, active composting, and tanks used to collect high BOD leachate (Box 12.7).

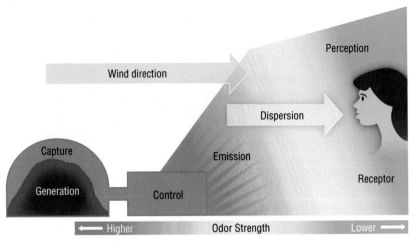

FIGURE 12.8

How capture and control fit into the odor sequence to reduce the chance of an odor incident.

Box 12.7 Capture and control efficiency

In air emissions management vernacular, the performance of each component of a capture and a control system is assigned an efficiency. The capture efficiency indicates the percentage of exhaust air that is captured. If a method has an efficiency of 95%, then the 5% of the exhaust air isn't captured represents fugitive emissions. The control efficiency refers to the percentage of the odorants or VOCs in the captured exhaust air that is destroyed or filtered from the exhaust air. So total OER is indicated by the following equations.

$$\text{Fugitive emissions} = \text{Odor flux} \times (1 - \text{Capture efficiency}) \tag{12.4}$$

$$\text{Controlled emissions} = \text{Odor flux} \times (\text{Capture efficiency}) \times (1 - \text{Control efficiency}) \tag{12.5}$$

$$\text{Total emissions} = \text{Fugitive emissions} + \text{Controlled emissions} \tag{12.6}$$

The "odor flux," is the measurement of the amount of odor emitted per unit of time by an odor source. It is usually expressed in odor units per second (ou/s or ou_e/s). Determining the odor flux of a given source requires knowing both odor concentration in, and the mass flow rate of, the air carrying the odor away from the source. It can be measured from a confined air stream by sampling. As described in Box 12.3, it is much more difficult to measure it from passive or area sources like an open window or pile.

---

10. Capture

Capturing odorants from any source at a composting facility involves either negative ventilation and/or adding a barrier to reduce fugitive emissions. When ventilation is used, the captured air is directed to a "control" technology (i.e., treatment). The simplest methods rely on the outer layer of the compost pile as a barrier and take advantage of the odor-controlling effect of its cool and moist environment.

10.1 Pile surface irrigation

Intermittent and timely irrigation moistens and cools the outer layer of the pile to better absorb odorants. This simple method can reduce odor emissions from either active composting or curing piles. Many methods have been devised to irrigate the top of a pile including water trucks equipped with side sprayers, sprinklers spread across the top of a pile, and high-volume traveling gun sprinklers (Chapter 9). A goal should be to minimize the amount of water applied to aisles, and thus avoid additional process water.

It doesn't take a lot of water to moisten the outside of a pile in the short-term. How much and how often water needs to be applied depends on ambient conditions and the heat generated by the composting process. The general rule is to apply surface irrigation intermittently during the hotter times of the day to get most of the benefit. The pile does not have to look moist all of the time, but it should feel moist below the surface, about 15 cm (6 in.) down. If the weather is dry, hot, and windy, water may need to be applied every 2—4 h during the heat of the day to maximize odor control efficiency (Cordova et al., 2015). On the other hand, when the weather is cold, applying water is not necessary. Condensation from hot moist air from the interior of the pile can keep the surface moist.

10.2 Applied biocover

A biocover is an exterior layer of relatively stable organic material added to the top of a pile to capture and contain odorants (Fig. 12.9). Biocovers are generally made up of locally available, relatively stable, organic materials that have either been through sanitization or are not considered pathogenic to begin with. Compost and screen overs are logical candidates because they are readily available. Wood chips are another common choice. The biocover should be 15 cm (6 in.) thick, at a minimum, although 30 cm (12 in.) is preferable and commonly required by regulations for sanitization of static piles.

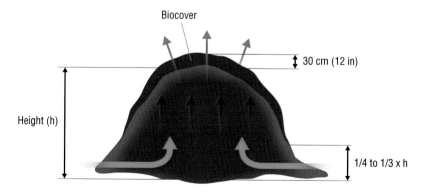

FIGURE 12.9

Placement and dimensions of a biocover.

Biocovers are an inexpensive capture and control option. As exiting air passes through the moist and cooler biocover, odorants become absorbed in the condensing vapor and adsorbed onto particle surfaces. The cooler more aerobic environment within the biocover is conducive to the decomposition of the captured odorants. Biocovers are generally made of stable materials that neither generate heat nor add to air emissions. In a study commissioned by CalRecycle (Büyüksönmez and Yucel, 2006; Büyüksönmez et al., 2012), the addition of a biocover was shown to significantly decrease the emissions for 24 h of certain odorant chemicals. Adding intermittent irrigation to the biocover can greatly enhance control efficiency. In another study commission by CalRecycle (Horowitz, 2013), combining a biocover with very heavy irrigation reduced air emissions by 98% compared to a windrow with the same feedstock. A more regularly attainable control efficiency for an irrigated biocover is in the range of 80%−90%. The combination of using semioptimized composting conditions to reduce the generation of odorants and VOCs plus the addition of a modestly irrigated biocover generally yields very low air emissions.

10.3 Fabric covers

Compost system designers have applied a wide range of fabric covers to the active composting process to provide a barrier to limit odor emissions. Three different types of covers standout − oxygen-permeable macroporous covers (see Chapter 5), microporous covers, and impermeable polyethylene bags (Chapter 6). Macroporous covers, also known as "fleece," are simple open-weave fabrics that are used primarily to exclude precipitation from windrows and piles. They form a slight barrier to odors by reducing passive emissions (Paré et al., 2000). Microporous covers do form an odor barrier. Microporous covers are a multilaminate assemblage of fabrics that include expanded polytetrafluoroethylene. These covers are used to blanket ASPs. In positive ASPs, as hot air leaving the pile hits the cover, water condenses on the underside of the cover. Odorants are absorbed in the water and return to the composting pile as the water drips back. Polyethylene bags (or pods) provide a high degree of containment. Air is exhausted only through numerous small ports along the bags. Moisture condenses readily on the inner surface of bags, which captures odorants but causes operational challenges.

In many cases, covers have helped facilities obtain permits because they could claim an enclosed process, and so afforded regulators a level of comfort. But covers do not protect against deficient process conditions. At some point, the material must be taken out of the cover or bag. Covering, without following composting fundamentals, can lead to emission of more odors than would have otherwise occurred.

10.4 Negative aeration

Negatively ASPs have a negative pressure, or suction, at the aeration floor and gentle pressure gradient to ambient pressure near the surface of the pile. Assuming the aeration system provides uniform and consistent aeration rates in excess of 2 m^3/h per m^3 of pile or about 1 ft^3/min per yd^3, this pressure gradient draws the vast majority of the pore space air down through the pile and into the aeration floor. A small

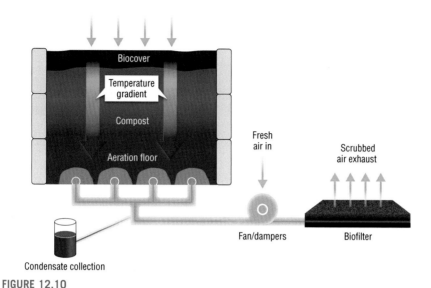

FIGURE 12.10

Diagram of a negatively aerated static pile.

portion of pore space air near the surface might escape, but these regions of the pile tend to have low odorant concentrations as they are cool and highly oxygenated. With the addition of a biocover, the fugitive emission of odorants is even further reduced.[7]

As the air flows from the surface deeper into the pile, it becomes hotter, and its oxygen level drops. In this case, the highest temperatures and lowest oxygen levels occur at the aeration floor. The exhaust air collected in the aeration floor is thus hot, very humid (100% RH), and much more odorous than the air leaving the top of a positively ASP. There is no "outer layer" cooling and oxygenating effect. In all cases, this captured air should be directed to an emissions control device that is generally connected to the outlet side of the aeration fan (Fig. 12.10).

10.5 Buildings

The need to enclose a processing area is often driven by permit requirements, and less often from the results of an odor study or actual odor complaints. Regulators are often quick to ask facility developers to locate processes in buildings to capture fugitive emissions. For a building to capture odor emissions, it must be nearly sealed most of the time, and be maintained under a negative pressure using large capacity exhaust fans. This generally calls for rapid opening/closing truck doors and careful

---

[7] The author has measured surface odor fluxes on at the surface of several negatively aerated static piles and found them to be at background levels.

control of air flow rates and interior pressures. While this practice has been successfully implemented at a number of locations, the full costs are not always appreciated.

Practical drivers for enclosing these processes are first, a highly odor sensitive site, and second, the nature and the volume of the incoming feedstocks. The most common process to be enclosed in a building is the receiving and preprocessing area (tipping floor). If the incoming feedstocks are hot and odorous, this investment can be worthwhile. But compared to active composting, the volume of feedstocks in preprocessing at a given time is generally relatively low, and the retention time is relatively short. Following the recommendations of quick processing, cleaning, and maintenance adequately manages odors without enclosing.

Once an interior space is fully enclosed, certain air quality standards must be maintained to protect the human occupants. Most trucks and loaders use internal combustion engines that exhaust carbon monoxide and other gasses of concern. Worker safety regulations require that minimum air change rates be maintained to avoid the buildup of these gases. The ventilation rate in a building with internal combustion engines (trucks, loaders, etc.) is usually required to be at least equal to 3 air changes per hour (ACH), and sometimes as high as 10 ACH. Most preprocessing buildings have high eves to accommodate tipping trucks. The combination of high air change requirements and big building volumes results in a large air handling and air scrubbing system, and the consumption of a significant amount of electrical power. In a cold climate, the energy requirements can skyrocket as it is often necessary to preheat the make-up air to prevent freezing and to enable the control mechanism to work (as most involve water).

Since trucks drive in/out the receiving building, large doors are needed. Odorous air can escape from the building through the doors. If the resulting odors are a problem, fast-acting doors can limit the amount of open time, but they do not eliminate the outflow of building air.

When active composting is located inside a building, new complexities arise due mostly to the warm humid nature of the air leaving the process. The insides of buildings must be highly corrosion resistant, which is difficult to achieve at large scale. This was demonstrated at a large compost process building in Edmonton Canada that was lined with stainless steel. The building was condemned some years before its design life due to corrosion of structural members. Despite high ventilation rates, low visibility inside compost buildings during cool weather is a significant hazard. Fig. 12.11 shows the typical visibility inside a large ASP composting building.

A direct effect of capturing a large air flow from a building is that the odor scrubbing device(s) must also be correspondingly large. In most cases, building exhaust air has relatively low odor concentrations since it is so diluted with ventilation air. For this reason, it is often desirable to capture and control the concentrated odors, but not to enclose the entire process.

FIGURE 12.11

Fog inside the building enclosed composting facility. This particular building has over 220 kW (300 hp) of continuously operating exhaust ventilation.

Source: ECS.

10.6 In-vessel composting

In-vessel composting systems offer the ability to contain and capture odor emissions via ventilation (Chapter 7). In many cases, the aeration system and the ventilation system are one in the same. In most cases, the exhaust from the vessel is directed to a biofilter (Fig. 12.12).

A typical in-vessel system generally has structural walls, a sealable door, an insulated shell, and relatively low headspace above the compost. One clear advantage of using a vessel instead of a building is that there is no requirement to exchange the air in the space above the composting materials in the vessel; the airflow rate is only dictated by the process. This results in a significantly smaller volume, but of higher

FIGURE 12.12

Exhaust air from a pair of vessels is directed to a small-scale biofilter (A), waiting to be loaded with media (B). The fans above the biofilter draw air from the vessels. The purpose of green drums is to capture condensate before the fan.

odor concentration, airflow than in compost building exhaust air. Like buildings, in-vessel systems are typically designed to operate with a modest negative pressure behind the loading doors such that all leaks are inwards.

10.7 Control methods

Once odorous exhaust air has been captured in a duct, the chemical concentrations of odorants must almost always be reduced, or "controlled," before release. Many smart scientist and engineers have dedicated themselves to developing and manufacturing equipment to do that. Biofiltration is the predominate odor control technology used at composting facilities. It is covered in the following section. This section briefly reviews other physical/chemical control methods, namely scrubbers, stacks, and nonthermal plasma. Although these methods are potentially viable for composting situations, they are seldom used in practice.

10.7.1 Wet scrubbers

There are a wide range of industrial air scrubbing technologies that can be applied to reduce odors and VOCs in exhaust air. To date they have found limited application at composting facilities due to their expense and inherent complexity. Wet scrubbers are among the more common technologies used for odor control in many industries.

Wet scrubbers remove odorants and VOCs from the air by passing the air through a solution of water, usually combined with reactive chemicals like sodium hydroxide or sulfuric acid (Coker, 2012d). The water absorbs water-soluble odors while the chemicals quickly react with target odorants like NH_3 (reacts with H_2SO_4) and hydrogen sulfide (reacts with NaOH). As an example, a scrubber might be used to reduce NH_3 concentrations before biofiltration. The air is exposed to the water so-lution by blowing the air through a reactor. Within the reactor, the water solution is either sprayed as a mist or it trickles through a matrix of media, usually plastic shapes with high surface area and some porosity to allow the air to flow. With the proper design and size, wet scrubbers can be effective at removing their target chem-icals, but they require steady supplies of the reactant chemicals and produce a liquid waste stream.

10.7.2 Stacks

One option for diluting captured odorous air is to exhaust it high in the air; by the time the odorants have returned to the ground, they will have mixed with ambient air and become less concentrated. This practice is rare, and mostly applied in juris-dictions that have the very restrictive odor limits limit at the property line. Also, stacks are of no help in managing the emission of regulated compounds such as VOCs or NH_3 where the actual mass emitted is important, and not just the concen-tration reaching a receptor. Tall stacks, of course, are expensive and require high airflow velocities generated by powerful fans.

10.7.3 Nonthermal plasma

Nonthermal plasma (NTP) creates an ionized gas within a relatively low-temperature reactor by applying an electric field (Sholtz et al., 2015). The free electrons associated with the ions react with chemicals, such as odorants, and convert them to oxidized forms. As an example, NTP converts molecular oxygen (O_2) to ozone (O_3), which a strong oxidant that breaks down most odorants and VOCs (Coker, 2012d). The air to be treated is passed through a catalyst within the NTP reactor. The treatment time is short, but, for composting situations, the reactor(s) must accommodate a high air flow rate due to the large volume of air that requires treatment. NTP may find niche applications in odor control, but its primary potential appears to be for disinfection in food and medical applications (Sholtz et al., 2015).

10.8 Biofilters

Biofilters are broadly used around the world to reduce the concentration of odorants and VOCs in exhaust air. The rule of thumb is that a properly constructed and maintained biofilter can be relied on to reduce emissions of odorants and VOCs by a factor of 10, or by 90%. Biofilters are remarkably tolerant and, unlike most air emission control devices, do not generally require special materials or manufactured elements. A simple "home-made" biofilter, that employs locally available coarse wood as the filter media, often suffices for many small applications. This type of biofilter can be assembled from parts common to basic ASPs (Fig. 12.13). But once the airflows grow large, or when specific performance metrics must be maintained, a more fully engineered approach is called for (Fig. 12.14). Because biofilters eventually need to be taken off-line for maintenance or replacement, it is helpful to have either redundant biofilters or redundant sections within a large biofilter. It should be possible to isolate each section so that any section can be "turned off" while remainder of the biofilter continues to function.

The operating principle of a biofilter is that the odorant chemicals are absorbed into the water film layer that surrounds the media particles. Once the odorous gasses are in solution, aerobic bacteria biooxidize the organic odorants to reduce their

FIGURE 12.13

Modest-sized biofilter with pipe-on-grade aeration floor.

Source: ECS.

FIGURE 12.14

Large biofilter for treating exhaust air from a large, enclosed composting facility.

Source: ECS.

concentrations. Ammonia is also absorbed and converted to less odorous nitrogenous compounds by a somewhat complicated set of pathways (Joshi et al., 1998) and with somewhat lower efficiencies than the classes of compounds.

Many of the same process consideration that determine composting efficiency—namely porosity, temperature, and pH—also are in play in biofiltration. Table 12.3 offers the generally recommended operating conditions for a compost process biofilter. However, these are only guidelines. Numerous biofilters have operated outside of these ranges and have achieved measured VOC reduction efficiencies above 80%.

Table 12.3 Guidelines for biofilter operating conditions.

Operating condition	Value
Media screen size	7–8 cm (3–4 in.)
Empty bed residence time (EBRT)	10–60 s
Media depth	1–2 m (3–7 ft)
Media temperature	4–50°C (40–120°F)
Media pH	5–9
Media moisture content	>50%
Initial pressure drops through fresh media	<17 mm wc per m of depth <0.2 in. wc per ft of depth
Maximum pressure drops through aged media	<67 mm wc per m of depth <0.8 in. wc/foot of depth

10.8.1 Media

An effective biofilter media is resilient (degrades slowly), porous, and has abundant wettable surface area to support microbial growth. Coarse shredded wood is the most common media found in the compost applications that meets these criteria (Fig. 12.15). Other medias include coarse bark, screen overs, and, much more rarely, engineered ceramic substrates found in special proprietary biofiltration systems. Wood-based media retains its porosity longer, if it is a resilient wood type, ideally sourced from stumps and not from branches, and only using the coarse fraction from shredding or grinding.

A small amount of compost added to the media is known to help the correct biome form more quickly. One common preparation sequence is to: (1) coarsely grind the woody substrate, mixing in a small amount of finished compost (2%–5% by mass); (2) then send the mix over a 50–100 mm (2–4 in.) screen; (3) light spray water on the compost-dusted overs as they come off the screen; and (4) use the overs for the media.

10.8.2 Pile depth and empty bed residence time

More odorants are absorbed into moist biofilter media as the time of contact increases. The empty bed residence time (EBRT) is a common metric that infers the length of that retention time. This metric assumes an "empty bed" (i.e., no media) because measuring the porosity of the media, is very difficult, even though it plays a big role in how fast the air travels through the bed (media quality is gauged by other metrics discussed in this section). Calculating the EBRT only requires the

FIGURE 12.15

Typical wood-based biofilter media; with a course, porous, and durable mix of particles. Note the drip irrigation tubing to keep the surface moist.

Source: ECS.

depth (d) in m or ft and footprint area of the media (A) in m^2 of ft^2, and the flow rate (Q) in m^3/s or ft^3/s.

$$EBRT(Seconds) = A \times d/Q \tag{12.7}$$

The calculation suggests that simply increasing the media depth is the answer to improving EBRT. This step, however, causes more compression which reduces porosity and shows up as increasing back pressure for given inlet airflow rate. The deeper the media, the more resilient and porous it must be to perform well. Most biofilters are constructed with an initial media depth of 1−2 m (3−6 ft).

A short EBRT usually suffices for exhaust air from an enclosed feedstock-receiving building, which is almost always lightly loaded with odorant chemicals. On the other hand, the exhaust air stream from poorly managed compost process, with high concentrations of odorants, requires a long EBRT, on the high side of the spectrum listed in Table 12.3.

10.8.3 Media temperature

While mesophilic temperatures are generally favored for degradation of odorants, numerous studies have shown effective biofiltration in the low to medium thermophilic range (<56°C or 133°F) (Moussavi et al., 2009). One drawback of allowing organic media to frequently operate at temperatures above roughly 45°C (113°F) is that the media particles degrade much faster and require replacement sooner. This degradation will show up as reduced media depth and increased back pressure.

The most common method of cooling the media is to mix the hot exhaust air from an active composting process with ambient air. If a large temperature drop is required, greater than 10°C or 20°F, the high energy content of the wet exhaust air requires the addition of a substantial amount of drier ambient air. This increase in total airflow either reduces the EBRT or requires a greater volume of media. Adding cooling air does decrease the concentration of odorants, which generally offsets effect on reducing the EBRT.

Too low of a temperature is not a problem for biofilters handling primarily composting process exhaust air. But it can be a problem when the air volume is made up of mostly cool building exhaust air in a cold climate. As mentioned above, biofilters require moisture to operate. If the building exhaust air temperature is low (e.g., below 5°C or 41°F) and the ambient temperature is well below freezing, the biofilter media can freeze. The most cost-effective method to heat such a cool air stream is to combine it with enough warm compost exhaust air. If this isn't possible, supplementary steam heating may be required.

10.8.4 Media pH

The pH of biofilter media impacts the removal efficiency of various compounds differently. Ammonia is most efficiently scrubbed in an acidic media. Many common odorants are most efficiently scrubbed at near neutral pH, but some are more effectively removed at either acidic or even alkaline pH (Liu et al., 2008). In general, biofilters can effectively scrub odors from compost exhaust air over a broad pH range.

When a biofilter loses removal efficiency due to pH, it is most often a result of acidic conditions (low pH). Acidic conditions develop by overloading the media with organic acids and/or excessive temperatures, especially when acidic water droplets are entrained in the exhaust air. Consistently acidic (pH <6) exhaust air, measurable only in condensate, is an indicator of poor composting conditions. If the process cannot be corrected, a wet scrubber, dosed with pH adjusting chemicals, can be used to neutralize the pH of the exhaust air and remove certain chemical odorants.

10.8.5 Media oxygen content

The aerobes that break down the odorants rely on the availability of oxygen in the water film layer.Low oxygen levels in the pore spaces of the biofilter media is rarely a problem. Compost aeration systems that provide enough airflow to affect even a modest amount cooling supply excess oxygen, (generally measuring >15% in the exhaust air). The addition of ambient air further raises the oxygen levels. For these reasons, regular measurement of oxygen levels in a compost biofilter is seldom of any value.

10.8.6 Pressure drop through media

Fortunately, most well-structured organic media, operated at reasonable conditions, lasts two or more years. However, over time all biofilter media loses porosity. Common causes include loss of structure through biological degradation, growth of biomass (the aerobes doing the scrubbing are being fed and are multiplying), and the accumulation of dust. For a given airflow rate, reducing the pore space in the media causes the air to flow through the bed faster and thus the back pressure at the inlet increases. If the pressure increases enough, the air flow finds channels of least resistance through the bed and does not flow uniformly. This effectively reduces the media surface area and the effectiveness of the biofilter.

When considering pressure drop across the media, it is important to separate the pressure drop caused by the ducting and aeration floor from the pressure drop through the media itself. The first step is to measure and record the inlet pressure at the design flow rate of the biofilter system prior to the addition of media. It is also a good idea to place a layer of very coarse material immediately over the aeration floor to minimize plugging at the orifices. Once the moist biofilter media has been placed, the inlet pressure at the design flow should be measured again. These measurements should be repeated roughly every 3—4 months as the change in porosity is slow to occur. As the back pressure approaches the upper limit given in Table 12.3, a plan to replace the media should be developed.

10.8.7 Media moisture

The media needs to remain moist in order to have the water film layer present to absorb the odorants and support the aerobes that biooxidize them. When the air stream to be scrubbed originates directly from an active composting process, the air is almost always fully saturated and warm; it is carrying lots of water. As that air travels through the media and toward the surface of the biofilter it cools, and significant amounts of water condense on to the media and even drip out the bottom (plan for drainage around a biofilter). The surface of the biofilter may look dry, but generally digging a few inches down will reveal moist media. If the media appears dry for deeper than roughly a foot below the surface, irrigation should be considered.

For dryer airstreams, such as those from a building, steps must be taken to avoid over drying the media. The most effective wetting mechanism is to saturate the incoming airstream. This requires either a wet scrubber or an in-line misting system. Achieving truly saturated air is tricky and best done with expert advice. The most common method of maintaining media moisture is irrigation. Surface irrigation is easy to apply but does not tend to wet evenly at depths over a half-meter (1—2 ft) from the surface. Some designers have used buried drip line laid out on a tight grid (12—16″). This appears to work well but complicates placing and removing the media.

10.8.8 Biofilter monitoring and maintenance

Once a biofilter is in place, it generally does not require much effort to operate and maintain. It is good practice to measure temperature and visually inspect the media every few weeks. Visual media inspection should include estimating depth, looking for surface cracks, assessing moisture, and looking for plant growth, which should be removed upon discovery. Every 3—4 months, the backpressure should be measured at the biofilter inlet when air is flowing at the design rate. The media pH should be checked with pH paper at the same frequency (Chapter 11).

If the media depth has fallen below the minimum, or surface cracks are apparent but surface back pressure is still acceptable, an additional layer of biofilter media can be added to the surface. This will extend the time before the entire media requires removal and replacement.

When the backpressure is approaching the design maximum, new biofilter media should be collected. If possible, the old media can be processed on a course screen (75—100 mm or 3—4 in.) and the overs can be mixed into the new media. This has the dual benefits of decreasing the amount of new media required and preinoculating the media so that it is effective at biooxidizing the effluent chemicals more quickly. Otherwise, the old media can be mixed with feedstock and composted.

11. Odor migration and dispersal

Odor migration is the critical link between odor emissions at the composting facility and odor perception by a receptor. Most odorants emitted into the air are destined to disappear through chemical reactions within the atmosphere and/or they return to the earth with precipitation and other forms of atmospheric deposition. However, much sooner, odorants *disperse*. They mix with, and become diluted by, the air within the atmosphere. In the meantime, the odor-carrying air volume remains somewhat intact. As the air volume travels, it gradually loses its integrity and the odorant concentrations decline.

Odorous air does not always go up. If a moving volume of odorous air is cooler than the air above, it moves horizontally or downwards. If the air hovering atop a composting facility is cooler than the air above, it stays near the ground, and it can "drain" downhill. This phenomenon, called *air drainage*, greatly depends on atmospheric conditions and the local topography.

How quickly odorants disperse, and how far they travel, largely depends on the atmospheric and weather conditions. The properties of the odorant also matter. Chemicals like NH_3 and hydrogen sulfide (H_2S) tend to dissipate rapidly while VOCs tend to persist. Topography and landscape are additional factors. The distance and direction to neighbors are very important in determining whether or not odors sufficiently disperse before they arrive.

11.1 Plumes and puffs

When odorants, or any other pollutants, are released into the air, the surrounding air assimilates them. The odorants do not move independently but travel within the assimilating air parcel and only gradually diffuse, decompose, or wash out. The moving parcel of air retains its identity (and embedded odors), but its composition dilutes over time and distance. How quickly that dilution occurs depends on the power of the forces that tend to diffuse, dilute, disperse, and chemically alter the compounds within the air. With little turbulence, wind, or other disturbing forces, the air travels intact for some distance while gradually mixing and spreading into the surrounding ambient air.

In air pollution terms, this intact but spreading contaminated air stream is called a plume. The vapor plume from the exhaust stack of a power plant is a good visual example. However, most emissions from a composting facility occur over a broad surface, rather than a stack. Most composting odor sources are considered "area sources." An area sources can be thought of as a plume that has already spread to the width of the area (Liu et al., 1997).

The wind, atmospheric temperature profile, and also the surrounding landscape determine the plume's direction, shape, and longevity. If the plume remains warmer than the surrounding air, the plume continues to rise. If it cools below the atmosphere around it, it may sink. Plumes disperse by diffusion, wind, and rising air currents. Because wind and air currents change constantly and instantaneous, so do the plumes. Under relatively calm conditions, plumes tend to disperse gradually, primarily by diffusion (de Nevers, 2000). The diffusion causes the plume to spread horizontally and vertically along its path of travel. At any moment, turbulence can cause the plume to disperse faster and take on a disorderly shape.

The way in which a plume spreads—its spreading shape—can be represented by mathematical formulas (e.g., Gaussian distribution). To anticipate how a plume might behave, air pollution professionals use these formulas in dispersion models that define the plume's spreading pattern under assumed conditions (Box 12.8). The spreading pattern in turn correlates with the movement, concentrations, and dispersal of pollutants within the plume. Dispersion models attempt to predict pollutant concentrations, like odor, at various directions and distances from the source.

A plume represents odor emissions in a simple manner—as a continuous stream of emissions at a constant strength and rate. At most composting facilities, most odor sources emit odor intermittently, at varying rates and concentrations. A good example is the odors emitted upon turning. In such cases, odor emissions can be modeled as "puffs" rather than plumes. Puffs are a discrete and brief burst of odor release. Each puff can differ in odor concentration and character.

Box 12.8 Atmospheric dispersion modeling

Atmospheric dispersion models attempt to predict the impact of a source of odors, or specific air pollutant, on the surrounding environment. The models make these predictions by assuming various atmospheric conditions (e.g., wind speed and direction) for the area in question and then calculating the dispersal and dilution of the odor emissions in one or more directions (Fig. 12.16). The models

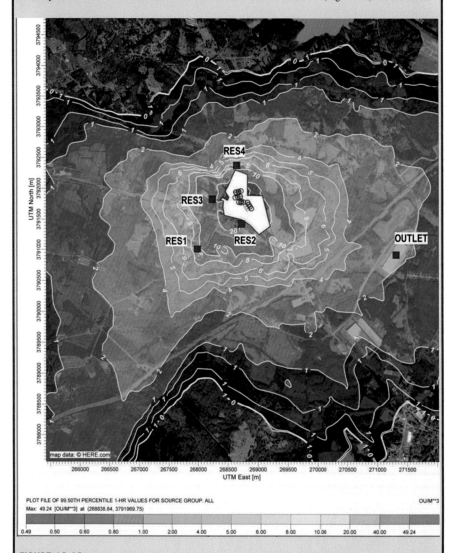

PLOT FILE OF 99.50TH PERCENTILE 1-HR VALUES FOR SOURCE GROUP: ALL

Max: 49.24 [OU/M**3] at (268838.84, 3791969.75)

OU/M**3

0.49   0.50   0.60   0.80   1.00   2.00   4.00   5.00   6.00   8.00   10.00   20.00   40.00   49.24

FIGURE 12.16

Example of the results from an atmospheric dispersion model. These results show the dilution of odor concentrations (in ou/.m^3) in all directions and various distances from the composting facility (white figure at center). The colors and lines indicate different ou concentrations.

> ### Box 12.8 Atmospheric dispersion modeling—cont'd
>
> employ sophisticated calculations, and their accuracy depends on having good input data. Those input data include: the typical range of atmospheric conditions for the site; land features, including topography, location and size of buildings, and vegetation; location of possible odor sources; and the OER or odor flux from sources on the site. Most odor sources at most composting sites are area sources. Measuring an OER from area sources is difficult and quantifying it over the range of changing conditions is inexact. Validating the quantitative results of odor dispersion models requires sampling and/or measurement of odor levels in the field, which is not easily done (Capelli et al., 2013). Therefore, atmospheric dispersion modeling for odors is an exercise in approximation, at best.
>
> Despite the challenges, atmospheric dispersion models are still among the most useful tools for gauging, in advance, how a particular composting facility might impact a particular neighborhood. In fact, in some places, dispersion models are used to establish regulatory criteria, such setback distances and/or odor concentration limits at the property line. These models are particularly useful for "what if" type of analysis, such as, "what if the wind is blowing only slightly to the southeast?" or "what if the biocover is used on new windrows?."
>
> Additional information discussing atmospheric dispersion models and their benefits and limitations for odors can be found in Balch et al. (2019) and Capelli et al. (2013).

11.2 Atmospheric factors

The atmosphere primarily determines the fate of an odorous air plume (or puff). The desired effect is rapid and substantial dilution, which is achieved by rapid and substantial mixing of the plume with the surrounding air. Mixing, in turn, is enhanced by atmospheric turbulence. Turbulence occurs horizontally (i.e., east, west, north, south) and vertically (up and down). Composters should hope for both. Short of dilution, a second desired effect is movement of the plume (or puff) in the direction of no one, or at least no one nearby. The best direction is straight up into a turbulent zone of the atmosphere. The next best direction is one that is opposite from objecting neighbors, if such a direction exists.

Wind and temperature are the atmospheric factors that chiefly decide the dilution and movement of an air plume (or puff). Wind drives the mixing and movement of air in the horizontal plane. Temperature plays a larger role when winds are calm. Temperature also determines the vertical stability of the atmosphere, which affects both air movement and turbulence. Wind and temperature are not only important at ground level but also at various higher altitudes in the atmosphere (e.g., up to about 2 km or about 1 mile).

11.2.1 Turbulence in the horizontal plane: wind speed and direction

Wind has positive and negative consequences. It disperses and dilutes odor-laden air, but it also carries odors to the neighbors. The net good or bad lies in the wind's strength and direction. The worst case is a slight wind (<8 km/h or 5 miles/h) in the direction of the most sensitive neighbors. The best case is a strong turbulent wind (>16 km/h or 10 miles/h) that frequently changes direction. In general, a brisk wind is usually good. However, the specifics of the site and situation determine whether that statement holds true. It is risky to relying on the wind as insurance against odor complaints because it is seldom consistent (Box 12.9).

Composters are generally advised to avoid site locations where prevailing winds blow toward sensitive neighbors. This advice is sound, but often difficult to follow. A favorable prevailing wind does not in itself preclude odor impacts in other directions. First, it is not unusual to find sensitive neighbors in all directions around the facility. Second, winds blow in the prevailing direction only sometimes, and not consistently so. The direction of prevailing winds is influenced by local topography including valleys and canyons, mountains, and large water bodies. Where these features exist, the prevailing winds are more predictable. Where they are absent, such as on broad inland plains, wind direction tends to be more variable.

For a given location, wind direction patterns can be ascertained from wind data collected at nearby weather stations. Wind data for a particular location is often presented as a *wind rose*. Fig. 12.17 presents a wind rose based on wind data collected from April 2015 to April 2016. in Binghamton, NY. A wind rose is a circular graph with 16 wind directional divisions and "spokes" radiating from the center. The spokes show the direction that the wind is blowing *from*. The length of each spoke represents the percentage of time the wind comes from the direction of the spokes.

Box 12.9 Wind basics

Wind is created by pressure differences in the atmosphere, primarily due to uneven solar heating and uneven terrestrial and radiational cooling. In general, wind direction and speed are influenced by the following factors (CalRecycle, 2007, after de Nevers, 2000):

1. *Storms and weather fronts*: The relative proximity and difference between high- and low-pressure centers primarily determine wind directions and speed, over and above other factors. Larger pressure differences and steep pressure gradients create stronger winds. A weather map showing isobars (lines of equal pressure) indicates the relative wind strength and direction.

2. *Land features*: In absence of overriding influences from storms and weather fronts, winds tend to follow valleys and canyons and locally blow up and down mountain slopes as the land surface heats and cools. In these situations, the wind typically reverses direction from morning to night, and between cold and warm seasons. The effect is greater with deeper or steeper features.

3. *Onshore and offshore breezes*: Winds blow on- and off-shore of water bodies, with the direction depending on the relative temperatures of the land and water. When the water body is cooler (summer and/or afternoons), the breeze blows onshore, replacing warm air over the land that has heated and risen. When the land surface is cooler (winter and/or at night), breezes move toward the water (offshore).

4. *General wind patterns*: When the foregoing effects are negligible, winds tend to be light and follow the dominant wind patterns on the earth's surface due to general atmospheric air circulation (e.g., trade winds, westerly's).

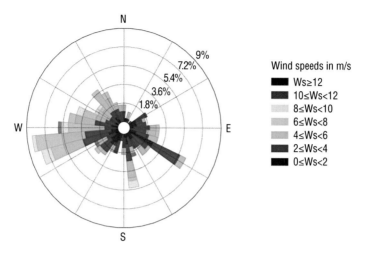

FIGURE 12.17

Wind rose for surface station at Binghamton, NY airport for the period April 2015–April 2016. Binghamton is an inland city in central New York, situated in a river valley among low mountains (1 m/s = 3.6 km/h = 2.24 miles/h.).

Source: NRCS, 2021.

The color bands on the spokes, and their thicknesses, correspond to the relative distribution of wind speeds. For example, Fig. 12.17 shows that the wind came from a direction slightly south of west more often than any other direction, and that the wind speed rarely exceeded 8 m/s (18 miles/h). The wind came directly from the west slightly more than 5.4% of the year. The wind rose also shows that the generally western prevailing wind prevails less than a third of the time. The direction of light winds may be the more important segments of the spokes (darker blues). Although the wind rose shows light winds came from the southeast most often, lights winds originated from almost all directions.

Wind roses can be generated for different locations and time periods by businesses, organizations, and government agencies that have interests in weather, climate, and/or the environment. In the United States, wind roses for many weather stations are available from the Natural Resources Conservation Service (NRCS, 2021). Some universities provide this service for in state or regional locations.

Wind roses provide a good visual representation of the wind characteristics for a given location. However, wind roses should be used with some caution. Most wind roses are developed from data collected at established weather stations. Wind characteristics can vary greatly among specific sites, even sites that are relatively close together. The differences can be due to landscape features, buildings, topography, and simply local variations in the weather conditions. Also, the data covered

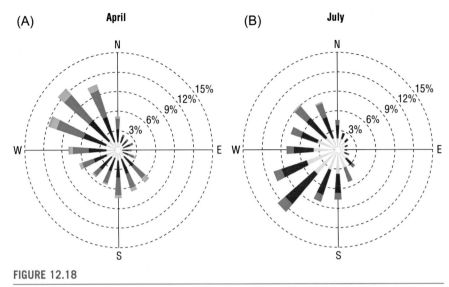

FIGURE 12.18

Two wind rose for Binghamton NY based on data collected in April 1961 (A) and July 1961 (B).

Source: NRCS, 2021.

by a wind rose deserve scrutiny. For instance, Fig. 12.18 presents two wind roses that exhibit different wind patterns, yet both are from Binghamton, NY (the same location as Fig. 12.17). The wind rose on the left (A) is for the month of April and the one on the right (B) is for July (1961 in both cases). A composter in Binghamton should likely pay more attention to the July wind rose since odors are more likely to be an issue then. Wind roses and other weather data from nearby weather stations remain good resources for planning, managing, and troubleshooting composting sites. Nevertheless, it is advisable for a composting facility to have, and properly install, a weather station of its own on-site (Fig. 12.19).

11.2.2 Turbulence in the vertical plane: temperature and atmospheric instability

Vertical turbulence and mixing occur as air rises and falls, largely due to temperature differences at difference altitudes. Vertical mixing happens when the atmosphere is *unstable*. When conditions hinder vertical mixing, the atmosphere is stable. An unstable atmosphere is good for composters because the plumes and puffs of odors are carried upwards, where they are very unlikely to encounter an objecting human nose. When air rises, it is also more likely to be diluted and dispersed by the more vigorous winds aloft. Table 12.4 lists conditions that lead to various levels of instability in the atmosphere.

FIGURE 12.19

Weather station mounted on the office roof at a composting facility.

Table 12.4 Atmospheric stability categories. See Haug (1993) for more detailed explanation of the solar and cloud cover conditions.

Surface wind speed (m/s); measured at a height of 10 m	Day Incoming solar radiation level			Night Cloud cover	
	Strong	**Moderate**	**Slight**	**Thinly overcast or ≥ 4/8 low cloud cover**	**≤3/8 cloud cover (≤3/8)**
<2	A	A–B	B	–	–
2–3	A–B	B	C	E	F
3–5	B	B–C	C	D	E
5–6	C	C–D	D	D	D
>6	C	D	D	D	D
	Assume neutral conditions (D) whenever overcast, either day or night				

A = extremely unstable; B = moderately unstable; C = slightly unstable; D = neutral; E = slightly stable; F = moderately stable.

Normally, the ambient air temperature decreases from the ground upward. Air near the ground surface is heated by the sun and becomes less dense compared to the cooler air above. The warm air below rises and the cool air above falls, creating vertical mixing. However, if the air near the ground (e.g., at the composting facility) is cooler than the air aloft, it rises only slightly, or not at all. This situation is known as a temperature inversion (Fig. 12.20).

The foregoing description is a simplification. A rising plume of air can be blocked even when inversions are not present. Because a plume of air expands and subsequently cools as it rises, its ascent also depends on its rate of cooling. To continue its upward trajectory, the rising plume must remain warmer than the surrounding air along the way up. The plume's rise is at the mercy of its rate of cooling, referred to as its "lapse" rate, and the vertical temperature profile of the ambient air.

For composters, temperature inversions are unwelcome events. They dampen vertical mixing and keep potentially odor-laden air near the ground, where it might travel to neighbors at ground level, carried by light winds or air drainage (Section 12.2.3). Temperature inversions are correlated with numerous odor complaints and severe odor complaints (Haug, 1993). Therefore, composting facilities should recognize and anticipate the conditions that lead to inversions.

Temperature inversions occur regularly due to diurnal heating and cooling when winds are calm. After a sunny day, during which the atmosphere and ground warm, radiational cooling reduces the air temperature near the earth's surface, while the air above remains warmer. A temperature inversion begins to set up, typically starting a few hours before sunset. If there is little wind and few clouds through the night, the inversion strengthens and heightens until the sun rises. A few hours after sunrise, the ground and the adjacent air become warm enough to generate circulating air currents

Normal atmospheric condition | **Atmospheric temperature inversion**

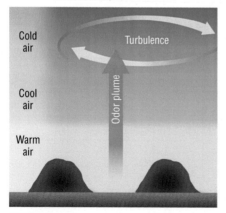 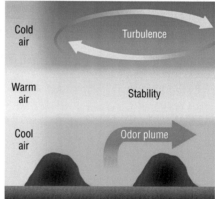

FIGURE 12.20

A temperature inversion occurs when a layer of warm air overrides cooler air near ground level.

and the inversion falls apart. This diurnal pattern explains why many composting facilities receive more odor complaints in the early morning and early evening. In such cases, it might be prudent to restrict odor-generating activities, like turning, to a mid-day window of time.

Inversions also brought on by other meteorological and/or geographic situations. When a warm front meets cool air, or cold air meets warm air, the cool air plows wedges under the warm air, creating an inversion that could last for days. Similarly, air cooled by an ocean or large lake pushes warm air upward as winds carry it onshore. A sinking upper-level high pressure system can lead to an inversion. The air sinks to a lower altitude, it becomes denser and subsequently wars, creating an inversion layer.

11.2.3 Cool air drainage

When the air near the earth's surface cools due to radiational cooling, the cooler low-level air hugs the ground. If a parcel of air is colder, and consequently denser, than the air at lower elevations, the parcel flows down slope. Only a slight slope is necessary. If winds are light, cool air tends to follow the "lay of the land" and collect in low spots. The draining air carries with it whatever odorous compounds are embedded.

Cooling occurs faster in wide-open expanses with a clear "view" of a cloudless night sky. In contrast, some areas remain warmer at night, shielded from the sky by trees, steep slopes, canyon walls, or buildings. These radiation-blocking elements are often common to lower elevations (e.g., river valleys). In developed landscapes, the air remains warmer due to the "urban heat island" effect, due to the thermal mass of features like asphalt parking lots and concrete buildings. Unfortunately, urban and suburban areas are prime air-drainage targets because of the heat island phenomenon and because cities and towns develop in river valleys.

A composting site is typically an expansive area of open land, exposed to radiational cooling. Sites that are located uphill of development may experience early morning odor complaints from downhill neighbors after clear relatively calm nights. Neighbors located in a nearby low spot, or bowl are most susceptible. A facility that is located in a natural air drainage "channel" should be considered with caution. As the next section describes, physical barriers and fluctuating topography disturb air drainage patterns, increasing turbulence and dispersal of odors.

11.3 Topography

The topography surrounding a site can greatly affect air movement in a number of ways. Topography can provide barriers to air flow (e.g., wind breaks), establish channels for air movement (e.g., drainages; utility line clearings), increase turbulence (e.g., trees) and generally creating microscale air circulation patterns.

A facility can use topography to its advantage. Retaining or planting a row of trees around the site can interfere with air drainage and create turbulent eddies when the wind is blowing (Chastain and Wolnak, 2000). A soil berm, tall fence or hedge can have a similar but lesser effect. On a tree-less site, fast-growing tree species can be planted along the perimeter for the longer term. In addition to the enhanced turbulence and odor dispersal, trees, fences, or soils berms along the perimeter provide a visible barrier that shields the facility from the public, and vice versa.

11.4 Distance

There is no better insurance against odor complaints than an isolated site, far from discerning neighbors. A long distance between an odor source and its nearest windward receptor provides time for the dispersal mechanisms to work.

The traditional way for composting facilities to establish "safe" separation distances is to observe recommended, or required, "setbacks." Setbacks are simply the distance established to create a buffer between the facility and nearby features, like wetlands, wells, and residences. They are also called buffer zones or buffer distances. Regulations commonly dictate minimum setbacks (Chapter 10). Before 2000, recommended setback distances between composting facilities and residences initially ranged from 50 to 150 m (about 200–500 ft). In many jurisdictions, they remain within this range (especially in the United States). Overtime, however, setback distances have tended to increase, surpassing 500 m (1640 ft) in some instances.

In some jurisdictions, setback requirements have become conditional on the situation at hand including facility size, feedstocks, composting process technology, and surrounding land uses. This approach is demonstrated by Table 12.5, which summarizes the current situation in Australia (Balch et al., 2019). As the table shows, the approach to establishing minimum setback distances vary among the states. In South Australia, setbacks depend only on facility size while the system in Victoria considers facility size, feedstock, and technology.

Blanket setback distances are increasingly being replaced by atmospheric dispersion models that predict odor impacts—based expected OERs at the facility and specific local conditions. Acceptable separation distances are determined by comparing results from the models with OIC, established by regulations and/or regulatory judgment.

It is worth emphasizing the nature, topography, and management of a specific facility, plus the attitude and expectation of its neighbors, determine the frequency and distance of odor complaints. Some composting facilities operate successfully within a few city blocks of residential neighborhoods. Some facilities in seemingly remote locations have been plagued by odor complaints from distant neighbors. Although setting a minimum setback distance provides some degree of protection from odor impacts, it is very difficult to establish a good general standard, even with atmospheric modeling. Models require good input data and obtaining good input data is challenging. Finally, setback distances cannot protect a facility from the incessant march of urban and suburban development.

Table 12.5 Examples of the criteria used to establish recommended setback distances among selected Australian states.

State	Technology	Types of feedstocks	Size of facility (tonnes/yr)	Recommended separation distance (meters)
Victoria	Open air receival, enclosed aerobic composting with secondary odor capture, and treatment, open air maturation	Green waste, vegetable organics, grease inceptor trap waste	1200 14,000 36,000 55,000 75,000 90,000	>300 >500 >800 >1000 >1200 >1400
	Open air receival, open turned windrow, open air maturation	Green wastes	1200 14,000 36,000 50,000	>600 >1100 >2000 >2000
Western Australia	Outdoor covered, turned windrows	Manures, mixed food/putrescible, and vegetative food waste	Not specified (NS)	750
		Biosolids	NS	250
		Green waste	NS	150
	Outdoor covered windrows with continuous aeration	Manures, mixed food/putrescible, and vegetative food waste	NS	500
		Biosolids	NS	250
		Green waste	NS	150
	Enclosed windrows with odor control	Manures, mixed food/putrescible, and vegetative food waste	NS	250
		Biosolids	NS	150
	In-vessel composting with odor control	Manures, mixed food/ putrescible, and vegetative food waste	NS	150
		Biosolids		150
South Australia	All technologies	All feedstocks	>200 >20 and < 200 <20	1000 300 100

Adapted from Balch et al., 2019.

12. Neighbor and community relations—complaints and more

With any waste or odor-generating activity, public scrutiny and opposition are inevitable. Objections most often arise from odors or fear of odors. However, other objections also disquiet the public including traffic, changes to the landscape, property values, or general fear and mistrust of not knowing what goes on "over there" (Goldstein and Beecher, 2005; Coker, 2012e). Regardless of the reasons, people have legitimate concerns that can cause them to oppose composting in their neighborhood or community.

It has become apparent that composters have a better chance of operating peacefully and successfully if they preemptively seek out neighbors and community stakeholders, rather than shun the public until a controversy erupts. As the work by Beecher et al. (2005) reveals, it is a mistake for a composting facility to try to operate "under the radar." Once an odor complaint occurs, the radar will zoom-in on the facility anyway.

12.1 Neighborhood and community serenity commandments

While it is often extremely vexing, the best way to interact with neighbors, and the community in general (including regulators), is with transparency, honesty, respect, and friendliness. It is the winning strategy in the long run in most situations. Some opponents of the facility will not be appeased by such cooperation, but not everyone has to be appeased. The goal is to show the majority of the public, and hopefully the most influential stakeholders, that the facility is more of an asset to the community than a liability (Coker, 2012e).

There are several good publications and web pages that provide information about public relations, starting with Beecher and Goldstein (2005). Good information is also available from agricultural sources because farmers also face complaints from neighbors unaccustomed to farming practices (e.g., manure spreading). The advice from these sources can read like moral commandments, so that is how it is presented here (there are more than 10, and "Thou shalt" and "Thou shalt not" are implied). Specific mini commandments are presented in Table 12.6. The key is to adopt these practices with an attitude of sincerity.

1. *Develop a public relations program:* Called a public "relationship" program by Beecher et al. (2005), establish a formal dialog with neighbors and members of the community to listen to their concerns, suggestions, ideas, and feedback. It provides a channel to inform the public about the composting facility and composting generally.
2. *Establish a neighborhood advisory or liaison committee:* The committee should be comprised of a cross section of neighbors and other stakeholders that can provide constructive information to the facility (including when odors are detected). It also provides a means for you to communicate to neighbors.

Table 12.6 Neighbor and community relations mini commandments. Some of these commandments apply always; some only in specific situations.

Preemptively engage	Be nice	Respect regulators
• Identify key community stakeholders, seek their input. • Cultivate friends, many friends, the right friends. • Provide information to anyone readily and regularly. • Hold tours; let people see the facility and the compost. • Don't shy away from conflict, address it calmly. • Use positive language/words ("feedstocks," not "waste"). • Show off your environmental credentials and benefits.	• Be polite, sincere, cooperative, transparent, and honest. • Don't complain back to complainers; if you feel harassed, tell your regulator. • Take neighbor complaints seriously (they may know something that you don't). • Respond to complaints promptly. • Don't hold back apologies. • Don't hold a grudge (grudges makes matters worse).	• Be personable, cooperative, polite, honest. • Heed the rules. • Don't be late to meetings. • Write everything down (including names and titles). • Request decisions and opinions in writing. • Keep very good records at the facility. • Don't violate your permit. • Don't let small problems escalate.

Operate well	Be an asset to the community	Be an asset to the neighbors
• Follow good composting practice. • Avoid odorous feedstocks if possible. • Stay small if necessary. • Make the sight attractive. • Pick up trash and debris (on-site and beyond). • Control dust and minimize noise. • Keep good records; keep a daily journal. • Plant trees; don't cut down trees. • Preserve open space around the facility. • Allow hiking or biking on undeveloped areas of the site.	• Inform the community members of facility activities that might impact them. • Remain aware of events in the community (e.g., fairs). • Be sensitive to traffic; make your customers sensitive to traffic. • Serve on local advisory boards and committees. • Build a community garden. • Support a local sports team. • Interact with the garden club; give them a facility tour. • Support backyard composters; help with bin give-away programs. • Give away compost. • Donate to political campaigns. • Live in the community.	• Get to know your neighbors; you will meet them eventually. • Inform neighbors of facility activities that might impact them. • Remain aware of events in the neighborhood (e.g., picnics). • Establish an odor monitoring network in the neighborhood. • Wave and smile to neighbors. • Help neighbors with tasks; use your equipment, if appropriate. • Plow snow from a few driveways; mow a few lawns. • Give away compost. • Plant a vegetable garden on-site for the neighborhood. • Live in the neighborhood.

The committee provides two-way communication and builds trust and goodwill. The committee should meet regularly, on a time frame dictated by the specific situation.

3. *Invite the public in:* It can help for neighbors and the community to see what the facility looks like and how composting takes place. Hold a facility tour or an open house to help counter uncertainty, mistrust, and misinformation. With these events, provide a few perks for the public, like food and/or free compost. Another option is to host classes on the site; classes about backyard compost, using compost in the garden or just gardening topics.

4. *Establish official lines of communication*: Create a dedicated phone, email, or internet "hot line" for odor complaints and other issues. It is better if neighbor complaints come to you rather than the authorities. It should be easy and comfortable for the public to connect with you. Any phone or email messages left by neighbors should be promptly addressed and replied back.

5. *Keep up appearances:* People want and appreciate an attractive neighborhood. A composting site can be a disorganized, unsightly place, but it does not have to be. Whatever you can do to improve the appearance of the site, also improves the prospects acceptance into the neighborhood. To whatever degree possible, maintain the perimeter with trees, shrubs, and other vegetation. Keep fences intact and painted. Make the entrance to the facility attractive and well-landscaped, using the compost produced to enhance the landscape (which also serve to promote compost use). Remove from the public's view litter (regularly), debris and rusting equipment.

6. *Accentuate the positive*[8]: Composting provides a public service (recycling) and environmental benefits (soil improvement, carbon conservation). The neighbors and the community at large are likely aware of composting's advantages. Still, during discourse about the facility, it only helps to remind stakeholders of the benefits of composting. Reference state or local policies and programs that encourage composting and compost use.

7. *Eliminate the negative*[9]: It is instinctive to react negatively, even with anger, to complaints, criticism, and accusations hurled by neighbors, political figures, and the media. It is important to squelch this instinct, no matter how difficult and unfair. In fact, the opposite attitude should be adopted, no matter how difficult and unjust. Comments and complaints from neighbors and public officials should be treated with respect, presumed to be legitimate and sincere, and accepted as positive feedback. Sincerely listen to what is being said, without prejudice. Your response should be honest, transparent, and empathetic, as if you are working on behalf of the community. Investigate and resolve any problems that the neighbors and community, with the assistance of the regulatory agencies as necessary.

---

[8] With respect and apologies to Johnny Mercer.
[9] Ibid.

8. *Prepare the staff:* At least one, if not all, members of the facility staff should be trained in conflict resolution and methods of dealing with the public. Have at least one staff member proficient in odor science and measurement.

9. *Don't shove science down their throats*: When confronted with accusations of wrong doings, one's first instinct is to flaunt the science, brandishing publications supported by research, and expert consensus. This strategy can be off-putting and ineffective with upset neighbors. First, they may not understand, nor care about, the science. They know what they smell. Second, the scientific minutia maybe too detailed for their needs. Third, if they already question your motives, they will also question "your" science. It is wise to bring up the body of research and science that supports the facility's practices. However, don't overplay it; and don't expect all of the neighbors to buy it. The better audience for scientific evidence is the group of stakeholders who are science-driven and more objective, including regulatory agents, arbitrators, government professionals, and influential members of the community who are technically oriented. Good science is a good asset, it just takes time to have its effect.

10. *Earn credibility and trust:* Most composters start with a credibility problem, just because people are naturally skeptical, especially of profit-minded organizations. You have to work to build credibility. Once credibility is established, you can work on building trust. The keys to earning credibility are objectivity, honesty, and transparency (Coker, 2012e). Relying on, or being accepting of, independent stakeholders, like government agencies, also helps, both with credibility and trust.

11. *Join the community:* It is easier to be accepted by a community if you are part of the community. There are many opportunities to join—serve on committees, support public causes (e.g., library, public garden), attend meetings, etc.

12. *Treat regulators with respect:* The people responsible for regulating the facility should be looked on as allies. They generally want to be your allies. Treat them with respect, talk with them as mutual professionals, give information freely and voluntarily, and admire them for the difficult jobs that they do. Avoid anger and angry exchanges. Accept regulators' suggestions with an open mind, not defensively. They can be persuaded by sound science and research. However, regulators also need to be kept somewhat "at arm's length" to maintain impartiality.

13. *Don't fight change (immediately):* In an attempt to resolve an odor issue, operational or facility changes are often suggested by consultants, regulators, or neighbors. Avoid meeting these suggestions with immediate scorn. Instead, consider them as opportunities for possible improvements. Although suggested changes often require effort and investment, they also can bring advantages, some which can save money or make composting easer (e.g., putting a roof over the feedstock storage bins). When a change has been proposed, evaluate it objectively and act on it accordingly.

14. *Write everything down:* Take judicious notes whenever communicating with neighbors, community members, regulators, and the media. Good notes help

when remembering who's who, what was said, and who said it. It also protects you when questions arise later, especially if someone else misremembers or presents "alternative facts." Write down names, titles, dates, times, notable quotes, and other salient information.

15. *Be a good neighbor:* Community relationships may rest on this commandment. Being a good neighbor means more than avoiding problems. It also means fitting into the neighborhood, getting to know the neighbors, helping fellow neighbors, making friends, building good fences, and tearing down bad ones. As much as possible, strive to be the kind of neighbor that you would like to have "next-door."

16. *Start early:* Start practicing the foregoing commandments as soon as possible; before the opening, in the case of new facilities; right now, in the case of existing facilities. Credibility, trust and goodwill will come sooner.

12.2 Odor complaint response

An odor complaint may come to a composter directly from a neighbor (i.e., receptor) or indirectly through the local government or regulatory agent. In the first case, the composting facility is more likely to have a primary role in investigating and resolving the odor complaint. In the second case, that role may be obligated to the regulatory authority. In either case, the composter's response should be one of concern, not annoyance, denial, or anger. If a problem existed, or exists, the best response is to be honest about it and describe how it was or will be resolved. This approach builds credibility. All situations are best handled when the composter is calm, respectful, sincere, and methodical.

The following suggestions provide some guidance for handling odor complaints. CASA (2015) provides detailed advice for interacting with a caller reporting an odor and for determining the steps to take following a complaint. This publication includes a sample odor complaint form and decision tree for prioritizing the complaint.

1. Set up a system for receiving complaints (e.g., telephone, email, webpage).
2. Designate at least one staff member to handle complaints (usually a facility manager). Staff should be trained in the art and science of dealing with the public and fielding complaints.
3. Gather the information from the "caller." CASA (2015) offers the following advice:
   * Remain calm and respectful when interacting with the caller,
   * Listen without interrupting or judgment,
   * Use a form that helps gather the complaint information,
   * Answer questions about the information being gathered,
   * Understand what can be and cannot be said to callers, and,
   * Explain the next steps in the process.
4. Every odor complaint that the facility receives should be recorded in a logbook (or equivalent computer file). The record should include as much information

as possible to help resolve the odor complaint (person's name time, date, location, duration, odor character, weather conditions, etc.). Odor complaint forms can be filled out electronically or on paper. In the latter cases, the forms should be kept in a notebook or file or scanned and stored digitally.

5. Investigate possible sources and causes of the odor as soon as possible.
6. Promptly correct any problems identified during the investigation.
7. Record in the logbook the investigation results and what was done to resolve the problem.
8. Contact the person who made the complaint and report what was done. If the investigation requires more than two days, contact the person to report its status.
9. As appropriate, contact the regulatory agency and inform them about the complaint and its resolution. Request their assistance as necessary.
10. Report complaints to your regulator as required by statutes and permits.

13. Odor characterization and measurement

Odors are perceptions, perceptions of one or more chemicals present at remarkably low concentrations. Odors are perceived differently by different people, and differently under differing circumstances. Therefore, it is challenging to objectively characterize and quantify odors. Nevertheless, scientists and practitioners have developed special techniques and terminology to gain a measure of the qualities and quantities of odors (St. Croix Sensory, 2005). As it turns out, the human element cannot be entirely removed from the process.

13.1 Parameters used to characterize odors

Because of the subjective nature of odors, both qualitative and quantitative descriptors are necessary. An odor may be described and measured according to its threshold concentration, pervasiveness, descriptive quality, degree of pleasant or unpleasantness, and the concentration of the odor-causing chemicals present.

13.1.1 Chemical concentrations

Through chemistry, the concentrations of odorants in a sample of air, are relatively easy to measure. Therefore, it is possible to measure and monitor for odors by sampling the air and analyzing the samples in a laboratory for the concentrations of specific individual odorants. The results are expressed in typical units of concentration of individual compounds such as parts per million (ppm) of ammonia and or parts per billion (ppb) of methyl mercaptan. These concentrations can then be compared to data that indicate the concentrations at which humans detect, recognize, and/or otherwise react to the odor of the chemical in question. Tracking the presence of specific compounds can help identify the sources and root causes of odors and to monitor the effects of odor mitigation practices.

This method of odor characterization has major flaws. First, the strength or quality of an odor cannot truly be represented by measuring the concentrations of a single compound, or even a few compounds. Second, humans can detect odorant concentrations of odorants well below the concentrations detectable by conventional analytical techniques. The human nose remains the better tool for detecting and measuring odors.

13.1.2 Threshold concentrations

Threshold concentrations refer to the D/T concentrations at which an odor stimulates a reaction from humans. The "detection" threshold is the minimum concentration at which humans detect an odor, essentially the D/T value. It is also called the "perception" threshold. The "recognition" threshold is the minimum concentration at which humans can recognize or identify the odor. The recognition threshold is usually many times higher than the detection threshold. Threshold concentrations are set at the point where some percentage of the people in an odor panel detect or recognize the odor, usually 50%. Other reactions, such as annoyance or irritation, can also be the basis for threshold concentrations. Threshold concentrations can be established for overall odors and expressed as D/T50. Also, threshold concentrations can apply to individual chemicals using conventional units of concentration like ppm (i.e., the concentration at which 50% of the panel members detect the chemical's odor).

13.1.3 Sniffing units (su)

Sniffing units are similar to odor units with a few important exceptions. Sniffing units are used when odors are measured by field olfactometry to distinguish between different methodologies of measuring odor in the field versus the lab. Also sniffing units relate to the number of dilutions needed to reach a recognition threshold, while odor units are associated with detection thresholds.

13.1.4 Odor intensity

Odor intensity is the perceived strength or pungency of an odor at a prescribed concentration (CASA, 2015). It greatly influences the tolerance that humans have for an odor. Odor intensity is quantified using a human odor panel to compare the intensity of a sample odor with the intensity of a standard odorant at various concentrations, commonly butanol. The intensity is reported in terms of the corresponding butanol concentration (e.g., ppm), or by rating the sample with a number scale (Balch et al., 2019). An overpowering odor would be rated 6; no odor would be 0.

Odor intensity and concentration are different but related parameters. Intensity (I) and concentrated (C) can be quantitatively correlated by either the Weber-

Fletcher Law or Stevens' Psychophysical Power law (Eqs. 12.8 and 12.9). The constants a, b, k, and n are specific to particular odorants.

$$\text{Weber-Fletcher Law: } I = a \log (C) + b \qquad (12.8)$$

$$\text{Steven's Law: } I = k \, C^n \qquad (12.9)$$

The equations are valid for either an overall odor or an individual odorant. For an overall odor, the threshold concentration would be used, expressed in ou_e or D/T. For a particular odorant, the actual concentration applies.

13.1.5 Odor pervasiveness

Pervasiveness refers to the staying power of an odorant. An odor is "pervasive" if its intensity decreases slowly even as it is diluted with fresh air. For example, in Eq. (12.9), the constant n indicates how easily the odor intensity decreases when odorous air is diluted. The value of n ranges from 0.2 to 0.8 for most compounds (Das, 2000). When n is large, intensity falls substantially as the concentration decreases. The odor is nonpervasive. Ammonia and aldehydes are easily diluted—they have high values of n. Amines have low n values—their odors disappear slowly as they are diluted (Haug, 1993).

13.1.6 Odor quality or character

Odor quality, or character, is what someone would say an odor smells like. It can only be identified by relating it to another odor that is generally recognizable. Lists of standard descriptors have been developed by several industries: the beer, wine, and food industries among them. Odor descriptors for waste industries have also been compiled that can be used by one person to convey the quality of an odor to another person (Table 12.7). Multiple descriptors can be used, and even weighted, for any given odor.

13.1.7 Hedonic tone

The hedonic tone is the degree of the pleasantness or unpleasantness of an odor. Although it is subjective, hedonic tone can be averaged for a group of individuals evaluating an odor. Hedonic tone can be rated on a scale from -10 for very unpleasant to $+10$ for pleasant (Chiumenti et al., 2005). Ratings scales can be "standardized" by relating the scale to reference odorants, such as isovaleric acid, as a model for a very unpleasant odor, and vanillin for a very pleasant odor (Haug, 1993). Hedonic tone is influenced by odor parameters like odor intensity, concentration, duration, and frequency of exposure, as well as the perceptions and associations of the individual.

Table 12.7 Standard categories and descriptors of odor character.

Earthy	Floral	Fruity	Offensive	Chemical	Medicinal
Ashes	Almond	Apple	Blood	Burnt plastic	Alcohol
Burntwood	Cinnamon	Cherry	Burnt	Car exhaust	Ammonia
Chalk	Coconut	Citrus	Burnt rubber	Cleaning fluid	Anesthetic
Coffee	Eucalyptus	Cloves	Decay	Coal	Camphor
Grain silage	Fragrant	Grapes	Fecal	Creosote	Chlorine
Grassy	Herbal	Lemon	Refuse	Diesel	Disinfectant
Mold	Lavender	Maple	Landfill leachate	Petrol	Menthol
Mouse-like	Licorice	Melon	Manure	Grease	Soap
Mushroom	Marigolds	Minty	Mercaptan	Foundry	Vinegar
Musky	Perfumy	Orange	Putrid	Kerosene	
Musty	Roses	Strawberry	Rancid	Molasses	
Peat	Spicy	Sweet	Raw meat	Mothballs	
Pine	Vanilla		Rotten eggs	Oil	
Smoky		**Vegetable**	Septic	Paint	
Stale		Celery	Sewer	Petroleum	
Swampy		Corn	Sour	Plastic	
Woody		Cucumber	Spoiled milk	Resins	
Yeast		Dill	Urine	Rubber	
		Garlic	Vomit	Solvent	
		Green pepper		Styrene	
		Nutty	**Fishy**	Sulfur	
		Potato	Amine	Tar/asphalt	
		Tomato	Dead fish	Turpentine	
		Onion	Perm solution	Varnish	
				Vinyl	

13.2 Odor measurement

In their review of odor measurement techniques, Bax et al. (2020) identify the following methods as "most important" in odor measurement. These methods apply to all applications of odor measurement, not just composting or waste management.

1. Dynamic olfactometry,
2. Chemical analysis (total, specific gases, single gas),
3. Gas chromatography-olfactometry (GC-O),
4. Electronic-nose (E-nose),

5. Field inspection,
6. Field olfactometry, and,
7. Citizen science.

Of these items, "dynamic olfactometry" is the principal method used, especially when reliable and defensible odor measurements are needed. Olfactometry involves the use of human odor panels, in a laboratory situation, that comply with standardized conditions and procedures. Items 2 and 3 in the list employ established analytical techniques of chemistry to determine concentrations or odorants, specific and nonspecific.

Items 4 through 7 attempt to identify and/or quantify odors where they occur— either at the locations of the odor sources or the odor receptors (e.g., impacted neighbors) or both. Ideally, odors should be measured in the field, where they are being either emitted or experienced. Because it is impractical to bring odor panels to the field, sampling and dynamic olfactometry are seldom used. Human noses still make odor evaluations in the field, but procedures tend to be less rigorous.

13.2.1 Dynamic olfactometry—standardized smelling

Despite their subjectivity, humans are still the most accepted instrument for detecting and distinguishing odors. Dynamic olfactometry is "dynamic" because a series of odor samples are presented to the evaluators in successively increasing concentrations (i.e., most diluted to least diluted). Samples of odorous gas are collected in inert containers at the odor sources, locations of the receptors, and/or other predetermined locations of interest. The samples are transported to the lab, diluted with odorless air in a controlled manner, and presented to the panel members for evaluation. The samples are diluted and presented with the aid of an air-handling device called an olfactometer. Each panel member discretely indicates when she/he detects an odor. The number of dilutions (i.e., concentration) at which half the panel members have detected an odor is the number of odor units (ou_e) or the D/T50 for the odorant in question. The odorant might be a specific chemical compound or an overall odorant from a source containing a mix of compounds. Similar procedures are used to identify recognition thresholds. This fundamental laboratory odor panel approach is used also to characterize odor intensity and hedonic tone.

The procedures for dynamic olfactometry are governed by European Standard EN 13725 (CEN, 2003). EN 13725 prescribes a full range of procedures from sampling practices to criteria for selecting the panel members. Although not universal, this standard has been widely adopted within and outside Europe and is the primary guidance document used for olfactometry worldwide (St. Croix Sensory, 2005; van Harreveld, 2021). An updated and revised version of EN 13725 is anticipated in 2021 (van Harreveld, 2021). While many laboratories and researchers in the United States follow EN13725 procedures, many US investigators use other standards such as ASTM E-670-04 (ASTM International, 2004; St. Croix Sensory, 2005). As a result, odor concentrations are more likely to be reported in D/T in the United States than elsewhere.

With the procedure well-standardized, and with long-standing experience, dynamic olfactometry is a well-accepted and scientifically sound method of odor measure. Its main disadvantages are: (1) that it evaluates a discreet sample of material taken at a single point in time; (2) samples must be transported and stored for a time, possibly allowing some odor constituents to change; and (3) it is a relatively time-consuming and expensive measurement technique. Nevertheless, no chemical/electronic instrument can yet replace humans in regard to their reliable and sensitivity to odors (Bax et al., 2020).

13.2.2 Analytical methods

Analytical measurements quantify the amount of concentration of one or more specific chemicals using various chemical, electrical, and physical techniques. The analytical approach is practical when one or a few specific and identifiable compounds dominate the odor. Analytical measurements also can be useful when monitoring the effects of changing situations. For example, analytical methods can track how hydrogen sulfide concentrations change in response to some odor mitigating practice, like adding an amendment to improve aeration. Even if hydrogen sulfide is not the offending odor, its presence and concentration may indicate the effect on odors.

Numerous analytical techniques and instruments are commonly used to identify specific chemicals or a broad category of chemicals. The target chemical and its concentrations typically determine what techniques are feasible. For specific compounds, the combined use of gas chromatographs (GCs) and mass spectrometers (MSs) is especially versatile, accurate, and popular. The GC device separates individual compounds while the MS identifies the compounds by their unique signals and quantifies them by the magnitude of the signals.

NH_3 and H_2S concentrations are sometimes measured as odor indicators (e.g., tracers), when they are not the primary odorants themselves. Specific analytical devices are available for each of these chemicals, some of which are relatively inexpensive. For example, a chemiluminescence analyzer is used for NH_3 and a gold leaf analyzer is used for H_2S (Bax et al., 2020). In the field, a detection tube can measure either NH_3 or H_2S concentration and provide an almost immediate result. The detection tube contains a chemical that reacts with the target odorant compound and changes color, expands or changes in some fashion in proportion to the amount of odorant present. The scale on the tube is calibrated to display the concentration of the odorant.

Rather than identify and measure individual chemicals, an alternative strategy is to measure a broad selection of chemicals without identification. For example, total VOC concentrations can be indicative of emissions levels, if not the general odor quality. Total VOCs can be inexpensively measured in the field with instruments like photo- and flame-ionizers (Bax et al., 2020).

Measuring specific or indicator chemicals has limitations for nuisance odor evaluation. First, the strength or quality of an odor cannot truly be represented by measuring the concentrations of a single compound, or even a few compounds. Second, humans can detect odorant concentrations of odorants well below the concentrations detectable by conventional analytical techniques.

To partially compensate for these limitations, GC/MS analysis is sometimes combined with human olfactometry (Conti et al., 2019). In this case, the GC/MS system includes a detection port for presenting the component gaseous compounds to human evaluators in order to assign a dilution threshold or odor intensity to each component gas. However, this method does not remove all of the difficulties with chemical analysis (Bratolli et al., 2011). In addition to being time consuming, the contribution of individual odorants with a general odorous gas is not additive. Some compounds may dominate or mask other compounds. Although the GC plus olfactometry approach can provide additional information about a given odor, it does not substitute for dynamic olfactometry.

13.2.3 Electronic nose

An electronic-nose, or E-nose, is a devise with a sample collection system, multiple sensors to measure selected chemicals, and a computer to process the sensor data. Each sensor detects and quantifies a specific chemical, or class of chemicals, and sends a corresponding electronic signal to the computer. Some E-nose systems feature artificial intelligence (AI) software, trained to recognize the pattern of signals generated by sensor arrays. Like a simplified version of a human nose, the sensors detect a combination of gaseous chemical compounds, at varying concentrations, and electronically inform the computer, which recognizes the signal patterns and attempts to describe the odor (Brattoli et al., 2011). An E-nose with AI software is intended to be trained, both by its own experience and by correlating E-nose interpretations with those of dynamic olfactometry panels.

E-noses have been used for years in a variety of industries, including in the food and beverage industry. Experience with E-noses has also been building in the environmental field, including investigations at composting facilities and at locations neighboring composting facilities (Brattoli et al., 2001; Bax et al., 2020). They offer several advantages including the ability for continuous monitoring at the odor source (for process management), continuous monitoring at possible receptor locations, and the potential to reduce the reliance on odor panels.

Odor chemistry is complicated, and localized gaseous diffusion adds an element of randomness. E-noses have not gained wide acceptance at composting facilities. However, with experience and technological advances, E-noses are expected to improve in their ability to detect and measure complex odors. Consequentially, they are expected to find increasing use in composting situations, both for odor management at larger facilities and for monitoring of impacts beyond facilities. Perhaps the most attractive application of E-nose devices is their potential to continually monitor odor emissions and trigger corresponding process changes, either by notifying operators of pending problems or by adjusting automatically.

13.2.4 Field inspections

Two recognized methods of field inspection exist that are defined by European standards (Bax et al., 2020). Both methods employ teams of trained evaluators, situated at defined locations within an area potentially impacted by odors. Although labor-intensive, these field methods provide qualitative information about odors beyond the source.

Standard EN16841-1 covers the "grid method." The grid method seeks to determine the level of annoyance that a particular odor source is imposing on its neighbors. In the grid method, evaluators are located in a grid pattern in an area surrounding or near an odor source. They sniff for odors at specified time intervals within a large, specified time span. The observations at the corners of each grid cell are combined to determine the percentage of time that an odor is observed. The assessments are supposed to be conducted under representative weather conditions.

The "plume method," covered by standard EN 16841-2, attempts to define the boundaries of a plume of odorous air moving downwind of an odor source. Within the plume, the odor in question can be detected and recognized. Evaluators make several assessments within a line that runs perpendicular to the expected plume direction. Assessments are made in a series of lines located at increasing distances downwind from the odor source. The nexus of assessment locations where odors are absent roughly define the envelope of the plume. This method is used for delineating areas being impacted under various meteorological conditions and for verifying dispersion models.

On a less formal level, field evaluations are made to anticipate and prevent odor problems, improve process management, verify a complaint and/or spot check for regulatory compliance or dispersion model verification. At its simplest, a field inspection is a casual survey of odors on and around the composting facility performed by a designated individual or group of individuals. The designated evaluators can be members of the composting staff (e.g., facility manager), members of the local community, or a third party hired to conduct the survey. In the process, field inspectors might make use of instruments like field olfactometers, detection tubes, or other portable instruments. In the best case, the evaluators are trained in the art and science of odor assessment. In the worst case, they are simply individuals sniffing the air. In all cases, survey forms should be used to provide a written record of observations, their locations, weather conditions, time, and date. Fig. 12.21 is an example of a field observation form, adapted from Sheffield and Ndegwa (2008).

Odor Field Observation Sheet

Date: _____ Time: _____ am / pm

Odor Panelist name or ID: _____

Sampling Location: _____

Purported source: _____ Distance from source: _____

General comments: _____

_____

Weather

Wind speed: _____ Direction: _____ Temp: _____ Humidity: _____

General weather conditions: _____

Odor Quality/Character

Primary description (from odor wheel or table): _____

Secondary description (your own words): _____

Odor offensiveness (hedonic tone): On a scale of 1 to 10, 10 being the worst smell that you have ever experienced or could imagine, this odor is a _____ .

Measurements (if applicable):

Odorant concentration. Odorant: _____

 Measured concentration(s): _____

 Measuring device: _____

Field olfactometry. Odor concentrations (D/T or ou): _____

 Measuring device: _____

Other Observations/Comments

_____

_____

_____

FIGURE 12.21

Example of an odor observation form.

Adapted from Sheffield and Ndegwa, 2008.

13.2.5 Field olfactometry

Several portable devices have been developed for controlled olfactory measurements of odors in the field. Generally, these field olfactometers collect a sample of air, dilute it with carbon-filtered air, and then guide it to the evaluator's nose, which is isolated from the surrounding ambient air by a mask or cup covering the nose. Field olfactometry began with the "Scentometer," in 1960 (originally manufactured by Bernebey Sutcliffe) and progressed with the development of the Nasal Ranger field olfactometer (St. Croix Sensory, 2005). Since then, other field olfactometers have emerged (Sheffield and Ndegwa, 2008; Henry et al., 2011a,b; Damuchali and Huiqing, 2019).

These devices work in a similar fashion. They dilute odorous samples with carbon-filter air applying a series fixed dilution ratios (e.g., 2, 4, 7, 15, 30, 60), although the series can be altered by exchanging instrument components (St Croix Sensory, 2005). The odor evaluator can shift among the dilution ratios adjusting a valve. With these instruments, field measures can estimate threshold concentrations in D/Ts.

A major challenge in field olfactometry is that the user, or "sniffer," is not in a neutral situation. The sniffer is exposed to both visual and olfactory signals that might influence her/his perception of the odor. For example, after spending an hour or more near an odor source, the sniffer becomes somewhat desensitized to the odor, a condition known as "odor fatigue." Afterward, the sniffer may not detect the odor as readily when the odor is diluted by the field olfactometer. Also, the sniffer's reaction to the appearance of the odor source can alter the way he/she perceives the odor.

The usefulness of field olfactometers has been reported to be generally favorable in the collective literature (Brant and Eliot, 2004; Henry et al., 2011a,b; Bratolli et al., 2011; Coker, 2013c; Damuchali and Huiqing, 2019; Bax et al., 2020). Agreement between field olfactometry results and dynamic olfactometry, the accepted standard of measure, has been mixed. Some authors report good agreement, some poor agreement, and in other cases the agreement depends on the nature of the odor. However, field olfactometry appears to positively identify the presence of odors and their general strength. Where dynamic olfactometry is not an option, due to its cost or availability, field olfactometry may be an acceptable alternative, even for establishing odor concentrations in D/T units and for regulatory compliance (CPCB, 2017).

13.2.6 Citizen science

Bax et al. (2020) describe citizen science as "the practice of public participation and collaboration in scientific research," in which "people share and contribute to data monitoring and collection programs." This approach may be referred to by other names including neighborhood monitoring or community studies (St. Croix Sensory, 2005). In regard to odor measurement, citizen science depends on members of the community-at-large to notice and record odors in regard to their quality, intensity, duration, frequency, suspected source, and other factors like weather conditions.

A desired goal is to create a map of odor impacts in the area of interest. St. Croix Sensory (2005) list several possible tools for citizen scientist to communicate their observations including a citizen odor hotline, surveys of recruited citizens, and citizen odor logbooks. Some of these tools can be replaced or improved by smartphone applications.

The citizen science or community-involvement strategy introduces outside and potentially innovative ideas into the situation (from the larger community) and fosters widespread acceptance of solutions because the community is involved, and the process is transparent (Bax et al., 2020). These advantages are contrasted by the considerable effort required to organize and manage the system and also the lack of knowledge and objectivity of some community members. The latter problem can be minimized by enlisting a large number of participants with diverse backgrounds and by providing training about odors and odor assessment. However, the potential for observer bias is difficult to eliminate.

References

Cited references

ARB, 2015. Emissions Inventory Methodology for Composting Facilities. California Air Pollution Control Board.

ASTM International, 2004. E679-04: Standard Practice for Determination of Odor and Taste Threshold by a Forced-Choice Ascending Concentration Series Method of Limits. Philadelphia, PA, USA.

Balch, A., Wilkinson, K., Schliebs, D., Hazell, L., 2019. Critical Evaluation of Composting Operations and Feedstock Suitability Critical Evaluation of Composting Operations and Feedstock Suitability, Phase 1 Report — Odour Issues. Arcadis Australia Pacific Pty Limited; prepared for the Department of Environment and Science (DES Queensland). https://environment.des.qld.gov.au/__data/assets/pdf_file/0024/226293/phase-1-composting-study-report.pdf.

Barthod, J., Rumpel, C., Dignac, M.F., 2018. Composting with additives to improve organic amendments. A review. Agron. Sustain. Dev. 38, 17. https://doi.org/10.1007/s13593-018-0491-9.

Bax, C., Sironi, S., Capelli, L., 2020. How can odors be measured? An overview of methods and their applications. Atmosphere 11, 92.

Beecher, N., Goldstein, N., 2005. Public perceptions of biosolids recycling. BioCycle 46 (4), 34. https://www.biocycle.net/public-perceptions-of-biosolids-recycling/.

Beecher, N., Harrison, E., Goldstein, N., McDaniel, M., Field, P., Susskind, L., 2005. Risk perception, risk communication, and stakeholder involvement for biosolids management and research. J. Environ. Qual. 34 (1), 122−128. PMID: 15647541.

Bokowa, A., Diaz, C., Koziel, J.A., McGinley, M., Barclay, J., Schauberger, G., Guillot, J.-M., Sneath, R., Capelli, L., Zorich, V., et al., 2021. Summary and overview of the odour regulations worldwide. Atmosphere 2021 (12), 206. https://doi.org/10.3390/atmos12020206.

Brant, R.C., Elliott, H.A., 2004. Odor Management in Agriculture and Food Processing in Pennsylvania. Pennsylvania State University, University Park, PA. https://www.agriculture.pa.gov/Plants_Land_Water/StateConservationCommission/OdorManagementProgram/Pages/default.aspx.

Brattoli, M., de Gennaro, G., de Pinto, V., Loiotile, A.D., Lovascio, S., Penza, M., 2011. Odour detection methods: olfactometry and chemical sensors. Sensors (Basel) 11 (5), 5290–5322. https://doi.org/10.3390/s110505290. Epub May 16, 2011. PMID: 22163901; PMCID: PMC3231359.

Buckner, S.C., 2002a. Effects of turning frequency and mixture composition on process conditions and odor concentrations during grass composting. In: Proceeding of the 2002 International Symposium, Columbus Ohio, May 2002.

Büyüksönmez, F., Yucel, A., 2006. Odor mitigation alternatives. In: 14th Annual U.S. Composting Council Conference. Albuquerque, TX.

Büyüksönmez, F., Rynk, R., Yucel, A., Cotton, M., 2012. Mitigation of odor causing emissions—bench-scale investigation. J. Air Waste Manag. Assoc. 62 (12), 1423–1430. https://doi.org/10.1080/10962247.2012.716808.

CalRecycle, 2007. Comprehensive Compost Odor Response Project (C-CORP). California Department of Resources Recycling and Recovery (CalRecycle). https://www2.calrecycle.ca.gov/Publications/Details/1241.

Capelli, L., Sironi, S., Del Rosso, R., Guillot, J.M., 2013. Measuring odours in the environment vs. dispersion modelling: a review. Atmos. Environ. 2013 (79), 731–743.

Carpenter, A., Brown, S., Souther, L., 2020. A Literature Review and Analysis of Compost Aeration Emissions Measurement Methods. Compost Research and Education Foundation. https://www.compostfoundation.org/Air-Emissions.

CASA, 2015. Good Practices Guide for Odour Management in Alberta Clean Air Strategic Alliance, Edmonton, AB. https://www.casahome.org/uploads/source/PDF/CASA_GPG_webversion_V3.pdf.

CEN, 2003. EN 13725:2003, Air Quality. Determination of Odour Concentration by Dynamic Olfactometry. European Committee for Standardization (CEN), Brussels, Belgium.

Charles, J., 2015. Environmental Odors and Public Nuisance Law: A Research Anthology. U.S. Centers for Disease Control and Prevention, Office for State, Tribal, Local and Territorial Support. https://www.cdc.gov/phlp/docs/anthology-environmentalodors.pdf.

Chastain, J.P., Wolak, F.J., 2000. Application of a Gaussian plume model of odor dispersion to select a site for livestock facilities. In: Proceedings of the Odors and VOC Emissions 2000 Conference, Sponsored by the Water Environment Federation, April 16–19, Cincinnati, OH.

Chiumenti, A., Chiumenti, R., Diaz, L.F., Savage, G.C., Eggerth, L.L., Goldstein, N., 2005. Modern Composting Technologies. The JG Press, Inc, Emmaus, PA.

Coker, C., 2012a. Part 1 Managing odors in organics recycling. BioCycle 53 (4), 25. https://www.biocycle.net/managing-odors-in-organics-recycling/.

Coker, C., 2012b. Part 2 Odor defense strategy. BioCycle 53 (5), 35. https://www.biocycle.net/odor-defense-strategy/.

Coker, C., 2012c. Part 3 "Going on offense against odors". BioCycle 53 (6), 25.

Coker, C., 2012d. Part 4 Odor treatment at composting facilities. BioCycle 53 (8), 21. https://www.biocycle.net/odor-treatment-at-composting-facilities/.

Coker, C., 2012e. Part 5 Resolving odor challenges. BioCycle 53 (11), 22. https://www.biocycle.net/resolving-odor-challenges.

Coker, C., 2013a. Part 6 Odor dispersion fundamentals. BioCycle 54 (6), 26. https://www.biocycle.net/odor-dispersion-fundamentals/.

Coker, C., 2013b. Part 7 The art and science of odor modeling. BioCycle. https://www.biocycle.net/the-art-and-science-of-odor-modeling/.

Coker, C., 2013c. Part 8 Odor monitoring and detection tools. BioCycle 54 (10), 16. https://www.biocycle.net/odor-monitoring-and-detection-tools/.

Conti, C., Guarino, M., Bacenetti, J., 2019. Measurement's techniques and models to assess odor annoyance: a review. Environ. Int. 134, 105261.

Cordova, T., Goodwin, J., Card, T., Schmidt, C.E., 2015. Positive air, biofilter cover layer control biosolids composting emissions. BioCycle 56 (3), 49. https://www.biocycle.net/positive-air-biofilter-cover-layer-control-biosolids-composting-emissions/.

CPCB, 2017. Odour Monitoring & Management in Urban MSW Landfill Sites. Central Pollution Control Board, Ministry of Environment, Forest and Climate Change, Govt. of India. https://cpcb.nic.in/archivereport.php.

St. Croix Sensory, Inc., 2005. A Review of the Science and Technology of Odor Measurement. Air Quality Bureau of the Iowa Department of Natural Resources, Des Moines, IA. http://publications.iowa.gov/34888/.

Dalton, P., 2003. How people sense, perceive and react to odors. BioCycle 44 (11), 26.

Damuchali, A.M., Huiqing, G., 2019. Evaluation of a field olfactometer in odour concentration measurement. Biosyst. Eng. ISSN: 1537-5110 187, 239–246. https://doi.org/10.1016/j.biosystemseng.2019.09.007.

Das, K.C., 2000. Odor related issues in commercial composting. In: Workshop Presented at the Y2K Composting in the Southeast Conference & Expo, October 2000, Charlottesville, VA.

de Nevers, N., 2000. Air Pollution Control Engineering. The McGraw-Hill Companies, Inc., Boston. MA.

Engineering Toolbox, 2006. Vapor Pressures Common Liquids (online) Available at: https://www.engineeringtoolbox.com/vapor-pressure-d_312.html. (Accessed 26 September 2021).

Epstein, E., 1997. The Science of Composting. Technomic Publishing Company, Penn.

Gage, J., 2020. Windrow odor management. In: Presentation at the U.S. Composting Council Annual Conference. Charleston, SC.

Guo, R., Li, G., Jiang, T., Schuchardt, F., Chen, T., Zhao, Y., Shen, Y., 2012. Effect of aeration rate, C/N ratio and moisture content on the stability and maturity of compost. Bioresour. Technol. 112, 171–178.

Haug, R.T., 1993. The Practical Handbook of Compost Engineering. Lewis Publishers, CRC Press, Boca Raton, Florida.

Haug, R.T., 2020. Elements of odor management. BioCycle. https://www.biocycle.net/elements-of-odor-management/.

Henry, C., Meyer, G., Schulte, D., Stowell, R., Parkhurst, A., Sheffield, R., 2011a. Mask scentometer for assessing ambient odors. Trans ASABE 54, 609–615.

Henry, C.G., Schulte, D.D., Hoff, S.J., Jacobson, L.D., Parkhurst, A.M., 2011b. Comparison of ambient odor assessment techniques in a controlled environment. Agric. Biosyst. Eng. Publ. 356. https://lib.dr.iastate.edu/abe_eng_pubs/356.

Horowitz, R., Barnes, K., Jones, J., Moon, P., Card, T., Schmidt, C.E., Noble, D., 2013. California composting trial in impacted air shed. BioCycle 54 (10), 33. https://www.biocycle.net/california-composting-trial-in-impacted-air-shed/.

Japan Ministry of the Environment, 2003. Odor Index Regulation and Triangular Odor Bag Method. Office of Odor, Noise and Vibration, Environmental Management Bureau,

Ministry of the Environment, Government of Japan. https://www.env.go.jp/en/air/odor/regulation/all.pdf.

Joshi, J., Cowan, R., Hogan, J., Strom, P., Finstein, M.S., 1998. Gaseous Ammonia Removal in Biofilters: Effect of Biofilter Media on Products of Nitrification. SAE Technical Paper 981613. https://doi.org/10.4271/981613.

Krzymien, M., Day, M., Shaw, K., Zaremba, L., 1999. An investigation of odors and volatile organic compounds released during composting. J. Air Waste Manag. Assoc. 49 (7), 804–813. https://doi.org/10.1080/10473289.1999.10463845.

Liu, D.H.F., Liptak, B.G., Bouis, P.A. (Eds.), 1997. Environmental Engineers' Handbook. Lews Publishers, New York, NY.

Liu, J., Liu, J., Li, L., 2008. Performance of two biofilters with neutral and low pH treating off-gases. J. Environ. Sci. (China) 20 (12), 1409–1414. https://doi.org/10.1016/s1001-0742(08)62541-3.

Lystad, H., Hammer, J.P., Sørhein, R., Molland, O., Bergersen, O., Smits, P., 2002. Using lime to reduce odors from biowaste composting. BioCycle 43 (9), 38–41.

Ma, J., Wilson, K., Zhao, Q., Yorgey, G., Frear, C., 2013. Odor in Commercial Scale Compost: Literature Review and Critical Analysis. Washington State Department of Ecology. https://apps.ecology.wa.gov/publications/documents/1307066.pdf.

McGinley, C., McGinley, M., 2006. An odor index scale for policy and decision making using ambient & source odor concentrations. In: Proceedings of the Water Environment Federation, vol. 2006, pp. 244–250. https://doi.org/10.2175/193864706783791696.

Michel Jr., F.C., Reddy, C.A., 1998. Effect of oxygenation level on yard trimmings composting rate, odor production, and compost quality in bench-scale reactors. Compost Sci. Util. 6 (4), 6–14. https://doi.org/10.1080/1065657X.1998.10701936.

Miller, F.C., 1993. Minimizing odor generation. In: Science and Engineering of Composting. Ohio State University, p. 219.

Moussavi, G., Bahadori, M.B., Farzadkia, M., Yazdanbakhsh, A., Mohseni, M., 2009. Performance evaluation of a thermophilic biofilter for the removal of MTBE from waste air stream: effects of inlet concentration and EBRT. Biochem. Eng. J. 45, 152–156. https://doi.org/10.1016/j.bej.2009.03.008.

Nordic Counsel of Ministers, 2009. Minimisation of Odour from Composting of Food Waste Through Process Optimisation : A Nordic Collaboration Project. https://doi.org/10.6027/TN2009-561.

NRCS, 2021. Wind Rose Resources. Natural Resource Conservation Service. U.S. Department of Agriculture. https://www.nrcs.usda.gov/wps/portal/wcc/home/climateSupport/windRoseResources/.

Paré, M., Paulitz, T., Stewart, K.A., 2000. Composting of crucifer wastes using geotextile covers. Compost Sci. Util. 8, 36–45.

Pelosi, R., 2003. Clearing the Air. American City & County. https://www.americancityandcounty.com/2003/10/01/clearing-the-air/.

Pinasseau, A., Zerger, B., Roth, J., Canova, M., Roudier, S., 2018. Best Available Techniques (BAT) Reference Document for Waste Treatment Industrial Emissions Directive 2010/75/EU (Integrated Pollution Prevention and Control). Publications Office of the European Union.

Qu, G., Edeogu, I.E., Fedde, J.J.R., 2010. Odor index: an integration of odor parameters for swine slurry odors. Trans. ASABE 53 (1), 219–223. https://doi.org/10.13031/2013.29497.

Sauer, N., 2012. Oxygen Solubility in Compost. UK Environment Agency. http://organics-recycling.org.uk/uploads/article2430/Tech%20Guide%203%20O2%20Solubility%20A.pdf.

Sauer, N., Crouch, E., 2013. Measuring oxygen in compost. BioCycle. https://www.biocycle.net/measuring-oxygen-in-compost/.

Schiffman, S.S., Williams, C.M., 2005. Science of odor as a potential health issue. J. Environ. Qual. 34 (1), 129−138. PMID: 15647542.

Scholtz, V., Pazlarova, J., Souskova, H., Khun, J., Julak, J., 2015. Nonthermal plasma–A tool for decontamination and disinfection. Biotechnol. Adv. 33 (6 Pt 2), 1108−1119. https://doi.org/10.1016/j.biotechadv.2015.01.002. Epub January 13, 2015. PMID: 25595663.

Sheffield, R.E., Ndegwa, P., 2008. Sampling Agricultural Odors. University of Idaho. https://www.extension.uidaho.edu/publishing/pdf/pnw/pnw595.pdf.

Sunberg, C., 2005. Improving Compost Process Efficiency by Controlling Aeration, Temperature, and pH (Doctoral thesis). Swedish University of Agricultural Sciences, Uppsala, ISBN 91-576-6902-3. ISSN 1652-6880.

Sundberg, C., Jönsson, H., 2008. Higher pH and faster decomposition in biowaste composting by increased aeration. Waste Manag. ISSN: 0956-053X 28 (3), 518−526. https://doi.org/10.1016/j.wasman.2007.01.011.

Sundberg, C., Yu, D., Franke-Whittle, I., Kauppi, S., Smårs, S., Insam, H., Romantschuk, M., Jönsson, H., 2013. Effects of pH and microbial composition on odour in food waste composting. Waste Manag. 33, 204−211.

TCEQ, 2016. Odor Complaint Investigation Procedures. Texas Commission on Environmental Quality. https://www.pearlandtx.gov/home/showdocument?id=22227.

van Harreveld, A., 2021. Update on the revised EN 13725:2021. Chem. Eng. Trans. 85.

Ward, H., Wiems, M., 2018. Odour From a Compost Facility. National Collaborating Centre for Environmental Health 2018, 200-601 West Broadway, Vancouver, BC V5Z 4C2.

Willmink, T.R., Diener, R.G., 2001. Handbook for Commercial and Municipal Composting in West Virginia.

Zhu, P., Shen, Y., Pan, X., Dong, B., Zhou, J., Zhang, W., Li, X., 2021. Reducing odor emissions from feces aerobic composting: additives. RSC Adv. 11, 15977−15988. https://doi.org/10.1039/D1RA00355K.

Consulted and suggested references

Bidlingmaier, W., Müsken, J., 2007. Chapter 11 Odor emissions from composting plants. In: Diaz, L.F., et al. (Eds.), Waste Manage Series. Elsevier London, pp. 215−324.

Buckner, S.C., 2002b. Controlling odors during grass composting. BioCycle 43 (9), 42−47.

Campbell, A.G., Folk, R.L., Tripepi, R.R., 1997. Wood ash as an amendment in municipal sewage sludge and yard waste composting processes. Compost Sci. Util. 5 (1), 62−73.

Capelli, L., Bax, C., Diaz, C., Izquierdo, C., Arias, R., Salas Seoane, N., 2019. Review on Odour Pollution, Odour Measurement, Abatement Techniques. D-NOSES. H2020-SwafS-23-2017-789315.

Coker, C., 2016. Controlling composting odors. BioCycle. https://www.biocycle.net/controlling-composting-odors/.

Dalton, P., Caraway, E.A., Gibb, H., Fulcher, K., 2011. A multi-year field olfactometry study near a concentrated animal feeding operation. J. Air Waste Manag. Assoc. 61 (12), 1398−1408. https://doi.org/10.1080/10473289.2011.624256.

Eijrond, V., Claassen, L., van der Giessen, J., Timmermans, D., 2019. Intensive livestock farming and residential health: experts' views. Int. J. Environ. Res. Publ. Health 16 (19), 3625. https://doi.org/10.3390/ijerph16193625. PMID: 31569632; PMCID: PMC6801788.

Enz, J.W., Hofman, V., Thostenson, A., 2019. Air Temperature Inversions. Causes, Characteristics and Potential Effects on Pesticide Spray Drift. North Dakota State University Extension. https://www.ag.ndsu.edu/publications/crops/air-temperature-inversions-causes-characteristics-and-potential-effects-on-pesticide-spray-drift.

Fletcher, L.A., Jones, N., Warren, L., Stentiford, E.I., 2014. Understanding Biofilter Performance and Determining Emission Concentrations under Operational Conditions. http://organics-recycling.org.uk/uploads/article2834/ER36%20Final%20Report%20for%20publication.pdf.

Gamble, S., 2018. Compost Facility Operator Study Guide. Minister of Alberta Environment and Parks, ISBN 978-1-4601-4130-4 (PDF online).

Goldstein, N., 2006a. Odor management strategy meets neighbor approval. BioCycle 45 (9), 50.

Goldstein, N., 2006b. Neighbor-friendly odor management. BioCycle 47 (3).

Goldstein, N., 2020. Odors and human responses. BioCycle. https://www.biocycle.net/odors-and-human-responses/.

Gutiérrez, M.C., Chica, A.F., Martín, M.A., et al., 2014. Compost pile monitoring using different approaches: GC—MS, E-nose and dynamic olfactometry. Waste Biomass Valor 5, 469—479. https://doi.org/10.1007/s12649-013-9240-0.

Janni, K.A., Nicolai, R.K., Stenglein, R.M., 2012. Biofilters for Odour and Air Pollution Mitigation. The Poultry Site, Gl;obal Ag Media. https://www.thepoultrysite.com/articles/biofilters-for-odour-and-air-pollution-mitigation.

Kelsey, T., Abdalla, C.W., 2017. Finding the Common Ground: Good Neighbor Relations: Advice and Tips From Farmers. Penn State Extension. https://extension.psu.edu/finding-the-common-ground-good-neighbor-relations-advice-and-tips-from-farmers.

Lucernoni, F., Capelli, L., Sironi, S., 2016. Odour sampling on passive area sources: principles and methods. Chem. Eng. Trans. 54, 55—60.

Mendrey, K., 2014. Legal Perspectives on EqOdor Impacts BioCycle. https://www.biocycle.net/legal-perspectives-on-odor-impacts/.

Michel, F.C., 1999. Managing compost piles to maximize natural aeration. BioCycle 40 (3), 56—58.

Michel, F.C., Forney, L.J., Huang, A.J.F., Drew, S., Czuprenski, M., Lindeberg, J.D., Reddy, C.A., 1996. Effects of turning frequency, leaves to grass mix ratio and windrow vs. Pile configuration on the composting of yard trimmings. Compost Sci. Util. 4 (1).

Missouri, N.R.C.S., 2004. Using Windbreaks to Reduce Odors Associated With Livestock Production Facilities. Natural Resource Conservation Service. U.S. Department of Agriculture. https://vdocument.in/windbreakshelterbelt-odor-control-windbreakshelterbeltaodor-control-conservation.html.

Parker, C., Gibson, N., 2009. Good Practice and Regulatory Guidance on Composting and Odour Control for Local Authorities. Department for Environment, Food & Rural Affairs (defra). www.defra.gov.uk.

Sashikala, M.P., Ong, H.K., 2015. Analytical Techniques for Odour Assessment. Malaysian Agricultural Research and Development Institute (MARDI), Kuala Lumpur, Malaysia. http://jtafs.mardi.gov.my/MARDIreport/MR217.pdf.

Sattler, M., Devanathan, S., 2007. Which meteorological conditions produce worst-case odors from area sources? J. Air Waste Manag. Assoc. 57 (11), 1296−1306. https://doi.org/10.3155/1047-3289.57.11.1296.

Schiffman, S.S., Walker, J.M., Dalton, P., Lorig, T.S., Raymer, J.H., Shusterman, D., Williams, C.M., 2004. Potential health effects of odor from animal operations, wastewater treatment, and recycling of byproducts. J. Agromed. 9 (2), 397−403. PMID: 19785232.

Schmidt, D., Jacobson, L., Nicolai, R., 2020. Biofilter Design Information. University of Minnesota Extension. https://conservancy.umn.edu/handle/11299/212362.

Western Australia DEP, 2002. Odour Methodology Guideline. Department of Environmental Protection Perth, Government of Western Australia. https://www.der.wa.gov.au/images/documents/your-environment/air/publications/odour-methodology-guidelines.pdf.

Williams, T.O., Miller, F.C., 1993. Composting facility odor control using biofilters: minimizing odor generation. In: Science and Engineering of Composting. Ohio State University.

Safety and health principles and practices for composting facilities*

13

Authors: Nellie J. Brown[1]

[1]*Workplace Health & Safety Program, ILR, Cornell University, Buffalo, NY, United States*

Contributors: Jane Gilbert[2], Ginny Black[3], Robert Rynk[4]

[2]*Carbon Clarity, Rushden, Northamptonshire, United Kingdom;* [3]*Compost Research and Education Foundation (CREF), Raleigh, NC, United States;* [4]*SUNY Cobleskill, Cobleskill, NY, United States*

1. Introduction

Composting is rugged work. It involves mechanical equipment, physical labor, and handling of diverse biological materials. It is usually practiced outdoors for long hours, in all types of weather. Even when composting takes place indoors, the environment can be difficult for workers. By its nature, composting exposes operators to assorted microorganisms (e.g., molds and bacteria), dust, gases, vapors, noise, sharp objects, heavy objects, fog, sunlight, heat, extreme cold, strain, fatigue, and mechanical and electrical machinery. Thus, composting inherently entails safety and health hazards. Even when composting facilities employ sound practices, there are always risks associated with day-to-day operations, and occasional accidents. However, awareness of the hazards, prevention, and preparedness keep the risks from becoming safety incidents and health problems.

Before considering the details of composting safety, health hazards, and risks, it is necessary to be clear about what these terms actually mean. *Safety* is associated with physical trauma, such as an equipment accident, a fall, or impact from a projectile. *Health* refers to physiological injury or illness to a person, usually from continued or repeated exposure to a hazard. A *hazard* is something that has the potential to cause harm, (even if the harm is made unlikely due to the proper controls) such as a machine, electricity, dust, or sunlight. *Risk* is the chance or likelihood of somebody being harmed by a hazard, combined with an indication of how serious the harm could be. Quantitatively, risk is the estimated cost of a hazard multiplied by the probability that it could occur. *Safety* is a judgment about the level of risk we are willing to accept.

---

* An early version of this chapter was issued by the Cornell Waste Management Institute under the title, "Composting Safety and Health" (Brown, 2016).

> **Box 13.1 Universal rules for reducing safety and health hazards at composting operations**
>
> - Have an educated workforce: trained on process control, process hazards, hygiene practices, signs and symptoms of overexposure or overexertion, and emergency response.
> - Exercise good process control, recognize the beginnings of trends that lead to process failures, and take prompt corrective action. Better process control will produce more consistent and predictable exposures and hazards and probably overall lower risks.
> - Meet occupational safety and health regulations where they exist; follow well-respected recommendations even where there are no regulations.
> - Practice good personal hygiene to protect the worker and his/her family.
> - Provide proper protective equipment to match the hazards of the specific site.

Although there are a few universal safety and health tenets (Box 13.1), different composting operations face different sets of hazards. Facilities vary in scale, the feedstocks handled, composting methods, types of equipment, climate, worker skills and training, hours of operation, and level of management. The feedstocks handled, methods employed, the equipment used, and the work practices followed strongly affect the specific safety and health hazards encountered and the associated levels of risk. Tables 13.1 and 13.2 provide an overview of the general hazards that composters may encounter in different facets of an operation.

This chapter describes safety and health issues related to composting and the practices that minimize the associated risks. While safety and health are not inseparable, in this chapter, safety is informally associated with physical trauma, such as an equipment accident, a fall or impact from a projectile. In contrast, health risks are loosely linked with physiological injury or illness to a person, and could involve acute exposure or continued or repeated exposure to a hazard.

Although health and safety principles are universal, specific practices, standards, laws and regulations vary among jurisdictions, and also over time. The general principles and practices discussed in this chapter are global. However, many details relate to conditions in the United States of America (US), the home country of the author and this book. US-specific details are identified within the chapter. Readers in other jurisdictions should become familiar with, and follow, local and regional situations, especially legislation. Appendix J lists internet links to several health and safety-related organization and web pages within and beyond the US. "Health and Safety at Composting Sites" (Gladding, 2012) provides guidance for conditions in the UK, and excellent safety and health information generally.

2. Hierarchy of controls

Safety and health hazards should be addressed using a hierarchy of the control measures shown in Fig. 13.1. Ideally, the hazard should be eliminated first; however, in practice, this can be difficult. Secondly, the hazard can be substituted for one that presents a lower risk, such as replacing machinery or using different equipment altogether.

Table 13.1 Relative risk from safety and health hazards, at a glance (see Table 13.2 for specific descriptions).

Hazard	Harm that may result	Receiving-sorting	Grinding	Mixing and pile forming	Turning	Screening	Bagging-shipping	Materials handling	Site management and monitoring
Machinery	Physical injury	✔✔✔	✔✔✔	✔✔	✔✔	✔	✔✔	✔✔	✔
Projectiles	Physical injury		✔✔✔	✔	✔✔	✔✔		✔	✔
Vehicles	Accidental collisions	✔✔✔	✔✔	✔✔✔	✔✔✔	✔✔	✔	✔✔✔	✔✔
Electricity	Electrocution			✔	✔✔		✔	✔✔	
Stockpiled materials	Engulfment	✔		✔✔	✔✔	✔	✔✔	✔✔	✔✔
Noise	Hearing damage	✔	✔✔✔	✔	✔✔	✔✔	✔	✔✔	✔
Bacteria, viruses, and other pathogens	Infection and allergic reaction	✔✔	✔✔	✔✔✔	✔✔✔	✔✔✔	✔✔	✔✔	✔✔
Dust and bioaerosols	Infection or respiratory illness; allergy	✔✔	✔✔	✔✔✔	✔✔	✔✔✔	✔✔	✔✔	✔
Volatile chemicals	Respiratory illness	✔	✔	✔	✔✔	✔✔	✔	✔	✔
Weather (ambient temperature, sunlight)	Heat stress, cold stress, extended sun exposure, dehydration	✔	✔	✔	✔	✔	✔✔	✔✔	✔✔✔
Heavy items	Physical overexertion (sprains, strains)	✔✔	✔	✔	✔	✔	✔✔	✔✔	✔

(✔) Relatively no to little risk; (✔✔) Little to moderate risk; (✔✔✔) Moderate to high risk.

Table 13.2 Summary of the potential hazards associated with composting.

Potential hazard	Harm that may result	Comments	Relevant operations and activities
Machinery	Physical injury	Accidents ranging from pinches, cuts, and scratches to loss of fingers and limbs, even strangulation.	All operations involving rotating and moving parts, such as conveyors, grinders, chippers, power take-off drives, bagging lines, motors and engines, etc.
Projectiles	Physical injury	Examples include rocks, fragments of wood, and whole pieces of fruit ejected during turning, hammers and bolts thrown from grinders.	Windrow turning; grinding of feedstock; screening.
Vehicles	Accidental collisions	Between vehicles or between vehicles and individuals. Vehicle operators' vision can be obscured by steam, fog, or piles. Also, loaders or other materials handling equipment can inadvertently entrain unseen workers.	Any activity that includes moving equipment; simply walking the site.
Electricity	Electrocution	From electric power supply wires or equipment.	Any operation near, under, or above high voltage lines (digging near buried cable); welding; and maintenance.
Stockpiled materials	Engulfment	Engulfment in piles, hoppers, bins, and other enclosures may occur from the collapse of piles or supporting materials underfoot.	Maintenance, process monitoring, sample collection, sorting, dislodging objects wrenched in machinery, simply walking the site.
Noise	Hearing damage and stress from noise	Risk depends on noise level, proximity to source and ear protection. Electric motors produce less noise than combustion engines of similar power.	Especially grinding, but all operations with equipment employing engines, vibrations, and abrasive action.

Table 13.2 Summary of the potential hazards associated with composting.—*cont'd*

Potential hazard	Harm that may result	Comments	Relevant operations and activities
Bacteria, viruses and other pathogens	Infection and allergic reaction	Infection can occur through inhalation, skin contact, hand-to-mouth contact, and especially through open wounds. Each type and source of feedstock brings its own mixed population of bacteria and fungi. Risks are greater prior to high temperature composting. The risk presented is dependent on an individual's susceptibility (e.g., allergies, immune system).	All operations involving materials handling, including receiving (feedstocks and bulking agents), grinding, mixing, pile windrow formation, turning, screening, compost handling and application, sampling and monitoring.
Dust and bioaerosols (also note above information on bacteria, viruses, and other pathogens)	Infection or respiratory illness	The primary pathway is via inhalation of particles. A constant factor, especially with dry materials and under dry conditions. Applying moisture with mists or sprays suppresses dust.	Vehicle movement on the site. All operations involving handling dry materials, including turning, screening, compost handling and application, bagging.
Volatile chemicals	Acute or chronic injury to body's target organs	The primary pathway is via inhalation of gases. Ammonia from the composting process is the primary risk.	Turning; receiving and handling of raw feedstock from storage; handling, pumping, transferring leachate.
Weather (air temperature/sunlight)	Heat stress (including dehydration), cold stress, extended sun exposure	Includes fatigue, extreme cold or heat, extended sun exposure, dehydration.	All operations, especially those occurring outdoors.
Heavy items	Physical overexertion (sprains, strains)	Includes heavy lifting, repetitive movement, general overexertion, sprains, strains.	All operations, including materials handling and maintenance tasks.

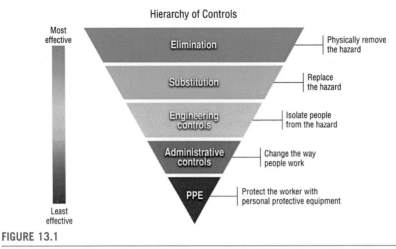

FIGURE 13.1

Hierarchy of safety and health hazard control (NIOSH, 2015).

Thirdly, introducing engineering controls, for instance, by altering the process, redesigning equipment, changing tools, installing ventilation, isolating the machine, or adding machine guards. Fourthly, the hazard could be controlled through management practices like improving working procedures or establishing administrative controls, such as job rotation or reduced work time. To guard against any potential hazards that remain, the final step is to introduce adequate protection through the use of personal protective equipment (PPE). To illustrate the distinction between control measures, Table 13.3 list a few steps within the hierarchy that are examples specific to composting.

Table 13.3 Examples of possible steps in the hierarchy of controls for improving safety and health at a composting facility.

Type of control	Example steps
1. Elimination of the hazard	Purchase wood chip amendments rather than grind on site Build heated maintenance shed to protect technicians from extremely cold weather
2. Substitution of the hazard	Replace tub grinder with horizontal grinder where projectiles present a high risk of injury Perform all welding at a workshop away from composting site to reduce fire risk
3. Engineering controls	Install air filtration on air intake for vehicle/equipment cabs Install hood and misting nozzles on screen conveyor to reduce dust Temporarily fence off windrow area while turning
4. Administrative controls	Require and fund annual physical examinations for employees Alternate employees to operate grinder to limit noise exposure of each employee
5. Personal protective equipment	Require hard hats, hearing protection, gloves, and reflective vests for workers in critical areas

Sources: N. Brown, R. Rynk.

3. Safety and health regulations

Well-established and legally binding safety and health regulations protect workers and govern employers at workplaces in most countries around the world. In the US, the Occupational Safety and Health Administration (OSHA) oversees and establishes occupational health and safety regulations on a *federal* level. OSHA regulations apply to private enterprises and federal government facilities (Box 13.2) In addition, individual states establish their own worker safety regulations, although most states tend follow OSHA's lead and adopt most of the federal requirements.

In countries other than the US, health and safety regulations vary greatly and may be more or less restrictive than the US regulations. Compared to the US, some other industrialized countries have more prescriptive regulations or approaches to risk assessment. Countries with developing economies tend to have fewer and more general regulations. Readers in other countries, and in specific states and provinces, are encouraged to consult the regulations of their own country and local jurisdictions.

4. Safety concerns at composting sites

Composting involves mechanical equipment, typically large machinery and typically several types. Equipment ordinarily employed in composting include bucket loaders, skid loaders, tractors, turners, trucks, excavators, grinders, chippers, turners, mixers, screens, conveyors, forklifts, and bagging devices. Any piece of machinery presents general safety risks and therefore deserves respect, usually in proportion to the size and power of the machine. Some machines are mobile (i.e., vehicles), which carry a separate set of safety considerations. In addition, specific operations employ specific pieces of equipment that pose specific hazards. This section covers the safety hazards and practices associated with composting equipment, starting with general equipment safety, and moving on to safety issues related to specific operations.

4.1 General equipment and site safety

All types of equipment have at least two common features—moving parts and power units that move those parts. Several common safety concerns arise from the operation and maintenance of mechanical devices.

Box 13.2 How OSHA regulations apply to public sector and agricultural compost facilities?

OSHA regulations generally apply to a specific hazard wherever it is encountered in the workplace. Specific US regulations that may apply to composting situations are listed in Appendix K. The federal Occupational Safety and Health Act of 1970 was initially intended for, and applies to, the private sector. By Executive Order of the President, it applies also to workplaces of the federal government; However, it may or may not apply to other public sector facilities, depending upon the individual state. So, for a private industry or commercial composting facility, the OSHA standards apply. For a public (city, town, county, or state) entity, other than the federal government, the state may have developed its own safety regulations or may have adopted Federal OSHA standards. To determine which regulations apply to a given public facility, check with the state's Department of Labor or the OSHA's website (www.osha.gov).

Although some safety and health regulations are written specifically for the agricultural sector, most of the general regulations do not apply to agriculture. However, even when a given regulation does not apply to agriculture, the regulation should be seriously considered, nonetheless. It is important to recognize that these regulations provide excellent health and safety guidance, including warnings, advice, and hazard reduction. Following such recommendations is prudent, even when not required.

For workplaces under its jurisdiction, OSHA can cite and/or fine employers for failure to comply with regulations. Where a state has assumed this responsibility, the state can cite or fine workplaces for noncompliance. If a specific health and safety regulation does not exist, a citation or fine may be issued for noncompliance with the "general duty clause" of the OSHA Act.

The "general duty clause" of the OSHA Act applies if a health and safety hazard does not have its own regulation. The general duty clause states:

SEC. 5. (a) Each employer:

(1) *shall furnish to each of his employees' employment and a place of employment which are free from recognized hazards that are causing or are likely to cause, death or serious physical harm to his employees,*

(2) *shall comply with occupational safety and health standards promulgated under this Act.*

Teen workers are a special case because they tend to suffer a higher rate of injury than their adult counterparts in similar jobs. US child labor laws prohibit teenagers from performing certain kinds of work, depending upon their age. In the US, a youth of 14 or 15 years old can work in agriculture, on any farm, but only in *nonhazardous* jobs (with limited exceptions[1]). For example, the following composting-related jobs would *not* be permitted for a youth under 16 working on a farm:

- operating a tractor of over 20 power take-off (PTO) horsepower, or connecting or disconnecting implements or parts to such a tractor,
- operating or helping to operate any of the following machines: feed grinder, auger conveyor, and earthmoving equipment. (Although woodchippers are not specifically listed in the law, the National Institute for Occupational Safety and Health (NIOSH), which conducts research and advises OSHA, believes that woodchippers are unsafe for operation by youths under the age of 18 years.), and,
- a list of other hazardous jobs can be found on the OSHA website (https://www.osha.gov/SLTC/youth/agriculture/machinery.html).

---

[1] In the US, a 14- or 15-year-old teen who has completed a youth training program (i.e., 4-H) for tractor operation and/or machine operation may conditionally work in occupations for which they have been trained. Consult a local Cooperative Extension office for details.

4.1.1 Moving parts

The moving parts of equipment can grab or entangle hair, clothing, and people during operations or maintenance. The moving parts require guards, shielding, and good work practices. To prevent clothing and extremities from being caught by moving parts, guards or shields should be installed and maintained wherever pinch points, nip points, or scissors points exist (Fig. 13.2). Loose clothing, long unbound hair, and hanging jewelry should not be worn when operating equipment with exposed shafts, chains, gears, and other moving parts that can grab.

Unprotected power-drive lines, such as on augers or the power-take-off (PTO) shafts, are potential hazards (Fig. 13.3). Injuries resulting from entanglement in an open PTO shaft include amputations, scalping, severe lacerations, multiple fractures of limbs, spine and neck injuries, or complete body destruction. PTO shafts have historically had little or no shielding. Workers often remove shields because they can get "in the way" or become bent or broken. Recent models have totally-shielded shafts, which are less likely to be removed. However, there continue to be a large number of inadequately shielded drives in use. Both open drive lines or partially-covered drive lines (e.g., a "U"-shaped shield) should be replaced by totally-shielded shafts (Box 13.3).

Some composting equipment—screeners, grinders, chippers, windrow turners— can throw objects. Once these objects become airborne, they are called projectiles. Eye protection such as shatter-resistant glasses with side shields or goggles (either may be further assisted with a faceshield) are needed to protect against potential impact and injury. In some cases, moveable screens or barriers can be effective in intercepting projectiles before they endanger bystanders.

(A)   **(B)**

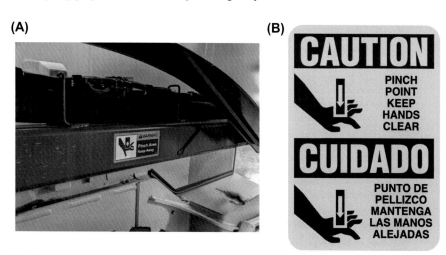

FIGURE 13.2

Pinch points, or nip points, are numerous on equipment commonly used for composting. Warning signs provide important reminders for workers.

Source: (A) James Hardin, (B) OSHA.

FIGURE 13.3

The power take-off (PTO) shaft of a tractor (left) and its coupling to the power shaft of a windrow turner (right). Both the shaft and the coupling are potential safety hazards and should be shielded. (A PTO or "power-take-off" is a mechanical connection for transferring power and mechanical energy from a power source, like a tractor or front-end loader, to another machine, like pump, woodchipper, or windrow turner.)

Box 13.3 What to do if someone is trapped by a PTO shaft?

1. If a victim is trapped against a PTO shaft, the very first step is shutting off the tractor and making sure it will not restart.
2. Next, chock the tractor wheels so that the tractor cannot move. It may be possible to extricate the victim by disconnecting the PTO shaft from the tractor or by putting the drive unit in neutral and turning the shaft counterclockwise to unwrap the victim's clothing or hair.
3. If the shaft is solid, rescuers may have to cut it with a cutting device such as a portable power grinder, hacksaw, or oxyacetylene torch. If there are combustible materials in the area, rescuers should be extremely careful when using any type of flame-producing equipment, or even portable grinders that produce sparks.
4. If a part of the body must be amputated, it should be located, covered, or wrapped cleanly and placed in ice for possible reattachment.
   For more details, see Farm Safety Association (2002).

Foreign objects delivered with some composting feedstocks are damaging to machinery and potentially hazardous to human operators. Examples include gas cylinders, batteries, containers with chemicals, cables, chains, rope, metal strapping, and large or sharp metal items. Policies for accepting feedstocks should prohibit such damaging items from deliveries. However, procedures should be in place for sorting and discarding these unwanted items before they come into contact with equipment, especially shredders, grinders, mixers, turners, and screens.

Equipment jams are also a regular occurrence at composting facilities, in part, because of the diverse and difficult materials encountered. Conveyors, grinders, shredders, chippers, mixers, screens, and even turners are each prone to jams. Operators have been known to climb inside the tub of a grinder with a sledgehammer and pickax to break up the jam. Climbing into such equipment should be avoided; wherever possible, the task should be done from outside.

When removing a jam from any equipment, the equipment first should be disconnected from the power source. If the unit is electric, power should come from an outlet equipped with a ground-fault interrupter to stop electrical current if a short occurs in the system. First, the equipment should be appropriately "locked-out" and "tagged out," and then the moving parts should be blocked and chocked. Even after power is shut off, it is possible that parts may still be able to move under their own weight once the jam is removed. Some equipment may have residual energy that is held back by the jam. After the jam is removed, the mechanism may thrust forward or blades, augers and paddles could rotate. A block-and-chock helps to prevent injury occurring in this way. If the jammed mechanism is driven by a PTO, the PTO should be disconnected first.

Before any type of powered equipment is serviced, that equipment must be fully shut down and "locked out" to ensure that the equipment is not inadvertently restarted. A lockout/tagout procedure should be established and always followed. Lockout refers to a procedure that physically prevents a machine from becoming re-energized. The procedure includes blocking to prevent the release of stored energy. A lock, or some other mechanism, is placed on the circuit breaker, switch, value, or other device that isolates the equipment from its energy source (Fig. 13.4). Only the person servicing the equipment can remove the lock. OSHA has a standard titled Control of Hazardous Energy (OSHA 29 CFR 1910.147) that provides more information (OSHA, 2021a). Tagout is used only when lockout is not an option. Tagout involves placing a suitably durable sign or tag at the point where the equipment's energy is shut off. The tag includes information about the service taking place, including the name of the service person. Lockout and, especially, tagout procedures require good communication, training, and management oversight (Flower, 2016).

4.1.2 Power lines

Overhead and underground power lines present the possibility of electrical shock. When power lines are located over composting operations such as windrows and storage piles, the elevated part of the vehicle should be kept a distance of at least 3 m (10 ft) from the power lines to prevent the equipment from becoming energized and shocking the operator.

It is also important to identify and permanently mark the location of underground power electrical lines and other utilities (gas, water, cable). The utility company needs to be notified when digging or trenching on site, especially in the vicinity of known underground utilities.

4.1.3 Site and equipment cleanliness

Wet and oily feedstocks, wet weather, or ponded leachate can create slippery conditions at a composting facility. These conditions can occur on the ground surfaces and

FIGURE 13.4

A lockout/tagout center at a commercial composting facility in California.

Source: C. Oshins.

on equipment platforms. Nonskid shoes or boots should be worn to reduce this problem, in addition to good site management practices to keep the site as clean and tidy as possible.

Workers should also be appropriately protected from dust. Dust can be a potential fire hazard and should be kept from accumulating in welding areas, on engine manifolds, mufflers, and on other equipment components that get hot (Chapter 12 dust control practices).

4.1.4 Vehicle safety

A composting site can be an active place with several vehicles (trucks, loaders, turners) moving amid each other and amid people on foot. Furthermore, the vehicle operator's view is occasionally obstructed by steam rising from windrows, fog, and piles of material (Fig. 13.5). The situation demands extra caution and attention to safety. Operators of bucket loaders and skid loaders, and those working near loaders, must be especially diligent as loaders maneuver at a rapid pace and frequently change direction.

To prevent an accident occurring, operators must make sure that co-workers and visitors are clearly out of the way. Each day an operator should start with an understanding of the general whereabouts of coworkers and visitors on that day and presume that they will remain there. Similarly, pedestrians walking in the vicinity of mobile equipment must make their presence known to the operators, and that they can actually be seen by the operators. Brightly colored clothing (e.g., reflective vests) is highly recommended. As a rule, do not approach an operating vehicle until you first make eye contact with the operator.

FIGURE 13.5

Visibility is often restricted at composting sites. Operators must be alert and know the whereabouts of co-workers and visitors.

All large mobile equipment (e.g., loaders and trucks) should have an audible back up (reversing) alarm. If the view to the rear is obstructed, a vehicle must not be used in reverse unless there is a reverse signal alarm or a coworker signals that it is safe to do so. If a tractor or front-end loader needs to travel on public roads it should be fitted with a slow-moving vehicle sign, flashing lights, or similar warning device (Fig. 13.6). This practice, which is required by law in many countries, applies to vehicles that travel at slow speeds, typically less than 16 km/h (25 miles/h). In the United States, slow-moving vehicle emblem requirements are defined OSHA regulations. They are based on American Society of Agricultural and Biological Engineers (ASABE) standard ASAE S276 .6 (ASABE, 2016).

As a basic rule, vehicle operators should wear seatbelts. When driving a tractor or loader, the driver should tighten the seatbelt sufficiently to keep him/herself within the area provided by the roll-over protection structure. The seatbelt should be inspected regularly to make sure it has not become worn or brittle—in which case, it should be replaced.

Mobile equipment, especially tractors, could upset or roll-over. Tractor roll-over continues to cause more fatalities than any other type of farm accident, despite increased emphasis on safer design and roll-over protection. Some older tractors have profiles, especially narrow wheel spacing that can easily upset the vehicle, causing injury or death to the operator. As 85% of all tractor overturns are to the side, the typical injuries tend to include a broken or crushed pelvis.

In many industrialized countries, regulations require that modern tractors must have roll-overprotective structures (ROPSs), seatbelts, and protection from spillage of batteries, fuel tanks, oil reservoirs, and coolant systems. Also, all sharp edges and corners at an operator's station have to be designed to minimize operator injury in the event of an upset. In the US, OSHA regulations require these features for tractors manufactured after October 1976. Older tractors need to be modified with roll-over protection and seatbelts. If a tractor without ROPS rolls over, it may be easier

FIGURE 13.6

Slow moving vehicle emblem on a tractor-pulled batch mixer. The sign is necessary for travel on public roads (not necessary for building windrows).

to rescue the victim from under the tractor by digging, after first stabilizing the tractor with cribbing and chocking the rear wheels so that they cannot turn (Rose and Baker, 2012).

In addition, moving equipment can also generate dust. In vehicles, exposure is best reduced using an enclosed cab with well-sealed, air-conditioned, and heated cab with filtered air intakes. Filters should be removable and washable or replaceable, and employees should be instructed to remove and clean or exchange filters frequently. If the filtration is inadequate, the operator can wear a particulate respirator (this requires the employer to meet the OSHA regulation on respiratory protection 29 CFR 1910.134).

4.2 Safety provisions for specific operations

All operations at a composting facility—receiving, sorting, grinding, turning, screening, shipping, materials handling, and maintenance—involve large equipment. Most operations also rely on vehicles as materials are transferred between operations. Thus, every operation demands respect for, and attention to, the general hazards associated with operating any mechanical device and vehicle, as previously discussed. In addition, specific operations and equipment bring specific hazards and safety concerns.

4.2.1 Site monitoring, process management—engulfment

Operators and managers frequently need to walk around the composting site to perform a variety of management tasks including monitoring the composting process, taking temperatures, recording data, inspecting feedstocks, performing maintenance, sampling materials, and generally assessing operations. Simply walking onsite exposes workers to several often-overlooked hazards, from accidents with vehicles (see previous section) to engulfment in piles.

It is possible for workers walking on windrows, piles, or bins to drop down into the material and be engulfed—trapped or buried alive. This problem is particularly risky when tall piles are loosely packed or have sufficient surface moisture to experience "crusting"—causing them to mistakenly look and feel solid underfoot. The interior of such piles can be very hot and cause burns, even if a person is only partially immersed (Fig. 13.7). This situation has occurred on farms with bins or silos of corn or other grain, as well as in other industries. Furthermore, the weight of the compost and resulting chest constriction can restrict breathing and produce asphyxiation as can the resulting relatively high concentrations of carbon dioxide and low oxygen levels.

On a composting site, relatively free-flowing bulk materials like wood chips and dry compost have the potential to collapse and partially engulf a person. Semisolid materials (e.g., manure, fruit pomace) stored in lagoons or depressions in the ground also pose this hazard to a person who attempts to walk over an apparently solid crust.

It is possible to eliminate this hazard altogether by not walking on lagoons, or climbing onto windrows, piles, vessels, or bins. If there is no alternative, workers should not work alone but with a co-worker observing from a safe stable point (i.e., not also on top of the windrow). Preferably a body harness should be worn with retrieval line attached to a fixed point. Even better protection is provided

FIGURE 13.7

A seemingly solid pile of material, like this vegetable processing waste, can be deceivingly fluid and dangerously hot inside. Avoid walking on top of piles and crusted lagoons.

with the retrieval line attached to a mechanical retrieval device with at least a 4:1 mechanical advantage (e.g., winch or come-along). Also, before climbing onto piles or windrows, check temperatures of those that could generate or retain heat.

Piles and bins that are burning internally (i.e., through spontaneous combustion) can collapse due to the voids created inside as the material burns. A pile that is smoldering, or suspected of burning, should never be walked on or driven on for any purpose, whether it is simply checking temperatures or extinguishing the fire.

As noted earlier, workers walking on-site and operators of equipment must be conscious of the whereabouts of co-workers, and continually look out for them. A worker standing near a pile, checking temperatures, or taking samples, for example, can go unnoticed by an equipment operator and inadvertently get run over, scooped up or hit by a loader or other equipment. Such accidents have happened at composting facilities. When working on the site, notify equipment operators and wear reflective clothing.

4.2.2 Conveyors

Conveyor belts are used for many tasks at composting facilities, by themselves and as integral components of other equipment such as grinders, screens, and turners. Conveyor belt accidents could involve stepping onto the belt, falling onto the belt, or otherwise having clothing or limbs caught or crushed by drive rollers, idlers, or belts. Conveyor chute openings should have covers or guard rails to protect workers from falls. Guards need to be in place to protect from pinch and shear points such as terminals, drives, take-ups, pulleys, and snub rollers. Critical points occur where a belt changes direction including where belts wrap around pulleys, at the discharge end of the belt, on transfer and deflectors, and at take-ups. Warning signs should be posted around conveyor belts—especially if guards are not practical. Operators should be trained on general rules for working around conveyors including:

- only permit authorized maintenance personnel to adjust conveyors,
- always ensure that the conveyor is locked out before working on it; this includes shutting off or disconnecting its electrical source and blocking or chocking the belt to prevent its movement,
- always make sure that the emergency stop is reachable, secured and in working order,
- always face the opposite direction from the belt movement when shoveling on a belt, and,
- ensure that no one can ride conveyors.

4.2.3 Size reduction—grinding, shredding, and chipping

Size reduction is an especially aggressive operation involving high-powered machinery and very fast rotational speeds. Therefore, size reduction deserves respect with regard to safety. Grinders, chippers, and shredders present common hazards but also some that are specific to the type of equipment at work (Chapter 9). Like screening, size reduction is an active operation with multiple vehicles and mechanisms working simultaneously. Operators need to remain alert to the whereabouts of other equipment, workers, and bystanders.

FIGURE 13.8

Open hopper of a slow speed shredder with exposed augers.

Hoppers and loading chutes for grinders and shredders are often open on the top or sides (Fig. 13.8). Guards or shields are often impractical because the operator needs to continually load feed material. Because they potentially can be pulled into moving parts, operators should stay clear of open top hoppers and feed chutes. As much as possible, avoid manual feeding of material into hoppers and chutes. It is important to recognize that there are feed mechanisms (e.g., rollers, chains, conveyors) operating behind hanging curtain deflectors.

Size reduction equipment is subject to jams from especially hard, pliable, or stringy material. Presorting troublesome items from the feed material will help reduce a jam occurring. Observing proper maintenance procedures when clearing jams should be carried out, as previously discussed.

Size reduction is an especially loud operation, therefore hearing protection is needed. Noise levels can vary with the model and whether it is powered by diesel or electricity, although machines with electric motors are substantially quieter.

Because grinders and shredders typically are open on the top or sides, flying debris is possible. Nearly every type of size reduction equipment can throw debris from the grinding chamber, but the risk of projectiles is greatest with tub grinders because of the often-uncovered tub that opens into the hammermill chamber.

Projectiles from tub grinders have been known to fly hundreds of meters. Flying objects can include hammer bolts, wood chips, stones, broken hammer tips, nails, and other metal objects in the feed material. Several precautions can be taken to reduce the risk of accidents from projectiles:

- Operators should wear either safety glasses with side shields or goggles to protect the eyes from flying objects, as well as a hardhat. A faceshield could be added as face protection (but not as an alternative to safety glasses or goggles; or a faceshield/goggle combination could be used).
- Keep the tub loaded with feed material when the grinder is operating to intercept items ejected from the hammermill chamber.
- Establish a zone around the grinder where only the grinder/loader operator is allowed. The radius of the zone depends on the grinder model and should be based on the manufacturer's recommendations. *Ask the manufacturer to specify the grinder's* "thrown object zone."
- Erect a tall screen to catch flying debris. Objects can be ejected from tub grinders in any direction, but they predominantly fly toward the direction that the hammers rotate at the top of the chamber (Fig. 13.9).
- Remove metals, rocks, and other hard items from the feed material prior to loading in the grinder.
- Install covers, partial covers, deflectors, or other safety devices designed to contain thrown objects.

FIGURE 13.9

Screen installed to intercept projectiles from a tub grinder. The screen is located such that the grinder rotates toward the screen at the base of the tub.

- Minimize the use of tub grinders near areas frequented by pedestrians and cars. In such areas, preferably use another type of grinder, such as a horizontal feed grinder, which is much less likely to eject objects.
- Never, at any time, step into a tub grinder, such as to loosen a jam!

Wood chipping equipment presents specific hazards because many models are fed manually. Both fatal and nonfatal injuries have resulted from working with chippers. Nonfatal injuries most commonly involve an injury to an upper extremity, including amputation. Operators can become caught in the feed mechanism and pulled into the rotating chipper knives. Chipping equipment also can produce flying objects, including a loose hood from the chipper itself. If the machine is being opened or closed with the knives still rotating, the hood can fly off if it contacts the rotating blades. The following safety precautions should be observed with chipping equipment:

- The chipper should be thoroughly inspected each day before start-up. The hood should completely cover the chipper knives, and workers should ensure that knives come to a complete stop before opening the hood.
- The area around the chipper should be kept clear to reduce tripping hazards.
- A long branch should be used as a push stick to feed shorter material into the chipper. If shredding leaves, use the tamper on the unit to push leaves into the shredding chamber.
- If purchasing a chipper, it is important to consider those with an interlock system where the chipper hood cannot be opened while the cutter disk is turning. Chipper shredders should have a certification symbol of safety from the Outdoor Power Equipment Institute (OPEI) and the American National Standards Institute (ANSI). Similar standards exist in other countries, notably the Machinery Directive (2006/42/EC) in the European Union, where conformance is declared through a CE mark (European Commission, 2021).
- Personal protective equipment (PPE) is mandatory when operating chippers due to the close proximity of workers to chippers and the danger from flying debris, blowback from the hopper, and the potential for entanglement of clothing in moving parts. The PPE recommended includes hard hat, eye protection (either safety glasses with side-shields or goggles; may be augmented with a faceshield or use a goggle/faceshield combination), hearing protection, safety boots, and close-fitting outer clothing. Gloves help prevent cuts if a tree limb is pulled from the operator's grasp, as well as reducing the effects of limb vibration as items are fed into the chute.
- Training for workers operating chippers should emphasize: (1) correct operation of safety devices and controls consistent with the recommendations of the manufacturer; (2) the need to keep hands and feet away from the feed chute; and (3) proper procedures for feeding items into the feed chute (including standing to the side in reach of the emergency shut-off when feeding items).

4.2.4 Mixing

Many types of mixing equipment involve rotating paddles or augers and hoppers with open tops. Falling into an operating mixer can be fatal. Operators should therefore stay clear of the top of the mixer when it is being loaded or otherwise operating. Material is preferably loaded using a front-end loader, grapple or conveyor, rather than by manual feeding.

Mixers are also prone to jamming from dense, rigid, and stringy feedstocks or foreign objects. Presorting troublesome items from incoming feedstocks can prevent jams from occurring. When clearing jams, observe the general safety precautions previously described.

4.2.5 Windrow turning

Turning causes the release of heat, water vapor, and other gases from piles and windrows. Thus, the immediate working conditions can be especially harsh. Operators are potentially exposed to concentrated levels of moisture, ammonia, and other gaseous products of decomposition. The abundant moisture released can literally fog the operator's vision. This situation is worse when the operator is situated at, or behind, the point of turning (for example, when using bucket loaders, some straddle turners and tractor-driven turners that are pushed). Excavators and tow-behind turners allow the operator more distance. In any case, enclosed cabs with air filters are a necessity. Operators should also be aware of slippery footing on equipment platforms due to condensation or freezing of the water vapor.

Most turners that employ a rotating shaft with paddles can fling objects behind them at high speed when they start moving into a windrow (Fig. 13.10). The flying objects, which can be anything from rocks to tennis balls, represent a potential accident. Once the turner housing is fully enclosed by the windrow, it no longer ejects material, until it reaches the end of the windrow. Bystanders should never stand

FIGURE 13.10

When entering a windrow the turner can throw objects backward at high speed. Anyone on the site should not be in the back of turners, especially at this point.

behind a windrow turner while it is entering a windrow (even to take photographs). Operators should not begin turning a windrow without first making sure that no one is behind the turner.

When debris needs to be cleared from the shafts and belts of windrow turners, operators should wear safety gloves, safety boots, eye protection, and a hard hat. Items wedged in the turning mechanism may retain some energy and spring outward when pulled loose. When working near the turner housing, operators need to be aware of projecting structural and mechanical elements, as well as slippery footing.

4.2.6 Screening

Screening can be a busy operation with multiple conveyors and one or more vehicles concurrently loading the screen and removing different piles of screened materials. Attention to the movement of co-workers, vehicles, and bystanders is crucial.

As screens rotate or vibrate, sharp objects, small particles, and dusts may become airborne, causing impact to the eyes and body and/or inhalation. Dust is a particular problem because materials tend to be relatively dry when screened. Loader cabs should have a properly operating air filter. When working outside of cabs, operators should wear safety glasses with side-shields or goggles and an N95 dust respirator (one that is designed to remove 95% of particles with a diameter of 0.3 microns and above) or a respirator with particulate (high-efficiency particulate air [HEPA]) cartridges.

It is important to clear dust and debris from engine manifolds and other components that can get hot, as this may lead to a fire. As sharp or pointed objects can be present in the screened material, especially the overs pile, appropriate safety gloves and safety boots fitted with protective midsole are necessary when sorting material by hand.

4.2.7 Welding

Although not normally considered to be a composting operation, the arc welder can be the busiest piece of equipment at a composting facility. Repairs done by welding or brazing involve simultaneous safety concerns—intense light, which can damage eyes and skin; burns from sparks or hot metal; fires from sparks igniting nearby oil spills, wood chips, or dry compost; and fumes that can damage lungs and other organs or induce cancer. The concentration of hazardous substances from gases and fumes increases in confined spaces.

Light

- In addition to the intense visible light, the invisible ultraviolet (UV) light and infrared light generated by a welding arc can damage the eyes; while UV light can also damage the skin. Barriers should be used to protect bystanders in the area from viewing the welding arc.
- All welding lenses have a shade number rating on a scale of 1 to 14. This provides an indication of the darkness of the lens (14 is the darkest; 1 is the lightest) and should be selected depending upon the type of welding and the brightness of the light. All welding lenses should screen out all UV and infrared light; a feature that is not dependent upon the shade rating.

- For comfort, the welder needs to wear a faceshield with the highest level of shade (shade level #10 to 14) compatible with viewing the work adequately. If the arc is hidden by the work piece, lower shade levels may be possible, with a minimum shade level of 7. Lenses may also be auto-darkening.

Sparks, burns, and fire

- In addition to a faceshield, necessary PPE includes gloves, head cap, hard-toed shoes, button-down shirt pockets, long-sleeved shirts, and cuff-less pants (trousers without any turn up at the bottom). Clothing must not contain synthetic fibers as these can melt and produce skin burns. For heavy-duty welding, leather gauntlet gloves, jacket, apron, and shoe covers may also be needed. For overhead or vertical welding, a cape or other shoulder protection may be needed. Ear plugs can protect from flying sparks as well as from the noise of a noisy welding operation.
- Although arc welding is usually done with low voltage (less than 100 V), there is always the hazard of electrical shock—especially when hot weather and dampness are present. Welders should be aware of the location of electrical contacts and replace any wires that have cracked or worn insulation.
- Welding should take place away from potential fuels such as oil spills, wood chips, or compost. A water supply or Class A portable fire extinguisher kept nearby enables rapid response for fires that start from wayward sparks (see Chapter 14, Box 14.5).

Gases and fumes

- Welding gases and fumes are a complex mixture which may include ozone and nitrogen oxides; metal vapors and other gases from the substrate being welded or from the welding rod; and carbon monoxide, carbon dioxide, and other gases from coatings and oils on the substrate (carbon dioxide may also be used as a shielding gas).
- When welding inside a building, sufficient ventilation is extremely important to dilute and remove gases and fumes. At the benchtop, the best arrangement is a slot hood connected to a blower that draws the gases and fumes to the outside. General building ventilation is not as effective. Welding curtains and other barriers can help to reduce drafts and improve the capture of air contaminants. If sufficient ventilation is not possible, a respirator with cartridges for welding fume and oxidizing gases could be worn.
- Cylinders of welding gases should be chained to a wall or bench and transported by fastening to a cylinder cart. The valve should be protected by a valve cover during transport or when the cylinder is not in use.

Confined spaces

- The OSHA standard on confined spaces provides excellent work practices for any confined workspace (e.g., storage bin, container). If welding is done in a confined space, gases can collect and increase to hazardous levels. Ventilation must be provided in the confined space to dilute and flush out these air contaminants and to provide sufficient oxygen for the welder to replace that consumed by burning the welding gases. A blower and hose can be used, with the blower kept outside the space in an area of clean air and the hose dropped into the confined space to provide good air overturn.

- It is important to inspect welding gas lines to prevent leaking gas lines. Gases such as acetylene or oxygen that leak into a confined space could reach explosive levels or lead to a rapid, intense fire.
- To determine if the atmosphere in a confined space is safe before and during a welding task, use an air tester with a direct readout that checks the air for levels of oxygen, explosive gases, and toxic gases. Toxic gas sensors used should reflect the gases expected to be produced by the specific welding operation (typically hydrogen sulfide and carbon monoxide). By monitoring the air during the task, an alarm can alert workers to exit the space if the atmosphere becomes unacceptable.

5. Physiological health concerns

Composting workers are subject to the hazards generally associated with performing physical labor near large equipment in an outdoor environment including noise, physiological stress, extreme heat, cold and fatigue.

5.1 Noise

Heavy equipment and processes of composting can be noisy, to the point of being potentially damaging. As most loud noise does not produce pain, hearing damage typically happens so gradually that there is no warning or indication that injury is taking place. The OSHA standard on noise, which does not apply to agriculture, endeavors to reduce exposure to levels below damaging levels. The level of noise in the regulation is an eight-hour time-weighted average of 90 decibels (dBA). A decibel, or specifically a dBA, is a measure of noise intensity, weighted on the "A" scale, which puts emphasis on exposure to higher-pitched noises. For levels of noise higher than 90 dBA, the time of exposure allowed for an employee decreases. At noise levels of 115 dBA, the amount of exposure allowed per day drops to just 15 min. OSHA's website notes that recent research shows that 90 dBA is not sufficiently protective and that an eight-hour time-weighted average of 85 dBA should be used, although the regulation has not yet been changed to reflect this.

If noise exposures equal or exceed an eight-hour time-weighted-average of 85 dBA, the employer must institute a hearing conservation program, which requires monitoring of noise levels. Equipment used at composting sites for wood chipping, shredding, grinding, turning piles, or other tasks have been reported as exceeding 90 dBA, with shredders as high as 98 dBA. Thus, the compost site's equipment should be evaluated to determine the exposure of the employees, along with the length of time the equipment is typically used.

A hearing conservation program requires audiometric testing of employees' hearing to establish a baseline level for each employee and note trends that indicate hearing damage. The program should also reduce exposure using administrative or engineering controls or, if not feasible, PPE. For some equipment, an enclosed cab may serve as an engineering control that sufficiently reduces the noise exposure for the operator. When acquiring equipment, noise hoods and mufflers should be specified and then these should be properly maintained. For people running chippers and grinders, PPE such as ear plugs, or earmuffs are needed. If staffing levels permit, it may be possible to reduce exposure by an administrative control—such as rotating workers among a variety of tasks so that the actual duration of exposure per day does not exceed what is allowed by either regulations or prudence (e.g., 85 dBA over 8 h).

5.2 Physiological demands: overexertion, sprains, or strains

The daily tasks involved in operating a composting facility can be physically demanding. Frequently, the work requires heavy lifting and prolonged postures (such as operating heavy equipment). These stressors can be compounded by exposure to heat or cold and UV light from the sun.

Ergonomics refers to the interaction of humans with their working and living environments. Ergonomic injuries such as overexertion and fatigue can occur when operating heavy equipment with repetitive use of machine controls and prolonged sitting. These tasks could involve acute and/or repetitive injury to muscles and joints. Musculoskeletal injury can be triggered by direct trauma (e.g., falls, impact, and bruising) or a single overexertion (e.g., pulling a muscle while lifting). Repetitive strain can result from static or dynamic work including prolonged sitting, holding the muscles in a tense, fixed position for extended periods, such as carrying something over a long distance; holding tools continuously, bending the wrists to hold or repeatedly move a tool, or prolonged exposure to vibrations. Typical symptoms of repetitive strain include soreness, pain, discomfort, redness and swelling, limited range of motion, stiffness in joints, weakness and clumsiness, numbing, tingling sensations ("pins and needles"), popping and cracking noises in the joints, and "burning" sensations. The symptoms occur more readily as people age. Repetitive trauma does not allow for complete repair of the tissue during rest. In the worst stage, pain persists even at rest, and sleep is often disturbed. Severe pain, limited mobility, or muscle weakness may make it impossible to perform most tasks.

Preventing ergonomic injuries involves evaluating the job for risk factors and obtaining symptom reports and observations from workers. Often the best ideas and solutions come from the people doing the work, especially if they know the risk factors. Table 13.4 list some general risk factors that reflect the body's limitations, along with several suggested solutions. Job analysis tools and checklists are available on OSHA's ergonomics website (OSHA, 2021b).

5.3 Heat stress

Heat stress occurs when the body is subjected to temperatures that cause the core temperature to stay above 38°C (100.4°F), over the course of the workday. Sweating

Table 13.4 Ergonomic risk factors and solutions.

Risk factors	Some solutions to consider
Joints (any joints including the back) are bent rather than neutral or relaxed; sometimes some tasks involve joints being bent as far as they can go—this stretches the surrounding muscles, making them easier to injure.	Use better designed tools and equipment; for example, modern payloaders have better and more adjustable seats. Use ergonomically designed shovels and tools which put the bend in the tool (rather than a bend in the user).
The work or load is too far out from the body, placing considerable strain on the lower back.	For any lifting job, bring the load as close to the body as possible. Use a tool to reach the distance, rather than extending the spine. Provide instruction on proper lifting techniques—if they are not or cannot be used for the job, figure out why not. If a load cannot be brought close to the body, such as a wide object that will not fit between the knees, buddy-lift or use a machine to perform the lift.
Body bent forward—the surrounding muscles are stretched and weaker, making them easier to injure	Lift using a squat position rather than a bent back. Buddy-lift or use a machine to perform the lift.
Twisted trunk—twisting the back places strain on muscles, tendons, and discs; lifting and twisting are a combination with a high risk of back injury.	Avoid any twisting. If lifting and carrying an object, lift while facing forward and then take steps to make the turn.
Sudden movements and forces—never swing things, especially as a way to lift heavy objects; the muscles can be forced to stretch faster than they are able to respond, producing over stretching or tearing and injury.	Avoid swinging motions. If an item is too heavy to lift alone, get help or use a machine.
Posture or movement maintained for a long period of time—this is the prolonged static and/or dynamic work described above. Also keep in mind prolonged contact stress such as pressing or leaning the body, hands, or wrists against a hard or sharp edge like the edge of a table.	Plan for breaks and recovery/rest periods. Divide the jobs among several people, so no one person does it for very long. Put padding on sharp edges.
Continuous stress on certain muscles producing localized muscle fatigue—sometimes a job or task involves overworking just a few muscles, such as sitting in place but using the hands and bending the wrists over and over again.	Modern heavy equipment tends to have controls which conform to the hands and produce better positions for the arms, wrists, and shoulders.

Continued

Table 13.4 Ergonomic risk factors and solutions.—*cont'd*

Risk factors	Some solutions to consider
Working to the point of exhaustion—as muscles become tired, they can suffer from insufficient oxygen supply and a build-up of waste products; once exhaustion occurs, muscle injury is more likely. Also, when workers become exhausted yet must continue working, they do the job any way that they can get it done—often with swinging of a lifted object, twisting of the back—anything to keep on going, this further increases the risk of injury.	Plan for breaks and recovery/rest periods. Frequent short rest periods reduce cumulative fatigue better than a few long breaks. The worst procedure is to let the worker work through breaks and go home early, exhausted. Divide tasks among several people; do a variety of tasks to use different postures and muscles.
Vibration—using vibrating tools, handling vibrating machine controls, or sitting on vibrating equipment. Prolonged vibration can produce damage to tiny nerves and blood vessels.	Padded or gel-filled gloves for vibrating tools or machine controls; good seat cushions and proper seat shape to protect the lower back. Plan for breaks and recovery/rest periods.

and the surrounding ambient air simply cannot cool the body sufficiently. Wearing protective clothing or a respirator can add to the risk. Workers may notice symptoms of irritability, low morale, increased numbers of errors or increased frequency of unsafe behavior. Usually, adaptation to heat exposure takes about 5–7 days; so abrupt changes in the weather can produce more discomfort than a gradual change in air temperature, which gives the body time to adjust its temperature and pulse rate. People with chronic illnesses of the heart, lung, kidney, or liver tend to have lower heat tolerance and may be at greater risk. Some drugs, prescription or illegal, can produce greater risk of heat injury. In this situation, seek the advice of a physician.

Prolonged heat stress can lead to related disorders such as fainting, prickly heat rash, heat exhaustion, and heat stroke. Hygiene practices can relieve some of the heat stress or assist in acclimation to abrupt changes in the weather. Such practices include:

- Fluid replacement—Drink small quantities frequently (150–200 mL or 5–7 fluid ounces every 15–20 min). Do not depend upon thirst as a warning. Salt intake in a normal diet is usually sufficient to meet salt demand. Workers on salt-restricted diet should never use salt tablets without consulting a physician.
- Training and self-determination—By providing accurate verbal and written instructions and training on heat stress, workers can limit heat stress and recognize symptoms in themselves and others. Managers and workers themselves can adapt to hot days by leveling out the work effort over the allocated time, rotate jobs, and take more frequent breaks. People who have been away from work for three or more days may require acclimation. Scheduling tasks to the cooler parts of the day and avoiding the hottest time of day should also be considered. It is also import to build in rest breaks spent in the shade or in a building with air conditioning.

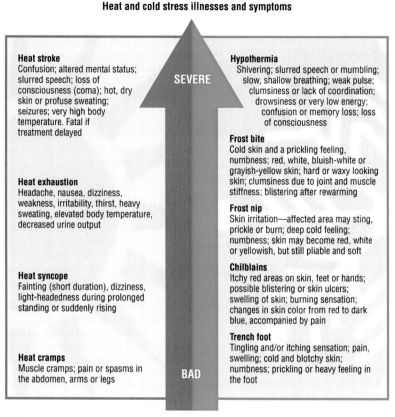

Heat and cold stress illnesses and symptoms

SEVERE

Heat stroke
Confusion; altered mental status; slurred speech; loss of consciousness (coma); hot, dry skin or profuse sweating; seizures; very high body temperature. Fatal if treatment delayed

Hypothermia
Shivering; slurred speech or mumbling; slow, shallow breathing; weak pulse; clumsiness or lack of coordination; drowsiness or very low energy; confusion or memory loss; loss of consciousness

Frost bite
Cold skin and a prickling feeling, numbness; red, white, bluish-white or grayish-yellow skin; hard or waxy looking skin; clumsiness due to joint and muscle stiffness; blistering after rewarming

Heat exhaustion
Headache, nausea, dizziness, weakness, irritability, thirst, heavy sweating, elevated body temperature, decreased urine output

Frost nip
Skin irritation—affected area may sting, prickle or burn; deep cold feeling; numbness; skin may become red, white or yellowish, but still pliable and soft

Chilblains
Itchy red areas on skin, feet or hands; possible blistering or skin ulcers; swelling of skin; burning sensation; changes in skin color from red to dark blue, accompanied by pain

Heat syncope
Fainting (short duration), dizziness, light-headedness during prolonged standing or suddenly rising

Heat cramps
Muscle cramps; pain or spasms in the abdomen, arms or legs

BAD

Trench foot
Tingling and/or itching sensation; pain, swelling; cold and blotchy skin; numbness; prickling or heavy feeling in the foot

FIGURE 13.11

Stages of heat and cold stress illnesses.

- Diet and life-style—Avoid large meals during work breaks because they increase circulatory load and metabolic rate. Adequate sleep, good diet, and regular exercise reduce the risks of heat stroke. Abuse of alcohol or drugs increases the risks.

There are varying levels of heat stress that demand different levels of action (Fig. 13.11). *Heat stroke* is an especially serious level of heat stress. It requires immediate medical attention to prevent loss of life. A person experiencing heat stoke is affected mentally, exhibiting confusion, loss of consciousness, and/or convulsions. *Heat exhaustion* is also a serious situation and typically precedes heat stroke. Heat exhaustion affects a person physiologically. It should be suspected if a worker appears unduly fatigued or disoriented or experiences inexplicable irritability, malaise, dizziness, lightheadedness, headaches, upset stomach, vomiting, decreased or dark-colored urine, fainting or passing out, or pale, clammy skin.

If symptoms of heat exhaustion appear, the worker should be moved to a cool location with rapidly circulating air for rest. Loosen and remove any heavy clothing. Have the person drink cool water (about a cup full every 15 min) unless he/she is sick to the stomach. Cool the person's body by fanning and spraying with a cool mist of water or applying a wet cloth to the person's skin. Keep the person under observation. *Call for emergency help if the person does not feel better within a few minutes.* If sweating stops and skin becomes hot and dry, the person may be experiencing *heat stroke* and immediate emergency care with hospitalization is essential. It is important to note that workers typically experience exertional heat stroke and thus may still be sweating (unlike typical heat stroke, such as in the elderly, for whom the skin is dry).

Before a worker experiences heat exhaustion, he/she may experience dizziness or fainting, which are signs of heat syncope. *Heat syncope* primarily occurs from dehydration or poor acclimation. Muscle pain or muscle spasms, without the other symptoms of heat exhaustion, signal *heat cramps*, which is normally cause by dehydration and low salt levels due to prolonged sweating. These less-severe heat-related illnesses are precursors to heat exhaustion and should be treated and taken seriously.

5.4 Cold stress

With extreme cold temperatures, workers face the risk of hypothermia, even to the point of frostbite. Cold temperatures are aggravated by wet clothing, wind, or contact with metal machine controls. Older workers or workers with circulatory problems require special precautionary protection against cold injury (e.g., extra insulating clothing, reduction in the exposure period). Some diseases and some medications reduce tolerance to work in cold environments. A physician should be consulted for workers who are at an elevated risk of cold stress.

As with heat stress, cold-related illnesses vary with the level of exposure and risk of permanent damage (Fig. 13.11). Hypothermia is of the greatest concern. It occurs as the body chills and its core temperature drops. It generally reduces mental alertness, rational decision-making, and manual dexterity—each of which can lead to accidents or unsafe behaviors. As body temperature lowers further, loss of consciousness is possible with the threat of fatality. Symptoms of hypothermia include shivering, frostnip, minor frostbite, the feeling of excessive fatigue, drowsiness, irritability, or euphoria. If these symptoms arise, the person should be moved to a warm area and wet clothing removed. Modest external warming should be provided using external heat packs or blankets. If the person is conscious, he/she should drink warm, sweet fluids. The person should be transported to the hospital. A person experiencing hypothermia should not be given drinks with alcohol or caffeine (coffee, tea, or hot chocolate).

Frostbite involves freezing into deep layers of skin and tissue. It can lead to permanent damage. Typically, the fingers, hands, toes, feet, ears, and nose are at the greatest risk. The skin appears waxy, unusually pale or dark and becomes hard and numb, or blisters. With frostbite, it is necessary to address freezing of damaged tissues. The tissue should be treated as a burn and the affected area should not be rubbed.

Other cold stress conditions include frost nip, trench foot, and chilblains. *Frostnip* is superficial freezing of skin surfaces. It does not result in permanent tissue damage if the skin is promptly warmed. Trench foot and chilblains occur from cold temperatures that may be above freezing. *Trench foot* damages the skin tissue of feet after prolonged exposure to cold and wet conditions. With *chilblains*, damage occurs to small blood vessels in exposed skin (e.g., cheeks, fingers, toes, ears). Damaged areas become red and itchy, sometimes permanently.

Practices that reduce the risk of cold stress, hypothermia, and frostbite include:

- Wear outer clothing that is impermeable to water where clothing may become wet on the job site during light work. For heavier work, the outer layer should be water repellent and changed as it becomes wetted.
- Introduce more frequent indoor breaks when work is performed continuously in the cold, e.g., below $-7°C$ ($19.4°F$), or equivalent wind chill temperature.
- Plan work to minimize sitting still or standing still for long periods.
- Protect the worker from drafts to the greatest extent possible. The cooling effect of the wind should be reduced by shielding the work area or by wearing an easily removable windbreak garment.
- Avoid unprotected metal chair seats. When working outside and using tools or machine controls, to prevent contact frostbite, workers should wear anticontact gloves. If the air temperature is $-18°C$ ($0°F$) or below, wear mittens or gloves. Machine controls and tools should be designed for handling without needing to remove the mittens.
- Ensure that workers regularly consume warm sweet drinks and soups to provide caloric intake and fluids. This is because dehydration can occur in a cold environment and may increase the susceptibility of the worker to injury due to a change in blood flow to the extremities. The consumption of coffee and other caffeinated beverages should be limited.

5.5 Exposure to sunlight—ultraviolet radiation

Exposure to sunlight is a health concern because UV radiation increases the risks of skin cancer. People who burn easily should be particularly careful in the sun. Employees who regularly work outside should frequently check their body for early signs of skin cancer, especially for a spot on the skin that is changing in size, shape, or color during a period of one month to two years. If such symptoms appear, a health care professional should be seen immediately as skin cancers detected early can be cured.

When working outside, to protect against UV rays of the sun, workers should:

- wear tightly woven clothing,
- wear a sunscreen with SPF of at least 15,
- wear a wide brim hat that shades the back of the neck (not a baseball cap),
- wear UV-absorbent sunglasses to block 99%−100% of UVA and UVB radiation,
- limit exposure to the sun between 10 a.m. and 4 p.m., and,
- wear special safety goggles or UV-absorbent sunglasses to protect the eyes, including from UV radiation reflected from snow in the winter (to prevent snow blindness).

6. Biological and chemical health concerns

Composters are exposed to the resident organisms involved in decomposition and their components (e.g., spores and endotoxins), plus the volatile compounds and dusts generated during composting. When these biological and chemical elements become airborne (usually with dust particles and mists), they are referred to as *bioaerosols*. The bad news is that both acute (i.e., immediate) and chronic (i.e., long-acting) adverse health effects could potentially arise from these biological and chemical agents if ordinary safety practices are ignored. The possible health effects range from short-term symptoms, like skin or eye irritations, to more worrisome allergic reactions and illnesses. The good news is that workers, compost users, or neighbors rarely experience adverse health effects, in part because these biological elements are widely present everywhere in the environment.

While these potential health concerns are real and merit attention, it is unclear if they make composting a "risky" occupation. Compost workers are an under-researched group and research results to-date are inconclusive, if not contradictory. Some of the potential adverse effects depend to a great extent on the susceptibility of the individual. Again, there is little evidence to suggest that compost workers—the group most exposed to the hazards—have any greater incidence of health problems than the population at large. Ordinary safety practices and the body's inherent defenses are apparently effective in keeping the risks low.

6.1 Routes of entry

The first step in minimizing the risks from biological and chemical substances is understanding their potential routes of entry into the body. Chemicals and biological agents enter or make contact with the body through eye contact, skin absorption, injection through the skin, or ingestion and inhalation. Common sense, good work practices and process control, hygiene, and PPE (see following section) block these routes of entry (Box 13.4).

- *Eye contact:* When present in high concentrations, water-soluble gases such as ammonia and hydrogen sulfide, volatile fatty acids and aldehydes can dissolve in eye moisture and produce eye irritation. Dusts and aerosols can also enter via eyes and produce irritation, while microorganisms in the eyes could produce infection. Defense mechanisms for eyes, including tears and blinking, exclude or remove most airborne hazards. However, eye protection is still recommended to lower the risks from bioaerosols, dust, and chemicals (in high concentrations) as well as flying particles. Hand-to-eye contact (e.g., rubbing eyes) is another pathway for irritation and infection. Mucous membranes in the eyes and respiratory tract are connected. Tears can drain into the nose, as they do when you cry, so infectious agents that land in the eye can be transferred to the nose and then reach the lungs. This is can be minimized by wearing goggles, as well as by frequent hand-washing and wearing gloves.

Box 13.4 Of particulates, PMs and particulate protection

Dust, mist, smoke, spores, bioaerosols, and other airborne particles all fall into the category of particulate matter, or just *particulates*. Particulates are ubiquitous in the environment. When the wind blows, particulates take to the air. When the family dog shakes itself awake, pet particulates fly. The fact that particulates are in the air all around is not in itself a cause for concern. Most particles in the air are harmless, and the human body has effective defenses against breathing in those particles that might cause harm. However, a concern for airborne particles grows when the level of particulates in the air becomes abnormally high and when the particulates become very small. Both of these factors can overwhelm the body's defenses. Also, some particular elements in the air can pose problems for some individuals, including those with allergies or whose defense mechanisms have been compromised. Therefore, it is sometimes necessary to provide protection against inhaling particulates.

Being small is one of the ways that particles defy the body's respiratory defenses. For this reason, particulates are classified by their size. The common unit of measure is the micron, which is one-millionth of a meter. Regarding their environmental and health impacts, particulates are rated according to the concentration of micron-sized particles in the air. The rating system uses the abbreviation PM, for particulate matter, followed by a number that represents a size fraction. PM_{10} relates to all particles smaller than 10 microns. PM_5, $PM_{2.5}$, and PM_1 relate to particles smaller than 5, 2.5, and 1 microns, respectively. Smaller numbers imply potentially more harmful particles.

Particle size is also relevant to rating the effectiveness of respiratory protective equipment such as respirators and filters. Respirators are rated with a letter followed by a number. The letter refers to the respirator's resistance to oil. The number signifies the percentage of particles of 0.3 microns in diameter that the respirator is rated to remove.

Respirator rating letter	Respirator rating number
N—not oil resistant	95—removes 95% of all particles that are 0.3 microns in diameter.
R—resistant to oil	99—removes 99% of particles that are 0.3 microns in diameter.
P—oil proof	100—removes 99.97% of all particles that are 0.3 microns in diameter.

Hence an N95 respirator is designed to remove 95% of particles with a diameter of at least 0.3 microns. Ventilation filters use similar criteria. For instance, HEPA or HE filters remove 99.97% of all particles that are 0.3 microns in diameter.

- *Skin contact or absorption:* The skin is an incredibly effective barrier to microorganisms and most chemicals. While rare, irritation or burns may result from skin contact with a few abrasive chemicals that may be used at some composting facilities, such as cleaning acids or caustics or laboratory chemicals. Skin reactions to plant toxins, such as poison ivy or poison oak, are more likely. Protection practices include wearing of appropriate clothing and gloves and awareness of the hazardous chemicals and plants that might be present. Special wipes are also available for both pre- and post-contact with poison ivy and poison oak.
- *Injection:* Chemicals and organisms are much more likely to cause infection or injury once they penetrate the fortress of the skin. Damaged skin (punctures, cuts, abrasions) can become infected, as well as serve as an entrance to the bloodstream. There are numerous opportunities for composting workers to acquire cuts and scrapes while walking near or handling materials containing sharps (e.g., glass fragments), scrap metal and wood with protruding nails and while working with equipment and tools. Cuts and scrapes should be cleaned

and treated immediately and then kept protected with bandages and washed frequently. Existing wounds must remain protected and cleaned. Appropriate gloves and footwear (e.g., steel soles) should be worn, and sharp, protruding objects should be removed promptly from harm's way.

- *Ingestion:* Dusts and particles carrying microorganisms can land on lips and can be inadvertently swallowed. In addition, hand-to-mouth transfer can occur by eating food or smoking cigarettes/pipe/vaping without first washing hands. If the microorganisms proceeded to the digestive tract, some may be killed by stomach acid and some may survive to infect the intestine or other body systems. Good hygiene greatly reduces hand-to-mouth ingestion. Respirators and masks worn in dusty situations minimize inadvertent ingestion.

- *Inhalation:* The atmosphere carries small to microscopic biological particles that people routinely inhale including dusts, mists, bacteria, fungal spores, viruses, and protozoa. These biological particles that are suspended in the air as an aerosol are termed "bioaerosols." At composting sites (and other locations where organic materials are disturbed e.g., raking leaves), bioaerosols are present at concentrations that are much higher than normal and thus present greater health risks. Most bioaerosols are removed before reaching the lungs by fluids and tiny hair-like structures in the nose, throat, and bronchial tubes (Fig. 13.12). Particles that are not water-soluble or are small are more likely to be carried into the air sacs of the lungs. Here white blood cells, called

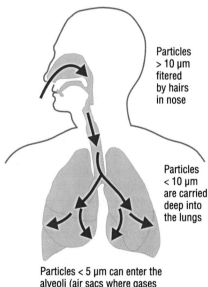

Particles
> 10 μm
fitered
by hairs
in nose

Particles
< 10 μm
are carried
deep into
the lungs

Particles < 5 μm can enter the
alveoli (air sacs where gases
are exchanged

FIGURE 13.12

Inhalation and fate of particles and gases in the lung.

Adapted from Gladding, T. (Ed.), 2012.

macrophages, engulf and destroy many bioaerosols and guard against infection. As with ingestion, most organisms that are inhaled die inside the body, while some may persist. The wearing of respirators or the use of air filters in the cabs of equipment should be used in dusty situations to reduce inhalation of bioaerosols. In addition, fewer bioaerosols become airborne when materials are kept moist.

6.2 Chemical hazards

In general, chemicals are not used widely at most composting facilities. The primary potential chemical hazards are due to gaseous compounds released during decomposition including carbon dioxide, ammonia, nitrous oxide, methane, hydrogen sulfide, and carbon disulfide. These gases are health hazards when they are present in high concentrations or displace fresh air needed for proper breathing. In the normal environments of a composting site, these gases are generated gradually and either further decompose or dissipate well before accumulating to hazardous levels. However, high concentrations can occur in certain situations, such as inside composting vessels, enclosed storage bins, or when an actively decomposing pile is opened. In these situations, the fumes can potentially overcome a worker. For example, ammonia has been observed to exceed regulatory limits in enclosed composting facilities. In poorly aerated areas of buildings or in a confined space (such as in-vessel composting), carbon dioxide may reach levels which could affect those with preexisting heart conditions.

To prevent accidents from exposure to chemical vapors, it is important to ventilate enclosed spaces before entering. Safety protocols for working in confined spaces should be observed when entering a composting reactor. Loaders and turners that are opening piles should have enclosed cabs with filtration on the intake air. While it may be possible to provide both particulate filtration and gas (such as ammonia) scrubbing for the intake air, it is also possible for the operator to wear respiratory protection which provides protection for both gases and particulates.

The only way to know about the level of exposure by inhalation of chemical vapors is to analyze air samples collected in breathing zone. However, serious overexposure can be indicated by symptoms listed in Table 13.5. These symptoms are typical for acute exposures; chronic exposures may produce subtle damage or effects without such obvious signs (see section below on medical surveillance). *It is important for workers to report any and all symptoms, injuries, and illnesses to their employer and/or regulatory officials so that hazards and overexposures can be addressed.*

6.3 Dust

Dust is a common hazard at composting sites. It is generated under dry conditions from materials handling, processing operations, and movement of vehicles on the site. When inhaled in large concentrations, or consistently over long periods, dust can interfere with a person's respiratory functions and can eventually lead to damage of the lungs

Table 13.5 Symptoms of overexposure to gaseous chemical compounds.

Structure or function affected	Possible effects
Central nervous system (brain)	Headache, dizziness, lightheadedness, euphoria, drunkenness, slowed response time, lack of coordination
Respiratory system (lungs)	Changes in rate or depth of breathing, chest tightness, irritation, difficulty breathing, feeling of "warmth" in the chest
Eyes	Tears, irritation, "burning" feeling, blurred vision, sensitivity to light
Heart or circulatory system	Heartbeat is rapid, slowed, or irregular; change in electrical activity of the heart, change in blood pressure, fainting
Digestive system	Vomiting, nausea, malaise, diarrhea, constipation
Skin	Swelling, redness, rashes, irritation, bumps, increase or decrease in pigmentation

and other organs. In addition, dust can be a bioaerosol as it carries biological elements including fungi, bacteria, and other particles (see following section).

Dust is minimized by keeping materials and surfaces moist for dust control practices. Exposure to equipment operators is best reduced by the use of an enclosed cab on the vehicle with well-sealed and air-conditioned/heated cabs with filtered air intakes. Filters should be removable and washable or replaceable with the best filtration rate possible, preferably HEPA efficiency; employees should be instructed to remove and clean or exchange filters frequently. Workers outside of cabs should use respirators in situations where dust is generated.

6.4 Biological hazards

Biological health concerns include potential exposures to bacteria, endotoxins, fungi (molds and yeast), parasites (protozoa), worm cysts, and viruses. While all these biological agents exist in the environment, they are likely to be present in higher concentrations at composting sites and farms, due to the nature of the feedstocks and the fact that composting fosters biological decomposition. A compost facility is a source of bioaerosols—airborne particles including fungi, bacteria, and endotoxin (Fig. 13.13). Elevated levels, sufficiently high to cause potential harm to workers, occur both upwind and downwind of activity areas, within about 25 m (90 ft) of the source.

Exposure to organic dust and bioaerosols at composting facilities has been associated with increased effects to the eyes and respiratory system in the form of irritation, chronic bronchitis, and an accelerated decline in lung capacity. These are believed to be the result of exposure to organic dust, possibly due to high concentrations of thermotolerant/thermophilic actinomycetes and filamentous fungi.

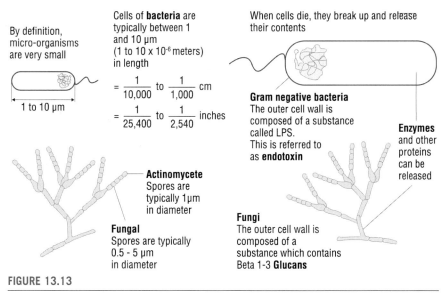

By definition, micro-organisms are very small

1 to 10 μm

Cells of **bacteria** are typically between 1 and 10 μm (1 to 10 x 10⁻⁶ meters) in length

$$= \frac{1}{10,000} \text{ to } \frac{1}{1,000} \text{ cm}$$

$$= \frac{1}{25,400} \text{ to } \frac{1}{2,540} \text{ inches}$$

When cells die, they break up and release their contents

Gram negative bacteria
The outer cell wall is composed of a substance called LPS. This is referred to as **endotoxin**

Enzymes and other proteins can be released

Actinomycete
Spores are typically 1μm in diameter

Fungal
Spores are typically 0.5 - 5 μm in diameter

Fungi
The outer cell wall is composed of a substance which contains Beta 1-3 **Glucans**

FIGURE 13.13

Biological elements prominent at composting sites and beyond. (A) bacteria; (B) endotoxins; (C) fungi and actinomycetes (i.e., actinobacteria).

Adapted from Gladding, T. (Ed.), 2012.

Disease-causing organisms, or *pathogens*, represent only a very small fraction of the microbial community in compost piles, and they are effectively destroyed by the high temperatures and antagonistic environment of the compost pile. However, this sanitation should not be a cause for overconfidence. It occurs only after the feedstocks have been mixed and the composting is well underway (1−3 days). Because new feedstocks are continually coming on the site, a resident population of pathogenic microorganisms inhabits the site to some degree, again depending on the feedstocks and their handling. In addition, some biological hazards are associated with agents that are not ordinarily considered pathogens (e.g., organic dust, fungal allergens) and that persist through the composting process. In short, despite the capabilities of the composting process and methods, potential biological hazards deserve respect and precautions that include minimizing exposure such as using air filtration (especially HEPA filters) on the air intake for vehicle/equipment cabs, sensible hygiene practices, PPE, and health monitoring. Currently, immunization recommendations for compost workers are not different from the general population. Due to the potential for cuts and puncture wounds, compost workers should consider keeping their tetanus vaccination up-to-date. If sewage sludge is the compost feedstock, immunization for hepatitis A and hepatitis B should be considered; this vaccine is also available as a combined hepatitis A and B vaccination.

The type and number of organisms present, and the predominant routes of entry, depend largely on the feedstock but the processing activities are also a factor. The primary groups of concern are summarized below.

6.4.1 Bacteria

Bacteria are simple single cell organisms present in composting feedstocks in tremendous variety and number. Bacteria are the dominant composting organisms. Although only a small fraction pose health risks to humans or animals, bacterial infections are responsible for numerous diseases and general infections. Symptoms differ among the various diseases and organisms. Infections can occur by any of the routes of entry discussed earlier. The bacterial species *Salmonella* and fecal coliform bacteria, including *Escherichia coli* (*E. coli*), are commonly used as indicators for the presence of pathogenic bacteria generally. Materials and environments are tested for these as representative groups, rather than each potentially pathogenic species of bacteria, since they are typically hardier and, if they are killed, it is assumed that the pathogens would have been killed as well.

The greatest risk of bacterial infection appears to be in the initial stages of composting, and from raw feedstocks like manure, biosolids, and postconsumer food scraps. There is little risk in the final composting stages because pathogenic forms die at the high temperatures of composting. Potentially-pathogenic bacteria may better survive low temperature processes but other factors, including microbial antagonism and competition, tend to contain pathogenic populations. The risks associated with pathogenic bacteria are reduced by maintaining high temperatures for an extended time. Regulations require that biosolids must be composted at specified temperatures for a specified time; that is, 55°C (131°F) for 3—15 days, depending on the composting method. Some jurisdictions have extended this requirement to all composting feedstocks. Testing for *Salmonella* and fecal coliforms is frequently performed to verify the destruction of bacteria. To avoid infection, it is equally important to minimize exposure and hinder the routes of entry with appropriate protection (e.g., gloves and dust respirators) and sensible hygiene.

6.4.2 Endotoxin

In addition to exposure to live bacteria, handling organic materials can exposure workers to pieces of dead bacteria, including endotoxin. Endotoxin is the outer cell wall of gram-negative bacteria (Fig. 13.13). At high levels of exposure, endotoxin could decrease lung function by producing inflammation. Dust containing endotoxin can be mechanically irritating to the mucous membranes of the eyes, nose, and throat and to the skin. The health effects of endotoxin exposure may include fatigue, fever (organic dust fever or organic dust toxic syndrome), chills, headache, malaise, cough, chest-tightness, and shortness of breath. In some cases, sweating, nausea, abdominal discomfort, and sometimes vomiting may occur. Endotoxin symptoms may develop from one to eight hours after exposure and usually last for up to 24 h, although the individual may feel unwell for a longer period of time. Repeated, prolonged exposure may lead to persistent airway disease, such as bronchitis and bronchiolitis.

Endotoxin is potentially present in nearly any feedstocks and the composts made from them. They are heat-stable and not substantially inactivated by composting temperatures. They are most often associated with dust. This association, plus the fact that endotoxins affect the respiratory system, suggests that inhalation is the primary entry route of concern. Dust suppression and respirators minimize exposure.

6.4.3 Fungi

Composting fosters the growth of fungi, especially with woody feedstocks and in the latter stages of the process. Up to 100 species of fungi have been found in compost. While predominant fungal species vary, common organisms include various species of *Aspergillus* and *Penicillium*—organisms that produce tiny, easily airborne spores. *Aspergillus fumigatus* is one species that has received particular attention in regard to composting because it is common and a well-known allergen (Box 13.5).

Typical reactions to fungal spores are irritation, allergy, or infection in susceptible people. However, the effects differ among individuals. Individuals who are not allergic and have a healthy immune system are less likely to be affected. People with specific allergies can exhibit immediate allergic reactions that may include irritation of the eyes, nose, and throat, lethargy, and headache.

Most fungi commonly encountered in the environment are unable to cause infectious disease. However, some species of fungi are considered opportunistic—they can cause infection in people whose immune system is compromised (i.e., *immunodeficient*) by other diseases or treatments for those diseases. Examples include: diabetes, cancer (especially leukemia), cystic fibrosis, alcoholism, inherited immune deficiency, acquired immune deficiency, invasive medical procedures, and certain medications (e.g., antibiotics, immunosuppressive drugs). Persons with weakened immune systems, fungal allergy, or medical conditions that compromise the body's ability to fight infection should use caution when handling compost. Recent attention has turned to substances called *mycotoxins,* which are fungal compounds that can cause harm to animals and humans if inhaled or ingested. Mycotoxin-contaminated dust inhalation in farm workers has been linked to liver cancer, fungal irritations, inflammations, and infections of the lung (mycotoxicoses). It is suggested that compost workers face similar, or greater, levels of exposure, although long-term epidemiological studies have not yet been reported for composting facilities.

Box 13.5 *Aspergillus fumigatus*

Aspergillus fumigatus is a fungus that is found in most parts of the world and is sometimes abbreviated to *A. fumigatus*, or even AF, in the literature. It is an adaptable organism that is able to degrade cellulose, hemicellulose, and lignin, which are the structural components of woody plants. AF is also able to tolerate relatively high temperatures (50°C or 122°F), so it is widely found in hot, composting materials, where it performs a useful role in helping to biodegrade the composting feedstocks and convert it into compost.

Being a fungus, *A. fumigatus* also produces tiny spores, which may survive temperatures as high as 70°C or 158°F. The spores are very tiny and often become airborne during the composting process, forming a bioaerosol. Due to their small size, the spores of AF can be inhaled and penetrate deep into the lungs where they may cause problems in some people.

As the fungus is so widespread, the average person inhales a few hundred spores every day without experiencing any undue problems. However, some individuals can develop an allergic reaction, or in extreme cases, become infected. Infection with AF is normally only associated with those individuals who have compromised immune systems, such as those with HIV or being treated with certain drugs (such as antirejection medicine following an organ transplant).

As composting exposes workers to higher concentrations of spores and for longer periods of time than the average person, care needs to be exercised to reduce exposure by implementing suitable control measures.

Fungi and fungal products (e.g., spores, glucans, mycotoxins) are prominent bioaerosols. High levels of fungal spores and mycelium fragments in the air can be generated from composting piles and activities that produce dust. The principal routes of entry for compost workers are inhalation and skin/eye contact.

6.4.4 Viruses

Viruses are very small pieces of genetic material wrapped within a protein cover. As a group, viruses cause a variety of diseases, from the common cold to polio. In general, viruses do not survive well beyond their host environment and appear to be inactivated readily during composting. Also, viruses tend to be retained well in the pile and are not readily airborne. The type of virus present primarily depends on the feedstocks and their sources. Those of greatest concern are those found in manure, biosolids, and food. Insect and tick bites can spread several viruses including West Nile virus. At a composting site, control of West Nile virus involves management of the composting site to reduce mosquito breeding, rather than process or feedstock management (Box 13.6).

Box 13.6 Protecting against West Nile virus and other mosquito-borne illnesses

West Nile virus is a virus that is mosquito-spread. It causes West Nile encephalitis, an inflammation of the central nervous system. Because many composting sites contain, or abut, reservoirs of water (even containers and puddles), they can also harbor mosquitoes that carry the disease. Prior to 1999, West Nile virus was found only in Africa, Eastern Europe, and West Asia. In August of 1999, it was identified in the United States and has spread across most of the country since then. There is no specific treatment for infection in humans. Elderly people are at the greatest risk of developing severe symptoms. In severe cases, hospitalization and intensive supportive therapy may be needed.

Certain species of birds act as reservoirs for the virus, and certain species of mosquitoes act as vectors. Mosquitoes pick up the virus when they bite, or take a blood meal, from infected birds and then, with the next blood meal, transmit the virus to people and other animals.

The best method of reducing the risk of West Nile virus is to eliminate mosquito breeding sites. Mosquitoes lay eggs and develop in stagnant water, so reduction of these sites involves eliminating stagnant water sources. Where water is a permanent feature, using chemicals to kill mosquito lava (larvicides) may be possible. Other control measures include:

- Using landscaping to eliminate standing water that collects on the composting site. Mosquitoes may breed in any puddle that lasts for more than 4 days. It may also help to grade (i.e., level) the site regularly to minimize puddles and ruts.
- Drilling holes in the *bottom* of containers left outdoors. Containers with drainage holes located only on the sides collect enough water to act as mosquito breeding sites.
- Removing old tires from the site. Old tires provide a place for water to collect and thus a breeding ground for mosquitoes. Tires that are used to hold down tarps or for other purposes should be sliced, quartered, drilled with drainage holes, or filled with foam plastic to exclude water.
- Using larvicides where water reservoirs cannot be eliminated; however, it is best to consult with the local pesticide regulation authorities before going down this route. Usually, larvicides must be applied by a certified pesticide applicator.
- Periodically inspecting the site for dead birds, such as crows. Any suspicious birds should be reported to the local health authorities. Protective gloves will need to be worn to handle dead birds, which should be placed in plastic bags, as directed by the health authority.

Viruses of veterinary importance include foot and mouth disease and avian influenza (AI). Studies of animal mortality composting have demonstrated that these viruses are readily destroyed by the composting process. However, the viruses remain viable in the raw feedstocks, which is one reason why European countries have generally not embraced composting as a means of disposal of dead animals (Chapter 8). AI or bird flu is a growing concern as a human health issue because the virus has infected humans. Infected poultry have been successfully composted as a means of controlled disposal. However, as of this writing, the topic continues to receive much attention and policies enacted in the near future may impact composting of AI-infected birds and manure.

6.4.5 Parasites, protozoa and worms

A number of parasitic species of protozoa and worms (i.e., helminths) can inhabit composting feedstocks, especially manure and biosolids. For example, workers at a biosolids composting project apparently were infected with *Giardia* from handling raw biosolids, probably via accidental ingestion. Protozoa such as *Cryptosporidium* are reportedly killed by composting and *Giardia lamblia* appears to be even less hardy than *Cryptosporidium*. However, little is known about the survival of most parasites during composting. Most likely, the harsh conditions during composting substantially destroy parasitic organisms. In any case, good hygiene is strongly recommended to avoid possible infection, especially when handling raw feedstocks.

6.4.6 Infectious prions

Infectious prions are the suspected cause of diseases such as mad cow disease, scrapie (affecting sheep), chronic wasting disease (affecting mainly deer and elk), and Creutzfeldt-Jakob disease (in humans).

A prion is not a living organism, nor a virus, but a protein with an abnormal conformation (i.e., shape) that triggers other proteins in the brain and spinal tissue to also change shape (see Prion Primer in Chapter 8). The possibility of prion-infected material coming to a composting site is extremely small, and that possibility applies almost exclusively to animal carcasses (e.g., farm mortalities and deer). In infected animal carcasses, the prions are concentrated in the brain and spinal tissue. It is best to keep potentially infected carcasses out of compost piles where there is any intention to use the resulting compost. Prions can conceivably find their way into municipal biosolids, MSW, and some food wastes but only at extremely low concentrations. The effect of composting on prion survival is unknown. Until research demonstrates composting can destroy prions, it should be presumed that compost made from infected material also carries the prions. Stringent rules preventing the composting of spinal cords is in place across the European Union for this reason.

7. Prevention and preparedness

Prevention begins with the goal of removing a hazard and finishes with protecting workers from those hazards that cannot be eliminated. It includes the control measures previously described, and includes sensible hygiene, PPE, and medical monitoring. With the proper control measures, even "higher-risk" individuals can work at a composting facility (Box 13.7). However, while prevention practices reduce the risks and number of incidents, there are no guarantees that accidents and illnesses will never occur. In fact, it is a good policy to expect and anticipate safety and health incidents and be prepared to react to and treat them.

Proper training of workers is a powerful means to reduce the safety and health risks associated with composting. Training starts with adequate instruction for the workers on how to perform their jobs and operate the required equipment. The work environment becomes even safer when workers are further trained about potential hazards, hazard reduction practices, emergency preparedness, and first aid. With training, workers can contribute valuable input and service to prevention and preparedness.

Box 13.7 Can someone with a health disability, such as asthma, work at a composting facility?

Given the bounty of biological stimuli and bioaerosols at composting sites (and farms), a standard hiring policy might be to immediately exclude individuals with certain health impairments, such as allergies to fungal spores or respiratory disorders. However, such a policy is not only unfair, it may also be illegal and unnecessary. Many governments have regulations that protect workers with disabilities from discrimination. For instance, in the US, the "Americans with Disabilities Act" requires the employer to provide a reasonable accommodation for the employee unless he/she constitutes a direct threat to the safety of him/herself or others. "Direct threat" is determined by:

- the likelihood (probability) that an injury will occur; this is high probability not just elevated risk or a remote or speculative risk; and
- the certainty (predictability) that an injury will occur; especially based upon individual factors.

The employer in conjunction with the employee and the occupational physician can determine if there is a significant risk of substantial harm by evaluating the individual employee's medical status and prognosis in relation to:

- probability—the statistical likelihood of the harm occurring
- severity—the nature and severity of the potential harm
- imminence—the time frame in which the harm is likely to occur
- duration—how long the risk is likely to be present.

The employer must determine that all reasonable accommodation cannot reduce the employee's risk to acceptable levels. Appropriate process management, engineering controls (where possible), work practices, hygiene practices, and PPE (respirator, powered air-purifying respirator) are often able to satisfactorily reduce the level of exposure for most people. This approach could afford more and better opportunities for employment for persons with disabilities. However, for some individuals, exposure still cannot be reduced sufficiently, even with the best practices and PPE. For example, someone with occupational asthma may still respond to very low-level exposures, even when wearing a full-face respirator. If, after this evaluation, the person with a disability truly cannot do the job without significant risk to him/herself or others, then the decision not to hire or retain such a person must be made on a sound scientific basis.

7.1 Reducing exposure to safety and health hazards

Safety and health risk can and should be a addressed via a hierarchy of controls that seeks to first eliminate a hazard (e.g., process changes), substitute for a lesser hazard, reduce exposure to the hazard (e.g., engineering controls or, if not sufficient or appropriate, then using administrative controls), and then protect workers from hazards that remain (Fig. 13.1). Items used from different rungs on the hierarchy of controls can work with each other for further hazard reduction.

Process changes seek to eliminate the hazard altogether by selecting alternative methods. Examples include paving roads and other surfaces to reduce dust and standing water. Engineering approaches would include providing enclosed, environmentally-controlled cabs for equipment. Many safety-conscious engineering controls have been suggested in previous sections of this chapter.

Administrative controls rely on the actions of people—these are more procedural in nature and can affect process and site management as well as worker activities. Training, work rules, standard job operating procedures, job rotation, and preventive maintenance procedures are examples of administrative controls. Limiting employee work hours to reduce fatigue is an administrative control that can reduce hazards in different ways. It is important to establish and encourage facilities and practices that encourage good hygiene (Table 13.6). It is highly desirable for any composting operation to have hygiene facilities—restrooms, change rooms, showers, clean-up facilities, lunchroom—to reduce the potential for exposures to chemicals and diseases.

Improvements in safety and health of any facility can be identified through hazard evaluation tools such as job hazard analysis, vulnerability analysis, and process hazards analysis. Specific programs include "Hazard and Operability Analysis" and "Hazard Analysis and Critical Control Point" (HACCP) planning. HACCP, for example, systematically identifies the critical points in a production system responsible for potential hazards, sequentially corrects the problems at those points, and then continues to monitor, test, and improve the system at new critical points (for more information, see Evans, 2003).

7.2 Medical monitoring: to establish a health baseline and to evaluate health

Regular monitoring of a worker's health status, beginning before the employee's first day on the job, can signal whether site conditions may be contributing to health problems. Monitoring is especially important for chronic ailments and those that may be due to cumulative or repeated exposure. It can also help to identify health issues that may have been caused by previous employment. The following recommendations for medical surveillance are adapted from those recommended for wastewater workers, considering the potential health effects and exposures associated with composting:

• A preplacement examination that should include a comprehensive physical examination, liver and kidney function tests, hematologic function tests, lung function tests.

Table 13.6 Elements of good hygiene at composting facilities.

Water and facilities

- Providing a source of potable water for drinking, hand/body washing, cooking, food and food preparation, eating, utensils washing, and personal service rooms. Drinking water should be provided in single-use cups or by fountains. Nonpotable water, such as that used for firefighting, should be clearly identified and distinguished from potable water. Nonpotable water sources and outlets should also be labeled as unsafe for drinking.
- Providing toilet facilities with toilet rooms, preferably separate ones for each gender. If the toilet room can only be occupied by one person at a time, it should be lockable from the inside. Toilets should have lids and signage should remind the user to close the lid before flushing.
- Providing changing rooms so that employees can wash and change into street clothes before leaving work. Showers are also highly desirable.
- Providing a lunchroom so that employees do not consume nor store food or beverages in any area exposed to potentially hazardous materials.
- Prohibiting eating, drinking, and smoking on the site except in designated locations (note that this practice is for fire, as well as disease, prevention.

Personal hygiene

It is good practice for employees to:
- Wash hands with soap and water before eating or smoking or whenever hands come into contact with compost or feed stocks.
- Shower at work and change into clean clothes and shoes.
- Wash hands before and after using the bathroom.
- Remove excess contaminants from footgear prior to entering a vehicle or a building.
- Not wear work clothes home or outside the work environment. Employees should remove contaminated clothing at the end of the shift.
- Avoid laundering work clothes at home. If home laundering is necessary, work clothes should be placed in a bag and kept bagged until they are actually placed in the washing machine. Clothing should be washed separately from other clothing using the hot water cycle and chlorine bleach (if appropriate).

Injuries and illness

Good practice measures include:
- Thoroughly but gently flushing eyes with water if contaminants contact the eyes.
- Caring for cuts and abrasions promptly, keeping wounds covered with clean, dry bandages.
- Promptly reporting injuries, illness, and symptoms.
- Keeping current vaccinations recommended by the relevant health department (i.e., for the general population) up to date, including tetanus. If biosolids are handled, vaccination against hepatitis A and B should also be considered.

- Yearly periodic health assessment using the same items as the preplacement examination and, including: a review of systems for symptoms suggestive of diseases (such as hepatitis, intestinal infections, respiratory infections) and for reactions to exposure to toxic gases (such as ammonia); update of immunizations (e.g., influenza, tetanus, diphtheria and polio). The US Centers for Disease Control suggests that the immunizations for wastewater workers should be the same as those for the general population. Operators that handle biosolids should consult a doctor regarding the need for vaccines for Hepatitis A and Hepatitis B.

- On-going health monitoring: prompt reporting of illnesses lasting longer than 2 days to employer or safety director; evaluation of suspected work-related illnesses as needed.

7.3 Personal protective equipment

Given the nature of composting, some potential hazards remain despite the best process, engineering, and management practices. In situations where workers are exposed to such hazards, PPE provides an important line of defense. PPE includes high visibility clothing, hard hats; eye and ear protection, dust respirators (disposable), and cartridge respirators. In addition, workers should wear appropriate clothing, gloves, and footwear (e.g., steel toe-capped boots with a reinforced midsole).

- *Overall visibility:* To ensure that they can be clearly seen by equipment operators, workers should wear brightly colored (e.g., orange or yellow) clothing or vests. Reflective tape or strips on clothing are also helpful.
- *Head protection:* Unless one is working in an enclosed cab, a hard hat should be used to protect the head from impact for any tasks involving flying or falling objects including mixing, turning, grinding, and wood chipping.
- *Eye protection*: Protecting the eyes is necessary from dusts and bioaerosols and from potential impact from projectiles. Unvented goggles give the best protection for gases and vapors, while still providing impact protection. Antifog wipes or sprays can be used to keep goggles from fogging. A faceshield could be used over goggles and is available in a style that attaches to a hard hat. Goggle-faceshield combinations are also available. A full-face respirator provides the same protection as goggles and a faceshield.
- *Hearing protection:* Earmuffs, ear plugs, and ear caps are available and must be worn (described above) during exposure to noisy work or equipment. Hearing protection devices must reduce the worker's exposure below the regulatory limit. In the US, these items are labeled with the manufacturer's assessment of their noise reduction rating (NRR). However, the US OSHA requires that the NRR rating be adjusted downward to account for uncertainty and a safety factor. The adjustment is made as follows: The value of 7 dBA is subtracted from the manufacturer's NRR to adjust for uncertainty regarding the frequency of the noise (the actual noise could be high-pitched to low-pitched, depending on that task or equipment). Then, the adjusted NRR can be divided by 2 to produce a 50% safety factor. If two noise reduction devices are used (such as ear plugs plus earmuffs), the NRR of the most effective protective device is used, then 5 dBA

is added for the second piece of protective equipment. The actual noise level of the noise source minus the adjusted NRR must be below the regulatory limit.

- *Respirator:* Respiratory protection should be worn for all the dust-generating tasks, unless working inside an enclosed cab with its own air filtration equipment. In the US, any respirator used must be marked as approved by NIOSH and must show a rating indicating its filtration level. The minimum level of protection from dusts and bioaerosols is an approved disposable N95 respirator. This respirator is able to filter 95% of particles at 0.3 microns in diameter—thus removing mold spores and other fine dusts. A disposable particulate respirator can be worn repeatedly until its filtering capacity has been used up—at this point, the wearer will notice that it is difficult to breathe through the respirator—or if it becomes damaged or grungy. Disposable dust respirators can be purchased with or without an exhalation valve, which more effectively removes exhaled moisture.

If protection is needed from gases or vapors, a half-face or full-face respirator is necessary. This type of respirator uses disposable cartridges that can remove a specific gas (such as a cartridge for ammonia gas). A dust cartridge can be layered over the gas cartridge to provide simultaneous protection from both types of air contaminants. If the worker is unable to tolerate the stress of using these types of negative-pressure respirators, it may be possible for the person to use a powered air-purifying respirator with the appropriate cartridges. This type of respirator uses batteries to provide the power to draw in the air and deliver it to the wearer's breathing zone.

Every time a respirator is put on, the wearer must use a fit check, as described in the manufacturer's instructions, to make sure that it fits tightly to the face and the seal does not leak. The OSHA respiratory protection standard (not applicable to agriculture) requires that workers must be physically able to wear a respirator, must be fit-tested so that their respirator makes a good seal on the face, must be clean-shaven where the respirator seal touches the face, and must be trained on donning, inspecting, using, and replacing the respirator. Even one day's beard growth can cause substantial leakage.

7.4 Emergency action planning and fire response

Regardless of the safety precautions implemented, it is always important to consider, well in advance, what potential emergencies your facility might encounter—accidents, fire, tornadoes, blizzards, hurricanes, floods, earthquakes, and any emergencies unique to a particular area (see Box 14.1 in Chapter 14). A *written* emergency plan should be in place, and employees should be trained on what they are expected to do in an emergency, including:

- how to report the emergency and how to raise the general alarm (the alarm should be a loud and distinguishable noise; flashing lights or other means can be used to accommodate workers with hearing impairment),
- roles and responsibilities in the event of a fire or other emergency,
- who is expected to leave the area and who remains to perform critical functions,

- what rescue and medical duties are expected of employees, and
- where people should gather after evacuation; what is the refuge or safe area (parking lot, open field, or street), and how they are all to be accounted for.

7.5 Fire preparation

Before or soon after the composting enterprise begins operating, the local fire department should be contacted and consulted so that, in the event of an emergency requiring their assistance, they will have already visited the facility, know its layout and potential problems. A local fire company may not be familiar with fires at composting facilities, especially spontaneous combustion. If they are receptive, fire authorities can be provided information about the nature of compost fires and recommended procedures for fighting compost fires Chapter 14 provides such information.

In order to respond to a fire on site, selected (or all) workers should have access to water or other extinguishing methods and should be trained to use them. For a fire in a building, Class A portable fire extinguishers, which deal with wood and paper, should be located within 23 m (75 ft) for rapid access. For a fire involving a vehicle (oils, fuels, etc.), a Class B portable fire extinguisher should be located on the vehicle or within 15 m (50 ft) of travel distance (Fig. 13.14). In addition, it is a good idea to have emergency fire-fighting equipment on site that contains the necessary equipment and materials such as a fire extinguisher, fire hose and couplings, instructions and keys to front-end loaders and access gates (Fig. 13.15).

FIGURE 13.14

Class ABC fire extinguisher strapped to a tub grinder.

Source: R. Rynk.

FIGURE 13.15

A fire preparedness kit placed at strategic location on a composting facility. It contains fire-fighting equipment, instructions and keys to vehicles, storage rooms and gates as needed.

Source: Matthew Cotton.

Water systems used for the composting process should be available to suppress a fire in an emergency. The recommended amount of water reserved for short term firefighting is 11 to 12 cubic meters (about 3000 gallons), enough to supply water at 380 L per minute (or 100 gallons per minute) for 30 min. The soft hose should be at least 90 mm (3.5 inches) in diameter to accommodate this flow rate.

With an enclosed composting facility, the building could be constructed with an automatic sprinkler system. For other enclosed structures on the site, local fire codes determine the requirements for building sprinkler systems (an enclosed structure means a structure with a roof or ceiling and at least two walls).

7.6 Means of exiting from a building

For any buildings on the composting site, workers should be able to exit or escape during an emergency, such as a fire. At least two exits should be available that provide free and unobstructed exit—with no locks, chains, or fastenings to prevent free escape from the inside, and free of obstructions or impediments. Exit doors should swing outward and all the exit door hardware should be functional and operational. The exit should be clearly visible and the route to it conspicuously indicated so everyone readily knows the direction of escape from any point.

Signs should be internally illuminated or self-luminous reading "EXIT" with letters at least 15 cm (6 inches) high. Clearly visible signs and arrows should be used if the way to the exit is not readily apparent. A sign reading "NOT AN EXIT" should identify any door, passageway, or stairway that is not an exit or a path to an exit. Exits should discharge directly onto a street, yard, court, or other open space that gives safe access to a public way.

7.7 First aid

It is unrealistic to expect that accidents will not happen. Therefore, employers must provide medical and first aid personnel and supplies commensurate with the hazards of the workplace. Proper first aid training is more important than the best first aid kit, as a first aid kit is useless if a person does not know how to use it.

For workplaces that are not located in proximity to a hospital or medical facility, a person or persons must be adequately trained to render first aid. First aid refers to medical attention that is usually administered immediately after the injury occurs and at the location where it occurred. It often consists of one-time, short-term treatment requiring little technology or training to administer—such as cleaning minor cuts, scrapes, or scratches; treating a minor burn; applying bandages and dressings; use of nonprescription medicine; draining blisters; removing debris from the eyes; massage; and drinking fluids to relieve heat stress.

Every workplace should have a first aid kit, stocked with the necessary supplies (Box 13.8). The contents need to be restocked as they are used from the kit. *It is important to ensure that the emergency kit contains personal medical information and supplies for those with special medical conditions.* For example, a sting to someone who's allergic to bee venom could be life-threatening, so appropriate antitoxins must be included. The name and telephone number of a family doctor for everyone who might be involved in a medical emergency also should be included. The kit should be clearly labeled and placed in an easily noticed and reachable easy-access location. It is worth considering having a first aid kit in all vehicles and buildings. Have a tickler file that notes the expiration dates of any first aid kit items that have them, such as bee sting epinephrine pens or blood coagulation bandages, so that they can be replaced at the appropriate intervals. Compost workers with serious medical conditions or allergic reactions would be will-advised to wear identifying bracelets or other medical alerts. It is also helpful if workers were apprised of their co-workers' special needs in an emergency (if they are willing to share such information).

Eyewashes and emergency (drench) showers are important for rapid response to chemical exposures. OSHA requires nonagricultural workplaces to have eyewashes and drench showers where corrosive materials are present. These devices also are useful for fuel accidents or for the rinsing of dusts from the eyes. Table 13.7 provides some recommended features of eyewashes and drench showers. Be sure to remove contact lenses and flush the eyes well. Although 15 min of flushing is typically recommended for chemical exposures, field studies of eye incidents indicate that at least 20 min are needed for chemicals which are alkaline/caustic in nature (some scientific literature suggests 30 min of rinsing). In any case, medical attention should be sought after any eye injury.

Box 13.8 Contents of a first-aid kit—US examples

The following list sets forth the minimally acceptable number and type of first-aid supplies for first-aid kits in the US, according to the OSHA standard. It includes additional items recommended for first aid kits for tractors and other farm machinery. The contents of the first-aid kit listed should be adequate for small work sites, consisting of approximately two to three employees. For operations with more employees, the supply quantities should be increased, or additional kits provided. Specifications: Minimum Contents per OSHA standard 29 **CFR 1910**.266 (OSHA, 2021a).

- Gauze pads (at least 4 × 4 inches)
- Two large gauze pads (at least 8 × 10 inches)
- One box adhesive bandages ("band-aids")
- One package gauze roller bandage at least 2 inches wide
- Two triangular bandages
- Wound cleaning agent such as sealed moistened towelettes
- Scissors (strong enough to cut through denim; e.g., stainless-steel bandage scissors)
- At least one blanket
- Tweezers
- Adhesive tape
- Latex gloves
- Resuscitation equipment such as resuscitation bag, airway, or pocket mask
- Two elastic wraps
- Splint
- Directions for requesting emergency assistance
- Additional items recommended for tractor accidents
- Several quarters 25 cents taped to the carrying case to make an emergency phone call
- A basic first aid manual
- Two triangular bandages with 36 in. sides (e.g., made from bed sheets)
- Spray antiseptic (not a pressurized can)
- Sterile saline solution
- Twelve adhesive bandages and four safety pins
- Eye goggles
- Three small packages of sugar
- Cold pack
- Amputation preservation kit (set of plastic bags: one large garbage bag, four kitchen-sized and two bread bags).

Table 13.7 Recommended features for emergency eyewash and showers.

Eyewashes
- Aerated potable water
- Deliver tepid flushing fluid.
- Temperature range—above 16°C (60°F) and below 38°C (100°F).
- Copious and gentle flow; 1.5 L per minute or 0.4 gallons per minute minimum for 15 min of water flow
- Hands should not be required to maintain the water flow
- Test on a regular basis; keep a record of the testing; for plumbed installations, flush the lines once per week
- Locate shower units no more than 10 s in time nor greater than 30 m (100 ft) in distance from the hazard
- For a disabled persons' use, the hand-held spray on a hose is the recommended unit

Emergency showers
- Chain pulls should be provided with a large ring or with a double ring at right angles; pull ring should not exceed 2 m (77 inches) from floor; for disabled persons, the maximum height should be determined functionally
- Shower head should be at least 2.1 m (84 inches) from the floor
- The horizontal distance from the center of the shower head to the pull bar should not be greater than 58 cm (23 inches)
- Valve should open readily and remain open until intentionally closed
- Water flow must be sufficient to drench person rapidly; shower must deliver minimum of 75 L (20 gallons) for 15 min and provide a column of water 50 cm (20 in.) wide at 152 cm (60 in.) above the surface floor of user and be 40 cm (16 in.) from any obstruction. (C
- Locate eyewash units no more than 10 s in time or greater than 30 m (100 ft) in distance from the hazard
- Deliver tepid flushing fluid.
- Temperature range—above 16°C (60°F) and below 38°C (100°F).
- Flow should accommodate more than one person, if necessary
- Label the location
- Shower area should be kept free of obstructions
- Label the location
- An associated floor drain is desirable, but its absence should not prohibit installation of a safety shower
- Test on a regular basis and keep a record of the tests

References

Cited references

Brown, N., 2016. Composting Health and Safety. Cornell Waste Management Institute. https://ecommons.cornell.edu/handle/1813/44632.

European Commission, 2021. EU Machinery Legislation. https://ec.europa.eu/growth/sectors/mechanical-engineering/machinery_en.

Evans, T., 2003. Hazard Analysis and Critical Control Point (HACCP) for Composters. The Composting Association, Wellingborough, England.

Farm Safety Association, 2002. Power Take Off Safety. Farm Safety Association Inc., Guelph, Ontario. Available from: https://nasdonline.org/49/d001617/power-take-off-safety.html.

Flower, W., November 1, 2016. Lock Out — Tag Out: A Fundamental Rule of Safety. Waste Advantage Magazine. https://wasteadvantagemag.com/lock-out-tag-out-a-fundamental-rule-of-safety/.

Gladding, T. (Ed.), 2012. Health & Safety at Composting Sites: A Guide for Site Managers, third ed. The Association for Organics Recycling, Wellingborough.

NIOSH, 2015. Hierarchy of Controls. U.S. National Institute of Safety and Health. https://www.cdc.gov/niosh/topics/hierarchy/.

OSHA, 2021a. Occupational Safety and Health Standards. https://www.osha.gov/laws-regs/regulations/standardnumber/1910.

OSHA, 2021b. Occupational Safety and Health Administration (U.S.). https://www.osha.gov/.

Rose, M., Baker, M., 2012. Tractor rollover rescue. Fire Eng. 165 (9). https://www.fireengineering.com/apparatus-equipment/tractor-rollover-rescue/#gref.

Consulted and suggested references

ACGIH, 1999. Bioaerosols: Assessment and Control. American Conference of Governmental Industrial Hygienists, Cincinnati, OH ("Endotoxin and other bacterial cell wall components").

ACGIH, 2003a. TLVs and BEIs based on the documentation of the threshold limit values and biological exposure indices. In: American Conference of Governmental Industrial Hygienists. Cincinnati, OH ("Biologically derived airborne contaminants, assayable biological contaminants").

ACGIH, 2003c. TLVs and BEIs. Based upon: documentation of the threshold limit values for chemical substances and physical agents. In: American Conference of Governmental Industrial Hygienists. Cincinnati, OH ("Thermal Stress").

Alexis, N.E., et al., 2003. Effect of inhaled endotoxin on airway and circulating inflammatory cell phagocytosis and CD11b expression in atopic asthmatic subjects. J. Allergy Clin. Immunol. 112 (2), 353.

American Conference of Governmental Industrial Hygienists, 1999. Bioaerosols: Assessment and Control. American Conference of Governmental Industrial Hygienists, Cincinnati, OH.

ANSI Z358.1-1990. Emergency Eyewash and Shower Equipment, American National Standards Institute, NY. https://ansi.org/.

ANSI, 2021. American National Standards Institute, NY. https://ansi.org/.

ASABE, 2016. Slow-Moving Vehicle Identification Emblem (SMV Emblem) ANSI/ASAE S276.8 (R2020. American Society of Agricultural and Biological Engineers, St. Joseph, Michigan. https://elibrary.asabe.org/abstract.asp?aid=46638.

AWS, 2021. Safety and Health Fact Sheets. American Welding Society. https://www.aws.org/standards/page/safety-health-fact-sheets.

Benfeldt, E., 2005. In-house composting of Turkey mortalities as rapid response to catastrophic losses. In: Presented at Symposium on Composting Mortalities and Slaughterhouse Residuals. May 24—25, 2005. Portland, Maine.

Benjamin, G.S., 1996. The eyes. In: Plog, B.A., et al. (Eds.), Fundamentals of Industrial Hygiene. National Safety Council, Itasca, IL ("Eye protection for welding," "Irradiation burns").

Bernard, T.E., 1996. Thermal stress. In: Plog, B.A., et al. (Eds.), Fundamentals of Industrial Hygiene. National Safety Council, Itasca, IL.

Boden, L.I., et al., 1995. Company characteristics and workplace medical testing. Am. J. Publ. Health 85 (8), 1070.

Bohm, R., 2002. Hygienic aspects of composting biowastes — legal situation and actual experiences in Germany. In: Proceedings of 2002 International Symposium on Composting and Compost Utilization. May 6-8, 2002. Columbus, OH.

Bohnel, H., et al., 2000. *Clostridium botulinum* and bio-compost. A contribution to the analysis of potential health hazards caused by bio-waste recycling. J. Vet. Med. B. Infect. Dis. Vet. Public Health 47 (10), 785.

Bonhotal, J., 2005. Cornell's natural rendering method. In: Presented at Symposium on Composting Mortalities and Slaughterhouse Residuals. May 24-25, 2005. Portland, Maine. Also, personal communication with the author following this presentation.

Brown, N.J., 1997. (revised). Health Hazards Manual for Water and Wastewater Treatment Plant Workers: Exposures to Chemical Hazards and Biohazards. Cornell University. Workplace Health and Safety Program (formerly Chemical Hazard Information Program).

Bünger, J., Antlauf-Lammers, M., Schulz, T.G., Westphal, G.A., Müller, M.M., Ruhnau, P., Hallier, E., 2000. Health complaints and immunological markers of exposure to bio-aerosols among biowaste collectors and compost workers. Occup. Environ. Med. 57 (7), 458—464. https://doi.org/10.1136/oem.57.7.458.

Bunger, J., et al., 2007. A 5-year follow-up study on respiratory disorders and lung function in workers exposed to organic dust from composting plants. Int. Arch. Occup. Environ. Health 80 (2007), 306—312. https://doi.org/10.1007/s00420-006-0135-2.

Burns, F.R., 1989. Prompt irrigation of chemical eye injuries may avert severe damage. Occup. Health Saf. (April), 33.

Burrell, R., 1994. Human responses to bacterial endotoxin. Circ. Shock 43, 137.

CalRecycle, 2020. Aspergillus, Aspergillosis, and Composting Operations in California. https://calrecycle.ca.gov/LEA/Advisories/06/.

Cramp, G.J., Harte, D., Douglas, N.M., Graham, F., Schousboe, M., Sykes, K., 2010. An outbreak of Pontiac fever due to Legionella longbeachae serogroup 2 found in potting mix in a horticultural nursery in New Zealand. Epidemiol. Infect. 138 (1), 15—20. https://doi.org/10.1017/S0950268809990835.

Curtis, L., et al., 1999. Characterization of bioaerosol emissions from a suburban yard waste composting facility. In: Johanning, E. (Ed.), Bioaerosols, Fungi and Mycotoxins: Health Effects, Assessment, Prevention and Control. Eastern NY Occupational and Environmental Health Center. Albany, NY, Proceedings of 4th International Conference on Bioaerosols, Fungi, Bacteria, Mycotoxins, and Human Health. Saratoga Springs. NY. September 23 — 25, 1998.

CWMI, 2004. Hygienic implications of small-scale composting in New York State. In: Harrison, E.Z. (Ed.), Final Report of the Cold Compost Project. Cornell University, Cornell Waste Management Institute, Ithaca, NY.

Cyr, D.L., et al., 2002a. Basic First Aid. National Ag Safety Database. http://www.cdc.gov/nasd/.

Cyr, D.L., et al., 2002b. First Aid Kits for the Farm and Home. National Ag Safety Database. http://www.cdc.gov/nasd/.

Cyr, D.L., et al., 2002c. First Response to Farm Accidents. National Ag Safety Database. http://www.cdc.gov/nasd/.

Deportes, I., et al., 1995. Hazard to man and the environment posed by the use of urban waste compost: a review. Sci. Total Environ. 172 (2–3), 197.

Droffner, M.L., et al., 1994. Survival of *E. coli* and *Salmonella* populations in aerobic thermophilic composts as measured with DNA gene probes. Zb. Hyg. 197, 387.

Druley, K., 2019. First aid kit requirements. Saf. Health. https://www.safetyandhealthmagazine.com/articles/19019-first-aid-requirements.

Dumontet, S., et al., 2001. Importance of pathogenic organisms in sewage and sewage sludge. J. Air Waste Manag. Assoc. 51, 848.

Eldridge, L., 2019. Wood Dust Exposure and Lung Cancer Risk. Very Well Health. www.verywellhealth.com/wood-dust-and-lung-cancer-whos-at-risk-3971878#:~:text=While%20wood%20dust%20is%20more,elevated%20rates%20of%20lung%20cancer.

Ellis, J.N., 1994. Suspenseful suspension. Occup. Health Saf. 63 (3), 38.

Epstein, E., 1996. Protecting workers at composting facilities. BioCycle (Sept.), 69.

Epstein, E., 1997. The Science of Composting. Technomic Publications.

Epstein, E., et al., 2000. Odors and Volatile Organic Compound Emissions from Composting Facilities. Odors and VOC Emissions 200 Conference. Water Environment Federation.

Fatal and nonfatal occupational injuries involving wood chippers — United States, 1992 — 2002. MMWR (Morb. Mortal. Wkly. Rep.) 53 (48), 2004, 1130.

Fink, J., 1986. Hypersensitivity pneumonitis. In: Merchand, J.A. (Ed.), Occupational Respiratory Diseases. USDHHS/CDC/NIOSH. Superintendent of Documents. U.S. Government Printing Office, Washington, D.C.

Fischer, G., et al., 1998. Airborne fungi and their secondary metabolites in working places in a compost facility. Mycoses 41 (9–10), 383.

Fischer, G., et al., 1999a. Mycotoxins of *Aspergillus fumigatus* in pure culture and in native bioaerosols from compost facilities. Chemosphere 38 (8), 1745.

Fischer, G., et al., 1999b. Species-specific production of microbial volatile organic compounds (MVOC) by airborne fungi from a compost facility. Chemosphere 39 (5), 795.

Fischer, G., et al., 2000. Exposure to airborne fungi, MVOC, and mycotoxins in biowaste-handling facilities. Int. J. Hyg Environ. Health 203 (2), 97.

Glanville, T., 2005. Evaluation of composting for emergency disposal of cattle mortalities in Iowa. In: Presented at Symposium on Composting Mortalities and Slaughterhouse Residuals. May 24-25, 2005. Portland, Maine.

Gordon, R., 2005. On-farm mortality management initiatives for Nova Scotia's agricultural sector. In: Presented at Symposium on Composting Mortalities and Slaughterhouse Residuals. May 24-25, 2005. Portland, Maine.

Gould, M., et al., 2002. Comparing the effectiveness of positive and negative aeration in controlling emissions from composting processes. In: 14th Annual Residuals and Biosolids Management Conference; February/March 2000. Water Environment Federation.

Gutierrez, S., 2005. Why a workshop on emerging infectious disease agents and issues associated with animal manures, biosolids and other similar by-products? In: Smith Jr., J.E., et al. (Eds.), Contemporary Perspectives on Infectious Disease Agents in Sewage Sludge and Manure. The JG Press, Inc. Emmaus, PA.

Hambach, R., Droste, J., François, G., Weyler, J., Van Soom, U., De Schryver, A., Vanoeteren, J., van Sprundel, M., 2012. Work-related health symptoms among compost facility workers: a cross-sectional study. Arch. Publ. Health (Archives belges de sante publique) 70 (1), 13. https://doi.org/10.1186/0778-7367-70-13.

Harber, P., et al., 1993. Accommodating respiratory handicap. Semin. Respir. Med. 14 (3), 240.

Harber, P., et al., 1994. Work placement and worker fitness: implications of the Americans with Disabilities Act for pulmonary medicine. Chest 105, 1564.

Harding, A.L., et al., 2012. Biological hazards. In: Plog, B.A., et al. (Eds.), Fundamentals of Industrial Hygiene. National Safety Council, Itasca, IL.

Heldal, K.K., Madsø, L., Eduard, W., 2015. Airway inflammation among compost workers exposed to actinomycetes spores. Ann. Agric. Environ. Med. 22 (2), 253−258. https://doi.org/10.5604/12321966.1152076.

Herr, C.E.W., et al., 2003. Individual exposure assessment in residents near large scale composting sites. In: Proceedings of 5th International Conference on Bioaerosols, Fungi, Bacteria, Mycotoxins, and Human Health. Saratoga Springs, NY. September 10−12, 2003.

Hoenig, D., 2005. Maine's emergency disease response plan: 'Are we ready?'. In: Presented at Symposium on Composting Mortalities and Slaughterhouse Residuals. May 24-25, 2005. Portland, Maine.

Hoffmeyer, F., van Kampen, V., Taeger, D., Deckert, A., Rosenkranz, N., Kaßen, M., Schantora, A.L., Brüning, T., Raulf, M., Bünger, J., 2014. Prevalence of and relationship between rhinoconjunctivitis and lower airway diseases in compost workers with current or former exposure to organic dust. Ann. Agric. Environ. Med. 21 (4), 705−711. https://doi.org/10.5604/12321966.1129919.

Hogan, T.J., 1996. Particulates. In: Plog, B.A., et al. (Eds.), Fundamentals of Industrial Hygiene. National Safety Council, Itasca, IL ("Welding fumes").

Huffman, L.J., et al., 2004. Enhanced pulmonary inflammatory response to inhaled endotoxin in pregnant rats. J. Toxicol. Environ. Health A 67, 125.

Hutchinson, M., 2005. Managing bones from slaughterhouse residuals during composting. In: Presented at Symposium on Composting Mortalities and Slaughterhouse Residuals. May 24-25, 2005. Portland, Maine. Also, personal communication with the author following this presentation.

Hygienic Implications of Small-Scale Composting in New York State Report, Cornell Waste Management Institute. http://cwmi.css.cornell.edu.

Industries, C.C.P., 2016. Get to Know the New First Aid Ansi Z308.1-2015 Standard. https://ccpind.com/assets/user/documents/ANSI%20Information.pdf.

Johanning, E., 1999. An overview of waste management in the United States and recent research activities about composting related occupational health risk. Schriftenr ver Wassen Boden Lufthyg 104, 127.

Kim, J.Y., et al., 1995. Survey of volatile organic compounds at a municipal solid waste composting facility. Water Environ. Res. 67 (7), 1044.

King, M., 2005. Observations of large animal carcass composting with different media. In: Presented at Symposium on Composting Mortalities and Slaughterhouse Residuals. May 24-25, 2005. Portland, Maine. Also, personal communication with the author following this presentation.

Klaassen, C.D. (Ed.), 2001. Casarett and Doull's Toxicology, the Basic Science of Poisons. McGraw-Hill, New York ("endotoxin").

Kontogianni, S., et al., 2017. Investigation of the occupational health and safety conditions in Hellenic solid waste management facilities and assessment of the in-situ hazard level. Saf. Sci. 96, 192–197.

Kothary, M.H., Chase, T., Macmillan, J.D., 1984. Levels of Aspergillus fumigatus in air and in compost at a sewage sludge composting site. Environ. Pollut. Ecol. Biol. ISSN: 0143-1471 34 (1), 1–14. https://doi.org/10.1016/0143-1471(84)90084-9.

Kroemer, K.H.E., 1996. Ergonomics. In: Plog, B.A., et al. (Eds.), Fundamentals of Industrial Hygiene. National Safety Council, Itasca, IL.

Lavoie, J., Dunkerley, C.J., Kosatsky, T., Dufresne, A., 2006. Exposure to aerosolized bacteria and fungi among collectors of commercial, mixed residential, recyclable, and compostable waste. Sci. Total Environ. 370 (1), 23–28. https://doi.org/10.1016/j.scitotenv.2006.05.016.

Lee, R., et al., 1993. Chipper-shredders. Dept. of Agricultural Engineering. University of Missouri – Extension, Columbia, MO 65211.

Linn, R., 2002a. Tractor Accident Victim Rescue. National Ag Safety Database. http://www.cdc.gov/nasd/.

Linn, R., 2002b. Power Take-Off Accident Victim Rescue. National Ag Safety Database. http://www.cdc.gov/nasd/.

MacLeod, D., 1995. Ergonomics Edge: Improving Safety, Quality, and Productivity. Van Nostrand Reinhold, New York.

Mahin, T., et al., 2000. When Is a Smell a Nuisance? Water Environment and Technology, p. 52. May.

Marsh, P.B., et al., 1979. A guide to the recent literature on Aspergillosis as caused by Aspergillus fumigatus, a fungus frequently found in self-heating organic matter. Mycopathologia 69 (1–2), 67.

McCunney, R.J., 1986. Health effects of work at wastewater treatment plants: a review of the literature with guidelines for medical surveillance. Am. J. Ind. Med. 9, 271.

Miller, G., 1996. Nonionizing radiation. In: Plog, B.A. (Ed.), Fundamentals of Industrial Hygiene. National Safety Council, Itasca, IL.

Millner, P.D., et al., 2005. Animal manure: bacterial pathogens and disinfection technologies. In: Smith Jr., J.E., et al. (Eds.), Contemporary Perspectives on Infectious Disease Agents in Sewage Sludge and Manure. The JG Press, Inc. Emmaus, PA.

MNOSHA, 2015. Contents of a First-Aid Kit. Minnesota Department of Labor and Industry. https://cdn.ymaws.com/www.mnta.org/resource/resmgr/files/Contents_of_a_first-aid_kit.pdf.

Naylor, L.M., 2000. A decade of in-vessel biosolids composting: survival of the fittest?. In: 14th Annual Residuals and Biosolids Management Conference; February/March 2000. Water Environment Federation.

NFPA. Standard 10 – Standard for Portable Fire Extinguishers, National Fire Protection Association, Quincy, MA.

Norkin, C.C., et al., 1992. Joint Structure and Function: A Comprehensive Analysis. F. A. Davis Co., Philadelphia.

OPEI, 2021. Outdoor Power Equipment Institute, Alexandria, VA; Vancouver, BC. https://www.opei.org/and https://www.opeic.ca/.

OSHA, 2002. Job Hazard Analysis. Occupational Safety and Health Administration (U.S.). https://www.osha.gov/Publications/osha3071.pdf.

OSHA, 2021c. Ergonomics. Occupational Safety and Health Administration (U.S.). https://www.osha.gov/SLTC/ergonomics/identifyprobs.html.

OSHA, 2021d. Youth in Agriculture. Occupational Safety and Health Administration (U.S.). https://www.osha.gov/SLTC/youth/agriculture/machinery.html.

OSHA, 2021e. 1910.266 App B - First-Aid and CPR Training (Mandatory). Occupational Safety and Health Administration (U.S.). https://www.osha.gov/laws-regs/regulations/standardnumber/1910/1910.266AppB.

OSHA, 2021g. Safety and Health Regulations for Construction. https://www.osha.gov/laws-regs/regulations/standardnumber/1926.

OSHA, 2021f. Occupational Safety and Health Act of 1970. Public Law 91-596; 84 Stat. 1590. (12/29/70). https://www.osha.gov/laws-regs/oshact/section5-duties.

OSHA, 2021h. Occupational Safety and Health Standards for Agriculture. https://www.osha.gov/laws-regs/regulations/standardnumber/1928.

Personal Communication. 06/24/04. Compost Worker Described an Accident Which Happened to Himself. He Was Walking on Top of a Windrow and Dropped Down into the Pile and Was Buried up to the Middle of His Chest. He yelled for help until he was found and rescued by a co-worker about 2 hours later.

Peterson, M.K., et al., 2000. Characterization of Emissions from Two Yard-Waste Composting Facilities. In: Odors and VOC Emissions 2000 Conference; April 2000. Water Environment Federation.

Piontek, M., et al., 2000. Effect of sewage sludge composting on the quantitative state of some groups of bacteria and fungi. Acta Microbiol. Pol. 49 (1), 83.

Poulsen, O.M., et al., 1995. Sorting and recycling of domestic waste. Review of occupational health problems and their possible causes. Sci. Total Environ. 168, 33.

Rautiala, S., et al., 2003. Farmers' exposure to airborne microorganisms in composting swine confinement buildings. Am. Ind. Hyg. Assoc. J. 64, 673.

Rodgers, S.H., et al., 1986. Ergonomic Design for People at Work, vol. 2. Van Nostrand Reinhold Co., New York.

Rothstein, M.A., 1996. Legal and ethical aspects of medical screening. Occup. Med. - State of the Art Rev. 11 (1), 31.

Rozman, K.K., et al., 2001. Absorption, distribution, and excretion of toxicants. In: Casarett and Doull's Toxicology, the Basic Science of Poisons. McGraw-Hill, New York.

Saif Corporation, 2015. Welding Health and Safety. https://www.saif.com/safety-and-health/topics/industry-specific-topics/welding.html.

Saldarriaga, J.F., et al., 2014. Assessment of VOC emissions from municipal solid waste composting. Environ. Eng. Sci. 31 (6), 300–307. https://doi.org/10.1089/ees.2013.0475.

Samadi, S., van Eerdenburg, F.J., Jamshidifard, A.R., Otten, G.P., Droppert, M., Heederik, D.J., Wouters, I.M., 2012. The influence of bedding materials on bio-aerosol exposure in dairy barns. J. Expo. Sci. Environ. Epidemiol. 22 (4), 361–368. https://doi.org/10.1038/jes.2012.257.

Santiago, A.V., et al., 1989. Angioedema, rhinitis and asthma provoked by fishing bait (Eisenia foetida). Allergol. Immunopathol. 17 (6), 331–335.

Schiffman, S.S., 2000. Potential health effects of odor from animal operations, wastewater treatment, and recycling of byproducts. J. Agromed. 7 (1), 7.

Schlosser, O., et al., 2012. Protection of the vehicle cab environment against bacteria, fungi, and endotoxins in composting facilities. Waste Manag. 32, 1106–1115.

Shih, J., 2005. Discoveries of prion degradation and a safe prion surrogate protein. In: Presented at Symposium on Composting Mortalities and Slaughterhouse Residuals. May 24–25, 2005. Portland, Maine.

Sigsgaard, T., 1999. Health hazards to waste management workers in Denmark. Schriftenr ver Wassen Boden Lufthyg 104, 563.

Sigsgaard, T., et al., 1994a. Lung function changes among recycling workers exposed to organic dust. Am. J. Ind. Med. 25, 69.

Sigsgaard, T., et al., 1994b. Respiratory disorders and atopy in Danish refuse workers. Am. J. Respir. Crit. Care Med. 149, 1407.

Sigsgaard, T., et al., 2000. Cytokine release from the nasal mucosa and whole blood after experimental exposures to organic dusts. Eur. Respir. J. 16, 140.

Slavin, R.G., et al., 1977. Epidemiologic aspects of allergic aspergillosis. Ann. Allergy 38 (3), 215.

Smith, L.W., 2001. Incidence of potentially pathogenic bacteria in liquor from selected wormfarms. Biol. Fertil. Soils 34, 215–217.

Smith Jr., J.E., Millner, P.D., Jakubowski, W., Goldstein, N., Rynk, R. (Eds.), 2005. Contemporary Perspectives on Infectious Disease Agents in Sewage Sludge and Manure. The JG Press Incorporated, Emmaus, PA.

Sobsey, M.D., 2005. Viruses in animal manures. In: Smith Jr., J.E., et al. (Eds.), Contemporary Perspectives on Infectious Disease Agents in Sewage Sludge and Manure. The JG Press, Inc, Emmaus, PA.

St. Clair, S., et al., 1992. Americans with Disabilities Act: considerations for the practice of occupational medicine. J. Occup. Environ. Med. 34 (5), 510.

Standard, J.J., 1996. Industrial noise. In: Plog, B.A., et al. (Eds.), Fundamentals of Industrial Hygiene. National Safety Council, Itasca, IL.

University of Tennessee, 1997. A Practical Safety Manual for the Composting and Mulching Industry. University of Tennessee Center for Industrial Services.

USDHHS. CDC/NIOSH, 1999. Injury Associated with Working Near or Operating Wood-chippers. HID 8. DHHS (NIOSH) Publication No. 99-145. Available at: http://www.cdc.gov/niosh/hid8.html.

USDHHS. CDC/NIOSH, 2002. National Institute for Occupational Safety and Health Recommendations to the U.S. Department of Labor for Changes to Hazardous Orders. Available at: http://www.cdc.gov/niosh/docs/nioshrecsdolhaz/pdfs/dol-recomm.pdf.

USDHHS/CDC/NIOSH, 1994. NIOSH Warns of Agricultural Hazards: Organic Dust Toxic Syndrome (07/19/94).

USDHHS/CDC/NIOSH, 1996. Independent Contractor Dies when Struck by Protective Hood from Chipper/shredder at Waste Management Facility in Massachusetts. Massachusetts FACE Report No. 96MA037.

USDHHS/CDC/NIOSH, 1999. Worker Dies Due to a Fall from a Conveyor Belt. Ohio FACE Report No. 99OH02301.

USDHHS/CDC/NIOSH, 2002. Guidance for Controlling Potential Risks to Workers Exposed to Class B Biosolids. DHHS (NIOSH) Publication No. 2002-149.

USDHHS/NIOSH, 1992. Working in Hot Environments. Superintendent of Documents. U.S. Government Printing Office, Washington, DC, 20402.

USDHHS.NIOSH, 1992. Working in Hot Environments. http://www.osha.gov/SLTC/heastress/recognition.html.

USDOL, 2005a. Fair Labor Standards Act Advisor: Exemptions from Child Labor Rules in Non-agriculture. www.dol.gov/elaws.

USDOL, 2005b. Fair Labor Standards Act Advisor: Prohibited Occupations for Agricultural Employees. www.dol.gov/elaws.

USDOL, 2005c. Fair Labor Standards Act Advisor: Prohibited Occupations for Non-agricultural Employees. www.dol.gov/elaws.

USDOL. OSHA. Cold Stress Equation. http://www.osha.gov/Publications/coldcard/coldcard. html.

USDOL. OSHA. Effective Ergonomics: Strategy for Success. http://www.osha.gov/SLTC/ ergonomics/index.html.

USDOL. OSHA. Heat stress. http://www.osha.gov/SLTC/heastress.html. ("Standards," "Hazards and possible solutions," "Heat stress card," Protecting yourself in the sun").

USDOL. OSHA, 1994. Preamble to the Final Rule for 29 CFR 1910.266 — Logging Operations. Available at: http://www.osha.gov/pls/oshaweb/owadisp.show_document?p_table=PREAMBLES&p_id+965. "employers in the logging industry must provide employees protection against occupational noise exposure by meeting the requirements of 29 CFR 1910. 95.

USDOL/OSHA, 2005. Medical and First Aid. http://www.osha.gov.

USDOL/OSHA, 29 CFR 1910. 151 — Medical Services and First Aid.

USEPA, 1994. Composting Yard Trimmings and Municipal Solid Waste. EPA 530-R-94-003.

USEPA, 2001. Mold Remediation in Schools and Commercial Buildings. EPA 402-K-01-001. http://www.epa.gov.

USEPA, 2004. Recommended Interim Practices for Disposal of Potentially Contaminated Chronic Wasting Disease Carcasses and Wastes. Springer. R. April 6, 2004. Office of Solid Waste. U.S. Environmental Protection Agency.

USEPA, 2005. New EPA Prion Studies Could Hinder POTW Push to Land-Apply Biosolids. InsideEPA.com (May 30, 2005).

Van den Bogart, H.G., Van den Ende, G., Van Loon, P.C., Van Griensven, L.J., 1993. Mushroom worker's lung: serologic reactions to thermophilic actinomycetes present in the air of compost tunnels. Mycopathologia 122 (1), 21—28. https://doi.org/10.1007/ BF01103705.

van Kampen, V., Hoffmeyer, F., Deckert, A., Kendzia, B., Casjens, S., Neumann, H.D., Buxtrup, M., Willer, E., Felten, C., Schöneich, R., Brüning, T., Raulf, M., Bünger, J., 2016. Effects of bioaerosol exposure on respiratory health in compost workers: a 13-year follow-up study. Occup. Environ. Med. 73 (12), 829—837. https://doi.org/10.1136/ oemed-2016-103692.

Van Tongeren, M., et al., 1997. Exposure to organic dusts, endotoxins, and microorganisms in the municipal waste industry. Int. J. Occup. Environ. Health 3 (1), 30.

Watanabe, T., Sano, D., Omura, T., 2002. Risk evaluation for pathogenic bacteria and viruses in sewage sludge compost. Water Sci. Technol. J. Int. Assoc. Water Pollut. Res. 46 (11—12), 325—330.

Waters, T.R., et al., 1994. Applications Manual for The Revised NIOSH Lifting Equation. USDHHS.NIOSH. No. 94-110.

Youngstom, R.A., 1981. Welding: Hazards and Controls. IUE Local 201 Health & Safety Committee, Inc.

Facility management

Authors: Robert Rynk[1], Nanci Koerting[2], Jeff Ziegenbein[3], James Hardin[1]

[1]*SUNY Cobleskill, Cobleskill, NY, United States;* [2]*Environmental Compliance, Grant County Mulch, Boonsboro, MD, United States;* [3]*Regional Compost Operations, Inland Empire Utilities Agency, Chino, CA, United States*

Contributors: Cary Oshins[4], Nellie J. Brown[5], Nancy J. Lampen[6], Dan Lilkas-Rain[7]

[4]*U.S. Composting Council, Raleigh, NC, United States;* [5]*Workplace Health & Safety Program, ILR, Cornell University, Buffalo, NY, United States;* [6]*Associates for Human Resource Development, Pittsford, NY, United States;* [7]*Growing Media, Town of Bethlehem, Bethlehem, NY, United States*

1. Introduction

With the composting methods, operations and site in place, the job becomes one of managing the composting facility with sensitivity to the surrounding conditions and environment (Fig. 14.1). Management can be the difference between a profitable efficiently operating facility, and one beset by frequent troubles with neighbors, regulators, the environment, and compost customers. In managing the facility, a composter should always keep in mind the goals of the facility—compost production, specific compost markets, desired compost quality, environmental benefits and its purpose within the community, farm, or overall organization.

Facility management tasks include administrative functions, employee management and training, process management, controlling odors, public relations, health and safety, site management, materials management, handling nuisances, weather/climate, and fire prevention. Process management, odors, site management, and health and safety are discussed in other chapters. This chapter focuses on the remaining issues.

An important management responsibility is guiding the enterprise through a crisis. Any crisis, large or small, is best managed in advance. Imagining a crisis that has not happened yet is a powerful way to handle that crisis when it does happen. Contingency planning anticipates serious or damaging events that might impact the enterprise. This exercise is not about a particular crisis, but every crisis that can potentially threaten the enterprise, within a reasonable probability. Box 14.1 is a guide for conducting contingency planning for composting facilities. Other resources for contingency planning are easily found via an internet search.

The Composting Handbook. https://doi.org/10.1016/B978-0-323-85602-7.00017-0

FIGURE 14.1

A well-managed composting facility looks well-managed—clean, organized, and compatible with the environment and the community.

Box 14.1 Thinking about the unthinkable—crisis and contingency planning

Authors: Nellie J. Brown and Nancy J. Lampen.

A crisis is a hazard or threat that you do not have the resources to cope with, and that will have negative effects if not addressed. It is not a question of whether a crisis will happen in an organization: it is only a matter of when, which type, and how it will occur. Managers and employees need to think about the unthinkable, *in advance of the unthinkable*, as a way of protecting themselves and their facilities. The better you can anticipate a crisis, the better you can manage it. This section provides templates and guidelines to help you to examine your organization's vulnerabilities and to develop plans for recovery. We will start by having you examine the range of crises that your facility could experience and then create an action plan for recovery by assuming that each crisis has already occurred and what you would need in people, equipment, and other resources to resume operation.

The following questions are intended to stimulate your thinking about some types of crises that could happen to your organization. Consider the extent of damage these crises could cause, and the effect they would have on employees, the organization, clients, neighbors, customers, animals, plants, others, and the environment.

1. What are three serious crises your organization could experience?
2. How well prepared is your organization to manage the crises listed above?
3. How would your organization manage if more than one of the above crises happened at the same time?
4. How would your organization cope if one or more of these crises occurred at the same time as a natural disaster?

Below are listed crisis categories, along with some possible examples, to help stimulate your thinking about possible vulnerabilities. See if you can come up with two or three crises in *each*

Box 14.1 Thinking about the unthinkable—crisis and contingency planning—cont'd

of these areas that your facility might be vulnerable to. We often first think of things that have already happened to us or to organizations that we are familiar with; but to be fully prepared for crises, it is necessary to examine vulnerabilities in each of the crisis areas.

- Terrorist/criminal attacks: such as arson, deliberate contamination of feedstock, or compost,
- Economic threats: such as privatization of a public facility, eminent domain by a government entity,
- Loss of proprietary information: such as identity theft, posting of employee personnel information on the web,
- Industrial disasters: such as a fire or accidental contamination of a feedstock,
- Natural disasters: such as tornados, hurricanes, and floods,
- Breaks in equipment: such as a broken conveyor belt, replacement parts are not available or are no longer manufactured,
- Legal: such as a lawsuit over odors or a compost fire,
- Reputation/perception: such as process is "stinky" or the compost is contaminated,
- Human resources/occupational: such as false credentialing of process operator,
- Health: such as pathogen survival through the process; fatality of an employee who climbed into tub grinder, and,
- Regulatory: such as final product does not meet maturity or disease kill requirements.

For each crisis type, imagine a scenario for the "crisis most likely to occur" or for the "crisis that would be most damaging if it occurred." Assume that such a crisis has already occurred—what is needed to recover from that crisis? For example:

- What are the minimum services/operations needed for the organization's functioning? Be sure to consider the needs of people and animals.
- What resources are needed to support a minimum resumption of services? How will you obtain these?
- What redundancies or backup do you have for the organization's most critical operations?
- What people or organizations will you need to recover? Identify the stakeholders most important to recovery.
- How will you communicate with stakeholders?
- How will you implement a critical incident response when needed?
- Are you prepared to interact with the media?

The following worksheet is an example of what can be used to identify actions to recover from the anticipated crisis. After working through this exercise, preferably with employees, you will be in a much better position to endure the crisis. In additional, it is helpful to ask which of these tasks can be accomplished *prior to the crisis*? (Start planning these now.) During the crisis? And after the crisis? Finally, are there other specialties or functions needed to manage a crisis recovery?

Tasks to achieve recovery	Capabilities needed by employees	Who will perform it?	What training is needed?

Anticipating potential crisis in a deliberate and organized manner is a valuable use of your time and your staff's time. Make an organizational effort to think through possible vulnerabilities because being proactive around crisis management is the key to successful recovery. Please don't trust it to luck—because you believe that "we'll just handle it when it happens" or that "we'll be able to think on our feet" in the midst of a bad situation. At such times, "we" are *not* at our best and "we" may be emotionally involved or even incapacitated. We will not have the luxury of time, resources, or capability.

2. Administrative functions

Apart from managing the composting process and products, operating a composting facility requires attention to those functions that administer to the business side of the enterprise, including complying with regulations. Like any agricultural or other commercial enterprise, there are a host of conventional business functions like accounting, payroll, billing, employee hiring and training, and the like. These functions are well beyond the scope of this chapter. Still, there are a few administrative duties that distinctly relate to operating a composting facility such as compulsory regulatory tasks, employee training, and maintenance of compost-specific documents and records.

2.1 Regulatory requirements

A variety of regulations and permits might apply to a composting site. In addition to designing the site to prevent pollution, regulations often require sites to perform, and document mandated regulatory tasks. Examples include employee training, self-inspections or third-party inspections, verification of pathogen reduction, process monitoring, verification of incoming and outgoing materials (both type and amount), stormwater runoff and recording odor episodes, and other complaints. It is best to be transparent and proactive with the regulators that oversee these conditions. A good manager develops procedures to carry out compliance tasks and document all conditions in a professional and organized manner. A good manager makes those records easily available for inspection by the regulators. Depending on the regulatory status of the facility, it might be necessary to submit periodic reports, notifications, and records. The required reports should always be submitted on time and managers should be prepared to discuss any of the details included. Neat and accurate records build trust with the regulators who will be far easier to work with and who will be more supportive of the facility. For many composting facilities, especially on-farm facilities, this regulatory burden is modest, if it exists at all. However, it is prudent to identify the management tasks that regulations compel the facility to perform, under what circumstances and how often.

2.2 Tips for employee hiring and management

The success of a composting facility rests on the skill, knowledge, and work ethic of its employees. Hiring good, committed, "in-it-for-long-term" candidates can be a challenge especially in an area with competitive labor market and/or high prevailing wages. Depending on the position level, salary, and benefits, it can be difficult to attract and keep longer term employees. A flexible and team-oriented work environment, casual dress, opportunity for advancement, and on-going training can be used to recruit especially in a business environment with higher salaries. The positive impacts of composting can be a hiring advantage. Many operators at composting facilities take pride in the positive environmental and local business benefits of their work. Also, composting and agriculture are very compatible in their activities, practices, knowledge base, and working conditions. Therefore, farm workers often make excellent employees for the composting industry.

When developing job descriptions, thoroughly research position responsibilities and build interview questions carefully around them. For example, will the position require a commercial-vehicle driver's license (e.g., CDL), or benefit from it? Will obtaining continuing education credits be a requirement for the job? Is training necessary to begin employment, or is it something that can be earned after the employee begins work? For positions that require employees to operate equipment, operating the equipment can be a component of the hiring process.

For new employees, it is common to institute a six-month probationary period with monthly meetings between the new employee and his/her supervisor. The meetings should include candid and documented evaluations, addressing topics like employee attendance and on-time performance, dependability, position progress, and honest concerns on either side of the table. A good strategy is to pair new employees, or newly promoted employees, with a seasoned staff member, at least for their first week in the position. Officially designating a seasoned team member as a mentor, and encouraging mentorship, can improve employee's comfort level and greatly help with their transitions. If a mentor is not available internally, such as in a small business, industry organizations can provide outside mentors. For example, the USCC has mentoring program for young professionals in the composting industry (USCC, 2021a).

Cross training employees for multiple jobs within a business boosts esteem and productivity among staff. It also eliminates the need for extra "back up people" in the event of absences, thereby reducing the overall number of employees necessary to run the site on a day-to-day basis. Field staff should be trained to understand the nuts and bolts of all phases of operations, include the site permits, site environmental responsibilities, equipment maintenance, and process management. Fully informed employees are the difference between a mediocre site and a model site.

It is helpful for employees at all levels of the organization to have access to written procedures, including administrative, safety, and operational procedures. These documents can be placed together or in separate notebooks at key locations within the operation where employees can review them at any time (e.g., office, shop, site trailer). The written procedures should change and grow with the business and should be reviewed and updated periodically such as at the beginning of each calendar year. Also, employees should have the opportunity to suggest changes to the procedures. Field operators have first-hand knowledge about what is happening on the site and with the process. They have valuable knowledge and experience that can only increase operational efficiency. In addition, including their input gives them additional pride and ownership of their contributions to the process.

Regular meetings among staff members, including management and site personnel, help maintain lines of communication and morale. Such meetings help employees to feel appreciated, and keeps valuable information flowing up *and* down the organization. Employees speak up and contribute valuable ideas, if given a nonthreatening opportunity, and if they feel that their suggestions and complaints are sincerely considered. There is much truth in the adage, *happy employees are productive employees.*

2.3 Training

A knowledgeable well-trained staff is the key factor in achieving a smooth-running trouble-free composting facility, even if it is a staff of only one person. Important topics of training include, but are not limited to, safety, operation of specific equipment and its maintenance, composting process fundamentals, composting methods and practices, process monitoring, record keeping, control and mitigation of odors and other nuisances, public relations, regulatory requirements, fire prevention, fire response, and general emergency response.

The most important subject of training is health and safety. Health and safety training should start with an initial training for new employees followed by frequent and regular safety meetings for all employees. At the best managed facilities, staff start the day and week with brief safety meetings. Every operation should have someone designated as the safety "coordinator," who is in charge of knowing the safety regulations, conducting the safety meetings, and maintaining safety records. Managers should be aware that in the event of a serious safety incident, there will likely be an investigation by the governing workplace safety authority (e.g., US Occupational Safety and Health Administration). The investigator(s) usually ask to review training and maintenance documents.

Some states, provinces, and/or countries require composting managers and operators to complete a basic composting course and then continue to attend additional educational forums. Often, this training must be certified, documented, and tracked (e.g., by accumulated "continuing education units"). Fortunately, the composting industry has spawned many good training opportunities for facility operators, managers, marketing staff, and compost users. Such training events occur annually or semiannually on a variety of topics ranging from general sessions about the science of composting to specific topics like contingency planning, forced aeration, and compost use in horticulture. Many training sessions are half-day to one-day workshops that take place in a classroom setting at conferences. Some workshops are stand-alone multiday events that include hands-on learning segments. The multiday training workshops tend to be comprehensive and broad in scope. Many provide hands-on learning activities (Fig. 14.2). While these sessions are often designed with the novice composter in mind, they are valuable to veteran composters as well. Much of the value in training events lies in the networking and information exchange that takes place among participants. While in-person training is generally preferred, it is also expensive. Therefore, training courses are increasingly offered on-line via webinars and other forums and often include breakout sessions to allow for networking and peer-to-peer interaction.

Training sessions are sponsored by university extension programs, industry organizations, composting trade associations, state and provincial agencies, and commercial equipment and product suppliers. In addition, some organizations offer certification and/or credentials to professionals who complete an educational program and then pass an exam. For example, the USCC offers two certifications with credentials, "Certified Compost Operations Manager" and "Certified Composting Professional" (USCC, 2021b). A list of current composting training and certification opportunities in the US is included in Appendix L.

Training should not be limited to outside organizations and locations, however. It should be a regular activity within the facility. For example, equipment dealers can

FIGURE 14.2

Five-day Compost Operators Training Course (COTC) offered by CREF. Participants work in teams to create, measure, and monitor composting batches from selected feedstocks.

Source: J.A. Biernbaum.

provide on-site instruction about operating the specific equipment that they supply (e.g., windrow turners, wheel loaders, and screens). In general, training should also take place via regular gatherings of staff at which training occurs via updates and discussions, perhaps lead by more experienced personal or staff members who attended an outside training event. A powerful training tool that is often overlooked is providing opportunities for operators, and managers, to tour other composting sites.

2.4 Documents and records

There is paperwork in managing any enterprise, composting included. Table 14.1 lists examples of documents and records that a composting facility might generate. The type and detail of the documents and records vary greatly, depending primarily on a facility's size and the governing regulations. Again, fewer requirements are imposed on operations handling small volumes and/or located on farms.

There are some documents that all facilities should keep, even when not required. Documents and records concerning safety, process control, and liability are especially important. For example, facilities in the US are required to keep safety data sheets for each chemical used and stored on the site. Most facilities have, and

Table 14.1 Example documents to create and keep at a composting facility. Compiled in part from composting process and compost product certification references (CQA, 2021; USCC, 2021c; BSI, 2018; Siebert and Auweele, 2018; CCME, 2005).

Documents	Records
Environmental and operational permits	Incoming feedstocks—date, time, source, amount, description, rejection
Facility construction plans, including buildings (design and "as-built")	Compost shipments or use—date, type, amount, destination
Notebook of safety data sheets (SDSs)	Process temperatures including sanitization/PFRP confirmation for each batch
Employee safety rules and regulations	Other process data—such as moisture and oxygen or carbon dioxide
Labor/employment rules and standards	Standard sampling protocol
Facility safety plan	Quality control samples and analysis
Standard Operating Procedure (SOP) (i.e., operation manual)	Feedstock and compost lab analysis
Odor management plan	Date and time-based log of complaints (especially odor complaints)
Fire management plan	Equipment maintenance, inspections, fluids and filter changes, lubrication
Stormwater management plan (including discharge permits if required)	Suspect or confirmed incidents of fires
Emergency response plan	Accidents and safety incidents
Accident/incident contingency plan	Records of incidents of relevant illnesses among employees and visitors (e.g., COVID-19, aspergillosis), as required, and within limits allowed by privacy laws.
Emergency contacts	Inspections and inspection results
Neighborhood contacts	Training and training certificates
Regulatory, technical, legal, and business contacts	Record of site visitors
Written copies of regulations	Remarkable incidents (power outage, wildfires, etc.)
Equipment maintenance plan and schedule	Weather conditions
Equipment manuals	Product inventory
Multiple copies of the "*The Composting Handbook*"	Equipment, tool, and parts inventory
Collection of resources and references	Hazard analysis report(s) (e.g., HACCP)

may be required to have, an operation manual, often referred to as an SOP (Standard Operating Procedure). In addition, it is good practice to develop and keep written plans for stormwater, odor, and fire management. Examples of other important operational documents include written safety procedures, written emergency procedures, operator certifications, and training records.

Perhaps the single best management adage is: "Keep good records of everything." Obviously, "everything" is an overly ambitious and impractical goal, but the underlying message is a valid one. As much as practical, records should be maintained for a variety of conditions and occurrences associated with the facility and business. Important records relate to verification of incoming feedstocks (amounts and types), confirmation of pathogen reduction requirements, employee health and safety incidents, training records, self-inspections, outside inspections, neighborhood complaints, compost sales, and inventories of feedstocks and products. Records help managers and operators learn about composting and compost quality. Records provide feedback about the quality of feedstocks and their source, the effects of turning and other process controls, and the influence of weather. Records also support defensible arguments when accusations arise regarding odors, permit violations, and other issues.

One or more employees at each site should be tasked with maintaining a written daily log that records date, weather, visitors, employees on site, and key events of each day. This log can be referred to in the event of complaints, odor issues, accidents, employee absences, possible fire-related issues, regulatory compliance, and so on. It can serve as the organization's collective memory.

3. Managing the carbon footprint

Composting and the use of compost generally have a net positive impact on climate change (USCC, 2021e). As explained in Chapters 1 and 16, the climate advantages of composting arise first from the diversion of feedstocks from more damaging alternatives, such as landfills or long-term manure storage. More climate advantages come from the benefits of compost to plants and soils, including carbon sequestration. The negative impacts come from the greenhouse gas (GHG) emitted during the manufacture and distribution of compost.

Composting contributes GHG to the atmosphere in primarily two ways: (1) by the direct release of Methane (CH_4) and nitrous oxide (N_2O) during composting and (2) by the use of fossil fuels for transportation and processing (Fig. 14.3). The CO_2 released from the composting pile does not count as a GHG. It is considered "biogenic" carbon,[1] derived from feedstocks that would have released the same

---

[1] "Biogenic carbon is the emissions related to the natural carbon cycle, as well as those resulting from the combustion, harvest, digestion, fermentation, decomposition or processing of biologically based materials." (UC Davis, 2021).

FIGURE 14.3

The carbon footprint of a composting facility depends on the process emissions and the fuel consumed by trucks, turners, and other equipment. The plumes of vapor shown in this picture carry primarily water and biogenic CO_2, which do not additionally contribute to climate change.

Source: Will Bakx.

amount of CO_2 (or less) if they were to decompose naturally. However, there is an uncounted benefit in conserving organic matter during composting so that less CO_2 is created (Chapter 11).

The GHG of composting that *do* count are CH_4 and N_2O. These gases trap much more heat in the atmosphere than CO_2 and they would not, presumably, occur via natural decomposition. In an anaerobic environment, CH_4 is created in place of some of the CO_2. Therefore, the goal should be to encourage aerobic conditions so that the carbon end-product of decomposition is CO_2. Aerobic conditions also discourage N_2O production but other factors like C:N ratio also are important. Practices that minimize the emissions of CH_4 and N_2O during composting are described in Chapter 11.

CO_2 generated from burning fossil fuels for electricity and engine power does count, and it is largely to blame for our current climate crisis. A lot of electricity and fuel is used between the collection of feedstocks and the end use of the compost.

Fuel and electricity are consumed by collection and delivery trucks, bucket loaders, mixers, turners, grinders, and screens (Brown, 2016; Linzer and Mostbauer, 2005). Whatever reduces fuel consumption, also reduces the carbon footprint of the composting facility (e.g., local feedstocks, local compost sales, efficient engines and engine-operation, using and/or generating electrical power from solar or wind). Composters seeking to reduce their carbon footprint can find thorough and solid guidance in *Greenhouse Gas Balance for Composting Operations* (Brown et al., 2008).

4. Weights and measures

Good management requires awareness of how much material is on the composting site and in particular piles. Such materials include feedstocks in staging areas, stored amendments, active composting windrows and piles, curing piles, and stored compost products. Measurements are made to first determine the volumes of materials on hand. But it is also useful, if not essential, to know the weights/masses of various windrows and piles. To calculate weights from volumes, and vice versa, one must know the average bulk density of the material in question. If the bulk density is not immediately known, it can be easily measured by the procedure outlined in Chapter 4.

Ordinarily, the volumes of piles and windrows are determined by measuring heights, widths, and lengths using standard "hand" tools and a bit of geometry. This approach works reasonably well in most cases, especially if measurements are made diligently (Keener et al., 2008a,b). For piles that are irregular in shape and dimensions, the geometry becomes complex and hand measurements are more laborious. For these situations, there are several electronic measuring options, including laser measuring devices and drones with cameras or scanning devices. Even with simple geometry, technical tools can provide results more rapidly and more accurately.

4.1 Traditional measurement tools

The traditional method of determining material volumes is to measure the dimensions of windrows and piles with standard measuring tools like a long measuring tape, distance measuring wheel (e.g., surveyors' wheel), a level pole, and perhaps a hand-held sight level (Campbell, 2014b). With these tools, pile dimensions can be measured. From those dimensions, and the proper geometric formula, the volumes can be calculated (Table 14.2). Except for conical piles, volumes are calculated by multiplying the cross-sectional area times the pile/windrow length. The cross-sectional shape of windrows and piles depends on how they are formed and worked. Piles that are wider than twice their height can be modeled as parabolas or

Table 14.2 Typical shapes assumed for piles and windrows and the geometric formulas for calculating volumes.

Assumed pile shape and dimensions	Cross-sectional area	Volume
Triangular	$A = 1/2 \, w \times h$	$V = A \times L$
Trapezoidal	$A = h \times (w + t)/2$ or $A = h \times (w - (h/\tan\theta)$ θ is the angle of repose. If $\theta = 45°$, $\tan 45° = 1$ then: $A = h \times (w - h)$	$V = A \times L$
Parabolic	$A = w \times h \times 2/3$	$V = A \times L$
Conical	$D = (d_1 + d_2)/2$ Base area: $A = \pi \times (D/2)^2$ $D = c/\pi$	$V = h \times A \times 1/3$

trapezoids. For narrower piles, triangular shapes are reasonable representations. Conical piles are more likely to be elliptical at their base rather than circular. Therefore, the width, or diameter, should be measured in two perpendicular directions and then averaged to estimate an "effective" diameter. It may be easier to measure the circumference of the conical base rather than the diameters.

It is difficult to obtain accurate measurements of pile dimensions using these traditional measuring tools. The biggest challenge is the irregular shapes and heights of piles. In such cases, the judicious approach is to make many measurements and average the measurements for each dimension. However, when manually estimating pile volumes and weights, it is impractical to measure with precision or to strive for a great deal accuracy. There are multiple places for error, including judging the level height of a pile, guessing its geometrical shape and determining its average bulk density. An accuracy of ±20% appears to be a reasonable expectation when making such measurements (Keener et al., 2008a).

4.2 Technical measuring tools

Recent and still-emerging technical tools are making it easier to obtain accurate measurements and manage of inventories of materials at composting sites and other facilities that handle bulk materials (Coker, 2016). Such tools include laser-based scanners and digital cameras mounted on unmanned aerial vehicles (UAVs), popularly known as "drones" (Fig. 14.4). In both cases, the equipment does only half the work, if not less. Computer software performs the magic of converting the electronic signals into useful information. The software might be purchased with the equipment and/or through a subscription fee. Users must be comfortable with computer technology and navigating the internet.

There is an initial cost in acquiring these tools, and a learning curve to master their use. However, they deliver results rapidly and with surprising ease. Depending on the scale of the facility, these tools can pay for themselves within a few years due

(A)   **(B)**

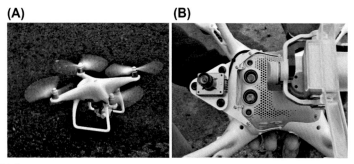

FIGURE 14.4

Unmanned aerial vehicle (A) equipped with sensors, communication instruments, and cameras (B). These "UAVs" can collect data that can be used to generate a three-dimensional image and estimate pile volumes.

to the timed saved in making measurements, improved accuracy, and better inventory management. Alternatively, the task can be hired out to companies that specialize in these technical services.

4.2.1 On-the-ground electronic tools

A variety of distance-measuring technologies are available that determine distance by the reflection of light, with and without the aid of Global Navigation Satellite System (GNSS) coordinates. The wavelengths of light used (visible and infrared), and the manner in which the reflected light is measured differ with the various technologies (Stockpile Reports, 2019). For example, a laser distance-measuring device emits and receives a narrow beam of light while a Lidar device sends and receives laser light in broad scanning swaths. Both measure the time interval of the reflected light to determine distance. Electronically measured distances can be used singly to determine a particular distance, as with surveyor's transit or a hand-held "range finder" for measuring dimensions. In addition, a set of many predetermined distance measurements also can be used to create three-dimensional (3D) models and images. From these 3D models, volumes can be calculated.

Simple straight-line measurements of a specific distance do not necessarily require expensive equipment. Hand-held range finders cost from a few hundred to a few thousand US dollars, depending on their capabilities. However, they have limited range and the measuring conditions need to be suitable; for instance, the target needs to be a good reflective surface.

Techniques and devices that can generate 3D models and calculate volumes range from hand-held laser-based range finders to scanning instruments mounted on tripods or vehicles. The range finders are relatively inexpensive and accurate, but they require many point-and-click measures to generate enough data to determine the shape of a pile. The scanning instruments are effective but expensive. Companies that use these instruments charge a fee for the service of determining pile dimensions and volumes for industries that deal with bulk materials including construction, mining, aggregates, and forestry.

Light sensing technologies can have other useful functions, in addition to measuring windrows and pile volumes. In particular, laser-based scanning systems are used to determine load volumes in trucks and containers (McEntee, 2019). Accurate volume measurements can complement or substitute for weight measurements from a weighbridge (i.e., truck scale). The volume scanning procedure works in a similar manner as a weighbridge. The scan from a loaded truck is compared to the reference scan from an empty truck to determine the loaded volume.

Sensor and measurement technology is still rapidly evolving. It is likely that new devices and technologies will make current technologies obsolete in a matter of years. Even smart phones can now be used with the appropriate application (Aggregates Business, 2017). Some have built-in Lidar capabilities and can generate 3D images. It is conceivable that smart phones may become the standard tool for measuring piles, with the right app (application).

4.2.2 Drones or unmanned aerial vehicles[2]

Measurement and image technology also can be sent skyward. UAVs with cameras, sensors, and automatic navigation can be used for rapid measurement of inventories at composting facilities. UAVs and associated image processing algorithms in many cases have displaced traditional methods using survey equipment for measuring stockpiles. With a modest investment and a moderate learning curve, UAV measurements can be done in-house by a facility employee. Regulations on UAV operations are evolving with the technology. In general, flights over people and near airports or other sensitive installations are either prohibited or require special permits. To operate UAVs for commercial purposes, operators should study the applicable regulations and may need to pass a test administered by an aviation agency (e.g., the FAA in the US). If costs or training requirements prove to be onerous, the task can be hired out to a company that specializes in UAV services.

There are many commercial software platforms that process UAV imagery into 3D models and allow for rapid analysis and reporting of the results. The most popular have easy to learn, intuitive browser interfaces to manage the process starting with planning the flight path, image upload and storage, processing, analysis, and reporting. A typical workflow involves the following four steps:

1. A flight plan for the UAV is configured with boundaries, altitude, image composition, and flight path. In most cases, this involves selecting the UAV from a list of popular models, superimposing a flight boundary onto an existing satellite image, and setting the flight altitude in the application. The application calculates a flight path with image acquisition points based on those preferences which the user can accept or modify (Fig. 14.5).
2. Prior to the flight, the operator downloads the flight plan to the UAV (flight path, altitude, and image acquisition points). During flight, the UAV follows the preset path using the Global Navigation Satellite System (GNSS) and collects images while the pilot monitors the process for safety (Fig. 14.6). Generally, the pilot only assumes flight control in case of problems.
3. Upon completion of the flight, images are uploaded to the cloud for processing into a model using photogrammetry algorithms. Photogrammetry uses trilateration[3] of features common to multiple images to build a 3D model of a surface by stitching together those same images. This approach is fundamentally different than other technologies such as Lidar, radar, and sonar which build 3D models from distances calculated from the time a signal takes to travel to and from an object (Fig. 14.7).

---

[2] This section was written by James Hardin with contributions from Dan Lilkas-Rain.
[3] Trilateration identifies a location by the confluence of three distance measurements. For an explanation of trilateration, see GIS Geography (2021).

FIGURE 14.5

A unmanned aerial vehicle (UAV) flight plans configures boundaries, altitude, image composition, and flight path.

FIGURE 14.6

From a preloaded flight plan site images are collected.

4. Once model processing has finished, measurements are made using the browser interface and reports are generated. To measure a stockpile volume the user simply draws a border around the stockpile and the volume is calculated from the elevation of the 3D model above the drawn border (Fig. 14.8).
Most platforms incorporate viewing of historical models to allow monitoring of changes over time.

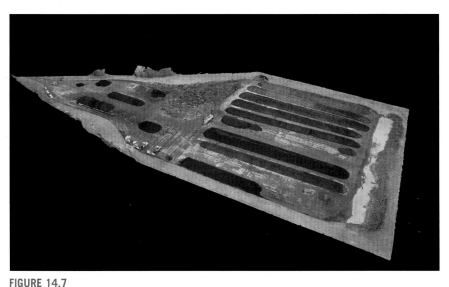

FIGURE 14.7

Photogrammetry uses trilateration of features common to multiple images to build a three-dimensional model.

Most of the underlying technology and mathematics is done by the software that users purchase, or purchase access to, with the UAV and its associated devices (e.g., camera). A composter does not need to understand trilateration or image processing algorithms to measure the sizes of piles. However, users do need to understand the features and capabilities of UAV technologies and how to access and manipulate them through on a computer.

UAVs costing from $1000 to $1500 are suitable for most composting installations. More expensive UAV systems with more sophisticated cameras, guidance, and flight characteristics are available, but are likely beyond the capabilities of non-professionals. Subscriptions to popular software platforms start at about $100/month, although there are free versions with limited support and functionality available. For example, the 3.2-hectare (8-acre) facility in the images above was surveyed using a DJI Phantom 4 UAV ($1200) and DroneDeploy software ($1200/yr). The flight took about 17 min at an altitude of 58 m (190 ft). The UAV camera acquired 261 images which were later uploaded to the web-based software for processing into the 3D model (Box 14.2). Absolute accuracy of volume measurements was not assessed, but ±5% is generally accepted. More sophisticated UAVs and software algorithms that reference imagery to precisely located ground control points can improve accuracy if it is needed.

FIGURE 14.8

The images from Figs 14.5—14.7 are used to estimate material volumes.

Box 14.2 Drone of Bethlehem

The UAV technology described in Section 4.2.2 was recently used to measure the volume of piles at the composting facility in the Town of Bethlehem, NY (Fig. 14.9). The images in Figs. 14.5—14.8 were generated by the UAV flyover in Bethlehem. On the day of the flyover, the facility contained several long piles of screened compost, unscreened leaf compost and piles of leaves that were still composting, plus a larger irregular-shaped pile of unprocessed yard trimmings (Fig. 14.10). To compare the UAV measurement technique, piles were also measured by traditional means, with measuring tape and a surveyor's wheel. Except for the unprocessed yard trimmings, piles were assumed to be parabolic in shape for the purpose of calculating volumes. The measurements, and the corresponding UAV results are shown in Table 14.3.

(A)

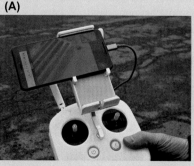

(B)

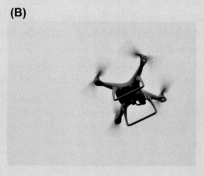

FIGURE 14.9

(A, B) Unmanned aerial vehicle (UAV) control panel and UAV in flight.

Box 14.2 Drone of Bethlehem—cont'd

FIGURE 14.10

UAV-taken photo of the Town of Bethlehem, NY, composting facility on the day of the UAV flyover. This is one of the 261 images used to build the three-dimensional model of the facility.

In this exercise, the volumes calculated by the UAV were lower than those estimated by manual measurements, but they were not terribly different. Because this was not a scientific study, it is uncertain which approach is more accurate. However, there are numerous factors in the manual measurement approach that can lead to inaccuracies. First, it is difficult to get accurate measurements of the heights and widths of large piles. Second, using a parabola to represent the shape of the piles is an assumption of convenience, and not necessarily accurate. For example, if the piles are assumed to be triangular instead of parabolic, the manually estimated volumes are 75% of those shown in Table 14.3. It would swing the comparison in the other direction; the UAV calculated volumes would be higher than the manually estimated volumes for the unscreened leaf compost and the composting leaves. Intuitively, it seems wise to bet on the drone.

Table 14.3 Comparison of manual and UAV measurements of pile volumes at the Town of Bethlehem, NY, composting facility.

| Pile type | Manual measurements[a] | | | Manually | UAV | |
	Est. Height	Est. Width	Combined length	Estimated volume[b]	Calculated volume	% Difference
Screened compost	4.3 m *14 ft*	12.2 m *40 ft*	39.7 m *130 ft*	1378 m³ *1798 yd³*	946 m³ *1236 yd³*	31% 31%
Unscreened leaf compost	2.7 m *9 ft*	9.8 m *32 ft*	52.2 m *171 ft*	932 m³ *1217 yd³*	777 m³ *1016 yd³*	17% 17%
Composting leaves	3.7 m *12 ft*	12.2 m *40 ft*	98.2 m *322 ft*	2925 m³ *3818 yd³*	2838 m³ *3710 yd³*	3% 3%
Yd. Trmgs. stockpile[c]				9000 m³ *11,750 yd³*	5599 m³ *7319 yd³*	38%

Notes:
[a] *Measurements were initially made in units of feet and converted to meters.*
[b] *Piles are assumed to be roughly parabolic in shape. Volume (V) was calculated from the estimated pile width (w), height (h), and the total length of similar piles (L) by the following formula:*

$$V = \tfrac{2}{3} \times w \times h \times L$$

[c] *The yard trimmings stockpile is an irregular shape. Its volume was visually estimated to be 11,500 to 12,000 cubic yards by two grinding professionals.*

5. Materials analysis

Most composters need to have feedstocks and compost products analyzed to determine their content and character, at least occasionally if not regularly. Feedstocks are analyzed primarily to develop recipes. Compost products are analyzed to determine their quality, value, and use. For process management, it is enlightening to analyze materials during the process.

Testing frequency is often tied to the size or output of the composting facility, as dictated by regulations or compost quality certification rules. For example, the USCC Seal of Testing Assurance (STA) program requires facilities that annually produce over 17,500 wet tons (imperial tons) must have compost samples analyzed monthly. Facilities that annually produce less than 6201 tons require analysis only once every three months and those in between must test every two months (USCC, 2021c). The Compost Quality Alliance voluntary program in Canada has similar sampling requirements—monthly samples if annual production exceeds 15,000 tonnes (t) and quarterly samples for facilities producing less than 5000 t (CQA, 2021). Such output-based testing obligations are common to many countries and quality-certification programs.

Table 14.4 lists the characteristics most often needing analysis. Although some data can be obtained from literature sources, it is usually necessary to conduct an analysis of an actual sample. Simple "in-house" testing can determine characteristics like moisture content, bulk density, and compost maturity. Only a professional laboratory (private or public) can provide a reliable analysis of nutrient concentrations, biological organisms, and chemical contaminants. In any case, a good representative sample is essential to the usefulness of the analysis.

The article, *Interpreting Compost Analyses* by Sullivan et al. (2018), provides a thorough yet concise summary of the analysis frequently performed for composting feedstocks and compost. Advice for selecting a professional laboratory is also included.

5.1 Professional laboratory analysis

Many of the important characteristics of feedstocks and compost products can only be ascertained through professional laboratory testing. Most feedstocks and compost products are highly variable. Values obtained from literature or the internet are rarely accurate enough to characterize a particular compost product or a particular feedstock from a particular source. In many situations, the analysis must be from a certified lab if compost products are required to meet quality standards, or to satisfy quality assurance programs (Chapter 15). Similarly, where applicable, regulations usually require analysis from independent professional and certified laboratories.

Professional laboratories possess the analytical equipment, technical skills, quality control, and knowledge of the appropriate methodologies. Many labs that serve composting facilities offer analytical "packages" that bundle useful tests together for

Table 14.4 Determining feedstock and compost characteristics for recipe development, process management, or compost quality and use.

Characteristic	Purpose	Source of information		
		Laboratory analysis	In-house testing	Literature
Moisture	Recipe Dev.	✔	✔✔	✔
	Process Mgt.		✔✔✔	
	Compost Qty.	✔✔	✔✔	
Total carbon	Recipe Dev.	✔✔		✔
	Compost Qty.	✔		
Total nitrogen	Recipe Dev.	✔✔		✔
	Compost Qty.	✔✔✔		✔
Bulk density	Recipe Dev.	✔	✔✔	✔
	Process Mgt.		✔✔✔	
	Compost Qty.	✔	✔✔	
Organic matter (or ash)	Recipe Dev.	✔	✔	✔
	Compost Qty.	✔✔	✔	✔
pH	Recipe Dev.	✔✔	✔	✔
	Process Mgt.		✔	
	Compost Qty.	✔✔✔	✔	✔
Pathogens (e.g., Fecal coliform)	Recipe Dev.			✔
	Process Mgt.	✔✔		
	Compost Qty.	✔✔✔		
Salinity (EC)	Recipe Dev.	✔✔	✔	✔
	Process Mgt.		✔	
	Compost Qty.	✔✔✔	✔	✔
Heavy metals	Recipe Dev.	✔		✔
	Compost Qty.	✔✔		✔
Herbicide	Recipe Dev.	✔	✔	
	Compost Qty.	✔	✔✔	
Specific N (Ammonia, nitrate)	Recipe Dev.	✔		✔
	Compost Qty.	✔✔✔		✔
Other elements (P, K, Ca, Mg, S)	Recipe Dev.	✔		✔
	Process Mgt.	✔		✔
	Compost Qty.	✔✔✔		✔
Inert contaminants (Plastic, glass)	Recipe Dev.	✔	✔	✔
	Process Mgt.			
	Compost Qty.	✔✔	✔	

✔✔✔ = *Necessary;* ✔✔ = *Recommended;* ✔ = *Useful or possible.*

one price. The number of labs certified to perform compost-specific analysis has increased since the 1990s, due to the availability of the testing standards specifically for compost and composting materials, and the associated quality control programs for laboratories.

The US Composting Council's "Test Methods for the Examination of Composting and Compost" (TMECC) specifies standard methods for nearly all analysis required for composting (USCC, 2012). It has been pivotal in creating consistent and predictable test results. Before TMECC, different labs would use a variety of methods for compost products meaning that the test results could vary widely even when testing the same product. TMECC has also been instrumental in expanding testing for the industry in North America, in part because the standards are compiled together in a single document (Duprey, 2019). While the TMECC is influential internationally, most countries and regional authorities follow testing procedures endorsed by standards organizations close to home, and/or they adopt international standards.

5.2 "In-house" testing

A few feedstock and compost characteristics can be determined "in-house" using simple procedures that require basic laboratory skills and inexpensive equipment. These characteristics include bulk density, moisture or solids content, pH, and soluble salts. They are essential qualities to know for process control, feedstock recipes, and compost product quality. A procedure for determining bulk density is described in Chapter 4 because it is relevant to recipe development. Methods to determine moisture content, soluble salts, and pH are explained in Chapter 11.

In general, accepted procedures for self-testing composting feedstocks and compost are thoroughly described in the TMECC. If feedstocks or compost are to be tested in-house, even occasionally, the operation should own a reference copy of the TMECC, or its equivalent outside of the US. At a minimum, a good scale is required, one that is able to read numbers which are at least 100th the size of the sample (for example, 1 g for a 100-gram sample or 1/8 ounce for a 1-pound sample). Scales that can read to 0.1 g are generally sufficient for most situations. Other equipment required depends on the specific test.

The in-house tests described in this book are not hazardous, but a few ordinary safety precautions need to be observed. When handling feedstocks, and compost, it is prudent to routinely wear a mask and disposable gloves, the type of gloves commonly used in food service and medical facilities. Insulated gloves should be worn when hot containers are handled. Safety glasses or goggles should also be used. Work areas should be well-vented. Observe appropriate equipment precautions. For example, do not use metal containers in a microwave oven and do not leave a microwave oven unattended while samples are being heated. After handling any organic material or soil, wash hands thoroughly, even if gloves were worn.

In-house testing can economically provide valuable information for internal management decisions and quality control. However, the results are generally not acceptable for quality certification programs or regulatory compliance.

5.3 Sample collection

Whether tests are conducted in-house or by a laboratory, collecting a good representative sample is an essential first step, followed by properly handling, preserving, and shipping of the sample. The techniques, frequency, sample sizes, and diligence required in sample collection depend on the materials being tested, the specific analytical tests being conducted, and on the purpose of the analysis. For instance, documentation of the "Chain of Custody" is important for samples taken to satisfy regulations, operating permits, and compost quality assurance programs (Fig. 14.11). In most cases, composters, regulators, certifiers, and compost users are interested in the average or overall characteristics of feedstocks or compost products. In this case, composite samples should be obtained.

Convenient tools for sample collection include a soil sampling probe (core), soil auger, shovel, spade, trowel, plastic bucket, sample bags with labels, permanent marker, meter/yard stick, a plastic tarp (about 2 m or 6 ft square), vinyl or nitrile gloves, a clip board for note taking, and most convenient, a wheel loader or skid steer with a bucket (Campbell, 2014a). Soil probes and augers are for taking core samples, augers being used for materials that are harder to penetrate. Shovels are for digging to the sampling point and withdrawing samples. Loader buckets are for exposing sections and cross-sections for withdrawing interior samples. Before sample collection begins, all tools and containers must be clean. In addition, particular analytical tests are inherently more sensitive to sample conditions and therefore demand more diligence in the choice of procedure, tools, and materials used in sample collection, and the handling of samples afterward. Examples include tests for biological organisms (e.g., pathogens), heavy metals, and foreign organic compounds (Box 14.3).

A good rule is to consult with and follow the recommendations of the laboratory conducting the analysis. Many laboratories provide customers with written procedures for collecting and handling samples.

Some of the most comprehensive sample collection guidelines can be found in the TMECC, which contains detailed protocol for gathering and handling samples (USCC, 2012). TMECC's section 02.01, *Field Sampling of Compost Materials*, is available via Appendix N. In addition, CREF has produced a pair of video programs and a written "companion guide" about sampling finished compost (CREF, 2020). The former Recycling Organics Unit (ROU) in Australia also created videos about sample collection in their CompostTV series (Campbell, 2012, 2014a). The preceding resources focus on sampling of compost products. The Vermont Department of Environmental Conservation has a fact sheet for sampling composting feedstocks, developed by the former Highfields Center for Composting (2015).

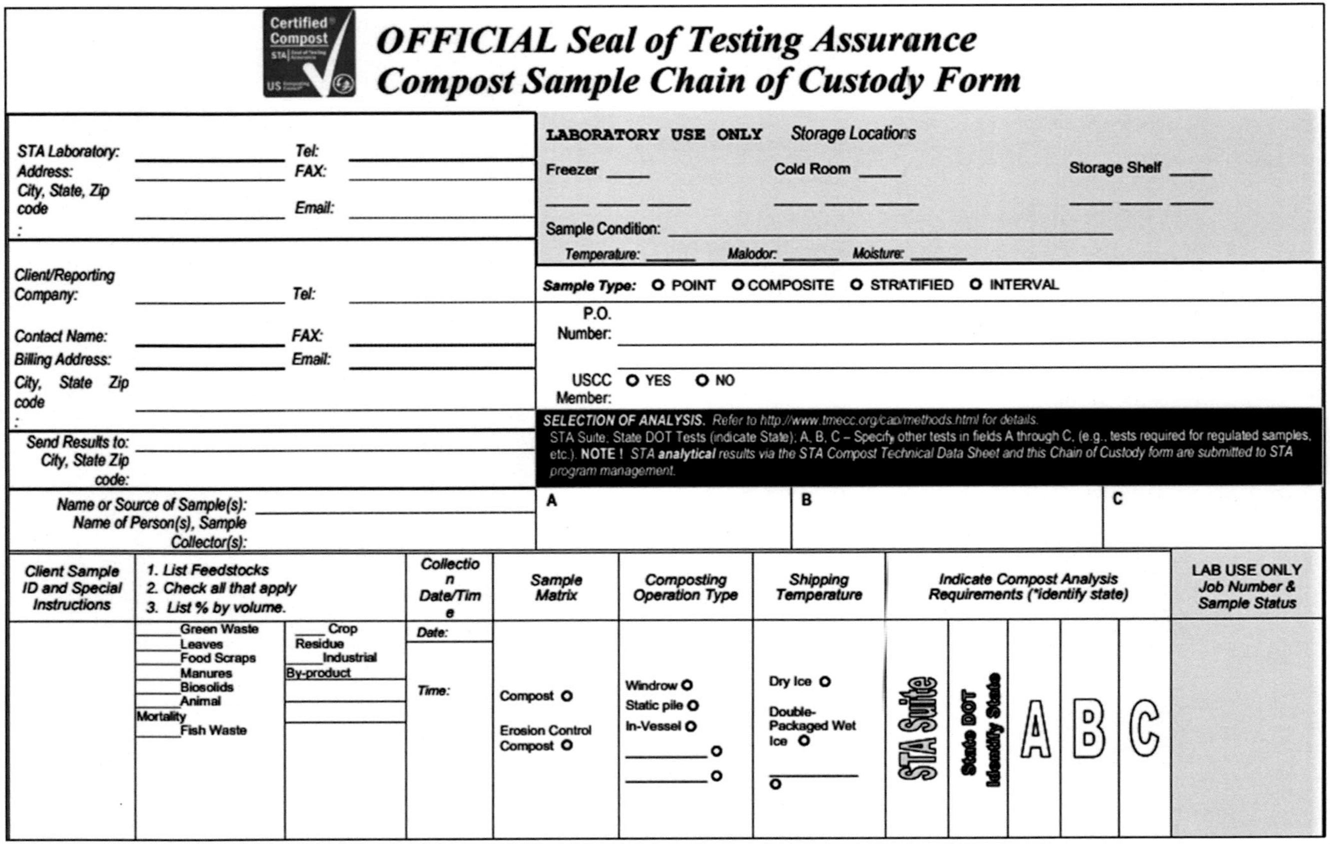

OFFICIAL Seal of Testing Assurance
Compost Sample Chain of Custody Form

STA Laboratory: ______ Tel: ______
Address: ______ FAX: ______
City, State, Zip code: ______ Email: ______

LABORATORY USE ONLY Storage Locations
Freezer ____ Cold Room ____ Storage Shelf ____
Sample Condition: ______
Temperature: ____ Malodor: ____ Moisture: ____

Client/Reporting Company: ______ Tel: ______
Contact Name: ______ FAX: ______
Billing Address: ______ Email: ______
City, State Zip code: ______

Sample Type: O POINT O COMPOSITE O STRATIFIED O INTERVAL
P.O. Number: ______
USCC Member: O YES O NO

Send Results to: City, State Zip code: ______

SELECTION OF ANALYSIS. Refer to http://www.tmecc.org/cap/methods.html for details.
STA Suite, State DOT Tests (indicate State); A, B, C – Specify other tests in fields A through C, (e.g., tests required for regulated samples, etc.). NOTE ! STA analytical results via the STA Compost Technical Data Sheet and this Chain of Custody form are submitted to STA program management.

Name or Source of Sample(s): Name of Person(s), Sample Collector(s): ______

A | B | C

Client Sample ID and Special Instructions	1. List Feedstocks 2. Check all that apply 3. List % by volume.	Collection Date/Time	Sample Matrix	Composting Operation Type	Shipping Temperature	Indicate Compost Analysis Requirements (*identify state)	LAB USE ONLY Job Number & Sample Status
	___ Green Waste ___ Leaves ___ Food Scraps ___ Manures ___ Biosolids ___ Animal Mortality ___ Fish Waste ___ Crop Residue ___ Industrial By-product	Date: Time:	Compost O Erosion Control Compost O	Windrow O Static pile O In-Vessel O ______ O ______ O	Dry Ice O Double-Packaged Wet Ice O ______ O	STA Suite | State DOT Identify State | A | B | C	

FIGURE 14.11

Example of a Chain of Custody document. This particular form is used for the USCC STA program.

Source: USCC, 2021d.

> **Box 14.3 Sample collection tips**
>
> The following tips are drawn from recommendations made by CREF (2020) and Campbell (2012):
>
> - Fill out laboratory sample forms completely, especially chain of custody forms. Dates are particularly important.
> - Prior to sampling, consult with the laboratory to determine the proper sample size to send. The sample volume required by the laboratory depends on the suite of tests to be performed. Three to four liters (about one gallon) is common. However, a laboratory may want three to four times the normal amount if they are testing for inert contamination.
> - Sample from both sides of a long pile or windrow and all around a square or circular pile.
> - The proper size of spot/grab samples depends on the particle size of the material. Normally a one liter or one quart volume is sufficient. For materials with large particles, larger samples may be required. Regulations or product standards may specify sample volume sizes.
> - If samples will be tested for *trace metals*, use plastic tools for removing samples. Scrape away materials from surfaces that are initially cut with loader buckets or metal shovels. Do not use galvanized tools.
> - If samples will be tested for certain *organic contaminants* (e.g., volatile organic compounds, pesticides), avoid plastic tools, buckets, and containers. Use thoroughly cleaned and rinsed (with deionized or distilled water) stainless steel shovels or trowels when removing samples. Place and ship samples in glass containers, amber colored if necessary. Rely on the laboratory for advice.
> - If samples will be tested for *pathogens or other microorganisms*, tools should be cleaned and sanitized before sample collection. Samples should be placed in a cold container immediately. Refrigerate, but do not freeze.

5.3.1 Collecting a composite sample

Composting feedstocks, and compost products, require a good composite sample—a blend of numerous random spot or grab subsamples taken from many locations in a pile or bin. When collecting and mixing a composite sample, it is worth remembering that the lab will use only a small fraction of the composite sample sent to them, generally 1 to 20 g for each test (Fig. 14.12). Therefore, a representative analysis requires a representative sample, obtained by thoroughly mixing many subsamples and then properly drawing the final sample(s) for the laboratory.

Proper sample collection is especially important for accurately characterizing a feedstock. Most organic materials are notoriously heterogeneous; that is, they are inconsistent in their texture and composition. Furthermore, coarse-textured materials tend to stratify, with the fines and moisture migrating toward the base of piles. A random sample from a pile of bedded livestock manure could yield a shovel full of mostly sawdust or a shovel full of mostly manure. The same situation could occur with a pile of mixed grass, leaves, and brush. Whether or not large pieces of feedstock should be picked out of the composite sample depends on the test being conducted. If unsure, consult the laboratory for advice. In some cases, the lab discards larger pieces from the sample before testing; in other cases, all sample material is finely ground up at the lab.

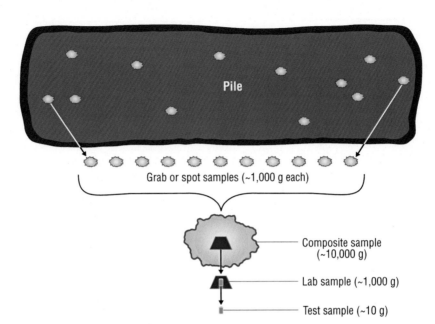

Grab or spot samples (~1,000 g each)

Composite sample
(~10,000 g)

Lab sample (~1,000 g)

Test sample (~10 g)

FIGURE 14.12

Portrait of the challenge of obtaining a representative sample for lab testing.

Adapted from Siebert, S., Jüngling, M.A., 2015.

The number of spot/grab sampling points depends on the size and configuration of the pile or windrow. As a general recommendation, material should be collected from *at least* five locations around the pile or windrow and from three depths at each location. Separate composite samples should be collected from different windrows or piles.

Determining a truly random spot sampling location takes more than a random effort, given the natural bias toward picking an easily accessible spot. In simple situations, or with a homogeneous pile of material, operators can indiscriminately pick locations that include diverse sections of the pile or windrow (throwing a lawn dart at the pile might be an entertaining tactic). In doing so, it should be recognized that the lower layers of a freestanding pile contain a much larger proportion of the material than the upper sections. Thus, a proportionally larger number of grab samples should be obtained from the base.

In general, and especially if the pile to be sampled is inconsistent, a random sampling procedure should be used to select sampling locations. There are several techniques for pinpointing random sampling locations (USEPA, 2002). One practical approach graphically overlays a grid onto a scaled sketch of the pile or windrow to be sampled. The grid is divided into cells that are numbered in sequence

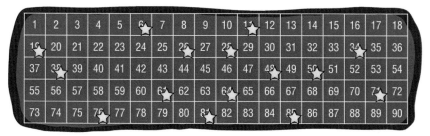

FIGURE 14.13

Numbered grid method of selecting random spot sample locations.

Source: Northern Tilth, LLC.

(Fig. 14.13). The locations to be sampled are identified by a set of random numbers, which can be generated from computer spreadsheets or programs available through the internet. Once the pile location is determined, most recommendations suggest taking three to five spot samples at different depths at the selected location. The numbered-grid approach can also be applied to identifying spot sample locations across the cross-section of a pile.

A challenge in collecting representative samples is accessing the interior of a pile or container. A bucket loader can be used to dig into the depths of the pile to sample the interior. The bucket loader can also take random cuts. Random samples can be drawn from the material removed by the loader. It is even more effective if the loader bucket draws random samples and the loader mixes these together (although composite subsamples should still be drawn in case the loader does not completely mix the samples.). An even better option is to collect random samples from conveyors or the discharge point of mixers. Windrows are best sampled after turning with a windrow turner.

Box 14.4 presents an example procedure for collecting a composite sample liberally based on recommendations from both the TMECC section 2.01 and The Pennsylvania State University Agricultural Analytical Services Laboratory (PSU, 2021). This procedure is presented to illustrate the steps and principals involved in gathering a composite sample. It is not a substitute for a site-specific sampling plan.

5.3.2 Sample handling

As a general rule, samples should be placed in a closed container and stored in a cool dry location, isolated from sunlight and moisture, and analyzed as soon as possible (e.g., within five days). This rule is sufficient in most situations, including analyzing feedstocks for recipe development and usually for analyzing compost for nutrient content, pH, salinity, and most other quality parameters. An exception to this rule

Box 14.4 Example procedure for collecting a composite sample

1. Tools needed:
   - a clean auger, soil probe, slotted tube, shovel, or spade,
   - a garden trowel or similar "scoop,"
   - a clean 16-liter or 20-liter bucket (4-gallon or 5-gallon)
   - a clean plastic tarp for mixing samples,
   - sample containers, 1 to 2 L or quarts (heavy plastic zip-loc bags, or wide mouth plastic bottles),
   - a permanent waterproof marker or printed moisture–resistant label,
   - disposable gloves (to protect hands from compost and compost from hand contamination) and/or a hand shove.

2. Using the marker or printed labels, label the sample containers that will be sent to the lab. Label the containers with the collection date, source, and type of material as well as contact information. Alternatively, the label can be a name or code that traces the date, source, and material.

3. Identify the sampling locations. Normally, at least five locations should be selected, and samples should be taken at three depths at each location. More sampling locations may be necessary for piles with diverse and randomly distributed materials (e.g., different batches of compost or feedstocks).

4. Proceed to the first sample collection location and collect approximately 1 L (1 quart) of material from upper section of the pile or windrow, but at least 20 cm (8 in.) beneath the surface. Place the samples in the clean bucket. Collect a second sample of the same size midway to the core of the pile and place that sample in the bucket. Collect the third sample of the same size near the core of the windrow. Mix the contents of the bucket.

5. Repeat this process at each of the sample collection locations in the pile, adding the samples to the same bucket.

6. After completing sample collection at all sampling locations, thoroughly mix the material in the bucket being careful to avoid stratification of the different particle sizes. Depending on the consistency and volume of the sampled material, mixing can be improved by dumping the bucket contents on a plastic tarp and mixing thoroughly to ensure homogeneity prior to taking a composite sample for analysis.

7. Collect your composite sample from the mixed material. The sample size needed by the lab, is typically 1 to 2 L (1–2 quarts), although 4-L (1-gallon) samples are also common. Ask the lab.

8. The mixed composite sample is usually substantially larger than the sample needed by the lab. In this case, the "cone and quarter" technique can be used to reduce the composite sample to lab size (Fig. 14.14). On the tarp, push up the mixed composite sample into a conical pile. Divide the conical pile into four equal parts and separate them into four small piles. If the separated piles are approximately lab size, select one to send to the lab. Retain at least a one of the other piles as a backup in case the lab needs a second sample. If each of the four separated piles are still substantially larger than lab size, discard two of the piles and then mix and recombine the remaining two into another conical pile on the tarp. Again, divide that pile into quarters. Repeat this divide-and-remix step until the quartered piles are slightly larger than the amount of sample material needed.

9. Place the lab sample in a suitable clean container for shipping. Label the container to identify the sample. Complete the compost sample analysis submission form, and the chain-of-custody form, if required. Pack the lab sample and form(s) in a container with frozen gel packs. Call the lab to let them know the when the sample will be shipped. Send the packaged sample to the lab by overnight delivery service.

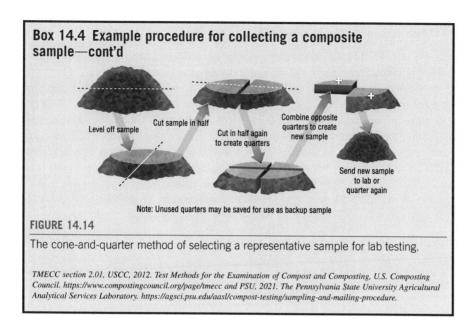

Box 14.4 Example procedure for collecting a composite sample—cont'd

Level off sample

Cut sample in half

Cut in half again to create quarters

Combine opposite quarters to create new sample

Send new sample to lab or quarter again

Note: Unused quarters may be saved for use as backup sample

FIGURE 14.14

The cone-and-quarter method of selecting a representative sample for lab testing.

TMECC section 2.01, USCC, 2012. Test Methods for the Examination of Compost and Composting, U.S. Composting Council. https://www.compostingcouncil.org/page/tmecc and PSU, 2021. The Pennsylvania State University Agricultural Analytical Services Laboratory. https://agsci.psu.edu/aasl/compost-testing/sampling-and-mailing-procedure.

is pathogen analyses (e.g., fecal coliform, salmonella), for which the sample needs to delivered to the lab within 24 h of sampling. Although compost and composting feedstocks are still biologically active, their physical composition and chemical characteristics do not change appreciably in the relatively short amount of time between sample gathering and analysis.

Some situations, however, call for more meticulous handling of samples. *Ideally*, samples should be stored between 0 and 4°C (32−39°F) and shipped to the laboratory by over-night delivery service in an insulated container packed with frozen gel packs. Regulations or quality control programs may prescribe specific and uniform sample handling conditions that lean toward this ideal. Also, some analytical parameters can change appreciably in a short amount of time, especially if handling conditions are less than ideal. Some of these are potential chemical contaminants that may be volatile or present in such low concentrations that even a small change becomes significant (Chapter 4). Some parameters are biological, such as tests for pathogens or other organisms. Test for specific pathogens, such as fecal coliform or *salmonella* sp., require samples to be continually kept near (but not below) freezing and analyzed promptly. Laboratory recommendations and regulatory specifications should be followed in all of these cases. Again, prior to sampling, it is prudent to call the laboratory that will perform the analysis to confirm proper containers, preservation techniques, and sample holding times.

6. Managing with the weather/seasons

Seasonal and weather variations often call for operational adjustments that compensate for, or take advantage of, the changing conditions (Table 14.5). Certain elements of the weather have benefits, under normal circumstances. Solar radiation helps to dry piles when they are too moist. The wind is also drying, and it contributes to passive aeration and the dilution of odors. Most importantly, moderate precipitation adds water that replaces moisture driven off by the composting process. However, when composters have to manage for the weather, it is more often in a negative mode. In dry climates, the concerns are lack of rain and hot temperatures that drive-off precious moisture. In wet climates, the primary problems are the added moisture from rain and snow and the lack of opportunity to dry out piles. In cold

Table 14.5 Season and Weather adjustments.

Weather condition	Possible course of action
Moderate rain	• Turn during or just after the rain • Flatten top of piles or create a furrow at the top
Heavy rain	• Turn piles after the rain unless the pad cannot be worked • Cover windrows/piles with compost fabric • Shape windrows/piles to shed precipitation • Clean out sediment basins and traps • Inspect or install filter berms and socks, replace and repair as needed
Occasional sun and/or favorable wind	• Turn wet windrows/piles. • Turn when wind is brisk and blowing away from sensitive neighbors • Add water as needed
Extended periods of dry weather (e.g., dry season)	• Reduce turning frequency • Reduce rate of forced aeration, if possible • Add water as needed • Apply water to roads and site surfaces to reduce dust
Extended period of wet weather (e.g., rainy season)	• Combine windrows/piles or otherwise increase their size • Cover windrows with compost fabric
Winter	• Combine windrows/piles or otherwise increase their size • Reduce turning frequency • Cover windrows with compost fabric • Add biocover to piles and windrows • Follow recommended cold-temperature worker-safety practices • Curtail or postpone site activities if and as needed
Extremely hot weather	• Follow recommended hot-temperature worker-safety practices. • Curtail site activities as needed

climates, subfreezing winter temperatures and precipitation are the main issues. Although extremely cold weather rarely stops the composting process, it does tend to retard it and can interfere with site activities (Lynch and Cherry, 1996). The lower temperatures reduce microbial activity, at least near the surface of the pile/windrow. The amount of heat generated decreases, worsening the situation. Small piles could freeze, halting composting temporarily.

With large-enough piles, composting can continue year-round, even in harshly cold climates, but it does not *have to* continue. For some composters, one possible adjustment is to give in to challenges of extreme weather. Operations can be suspended or slowed for the winter, or during the dry season, or even during extended bouts of bad weather. Since sales of compost can also be seasonal, an interruption in the operation may not harm the enterprise. This strategy is, however, at the mercy of the generation, delivery, and storage qualities of feedstocks.

6.1 Exposure to weather extremes

Managing the operation in response to weather and climate is essentially a matter of exposure (of the composting piles). To shield the operation from negative effects of the weather, adjustments can be made to reduce exposure. Examples include enlarging and insulating piles, reducing turnings, and operating under covers or inside a building. The opposite steps are taken to take advantage of favorable effects of weather, including making smaller or flatter piles and turning more often. Harsh conditions due to wind, rain, snow, and temperature extremes also present safety concerns for employees that also need to be addressed (Chapter 13).

In all climates, weather is primarily a concern to turned windrow composters, and generally outdoor operations managing small piles (<1.25 m or 4 ft). Aerated static pile and in-vessel operations are much less affected by the external environment. Weather affects large piles less than it does small piles. Large piles retain more heat and moisture. Therefore, one way to buffer the impacts of weather is to increase the pile size. In cold climates, at the onset of winter, it can be an advantage to build larger piles or to combine small piles into larger ones. A reasonable *minimum* winter-sized pile for cold climates is in the range of 1.5−2 m (about 4−6 ft). Still, piles that are started during the winter can still heat slowly and may not sustain pathogen-killing conditions until the spring or summer (McCartney and Eftoda, 2011). Piles that are at the tail end of active composting should be larger as less heat is being generated. Increasing pile size also can reduce moisture loss that comes with hot dry weather. It is worth noting that large piles are more likely to develop anaerobic sections that could be odorous when exposed.

Turning further opens the pile contents to the elements of weather, encouraging greater heat and moisture loss. During cold weather, and winter generally, reducing the frequency of turning, or halting it altogether, helps to cope with the severe cold. Similarly, more moisture is retained by limiting turnings during hot and dry weather.

The trade-off for the weather-buffering advantages of both large piles and less-frequent turning is reduced aeration and process control. Managers must decide which factor inhibits the process more—weather or aeration. In cold climates, at least, larger piles and/or reduced turning usually win the day (or the season).

It is possible to shield the operation, or key parts of it, from the weather even without altering pile size or the process. One approach is to add insulating layer of compost to the surface of the piles. Unscreened compost, wood chips, screen "overs" or other dry, coarse feedstocks can also be used. The drier the material is, the better it is as an insulator; the coarser the material is, the less it interferes with aeration. The thicker that the layer is, the better it insulates. The strategy works well for either passively aerated or forced aerated systems but only if the piles are not turned during the time that the insulation is needed. Otherwise, the insulating layer will need to be reapplied. When the piles are eventually turned or moved, the insulating material will be incorporated into the compost. A layer of compost on the surface of piles has the additional advantage of reducing odor and volatile gas emissions.

Another option is to use a removable macroporous fabric cover (Chapter 6). A number of flexible covers are available that exclude some of the effects of the weather from windrows, forced-aerated piles, and curing and storage piles. These covers range from selectively porous nonwoven fabrics to reinforced plastic tarps. The primary functions of these covers are to keep out precipitation. They also moderate the effects of wind, ambient temperature, and slow or eliminate moisture loss from the pile, depending on the type of cover used. Fig. 14.15 shows macroporous covers used to reduce the effects of cold weather and precipitation during the winter. In climates with long winters, turning and active windrow management usually are suspended until the spring because the covers freeze against the ground and are very difficult to remove. Synthetic covers can be used for both passively aerated and forced-aerated piles. In the case of forced aeration, the covers are usually microporous covers, an integral component of the composting system.

FIGURE 14.15

Fleece covers to prevent precipitation from saturating windrows. These covers pictured here will likely stay in place until the ground thaws.

FIGURE 14.16

Open roofed structure provides partial protection from precipitation for mature compost piles but otherwise interferes only slightly with site activities. Roofs on buildings should have drain gutters and drainpipes to collect and channel clean precipitation away from processing areas. The water collected from the roof can be handled as clean runoff, not leachate or process water.

Source: Jane Gilbert.

Perhaps the most effective strategy for coping with the weather is to isolate the operation from the weather altogether. One way to do so is to move the operation indoors or under cover (Fig. 14.16). Short of that, the operation, or part of it, can be covered by a building that is otherwise affected by the environment outside. Composting facilities are increasingly using simple unheated fabric-roofed buildings, often open at the ends. These partial enclosures exclude rain and snow, shelter the operation from wind, and moderate the temperature in times of freezing or hot conditions. However, a building also excludes the favorable effects of weather, including solar radiation, which helps to dry a pile when it is too moist; wind, which contributes to passive aeration; and especially beneficial, moisture-replacing rain. Ryndin and Tuuguu (2015) describe several composters that have adapted their operations to cold winter conditions.

6.2 Precipitation

Rain is welcome when it comes in moderation and the piles are dry or can otherwise use the added water. Water falling from the sky is the easiest way to moisten piles and dampen dust on the site. Composters can take better advantage of a favorable rainfall by manipulating the piles to absorb the water more effectively. For instance, moisture-starved piles can be turned during a rain shower, assuming the equipment allows for safe operation in the rain. In addition, piles can be reshaped or spread out to intercept more the rainwater (see below).

Heavy rain and snow are always unwelcome and can lead to a muddy, messy site. There is little that a manager can do in response other than rely on a well-designed, well-drained site. Where the rain is frequent, a paved composting site may be necessary. Rain can leave puddles of water on the site long after it is gone. The site should be graded as soon as possible after rain to eliminate the puddles. The puddles can become sources of odor. The puddles are not only a potential odor source but they are also unsightly and an easy target for site inspectors and unfriendly neighbors. Snow melts from the surfaces of actively composting piles and the water is absorbed within. However, the snow needs to be plowed from the path of equipment if operations are to continue.

Even moderate rain and snow are unwelcome when piles are already moist, and the season is cool. The unwanted water hinders the process, makes screening difficult, and makes soggy compost products. When rainwater falls on a composting pile, it tends to follow the contour of the piles, running down the sides, and gradually seeping through the surface. Some of it makes its way to the foot of the pile where it either runs off onto the site or gets absorbed at the base. The base can become saturated, at least temporarily. How much water seeps into the pile versus runs off depends in part on the feedstocks and their stage of decomposition. For example, fresh deciduous leaves tend to shed more water than shredded yard trimmings. The leaves shed less water after they are partially composted.

The shape of the pile is another factor. A pile with a wide, flat top intercepts a larger proportion of the rainwater before the rain has the chance to run down the sloping sides. The pile with a trough-shaped top does this even better (Fig. 14.17). Therefore, when it is advantageous to retain as much rainwater as possible, the pile can be shaped in such a manner in anticipation of an approaching storm. Windrows can be "leveled" by pushing the tops into the aisle space between the windrows. While this approach requires labor and equipment, it may still be the best option for thoroughly wetting dry windrows and piles.

The rain that is absorbed by a pile is not evenly distributed. After a light to moderate rain, the absorbed water tends to remain several inches below the surface while some of the runoff is soaked up at the foot of the pile. Ideally, piles should be turned soon after rain to disperse the water throughout. Also, if the opposite occurs, if the rain saturates the pile, the pile should also be turned to drive off as much of the excess water as possible.

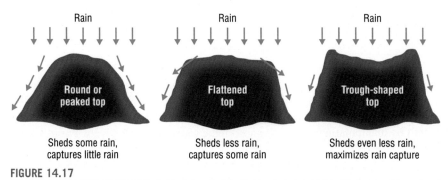

FIGURE 14.17

A pile with a wide flat top or a trough-shaped top retains more precipitation than a pile with a rounded or peaked top.

Compost that is approaching stability does not generate enough heat to drive off moisture added by precipitation. Nor does it require an abundant supply of oxygen. Therefore, storage piles of compost can be covered with plastic tarps to exclude the rain and snow melt. The tarp provides added protection against weed seeds. The plastic should be reinforced to increase the strength and reduce tearing. Because finished compost requires some oxygen, the tarp can lead to the accumulation of intermediate compounds of decomposition, some of which may be phytotoxic. For this reason, it is always wise to turn or otherwise "air out" compost before it is used.

7. Preventing and managing fires

A fire can happen anywhere large quantities of organic materials are handled, including compost- and mulch-producing facilities. The organic materials not only fuel the fire, they can also be the genesis of it.

At a composting facility, fire can start from careless disposal of cigarette butts, intense heat from equipment, sparks from welding, lightning strikes, wildfires, arson, and spontaneous combustion (SC). As the survey summarized in Table 14.6 attests, SC is the most frequent cause of fires at composting facilities; SC is the internal generation and accumulation of heat that leads to fire-instigating temperatures. It differs from the other causes of fires in that it does not require an external spark or heat source. It usually begins and grows unseen deep within a pile. Prevention and control of SC fires are different from that of the surface fires.

Table 14.6 Reported causes of fires at composting and mulch facilities from a survey of facilities known to have experienced fires. SC dominates the collective responses.[a]

Suspected cause of firm	Certain	Probable	Possible	Total
Spontaneous combustion (SC)	19	7	1	27
Cigarette	1	1	1	3
Spark from welding	0	0	0	0
Spark from equipment	0	0	0	0
Heat from equipment	1	0	0	1
Arson	1	1	1	3
Wildfire	0	0	0	0
Lightning	0	0	0	0
Other: "hot in-bound load"	1	0	0	1

[a] *The survey respondents were asked to identify the suspected cause(s) of the fire and whether the cause(s) was "certain," "probable," or "possible." If not certain, a respondent could provide more than one possible cause (Rynk and Buggeln, 2008)*

7.1 Surface fires

Fires caused by cigarettes, equipment, welding sparks, wildfires, and lightning can be considered "surface" fires because they start at an exterior location and spread along surfaces where oxygen is abundant. Prevention of surface fires enlist common fire-prevention practices, including:

- no-smoking or restricted-smoking policies, with receptacles for discarded cigarettes,
- proper maintenance of equipment,
- dust control,
- prudent welding and cutting procedures ("hot works"),
- clearing dry brush and wood debris from the site and its surroundings, and
- installation and upkeep of fire extinguishers.

Surface fires, regardless of cause, tend to be managed in the same way. They are contained and extinguished with conventional water, foam or chemicals. A good review of fire prevention and safety for surface fires, and SC, is the publication *Reducing Fire Risk at Waste Management Sites* from the Waste Industry Safety and Health Forum in the UK (WISH, 2017).

Equipment fires occur on equipment commonly found at composting facilities, including bucket loaders, grinders, and screens (Fig. 14.18). Composting activities generate a good deal of dust which is a common factor in equipment fires. The dust has several roles in the start of fire. First, it is dry, light, and easily combustible.

FIGURE 14.18

Smoke from a smoldering fire beneath the surface of pile. With time, the fire will migrate toward the surface where there is enough oxygen to support flames.

Source: M. Cotton.

It is the fuel that first ignites in an equipment fire. Second, the dust blankets equipment surfaces, forming a dry insulating layer that prevents heat from escaping hot components like mufflers and engine manifolds. As a result, the temperature of these components can build up to the temperature high enough to combust the dust. Third, the dust can be source of friction that leads to excess heat from mechanical parts. Prevention of equipment fires includes daily cleaning of the equipment components with high pressure air to remove dust, dried leaves, and anything else that might burn. Regular maintenance of equipment, including lubrication, reduces the heat generated by operating equipment. Every piece of equipment should have at least one multipurpose (dry chemical) fire extinguisher, mounted in an easily reachable location (Figure 13.14 in Chapter 13), This fire extinguisher should be a dry powder extinguisher, equivalent to a US Type ABC, European Type ABCF/D, or Australian Type ABCE. Depending on the vehicle size, it should have a capacity of 4–9 kg or 10–20 lbs.

Other possible causes of surface fires at composting facilities are sparks from welding and cutting, so-called "hot work" activities. Hot work encompasses any operation that produces sparks or a flame that can ignite combustible materials. Welding and cutting should only be conducted by employees or contractors well-trained in welding and welding safety (Chapter 13). Welding and cutting should be done in a work area suitable for hot work. If not, a hot works permit, or management program should be implemented (OSHA, 2021). A hot works management program relies on a trained responsible person to inspect, approve, and sign off on the hot work location in advance of the work. The program also should include a standard procedure for preparing the area and performing the work, preferably with the safety checklist (Box 14.5). The program should be implemented every time welding, cutting, or other hot work is performed outside of an approved area.

Box 14.5 Hot works site preparation

In addition to following established hot works procedures, the Canadian Center for Occupational Health and Safety offers the following recommendations for preparing the work area (CCOHS, 2017).

- Make sure that all equipment is in good operating order before work starts.
- Make sure that all appropriate personal protective devices are available at the site, and each worker has been trained on how to use, clean, and store them properly.
- Inspect the work area thoroughly before starting. Look for combustible materials in structures (partitions, walls, ceilings).
- Move all flammable and combustible materials away from the work area.
- If combustibles cannot be moved, cover them with fire resistant blankets or shields. Protect gas lines and equipment from falling sparks, hot materials, and objects.

Continued

Box 14.5 Hot works site preparation—cont'd

- Sweep clean any combustible materials on floors around the work zone. Combustible floors must be kept wet with water or covered with fire resistant blankets or damp sand.
- Use water *only* if electrical circuits that have been de-energized to prevent electrical shock.
- Remove any spilled grease, oil, or other combustible liquid.
- Vacuum away combustible debris from inside ventilation or other service duct openings to prevent ignition. Seal any cracks in ducts. Prevent sparks from entering into the duct work. Cover duct openings with a fire-resistant barrier and inspect the ducts after work has concluded.
- Make sure that appropriate fire extinguishers (e.g., ABC fire extinguishers) are available and easily accessible.
- Make sure that the first-aid boxes are available and easily accessible.
- Block off cracks between floorboards, along baseboards and walls, and under door openings, with a fire-resistant material. Close doors and windows.
- Cover wall or ceiling surfaces with a fire resistant and heat insulating material to prevent ignition and accumulation of heat.
- Secure, isolate, and vent pressurized vessels, piping and equipment as needed before beginning hot work.
- Inspect the area following work to ensure that wall surfaces, studs, wires, or dirt have not heated up.
- Post a trained fire watcher (i.e., "spotter") within the work area, including lower levels if sparks or slag may fall during welding, including during breaks, and for at least 60 min after work has stopped. Depending on the work done, the area may need to be monitored for longer (up to three or more hours) after the end of the hot work until fire hazards no longer exist.
- Eliminate explosive atmospheres (e.g., vapors or combustible dust) or do not allow hot work. Shut down any process that produces combustible atmospheres, and continuously monitor the area for accumulation of combustible gases before, during, and after hot work.
- If possible, schedule hot work during shutdown periods.
- Comply with the required legislation and standards applicable to your workplace.

7.2 Spontaneous combustion

A sure way to create a SC fire is to follow this recipe:

1. Create a large pile of biologically active materials (e.g., over 5 m or 16 ft high),
2. Fill the pile with materials that are relatively dry but not too dry (20%−40% moisture content), or allow a section of a moist pile to dry below 40%, or allow a section of a dry pile to remain or become moist, and,
3. Leave the pile undisturbed for a few months, and then open the pile, or wait for a brisk wind.

Given the foregoing recipe, it should not be surprising that SC is the leading cause of composting-related fires. Composting facilities hold large piles of heat-generating semimoist combustible materials that are often left undisturbed for months. These conditions are ideal for the start of an SC fire (Box 14.6). The way to avoid an SC fire is to eliminate at least one of the foregoing conditions.

Box 14.6 Spontaneous combustion explained

SC happens for one reason—more heat is generated within a pile than lost to the surroundings. Hence, the temperature rises, and rises, until it reaches a level that ignites the surrounding material. No external energy source or spark is needed.

The heat generated within the pile is the result of a chain reaction of several heat-generating biological and chemical processes (Buggeln and Rynk, 2002). At the start, the temperature rise is largely biological, due to the heat released by decomposition and possibly the respiration of surviving plant cells (biological oxidation). The biological heating can carry the temperature up to 80°C (176°F). At this point, the biological organisms die and stop producing more heat. However, a variety of chemical processes, including chemical oxidation and pyrolysis, are waiting to pick up the cause. The chemical reactions bridge the gap between biological self-heating and ignition point of the materials, roughly 200−300°C (400−570°F). The rising temperature is self-sustaining. As the temperature increases, it accelerates the rate of biological and chemical reactions. Assuming nothing interferes to increase heat loss, the temperature eventually becomes hot enough to the burn material in the pile.

Oxygen is a well-known fire factor, but oxygen is not necessary for SC. If ample oxygen is available, the hot material burns quickly and spreads fast, as a flaming fire. If oxygen is scarce, the hot material smolders and slowly spreads until it reaches an ample supply of oxygen. In a composting pile, the origin of smoldering fire is often a relatively small nucleus of material where the conditions are most favorable for the fateful temperature rise. From that nucleus, lacking sufficient oxygen, a smoldering fire often advances slowly, and in one or more narrow veins. The surrounding material is hardly affected. A vein of smoldering material may go undetected until the pile is opened, or the vein approaches the pile surface, and then it yields smoke or even flames.

Moisture plays a complex and conflicting role in the SC story. In essence, a small to moderate amount of moisture (20%−40% moisture content) favors SC while a plentiful amount of water suppresses it. On one hand, some amount of moisture is necessary for SC to occur because the initial biological activity requires water. This is why hay must be very dry before it is put in the barn. Without water, biological heating does not start (SC can start without the biology if there is a chemical or external heat source). Water also stimulates SC because it catalyzes chemical oxidation, and it transfers heat when and where it condenses. On the other hand, ample moisture is a huge factor in suppressing an SC fire. The primary means of heat loss in a composting pile is evaporation. When abundant water is available, the temperature rise is kept in check. As long as some water is present in the local environment, the temperature cannot rise above 212°F (100°C). The fire does not start until all of the free water is gone. In addition, water further discourages SC because it adds thermal mass and also increases heat loss via conduction.

7.2.1 A closer look at the spontaneous combustion recipe

The reason SC occurs in *large piles* is that large piles generate more heat than they lose. It does not help that it is difficult to monitor the conditions within a large pile. The pile can easily harbor the beginnings of a fire without it being recognized, even by an operator who diligently monitors temperatures. There is no critical size that guarantees that SC will or will not occur. If the amount of heat generated overruns the heat loss, a fire-starting temperature can develop even in relatively small piles (e.g., less than 2 m or 6 ft). Limiting the pile to some maximum height reduces the risk of SC, but it does not eliminate the possibility. In addition, there are other risk factors for SC, especially the amount of time that a pile remains undisturbed. Piles that are turned or moved with a month's time can be larger than piles that remain static for a longer period.

Some fire-prevention guidelines suggest that piles of feedstocks, active compost, and mulch, should be lower than 3 to 8 m (10–25 ft) and less than 10 to 45 m wide (30–150 ft). The range is wide because different jurisdictions, and different levels of government, impose different conditions (Benton, 2012). Some state, provincial, and local jurisdictions impose strict height and width limits for composting facilities, especially where a local community has experienced a damaging fire. In other jurisdictions, there are no regulations regarding pile dimensions.

Many jurisdictions follow the guidelines of the International Fire Code (IFC). The current version of the IFC (2018) sets maximum pile dimensions to 7.6 m (25 ft) in height, 45.7 m (150 ft) in width, and 76.2 m (250 ft) in length (ICC, 2017). These dimensions are rather large and liberal, in part because they also cover mulch producers, who tend to steadily remove or relocate the mulch piles. The IFC allows even larger piles if a facility has an approved fire prevention plan. However, the IFC does not reference the time that piles remain static. Guidance from WISH (2017) suggests a maximum period of three months for shredded combustible materials.

SC occurs in *dry-to-moderate moisture* conditions because there is enough water to sustain heat-generating biological activity but not enough water to evaporate away the heat that is generated. This critical moisture level applies to any part of a pile. Variations in moisture content within the pile create localized conditions that are well-suited to SC, even though the overall condition of the pile is not. The worst case is moderately moist pocket of material surrounded by dry material. The situation can occur as the exterior sections of a moist pile dry over time but one of more interior sections remain somewhat moist.

Leaving the pile undisturbed gives a smoldering fire time to get established and travel (Fig. 14.18). When the pile is opened, by turning for instance, the accumulated heat rapidly dissipates and interrupts the SC cycle. However, the hot areas of the pile need to cool completely, and also should be watered. If a hot spot is uncovered and then subsequently covered over again, the SC process resumes.

Any of the following conditions can hasten the development of an SC fire or encourage its growth:

- Feedstocks that decompose quickly and produce a great deal of heat over short time (e.g., grass clippings). These feedstocks give the SC heating cycle a quick start.
- Food and vegetative feedstocks that are abundant in oils (e.g., flaxseed, cottonseed). Oily compounds readily oxidize and generate heat.
- Channeling of forced aeration. When air flows through a pile in channels, it removes heat from sections exposed the air flow but not the pockets that the air flows around. The situation can allow the accumulation of heat in those pockets. When the smoldering fire grows toward the air channels it can become a flaming fire.

- Compaction of piles. Some composters believe that fires are prevented by compacting large piles (e.g., driving equipment on them) The flawed logic surmises that compaction drives out the oxygen-carrying air and a fire will not start without oxygen. As Box 14.6 explains, an abundance of oxygen is not necessary for an SC fire to start. In fact, evidence suggests that SC may be more likely in compacted piles (Rynk and Buggeln, 2009; Heller, 2011).

7.2.2 Spontaneous combustion monitoring and diligence

SC can occur with nearly any composting feedstock and at any stage in the production system—from feedstock storage to piles of finished compost. All other things being equal, raw feedstocks are more likely to develop a fire than finished compost simply because they generate more heat. However, the pile size, moisture, and agitation are equalizing factors. Moderately dry and degradable feedstocks are prime candidates for SC but only if they are stored in large stationary piles. Woodchips, which are slowly degradable, develop fires because they are moist and stored for months in large piles, storage bins, or silos. For the same reasons, curing piles and compost storage piles can also develop fires, despite the reduced heat generation.

Monitoring for SC fires should include regular temperature measurements for process control and routine inspections of piles for signs of fire. Large piles (greater than 4 m or 12 ft tall) merit more monitoring than small piles. Temperature measurements made for the purpose of process control should be examined for signs of potential fire. For instance, if a temperature reading above 170°F (77°C) is encountered, the hot section of the pile can be opened up for inspection. Although the chances of finding hotspots improve with additional temperature measurements, it is difficult to detect the presence of a developing fire, or even a smoldering fire, in a composting pile. The materials in a composting pile are very good insulators. The thermometer may read normal composting temperatures within a 15 cm (6 in.) of a vein of smoldering material (Buggeln and Rynk, 2002). Therefore, fire monitoring also requires diligence in looking for signs of fire. One such sign is a steam vent in a pile where the hottest air is escaping. Wet patches on the pile surface also indicate hotter sections. Steam vents and wet patches are good locations to probe with a long-stemmed thermometer. Other signs include charred or ash-colored sections of the pile, a burning or smoke smell, and smoke itself. The smoldering activity can reduce the internal sections to ash, leaving a void that creates an engulfment hazard. Operators and fire fighters should avoid standing on or driving over suspicious sections of piles.

New technical tools for detecting hot spots are emerging. Thermal photography, in particular, holds promise (Büyüksönmez, 2012). Drone technology may prove to be an important tool in detecting a nascent fire (Coker, 2016).

There are several situations in which a smoldering fire is more likely to grow into a flaming fire and merit more SC scrutiny on the composter's part. Composting operators often report that flames or smoke occur after a brief, nonsoaking, rain and at the onset of windy weather (Rynk and Buggeln, 2008). For instance, in the Western US, Chinook winds have been reported as the "cause" of a fire. However, neither the rain nor the wind causes the fire. Rather they intensify an embryonic fire that already exists in the pile. In the case of wind, it is the deeper penetration of oxygen. In the case of rain, the intensifying factor is more complicated. Luangwilai et al. (2012) found that a moderate amount of rain stimulates both biological and chemical oxidation in moderately dry piles. Small and large amounts of rain do not.

7.2.3 Fighting a spontaneous combustion fire

Once an SC fire takes hold, it can only be extinguished by removing the hot material from the pile and dousing it with water after it is removed. A bucket loader, excavator, or similar equipment has to be part of the solution. The burning sections of the pile must be persistently removed, spread out, and watered until all hot spots are removed. The equipment operator must work the pile from ground level because it is too dangerous to drive or walk on the pile. The fire may have already scoured out large void spaces that can cause the pile to collapse.

Conventional firefighting techniques do not work with an SC fire. Simply hosing a pile with water does not extinguish the fire. The material in the pile soaks up much of the applied water, preventing it from reaching the burning sections. Furthermore, by the time the fire shows itself, a tremendous amount of heat is stored within the pile. An impractical amount of water is required to adequately suppress the heat unless the water is confined to the hot material. Smothering a burning pile with a cover of soil or sand also is not a solution, at least not a permanent solution. While the cover material may quell the flames and smoke in short term, the fire continues to smolder beneath the cover. It can remain in that state for a long time, or it may burn through to the surface and revert into flaming fire.

7.3 Fire readiness

Most composting facilities experience a fire at some point, so it is prudent to accept the likely reality that a fire will eventually occur and prepare for it.

An early step should be a meeting with the local fire department to discuss what should happen if a fire develops at the facility, especially from SC. Once the fire department is called to a fire, the fire-fighters are in charge. They may or may not consult facility personnel for advice and information. Most fire departments are not familiar with the materials occupying a composting facility and how they behave in a fire. Often, a fire department's first reaction is to pull out the hoses and soak the fire with water. As mentioned above, this reaction is the wrong one for an SC fire in a pile of compost or mulch. A meeting with the fire department in advance of fire is an opportunity to provide them with information about fighting composting fires and impart an understanding of what facility operators can do to help (e.g., operate loaders to remove burning material from the pile). Some fire departments resent it as an intrusion but most welcome it. It is worth the effort in any case.

Facilities should be prepared to deal with fires by having appropriate fire extinguishers in place, a designated source of water and equipment to apply water to the fire. A fire preparation kit is a good idea, such as the one shown in Figure 13.15 in Chapter 13. However, the kit and its contents should be maintained intact and in good working order. For instance, if tools for opening hydrants are part of the fire preparation kit, those tools should not be "borrowed" for other purposes because they may not be returned to the kit. Facility staff should be trained about who to contact and what to do if a fire occurs. Each employee should have predetermined assignments (e.g., move equipment, contact authorities, operate the water tank truck). A fire of any sort is a prime candidate for contingency planning (Box 14.1).

Depending on the applicable fire codes, automatic sprinkler systems may be required in buildings. The sprinklers can help control surface fires, but they do little to extinguish a fire from SC. Nevertheless, sprinklers protect the building and its contents and can prevent even an SC fire from spreading.

The design of the composting site should also consider the potential implications and impacts of the fire. To start with, site should be large enough to handle incoming feedstocks without exceeding the fire-safe height of piles. Otherwise, the operation may be forced to live with large piles. Enough empty space should be available at any time to spread out the pile if an SC fire occurs. The space between windrows and piles is often sufficient for temporarily storing the material removed from a burning pile. The site should have adequate access for firefighting equipment, at least around the perimeter. Many jurisdictions conservatively require windrows and piles to be separated by a distance large enough for fire-fighting trucks to pass.[5] There should be access to an adequate supply of water plus plans for the drainage of the abundant water that might be applied during the fire. If that water runs off the site and into a stream, environmental authorities could impose fines, making a fire that much more of a disaster.

8. Preventing and managing nuisance conditions

Every composting operation is obligated to manage nuisances for the sake of employees, visitors, and neighbors. Nuisances include dust, noise, litter, traffic, plus flies, mosquitoes, birds, and other pests. A good defense against nuisances, generally, is a clean, well-landscaped site. Trees and shrubs serve as visual barriers, noise buffers, and wind breaks. These features have a tangible positive effect on nuisance conditions and a psychological impact as well.

Dust control is primarily about moisture. Dust becomes airborne, when it becomes dry. Practices to reduce dust include spraying water on dry feedstocks, wetting down dirt roads and working surfaces (Fig. 14.19), misting dry material on conveyors especially as it comes off screens and grinders, misting windrows while turning and maintaining adequate pile moisture. Dust can be further suppressed by installing windbreaks and hoods over conveyors. Controlling dust also controls the transport of bioaerosols.

---

[5] As stated previously, applying water directly to a windrow or pile during an SC fire is ineffective. Fire vehicles rarely need to drive up to burning windrows/piles.

FIGURE 14.19

Dust control is largely a matter of keeping surfaces moist, such as regular watering of unpaved working surfaces during dry weather.

Noise is a nuisance as well as a worker-protection issue (Chapter 13). In both cases, the impact of noise depends on its loudness and duration. As a nuisance, noise also depends on timing. Noise restrictions are rarely imposed on farm composting operations because farm and composting activities are similarly noisy. However commercial composting facilities are sometimes restricted to certain hours of operation by regulatory agencies, local codes, or by agreements with neighbors. Those hours tend to be in the 8 a.m. to 5 p.m. range, give or take an hour or two.

The sources of noise at composting operations are equipment—grinders, turners, loaders, generators, trucks, fans, compressors, etc. Grinders are the loudest machines. The level of noise, or its impact, from grinders can be reduced with noise barriers such as concrete blocks, enclosed wooden fences, trees, and other vegetative barriers. If it is essential to operate grinders at a low noise level, electric motors can be used as a power source instead of diesel engines. Excessive vibration amplifies noise, so the condition of equipment is an important noise factor. Backup alarms on vehicles are a common noise complaint. Because backup alarms are required safety features, there is little that can be done to eliminate this annoying source of noise. The sound volume of the alarms should be set as low as safely allowed. Otherwise, restricting operation to certain times of the day and establishing sound barriers around the facility are the only other options. Conceivably, alarms can be replaced with another type of warning device, such as a strobe light, but safety authorities are rarely willing to consider alternatives to backup alarms.

Traffic is a large concern of the community. The composting operation might increase traffic in the neighborhood, largely due to trucks transporting feedstocks and finished compost. On a farm, the composting operation might result in manure spreaders and trucks traveling on public roads. Careful selection of travel patterns and efficient site design minimize the impact of increased traffic on the roads adjacent to the facility. If traffic is a concern on a farm, it is best to use farm roads to connect the composting site with the barns or whatever locations the feedstocks are generated. If public streets are used, care should be taken to avoid spilling feedstocks and finished compost, as it is with the transport of any farm material. A common traffic-related complaint is the presence of trucks waiting on or along the side of the access road. The entrance to the site should permit the trucks to enter the site if they need to queue up or stop to make adjustments (e.g., remove the tarp). In general, all truck drivers, employed by the composting operation and others, should be made aware of the community's sensitivity to traffic. They should be well-informed, without ambiguity, about the allowable hours for deliveries and what practices are unacceptable and what practices are preferred.

Litter comes to a composting operation with the feedstocks, some more than others. Yard trimmings and food scraps can carry a bothersome amount of paper and plastic. Feedstocks that originate on-site are much less of a concern. Litter seems to accumulate as the feedstocks decompose into compost. The litter easily find its way to all corners of the site, blown about as feedstocks are unloaded, windrows are turned, and compost is screened. A chain-link fence surrounding the site greatly helps keep the litter on-site. The best way to manage litter is to frequently collect it from the site and then securely store and dispose of it. Paper items can be returned to the composting system, but plastics and metals have to be discarded or recycled. All composting sites should have one or more trash receptacles of adequate size and accessibility. In most cases, a small trash dumpster is sufficient.

Litter collection starts when feedstocks are delivered. Plastic and metal items should be pulled out at that point, as many as possible. Feedstock should not be shredded without first removing as much plastic as possible. The shredded plastic greatly increases the work in managing litter and contamination of the compost. After piles are formed, and as they continue to compost, plastics and metals should be routinely picked from the surface of the pile (large items tend to migrate to the surface). The screening area can be a critical point for litter control as a large amount of trash is uncovered by the screen. Much of the litter is in the overs pile but small pieces may also accompany the compost. Again, inorganic items should be picked off at this point. Trash can blow away and become litter when screens and conveyors toss the material into the air. Therefore, wind and litter screens around the screening area can help control litter (and dust).

Flies, mosquitos, birds, rodents, and other animal scavengers can find a composting site inviting because there is food and habitat for them. Not surprisingly critters are most attracted to those composting operations that handle food scraps or animal carcasses. The key to controlling most animal nuisances is containment of the food within the windrow, pile or vessel. An active compost pile is good at disguising the potential food within. Plus, the hot conditions inside the pile are inhospitable. Problems with critters occur when their food escapes the confines of a composting pile or container. Feedstocks with food value should be placed in the compost pile as soon as possible, if not immediately. In vessel and membrane-enclosed composting methods have the advantage of the physical barrier to exclude animal pests. Nevertheless, feedstocks still need to be handled properly.

Flies are attracted to the food and habitat offered by food scraps, dead animals, manure, and other rotting materials. Their interest is not confined to human food. Again, the primary way to control flies is to contain these attractive feedstocks within the pile. However, flies can find food and breed even in the drier but also cooler outer layer of an active composting pile of manure. During fly breeding season, flies can be controlled by turning the pile before larva hatch. In general, development time for most fly species of concern falls within the range of 5–7 days. Thus, complete fly control requires turning every five days. Such a frequent cycle is a challenge for many composting operations, especially those that rely on static piles. An alternative to turning is to blanket piles with a layer of finished compost or another material that does not attract flies (e.g., woodchips).

Seagulls, crows, and other birds are often a nuisance problem at composting facilities that handle food scraps (Fig. 14.20). Prompt handling of food-containing feedstocks and containment of the feedstocks within the pile are the main control measure. Areas on the site where food has been stored should be thoroughly cleaned and preferably paved for more effective cleaning. If these steps do not solve the problem, bird control services can be hired to establish a bird control program. Local landfills also may have tips to share.

Mosquitos are a nuisance and a potential health issue for employees, neighbors, and visitors. Chapter 13 discusses control measures in relation to mosquito-borne diseases like West Nile virus.

FIGURE 14.20

Seagulls are among the birds that regularly reside at composting facilities that handle food-containing feedstocks. Crows and vultures often join the party.

Source: Cary Oshins.

References

Cited references

Aggregates Business, 2017. New App to Measure Stockpiles Using iPhone. https://www.aggbusiness.com/ab1/news/new-app-measure-stockpiles-using-iphone.

Benton, J., April 22, 2012. Measuring Mulch Piles; Fire Codes Vary Significantly across US — Spontaneous Combustible Fires A Continuing Problem. EHS Safety News America. https://ehssafetynewsamerica.com/tag/international-code-council/.

Brown, S., Kruger, C., Subler, S., 2008. Greenhouse gas balance for composting operations. J. Environ. Qual. 37 (4), 1396—1410.

Brown, S., 2016. Greenhouse gas accounting for landfill diversion of food scraps and yard waste. Compost Sci. Util. 24 (1), 11—19. https://doi.org/10.1080/1065657X.2015.1026005.

BSI, 2018. Specification for Composted Materials. PAS 100:2018. British Standards Institution, London. http://bsigroup.com.

Buggeln, R., Rynk, R., 2002. Self-heating in yard trimmings: conditions leading to spontaneous combustion. Compost Sci. Util. 10 (2), 162—182. https://doi.org/10.1080/1065657X.2002.10702076.

Büyüksönmez, F., 2012. Full-scale VOC emissions from green and food waste windrow composting. Compost Sci. Util. 20, 57—62.

Campbell, A., 2012. Video: Compost Sampling_AS4454 Method B. Recycled Organics Unit (ROU), CompostTV. https://www.youtube.com/watch?v=Tqo-qK2xnm8.

Campbell, A., 2014a. Video: Compost Sampling AS4454 Method A. Recycled Organics Unit (ROU), CompostTV. https://www.youtube.com/watch?v=-G8RlcHyyC0.

Campbell, A., 2014b. Video: Ground Survey: How to Estimate the Volume of a Compost Pile. Recycled Organics Unit (ROU), CompostTV. https://www.youtube.com/watch?v=Tqo-qK2xnm8.

CCOHS, 2017. Canadian Centre for Occupational Health and Safety. http://www.ccohs.ca/oshanswers/safety_haz/welding/hotwork.html.

CCME, 2005. Guidelines for Compost Quality. Canadian Council of Ministers of the Environment, Winnipeg, Manitoba.

Coker, C., 2016. Using drones to measure compost piles. BioCycle 57 (6), 17.

CQA, 2021. The Compost Quality Alliance. Compost Council of Canada. http://www.compost.org/compost-quality-alliance/.

CREF, 2020. Compost Sampling Videos and Reference Guide. Compost Research & Education Foundation. https://www.compostfoundation.org/Education/Sampling-Videos.

Duprey, C., 2019. Comparison of compost laboratory results. BioCycle. https://www.biocycle.net/comparison-compost-laboratory-results/.

GIS Geography, 2021. How GPS Receivers Work — Trilateration vs Triangulation. https://gisgeography.com/trilateration-triangulation-gps/.

Heller, P.J., July/August 2011. Spontaneous Combustion. Hot Issue for Compost, Mulch Producers. Soil & Mulch Producers News. https://issuu.com/downingandassociates/docs/s_mp_jul-aug__11_final.

Highfields Center for Composting, 2015. Feedstock Sampling. Vermont Agency of Natural Resources, Department of Environmental Conservation, Solid Waste Program. https://dec.vermont.gov/sites/dec/files/wmp/SolidWaste/Documents/ANR%20Feedstock%20Sampling.pdf.

ICC, 2017. 2018 International Fire Code. International Code Council, Inc. https://codes.iccsafe.org/content/IFC2018

Keener, H.M., Wicks, M.H., Skora, J.A., 2008a. Measurement of volumes at Ohio class IV composting facilities. BioCycle 49 (6), 38. https://www.biocycle.net/measurement-of-volumes-at-ohio-class-iv-composting-facilities/.

Keener, H.M., Wicks, M.H., Skora, J.A., 2008b. Measuring material volumes at composting sites. BioCycle 49 (7), 42. https://www.biocycle.net/measuring-material-volumes-at-composting-sites/.

Linzer, R., Mostbauer, P., 2005. Composting and its impact on climate change with regard to process engineering and compost application-A case study in Vienna. In: Sardinia 2005 Tenth International Waste Management and Landfill Symposium, pp. 59–60.

Luangwilai, T., Sidhu, H.S., Nelson, M.I., 2012. Understanding the role of moisture in the self-heating process of compost piles. In: CHEMECA 2012: Australasian Chemical Engineering Conference. Engineers Australia, Australia, pp. 1–13.

Lynch, N.J., Cherry, R.S., 1996. Winter composting using the passively aerated windrow system. Compost Sci. Util. 4 (3), 44–52. https://doi.org/10.1080/1065657X.1996.10701839.

McCartney, D., Eftoda, G., 2011. Windrow composting of municipal biosolids in a cold climate. J. Environ. Eng. Sci. 4. https://doi.org/10.1139/s04-068.

McEntee, K., 2019. Improve load measurements using laser scanning technology. Soil Mulch Prod. News XIII (2). https://issuu.com/downingandassociates/docs/s_mp_mar-apr__19_final.

OSHA, 2021. Welding, Cutting and Brazing. U.S. Dept. of Labor Occupational Safety & Health Administration. https://www.osha.gov/welding-cutting-brazing.

PSU, 2021. The Pennsylvania State University Agricultural Analytical Services Laboratory (2021). In: https://agsci.psu.edu/aasl/compost-testing/sampling-and-mailing-procedure.

Ryndin, R., Tuuguu, E., 2015. Composting Manual for Cold Climate Countries. ACF (Action Contre le Faim), Ulaanbaatar, Mongolia. http://www.susana.org/en/resources/library/details/2389. (Accessed 29 March 2017).

Rynk, R., Buggeln, R., 2008. Composting and Mulch Fire Causes and Conditions (Unpublished survey of composting and mulch producers).

Rynk, R., Buggeln, R., March 2009. Fires in Mulch Piles — Advice and Experience From The Industry. Findings of a Preliminary Survey. Amerimulch Newsletter.

Siebert, S., Jüngling, M.A., 2015. Composting and Quality Assurance in Germany. Bundesgütegemeinschaft Kompost e.V. (BGK). Kättesaadav: http://www.kompost.de/uploads/media/Compost_Course_gesamt_01.pdf (viimati külastatud February 25, 2015).

Siebert, S., Auweele, W.V., 2018. European Quality Assurance Scheme for Compost and Digestate. European Compost Network. https://www.compostnetwork.info/ecn-qas/ecn-qas-manual/.

Stockpile Reports, September 26, 2019. Stockpile Measurement Methods that Work. https://www.stockpilereports.com/stockpile-measurement-methods-that-work/.

Sullivan, D., Bary, A., Miller, R., Brewer, L., 2018. Interpreting Compost Analyses. Oregon State University Extension Service. https://catalog.extension.oregonstate.edu/em9217.

UC Davis, 2021. Science & Climate Definitions. University of California at Davis. In: https://climatechange.ucdavis.edu/climate-change-definitions/biogenic-carbon/#:~:text=What%20is%20Biogenic%20Carbon%3F,processing%20of%20biologically%20based%20materials.

USCC, 2012. Test Methods for the Examination of Compost and Composting. U.S. Composting Council. https://www.compostingcouncil.org/page/tmecc.

USCC, 2021a. Get Mentored. U.S. Composting Council. USCC, 2021b. Combatting Climate Change, U.S. Composting Council. https://www.compostingcouncil.org/general/custom.asp?page=MentorYoungProfessionals.

USCC, 2021b. Welcome to the USCC's Professional Certification Program. U.S. Composting Council. https://certificationsuscc.org/.

USCC, 2021c. Seal of Testing Assurance (STA). U.S. Composting Council. https://www.compostingcouncil.org/general/custom.asp?page=SealofTestingAssuranceSTA.

USCC, 2021d. Seal of Testing Assurance Program Chain of Custody (COC) —Guide and Form. U.S. Compostiong Council. https://cdn.ymaws.com/www.compostingcouncil.org/resource/resmgr/images/Chain-of-Custody-Form.pdf.

USCC, 2021e. Combatting Climate Change. U.S. Composting Council. https://www.compostingcouncil.org/general/custom.asp?page=ClimateChangeBenefits.

USEPA, 2002. Guidance on Choosing a Sampling Design for Environmental Data Collection. U.S. Environmental Protection Agency. https://www.epa.gov/quality/guidance-choosing-sampling-design-environmental-data-collection-use-developing-quality.

WISH, 2017. Reducing Fire Risk at Waste Management Sites. Waste Industry Safety and Health Forum. https://www.360environmental.co.uk/documents/Revised%20WISH%20Mar%202017.pdf.

Consulted and suggested references

A & L Canada Laboratories, 2005. Compost Management Program. Compost Analysis for Available Nutrients and Soil Suitability Criteria and Evaluation. https://www.alcanada.com/pdf/technical/compost/Compost_Handbook.pdf.

Abdellah, Y., Li, C., 2020. Livestock manure composting in cold regions: challenges and solutions. Agriculture (Pol'nohospodárstvo) 66 (1), 1—14. https://doi.org/10.2478/agri-2020-0001.

AgroLab, 2021. Compost Analysis — STA Chain of Custody Form. https://www.agrolab.us/pdfs/chain_of_custody.pdf.

Al-Alawi, M., Szegi, T., Simon, B., Gulyás, M., 2019. Investigate biotransformation of green waste during composting by aerated static windrow with GORE (R) cover membrane technology. Jord. J. Eng. Chem. Ind. (JJECI) 2, 32—40. https://doi.org/10.48103/jjeci252019.

BC Ministry of the Environment, 2016. Summary of General Composting Best Management Practices. British Columbia Ministry of the Environment. https://www2.gov.bc.ca/assets/gov/environment/waste-management/organic-waste/compost-best-practice-info-notice.pdf.

Bilsen-Brolis, L., 2019. Community Composting Done Right: A Guide to Best Management Practices. Institute for Local Self-Reliance. https://ilsr.org/composting-bmp-guide/.

BioCycle, 2003. Preventing fires in grinding equipment. BioCycle 44 (11), 41.

Brinton, W., Bonhotal, J., Fiesinger, T., 2012. Compost sampling for nutrient and quality parameters: variability of sampler, timing and pile depth. Compost Sci. Util. 20 (3), 141—149. https://doi.org/10.1080/1065657X.2012.10737039.

Brooksbank, K., 2018. Composting to Avoid Methane Production. Department of Primary Industries and Regional Development, Government of Western Australia. https://www.agric.wa.gov.au/climate-change/composting-avoid-methane-production.

Brown, S., Cotton, M., Messner, S., Berry, F., Norem, D., 2009. Methane Avoidance from Composting. Climate Action Reserve. http://faculty.washington.edu/slb/docs/CCAR_Composting_issue_paper.pdf.

Burke, R., 2003. The phenomenon of spontaneous combustion. Firehouse.

Büyüksönmez, F., Evans, J., 2007. Biogenic emissions from green waste and comparison to the emissions resulting from composting Part II: volatile organic compounds (VOCs). Compost Sci. Util. 15, 191—199.

Büyüksönmez, F., Rynk, R., Yucel, A., Cotton, M., 2012. Mitigation of odor causing emissions—bench-scale investigation. J. Air Waste Manag. Assoc. 62 (12), 1423—1430. https://doi.org/10.1080/10962247.2012.716808.

CCC, 2016. Best Practices for Operating an Aerated Windrow Composting Facility. The Compost Council of Canada (for the Government of Manitoba Conservation and Water Stewardship). http://www.compost.org/.

Chou, C.-H., Büyüksönmez, F., 2006. Biogenic emissions from green waste and comparison to the emissions resulting from composting (Part 1: Ammonia). Compost Sci. Util. 14 (1), 16—22. https://doi.org/10.1080/1065657X.2006.10702258.

Clinton, C., 2011. Prevent Grinder Fires. Grinderinfor.com. https://earthsaverequipment.wordpress.com/2011/05/05/prevent-grinder-fires/.

Coker, C., 2016. Sampling plans for organics recycling facilities. BioCycle. https://www.biocycle.net/sampling-plans-organics-recycling-facilities/.

Coker, C., 2019. Managing compost and mulch fires. BioCycle. https://www.biocycle.net/managing-compost-mulch-fires/.

Coker, C., 2021. Using drones to measure compost piles. BioCycle. https://www.biocycle.net/using-drones-to-measure-compost-piles/.

Cook, K.L., Ritchey, E.L., Loughrin, J.H., Haley, M., Sistani, K.R., Bolster, C.H., 2015. Effect of Turning Frequency and Season on Composting Materials From Swine High-Rise Facilities. Publications from USDA-ARS/UNL Faculty, p. 1498. https://digitalcommons.unl.edu/usdaarsfacpub/1498.

Fey, M., Choate, J., Murphy, E., 2018. A Guide to Collecting Soil Samples. Oregon State University Extension Service. https://catalog.extension.oregonstate.edu/ec628.

Grainger, 2021. Grainger Catalog. W.W. Grainger, Inc. https://www.grainger.com/content/general-catalog.

Hao, X., Stanford, K.I.M., Xu, S., Larney, F.J., McAllister, T.A., 2010. Cattle mortality disposal via composting in a cold climate: feasibility, greenhouse gas emissions, and land application of the end product. In: Lethbridge, A.B. (Ed.), 47th Annual Alberta Soil Science Workshop, February 16—18, 2010. Canada, p. 19 (Workshop).

McSweeney, J., 2019. Community-Scale Composting Systems. Chelsea Green Publishing, White River Junction, VT.

NFPA, 2019. Standard for Fire Prevention During Welding, Cutting, and Other Hot Work, National Fire Protection Association. www.nfpa.org/codes-and-standards/all-codes-and-standards/list-of-codes-and-standards/detail?code=51BNova Scotia Agriculture. How to take a compost sample. https://novascotia.ca/agri/documents/lab-services/analytical-lab-manure-compost-sample.pdf.

Nutrient Advantage. Compost or Manure Sampling Procedure, 2021. Incitec Pivot Fertilisers. https://www.nutrientadvantage.com.au/~/media/Files/NA/Sampling/Sampling%20ProcedureCompostManure.pdf.

Oshins, C., 2020. Compost Operators Training Course Manual. Compost Research and Education Foundation.

Paré, M., Paulitz, T., Stewart, K.A., 2000. Composting of crucifer wastes using geotextile covers. Compost Sci. Util. 8, 36—45.

Parkinson, R., Gibbs, P., Burchett, S., Misselbrook, T., 2004. Effect of turning regime and seasonal weather conditions on nitrogen and phosphorus losses during aerobic composting of cattle manure. Bioresour. Technol. 91 (2), 171–178.

Pergola, M., Persiani, A., Maria Palese, A., Di Meo, V., Pastore, V., D'Adamo, C., Celano, G., 2018. Composting: the way for a sustainable agriculture. Appl. Soil Ecol. ISSN: 0929-1393 123, 744–750. https://doi.org/10.1016/j.apsoil.2017.10.016.

Platt, B., 2015. Neighborhood Soil Rebuilders' Capstone Project Compost Best Management Practices & Monitoring Guide. Institute for Local 98 Self-Reliance. https://ilsr.org/composting-best-practices/#Oxygen.

RISN, 2016. Organic Waste Processing Facilities Recommended Management Guidelines. Resource Innovation and Solutions Network. Arizona State University. In: https://www.maricopa.gov/DocumentCenter/View/20058/Organic-Waste-Processing-Facilities-Management-Guidelines-PDF.

Rynk, R., 2000a. Fires at composting facilities: causes and conditions. BioCycle.

Rynk, R., 2000b. Fires at composting facilities: handling and extinguishing fires. BioCycle.

Sall, P., 2016. On farm composting of fruit and vegetable waste from grocery stores: a case under cold climatic conditions of Eastern Canada. In: SUM 2016, Proceedings Third Symposium on Urban Mining and Circular Economy.

Saveyn, H., Eder, P., 2014. End-of-waste Criteria for Biodegradable Waste Subjected to Biological Treatment (Compost & Digestate): Technical Proposals. European Commission, EU Science Hub. https://ec.europa.eu/jrc/en/publication/eur-scientific-and-technical-research-reports/end-waste-criteria-biodegradable-waste-subjected-biological-treatment-compost-digestate.

Schreuder, H.T., Ernst, R., Ramirez-Maldonado, H., 2004. Statistical Techniques for Sampling and Monitoring Natural Resources. U.S. Department of Agriculture. https://www.fs.usda.gov/treesearch/pubs/6287.

SGS, 2013. Stockpile Measurement Services. SGS Management Group. https://www.sgs.com/-/media/global/documents/flyers-and-leaflets/sgs-min-wa097-stockpile-management-services-2-en-11.pdf.

WasteMINZ, 2009. Consent Guide for Composting Operations in New Zealand. Waste Management Institute of New Zealand, ISBN 978-0-473-13713-7. https://www.wasteminz.org.nz/wp-content/uploads/Compost-NZ-Consent-Guide.pdf.

WoodProductSfi, 2021. Fire Properties of Wood. www.woodproducts.fi/content/wood-a-material-4#:~:text=jpg,of%200.8%20mm%20per%20minute.

Compost characteristics and quality

Authors: Richard Stehouwer[1], Leslie Cooperband[2], Robert Rynk[3]

[1]*Pennsylvania State University, State College, PA, United States;* [2]*Praire Fruits Farm and Creamery, Champagne, IL, United States;* [3]*SUNY Cobleskill, Cobleskill, NY, United States*

Contributors: Johannes Biala[4], Jean Bonhotal[5], Susan Antler[6], Tera Lewandowski[7], Hilary Nichols[8]

[4]*Centre for Recycling of Organic Waste & Nutrients, The University of Queensland, Gatton, QLD, Australia;* [5]*Cornell Waste Management Institute, Cornell University, Ithaca, NY, United States;* [6]*Compost Council of Canada, Toronto, ON, Canada;* [7]*Growing Media, The Scotts Miracle-Gro Company, Marysville, OH, United States;* [8]*U.S. Composting Council, Raleigh, NC, United States*

1. Introduction

Compost is made from organic feedstocks by subjecting them to a managed aerobic decomposition process such that the original feedstocks are no longer recognizable, the organic materials have been stabilized, and plant toxicity has been reduced to benign levels. Several factors influence the quality of the resulting compost, including how they are processed and for how long (Bonhotal et al., 2008). However, more than any other factor, the feedstocks determine the compost product characteristics. Thus, the "qualities" of compost products are nearly as diverse as the feedstocks used to produce them.

It is important to note that the characteristics necessary for "quality" compost are dependent on the intended use for the compost and the soil conditions, specific types of plants, climate, annual rainfall, and other factors. These represent a wide range of potential conditions that make it difficult to predict the optimal finished compost to use in all situations. Since there is a great deal of latitude in the description of what compost quality means, many countries have developed compost quality standards that specify minimum and/or maximum levels for the primary quality parameters, especially those relating to safety. Elsewhere, notably in the United States, there has been resistance to the establishment of universal standards for performance parameters. The primary argument for such resistance is that compost quality depends so much on the end use of the product, and conditions vary widely. Other jurisdictions have addressed this barrier, in part, by creating different standards for different applications (e.g., container mixes, agricultural fields, home horticulture) or based on different feedstock materials (e.g., garden & food residues, and biosolids). There has also been reluctance to specify quality grades of compost

The Composting Handbook. https://doi.org/10.1016/B978-0-323-85602-7.00012-1

(A, B, C, or high, medium, low) since characteristics that are essential for some intensively managed systems may be irrelevant for other uses and the grades would indicate greater or lesser quality. This approach is used in many countries for categorizing biosolids products and their allowable uses. Compost standards and quality assurance and disclosure programs are discussed later in this chapter.

The characteristics that help define the quality of compost can be grouped into three realms: those related to *product performance*; those related to *product aesthetics*; and those related to *product safety*. Product performance characteristics include numerous parameters that relate to how compost affects plant growth and health, soil quality, and moisture. Aesthetic characteristics are primarily visual and olfactory and include color, odor, and the content of inert contaminants such as film plastics and glass. Safety refers primarily to substances in compost that could negatively impact human, animal, or environmental health. Prime examples are pathogens, potentially detrimental chemicals, and physical contaminants such as glass and plastic. Safety attributes form the basis of compost quality standards and regulations that are established by various public entities and compost industry organizations. While it is helpful to categorize quality characteristics in this way, there clearly is overlap with some compost characteristics affecting quality in more than one realm (Fig. 15.1).

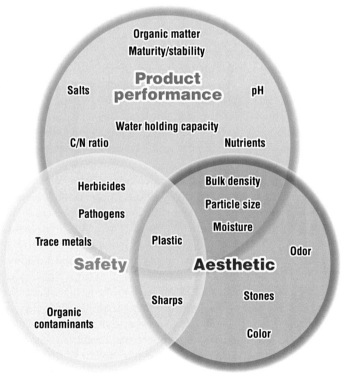

FIGURE 15.1

Overall compost quality is determined by product characteristics that fall into three intersecting realms: performance, aesthetics, and safety/risk. While some characteristics or measured parameters of compost may relate to only one of these realms, others may affect two or all three.

Compost quality is tied to the product's ability and capacity to perform its intended function. The better the compost is able to perform the intended function, the higher its quality. Because compost performs varying functions in different applications, the designation of quality depends on the ultimate use of the compost. While the ideal is to produce 'fit for purpose' compost products, a particular compost product often serves multiple purposes. This chapter presents compost qualities primarily based on the three categories discussed above—product performance, aesthetics, and safety. The compost characteristics suited to specific uses are discussed in Chapter 16. Chapter 17 covers the compost quality factors specific to disease suppression.

2. Typical, and typically variable, compost product qualities

Compost performance parameters are known to vary considerably in composts. For example, Fig. 15.2 shows three different products, called "compost." They visibly differ in color and texture, and likely differ in other respects including nutrients and organic matter (OM). These differences are largely due to the originating feedstocks, but processing also plays a role (e.g., screening, moisture).

Tables 15.1 and 15.2 present data from over 500 composts from commercial, municipal, and agricultural producers in the northeastern US. The samples are are grouped according to the primary feedstock type used in making the compost. All compost samples in Tables 15.1 and 15.2 were analyzed at the Pennsylvania State University Agricultural Analytical Services Laboratory using standard methods for compost analysis from the Test Methods for the Examination of Composting and Compost or "TMECC" (TMECC, 2012) (see Section 8). Appendix R presents additional data that updates Tables 15.1 and 15.2.

Tables 15.3 and 15.4 present similar data from a much smaller group of composts produced in the UK. Examination of these data shows a considerable

FIGURE 15.2

Three products, all called compost, have visibly different characteristics that may or may not determine their performance for a particular use.

Table 15.1 Range of physical and fertility parameters for composts produced from different primary feedstocks.

Parameter[a]	pH	Soluble salts	Solids	LOI	Total N	NH$_4$	OC	C/N ratio	P	K	Ca	Mg	S
Units		dS/m	%	%	%	mg/kg	%		%	%	%	%	%
Leaf and yard trimmings													
10th percentile	6.7	0.41	36.39	25.2	0.8	4.5	15	11.3	0.09	0.31	0.95	0.21	0.10
Median	7.5	1.76	53.30	46.5	1.5	6	27.1	15.5	0.24	0.53	2.05	0.49	0.25
Mean	7.4	2.00	53.17	46.6	1.7	366	26.5	19.5	0.48	0.70	3.14	0.66	0.31
90th percentile	8.0	3.33	70.92	67.0	2.6	1730	37.8	26.9	1.35	1.28	6.78	1.18	0.40
Number of samples	130	130	130	130	130	130	130	130	125	125	125	125	125
Food waste (with various co-feedstocks, such as yard trimmings)													
10th percentile	7.3	0.58	40.1	8.2	0.3	3.9	4.5	10.5	0.09	0.36	1.05	0.2	0.10
Median	8.0	1.86	58.10	35.6	1.1	5	19.6	15.1	0.20	0.68	2.98	0.39	0.18
Mean	7.9	3.41	57.91	36.5	1.5	140	19.9	15.8	0.37	0.90	3.93	0.53	0.23
90th percentile	8.5	7.79	77.10	67.0	2.9	253	34.4	22.4	0.63	1.74	8.37	0.97	0.41
Number of samples	99	99	99	99	99	99	99	99	88	88	88	88	88
Manure													
10th percentile	6.8	0.65	29.4	27.1	0.8	2.2	14.9	10.4	0.12	0.42	0.78	0.18	0.10
Median	7.7	2.64	50.20	55.3	1.6	17	29.8	16.1	0.37	1.00	2.74	0.68	0.26
Mean	7.6	3.96	54.22	56.5	1.8	351	30.8	29.3	0.59	1.18	3.80	0.71	0.31
90th percentile	8.4	10.01	82.10	85.5	2.9	805	47.1	37.0	1.48	2.10	8.91	1.16	0.62
Number of samples	201	201	201	201	201	201	201	201	182	182	182	182	182
Biosolids													
10th percentile	5.9	1.04	31.25	41.6	1.1	25.2	19.7	10	0.37	0.22	1.24	0.2	0.18
Median	7.1	2.68	49.85	58.3	2.2	1297	29.0	13.4	0.95	0.37	1.77	0.35	0.48
Mean	7.0	3.05	50.78	55.6	2.1	1727	28.2	14.4	0.94	0.45	1.93	0.37	0.48
90th percentile	7.7	5.37	70.85	68.7	3.0	3699	35.4	19.3	1.42	0.69	2.59	0.51	0.73
Number of samples	86	86	86	86	86	86	86	86	81	81	81	81	81

[a] The Mean is the arithmetic average of all samples. The Median is the value that half of tested samples were greater than and half were less than. The 10th percentile is the value that 10% of samples were less than, and the 90th percentile is the value that 90% of samples were less than.

Table 15.2 Range of metal and trace element contents from compost of various primary feedstocks.

Trace element[a]	Al	Fe	Mn	Na	As	Cd	Cu	Pb	Hg	Mo	Ni	Se	Zn
Unit							mg/kg						
Leaf and yard trimmings													
10th percentile	5224	5408	305	163	0.6	1.1	28.6	11.2	0.06	3.1	6.2	1	79.1
Median	12,375	11,395	639	405	2.6	2.1	52.4	33.0	0.21	3.3	11.9	1.1	178.5
Mean	14,108	12,276	715	532	4.2	2.0	74.4	48.2	0.33	3.9	12.8	1.7	235.8
90th percentile	24,323	18,607	1235	935	10.0	2.2	127.3	96.4	0.64	6.0	18.8	3.3	501.8
Number of samples	125	125	125	125	125	125	125	125	58	125	125	125	125
Food waste													
10th percentile	4650	4621	117	292	0.6	2	21.3	7.1	0.05	3.1	4.6	1	62.4
Median	8516	12,307	322	1077	3.9	2.1	36.9	18.8	0.07	3.2	7.8	1.1	96.4
Mean	9231	12,231	390	1971	3.6	2.1	40.3	24.3	0.09	3.3	8.6	1.3	124.4
90th percentile	13,701	19,781	626	4769	8.0	2.2	63.5	44.4	0.13	3.3	12.3	1.1	201.1
Number of samples	88	88	88	88	88	88	88	88	39	88	88	88	88
Manure													
10th percentile	1049	1193	146	187	0.6	1.1	18.8	4.2	0.02	3.1	2.4	1	46
Median	5419	6548	559	1025	0.7	2.1	45.3	8.8	0.05	3.3	8.3	1.1	127.0
Mean	8267	8667	620	1678	3.0	1.9	84.2	16.7	0.06	3.6	11.6	1.9	211.1
90th percentile	16,982	19,314	1192	4396	5.6	2.2	166.0	37.9	0.12	4.6	19.6	3.7	437.6
Number of samples	182	182	182	182	182	182	182	182	43	182	182	182	182
Biosolids													
10th percentile	7377	9984	293	446	0.6	1.2	83.4	17.5	0.11	3.2	6.8	1.1	203.4
Median	10,471	19,410	496	595	6.5	2.2	221.0	43.7	0.71	3.6	13.9	2.3	373.0
Mean	11,264	24,244	729	694	7.3	2.1	242.1	60.5	0.91	4.7	14.3	2.6	417.3
90th percentile	15,622	49,202	1343	948	13.1	2.6	422.5	127.9	2.34	5.8	21.0	3.7	660.7
Number of samples	81	81	81	81	81	81	81	81	51	81	81	81	81

[a] The Mean is the arithmetic average of all samples. The Median is the value that half of tested samples were greater than and half were less than. The 10th percentile is the value that 10% of samples were less than, and the 90th percentile is the value that 90% of samples were less than.

Table 15.3 Range of physical and fertility parameters for urban derived composts produced in the UK and analyzed in the ReMaDe project. Data presented are extracted from WRAP (2004) and additional compost analytical data can be found in this publication.

Parameter[a]	Bulk density	pH	Soluble salts	Solids	LOI	Total N	C/N	OC	P	K	Ca	Mg	S
Units	g/L		dS/m	%	%	%		%	%	%	mg/kg	mg/kg	mg/kg
Minimum	260	6.3	0.34	33	17	0.7	8	7	0.13	0.28	10,900	1550	1000
Median	483	8.2	0.98	64	35	1.2	12	14	0.23	0.89	22,278	2655	1834
Mean	492	8.1	0.96	63	40	1.3	14	17	0.25	0.85	23,333	2786	1970
90th percentile	611	8.7	1.39	78	57	1.5	23	28	0.27	1.13	30,893	3973	2556
Maximum	720	9.0	1.80	80	72	2.8	25	32	1.02	1.98	40,200	478	4390
Number of samples	33	33	33	33	33	33	33	33	28	28	28	2	28

[a] The Mean is the arithmetic average of all samples. The Median is the value that half of tested samples were greater than and half were less than. The 10th percentile is the value that 10% of samples were less than and the 90th percentile is the value that 90% of samples were less than.

Table 15.4 Metals and trace elements in urban derived composts produced in the UK and analyzed in the ReMaDe project. Data presented are extracted from WRAP (2004) and additional compost analytical data can be found in this publication.

Parameter[a]	Fe	Mn	Na	B	Cd	Cr	Cu	Pb	Hg	Ni	Zn
Units						mg/kg					
Minimum	5570	186	354	11	0.4	8	18	32.5	0.1	7.6	75
Median	14,750	317	660	27	0.6	22	58	92.0	0.2	15.2	188
90th %	18,779	426	1238	81	1.3	50	187	206	0.5	30.0	316
Maximum	34,800	593	1625	102	2.5	83	1160	301	13.3	44.3	572
Number of samples	28	28	28	28	21	21	28	21	19	21	28

[a] The Median is the value that half of tested samples were greater than and half were less than. The 90th percentile is the value that 90% of samples were less than.

range for many measured parameters, some of which are clearly influenced by the feedstocks. For example, the total nitrogen and OM concentrations of compost products can differ by a factor of 3 to 4. Values for pH range from slightly acidic to highly alkaline. Clearly, there is no such thing as "typical" composts.

Fortunately, there is now a wealth of data available about the characteristics of composts produced around the world. Appendix R updates and expands upon the data listed in Tables 15.1 and 15.2. Bhogal et al. (WRAP, 2015) present feedstock-specific data about composts (and anaerobic digestates) produced in the UK.

Every year, the German Association for Compost Quality Assurance publishes median values for a range of compost quality characteristics, including total and available plant nutrients, physical (bulk density, moisture, impurities), chemical (pH, EC) and biological (bioassay, weed seeds) characteristics, organic matter and liming value and heavy metal contents. The published median values represent all compost products that were analysed under their quality assurance program, which amounted to more than 3,800 in 2020 (BGK, 2021). The published data allow compost producers to compare and benchmark their products.

3. Compost performance characteristics

Composts are primarily used as soil additives to improve soil quality for various horticultural and agricultural uses and soil erosion control (see Chapter 16). There are numerous measurable parameters of compost that relate to the ability and capacity of compost to perform these functions, including pH, soluble salts, maturity, stability, OM content, nutrient content and form, C/N ratio, particle size, cation exchange capacity, and viable weed propagules. These parameters are tied integrally to the properties and processes of the soil and growing media in which the compost is applied. Therefore, it is important for compost producers and users to understand the basic properties of soils and soil OM. To that purpose, a "primer" on soil and soil OM interactions is presented in Appendix N.

3.1 pH

pH is the measure of the hydrogen ion activity (or active acidity) in compost. pH values are reported as the negative log of the hydrogen ion activity, thus the pH scale is logarithmic (a 1-unit change in pH means a 10-fold change in acidity). pH ranges from 0 to 14, with a pH of 7 indicating neutrality, less than 7 becoming more acidic and greater than 7 becoming more alkaline (see Chapter 3). It should be emphasized that pH is a measure of active acidity (acidity in solution) and provides no information on the reserve acidity or "buffering capacity" of the compost. Thus, a pH measurement alone can only predict the direction of change in soil pH that might result from compost application, but not how large the change will be. Moreover, soils vary widely in their capacities to buffer changes in pH. Soil pH buffering capacity tends to increase as clay and OM contents increase. Thus, high pH compost applied to a low pH sandy soil will likely increase soil pH, but the same compost applied at the same rate to a low pH heavy textured soil may result in little or no change in soil pH.

Compost (and soil) pH is extremely important for plant production because it has a great influence on many other soil parameters including macro- and micronutrient availability, trace element toxicity, soil microbial composition and activity, and cation exchange capacity. For most plant species, the pH range 6—7 provides adequate availability of most nutrients, although horticultural plant species with high iron demand benefit from somewhat lower soil pH. Most beneficial soil microorganisms also function well in the 6—7 pH range. Compost pH can be adjusted downward by adding acidifying agents such as elemental sulfur or upward by adding liming materials. In general, however, it is more effective to use these materials to adjust soil pH, rather than compost pH.

The pH of compost is influenced by a complex of chemical factors including the mix and concentrations of cations (positively charged ions) and anions (negatively charged ions) in solution. The median pH of composts in Tables 15.1 and 15.3 falls in the relatively narrow range of 7.1—8.2. Most composts have neutral to slightly alkaline pH. Compost pH above 8 is often associated with high sodium (Na) concentrations as can be seen in the 90th percentile pH and Na levels in food waste and manure composts (Tables 15.2 and 15.4).

3.2 Soluble salts

"Soluble salts" refer to the concentration of dissolved ions in a slurry of compost and water. These dissolved ions are the source of nutrients for root uptake, but the large osmotic potential resulting from excessively high concentrations can impede plant uptake of water. Specific dissolved ions, such as sodium (Na) or chloride (Cl), can be directly toxic to plants. Plant species vary widely in their tolerance to salinity and growers should familiarize themselves with the tolerance range of the species they produce, and also with salinity levels in their soil.

The concentration of soluble salts is typically estimated by measuring the ability of the solution to conduct an electrical current and is reported as electrical conductivity (EC) using units of dS/m, μS/cm, or mS/m (1 dS/m = 100 mS/m = 1000 μS/cm). Some laboratories may report EC using the unit mmhos/cm, which is numerically equivalent to dS/m. EC can be measured in a saturated paste or using a quantitative dilution procedure. A ratio of 1:5 compost (dry weight) to deionized water is the most common, but 1:10 also is used. Historically, the horticulture industry has used the saturated paste method to determine both pH and EC. It is important to know the method used to determine EC because each will produce very different values. Whatever method is used, it is critical to be consistent in using the same method to compare composts or track changes in EC over time.

The soluble salt levels of composts can span a wide range, typically from 0.2 to 16 dS/m in a 1:5 compost:water extract. However, as Tables 15.1 and 15.3 indicate, 50% of composts are likely to have soluble salt levels of less than 2.7 dS/m. Compost feedstocks are the source of salts, with manures and food residuals generally containing more salts than leaf and yard trimmings. However, soluble salts in the final compost also are influenced by composting management such as feedstock ratios, the extent of leaching, and whether or not leachates are returned to the compost.

Soils and other rooting media to which composts are added typically have lower soluble salt levels and often dilute or mitigate the compost salts. Because of the limited capacity for dilution, salt levels are important in composts destined for use in container media. It is critical that the combined soluble salt level in the blend of compost and soil or growing media remains low after compost application. Soluble salts can often be removed from soils or potting media by leaching with water. Because this level is a function of soluble salts in the compost and in the soil, the ratio of compost to soil, and the amount of leaching prior to planting, it is difficult to establish firm guidelines for compost soluble salt levels. However, quality specifications for blended growing media usually prescribe maximum salt levels. For example, guidelines issued by the European Compost Network (ECN) for compost used in growing media specify a maximum EC of 1.9 dS/m, using a 1:5 extraction (Siebert and Gilbert, 2018). The corresponding guidelines in the UK, set the maximum EC at 1.5 dS/m (WRAP, 2014). These levels are difficult targets for many compost products.

3.3 Organic matter and organic carbon

These are critical quality parameters because composts are frequently used to increase soil organic matter (OM). For such applications, composts with increased OM or organic carbon (OC) contents are desirable. OM in composts is most often measured using a loss on ignition (LOI) method. The LOI method determines the loss in mass of a dry compost sample that is heated to 550°C. The amount of mass lost is assumed to represent the organic fraction of the compost, and the remaining ash is the mineral fraction. OM and volatile solids content are nearly identical parameters. Both are determined by the LOI method.

OC is determined by directly measuring the *carbon* released by a compost sample during heating to greater than 1000°C. Even though carbon is the backbone of the molecules that make up OM, there is no accurate way to mathematically convert the results of an OM analysis to OC and vice versa. This is because the carbon content of compost OM is not constant but varies considerably among various composts (frequently from 45% to 60%). This variability may be associated with differing feedstocks and with compost maturity. Therefore, estimates of OC content that are derived from an OM analysis may be inaccurate along with subsequent calculations of carbon to nitrogen ratios. Similar inaccuracies occur when OM content is estimated from an OC analysis. OM to OC conversions is generally most accurate for leaf and yard trimmings composts and less accurate for food waste, manure, and biosolids composts. The median OM and OC contents of composts in Table 15.1 range from 35% to 58% and 20%–30%, respectively. But again, there is a significant range of values around the medians. Manure and biosolids composts tend to have the higher OM and OC values and food waste the lowest. Even composts made entirely of "organic" materials are not 100% OM due to the ash content of the organic materials. When composts have OM or OC levels much below the median, it generally indicates a significant amount of soil or other mineral material has been mixed into the compost, or that the compost contains a lot of stones or other inert contaminants.

Another important consideration related to measurements of either OM or OC is that not all material burned off at 550°C necessarily behaves like, or is readily converted to soil OM. Likewise, not all C measured as described above is contained in organic compounds that are readily available for microbial decomposition. Woody material is measured as OM or OC by these methods, but does not contribute significantly to cation exchange capacity or water holding capacity (WHC) while it remains wood. Similarly, the C in woody and other highly lignified materials is measured as OC but is quite resistant to microbial decomposition and thus is not part of the active C fraction of the compost.

3.4 Nutrient content

Compared to synthetically produced inorganic fertilizers, composts have very low nutrient contents and would be considered low analysis fertilizer materials. Composts are rarely marketed as fertilizers because to do so would require guaranteed nutrient content labeling under fertilizer regulations in most states, provinces, and/or countries. Nevertheless, composts are used to provide plant nutrients in many applications. For such uses, composts with larger nutrient contents would be desirable. By contrast, composts with low nutrient content may be more desirable for erosion control uses and certain potting mixes. Furthermore, for almost all compost uses, knowledge of nutrient contents and nutrient forms is needed for environmentally responsible nutrient management.

The most important plant macronutrients are nitrogen (N), phosphorus (P), and potassium (K) but composts can also be sources of other macronutrients including calcium (Ca), magnesium (Mg), and sulfur (S) as well as numerous micronutrients. Laboratory analyses usually measure the total amounts of these nutrients present (using strong acid digestion or combustion processes), but in composts these nutrients are present largely in organic forms (particularly N, S, and P) or they are held by organic molecules. Organic forms of nutrients are not available to plants until they are transformed to inorganic forms (mineralized) during decomposition of OM. Decomposition rates are strongly influenced by the soil texture and environmental conditions (temperature, moisture, aeration) under which the compost is used as well as by compost characteristics (stability, maturity, C/N/P ratios). As such, it is very difficult to predict the rate at which most compost nutrients will become available to plants.

Laboratories report the concentrations of nutrients in composts (and feedstocks) based on the dry weight of the compost. Using moisture content and bulk density determinations, laboratories often convert the dry weight basis results to wet weight and volume basis results. However, such values should deserve caution since the bulk density of the samples in the lab may not be representative of conditions at the composting operation and because moisture content can easily change during storage or from one windrow to another. It is better for compost producers and user make these conversions themselves, using *on-site* measurements of solids content and bulk density. Box 15.1 explains this easy procedure.

Box 15.1 Conversion of compost analytical results from dry weight to *as is* (wet) weight and to volume

Nutrient concentrations and other chemical parameters in composts are usually reported on a dry weight of compost basis. This is the standard practice because it gives the most consistent baseline for reporting and use. However, composts are usually sold and used based on volume and/or wet weight. Therefore, dry weight analytical results must be converted to a wet weight or volume basis. This conversion is simple to make if one knows the moisture content and bulk density of the compost. However, in practice, it is complicated by the fact that these factors can change over time. Hence, it is important to also determine bulk density and solids content as it exists at the point of use or sale. These conversions are most easily done by calculating conversion factors that can then be applied to the analytical results.

Converting from dry weight basis to *as is* (wet) weight basis.

The solids (dry matter) content of the compost must be known. Most analytical laboratories report either solids content or moisture content of the compost sample that was submitted for analysis. These values are usually reported on a percent basis. The laboratory results can be used if you are reasonably certain the moisture content of the sample received by the lab is representative of the bulk compost of concern.

$$\% \text{ Solids} = \frac{\text{Weight of sample after drying}}{\text{Weight of sample before drying}} \times 100\%$$

Many laboratories report only moisture content of compost and not solids. In that case, the solids content is the moisture content subtracted from 100%, e.g., 100%−40% moisture = 60% solids.

To convert analytical results from dry weight basis to "wet" or "as is" basis, simply multiply by the decimal form of the % solids result (% solids/100). For example, the conversion factor of compost with 60% solids would be 60/100 or 0.6. Thus, if the compost analysis reports 2.5% N, or 25,000 mg N per kg, on a dry weight basis, the N content of the moist compost would be:

2.5%N × 0.6 = 1.5% N on an as is basis

25,000 mg N/kg dry compost 0.6 = 15,000 mg N/kg of compost (as is).

Converting from *as is* (wet) weight basis to *as is* (wet) volume basis

To convert chemical analytical values from a wet *weight* basis to a wet *volume* basis, the bulk density must be known. For analytical results reported on a percent basis, multiply the % value by the bulk density and divide by 100. If the analytical results are reported as mg/kg, the concentration also must be divided by 1,000,000.

For example, assume that the bulk density of the compost is 540 kg/m^3 (900 lbs/yd^3) and the wet weight concentration of N is 1.5% (15,000 mg/kg). Then, the volume-based concentration of N is calculated as:

$$\frac{1.5\% \text{ N}}{100\%} \times 540 \text{ kg/m}^3 = 8.1 \text{ kg of N per m}^3 \text{ of compost (wet or as is), or}$$

$$\frac{1.5\% \text{ N}}{100\%} \times 910 \text{ lbs/yd}^3 = 13.7 \text{ lbs of N per yd}^3 \text{ of compost (wet or as is), or}$$

$$\frac{15,000 \text{ mg N/kg}}{1,000,000 \text{ mg/kg}} \times 540 \text{ kg/m}^3 = 8.1 \text{ kg of N per m}^3 \text{ of compost (wet or as is), or}$$

$$\frac{15,000 \text{ mg N/kg}}{1,000,000 \text{ mg/kg}} \times 910 \text{ lbs/yd}^3 = 13.7 \text{ lbs of N per yd}^3 \text{ of compost (wet or as is)}$$

3.4.1 Nitrogen

In most composts much of the N is present as organic N (typically 1% to 3% by dry weight) with very small amounts (usually less than 500 mg N/kg dry weight) present in the inorganic forms of ammonium (NH_4^+) and nitrate (NO_3^-). Nitrate can be readily taken up by plants. It can be assumed, under most situations, that 100% of this N will be immediately available to plants. Ammonium is also immediately available, although not all plant species have an affinity for ammonium. Also, plant damage and inhibited seed germination can result from a high concentration of ammonium, which is primarily a consequence of immature compost.

Organic N forms, however, must be mineralized to the inorganic N forms before most plants can utilize it. A conservative estimate for mature composts is that up to 10% of total N is present in mineral form and immediately plant available, and that 5%−10% of the organic N is mineralized in the first year (Box 15.2). Mineralization rates decline in subsequent years unless compost is added annually, in which case they increase. One-time compost applications tend to show low mineralization rates and nitrogen use efficiency, which improves markedly with repeated compost applications. Nitrogen mineralization rates may be greater for composts made from N-rich feedstocks such as manures and biosolids and less for composts derived from C-rich feedstocks such as yard trimmings or wood shavings. Soil environmental factors are also important; mineralization rates are faster with warm, moist conditions, and slower under hot and dry or cold and wet conditions.

3.4.2 Phosphorus, potassium, and other nutrients

For most composts, no more than 50% of total P will be available to plants during the first year after application, with a much larger proportion of the P becoming available over a three-year period. Many laboratories also offer a water-soluble P test for compost (as well as other nutrients). These tests are conducted either in a compost-water slurry (saturated media extract—most typical for the container horticulture industry), or a fixed weight to volume ratio of compost to water (often a 1-part compost to 5-parts water dilution). Such tests measure the amounts of nutrients that are immediately available for plant uptake, but do not indicate how much might become available during a growing season or crop rotation (1−3 years). When interpreting compost test results, it is important to distinguish between total and water-soluble measurements, and to request how the water-soluble measurements were determined (slurry vs. fixed ratio).

The K in composts is quite soluble. It is generally assumed that all K contained in compost will become available for plant uptake over a three-year period.

In the United States, fertilizer P and K must be reported on an oxide basis as P_2O_5 and K_2O, respectively, and many laboratories also report compost P and K on an oxide basis. Compost producers and users must pay attention to the form in which the results are reported. To convert P_2O_5 to P, divide by 2.29. To convert K_2O to K, divide by 1.2. Most composts are in the range 0.2%−1.3% P and 0.3%−2.0% K by dry weight.

Micronutrients including iron (Fe), zinc (Zn), copper (Cu), boron (B), manganese (Mn), Cobalt (Co), etc. are required by plants in very small quantities; nonetheless, they are essential for plant growth. Most composts contain a diverse array of micronutrients present in varying concentrations). Micronutrients are measured

Box 15.2 Nitrogen availability from composts—research from temperate climates

Author: Johannes Biala.

Amlinger et al. (2003) reviewed the results of 28 experiments that studied nitrogen mineralization from composts. The authors provided the following generic observations and conclusions:

- Nitrogen availability in field trials ranged between 2.6% and 10.7% (*average* of minimum and *average* of maximum values, respectively) during the first year after compost application, which led some authors to conclude that, on average, approximately 5% of nitrogen contained in compost can be considered plant available.
- Subsequent nitrogen availability (three and more years after compost application) depends largely on site and production specific conditions but is generally in the range of 2%−3% of added compost nitrogen.
- Use of mineral nitrogen fertilizer usually reduces the efficiency of plants in utilizing compost-derived nitrogen. Therefore, research results should be differentiated between those that use mineral fertilizer and those that do not.
- Nitrogen use efficiency tends to be higher for rotations that contain crops with high nitrogen demand.
- Nitrogen use efficiency increases as the growth period of crops increases (e.g., corn vs. wheat or leafy vegetables).
- Continuous compost use (e.g., annual applications) increases nitrogen mineralization in the initial year after compost application, with a maximum of 40% measured after 21 years of compost use.
- Mineralization from mature composts usually occurs faster than from fresh composts
- Fresh composts as well as those with high C/N ratio often incur temporary nitrogen immobilization or nitrogen-rob.

Similar nitrogen mineralization patterns are evident from research conducted at five sites in Germany after 9 and 12 year compost applications (Haber et al., 2008). Averaging results across all sites, years, and compost products, the results indicate that 5%−10% of nitrogen added with compost is utilized by crops in the first year after application (Haber et al., 2008). However, the relatively long and diverse compost application trials revealed nitrogen use efficiency data that are much more complex than what the above average suggests. The medium-term trials with three or four crop rotation cycles (corn—winter wheat—winter barley) showed very clearly that nitrogen use efficiency increased from less than 5% during the first crop cycle, to levels between 5% and 15% during subsequent cropping cycles.

When no additional mineral nitrogen was used, plant up-take of N averaged around 5% in the first crop cycle, and between 8% and 15% in subsequent crop cycles. However, nitrogen immobilization (−15%) was also witnessed in the first crop cycle. Nitrogen use efficiency decreased somewhat when compost was used together with mineral fertilizer. When 50% recommended mineral nitrogen was applied, compost nitrogen use efficiency dropped to 2%−3% annually across all sites during the first cropping cycle and to 5%−12% in subsequent cropping cycles. When 100% recommended mineral nitrogen was applied, and plants had "luxury" nitrogen supply, compost nitrogen use efficiency was minimal during the first cropping cycle, but still reached 5%−10% in subsequent cropping cycles. Observations suggest that "biological preconditioning" may result in greater nitrogen use efficiency from compost.

The medium-term trials in Germany showed further that higher nitrogen use efficiency occurs when compost is used in crop rotations that have high nitrogen demand compared to those that require lower nitrogen. For example, compost nitrogen-use efficiency was generally higher where corn was grown for silage; due to higher nitrogen demand for this crop (rotation) compared to when corn was grown for grain. The difference was particularly apparent where no mineral fertilizer was used, with compost nitrogen use efficiency averaging 13%−16% for high nitrogen demand rotations, and 7%−10% for crop cycles with lower nitrogen demand.

either as total concentrations (using acid digestion or combustion procedures similar to those described for total C and macronutrients) or in plant-available forms. Plant available forms are extracted using a chelating or binding agent like EDTA or DTPA. To assess the plant available forms, it is better to request a test for EDTA or DTPA extractable micronutrients than the standard analysis of total micronutrients or even a water extraction method.

3.5 C/N ratio

The ratio of OC to nitrogen (weight to weight) in compost is important primarily because it can be used to predict whether or not organic N in the compost will be mineralized to plant available inorganic forms and whether there is a risk of N immobilization. The general rule of thumb is that if the C/N ratio is less than 20, organic N will be mineralized. Composts with C/N ratios greater than 25 may not be stable. It is increasingly likely that a compost will immobilize (i.e., tie up) soil inorganic N as the ratio increases above 25. While many composts have C/N ratios in the range 10 to 20, some composts made from yard trimmings have C/N ratios above 20. In general, as C/N ratio decreases, N availability from compost application increases. However, the actual quantity and rate of inorganic N release from compost depends also on total N content and compost stability. This ratio holds for organic C that is susceptible to microbial decay. The validity of the ratio decreases as more and more recalcitrant or microbially unavailable C is included in the measurement of organic C (see the section above on organic C). Also, as previously discussed, estimates of organic C-based on measured OM are susceptible to significant error because of the variability in C content of OM. Thus C/N ratios determined from direct measurement of C and N are preferred.

3.6 Particle size

Maximum particle size, and in some cases the particle size distribution, is one of the more important compost characteristics affecting quality for various end uses, particularly when the compost is derived from woody feedstocks. To determine particle size, the compost sample is simply passed through a screen of a specified mesh size. The value is typically reported as the percent passing the specified mesh size. To determine particle size distribution, compost is passed through a series of screens with decreasing mesh sizes (Fig. 15.3). The percent of material retained on each screen is reported. These determinations are usually done on "as received" samples and the results expressed on a moist sample weight basis. However, this should be confirmed with the lab that is conducting the analysis.

For most uses as a soil amendment, a maximum particle size of 13 mm (1/2 in.) is satisfactory, although composts with particles up to 25 mm (1 in.) in diameter perform well in many circumstances.. For most potting mixes and topdressing applications (such as on established turf), finer material is required with maximum particle sizes of 7 to 10 mm (1/4 to 3/8 in.). For erosion control applications and use as mulch in vine and tree cropping, a much coarser material is desirable so that particles knit together and hold in place. Particles of up to 25 mm (1 in.) in size are usually acceptable.

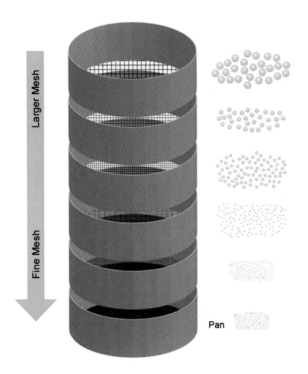

FIGURE 15.3

Illustration of sieves used to determine particle size distribution in composts, soils, and other particulate materials.

Source: PTL (Particle Technology Labs).

3.7 Bulk density

Compost bulk density (the weight of compost per unit volume) can vary greatly depending on moisture content and actions that might tend to compress or fluff the compost such as stacking height, depth in a pile, or time from turning, screening, or moving. Nevertheless, bulk density is a critical measure since most compost is marketed and applied on a volume basis. Many laboratories measure bulk density of compost if they are provided sufficient sample volume. The analysis is normally measured with samples "as received" meaning without any moisture adjustment, screening or milling, and is done simply by filling a known volume container with compost using a prescribed amount of packing and then determining the weight of the compost.

Bulk density is best measured on-site using a procedure that mimics the compaction of the compost in its storage location. The procedure described in Chapter 4 works as well, if not better, for compost as it does for feedstocks.

Determination of compost bulk density is easily done by packing a container of known volume with compost and measuring the weight of compost in the container. The resulting value is highly dependent on the extent of packing and compaction used when packing the container. Therefore, packing should be done to simulate the packing that would occur with loading equipment used for sale or spreading of the compost.

3.8 Water holding capacity

The amount of water compost can retain is particularly important when it is used in container media, but may also be a factor for soil uses if the intent is to increase soil WHC of coarse textured sandy soils. The WHC of compost refers to its ability to hold water under varying tensions or suction pressures. It is usually expressed as the percentage moisture of the total weight of the compost sample. The most common methods for determining WHC involve collecting a sample of finished compost, sieving the material, and then transferring it into a container of known volume (a bucket, a beaker or a graduated cylinder). The bulk density is determined and then the compost is saturated with water and excess water is drained. Changes in compost volume and mass, and the ratio of water retained relative to the amount of drained water provide a means for estimating compost bulk density, porosity/pore water volumes and free airspace, and WHC. The methods used are modified from those used to determine WHC of peat because the physical character of compost is similar to that of peat.

3.9 Stability and maturity

These are two very important compost quality parameters because they directly influence how compost affects plant growth and soil properties. However, the exact definition and meaning of the two terms, and more so, how they should be measured, has been and continues to be much debated. Stability is generally understood to mean the degree of resistance to further decomposition of the compost. Thus, it relates to the types of remaining organic compounds in the compost (how resistant are they to further microbial attack). It is indicated by diminished biological activity in the compost. Stability is important because of its impact on N mineralization or immobilization when compost is applied to soil or used in a container media, and to the potential for volume reduction of compost from further decomposition.

Laboratory measures of stability attempt to quickly measure biological activity in compost and most do so indirectly. One approach measures the temperature rise, due to self-heating, of a sample in an insulated container (the Dewar test). A second approach assesses microbial respiration by measuring oxygen consumption or carbon dioxide evolution under favorable temperature and moisture conditions. The TMECC decribes several methods of stability measurement (TMECC 2012).

Maturity refers to the degree of completion of the composting process. Compost that is not fully mature may contain phytotoxic compounds such as ammonia, organic acids, or other soluble compounds in quantities sufficient to adversely affect seed germination, seedling, and root development, or to cause odors. Maturity assessment is complex and is not determined by a single measurement (Wichuk and McCartney, 2013).

Maturity can be assessed using plant bioassays as well as direct measurements of potentially toxic compounds such as ammonia or organic acids. The main drawback to performing bioassays, which measure seed germination, seedling emergence, vigor, and survival, is that they are labor intensive and slow (12−14 days). The general approach used in these bioassays is to prepare mixtures of compost with peat or some other standard potting media at ratios varying from 0% to 100%. Seeds of a test species are planted in the various mixes and germination, emergence, and growth are measured at various time intervals. Seed species commonly used include cucumber, cress, lettuce, carrot, peas, Chinese cabbage, and amaranth (Emino and Warman, 2004). Fig. 15.4 shows plant bioassay tests using radish seeds. Performance in compost is compared to performance in the standard potting media. In another bioassay, used principally in Australia, radish seeds are grown in pure compost and root length is measured after four days. Root lengths of 60 mm or longer indicate that growth is not inhibited in any way by the compost.

Because maturity is really determined by several compost properties, another approach to maturity assessment is to utilize an index of several tests or measures of both stability and maturity. One of these is the compost maturity index developed by the California Compost Quality Council (Buchanan et al., 2001). This index first considers the C/N ratio. If it is greater than 25, the material is not considered to be compost (classified as very immature). If it is less than or equal to 25, it is then subjected to at least three further tests, at least one test from each of two groups of analyses (Table 15.5). Group A tests relate to stability indicators (CO_2 release, O_2 demand, self-heating) and Group B tests relate to toxicity (ammonium-to-nitrate ratio, ammonia concentration, volatile organic acid

FIGURE 15.4

Plant bioassay using radish seeds to compare treated pots with control points in terms of seedling emergence and on root condition.

Source: D. Neher.

concentration, plant bioassays). Composts are then assigned to one of three maturity levels based on these test results: immature, mature, and very mature.

Rapid maturity tests have been developed to provide a tool for quick maturity assessment. One of these is the Solvita maturity index, which is a four-hour colorimetric measurement of ammonia and carbon dioxide evolution from compost (Fig. 15.5). A maturity rating of 1 to 8 is assigned based on levels of both gasses. The Solvita test is widely used on-site in the composting industry, including analytical laboratories. It is a "do-it-yourself" test that can be performed on-site without the need to send samples to a laboratory (Solvita, 2021). Results are available within a day at a relatively moderate cost, with a basic test kit costing less than $200 (covering 6 samples). A digital color reading device is also available to more accurately interpret the colorimetric results.

Stable and mature composts are desired for many compost uses, in particular where applications are made to existing vegetation, such as turf topdressing, or where plants will be grown soon after mixing in significant amounts of composts such as potting mixes, plant bedding mixes, or other manufactured topsoil. For soil applications where there will be a significant time lag before planting, composts that are not fully mature may be satisfactory or even preferred. For example, fall applications to soil that will be used for horticultural or agronomic crop production the following spring. There is some evidence to suggest that the increased soil microbial activity resulting from application of immature compost has greater benefits for soil physical properties than mature compost, and the lag until planting allows ample time for degradation or dissipation of plant toxic compounds.

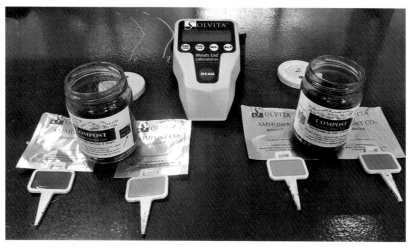

FIGURE 15.5

Solvita compost maturity test. Two paddles placed in the closed jars change color as they absorb ammonia and carbon dioxide, one paddle for each compound. The device in the background reads the color and indicates the associated stability/maturity level. Handheld color charts also are used.

Source: J. Biala.

Table 15.5 California Compost Quality Council (CCQC) compost quality index (Buchanan et al., 2001).

C/N Ratio	
Group A (stability)	**Group B (phytotoxicity)**

Respirometry tests
   Specific O_2 uptake rate (SOUR)
   CO_2 evolution
   Dewar self-heating test
   Solvita CO_2
	Ammonium (NH_4)
	 NH_4:N to NO_3:N ratio
	 Solvita NH_3
	 Volatile fatty acids (VFAs)
	 Biological assays
	   Emergence and seedling vigor
	   In-vitro germination and seedling elongation

4. Aesthetic characteristics

Compost users first judge the product using their senses, thus the appearance, smell, and feel of compost is important to successful marketing. These sensory characteristics generally go beyond aesthetics and can be indicative of the composting process and product quality/performance. An earthy aroma with no objectionable odors is not only pleasing but also indicates the composting process was well managed, while unpleasant odors indicate the opposite and suggest the product may not be stable or mature. Similarly, for almost all compost uses, a dark brown to nearly black product is preferred. Additionally, original feedstocks should not be recognizable in the finished product. Exceptions can be made for compost products derived from feedstock mixes that include wood chips, and are intended to be used as mulch, topdressing and erosion control. Small fragments of wood are likely to be present in these types of compost.

4.1 Physical contaminants (impurities)

Physical contaminants, also called "impurities," include any inert materials that should not be present in finished composts, but most commonly refers to stones, metal, glass, string, and the bane of many composting operations—film plastic. While such contaminants can find their way into many feedstocks, they are often present in feedstocks collected in urban and suburban areas (Fig. 15.6). Stones can enter the compost from composting operations conducted on nonpaved surfaces. These contaminants not only detract from the finished product quality but also can be a nuisance to the composting operation in the form of blowing trash and wrapping around turners and other equipment. Limits on physical contaminants in finished compost are included in the standards of most countries. These standards often distinguish between natural foreign matter (stones, clumps of clay), anthropogenic foreign matter (glass, plastic, metal), and some further distinguish heavy plastics and film plastics (Biala and Wilkinson, 2020). The standards are

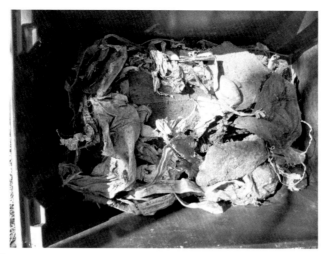

FIGURE 15.6

Contaminants that have been manually picked from compost at the point of application.

Source: J. Biala.

normally based on the weight fraction of contaminants (dry weight basis) present in a certain size fraction of the compost. However, because of the very low density of film and foam plastics, surface area standards also are being established (e.g., cm^2 of plastic per liter of compost).

Most standards for natural foreign matter specify less than 2%–5% in the greater than 5 mm size fraction. Standards for anthropogenic foreign matter mostly specify less than 0.5%–2% in the greater than 2 to 5 mm size fraction. The 2018 German Compost Quality Standards specify that the surface area of impurities must not exceed 15 cm^2 per liter of fresh compost while also meeting the 0.5% weight-to-weight limit (Biala and Wilkinson, 2020).

The presence of micro-plastics in composts are an increasing concern, as they are for all products that are used in the environment. However, micro-plastics are more of a chemical contaminant than a physical one.

4.2 Color, odor, and moisture

Although specifications for these characteristics are rarely included in compost quality standards, they are frequently mentioned in guidelines for compost use. The desired color for most applications is dark brown to black. This is what consumers expect composts to look like and, color may provide some indication of compost maturity. Composts should have a nonoffensive, earthy aroma. An unpleasant odor in compost is most often the result of inadequate aeration during the composting process but may also be a carryover from particularly odorous feedstocks. Ammonia odors may indicate the C/N ratio of the starting feedstock mix was too low.

Very wet or very dry compost does not feel right to many compost users. High moisture compost is difficult to screen and spread and users must haul a lot of water weight. Very dry compost presents problems with excessive dust formation and may exhibit hydrophobicity and thus is difficult to rewet. The moisture content of finished composts can certainly change during storage but may also indicate the composting process was conducted at nonideal moisture contents. Finished composts typically should have moisture contents between 35% and 45%. Certain customers may desire a drier product to decrease weight and make spreading easier. However, if moisture contents drop below 35%, there will be a lot of dust generated when these composts are transported or applied to fields.

5. Safety characteristics

Primarily because many compost products are derived from waste materials, they potentially carry the environmental and safety concerns associated with the parent materials—pathogens, elevated trace elements, potentially toxic organic compound, plastics, and sharp particles that can injure composting operators and compost users. These safety characteristics are the starting point for most compost quality standards and regulations.

The composting process removes many of the risks associated with the initial feedstocks. Pathogens are effectively eliminated by proper thermophilic composting. Many organic contaminants, like antibiotics, substantially decompose during composting. However, inorganic contaminants (e.g., trace elements, glass, plastic) remain in the final product since they are not broken down and do not appreciably volatilize (if at all).

The old computer adage "garbage in—garbage out" applies well to compost contaminants. The best way to ensure low contaminant levels in compost is to use consistently clean feedstocks that are kept clean and collected separately at the point of generation (rather than comingling). Again, the best cure is prevention—use feedstocks with no or little contamination.

5.1 Pathogens and pathogen inactivation (sanitization, pasteurization)

Nearly any feedstock has the potential to contain viable human, animal, or plant pathogens. However, some feedstocks pose greater pathogen risks, mainly manures, slaughterhouse wastes, diseased plant residues, biosolids, source-separated organics and postconsumer food residuals. When managed properly, pathogenic organisms that might be contained in feedstocks are inactivated during the composting process.

There is much agreement worldwide on requirements for pathogen inactivation of composts produced from feedstocks that may contain human or animal pathogens (often termed sanitization, pasteurization, or hygienization). Most existing standards specify some combination of time and high-temperature requirements together with turning requirements in the case of open windrow composting. Requirements generally allow for longer times at lower temperature and shorter times at higher temperatures. These time-temperature levels very nearly pasteurize the compost and achieves very large reductions and inactivation of human, animal, and plant pathogens.

In the United States, there are national standards for pathogen reduction, but only for composting of biosolids (see Chapter 3). These standards, known as the Process to Further Reduce Pathogens (PFRPs), are specified in the EPA Biosolids Rule, USEPA 40 CFR Part 503 (Walker et al., 1994). They specify time and temperature requirements. For windrow composting biosolids must remain at 55°C (131°F) or higher for a minimum of 15 days during which there must be at least five turnings. For in vessel systems or static aerated pile (ASP), the requirement is 55°C for at least 3 days. Many individual states have adopted regulations requiring PFRP standards for composting of other feedstocks, most commonly food wastes. A few states require PFRP standards to be met for all feedstocks, as does the USCC's Seal of Testing Assurance program (see Section 7). The USDA National Organic Program standard for composting uses the same time-temperature requirements for pathogen destruction as those described for composting of biosolids (USDA, 2021). These PFRP requirements are commonly extended to other feedstocks, where mandated by state and local regulations.

Time-temperature sanitization guidelines in the ECN Quality Assurance Scheme are slightly different, requiring 3 days at 60°C (140°F) for closed systems (e.g., ASP, in-vessel) and either 10 days at 55°C or three days at 65°C (149°F) for windrow systems. These requirements are generally applied to all composting feedstocks. European composters also have to comply with the stricter requirements of the European Union's Animal By-Products Regulation, EC No. 1069/2009 (Amlinger and Blytt, 2013). This regulation applies to many animal by-products, including manure and stomach contents (e.g., paunch manure), but they are most restrictive for feedstocks that include animal body parts and meat, including food wastes.

Attainment of these temperature/time requirements is easily achieved in a well-run composting operation and can be used to ensure effective sanitization of all materials commonly composted on farms. Though not necessarily required, composters using animal manure as a feedstock should consider keeping temperature and time records to document pathogen control in their composting process. The time/temperature and turning requirements for pathogen control also inactivate weed propagules and plant disease agents present in organic feedstocks.

In addition to, or sometimes in lieu of, demonstrated temperature/time performance standards, composts must be tested for pathogens by a certified laboratory, using accepted protocols. The tests signal and/or enumerate the presence of indicator organisms, usually Fecal Coliform, or *Escherichia coli*, and *Salmonella* spp. A common standard for Fecal Coliform is less than 1000 MPN (Most Probable Number) per gram of dry compost. Usually, *Salmonella* spp. is required to be absent in a specified compost sample size (e.g., 25 g dry weight, 50 g wet). The frequency of pathogen tested is determined by the feedstocks and the composting facility production level.

5.2 Chemical contaminants

The chemical contaminants of primary concern are a group of trace elements[1] (i.e., heavy metals) that, at high concentrations in the soil, can adversely impact human, animal, or plant health. Composts can also contain a myriad of other, mostly organic, chemical compounds that can potentially pose a risk to humans, animals, or the environment. There is increasing concern about micro-plastics and industrial compounds known by the acronym PFAS. These contaminants, which are largely feedstock-specific, are discussed in Chapter 4. Trace elements in composts are widely regulated; the other chemicals are not. In some countries, chemical contaminants are labeled "Potentially Toxic Elements" or PTEs.

Composts can also contain a myriad of other, mostly organic, chemical compounds that can potentially pose a risk to humans, animals, or the environment. PFOA and PFOS (Perfluorooctanoic acid and perfluorooctanesulfonic acid) are of increasing concern to the public, but currently there are no comprehensive lab test methods developed or standards programs measuring them. PFAS - www.compostingcouncil.org/PFASincompost.

5.2.1 Trace elements

Most composting operations should be able to utilize feedstocks with low trace element contaminant levels. Biosolids composts tend to have higher concentration ranges for some trace elements than other composts, notably for copper, lead, and zinc (Table 15.2). Still, most wastewater treatment facilities produce biosolids that easily meet regulatory limits. Certain manures also can contain elevated levels of some trace elements. For example, swine manure is often high in copper due to dietary supplements, and dairy manure can be high in copper if copper sulfate hoof baths are used. Plant residues normally have very low trace element concentrations as do leaf and yard debris and woody wastes. However, treated lumber has very high copper levels (in the United States, treated lumber produced prior to 2004 also likely contains arsenic and chromium). Care should be taken to keep this material from finding its way into woody feedstocks.

While there is general agreement around the world as to which trace elements in composts could be of human health or environmental concern, there is much less agreement on what the threshold levels should be. Most European countries and Australia have adopted standards with stringent limits intended to prevent any significant accumulation of these elements in the soils to which composts are applied (i.e., precautionary principle). In some cases, countries have adopted limits based (at least in part) on best available technology; that is, setting limits that are judged to be achievable after adopting currently available practices, such as a particular

---

[1] A trace element is an element that is present in soil or other media at concentrations below 0.1%. Heavy metals are elements with a density greater than 5 g per cubic centimeter. Because heavy metals are common trace elements in soils, they are often referred to as trace elements or, more specifically, trace metals.

separation technology (e.g., centrifugation). In the U.S., federal regulations do not impose trace element limits for composts other than those that contain biosolids. However, the biosolids limits are often used as guidelines for other composts. For example, all compost products accepted into the USCC's Seal of Testing Assurance program must meet the U.S. EPA's "exceptional quality" trace element requirements (U.S. EPA, 2021). The U.S. trace element limits are derived from a set of risk assessments that seek to determine the maximum level of trace elements that could be present without causing any adverse effect on human health or the environment. Using these different approaches, there are understandably very large differences in the established limits.

Table 15.6 provides examples of the limits used under the three approaches. The three approaches clearly result in very different thresholds. The ECN Quality Assurance Scheme (Siebert and Auweele, 2018) generally follows the "no net accumulation" approach. Its limits are similar to the upper range of soil concentrations of these elements. Model compost quality guidelines, developed by Canadian Council of Ministers of the Environment (CCME, 2005), consider both the "no net accumulation" and "best available technology" for each trace element. For composts with general use (Category A), the concentration limit selected is the higher concentration determined between the two approaches (Table 15.7). In addition, the CCME guidelines still allow uses for compost products that do not meet the Category A thresholds (e.g., field agriculture vs. lawns and garden). In addition, the cumulative addition of trace elements is capped for Category B composts. The CCME guidelines are more reflective of the upper range of trace element concentrations found in composts made from common feedstocks. The US standards clearly allow the use of composts with much greater trace element concentrations than normal soil concentrations and also allow substantial increases in soil concentrations of these elements.

Almost all trace elements in composts originate in the feedstocks. Composts made from clean source separated materials should have no problem achieving trace element concentrations below the CCME guidelines. Biosolids composts may exceed the Canadian standards, particularly for copper. If composts consistently exceed these limits, producers should analyze their feedstocks to determine the source of the contamination.

5.2.2 Organic chemical contaminants

The list of possible chemical contaminants in composts is long because feedstocks can harbor many chemical components. General categories include functional products like pesticides, cleaning products, personal care products, antibiotics, steroids, and other pharmaceuticals. Other chemicals are identified by their chemistry such as dioxins, PAHs, and PCBs. Most are organic compounds and decompose, to varying degrees, during composting. At the time of this writing, very few organic chemicals have been found to be troublesome enough to merit more than a general caution. The exception is a group of herbicides in the pyridine family. There have been cases of these herbicides surviving the composting process at concentrations large enough to cause harm to susceptible plants (Box 15.3).

Table 15.6 Compost trace element concentration limit values established by various countries.

Parameter	ECN Quality assurance[a] scheme	Canada—CCME guidelines[b] Category A (unrestricted use)	Category B	Max. Cumulative addition of B	US EPA Exceptional quality (EQ) biosolids[c]
	mg/kg dry weight	mg/kg dry weight		kg/ha	mg/kg dry weight
Arsenic	–	13	75	15	41
Cadmium	1.3	3	20	4	39
Chromium	60	210	1060[d]	210[d]	1200
Cobalt	–	34	150	30	–
Copper	300	400	757[d]	150[d]	1500
Lead	130	150	500	100	300
Mercury	0.45	0.8	5	1	17
Molybdenum	–	5	20	4	75
Nickel	40	62	180	36	420
Selenium	–	2	14	2.8	100
Zinc	600	700	1850	370	2800

[a] European Compost Network, Quality Assurance Scheme (QAS) for Compost and Digestate (Seibert and Auweele, 2018).
[b] Canadian Council of Ministers of the Environment (CCME) Guidelines for Compost Quality (CCME, 2005).
[c] A Plain English Guide to the EPA Part 503 Biosolids Rule, US Environmental Protection Agency (Walker et al., 1994).
[d] Category B limits for copper and chromium are not fully established. The values listed in Table 15.6 are based on calculated estimates. See CCME (2005) for details.

Table 15.7 Determination of the Canadian Category A compost trace element limits, considering both the "no net accumulation" and "best available technology" concepts. The values in bold italics are values designated as the "maximum trace element concentrations for Category A compost," as listed in Table 15.6.

Parameter	No net accumulation concept	Best achievable approach concept	Maximum trace element concentrations for Cat. A composts
	(mg/kg dry weight)		
Arsenic	10	*13*	13
Cadmium	*3*	2.6	3
Chromium	121	*210*	210
Cobalt	*34*	26	34
Copper	60	*400*	400
Lead	150	*150*	150
Mercury	0.15	*0.8*	0.8
Molybdenum	2	*5*	5
Nickel	*62*	50	62
Selenium	2	*2*	2
Zinc	500	*700*	700

Adapted from Appendix A in CCME (2005).

Organic chemicals are sporadically regulated in composts. A few countries have established limit values for very few chemicals, most consistently PCBs and PAH (Saveyn and Eder, 2015; Biala and Wilkinson, 2020). However, regulations are increasingly being considered for composts made from nonsegregated waste. Analysis of composts produced from clean, source-separated feedstocks have rarely found organic chemicals at concentrations that could cause any adverse effects.

Box 15.3 Compost's problem herbicides

Author: Robert Rynk.

As noted in Chapter 4, residues of most pesticides rarely create problems for compost users or producer, either because the pesticides are present in insignificant concentrations or because they decompose during composting. However, since the early 2000s, the pyridine family of herbicides has haunted the composting industry. Four pyridine products/compounds, in particular, have been found to cause problems in composts—clopyralid, aminopyralid, aminocyclopyrachlor, and picloram (USCC, 2021a). These products/compounds are used on a wide variety of agricultural and horticultural crops, including turf. One or more of these herbicides are ingredients in numerous commercial herbicide products.

At high-enough concentrations, pyridine residues in compost can damage susceptible plants, causing abnormal cupping of leaves, stunted growth, loss of compound leaves, and/or poor fruit set (Fig. 15.7). One particular problem with these herbicides is that the universe of susceptible

Continued

BOX 15.3 Compost's problem herbicides—cont'd

plants includes several favored families of crops such as nightshade/Solanaceae (tomatoes, pepper, potatoes), composites (sunflowers, daisy, asters), cucurbits (cucumber, squash), and legumes (peas, beans, clover) (Davis et al., 2020). A second problem is that "high-enough" concentrations are not high at all. These herbicides can damage susceptible plants at concentrations even lower than 10 parts per *billion*. The ability of pyridine herbicides to cause damage at very low concentrations is a challenge for composters. Even if these compounds are substantially diluted and degraded by composting, the residual concentrations may still be enough to cause damage if the compost is used to grow tomatoes, for example.

Many types of feedstocks can potentially bring pyridine herbicides to a composting facility. The list includes grass clippings, mixed yard trimmings (e.g., leaves with grass clippings), hay, straw, residues from miscellaneous crops, and livestock manure (e.g., cattle, horses). Manure is a difficult potential culprit because the original herbicide source can be livestock feed brought onto the farm without the farmer's knowledge. The herbicides pass through the animal largely unchanged and remain in the manure and bedding. Nevertheless, screening sources of feedstock is still one of the better ways to reduce the possibility of herbicide contamination (Goosen, 2014).

Because feedstock screening is far from perfect, composters should also test their compost products for potential contamination. Laboratory analysis is prohibitively expensive for routine testing (Coker, 2013). It is best used for investigation after a serious problem has emerged. However, composters can and should perform plant growth tests (i.e., "bioassay"), if herbicide contamination is even remotely possible. The general principle is to plant seeds of a sensitive plant (e.g., beans, peas, cucumber, tomato) in a mixture of the compost and common potting soil and evaluate the plants for signs of herbicide damage. The test should also include a control in which the seeds are grown in potting mix without compost added. Several plant bioassays have been developed for detecting the presence of pyridine herbicides in compost. Appendix O is the procedure promoted by the USCC (Whitt and Coker, 2015a).

FIGURE 15.7

Typical symptoms of pyridine herbicide damage including cupping of leaves, stunted growth, and loss of compound leaves.

Source: Courtesy of Washington State University, Whatcom County Extension.

5.3 Sharps

Physical contaminants are primarily an aesthetic issue with the exception of sharps—shards of metal or glass and needles. Again, the best cure is prevention by using feedstocks with no or little contamination. Physical contaminants often get reduced to smaller pieces during mixing and turning operations making them less visible, but also more difficult to remove by screening.

Sharps are are a safety issue for producers and users, as well as a liability concern for producers and marketers of compost. The presence of visible sharps in any compost is unacceptable. The presence of impurities such as glass or pieces of metal may make compost products unacceptable for growers of root crops in particular.

The major source of physical contaminants is poor source separation. Municipal solid waste (MSW) composts are most prone to unacceptable amounts of physical contaminants although composts made from source separated organics (SSOs) can also contain high levels. Such contaminants render the finished product unattractive to most end users; consequently, over the past decade, the number of facilities composting MSW in North America and Europe has declined to a handful. Yard trimmings and feedstocks generated on farms should be nearly free of such contaminants.

Limits for sharps are usually included in compost standards and regulations, either specifically identified as "sharps" or counted with large pieces of glass, metals, and rigid plastics (e.g., physical contaminants, impurities, or "inerts"). The limits are expressed either as a percentage of the dry weight of compost, typically <0.5%, or as a number of sharp items in a specified sample volume. For example, the Canadian compost quality guidelines specifies that a 500 mL sample of compost should contain no sharp items over 3 mm (about $1/8$ in.) in size to qualify for general use (CCME, 2005). Category B compost can have up to three sharp items per 500 mL sample but no sharp items over 12.5 mm (½ in.).

6. Compost quality standards

Compost quality standards are seen as a means to promote the industry, safeguard public health and the environment, encourage production of compost products meeting minimum requirements, and instill confidence in compost users that composts will not cause damage. Australia, New Zealand, most individual European countries as well as the European Union (EU), and Canada are among countries that have established compost quality standards or guidelines.

As an example, Table 15.8, presents the "minimum compost quality" specified in the 2018 PAS 100 "Specifications for composted materials" standard (British Standards Institute, 2018). The inclusion of tomato-based plant growth parameters is due to the possible presence of pyridine herbicides. Compost product performance characteristics are not qualitatively specified, but PAS-100 and many other compost standards require that the following parameters are tested and test results are declared: moisture content (or dry matter), OM, soluble salts (EC), pH and concentrations of N (total), P, K, Ca, Mg, and S. Depending on the intended compost use, values for total OC, C/N ratio and concentrations of Na and specified micronutrients may also have to be declared. PAS-100 is not a regulatory standard but a set of guidelines, or best practices, that apply broadly to the manufacturing of composts. Compost quality is just one component of the standard. It also covers allowable composting feedstocks, composting operational practices, monitoring, sampling, testing frequencies, test methods, and quality assurance procedures, among other topics.

Several of the European compost standards and regulations are very comprehensive. They go well beyond the product quality requirements, and could be seen and

Table 15.8 Minimum compost quality criteria within the 2018 PAS 100 standard.

Quality parameter	Allowable limit[a]
Pathogens	
Escherichia coli	1000 CFU/g fresh mass
Salmonella spp.	Absent in 25 g fresh mass
PTEs (trace elements)	
Cadmium	1.5 mg/kg dry matter
Chromium	100 mg/kg dry matter
Copper	200 mg/kg dry matter
Lead	200 mg/kg dry matter
Mercury	1.0 mg/kg dry matter
Nickel	50 mg/kg dry matter
Zinc	400 mg/kg dry matter
Physical contaminants	
Total glass, metal plastic and other nonstone fragments >2 mm	0.25% of the mass of an air-dried sample of which <0.12% is plastic
Stones >4 mm in grades of compost products other than mulch	8% of the mass of an air-dried sample (10% in mulch grades)
Stability/maturity/weeds	
Microbial respiration rate	16 mg CO_2/g organic matter per day
Seed/propagule germination	Zero in a liter sample of compost
Minimum plant response	
Tomato plant germination in compost/peat moss growing medium	80% germination, compared to germination in 100% peat moss control
Tomato plant mass in compost/peat moss growing medium	80% plant mass, compared to mass in 100% peat moss control
Tomato plant abnormalities in compost/peat growing medium	No abnormalities

[a] *Limits may vary within the UK. For example, in Scotland, the current allowable limit for plastics in compost is <0.6% of an air-dried sample.*
Adapted from BSI, 2018.

used as End-of-Waste codes. This case is made very clear in the European Compost Network's quality assurance framework, which specifically references End of Waste Criteria. The Compost Quality Protocol for England, Wales and Northern Ireland, in fact, "represents end of waste criteria for the production and use of quality compost from source-segregated biodegradable waste" (WRAP 2012).

In the United States, there is no comprehensive national standard for compost, despite much debate and discussion of the issue over many years. The US EPA has nationally applicable standards that affect the use of biosolids-based composts, in regards to heavy metals and pathogens, but these restrictions do not widely apply to other feedstocks. Some US states (CA, WA) have adopted compost standards, as have some state transportation departments (USCC, 2021c).

One major benefit of compost quality assessment is for marketing the product. With regular evaluation of key quality parameters, compost producers can be assured of their product's characteristics and consistency. Producers can also use compost quality assessment to monitor their composting process and to help identify and diagnose any problems with either process management or feedstocks. The following discussion of compost quality parameters will include consideration of both uses for product quality information.

Assessment against, and compliance with, compost quality standards is usually voluntary, unless public/environmental health and safety agencies adopt or cite them within regulations. Standards should ensure that the products meet requirements for soil, groundwater and environmental protection, as well as health and safety requirements. As such they could be incorporated into regulatory frameworks and should be considered as "minimum standards" that apply to composted products of all types. It can be argued that parameters relating to product performance and therefore end use are best included in voluntary product specifications. Voluntary specifications are typically developed at an industry (organization/association) level for classes of compost products intended for different uses such as agricultural soil incorporation, potting media, land rehabilitation, or mulching. They can also be developed by individual producers for specific products to more narrowly define product quality and ensure the desired performance for the intended end use, such as vegetable production, container media, and turf management. This concept of "fit-for-purpose" products and associated quality standards has been advanced to some degree in some European countries and is being pursued in Australia.

Several publications provide a comprehensive review of compost standards and their evolution. The most recent review, as of this writing, is International comparison of the Australian standard for composts, soil conditioners, and mulches (Biala and Wilkinson, 2020); the most detailed, at least for Europe, is end-of-waste criteria for biodegradable waste subjected to biological treatment (compost and digestate): Technical proposals (Saveyn and Eder, 2015). Other good reviews are listed in this chapter's references including Azim et al. (2018), Bernal et al. (2017), Hogg et al. (2002), and Brinton (2000).

7. Compost testing assurance

Assuring compost quality involves, at a minimum, (1) testing compost to determine many of the compost characteristics described in the previous sections, (2) a system to assure that compost samples are representative and that analytical procedures used produce reliable and accurate results, and (3) allows for the comparison of the results to established standards or other guidelines. Compost quality assurance efforts more commonly take on a broader scope that reaches beyond the compost itself. Various assurance programs also document that composting practices are sufficient to produce a quality product (e.g., sanitization specifications), and that quality control protocols are in place. Examples of such comprehensive quality assurance programs can be found in the UK's PAS 100 specifications (British Standards Institute, 2018) and the ECN Quality Assurance Scheme (Siebert and Auweele, 2018).

Internationally, there are several countries that have compost quality assurance programs (Saveyn and Eder, 2015). In addition to the UK and ECN, the German Compost Quality Assurance Association is probably the oldest and one of the largest and most successful quality assurance organizations. It currently certifies products generated by some 430 composting operations and 100 anaerobic digestion facilities. Their standards and reporting format cover both safe use of compost and critical information for end-users. Using certified compost provides agricultural users in Germany with some exemptions from strict EU/German regulations. In countries or regions with national standards for compost, quality assurance programs generally include voluntary certification that those standards have been attained. The Composting Council of Canada has a voluntary compost quality assurance program called The Compost Quality Alliance (CQA, 2021). The program's goals are to standardize testing and interpretation of quality results to engender consumer confidence in composts and their use.

In the U.S., composters can particip[ate in the USCC's Seal of Testing Assurance (STA) program, which is an information disclosure program (USCC, 2021b). The aim of this program is to provide compost customers a uniform and consistent basis to evaluate and compare compost products and understand their characteristics and use. The STA program allows participating compost producers to apply the STA logo to their product IF:

1. the product meets the USCC's definition of compost;
2. the facility complies with all federal state and local regulations, including limits for trace elements and pathogen destruction;
3. the compost was tested by a certified laboratory, using TMECC methods
4. compost products are tested at the required frequencies (based on the volume of compost produced); and
5. customers are provided the STA Compost Technical Data Sheet that discloses the compost test results, the parent feedstocks, and information about how to use the compost.

The STA logo certifies that the compost has been tested but does not certify that the compost meets any minimum quality or end-use based standards, except for pathogens and heavy metals. However, the USCC has a companion program known as the Consumer Compost Use (CCUP) program that includes product standards for three specific and common compost uses – lawns, trees and shrubs, and flowers and vegetables. Appendix Q provides more information about both the STA and CCUP programs with Internet links to individual program content pages.

8. Laboratory analysis of compost products

Having a laboratory analyze a compost product is essential, at least periodically. Compost users want to know the performance qualities of the compost that they are buying; regulators may want proof that the compost is safe for the environment and for users; and compost producers should want to know the characteristics of their composts (Sullivan et al., 2018). Third-party laboratory verification of test results is particularly important for customer confidence. In many cases, periodic testing of compost products is mandated by applicable regulations, compost standards, or compost testing assurance programs. The required frequency of testing typically depends on the feedstocks and amount of compost that a facility produces. Composts produced from biosolids, MSW, SSO, and other postconsumer food residues usually merit more inclusive and more frequent testing.

Although small- and moderate-scale compost producers may be exempted from compulsory testing, it is still in their interests to have compost products tested occasionally. Accurate analysis enables the composter to fine-tune the process and the products. Laboratory testing is most important when an operation is just beginning. Later when procedures change or when new feedstocks or a different source of material is being considered, additional analysis is prudent. Otherwise, laboratory analysis is needed for periodic quality control evaluations.

The most important compost quality parameters to have tested are those parameters that have been discussed in the preceding sections of this chapter and summarized in Table 15.9. Pathogens, nutrients and other chemical properties parameters must be tested by a professional laboratory. Most laboratories provide a package of standard tests, in addition to individual requested tests. The parameters that can be easily tested "in-house" are moisture/solids content, bulk density, pH, and soluble salts. Chapters 11 and 4 describe the procedures for conducting these tests. In addition, composters can assess compost maturity in-house using Solvita and/or plant bioassays.

Sullivan et al. (2018) provides a thorough discussion of the compost parameters commonly tested by laboratories and the methods used for testing. Similar information is often available from regional laboratories and universities (see Consulted and suggested references section).

Fig. 15.8 is a copy of the first page of a compost analysis report from Pennsylvania State University's Agricultural Analytical Services Laboratory (PSU AASL, 2021). The test results summarized in Fig. 15.8 are usually sufficient for the routine purposes of compost producers and users. However, a more comprehensive suite of analysis is often necessary to satisfy regulations, compost quality programs, and

Table 15.9 Compost quality parameters worthy of laboratory analysis.

General qualities	Nutrients	Environmental, safety and aesthetic qualities
Moisture or solids content[a]	Organic carbon	Heavy metals
pH[b]	Total nitrogen	Pathogens
Soluble salts (electrical conductivity)[b]	Organic nitrogen	Inert contaminants and sharps
Bulk density[a]	Inorganic nitrogen (ammonium and nitrate)	Persistent pesticide residues (bioassay)[a]
Organic matter or ash (LOI)	C/N ratio	Weed seed germination[a]
Stability (e.g., respiration)	Phosphorus and potassium	
Maturity[a] (e.g., bioassay, Solvita)	Calcium, magnesium, sulfur	
Particle size distribution	Micronutrients (Fe, Zn, Cu, B, Mn, Co)	
Water holding capacity	Sodium and chloride	

[a] Can be tested "in-house" (i.e., on-site) with commonly available supplies.
[b] Possibly can be tested "in-house" with appropriate equipment and training.

many compost users. Additional tests typically conducted include compost stability, maturity (i.e., bioassay), particle size distribution, pathogens, and trace elements. In addition, analytical laboratories often provide additional information to help clients interpret the test results, and may provide recommendations for use of the compost analyzed. As an example, Appendix P presents the full PSU AASL report associated with Fig. 15.8 It typifies the type of a report required by compost quality programs (e.g., STA).

Compost should be analyzed by methods approved specifically for compost and by laboratories that are certified for compost analysis. In any case, an accurate analysis depends on a good representative compost sample. There are many different methods for analyzing the parameters discussed in the previous section, but each different method is likely to give a different result. Furthermore, some laboratories use methods developed for soil, manure, or plant tissues, and these may not be appropriate for compost analysis without modification. Such methods not only give inconsistent and inaccurate results but also different laboratories would obtain very different results for the same compost.

In an attempt to address this inconsistency, most countries with a well-established composting industry have sanctioned standardized compost testing methods that are backed by standards-setting agencies and organizations (e.g., ISO, ASTM, BSI). Compost quality specifications stipulate which standard method(s) must be used when analyzing a particular parameter (e.g., soluble salts).

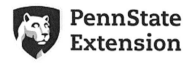

Agricultural Analytical Services Laboratory
The Pennsylvania State University
111 Ag Analytical Svcs Lab
University Park, PA 16802

(814) 863-0841   aaslab@psu.edu   www.aasl.psu.edu

Analysis Report For:	Copy To:
Compost Research and Education Foundation (CREF) PO Box 19246 Raleigh, NC 27619	

LAB ID:	SAMPLE ID:	REPORT DATE:	SAMPLE TYPE:	FEEDSTOCKS	COMPOSTING METHOD	COUNTY
C13133	Primed Compost	04/02/2021	Finished Compost	Spent mushroom substrate	Windrow	Wake

COMPOST ANALYSIS REPORT

Compost Test 3A

Analyte	Results (As is basis)	Results (Dry weight basis)
pH	8.7	——
Soluble Salts (1:5 w:w)	7.28 mmhos/cm	——
Solids	35.0 %	——
Moisture	65.0 %	——
Organic Matter	18.7 %	53.3 %
Total Nitrogen (N)	0.7 %	2.0 %
Organic Nitrogen[1]	0.7 %	2.0 %
Ammonium N (NH$_4$-N)	258.0 mg/kg or 0.0258 %	737.0 mg/kg or 0.0737 %
Carbon (C)	10.3 %	29.5 %
Carbon:Nitrogen (C:N) Ratio	14.40	14.40
Phosphorus (as P$_2$O$_5$)[2]	0.49 %	1.39 %
Potassium (as K$_2$O)[2]	0.78 %	2.22 %
Calcium (Ca)	2.33 %	6.66 %
Magnesium (Mg)	0.52 %	1.47 %
Particle size (< 9.5 mm)	81.57 %	——
Fecal Coliform[3]	4900 MPN per g	13998 MPN per g

[1] See comments on back of report .
[2] To convert phosphorus (as P$_2$O$_5$) into elemental phosphorus (P), divide by 2.29.   To convert potassium (as K$_2$O) into elemental potassium (K), divide by 1.20.
[3] Fecal Coliform subcontracted to Acme Labs, Altoona, Pa
Sample was overnighted and on ice. Received at 2°C, frozen

FIGURE 15.8

First page of a comprehensive compost laboratory analysis. This page contains the primary test results sought by compost producers and users. Appendix P presents the remainder of this report.

Source: PSU AASL (2021).

The USCC has compiled standardized methods for compost analysis within a single document known as Test Methods for the Examination of Composts and Composting (TMECC, 2012). In conjunction with TMECC, the council has also developed the Compost Analysis Proficiency proram (CAP) as a means of maintaining both the robustness of the TMECC methods and the ability of participating laboratories to achieve similar results. Other countries similarly carry out regular "Round Robin" testing for accredited laboratories.

With the proliferation of standardized testing, there are numerous analytical laboratories throughout the US, Europe, Asia, Australia, New Zealand, and other parts of the world that offer compost testing programs, using well-accepted standard methods, covering all of the parameters discussed above and more.

References

Cited references

Amlinger, F., Götz, B., Dreher, P., Geszti, J., Weissteiner, C., 2003. Nitrogen in biowaste and yard waste compost: dynamics of mobilisation and availability—a review. Eur. J. Soil Biol. 39, 107–116.

Amlinger, F., Line Diana Blytt, L.D., 2013. How to Comply with the EU Animal By-Products Regulations at Composing and Anaerobic Digestion Plants. European Compost Network. https://www.compostnetwork.info/download/good-practice-guide-comply-eu-animal-products-regulations-composting-anaerobic-digestion-plants/.

Azim, K., Soudi, B., Boukhari, S., et al., 2018. Composting parameters and compost quality: a literature review. Org. Agr. 8, 141–158. https://doi.org/10.1007/s13165-017-0180-z.

Bernal, M., Sommer, S., Chadwick, D., Qing, C., Guo-xue, L., Michel, F., 2017. Current approaches and future trends in compost quality criteria for agronomic, environmental, and human health benefits. Adv. Agron. 144, 143–233.

Bhogal, A., Taylor, M., Nicholson, F., Rollett, A., Williams, J., Newell Price, P., Chambers, B., Litterick, A., Whittingham, M., 2016. Field experiments for quality digestate and compost in agriculture.

BGK [Bundesguetegemeinschaft Kompost], 2021. Evaluation of analytical results of quality assured compost products 2020. https://www.kompost.de/fileadmin/user_upload/Dateien/Zahlen/Kompost_D_2020.pdf.

Biala, J., Wilkinson, K., 2020. International Comparison of the Australian Standard for Composts, Soil Conditioners and Mulches. Centre for Recycling of Organic Waste and Nutrients (CROWN), University of Queensland.

Bonhotal, J., Schwarz, M., Jack, A.L.H., Olmstead, D., Harrison, E.Z., 2008. Compost Quality Assessment for Use in Horticulture: Impact of the Composting Process. Cornell Waste Management Institute, Soil and Crop Sciences, Cornell University.

Brinton, W., 2000. Compost Quality Standards & Guidelines. Woods End Research Laboratory. Available from: http://compost.css.cornell.edu/Brinton.pdf.

BSI, 2018. Specification for Composted Materials. PAS 100:2018. British Standards Institution, London. http://bsigroup.com.

Buchanan, M., Brinton, W., Shields, F., Thompson, W., 2001. Compost Maturity Index. California Compost Quality Council, Nevada City, CA.

CCME, 2005. Guidelines for Compost Quality. Canadian Council of Ministers of the Environment, Winnipeg, Manitoba.

Coker, C., 2013. Composters defend against persistent herbicides. BioCycle 54 (8), 21. https://www.biocycle.net/composters-defend-against-persistent-herbicides/.

CQA, 2021. The Compost Quality Alliance. Compost Council of Canada. http://www.compost.org/compost-quality-alliance/.

Davis, J., Johnson, S.E., Jennings, K., 2020. Herbicide Carryover in Hay, Manure, Compost and Grass Clippings. North Carolina State Extension. https://content.ces.ncsu.edu/herbicide-carryover.

Emino, E.R., Warman, P.R., 2004. Biological assay for compost quality. Compost Sci. Util. 12 (4), 342−348. https://doi.org/10.1080/1065657X.2004.10702203.

Goossen, D., 2014. Persistent herbicides. In: Presentation at USCC Annual Conference. Dan Goossen, General Manager Green Mountain Compost, Williston, VT.

Haber, N., Kluge, R., Deller, B., Flaig, H., Schulz, E., Reinhold, J., 2008. Nachhaltige Kompostanwendung in der Landwirtschaft (Sustainable Use of Compost in Agriculture). Landwirtschaftliches Technologiezentrum Augustenberg, Karlsruhe, Germany.

Hogg, D., Barth, J., Favoino, E., Centemero, M., Caimi, V., Amlinger, F., Devliegher, W., Brinton, W., Antler, S., 2002. Comparison of Compost Standards within the EU, North America and Australasia. Waste and Resources Action Programme (WRAP).

PSU, A.A.S.L., 2021. Pennsylvania State University Agricultural Analytical Services Laboratory. https://agsci.psu.edu/aasl.

Saveyn, H., Eder, P., 2015. End-of-waste Criteria for Biodegradable Waste Subjected to Biological Treatment (Compost & Digestate): Technical Proposals. European Commission, EU Science Hub. https://ec.europa.eu/jrc/en/publication/eur-scientific-and-technical-research-reports/end-waste-criteria-biodegradable-waste-subjected-biological-treatment-compost-digestate.

Siebert, S., Auweele, W.V., 2018. European Quality Assurance Scheme for Compost and Digestate. European Compost Network. https://www.compostnetwork.info/ecn-qas/ecn-qas-manual/.

Siebert, S., Gilbert, J., 2018. Specifications for Use of Quality Compost in Growing Media. European Compost Network. https://www.compostnetwork.info/ecn-qas/ecn-qas-manual/.

Solvita, 2021. Compost Maturity Testing in Your Hands, On-Site. Solvita & Woods End Laboratories. https://solvita.com/compost/.

Sullivan, D., Bary, A., Miller, R., Brewer, L., 2018. Interpreting Compost Analyses. Oregon State University Extension Service. https://catalog.extension.oregonstate.edu/em9217.

TMECC, 2012. Test Methods for the Examination of Compost and Composting. U.S. Composting Council. https://www.compostingcouncil.org/page/tmecc.

USCC, 2021a. Persistent Herbicides FAQ. U.S. Composting Council. https://www.compostingcouncil.org/page/persistent-herbicides-faq?&hhsearchterms=%22herbicides%22.

USCC, 2021b. State Regulations. U.S. Composting Council. www.compostingcouncil.org/general/custom.asp?page¼StateRegulations.

USCC, 2021c. Seal of Testing Assurance (STA). U.S. Composting Council. https://www.compostingcouncil.org/page/SealofTestingAssuranceSTA.

USDA, 2021. National Organic Program (NOP). U.S. Department of Agriculture, Agricultural Marketing Service. https://www.ams.usda.gov/about-ams/programs-offices/national-organic-program.

Walker, J., Knight, L., Stein, L., 1994. A Plain English Guide to the EPA Part 503 Biosolids Rule. U.S. Environmental Protection Agency, Washington, D.C.

Whitt, M.B., Coker, C., 2015a. Implementing a Plant Growth Testing Program. U.S. Composting Council. https://cdn.ymaws.com/uscc.site-ym.com/resource/resmgr/images/USCC-PH-Fact-Sheet-3-for-web.pdf.

Wichuk, K.M., McCartney, D., 2013. Compost stability and maturity evaluation: a literature review. J. Environ. Eng. Sci. 37, 1505–1523.

WRAP, 2004. To support the development of standards for compost by investigating the benefits and efficacy of compost use in different applications. WRAP, Banbury, UK. www.wrap.org.uk.

WRAP, 2012. Quality Protocol Compost. WRAP, Banbury, UK. www.wrap.org.uk.

WRAP, 2015. DC-Agri; field Experiments for Quality Digestate and Compost in Agriculture – WP1 report, Prepared by Bhogal et al. WRAP, Banbury, UK. www.wrap.org.uk.

WRAP, 2014. Guidelines for the Specification of Quality Compost for Use in Growing Media. WRAP, Banbury, UK. www.wrap.org.uk.

Consulted and suggested references

Brinton, W., Bonhotal, J., Fiesinger, T., 2012. Compost sampling for nutrient and quality parameters: variability of sampler, timing and pile depth. Compost Sci. Util. 20, 141–149.

Campbell-Nelson, K., 2015. Compost Analysis and Interpretation. UMass Extension Vegetable Notes. https://ag.umass.edu/sites/ag.umass.edu/files/newsletters/november_5_2015_vegetable_notes.pdf.

Crohn, D.M., 2016. Assessing Compost Quality for Agriculture. https://doi.org/10.3733/ucanr.8514. Retrieved from: https://escholarship.org/uc/item/4v1576f8.

Dillon, J., 2013. Year-long investigation finds chemical herbicide in compost. Vermont Public Radio Transcript. https://www.vpr.org/post/year-long-investigation-finds-chemical-herbicide-compost#stream/0.

ECN, 2020. Soil Structure & Carbon Storage. European Compost Network. https://www.compostnetwork.info/download/soil-structure-carbon-storage/.

EUR-LEX, 2021. Regulation (EC) No 1069/2009 of the European Parliament and of the Council of 21 October 2009 Laying Down Health Rules as Regards Animal By-Products and Derived Products Not Intended for Human Consumption. https://eur-lex.europa.eu/eli/reg/2009/1069/oj.

Ge, R., McCartney, D., Zeb, J., 2006. Compost environmental protection standards in Canada. J. Environ. Eng. Sci. 5 (3), 221–234.

Gichon, Y., 2006. Compost Standard in the United States Department of Agriculture's National Organic Program Regulation: Implications and Limitations. Graduate Thesis. University of Florida, Gainsville.

Gilbert, J., Ricci-Jürgensen, M., Ramola, A., 2020. Benefits of Compost and Anaerobic Digestate when Applied to Soil. ISWA – International Solid Waste Association. https://www.iswa.org/media/publications/iswa-soils-project/#c8146.

Goldstein, N., 2013. Unraveling the maze of persistent herbicides in compost. BioCycle 54 (6), 17. https://www.biocycle.net/unraveling-the-maze-of-persistent-herbicides-in-compost/.

Heydarzadeh, N., Abdoli, M.A., 2009. Quality assessment of compost in Iran and the need for standards and quality assurance. J. Environ. Stud. 34 (48), 29−40.

Khater, E.S.G., 2015. Some physical and chemical properties of compost. Int. J. Waste Res. 05 (01). https://doi.org/10.4172/2252-5211.1000172.

McCauley, A., Jones, C., Olson-Ruiz, K., 2017. Soil pH and Organic Matter. Montana Stae University Extension. https://landresources.montana.edu/nm/documents/NM8.pdf.

Nova Scotia Agriculture, 2010. Understanding a Compost Test Report. https://novascotia.ca/agri/documents/lab-services/analytical-lab-manure-compost-report.pdf.

Ontario Ministry of the Environment, Conservation and Parks, 2019. Ontario Compost Quality Standards. https://www.ontario.ca/page/ontario-compost-quality-standards#section-2.

Pivato, A., Raga, R., Vanin, S., et al., 2014. Assessment of compost quality for its environmentally safe use by means of an ecotoxicological test on a soil organism. J. Mater. Cycles Waste Manag. 16, 763−774. https://doi.org/10.1007/s10163-013-0216-8.

Stenn, H., 2019. Specifying Compost: How to Get Quality Compost and Avoid Unnecessary Complications. Deeproot Blog. https://www.deeproot.com/blog/blog-entries/specifying-compost-how-to-get-quality-compost-and-avoid-unnecessary-complications.

University of Missouri Extension, 2020. Compost Analysis. https://extension.missouri.edu/programs/soil-and-plant-testing-laboratory/spl-compost-analysis.

van der Wurff, A.W.G., Fuchs, J.G., Raviv, M., Termorshuizen, A., 2016. Handbook for composting and compost use in organic horticulture. BioGreenhouse. https://doi.org/10.18174/375218.

WasteMinz, 2012. A Tool Kit for:NZS4454: 2005 the New Zealand Standard Conditioners and Mulches. https://www.wasteminz.org.nz/pubs/a-tool-kit-for-nzs4454-2005-the-new-zealand-standard-for-composts-soil-conditioners-and-mulches-2007/.

Whitt, M.B., Coker, C., 2015b. Strategies to Mitigate Persistent Herbicide Contamination at Your Composting Facility. U.S. Composting Council. https://cdn.ymaws.com/www.compostingcouncil.org/resource/resmgr/images/USCC-PH-Fact-Sheet-2-for-web.pdf.

Compost use

Authors: Monica Ozores-Hampton[1], Johannes Biala[2], Gregory Evanylo[3], Britt Faucette[4]

[1]*TerraNutri, LLC, Miami Beach, FL, United States;* [2]*Centre for Recycling of Organic Waste & Nutrients, The University of Queensland, Gatton, QLD, Australia;* [3]*School of Plant and Environmental Sciences, Virginia Polytechnic Institute and State University, Blacksburg, VA, United States;* [4]*Filtrexx International, Decatur, GA, United States*

Contributors: Leslie Cooperband[5], Nancy Roe[6], Jeffrey A. Creque[7], Dan Sullivan[8]

[5]*Praire Fruits Farm and Creamery, Champagne IL, United States;* [6]*Agricultural Consultant, Tucson, AZ, United States;* [7]*Rangeland and Agroecosystem Management, Carbon Cycle Institute, Petaluma, CA, United States;* [8]*Department of Crop and Soil Science, Oregon State University, Corvallis, OR, United States*

1. Introduction

Compost is a soil-amending product that improves the physical, chemical, and biological properties of the soils, resulting in increased crop productivity and enhanced environmental quality (Brown and Cotton, 2011). As such, compost is beneficial in nearly any plant-growing and land management application including landscape and turfgrass applications, field crop production (Fig. 16.1), grazing lands, fruit and vegetable production, organic agriculture, greenhouse and container plant production, remediation of brownfield and contaminated soils, erosion control, and even as a growing medium for roof-top gardens or green roofs. This variety of uses collectively entails a wide range of compost products with varying characteristics.

This chapter describes the practices and benefits of a wide array of compost uses in agricultural, horticultural, viticulture, forestry, urban landscape, and greenhouse systems and for erosion and stormwater control. Compost application practices are also discussed, both in general and specific to particular crops. It is important to realize that the effects of using composts depend on a range of variables, including the soil type and properties, the characteristics of the compost applied, the rate and frequency of application, and the intended use and benefits. The chapter is intended to give the reader an understanding of the many uses of compost and the factors that influence the type and rate of application. Sources of additional information about specific uses are cited throughout this chapter. For more comprehensive information about compost use generally, readers should consult "Compost Utilization in Production

FIGURE 16.1

Compost application to an agricultural field.

Source: M. Ozores-Hampton.

of Horticultural Crops" (Ozores-Hampton, 2021), "Handbook for Composting and Compost Use in Organic Horticulture" (Van der Wurff et al., 2016), "Organic Matter Management and Compost Use in Horticulture" (Biala et al., 2014), "Compost Use for Landscape and Environmental Enhancement" (Hartin and Crohn, 2007), and "Handbook Compost in Horticulture" (in German) (Zentralverband, 2002) and "Compost Utilization in Horticultural Cropping Systems" (Stofella and Kahn, 2001).[1]

2. General considerations for compost use

When considering the use of compost, it is useful to contrast compost characteristics with those of the raw or fresh organic materials that often serve as composting feedstocks. One difference between the properties of composted and uncomposted organic residuals is the increased biological stability of the composted organic matter. If the same amounts of organic matter are applied to soils, either from raw organic materials or as composts, the organic matter contained in uncomposted residuals decomposes faster, leaving about one third the amount of organic matter in the soil after 1 year, compared to that from compost (Fig. 16.2). A larger proportion of the nitrogen (N) in most raw organic residuals is water-soluble (i.e., mobile) and readily available to plants compared to N in the corresponding composts. This characteristic is a disadvantage for composts regarding supplying N to plants, but it is an environmental advantage in favor of composts as less N is potentially volatilized, leached from the soil, or lost in runoff. Additionally, uncomposted

---

[1] While several of these references are older publications, they remain among the more useful and comprehensive references about compost use.

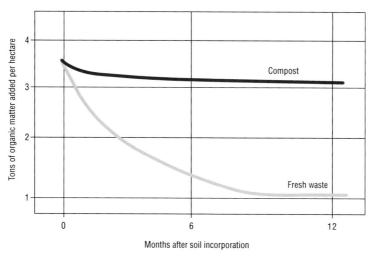

FIGURE 16.2

Degradation and residual organic matter quantities from fresh and composted materials.

Source: L. Cooperband.

organic materials are more likely than composts to contain pathogens of human, animal, and plant origins and viable weed seeds; thus, composting and compost use improves biosecurity and reduces food safety risks.

In making decisions about compost use, it is constructive to ask, "What is the intended purpose of using compost?" Is it to:

- build soil organic matter?
- supply plant nutrients?
- improve soil structure and water status (infiltration, availability)?
- enhance soil biological activity?
- suppress plant diseases?
- improve water and nutrient use efficiency?
- control weeds (e.g., mulch)?
- manage stormwater and reduce erosion?
- sequester carbon?
- overcome (sub)soil constraints?

In reality, many compost users do not have or realize that they have a specific purpose in mind. Most simply want to improve the soil. But the various purposes call for different combinations of compost characteristics and application practices. Compost use decisions become clearer after identifying major purposes. Still, it must never be forgotten that compost is a multipurpose product that can modify a wide range of chemical, physical, and biological soil properties and also deliver a range of benefits.

Depending on the intended use, compost quality parameters of likely interest include organic matter content, nutrient content, nutrient availability, C:N ratio, pH, soluble salts (EC) concentration (i.e., electrical conductivity), microbial activity, compost maturity (i.e., fitness for use and stability), contaminants (physical,

Table 16.1 Relative importance of compost characteristics to the purposes of compost use.

Compost use purpose	Sanitized[a]	Maturity	Organic matter	Nutrient content	C:N ratio	pH	Soluble salts	Microbial content	Particle size
Improve soil quality	✔✔	✔✔	✔✔✔✔	✔✔	✔	✔✔	✔✔	✔✔✔	✔
Control erosion	✔✔✔	✔✔	✔✔✔	✔	✔✔	✔	✔✔	✔✔	✔✔✔✔
Supply nutrients for plant growth	✔✔	✔✔	✔✔✔✔	✔✔✔✔	✔✔✔	✔✔	✔✔	✔✔	✔✔
Sequester carbon	✔✔	✔✔	✔✔✔✔	✔	✔✔	✔	✔	✔✔	✔✔
Conserve water and nutrients	✔✔	✔✔✔	✔✔✔✔	✔✔	✔	✔	✔	✔✔	✔
Suppress diseases	✔✔✔✔	✔✔✔	✔✔	✔✔	✔✔	✔	✔✔	✔✔✔	✔
Suppress weeds	✔✔✔✔	✔	✔✔	✔	✔	✔✔	✔	✔✔	✔✔✔

✔✔✔✔: most important; ✔✔✔: important; ✔✔: slightly important in many cases; ✔: not or rarely important.

[a] Sanitized = meeting sufficiently high temperatures for a prescribed time period, and number of turns, to destroy pathogens and weed seeds.

chemical, and biological), and particle size distribution (Table 16.1). Experienced compost producers and users recognize that composts derived from different feed-stocks possess variable characteristics that influence their suitability for various uses (Pittaway, 2004). The creation of "designer" composts has bolstered the concept of distinct value for various applications. Such designer composts are often fortified with products that enhance their properties and value for specified end-uses (e.g., lime, acidifiers, nutrients, biological control agents, chemical binding agents, microbial additives, and biochar).

Because the attributes of composts vary with feedstock composition and com-posting processing methods and duration, composts should be tested to determine their characteristics. Whether or not compost is used primarily as a nutrient source, its nutrient contribution should be calculated and accounted for within the crop fertility programs. Sound soil fertility management is necessary to prevent nutrient imbalances and protect surface water and groundwater from N and phosphorus (P) pollution.

Compost is most effective as a soil amendment when it is incorporated into the receiving soil, especially where heavy applications of compost are made to restore soils that are, for example, compacted or low in organic matter. Soil incorporation situates the compost within the soil matrix, where it functions best. It is important to apply the compost evenly so that changes in soil properties are uniform.

In many situations, it is not possible or practical to incorporate compost into the existing ground, as is the case with established perennial crops or in minimum-till farming. The compost can be applied to the surface soil as a top dress, side dress, or mulch. When applied at a sufficient depth (>25 mm or 1 in.), surface-applied compost functions as a mulch and provides similar benefits—moderates soil temper-ature fluctuations and reduces water evaporation, soil erosion, and weeds. After-ward, and over time, nutrients and organic matter from the compost enter the soil as particles are physically mixed into the soil, as microarthropods and earthworms carry organic matter downward, and as soluble inorganic and organic compounds trickle down.

It is important to use mature compost products for plants that exhibit sensitivity to unstable organic matter, high ammonia contents, and other plant growth-limiting factors, or where compost is not diluted with other components. Recently seeded fields, seedlings, and containerized plants are especially risk prone. Fig. 16.3 provides a conceptual view of how compost maturity changes during composting and how compost should be best used at various maturity stages.

2.1 Improving soils

Most compost users primarily want compost to improve soil condition; that is, build soil organic matter, improve soil tilth (structure), stimulate biological activity, reduce compaction, and/or improve water use efficiency. Such users are best served by stable composts with high organic matter content. All things being equal, compost with greater amounts of stable organic matter yields better results. Stable organic compounds boost the cation exchange capacity of soils, especially coarse (sandy) textured soils and soil-less growing media.

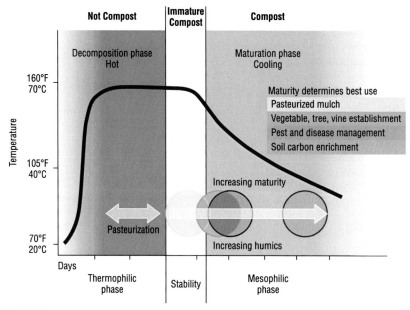

FIGURE 16.3

Conceptual view of the composting process and how it is best used at various maturity stages.

Adapted from Paulin, B., Western Australian Department of Agriculture.

Compost intended to improve the condition of poor-quality soils should be physically incorporated into the soil (Fig. 16.4). While surface applications eventually benefit the soils below, immediate incorporation greatly accelerates the effects. After the initial application of compost has been incorporated, further regular applications help to maintain soil quality and organic matter levels.

The pH of compost-amended soils tends to move in the direction of the compost pH, which is often slightly alkaline (Chapter 12). The pH change is faster and greater in soils that are poorly buffered (e.g., sandy soils with little organic matter) and low in clay content than in soils with higher clay content. Most composts have a small neutralizing value of 5% calcium carbonate equivalent in the dry matter (3% in fresh compost), compared with at least 80% for ground limestone or chalk. For example, the neutralizing value of 30 metric tons per hectare (t/ha) or 13.5 US tons/acre (tons/acre) of fresh compost is approximately equivalent to 2 t/ha (0.9 tons/acre) of agricultural lime. Regular compost application can maintain or even slightly increase the soil pH over time, but it cannot achieve major pH corrections.

Alkaline soils, whose pH is greater than 7, are common in arid and semiarid regions because little leaching and high evaporation cause ions to concentrate in the soil. Sodic soils, which are alkaline soils with excessive sodium (Na) concentrations create potential physical (dispersion, reduced hydraulic conductivity) and biological (phytotoxicity) problems. Compost products used to amend such soils should have low Na content and lower pH than most composts.

FIGURE 16.4

Compost is most effective when it is incorporated into the soil.

Source: K. Wilkinson.

Appendix N provides more detailed information about soil attributes and how soils are improved by organic matter, and by applying compost in particular (Box 16.1).

2.2 Supplying plant nutrients (it's mostly about N, but P and K deliver value too)

Macro and micronutrient concentrations in compost products vary widely, and rarely balance to match the N:P:K:Mg:Ca demand of agricultural and horticultural crops (Table 16.2). Therefore, other nutrient sources should be used to avoid long-term nutrient imbalances, deficiencies, or surpluses and to achieve optimum crop yield.

A grower relying on compost to supply N for crops should seek composts with a high N concentration and low to moderate C:N ratio. The C:N ratio of compost typically varies between 10:1 and 25:1, depending on the parent feedstocks and duration of the composting process. Compost with a C:N ratio higher than 25:1 can potentially temporarily immobilize N when applied to the soil. Finished compost typically has 0.5%−2.5% total N (of dry weight), depending on the parent feedstocks. However, only a small proportion of the N in mature compost is plant available, while most N is in an organic form that becomes available for plant uptake only after it has been mineralized.

Normally, more than 90% of the total N in compost occurs in an organic, slow-release form and 10% or less in readily available inorganic forms as nitrate-N (NO_3) and/or ammonium-N (NH_4). This aspect is very important when relying on compost as a major source of N for crop production. Although compost contains

---

Box 16.1 Deep placement of organic amendments for overcoming subsoil constraints

Most clay and sodic subsoils in the agricultural regions of southern Australia present a challenge to cropping. These dense and low macroporosity soils often contribute to surface and subsurface water-loging mainly during winter. The practice of subsoil manuring (deep placement of organic amendments) in the Victorian high rainfall zone has been evaluated since about 2003. The practice involves the placement of large amounts (usually between 10 and 20 t/ha or 4.5–9 tons/acre) of organic manure within the dense clay matrix in a single ripping operation. The soils studied so far are known to be very poor in their plant-available water capacity because of poor macro- and mesoporosity (and aeration) and therefore are a major impediment to root growth and proliferation. The following paragraphs, reproduced from Peries (2013), summarize results achieved in this field.

"Dense and impervious subsoils can be a major impediment to achieving potential water use efficiency from cereal and canola crops grown in high rainfall Southern Australia. The placement of large volumes of organic matter within and above the heavy clay layers of the soil in a single deep ripping operation, referred to as subsoil manuring, led to improvements in soil bulk density by 25% and a corresponding increase in soil macro-porosity. This resulted in nearly a 37% increase in plant-available water capacity that contributed to yield increases of between 27% and 96% across different soil types and rainfall regimes. The change in soil physical properties with deep placement of organic amendments, of the otherwise hostile subsoil, led to capture and storage of summer rainfall (with a near doubling of the summer fallow efficiency) and made it available to crops during grain filling in the following spring, when below-average rainfall conditions existed.

A collaborative research and extension program since 2007 looked at the technology, the resulting productivity increases, and the economics of the practice. With the magnitude of yield increases obtained, it was found possible to recover the costs of the initial operation in less than 4 years, and the effect of a single intervention showed no signs of diminishing after 4 years. The emerging technology represents a powerful adaptive strategy to a warming and drying climate with the potential to double the crop yield in the region and revolutionize the farming system."

---

comparatively low concentrations of soluble N, nitrogen loads can still be excessive if compost is applied at extremely high rates. Excess soluble N, regardless of its source, can move readily through the soil profile, resulting in reduced plant-available N and NO_3 enrichment of groundwater. Nutrient management regulations in many countries govern both mineral fertilizers and organic soil amendments. One of the advantages of using composts to build soil fertility is that it releases N slowly over a long period and therefore less likely to create an excess of N.

The amount of N from compost that becomes available to plants in the first year after application equals the inorganic N at the time of application plus the amount of organic N that becomes available due to mineralization. To convert, or "mineralize," organic N to plant-available N, the organic compounds in compost must biologically decompose.[2] The mineralization rate accounts for this conversion. N mineralization rates for composts reported by researchers range widely, from less than 0% to over 40% of the initial organic N content (Amlinger et al., 2003; Lee, 2016). Immature

---

[2] There is on-going debate and research about whether and to what extent plants can absorb and obtain nutrients from organic compounds (e.g., amino acids, peptides) without them having to be fully mineralized (Paungfoo-Lonhienne et al., 2012).

Table 16.2 Example of key characteristics and nutrient content of different compost products.

	Composted chicken layer manure	Composted cattle feedlot manure	Composted biosolids	Urban-derived compost[a]	Composted yard waste
pH	7.64	8.74	7.74	7.65	7.3
Electrical conductivity (dS/m)	10.95	11.24	3.69	4.25	2.2
Organic carbon (% dm)	24.5	20.2	20.1	16.8	46
Nitrogen (% dm)	4.32	2.60	2.18	1.58	0.93
Available nitrogen[b]	3003	1372	1503	403	14
Phosphorous (% dm)	1.75	1.19	0.94	0.33	0.24
Potassium (% dm)	2.55	3.03	0.98	1.01	0.50
Magnesium (% dm)	1.03	1.34	0.48	0.64	0.32
Calcium (% dm)	9.79	4.12	2.38	2.75	1.6

dm, dry matter (i.e. dry weight).
Notes:
[a] Urban-derived compost is a general term for compost made from mixtures of residential and commercial yard waste with lesser amounts of residential and commercial food waste.
[b] Ammonium (NH_4) + Nitrate (NO_3) in mg/kg.

Source: J. Biala, except for yard waste data, which is from Wilkinson, K., Paulin, R., Tee, E., O'Malley, P., 2002. Grappling with Compost Quality Down under. In: Michel Jr., S.C., Rynk, R.F., Hoitink, H.A.J. (Eds), Proceedings of the 2002 International Symposium on Composting and Compost Utilization. 6–8 May 2002, Ohio State University, Columbus, Ohio, pp. 527–539.

composts with a high C:N ratio can temporarily immobilize soil N, leading to negative mineralization rates, as low as -15%, which means that microbes use soil N making it unavailable for plants.

Considering all types of composts and soil conditions, a "typical" mineralization rate for compost in temperate climates falls in the range of $0\%-20\%$ for the first year after application. The higher ends of these ranges can be justified for composts that have a high N content, for tropical and subtropical climates, and where compost is applied frequently (e.g., annually). For instance, in regions with cold winters, rates in the range of $0\%-10\%$ are commonly reported. In regions with moderate winters, rates of $5\%-15\%$ are more common. However, there are additional factors that influence the availability of N from composts applied to soils, with soil moisture being a key factor in dryland farming systems (see Box 16.2). In the following 3 to 5 years after an application, it is reasonable to assume that $2\%-5\%$ of the initial organic N per year is further mineralized, bringing the total to between 20% and 40% of total organic N applied in compost. It is not certain how much of this mineralized N is used by plants. Yet, when considering that globally, N use efficiency (NUE) for fertilizer ranges between 35% and 50% (Raun and Johnson, 1999), NUE for compost doesn't look bad.[3]

Whenever possible, N mineralization estimates should be based on research, laboratory incubation, and/or experience for specific types of compost under local conditions. Also, testing the soil is recommended to obtain information about the potential release of N from soil organic matter, including compost applied in previous years. Efforts are currently underway in Australia to develop a web-based integrated organomineral nutrient calculator that allows farmers to estimate N, P, and K contributions from compost and manure and use mineral fertilizer only as a top-up to achieve the desired supply of plant-available nutrients for a crop (Fig. 16.5).

Phosphorus (P) contents in composts vary widely from less than 0.5% of dry weight to over 2% (Alexander, 2016). Composts generated from animal manures and biosolids usually contain higher concentrations of P than those generated from vegetative feedstocks. As with N, composts contain both immediate plant-available inorganic P and also organic forms of P that must be converted to soluble forms to become available to plants. As a rule of thumb, P-availability in the year of application ranges between 30% and 50%, and typically between 60% and 90% over 2 years. Virtually all P in compost becomes available to plants and needs to be accounted for in nutrient budgets and fertilization programs (Ebertseder and Gutser, 2003; Gagnon et al., 2012). If compost is applied at N-based application rates, P is normally oversupplied for plant needs. On land that is nearly saturated

---

[3] NUE refers to the amount of N actually used by the plants relative to the amount of N applied. NUE is not analogous to mineralization. Even though it is slowly released, mineralized N from compost, could be lost from the system due to asynchrony between the mineralized N and the plant N needs.

Box 16.2 Factors affecting the mineralization of organic nitrogen in composts

The range of N mineralization rates for compost is wide and difficult to accurately predict because mineralization is influenced by compost composition and soil, weather, and climatic conditions (Fig. 16.6). Relevant soil conditions include soil texture, moisture, temperature, and also the depth of incorporation, and the frequency/intensity of cultivation. Less mineralization takes place where microbial activity is suppressed by dry or cold soils or any soil conditions that reduce microbial activity. Conversely, once the soil microbiome is adapted after prolonged compost use, N release and NUE from compost are greatly enhanced compared to single and infrequent applications (Haber et al., 2008). Compost-related factors are compost maturity, N content, C:N ratio, and particle size distribution. Other relevant factors include the N demands of the crop and whether or not supplemental N fertilizer is used (Amlinger et al., 2003; Haber et al., 2008).

Compost N mineralization rate increases with N concentration and as C:N ratio decreases (Lee, 2016). It is a double benefit—there is more N and more of it becomes available. This benefit is largely due to increased microbial activity stimulated by available N. Similarly, when compost is applied annually, N mineralization increases over time due to organic N accumulation and increased soil microbial activity.

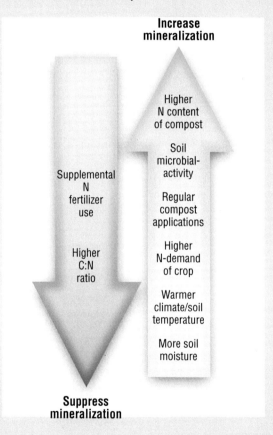

FIGURE 16.6

Factors affecting the rate of organic nitrogen (N) mineralization in soils.

Since compost characteristics are strongly tied to feedstocks, N mineralization rates are feedstock-dependent. The highest mineralization rates tend to be associated with composts made from N-rich feedstocks, such as animal manures, fish, biosolids, and mixed wastes that include substantial proportions of food. These N-rich composts can exhibit mineralization rates in the range of 15%−20% for the first year after application (Eghball, 2000). Yard trimmings (i.e., green waste) composts tend to have low to moderate N concentrations and relatively low mineralization rates, in the negative to 10% range. Negative mineralization rates occur when soil N is temporarily immobilized by composts with little N and high C:N ratios. It is especially true for composts with C:N ratios >20 and immature composts because more of the carbon remains readily available for degradation, for which microbes require N.

A related textbox in Chapter 14 presents a summary of selected research concerning N availability from composts.

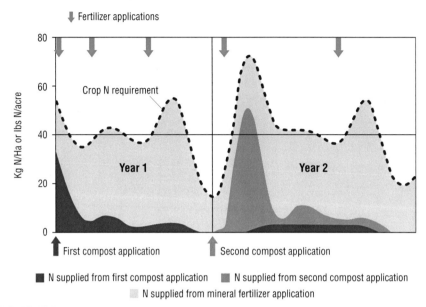

FIGURE 16.5

Model for integrated nitrogen (N) supply from organic soil amendments and mineral fertilizer in a horticultural cropping system.

Source: Adapted from Biala, J., 2017. Benefits of Accounting for Nutrients in Organic Soil Amendments, Vegetables Australia, pp. 24–25.

with P, compost applications should be based on crop P requirements, rather than N.

Most composts have potassium (K) contents in the range of 0.5%–1% of dry weight, although values as high as 3% and almost zero are possible. Nearly all of the K in compost is water-soluble and readily available to plants during the first year after application (Ebertseder and Gutser, 2003; Paulin et al., 2005).

When considering the economics of replacing mineral fertilizer with compost, P and K provide more value than N on a strict quantity basis. Table 16.3 provides an example where raw and composted animal manures were applied annually in an intensive horticultural production system. As the table shows, the replacement of P and K fertilizer provides much higher savings in fertilizer costs than the replacement of N. This aspect argues in favor of applying compost like a base fertilizer, that is, to first of all satisfy P and K crop demand. However, where the P and K requirements are low, the compost may fall far short of meeting the N demands of the crop, and therefore also fall short of its potential value.

Table 16.3 Quantity and value of mineral fertilizer replaced through raw and composted manure over two years in intensive horticulture.[a]

Products	Nitrogen	Phosphorous	Potassium	Total ($US)
	kg/$US			
Raw layer hen manure	152/$185	109/$450	249/$489	$1124
Composted layer manure	160/$195	139/$574	376/$738	$1507
Raw feedlot manure	286/$348	139/$574	376/$738	$1660
Composted feedlot manure	174/$212	139/$574	376/$738	$1524

[a] Note. P and K fertilizer replacement for composted layer manure and raw/composted feedlot manure are identical, representing standard fertiliser application, and indicating P and K supply above standard farm practice application rates.
Source: J. Biala.

2.3 Sequester carbon in soils[4]

A still-evolving use for compost is sequestration of carbon in the soil as a means to mitigate climate change. The addition of compost to soils in any application inherently increases soil organic matter and its carbon component. However, only a portion of this added carbon remains in the soil long enough to be considered "sequestered." Carbon sequestration does not refer to the short-term increases in soil carbon following a once-off application of compost, but the retention of some of the added carbon for periods of 25, 50, or 100 years. The magnitude of the sequestered portion is an elusive quantity because it is difficult to measure and depends on multiple factors. The potential for organic amendments to contribute to carbon sequestration depends on the quantity and quality of the amendment, the quality of the organic matter, the local soil, climatic conditions, and agricultural or land management practices. Nevertheless, compost applications are now being planned and studied with carbon sequestration as a prime purpose.

The Intergovernmental Panel on Climate Change (IPCC, 2019) and numerous government and nongovernment agencies consider soil carbon sequestration a key mechanism for countering increasing greenhouse gas (GHG) emissions (Teske, 2019). As noted by the French Ministry of Agriculture, even a small annual increase in photosynthetically derived soil carbon throughout the world's arable lands could significantly reduce atmospheric carbon dioxide (CO_2) levels (agriculture.gouv.fr/

---

[4] This section was contributed by Jeffrey A. Creque, Marin Carbon Project; Carbon Cycle Institute.

agriculture-et-foret/environnement-et-climat). Sequestration in soil of photosynthetically captured carbon offers the lowest cost, lowest risk option available for reducing GHG concentrations (Follett and Reed, 2010) while offering a host of ancillary ecological benefits. Yet only a handful of countries specifically address soil carbon sequestration and include them in their emissions reduction commitments to meet the Paris Agreement climate change goals. One of these countries is Australia, where the Emissions Reduction Fund provides a framework that enables farmers and other land managers to gain carbon credits for eligible activities, including the use of manures or composts in place of synthetic fertilizers (Clean Energy Regulator, 2021a).

Depleted and degraded soils have a high potential for sequestering soil carbon (Lal et al., 2015). Turfgrass, urban plots in need of renovation, and disturbed lands (e.g., mine lands) are common targets for such applications (Alvarez-Campo and Evanylo, 2019; Badzmierowski et al., 2015; Lal and Augustin, 2012).

Brown and Beecher (2019) modeled the carbon-benefits and carbon-costs of using biosolids compost for various urban uses in King's County, WA. Table 16.4 provides a summary of their model results. The most negative value provides the highest carbon saving.

Grasslands, or "rangelands," on which animals graze, also are considered a promising application, especially in arid regions. Rangelands are the most extensive land type on the planet, constituting roughly 50% of global land area (Holechek et al., 1998), and much of the world's agriculture is defined by forage-based livestock production on native, unimproved pastures. Rangelands have inherently low rates of natural soil organic carbon (SOC) stocks, primarily due to limited rainfall

Table 16.4 Modeled end uses for compost in an urban area with associated carbon benefits/costs. The most negative value provides the highest carbon saving.

End use[a]	Transport	Soil carbon	Fertilizer offset	Above-ground biomass	Balance
		(Tons CO_2 eq. per tons compost)			
Turf A homeowner with an established lawn drives their vehicle to pick up compost.[b] Uses compost as topdressing and adds additional fertilizer.	0.09	−0.036			0.054

Table 16.4 Modeled end uses for compost in an urban area with associated carbon benefits/costs. The most negative value provides the highest carbon saving.—*cont'd*

End use[a]	Transport	Soil carbon	Fertilizer offset	Above-ground biomass	Balance
	(Tons CO$_2$ eq. per tons compost)				
Turfgrass Homeowner/developer in a new home/community has compost delivered in bulk[c] to establish a lawn and does not use supplemental fertilizer.	0.005	−1.1	−0.09		−1.19
Trees Homeowner drives own vehicle to pick up compost[b] that is used as a topdressing on a mature tree	0.09	−0.036		−1.53	−1.66
Trees Homeowner/developer has compost delivered in bulk.[c] Compost is incorporated into existing soil to establish trees.	0.005	−1.1	−0.09	−4.58	−5.78
Highway Compost is tilled into surface soil for right-of-way to control erosion and establish a vegetative cover. Compost is delivered in bulk.[c]	0.005	−1.1	−0.09		−1.20
Home-garden A homeowner with an established lawn drives their vehicle to pick up compost.[b] Uses compost to grow vegetables on a portion of that lawn	0.09	−0.036	−0.09		−0.22

Continued

Table 16.4 Modeled end uses for compost in an urban area with associated carbon benefits/costs. The most negative value provides the highest carbon saving.—*cont'd*

End use[a]	Transport	Soil carbon	Fertilizer offset	Above-ground biomass	Balance
	(Tons CO_2 eq. per tons compost)				
Community garden A newly established community garden on derelict urban soil. Compost is delivered in bulk.[c]	0.005	−1.1	−0.09		−1.20

Notes:
[a] *See Brown and Beecher (2019) for details and assumptions.*
[b] *Homeowner obtain a 50 kg (110 lbs) load of compost, driving 20 km (12.5 miles) round trip. The fuel mileage rate is 6.4 km/L (15 miles/US gallon).*
[c] *Bulk delivery is made by 50 metric ton truck (55 US tons), driving a round trip distance of 20 km (12.5 miles). Fuel mileage rate is 4.25 km/L (10 miles/US gallon).*
Adapted from Brown, S., Beecher, N., 2019. Carbon accounting for compost use in urban areas. Compost Sci. Util. 27 (4), 227–239.

which restricts primary biomass production. The vastness of the world's rangelands offers large theoretical carbon storage potential. The Marin Carbon Project (MCP) in Marin County, California, has been investigating whether improved grassland management, including the use of compost, might measurably and durably increase SOC on rangelands (Box 16.3).

In MCP research trials, yard trimmings compost was broadcasted one time to a target depth of 13 mm (1/2 in.) over standing vegetation on grazed rangelands. This equates to an application rate of 130 m³/ha (about 70 yd³/acre). The control plots received no compost and no fertilizer (other than manure and urine from grazing cattle). This single application of compost had positive effects. Cattle preferentially grazed the treated plots (Fig. 16.7). Laboratory analysis revealed higher protein in the forage from those plots. After 1 year, compost-treated plots yielded 40% −70% more forage, while soil C also increased relative to controls (Fig. 16.8). Notably, greater plant productivity on treated plots resulted in net increases in total ecosystem carbon, above that contained in the applied compost, including soil carbon increases, resulting from plant photosynthetic additions, while control plots lost carbon (Ryals and Silver, 2013). This trend has continued through 10 years of data collection and at multiple additional trial sites across California (Mayer and Silver, 2018).

Box 16.3 The Marin Carbon Project (MCP)

Jeffrey A. Creque, Ph.D., Marin Carbon Project: Carbon Cycle Institute.

As stated on its website "The Marin Carbon Project (MCP) is a consortium of independent agricultural institutions in Marin County, including university researchers, county and federal agencies, and nonprofit institutions. We seek to enhance carbon sequestration in rangeland, agricultural, and forest soils through applied research, demonstration, and implementation in Marin County." (MCP, 2018).

The MCP began as a conversation around the question of marketing carbon credits derived from sequestration of atmospheric CO_2 on County rangelands. The carbon credits, embodied in grassland biomass and soil carbon, might be realized through improved rangeland management and other agricultural practices, under the general rubric of "carbon farming." This in turn highlighted the need for a scientifically rigorous soil carbon sequestration quantification protocol for verification and marketing purposes (Lal, 2007). To address this need, the fledgling MCP reached out to the University of California (UC) at Berkeley for assistance. Dr. Whendee Silver, an ecosystem ecologist and biogeochemist with extensive experience quantifying soil carbon, joined the Project as the lead scientist.

Drawing on the expertise of UC Extension, the Marin Agricultural Land Trust, the Marin Resource Conservation District, and local rangeland managers, a GIS-based sampling protocol was used to establish a carbon baseline for the majority of Marin's rangeland and pasture soils. The USDA Natural Resources Conservation Service (NRCS) and Marin Organic also recognized the potential for traditional conservation and organic agricultural practices to fall within a carbon sequestration framework and became active partners in the Project, providing administrative, technical, and outreach support. The MCP thus emerged, and continues, as a uniquely diverse and comprehensive coalition of researchers, producers, and Marin agricultural organizations and agencies, working toward the shared goal of effecting atmospheric GHG reductions through land management activities.

MCP's compost field trials began with applications of composted urban green waste to grazed rangeland sites, both on privately owned rangeland in Marin County and at the UC Sierra Foothills Research Center in the more arid Yuba County. The decision to use compost, rather than manure, a more common practice in the region, was prompted by the need to apply identical materials to experimental plots on both the Marin and Yuba County sites and the desire to avoid negative impacts often associated with manure applications.

FIGURE 16.7

Cattle distribution on experimental plots, showing a clear preference for the two compost treated plots (far left and second from right).

Source: John Wick.

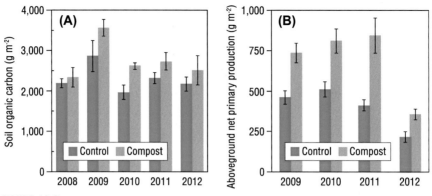

FIGURE 16.8

Marin Carbon Project results following the single 13 mm (½ in.) compost application in 2008 on California rangeland. (A) Soil organic carbon pool through 2012; (B) Aboveground forage production through 2012.

Based on data presented by: (A) Ryals, R., Silver, W.L., 2013.

Further information about soil carbon sequestration and other GHG mitigating effects of using compost is provided in *The benefits of using compost for mitigating climate change* (Biala, 2011), *Quantifying the benefits of applying quality compost to soil* (Gilbert et al., 2020) and *The potential for enhancing soil carbon levels through the use of organic soil amendments in Queensland* (Biala et al, 2021).

2.4 Using compost as mulch to manage soil erosion, water, and weeds

A mulch is a layer of material applied to the surface of the soil, and as such, compost products can be used as mulch to protect the soil, control weeds, reduce soil erosion, conserve soil moisture, moderate soil temperatures, and improve visual appeal. A typical mulch should be relatively coarse with a small proportion of fines so that water infiltration and gas exchange is not impeded. Mulch is comprised of coarse woody material that degrades slowly and lasts and functions much longer than mulch which is comprised primarily of fine materials. All recycled organic materials utilized as mulch in locations other than where the raw material originated from should be sanitized to ensure pathogens and viable weed seeds are not spread in the raw mulch. The "pathogen and weed-free" label is a major advantage composted mulches have over ground yard waste that is often given away as free mulch.

FIGURE 16.9

Composted screen overs applied as a coarse mulch to avocados.

Source: J. Biala.

Composted products applied to the soil surface as mulch, i.e., at a depth that is adequate to form a protective layer above the soil surface, buffer the impact of raindrops on the soil, lessening the likelihood that soil particles are dislodged and moved. Beneath the surface, organic matter in composted mulches binds with soil particles, increasing resistance to erosive forces form wind and water. Over time, composted mulches break down, adding nutrients and organic matter to the soil, potentially even aid in re-establishment of the soil's A-horizon (Biala and Chapman, 2017).

Coarse-textured compost, containing large woody particles, can be used as a surface mulch for vines, tree crops, and perennial berry crops and landscape plantings (Fig. 16.9). When used generally as surface mulch, composts can contain particles as large as 50—75 mm (2—3 in.). Compost used strictly for mulch can have a high C:N ratio (i.e. >25:1). Using a coarse mulch with a high C:N ratio limits the addition of N, something that may be required by viticulturists as fine compost mulch may increase canopy growth at the expense of grape/wine quality.

Compost used for mulching established trees and bushes needs to be pasteurized to ensure no weeds and diseases are introduced, but it does not need to be fully matured since it is applied on the soil surface and does not get into direct contact with new roots. One of the great success stories of compost use in Australia is the application of pasteurized compost as mulch under vines and other crops in South Australia. The economic analyses by the South Australian Centre for Economic Studies, based on two-year field trial results in vines, cherries, almonds, pear, and citrus suggested that on the whole, the quantitative benefits from lower rates of

FIGURE 16.10

Surface-applied blanket of compost used to reduce erosion and establish vegetation on a steep slope.

Source: B. Faucette.

compost applications outweigh the costs associated with the compost for most horticultural crops (SA Center for Economic Studies, 1999). Also, growers obtained considerable other benefits that were not quantified, including water savings, increased flexibility with water management, and soil improvements.

Immature compost only should be used as mulch on unplanted areas to discourage weed growth. The same processes that are undesirable for the growth of crops can be put to work to kill weed seeds and seedlings in the mulch (Ozores-Hampton et al, 2001, 2002).

More recently, surface-applied compost has found favor in environmental erosion and stormwater management applications, such as road construction (Fig. 16.10) and streambank protection. In environmental erosion control projects, compost particle size is important, and often specified by project standards, particularly where the product is applied with a blower unit. Compost products utilized for this kind of erosion control usually contain a mix of large and small particles, typically screened to <16 or 20 mm (5/8 or 3/4 in.). Composts contained in erosion control substrates to which seeds are added for establishing vegetation cover should meet the same quality requirements as composts used for horticulture (e.g., maturity, nutrients, and low salt). Compost to be used for erosion control and stormwater management should undergo the sanitation required to kill pathogens and weed seeds.

3. Compost application rates

Determining a meaningful and outcome-focused application rate for compost is seldom straightforward. Little or no data exist that relates plant growth or crop yield with rates of compost application. In lieu of such data, crop nutrient requirements or fertilizer recommendations provide the most objective means to estimate how much compost to apply to a given area. Other approaches to selecting application rates rely on empirical research results, user experiences, intuition, and/or practical and financial constraints.

All of the following approaches have been used to determine compost application rates. Only the first three have a mathematical basis for determining application rates. Meeting targets for organic matter and nutrients are covered in more detail in the following subsections:

1. Soil organic matter target—Armed with analysis of compost and soil characteristics, the compost user can determine the amount of compost needed to achieve the desired concentration of organic matter or carbon in the soil or soil mix.

2. Nutrient targets—Compost is applied at rates that meet some or all of the nutrient needs of the crop at an assumed yield. The target nutrient can be N, P or K. This approach is similar to that used for conventional fertilizer recommendations. It is likely to result in high application rates, in some cases higher than practical or supplying a surplus of nutrients, which risks water pollution.

3. Nutrient limits—Nutrients in the compost, usually P or N can constrain the rate of application if the soil is near its nutrient-loading capacity or if regulatory nutrient management requirements impose limits to annual nutrient loads. At a minimum, this approach requires knowledge of the concentrations of the nutrients in the compost, the existing concentrations in the receiving soils, and the concentration limits allowed on the land. The rate is determined mathematically by calculating how much compost can be added to a given depth of soil without significantly exceeding crop nutrient demand or regulatory nutrient supply limits allowable soil concentration of the nutrient in question.

4. Trace element and contaminant limits—Limits on trace elements or contaminants can restrict application rates if the compost is derived from feedstocks that are prone to such contamination (e.g., biosolids, MSW, etc.). This situation is rare in the United States and other countries with risk-based trace element limits, but it can come into play in countries with stricter limits (see Chapter 14). Regulations usually set limits based on the maximum allowable concentration of a contaminant that accumulates in the soil. The allowable application rate is calculated in much the same way as nutrient-based applications.

5. Local conventional practice—Compost application rates are based on historical use by growers in the local area. In some cases, these rates are developed and/or perpetuated by the local compost suppliers.

6. Local research results or growers' compost experience—Ideally, such recommendations should be specific to a given type of compost, crop, climate, and soil type or soil mix.

7. Cost of compost and its application—Rates are often constrained by how much compost a user can afford to apply in a given year. This approach typically yields lower-than-desired application rates, however, can have compound effects in a multiple-year annual application.

8. Availability of compost—Supplies of compost can be locally scarce. Application rates are determined by the amount of compost available and the area of land to be treated. Again, these rates typically fall short of soil organic matter or nutrient goals.

Regardless of how they are determined, compost application rates are variably expressed in terms of weight per unit area (e.g., kg/m^2, t/ha, Mg/ha, tons/acre),

Table 16.5 Formulas for converting application rates between weight, volume, and depth.

Weight $\leftrightarrow$ Volume:
General formulas:
(Weight per unit area $\div$ bulk density) $\times$ unit conversion factor $=$ volume per unit area
Volume per unit area $\times$ bulk density $\times$ unit conversion factor $=$ weight per unit area
Some specific "close-enough" formulas:

$(t/ha \div kg/m^3) \times 1000 = m^3/ha$	$m^3/ha \times kg/m^3 \times 0.001 = t/ha$
$(t/ha \div kg/m^3) \times 10 = m^3/100\text{-}m^2$	$m^3/100\text{-}m^2 \times kg/m^3 \times 0.1 = t/ha$
$(tons/acre \div lbs/yd^3) \times 2000 = yd^3/acre$	$yd^3/acre \times lbs/yd^3 \times 0.0005 = tons/acre$
$(tons/acre \div lbs/yd^3) \times 46 = yd^3/1000\text{-}ft^2$	$yd^3/1000\text{-}ft^2 \times lbs/yd^3 \times 0.022 = tons/acre$

Volume $\leftrightarrow$ depth:
General formulas:
Volume per unit area $\times$ area conversion factor $=$ depth
Depth $\times$ area conversion factor $=$ volume per unit area
Some specific "close-enough" formulas:

$m^3/100\text{-}m^2 \times 10 = mm$	$mm \times 0.1 = m^3/100\text{-}m^2$
$m^3/ha \times 0.1 = mm$	$mm \times 10 = m^3/ha$
$yd^3/1000\text{-}ft^2 \times 0.324 = in.$	$in. \times 3.1 = yd^3/1000\text{-}ft^2$
$yd^3/acre \times 0.007 = in.$	$in. \times 134 = yd^3/acre$

Convenient conversion factors

1 kg $=$ 2.2 lbs 1 lb $=$ 0.454 kg	
1 metric ton or tonne (t) $=$ 1000 kg $=$ 1.1 US. tons (ton)	1 US ton $=$ 2000 lbs
$= 0.907$ t $= 907$ kg	
1 meter (m) $=$ 100 cm $=$ 1000 mm $=$ 1.1 yards (yd)	1 yd $=$ 3 ft $=$ 0.91 m
1 inch (in.) $=$ 2.54 cm $=$ 25.4 mm	1 cm $=$ 10 mm $=$ 0.39 in.
1 hectare (ha) $=$ 10,000 m^2 $=$ 2.47 acres (acre)	1 acre $=$ 43,560 ft^2 $=$ 0.405 ha
1 square meter (m^2) $=$ 1.2 square yards (yd^2)	1 yd^2 $=$ 9 ft^2 $=$ 0.84 m^2
1 cubic meter (m^3) $=$ 1.31 cubic yards (yd^3)	1 yd^3 $=$ 27 ft^3 $=$ 0.765 m^3
1 m^3/ha $=$ 0.53 $yd^3/acre$	1 $yd^3/acre$ $=$ 1.89 m^3/ha
1 t/ha $=$ 0.45 tons/acre	1 ton/acre $=$ 2.25 t/ha

Table 16.6 Equivalent compost application rates for selected depths of application.

Depth of application in mm	Application rates per unit area			Approx. depth in inches	Application rates per unit area		
	m^3 per 100 m^2	m^3/ha	t/ha at 600 kg/m^3		yd^3 per 1000-ft^2	yd^3/ acre	Tons/ acre at 1000 lbs/yd^3
3	0.3	30	18	0.125	0.4	17	8
6	0.6	60	36	0.25	0.8	34	17
12	1.2	120	72	0.5	1.6	67	34
20	2	200	120	0.75	2.3	101	50
25	2.5	250	150	1	3.1	134	67
40	4	400	240	1.5	4.7	201	101
50	5	500	300	2	6.2	268	134
65	6.5	650	390	2.5	7.8	335	168
75	7.5	750	450	3	9.3	402	201
90	9	900	540	3.5	10.9	469	235
100	10	1000	600	4	12.4	536	268
125	12.5	1250	750	5	15.5	670	335
150	15	1500	900	6	18.6	804	402

volume per unit area (e.g., m^3/ha, yd^3/acre), and application depth (cm, mm or in.). Smaller-scale applications are sometimes related to areas of 100 square meters (m^2) or 1000 square feet (ft^2). Each of these terms expresses an amount of material for a given surface area. They can be converted from one to another using the formulas listed in Table 16.5. In Table 16.6 "does the math" for selected depths. As an example, for compost with a bulk density of 600 kg/m^3 (37.5 lb/ft^3or 1000 lbs/ yd^3), an application rate of 50 t/ha (22.3 tons/acre) equates to 83 m^3/ha (44 yd^3/ acre), which is only 8 mm (0.3 in.) in depth when evenly spread.

3.1 Target soil organic matter level

Organic matter level can be an objective criterion to measure the soil quality gained through the use of compost (Hinman, 2012; Stenn, 2018). It is, after all, the organic matter that carries most of the beneficial components of compost. In some urban land development situations, minimum soil organic matter levels are specified, either by regulations or by contract. For example, several municipal governments require minimum organic matter levels for the soil in new property developments.

With a specific soil organic matter target, a compost user can mathematically calculate how much compost to apply given the organic matter contents of both the compost and the receiving soil (Examples 16.1 and 16.2). The calculation is made on a dry weight basis assuming that the compost will be incorporated into the soil to a given depths (e.g., 15 cm or 6 in.). Eq. (16.7) calculates the required depth or thickness of compost to apply to achieve a target soil organic matter.

$$CT = D \times \frac{SBD_d}{CBD_d} \times \frac{(TOM - SOM)}{(COM - TOM)} \qquad (16.1)$$

where:

CT is the compost application depth thickness (cm or in.).

D is the depth of incorporation into the existing soil (cm or in.).

SBD_d is the *dry* bulk density of the soil (kg/m³ or lbs/yd³)*.

CBD_d is the *dry* bulk density of the compost (kg/m³ or lbs/yd³)*.

TOM is the target organic matter content of the amended soil (% or fraction).

SOM is the organic matter content of the existing soil (% or fraction).

COM is the organic matter content of the compost (% or fraction).

* Multiply the "as is" bulk density by the solids content to calculate the dry bulk density.

The equation calculates the application based on the dry bulk densities of the compost and soil. The dry bulk density can be determined by multiplying the "as is" or "wet" bulk density by the respective solids content as a fraction (Eq. 16.2). Also, the equation makes the reasonable but liberal assumption that the bulk density of the compost does not change after it is placed on the soil.

$$BD_d = BD_w \times SC = BD_w \times (1 - MC) \qquad (16.2)$$

where:

BD_d is the *dry* bulk density of the compost or soil.

BD_w is the *as-is* or *wet* bulk density.

SC is the *solids* content of the compost or soil (as a fraction).

MC is the *moisture* content of the compost or soil (as a fraction).

The TOM in Eq. (16.1) is the organic matter level of the soil just after mixing in the compost. From that point forward, some of the organic matter decomposes. The soil organic matter concentration will gradually fall unless more compost or other sources of organic matter are added. To achieve the target soil organic matter concentration over the long term, the compost needs to be applied repeatedly (e.g., annually). Alternatively, the target organic matter concentration can be set higher than the desired long-term goal, based on a "fudge factor" that accounts for organic matter depletion between applications.

Eq. (16.3) calculates an initial soil organic matter target (TOM_i) based on a constant annual organic matter depletion factor (k) and a final organic matter target (TOM_f) after a selected number of years:

$$TOM_i = TOM_f / (1 - k)^y \qquad (16.3)$$

where:

TOM_i is the initial target organic matter content of the amended soil just after the compost is applied (% or fraction).

TOM_f is the final target organic matter content of the amended soil a selected number of years after the compost is applied (% or fraction).

y is the number of years after the initial compost application.

k is the estimated annual rate of soil organic matter depletion (as a fraction).

Example 16.1 Compost application rate to meet an immediate soil organic matter target

Suppose that the local land development regulations require a minimum soil organic matter of 5% for all new housing developments, measured to a depth of 200 mm (about 8 in.). The developers have stripped the site of topsoil, leaving subsoil with and organic matter content of only 1% and a dry bulk density of roughly 1100 kgs per cubic meter (1850 lbs per cubic yard). The developer has the following two compost products available. How much of either compost will the developer need to apply?

Compost 1	Compost 2
Organic matter (COM1): 45%	Organic matter (COM2): 30%
"As is" bulk density: 600 kg/m^3	"As is" bulk density: 560 kg/m^3
Solids content: 50%	Solids content: 60%
Dry bulk density (CBD$_d$) = 0.5 × 600 kg/m^3	Dry bulk density (CBD$_d$) = 0.6 × 560 kg/m^3
= 300 kg/m^3	= 336 kg/m^3

For Compost 1:

$$CT = D \times \frac{SBD_d}{CBD_d} \times \frac{(TOM - SOM)}{(COM - TOM)} = 200 \times \frac{1100}{300} \times \frac{(5-1)}{(45-5)} = 73 \, mm (2.9 \, in.)$$

For Compost 2:

$$CT = D \times \frac{SBD_d}{CBD_d} \times \frac{(TOM - SOM)}{(COM - TOM)} = 200 \times \frac{1100}{360} \times \frac{(5-1)}{(30-5)} = 105 \, mm (4 \, in.)$$

All other factors equal (e.g., price, quality), compost 1 is the better choice.

3.2 Target nutrient applications rate

Compost application rates can be calculated to supply a selected amount of a plant nutrient. Usually, N is the desired nutrient. However, when it comes to the supply of nutrients from compost, it is often wiser to concentrate on P or K. It is rarely possible to supply the optimum concentrations of all nutrients with compost because all the nutrients are rarely in the right proportions for a given crop. Also, nutrient availability varies with temperature, moisture, and other factors that cannot be controlled. It is possible to grow acceptable crops with compost as the only added nutrient source (apart from legumes) if the receiving soil already has a large reserve of nutrients. Typically, nutrients supplied from compost supplement other sources of nutrients like commercial fertilizers, cover crops, and animal manure. It has often been reported that the best plant growth and yields come from compost combined with fertilizers to jointly supply the recommended amount of nutrients (De Rosa et al., 2017; Ozores-Hampton, 2012).

It is usually best to determine compost application rates based on the nutrient that requires the lowest application rate in order to avoid overapplication of nutrients. If an application rate is calculated to meet the N requirement of the crop, P might be oversupplied. In such cases, it is advisable to reduce the rate so that crop P demand is satisfied and rely on other sources to supply N, much of which is often supplied as side-dress fertilizer while the crop is growing.

The procedure to calculate nutrient-based compost application rates is the same for any nutrient. It is also used to calculate nutrient contributions from compost at any given application rate, regardless of whether the aim is to fully meet specific nutrient requirements or not. Nitrogen is used here to demonstrate the procedure. The steps are listed below and demonstrated in Example 16.3.

1. Determine the nutrient needs of the crop or planting from an authoritative source, like a crop consultant, agricultural university, or public agency. These values should be obtained in units of kg/ha or lbs/acre.

2. Estimate the anticipated nutrient contribution of the soil in kg/ha or lbs/acre. The best way to obtain this estimate is by sampling the soil to rooting depth (at least 0–30 cm or 0–12 in.) and having the sample analyzed by a laboratory. Convert laboratory results, usually provided in mg/kg or ppm, to nutrients per hectare (acre). The amounts of nutrients required from compost, plus supplemental fertilizer and other nutrient sources, equal crop nutrient requirements minus the estimated soil contribution.

3. Determine the total amount of N [organic-N, ammonium (NH_4-N), and nitrate (NO_3–N)] contained in the compost. Usually, and preferably, this step requires sampling the compost and having it tested by a laboratory (Chapter 12). Estimates can also be obtained from literature or other sources, which is a poor substitute for laboratory testing. If you are purchasing compost, the supplier should provide an up-to-date and comprehensive analytical test result.

4. Convert compost nutrient concentrations to values that make sense for land application—either kg per *dry* metric ton (kg/t) or pounds per *dry* US ton (lbs/ton). Generally, nutrient concentrations are reported as a percentage (%) of the *dry* weight. Lesser nutrients are commonly reported in units of milligrams per *dry* kilogram (mg/kg or ppm). Multiply % values by 10 to convert to kg/t or by 20 to convert to lbs/ton. Multiply mg/kg values by 0.001 to convert to kg/t or by 0.002 to convert to lbs/ton.

5. Estimate how much N will be available from the compost in the first year after application by multiplying the organic N by a mineralization rate and adding that to the amount of NH_4–N and NO_3–N. A reasonable N mineralization estimate for the initial year following application is 5%. If a higher rate of mineralization is expected, consider using a mineralization rate ranging between 5% and 20%; for example, if growing in a warm climate, when using certain plastic mulches (soil temperatures are higher), or using irrigation for high-N composts, like biosolids- and animal manure–based compost.

6. Determine the amount of compost to apply by dividing the amount of N *available* in the compost by the amount of N *needed* by the crop. The calculated application rate is in dry weight per unit area. Remember to calculate the rate in terms of wet weight, or "as is" weight, by dividing the dry weight application rate by the solids content (as a fraction). Remember that other sources of N (e.g., legumes, green manures, or fertilizers) might need to be added to make up the total N supply, which needs to match crop N demand.

7. Calculate the quantity of P and K supplied with the compost at this application rate. To calculate the correct application rate of compost, multiply by availability factors to obtain the amount of P and K that will be available to the crop from compost.[5] Where P and K requirements are given in terms of P_2O_5 and K_2O, multiply the total P by 2.29 and K by 1.2 to obtain P_2O_5 and K_2O values.

8. Compare the amount of nutrients supplied from compost with recommended nutrient application rates and consider if the chosen compost product and application rate are appropriate.

Example 16.2 Compost application rate to meet a long-range soil organic matter target

Suppose that the owners of the land in Example 16.1 want to establish a soil organic matter concentration that remains above 5% for five years after the initial application. No other applications will occur in the interim. The estimated soil organic matter depletion for the local area is 10% annually. What depth of compost should be applied if Compost 1 is used?

Given a 10% annual soil organic matter depletion over five years, the initial target soil organic matter content is:

$$TOM_i = TOM_f/(1 - k)^y = 5\%/(1 - 10)^5 = (5\%)/.59 = 8.5\%$$

Therefore, enough compost must be added to raise the soil organic matter initially to 10.2%. The required depth of application for Compost 1 is:

$$CT = D \times \frac{SBD_d}{CBD_d} \times \frac{(TOM - SOM)}{(COM - TOM)} = 200 \times \frac{1100}{300} \times \frac{(8.5 - 1)}{(45 - 8.5)} = 151 \text{ mm}(5.9 \text{ in.})$$

While 150 mm (6 in.) of compost may seem like a reasonable amount, it can be challenging to physically incorporate such a large application rate of compost into the soil.

Example 16.3 Application rates based on available N supply

A dairy-manure compost is to be applied to a strawberry field before planting. The compost has 1.48% total N, 628 mg/kg ammonium-N, and 2000 mg/kg nitrate-N. The P and K concentrations are, respectively, 1.78% and 0.37%. All concentrations are on a dry-weight basis. The compost moisture content is 43%. After considering the soil N and other sources, it is estimated that the strawberry plot needs an additional 34 kg/ha (30 lbs/acre) of N from the compost.

---

[5] Locally recommended P and K availability factors should be used. P availability factors can vary from 40%–50% in some regions to 90%–100% in others. K availability factors range from 80%–100%.

The estimated N mineralization rate is 5% in the first year after application. The recommended nutrient application rates for a 30 t/ha strawberry crop are 155 kg-N/ha (138 lbs-N/acre), 45 kg-P/ha (40 lbs-P/acre), and 236 kg-K/ha (210 lbs-K/acre).

Considering the results of this example, P is well oversupplied, even after considering that not all of the calculated P supply is available P. Therefore, depending on the P status of the receiving soil and the long-term cropping system, the compost application rate may need to be reduced to eliminate or reduce the P surplus.

Step 1: Nitrogen content of compost. Convert the organic and inorganic N levels to kg per dry metric ton (kg/t) or pounds per dry US tons (lbs/ton):

$$
\begin{aligned}
&\text{Total N} = 1.48\% && \times\, 10 = 14.8\ \text{kg/t} && \times\, 20 = 29.6\ \text{lbs/ton} \\
&\text{Ammonium-N} = 628\ \text{mg/kg} && \times\, 0.001 = 0.6\ \text{kg/t} && \times\, 0.002 = 1.3\ \text{lbs/ton} \\
&\text{Nitrate-N} = 2000\ \text{mg/kg} && \times\, 0.001 = 2\ \text{kg/t} && \times\, 0.002 = 4\ \text{lbs/ton} \\
&\text{Organic N} = \text{Total N} - \text{inorganic N} && = 14.8 - (0.6 + 2) = 12.2\ \text{kg/t (dry)} \\
& && = 29.6 - (1.3 + 4) = 24.3\ \text{lbs/ton (dry)}
\end{aligned}
$$

Step 2: Total plant-available N at 5% mineralization rate:

$$
\begin{aligned}
&\text{Organic N} = && 12.2 \times 0.05 = 0.6\ \text{kg/t} && 24.3 \times 0.05 = 1.2\ \text{lbs/ton} \\
&\text{Inorganic N} = && 0.6 + 2 = 2.6\ \text{kg/t} && 1.3 + 4 = 5.3\ \text{lbs/ton} \\
&\text{Total available N} && 0.6 + 2.6 = 3.2\ \text{kg/t (dry)} && 1.2 + 5.3 = 6.5\ \text{lbs-N/ton (dry)}
\end{aligned}
$$

Step 3: Amount of compost to apply for 34 kg/ha (30 lbs/acre) at 43% moisture (57% solids content).

$$
\begin{aligned}
\text{Dry weight} &= 34\ \text{kg-N/ha} \div 3.2\ \text{kg N/t} = 10.6\ \text{dry t/ha} \\
&= 30\ \text{lbs N/acre} \div 6.5\ \text{lbs-N/ton} = 4.6\ \text{dry tons/acre} \\
\text{Wet weight} &= 10.6\ \text{dry t/ha} \div 0.57 = 18.6\ \text{wet t/ha.} \\
&= 4.6\ \text{dry tons/acre} \div 0.57 = 8.1\ \text{wet tons/acre}
\end{aligned}
$$

Step 4: Total P and K applied.

P = 10.6 dry t/ha × 0.0178 = 0.189 t-P/ha = 189 kg-P/ha × 2.29 = 0.432 t-P_2O_5/ha = 432 kg-P_2O_5/ha
4.6 dry tons/acre × 0.0178 = 0.082 tons-P/acre = 164 lbs-P/acre × 2.29 = 0.188 tons P_2O_5/acre = 376 lbs P_2O_5/acre
K = 10.6 dry t/ha × 0.0037 = 0.039 t-K/ha = 39 kg-K/ha × 1.2 = 0.047 t-K_2O/ha = 47 kg-K_2O/ha
4.6 dry tons/acre × 0.0037 = 0.017 tons-K/acre = 34 lbs-K/acre × 1.2 = 0.020 tons-K_2O/acre = 41 lbs-K_2O/acre.

Step 5: Compare nutrient supply from compost with recommended application rates (= demand).

N	Demand	155 kg/ha	138 lbs/acre
	Supply from compost	34 kg/ha (22%)	30 lbs/acre (22%)
P	Demand	45 kg/ha	40 lbs/acre
	Supply from compost	189 kg/ha (420%)	164 lbs/acre (410%)
K	Demand	236 kg/ha	210 lbs/acre
	Supply from compost	39 kg/ha (17%)	34 lbs/acre (16%)

4. Equipment for spreading compost

Compost users have several different choices of equipment for spreading compost (Table 16.7). The selection depends primarily on the particular application, the rate of application (e.g., m^3/ha, tons/acre), and the compost product's physical characteristics (Alexander and Wagner). The rate of application, plus the area to be covered, influences the equipment size.

Table 16.7 Summary of compost spreading equipment and techniques.

Spreading device or technique	Suitable applications
Straight-drop brush	Precise application rates to golf courses, sports fields, vegetable beds, perennial field crops including small fruit, orchards, and nurseries. Longer application times due to the narrower breadth of the spread.
Beater box spreader	Best spreader for a wide range of compost moisture levels. Suitable to wet composts. Less precise but faster application of compost to mostly open fields. Commonly used for periodic applications of compost to agricultural fields, sports fields, and landscapes.
Spinner	Best spreader for most applications where compost is screened and dry (<45% moisture).
Side discharge broadcast	Convenient for areas with difficult vehicle access such as hillsides, over established vegetation, and among trees. It can also be used in open fields. Able to handle a wide range of composts, including moist compost.
Side discharge band/ mulch	Primarily used for concentrated applications to rows of crops, especially vegetable beds, vineyards, orchards, small fruit rows, and field nursery crops. Able to handle a wide range of composts, including moist compost.
Pile and grade	Efficient but imprecise spreading method where thick uniform applications (>25 mm or 1 in.) are to be made over large fields or relatively clear application area.
Pneumatic blower	A good option for spreading compost on difficult-to-reach areas like hillsides, forests, and other sites impeded by existing vegetated areas. Especially well-suited to erosion and stormwater management applications.
Small-scale spreaders	Spreaders are designed for small spaces, like gardens, commercial landscapes, and, small farms. Different models use different mechanical schemes (e.g., beaters, brushes, and spinners). They also differ in drive mechanisms and manner of propulsions (walk behind, ground-drive, tractor-pulled, and tractor-powered).
Fertilizer spreaders and air-seeders	Only suitable for pelletized and granulated compost products.

As described in previous sections, composts applied to agricultural fields are usually broadcast thinly and widely at relatively low application rates. Composts applied to many horticultural crops, like established vineyards and field-grown vegetables, tend to be placed at higher rates in narrow bands or concentrated in planting beds. Most turfgrass applications also tend to be widely broadcasts, but the application rates are much greater compared to farm fields. Erosion control applications vary from blankets to concentrated berms, but the application rates tend to be heavy and involve fibrous compost products with large particles. Residential applications are often best served by hand tools including shovels, rakes, and wheelbarrows. However, in larger landscape applications, mechanical spreaders are used.

In addition, the characteristics of the compost must be considered. Different mechanical spreading techniques are used depending on the physical characteristic of the compost. If composts are moist and coarse-textured, they may be difficult to apply. Compost that contains large particles, like branches and stones, can clog and even damage some types of spreaders. However, finely screened composts with moisture contents below 40% generally have fairly good "flow" characteristics and can be handled by commercial compost/manure spreading equipment. Pelletized and granulated compost are especially suitable for equipment that requires flowable material.

Seven basic mechanical designs are currently being used for spreading compost plus various small-scale units designed for low-volume applications (USCC, 2020a; Snohomish County Extension, 2020; Rahman and Widerholt, 2012). Each of these units has advantages and disadvantages depending on what the compost user wants to accomplish. There is generally a tradeoff between application uniformity and application time (i.e., efficiency). Faster machines are generally less accurate and consistent in placing compost evenly at the desired application rate.

- Straight-drop brush mechanism (Figs. 16.11A and 16.26A): Straight-drop brush spreaders discharge and disperse material directly out the back of the spreader box. The material is moved to the back of the box using a belt or chain conveyor. A rotating brush at the end of the box pushes the material out evenly in a swath equal to the width of the box. Deflectors can be installed to widen the swath further. This mechanism has been used for greens and tees by the golf industry for years. Larger versions have been used on athletic fields. The advantage is that it is the most precise method of application depth. The disadvantage is the slow application because the spread pattern equals only the width of the unit. Light application rates of finely screened compost are ideal for this mechanism.
- Beater box spreader (Fig. 16.11C and D): This spreader is similar to a beater-style solid manure box spreader, the type commonly found on farms. A conventional farm manure spreader is suitable for spreading moist and unscreened compost in most field applications. Clumps of compost are broken apart as they pass through the beater bars. The material travels to the back of the box via a belt or chain conveyor and then thrown outward through the action of one or more rows of horizontally or vertically positioned beaters at the end of the spreader box. Horizontal beaters throw materials only slightly wider than the

FIGURE 16.11

Various examples of compost spreading mechanisms: (A) straight-drop (B) spinner with dual spinner wheels, (C) truck-mounted horizontal beater box, (D) vertical beater box, (E) side discharge spreader in broadcast mode, and (F) spreading compost with a skid-steer bucket.

Source: (A, F) R. Alexander. (C, D) R. Rynk. (E) USCC.

box width. Spreaders with vertical beaters can produce a spread pattern about twice the width of the machine, which means fewer trips over the field compared to the straight drop brush mechanism. The uniformity of the application with beater mechanisms is less than that of straight drop brush spreaders or spinners. More compost is applied near the center of the machine, less toward the outsides. As with any other spreading operation, the uniformity should be improved by overlapping applications. With dry screened compost, box spreaders can produce an uneven spreading pattern because the compost tends to flow out between the beaters, particularly vertical beaters.

- Spinner mechanism—single or dual (Figs. 16.1 and 16.11B): Spinner mechanisms are commonly used to apply dryer compost materials, as well as fertilizer and lime. This mechanism is the best spreading method if the compost is screened (i.e., consistent in texture and without large particles) and relatively dry (e.g., well below 50% moisture) or granulated/pelletized. The material flows evenly to either the single or dual funnel that feeds the rotating spinner(s). Compost can be thrown 3—5 m (10—15 ft) to either side of the spreader so a dual spinner can easily cover a width of 10 m (30-ft) or more. The advantage of this mechanism is that it can cover a large width in one pass. The disadvantage is that many spreaders with spinners have relatively small loading capacity and that bulky and damp products are difficult to spread. Most commercially produced screened compost is spread with a spinner-type spreader.

- Combined beater box and spinner spreader (Figs. 16.12 and 16.13): To overcome shortcomings of applying compost with beater box spreaders (inaccurate and narrow spreading, not ideal for dry compost) and spreaders that employ spinners (low loading capacity, unsuitable for wet compost, tapered body results in bridging of moist and coarse products), some manufacturers have combined the two to create a new type of "universal" spreader that is equally suited for applying dry and wet compost, as well as manure and various other materials. The chain conveyor and horizontal beaters resemble a manure spreader, but the dislodged compost is caught in a box and drops down onto two spinners that distribute the material relatively evenly to a width of 10—12 m (33—39 ft).

- Side-discharge broadcasting spreaders (Fig. 16.11E): Side discharge spreaders unload compost to one side of the vehicle through a discharge conveyor. Augers within the storage hopper agitate the compost and move it toward the discharge point. These units are similar in appearance and design to horizontal mixers used for blending feedstocks. At the discharge point, rotating flails or paddles throw the compost outward at an upward angle, broadcasting it over a distance. The distribution is uneven as more materials fall closer to the spreader. The advantage is that the spreader can apply compost without driving through the application area. It is useful for applying compost broadly in difficult to access areas like hillsides and forests. It can handle a wide variety of compost products, including wet composts and coarse-textured composts.

(A)

(B)

FIGURE 16.12

Spreader with combined beater box and spinners.

Source: Ludwig Bergmann GmbH.

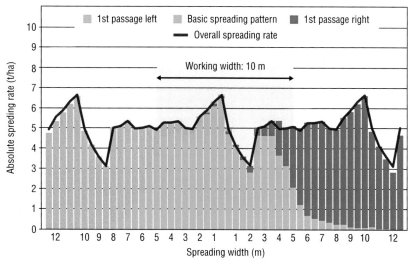

FIGURE 16.13

Spreading pattern produced by a combined beater box/spinner spreader for target compost application rate of 5 t/ha (2 tons/acre).

Source: Adapted from DLG, 2012.

- Side-discharge band/mulching spreaders (Figs 16.17 and 16.18): This type of spreader can place material in thick narrow bands or even piles. Up to the discharge point, they work in the same manner as broadcasting side-discharge spreaders. But at discharge, the compost is unloaded to the side through a chute that places the compost in a concentrated band. This type of spreader is well suited to applying compost to vegetable beds, orchards, vineyards, and berries where banding the compost along the crop is desired. This mechanism also suits other high-volume applications such as placing compost as mulch or laying down compost for manufacturing topsoil.
- Pile and grade (Fig. 16.11F): An efficient way to spread heavy applications of compost over a large area is to unload the compost in a pile and then evenly spread the compost over the field of application with a grading blade. The blade can mount on a tractor, wheel loader, bulldozer, or grader. The equipment operator simply pushes and pulls the compost from a pile over the application area with the blade held at the desired height. This approach is commonly used when application depths exceed 25 mm (1-in.). It is an imprecise way to apply compost, as the depth varies across the field, especially on uneven ground.
- Pneumatic blower truck mechanism (Chapter 19, Fig 19.7): This method of applying compost uses a blower to discharge compost through a manually directed large-diameter hose (minimum 10 cm or 4 in.). The advantage is that it can cover difficult areas and terrain that other mechanisms cannot access, such as tight corners, narrow landscapes, and steep slopes. The disadvantage is that the mechanism is very expensive and labor-intensive. An excellent use of this mechanism is for erosion control applications, especially filling filter socks that are used to filter stormwater runoff. It also allows for precise application around plants.

- Small-scale spreaders (Figs. 16.14B and 16.26): Small compost spreaders of varying designs and scale are available for low-volume and small-area applications like small farms, residential lawns, raised-bed vegetable plots, flower gardens, and small orchards. These spreaders use either spinners, beaters, or brushes to broadcast or top-dress compost onto lawns or gardens. Some of these spreaders are walk-behind machines that discharge compost through either a ground-driven or a motorized spreading device. Others are pulled by a small tractor or an all-terrain-vehicle. Some advanced equipment includes a control panel for precision rate application.

5. Specific agricultural, horticultural, and forestry applications

As a soil conditioner, compost offers a range of potential benefits for improving soil quality, and plant health, while also delivering environmental advantages. The most prevalent use of compost is in horticulture, the branch of agriculture concerned with the production and maintenance of vegetables, nuts, fruit, and ornamental crops. The functions and benefits of compost for horticulture have expanded to other types of agriculture. The use of compost should be tailored specifically to the cropping system and environmental conditions.

5.1 Field crops (small grains, maize, fiber crops, hay, and pasture)

Field crops include maize, sorghum, soybeans, wheat, barley, cotton, potatoes, sugar beets, sugar cane, hay, pasture, and many others. Traditionally, raw animal manures are used to supply nutrients and maintain favorable soil properties for these crops, which often cover vast areas of land. In many locations, compost application to field crops is uncommon because sources of raw animal manure are typically more abundant, and similarly suitable as a soil amendment. Turning raw animal manure into compost is costly, and farms are often not inclined to pay higher prices for compost. However, there is an increasing supply of composted animal manures, which can be co-composted with municipal or commercial organic residues. Composting of manure from intensive animal operations can provide marked benefits for both manure producers and compost users. However, there are situations where the land application of compost, rather than manure, makes sense for farms. Compost can be less costly to apply to remote fields than raw animal manure as it is a drier, lighter, and a more concentrated package of organic matter and nutrients compared to raw manure. If it is sufficiently dry, compost can be spread with broadcast-style spreaders (Fig. 16.14). Compost releases less odor and attracts fewer flies than uncomposted animal manure. Because of its reduced N availability, compost is well-suited to N-fixing crops like alfalfa and soybean.

Compost applied to field crops is commonly surface-applied and incorporated into the soil with tillage. Compost can be applied prior to planting, after harvest,

FIGURE 16.14

Manure-based compost applied to agricultural land at two different scales with (A) a large spinner-type spreader (at roughly 17 t/ha or 7 tons/acre), and (B) a small-scale box spreader with horizontal beaters.

Source: (A) J. Biala. (B) R. Rynk.

or to fallow fields. Because fields are not immediately seeded after application, the compost does not need to be fully mature and stable, which usually translates also to reduced purchase costs (Box 16.4).

Many farms apply compost to field crops primarily to replenish soil organic matter, which increases biological activity and improves physical soil properties, resulting in increased water infiltration and retention. In arid climates, improved water infiltration and retention benefits may return the greatest value. Improvement of plant available water through increased soil organic matter is greatest in sandy soil. For soil conditioning, application rates tend to follow local recommendations, adjusted by affordability. They commonly fall in the range of 5−20 t/ha (about 2−10 tons/acre), although compost is often not applied to the same field annually.

Box 16.4 Use of granulated compost in minimum-till farming

Granulated compost products provide an opportunity to efficiently use compost, without the need for special equipment, in situations where soil incorporation is impractical, such as some minimum-till and no-till farming systems. Granulated compost products can be delivered to the root zone of crops with subsurface fertilizer applicators or air-seeders during seeding operations. In these cases, compost application rates range between approximately 100 and 500 kg/ha or lbs/acre. This range is substantially lower than bulk applications of compost. However, the compost is delivered exactly to where plant roots grow, potentially improving nutrient use efficiency and amplifying the plant growth−stimulating effects of compost. The beneficial effects of compost use on soil organic matter and soil properties are modest due to low application rates. Therefore, the aim is to partially or fully replace mineral base fertilizer and boost germination and early plant growth and possibly trigger disease-suppressing factors. Additives such as micronutrients, lime, or gypsum can be incorporated into compost granules (Fig. 16.15) allowing the supply of crop and/or soil-specific products over long distances. Furthermore, the organic substrate in the compost granules can serve as a carrier for microbial agents that enhance crop growth (e.g., *Rhizobium*) and protect it (e.g., *Trichoderma* spp).

Continued

Box 16.4 Use of granulated compost in minimum-till farming—cont'd

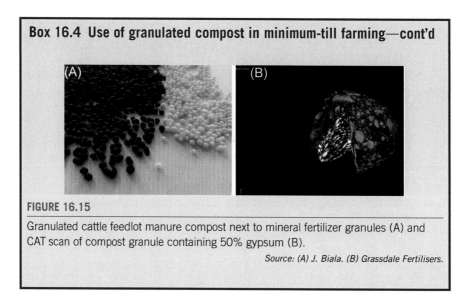

FIGURE 16.15

Granulated cattle feedlot manure compost next to mineral fertilizer granules (A) and CAT scan of compost granule containing 50% gypsum (B).

Source: (A) J. Biala. (B) Grassdale Fertilisers.

Where compost is a major source of nutrients, as in organic crop production, application rates may need to be heavier. The actual rate depends on the crop, compost nutrient content, organic N mineralization rate, preceding (legume) crop, existing soil N, and other nutrient sources used. Subsequent annual applications would gradually decrease due to the residual organic N in the soil. Applications rates can be lower if high plant available N-containing (e.g., animal manure—based) compost is used. Heavy application rates that speed up soil fertility building are costly and can increase the risk of surface water pollution due to rising soil P concentrations. They can also bump up to N or P application limits imposed by regulations, especially where fields have been historically manured (Box 15.2).

Nutrient balancing for individual fields and whole farms is an excellent tool for tracking nutrient inputs and outputs, and for recognizing nutrient deficiencies or surpluses that require corrective action, such as modifying the rate at which compost is applied or the type of product used.

In general, the best way to determine appropriate application rates and avoid excessive buildup of nutrients is by tracking crop nutrient removal, testing the compost and soil, keeping application rates below crop use levels, and applying the compost at times when nutrients will be taken up by crops (Box 16.5).

Box 16.5 Limits to field-applied nutrients

A significant increase in the use of fertilizer, combined with an equally significant increase in intensive animal production and generation of manure has resulted in a sharp increase in nutrient inputs on agricultural land. In Denmark, for example, N inputs on agricultural land increased from about 46 to 240 kg-N/ha/year (214 lbs/acre) between 1946 and 1990 (Hansen et al., 2017). The highest rates of N fertilizer application are found in the midwestern United States, western Europe, northern India, eastern China, Vietnam, Indonesia, and Egypt's Nile Delta.

Regulations now limit animal manure applications to some fields where concentrations of soil plant-available N or P exceed a critical level. For instance, the European Union's Nitrates Directive, which was introduced in 1991, limits application of N sources to maintain groundwater concentrations <50 mg NO_3 per liter (<10 mg NO_3–N per liter). The fact that the areas of England designated as nitrate vulnerable zones cover around 55% of agricultural land, demonstrates the significant impacts (Gov.UK, 2018a).

Farmers in Europe must meet quantitative and seasonal restrictions regarding N and P application to farmland. This covers both mineral and organic nutrient sources, including animal manures, compost, and any other organic soil amendments. As a result, depending on which crops are grown, farmers in the UK are allowed to apply up to between 120 and 370 kg-N/ha (107 and 330 lbs-N/acre), including mineral fertilizer and mineralized N from organic soil amendments, with a maximum of 250 kg-N_{tot}/ha (223 lbs-N_{tot}/acre) coming from organic sources in an individual field in any 12 months (Gov.UK, 2018b). Likewise, farmers in Germany can apply organic amendments at a rate of up to 170 kg-N_{tot}/ha (152 lbs-N_{tot}/acre), averaged over the entire agricultural area of the farm, but their nutrient plan and nutrient budget must also show surpluses of no more than 50 kg-N/ha (45 lbs-N/acre) and 4.5 kg-P/ha (4 lbs-P/acre) (Bundesamt für Justiz, 2017). Some countries recognize that most N in compost is organically bound and becomes available only over prolonged time periods, and therefore make special allowance for use of compost. In Germany, for example, total N applied with compost can amount to 510 kg/ha as an average over three years, provided no other organic amendments are used Schneider (2017) Because of prolonged winter periods, organic amendments cannot be land applied between 30 November and 15 February.

In other countries, including United States, Canada, and Australia, nutrient application limits are not as explicitly regulated, except for regulations in specific protected areas or watersheds, such as the Great Barrier Reef in Australia and the Chesapeake Bay watershed in the mid-Atlantic United States. Instead, applications of nutrients are governed by each farm's individual nutrient management plan. Periodic soil testing serves as a guide to adjust nutrient management plans as needed.

5.2 Vegetable crops

Vegetable crops are well suited to the use of compost. Most vegetables are intensively cultivated and require high levels of nutrients and water (Fig. 16.16). Moreover, many vegetables are produced on soils with low native organic matter. Composts can improve the overall soil fertility and available water status for vegetable production, particularly on sandy soils (Ozores-Hampton et al., 2011).

The effects of using composts on vegetables depend on the production systems and the compost characteristics (Ozores-Hampton, 2017a,b). It is often difficult to see yield increases when compost is added to a very intensive horticultural production system (e.g., plastic mulch, soil fumigation, irrigation, high fertilization rates, etc.). However, growers may find that the addition of compost allows them to reduce

FIGURE 16.16

Intensive vegetable production system. Tomato beds with compost-amended soils, plastic mulch, and drip irrigation.

Source: Monica Ozores-Hampton.

rates of inorganic synthetic fertilizers by as much as one-third to one-half and still produce comparable yields (Ozores-Hampton, 2012; De Rosa et al., 2017). There can also be changes in crop growth and quality, such as less fruit cracking and more even ripening, due to evening out of soil moisture levels. In low input and organic production systems, the addition of composts can be the main source of external nutrients and a critically important factor in ensuring satisfactory crop yields and quality.

An advantage of using composted rather than uncomposted animal manures as soil amendments in vegetable production systems is consumer safety and bio-security. Because compost undergoes sanitization under thermophilic temperatures, plant, animal, and human pathogens are largely destroyed and reduced to a level where they no longer pose a risk to public health (Millner, 2014). As such, vegetable producers have no time restriction on when compost can be applied to food crops, whereas raw or fresh inputs (particularly animal manures) must be applied well before the harvest date, which is 90−120 days in the United States (shorter and longer in other countries).

Compost can be broadcast throughout an entire field and then incorporated with discs, a plow, or other tillage equipment but it is more efficiently used when incorporated into the planting bed (Fig. 16.17). The product should be uniformly surface-applied on the bed and then incorporated to an approximate depth of 12−15 cm (5−6 in.) using a rototiller, or bed-former.

Using the compost as a mulch requires modification of seeding and/or transplant-ing methods. Whether mechanical or manual planting methods are used, planting

FIGURE 16.17

Compost application prior to incorporation and final bed-forming in large-scale field horticulture in Australia.

Source: C-Wise.

holes must be dug through the compost mulch, so that seeds or roots of transplants are placed directly into the soil. While compost mulch does help to control weeds, it is not as effective as polyethylene mulch. New plastic technology may allow the use of bio-based degradable plastic under compost. Compost mulches tend to moderate temperature changes, which is usually positive in warmer climates but may slow early production in a cooler climate.

Appropriate compost application rates in vegetable production are influenced by existing soil conditions, compost characteristics, and the nutrient requirements of the crop. In very sandy soils in tropical and subtropical climates, growers have used as much as 110 t/ha (about 50 tons/acre) for initial applications. More often, the first application ranges from 10 to 30 t/ha (about 5—15 tons/acre) with applications of 5—10 t/ha (about 2—5 tons/acre) in subsequent years. When high rates are used, N applied in the compost should not exceed crop requirements and the impacts on soil P levels must also be considered. Some specialty composts which are added mainly for beneficial microorganisms are used at rates less than 2 t/ha (1 ton/acre).

For growers who cultivate their fields intensively, and harvest several crops per year, it might be a challenge to find the right time to apply compost. A good time might be to apply it alongside other soil amendments (lime or gypsum) or with/instead of base fertilizer (if needed). If possible, it is best to apply the compost at the beginning of the growing season, a few weeks before sowing or planting. If the compost is mature and biologically stable it can be applied to beds immediately before sowing or planting. Alternatively, compost can also be applied before or after the growth of a cover crop or when the soil is fallow during the off-season, when more labor and equipment are available.

5.3 Small fruits

Strawberries, raspberries, blackberries, blueberries, currants, and gooseberries are all examples of small fruit crops. Because of the diversity of these botanically unrelated perennial crops, it is difficult to generalize about compost use on small fruit crops. Grape production, especially for wine, is a special case in itself and is presented separately in the following section.

Compost can be a valuable addition to all small fruit production. Compost use is integrated with other nutrient management and cultivation strategies including supplementary fertilizers, green manure, and cover crops, drip irrigation, and mulching (Fig. 16.18). The primary benefit of compost to small fruit production is that it improves soil properties, which is an important aspect considering small fruits are perennial crops with soil cultivation being limited to replanting (every 2–20 years). In addition, compost adds available and slow-release plant nutrients, which is a benefit that can be overdone.

FIGURE 16.18

Compost applied to blueberry plants as a mulch with drip irrigation at Oregon State University (Box 16.3).

Source: Bernadine Strik.

With small fruit crops, it is important to be aware of the amount of nutrients applied relative to the nutrient needs of the crop. Therefore, compost selection and application rates are important. Except for herbaceous perennials like strawberries, woody small fruits like raspberries, blackberries, currants, and gooseberries require very low amounts of N each year. Appropriate composts have low to moderately low N contents (e.g., <1%). Strawberries, on the other hand, are similar to annual crops in that they are sensitive to soil N status. At the early stages of growth, high N rates delay bud differentiation and promote too much vegetative growth. Too much N can lead to soft fruit, susceptibility to molds, powdery mildew, and lower yields. However, during reproductive growth, low N reduces flower quantity, root mass, and fruit size. Example 16.3 demonstrates the substantial nutrient input that can be delivered to a strawberry field before planting with composted dairy manure.

Compost analyses provides important information to guide compost selection. Some small fruit crops, like blueberry, prefer acidic soils and are sensitive to high salt (EC). Therefore, composts selected for these crops should maintain soil pH in an acceptable range and avoid plant-damaging levels of soluble salts.

Table 16.8 presents compost quality targets for blueberry production, based on recent research at Oregon State University (Sullivan, 2015). The C:N ratio of a compost can provide clues to its suitability. Composts with low C:N ratios (<12) tend to have N contents above 2% that supply too much N for blueberry production. In addition, low C:N ratio composts are often made from feedstocks that have unacceptably high pH, EC, and K, like mint distilling residuals, spent mushroom compost, and livestock and poultry manure. Most, if not all animal manure composts are not suitable for blueberry. Compost with C:N ratios in range of 12–25 can be acceptable for blueberry production if the pH, EC, and K values are not excessive. Yard waste compost is usually suitable for blueberry, but excess K may be a long-term problem.

Table 16.8 Compost quality targets for blueberry production.

Parameter	Unit	Preferable	Acceptable	Goal
pH	pH units	<6	<7.5	To maintain soil pH in the desired range of 5.0–5.5.
EC	dS/m (mmhos/cm)	<4 with the saturated media extract method <2 with the 1:5 method <1 with the 1:10 method		To avoid salt-related plant damage
Total K	% dry weight	<0.7%	<1.3%	To avoid excess accumulation of K in the plant
Total N	% dry weight	<1%	<2%	To avoid an excess supply of plant-available N

Source: Sullivan, D., 2015. Compost for Blueberry Plants: Testing and Tips. Oregon State University Extension Service. https://extension.oregonstate.edu/crop-production/berries/compostblueberry-plants-testing-tips.

When used before plants are established, composts may be tilled into the soil at low to moderate application rates. After the initial plant establishment, compost can be added as topdressing if and when required. Timing the application of compost in small fruit, as in all other crops, is different than for adding mineral fertilizers because nutrients, particularly N and to some degree P become available for plant uptake only over time. When applying compost, allowance has to be made for the mineralization and the associated release of nutrients, which is slow when soil temperatures are low. For example, June-bearing strawberries have a high nutrient demand in the previous fall as they produce flower buds for the crop the following season. Compost may need to be applied in late summer so it will have sufficient time to decompose and release nutrients in time to meet plant nutrient demand in fall. Applying compost at the wrong time can result in vigorous plant growth late in the season, which delays hardening off of the plants and can lead to winter injury.

Woody or fibrous composts can be used as mulch to control weeds, reduce soil loss from erosion, improve water infiltration and soil water retention, and moderate soil temperatures. Compost mulches keep the soil cool so bushes and vines may not "leaf out" or bloom as early as in bare soil, and in areas where spring freezes are common, later blooming can save fruit crops. However, this delay can be a disadvantage if growers are trying to get early crops. Mulches can also moderate temperature changes in the winter, helping to protect root systems, plant crowns, and graft unions. Mulch is particularly effective in dry areas, where there is no irrigation available as it reduces water loss through evaporation.

To discourage weed growth, the mulch should be at least 5−10 cm thick (2−4 in.) and it should be coarse enough to hold down weed growth. When mulch is applied to raspberry plantings, it is best to use a fine particle-size compost because primocanes have difficulty emerging through large clumps. However, fine mulch with appreciable nutrient content encourages weed growth (Box 16.3). This problem can be overcome by placing compost beneath plastic mulch (Box 16.6).

Box 16.6 Ten years of field research using yard trimmings compost as a mulch for blueberry production

Author: Dan Sullivan.

Oregon State University recently concluded a long-term organic systems research trial studying compost and other materials as mulch for blueberry plants (Strik et al., 2017, 2019). Three mulches were compared in a 10-year field trial: (1) sawdust; (2) yard-trimmings compost topped with sawdust (compost + sawdust); and (3) black, woven polyethylene groundcover (weed mat). The mulches were applied after planting in 2006, and were replenished in 2011, 2013, and 2015 (Fig. 16.8).

Results:

- Cumulative berry yields during the 10-year trial were, on average, about 5% greater in weed mat plots than with either the sawdust or the compost + sawdust mulches.
- The compost + sawdust mulch increased weed control costs substantially compared to sawdust alone.

Box 16.6 Ten years of field research using yard trimmings compost as a mulch for blueberry production—cont'd

- Compost applications supplied over 900 kg (2000 lb) of total N, 160 kg (350 lb) of P, and 900 kg (2000 lb) of K per acre during the trial. Most of these nutrients were stored in soil and will contribute to future crop nutrient needs.
- Compost increased soil organic matter. During the last five years of the trial, soil organic matter averaged 4.1% under compost + sawdust, 3.3% under sawdust, and 2.9% under weed mat,
- With compost, soil pH was maintained in optimum range for blueberry (4.5−5.5). During the last five years of the trial, soil pH averaged 5.2 with compost + sawdust, compared to 5.0 with sawdust alone.
- Exchangeable soil potassium (K) increased from 220 ppm with sawdust alone to 360 ppm with compost + sawdust mulch (Fig. 16.19).

FIGURE 16.19

Blueberry research plots receiving three mulch treatments: sawdust, compost + sawdust, and weed mat. Note: Fig. 16.18 shows the initial application of compost to these plots, before the sawdust was applied.

Source: Bernadine Strik.

5.4 Compost use in vineyards

Compost is used in vineyards primarily as a source of nutrients and for its soil improvement and water-saving benefits (Biala, 2000; Eleonora et al., 2014). Compost also helps to control weeds. Although it can be broadcast across the entire vineyard area, including inter-row areas, compost is typically applied to vines on the soil surface like mulch, concentrated in a band under the vines (Fig. 16.20). Compost

FIGURE 16.20

Pasteurised coarse compost applied in a concentrated band to vines using a side-discharge spreader.

Source: (A) J. Biala. (B) K. Wilkinson.

is not without potential drawbacks in grape production. In humid regions, wine-grape quality can suffer from the higher soil-water-holding capacity plus excess available-N late in the growing season. Grapes grown in such regions benefit from water stress and low N and K availability.

Different compost types, preferably with low N content, can be equally useful for application in vineyards. One of the unique considerations in vineyards is the opportunity to use composts produced from grape processing residues, primarily grape pomace or marc, but also vine trimmings. Most compost produced from grape pomace plus amendments have a moderately high N content, in the vicinity of 1%−2.5%, and relatively high available K levels in the range of 2%−3% (Patti et al., 2009). If several types of compost are available to the grower, then choices should be based on the content of N and other nutrients, salt content, microbial activity, pasteurization, hauling considerations, texture, and moisture levels which affect the ease of application.

Grapes do not require high levels of N and too much N or available N at the wrong time can be detrimental to fruit production and wine quality. As a general guide, annual compost application rates tend to be in the range of 5−30 t/ha (about 2−15 tons/acre) depending on the compost used and past applications. However, application rates and timing should be specific to each situation, and based on nutrient requirements of grapes as well as N concentrations and expected N supply after analysis of the soil, compost, and possibly plant tissue (see Section 2.2). If vines are growing well, with adequate N already in the soil, then yard compost, with little or no available N, may be most appropriate to avoid an oversupply of N. On the other hand, if N and soil organic matter are low and plants are growing poorly, then it is appropriate to use compost with a higher N content and mineralization rate, or a higher rate of compost with moderate N. The key is to know the N content of the compost through laboratory analysis and know how much N needs to be applied to the vineyard. Compost should be applied either in the spring before bud break or in the fall after the grape harvest, avoiding summer when the available N might encourage excessive vegetative growth.

Because of the high application rate and N load, compost is used as banded mulch under vine only once every 3—5 years. A 50 mm (2 in.) deep application of compost amounts to 160—170 m³/ha (2287 to 2430 cubic feet/acre), depending on vine row spacing (Port Phillip and Westernport CMA, 2020). For compost with a bulk density of 0.5 t/m³, these application rates equate to 85 t/ha (38 ton/acre) and 125 t/ha (55 ton/acre). Yet, as the area under vine covers only about one-third of the total vineyard area, the actual rate of application where mulch is used is three times the tonnage shown above. This is a critically important aspect when considering potential accumulation of nutrients and soluble salts.

Several references listed at the end of this chapter focus on the use of compost in vineyards, including Port Phillip and Westernport CMA (2020), Westover (2019), Travis et al. (2003), and Biala (2000). Australian Native Landscapes, Australia's largest compost manufacturer, provides a range of technical and research information about the use of compost and mulch in viticulture and tree crops on its website (ANL, 2021).

5.5 Tree crops

Since most tree crops should last for 10—20 years or more, it is beneficial to prepare the soil before planting trees. These initial rates of compost can be high, 50 to 100 t/ha (about 20—50 tons/acre) if broadcast over the entire area. With cover crops, 10 to 20 t/ha (about 5—10 ton/acre) is equally effective. Applications can be also considerably lower if compost is banded in future tree rows (Fig. 16.21) It is important to incorporate the compost well to encourage uniform distribution in the soil profile.

After initial tree establishment, compost can be added annually as mulch (Fig. 16.22). The mulch should be 50—100 mm thick (2—4 in.) and coarse enough to discourage weed growth. Otherwise, the nutrients in the fine compost may promote more weeds. Mulches are particularly effective in dry areas by reducing

FIGURE 16.21

Compost applied in a band where future row crop will be planted.

Source: C-Wise.

FIGURE 16.22

Compost applied under mature macadamia trees with side-delivery belt and spinner in Australia.

Source: J. Biala.

water loss through evaporation, also assisting in reducing irrigation water use (Uckoo et al., 2009). Growers have observed increased tree growth in the first few years when compost mulches were used. Some growers have reported that compost applications increase fruit size and enhance fruit flavor. Mulches can moderate temperature changes in the winter, helping to protect root systems, plant crowns, and graft unions. Compost mulches also keep the soil cool so trees may not "leaf out" or bloom as early as trees growing in bare soil. Since that can be a problem if growers are trying to get early crops, some growers wait until after bloom and soil warming to apply mulches. However, in areas where spring freezes are common, later blooming can save fruit crops.

Some growers chip their tree trimmings to use as mulch. Although the chips can be used directly, composting them first is preferred. Composting reduces pathogens, insects, and the tendency for fresh chips to immobilize N. Since woody trimmings are low in N, an additional N source is needed for the composting process. On a cautionary note, compost mulches applied right up to the trunk of the tree may increase small rodent infestations and damage to tree trunks. Such application strategies may also exacerbate crown fungal diseases. These problems can be avoided if mulches are kept away from tree trunks. For example, in citrus orchards grown in a subtropical humid environment such as Florida, compost mulches applied soon after planting helped to decrease root diseases. There is also evidence that compost use can discourage diseases in apple and cherry orchards (Granatstein and Dauer, 1999; Brown and Tworkowski, 2004; Dupont and Granatstein, 2018). In Australia, it was demonstrated that mulching of citrus trees with high rates of pasteurized yard trimmings compost, in particular, resulted in the reduction of citrus thrips and improved yields through increased numbers of predatory insects and other beneficial effects (Crisp et al., 2013). The use of mulch markedly improved the economic performance of the orange orchard over 6 years.

It is not only coarse composted mulch that is used in tree crops but also screened compost products at lower application rates, which aim to supply macro and micronutrients, improve soil properties, and re-establish topsoil (Ozores-Hampton et al., 2015). Usually, it takes a range of integrated management changes to achieve significant changes in soil properties, soil health, productivity, and profitability. For example, a California almond orchard increased yield by 10% and reduced pesticide and water use after implementing sustainable practices that included compost use (BioCycle, 2020). The benefits resulted in a 553% return on investment. In contrast, compost use in two macadamia nut orchards in Australia was financially beneficial in the low-input but not the high-input orchard (Biala and Chapman, 2017). Furthermore, the project re-established a topsoil horizon under trees and significantly reduced soil erosion.

Dupont and Granatstein (2018) and Baldi et al. (2021) provide more specific details about the beneficial uses of compost in orchards.

5.6 Forest crops

Just as it is used in orchards, compost can be used to establish forest tree seedlings and as a mulch to protect young trees. The use of compost is especially helpful on slopes where it can prevent erosion while protecting tender young tree roots.

Some studies have shown long-term benefits of applications of composts to forests (Rockwood et al., 2012). Slightly immature compost at rates over 100 Mg/ha (50 tons/acre) may be used in silviculture. Forestry offers an opportunity to use composted biosolids, which may have some restricted use in edible crops and urban landscapes. However, the compost must still be of acceptable quality for this use and without excessive contamination.

Application can be the most difficult part, since trees are not always in regular rows, and repeated trips with a spreader can damage tree roots and forest floor ecology. Side casting or "slinger" type spreaders and pneumatic blower trucks can apply compost between trees with less disturbance to the forest floor. More information about compost application to forest land can be found in "Compost Use in Forest Land Restoration," (Henry and Bergeron, 2005).

6. Nursery and greenhouse applications

Nursery and greenhouse operations differ from other agricultural enterprises largely by the production of higher-value ornamental crops and container-grown plants, although nursery crops are also grown in field soils, and greenhouse operations also produce vegetable crops. For these industries, compost is used primarily to manipulate the qualities of growing media, add micronutrients, and/or improve biological properties. The potential for disease suppression is an especially attractive benefit to these industries (Chapter 17). The nutrients supplied by compost is usually unimportant in greenhouse production, but it is beneficial for nursery crops, which need more time to grow to their market size.

In container media, compost is combined with other ingredients like peat moss, coconut coir, bark, perlite, rock wool, and sand. A strong incentive for using compost is to replace much or all of the peat moss conventionally used. The amount of compost typically used falls in the range of 20%−33% by volume, although it varies from 5% to 50% for various applications (Table 16.9). The proportion depends on the crop, its stage of growth, the qualities of the compost, and grower preferences. Plants grown in containers are sensitive to soluble salts, high ammonium-N levels, and pH, especially seedlings and ericaceous plants (e.g., rhododendrons, blueberry). Much more information about the qualities of compost for use in container media can be found in "Guidelines for the Specification of Quality Compost for use in Growing Media" (WRAP, 2014b) and "*Specifications for Use of Quality Compost in Growing Media*" (Siebert and Gilbert, 2018).

In countries where peat was not readily available, such as in Australia, growing media based on composted bark and sawdust were successfully developed and refined in the 1970s and 80s (Handreck, 2014). Extensive information on growing media is available in a book on "Growing Media for Ornamental Plants and Turf" (Handreck and Black, 2010).

Many greenhouses and nurseries can make composts using trimmings and discarded plant materials from their own operation. While this is a sensible way to recycle discards, trimmings, and growing media, the composting process must achieve pathogen-killing temperatures for a sufficient time. "Compost Production for use in Growing Media—a Good Practice Guide" (WRAP, 2014a) offers additional information about compost production practices specifically for container media.

Table 16.9 Suggested volume percentage of compost in container growing media.

Container media application[a]	Compost[b] by volume (%)
Seed mixes	5%−10%
Bedding plants	20%−25%
Pot plants	20%−25%
Nursery stock (general), excluding ericaceous species[c]	30%−35%
Nursery stock (vigorous), excluding ericaceous species[c]	35%−50%
Multipurpose growing media	20%−40%

[a] These recommendations are for composts mixed with low nutrient, low conductivity ingredients.
[b] For compost that meets quality standards for container growing media.
[c] For ericaceous plants, the percentages depend on the pH, soluble salts, and other qualities of the specific compost used.
Adapted from WRAP, 2014b. Guidelines for the Specification of Quality Compost for Use in Growing Media, WRAP, Banbury, UK. www.wrap.org.uk.

FIGURE 16.23

This nursery manufactures its own compost for use in its field and container production operation and also sells much of the compost produced.

6.1 Field nursery production

In field nurseries, compost can be incorporated into the soil before planting nursery crops, it can be used to fill holes from harvested plants before future cultivation and plantings, or it can be used as mulch (Fig. 16.23). While compost can improve the soil for nursery field stock, the stability of the growing plant and the final root ball must be considered. Using a light compost to fill holes after digging trees can result in wet soil and/or unstable trees. If plants are harvested from the field before being firmly established, composts may be too light to hold the root ball together. On the other hand, compost tends to encourage root growth, which helps to form a stable root ball.

Application depths for initial incorporation in tree rows or planting holes usually vary from 12 to 25 mm (1/2 to 1 in.) based on soil, compost cost, and the value of the crops to be planted. Rates at the high end of this range are used for poorer quality soils and/or higher value crops.

As with tree crops, a 30−50 mm (about 1−2 in.) layer of large-particle compost can be used as mulch, especially around trees that will be in the field for several years. If trees or shrubs are subject to crown or root rot diseases in a particular climate and soil, mulch should be kept several inches away from the tree trunk or main stem of the shrub.

6.2 Container nursery production

There are wide differences among various container-grown nursery crops in their responses to compost in their soil mix. The best approach is to obtain compost use recommendations for specific crop species from agricultural research and extension resources and/or nursery industry experts and organizations. In any case, a new mix should be tested on a small number of container plants before fully adopting a container mix that includes compost. For most plants, however, an appropriate container mix consists of 20%–35% compost by volume. Some growers use up to 60% compost in the mix, depending on the plant species and compost characteristics.

Composts derived from yard trimmings, and other low-nutrient feedstocks, rarely contain enough nutrients to meet the nutrient requirements for profitable growth of nursery crops. The best growth is usually obtained when a nutrient-rich compost is used and inorganic fertilizers are used to top up where compost nutrient supply is inadequate. Due to the slow release of N and P, compost may be able to supply more of the nutrient requirements for slow-growing species. Testing of the growing media, knowledge about crop nutrient demand, including temporal variation, and plant tissue testing provide information about potential nutrient deficiencies and help prevent them.

There are additional considerations beyond the growth of the plant in the container mix. The low bulk density of most composts may be positive because it makes the pots lighter to ship than soil-based media. However, pots that are too light can be unstable and easily blow over by the wind. Also, if landscape plants are to be planted in heavy clay soil, a very light container mix can result in a waterlogged root ball after planting.

6.3 Greenhouse production of seedlings and annual crops

Compost is used largely as a substitute for peat moss and pine bark in greenhouse container media. Like from peat moss, growers expect little or no fertility from compost. In fact, nutrients supplied by compost can complicate precise fertilization schedules. In planting beds, compost serves as a soil amendment and nutrient source in the same manner as it does for field vegetables, although the compost quality is more important in greenhouse beds (Fig. 16.24).

Greenhouse production requires the most mature, stable, and finely screened composts for inclusion in container mixes or incorporation into planting beds. Since many seedlings and annual crops are susceptible to elevated soluble salt levels, particularly sodium (Na) and chloride (Cl), composts used for these situations must be low in salts, which is measured as EC. The Australian Standard for Potting Mixes (AS3734-2003), for example, requires seedling mixes to have an EC of ≤ 1.5 dS/m and contain ≤ 60 mg/L of Na. Compost used in seedling mixes is usually screened to 6 mm (¼ inch), or even smaller for planting fine seeds. Growers also need to know that the compost has the same consistency each time they use it since crops are often grown using standardized nutrient and water application rates. Also, smooth operation of filling and planting equipment may depend on the texture of the container media used.

When used in a greenhouse container media, compost typically comprises 20%–30% of the mix, along with other ingredients such as perlite and/or vermiculite and peat moss or alternatives such as composted bark and sawdust or coco peat

FIGURE 16.24

Compost used as a soil amendment and nutrient source in greenhouse beds producing salad greens.

(Fig. 16.25). Compost is rarely used alone as a container medium. Most composts are too porous, have inadequate water-holding capacity, and the soluble salt levels are often high. If the compost is low in salts (<3 dS/m) and low in ammonium (<3 ppm NH_4-N), growers may use up to 50% in container mixes for larger vegetable transplants or outdoor container mixes (Box 16.7).

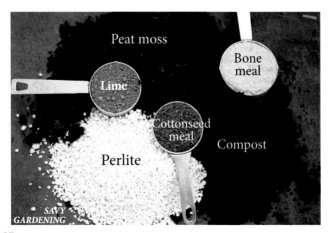

FIGURE 16.25

One example of a potting media that includes compost (Walliser, 2019).

Source: Jessica Walliser.

> **Box 16.7 Why compost quality matters?**
>
> Based on a study comparing horse manure (HM) compost with food residuals-yard trimmings (YT) compost for production of organic vegetable transplants (Clark and Cavigelli, 2005).
>
> Two composts, one made from YT and a second from HM, were separately used in a potting mix at 100% (i.e., only compost) and 50% in combination with a commercial potting media (bark, peat, and sand). At both proportions, the potting mix containing YT compost produced vegetable transplants of similar quality to those grown in the commercial potting mix alone. However, the HM compost was unacceptable for commercial production.
>
> The two composts had similar amounts of total N (approximately 2%), C:N ratios (15:1), and bulk densities (0.3 g/cm^3). Also, they were equally biologically stable, according to the Solvita Maturity Test. However, the pH and EC differed greatly between the two. The pH was 6.6 for the YT compost versus 8.0 for the HM compost. The ECs were 2.9 dS/m(YT) and 5.2 dS/m(HM), respectively.

7. Turf and landscape applications

In many localities, a major market for compost is the landscaping and amenity horticulture industries—lawns and sports turfgrass, gardens, green roof installations, and landscape plantings, including the planting of ornamental and shade trees. These industries are the primary users of compost in many countries, such as Australia and the United States, excluding California (Chapter 18). Landscaping and amenity horticulture are prominent green industries in urban and suburban settings where many composting feedstocks are generated, including yard trimmings, municipal and commercial food wastes, food processing residuals, and biosolids. Compost products sold at the retail level are primarily used for landscaping.

Composted mulches and soil conditioners are commonly used in a range of landscaping applications, such as in commercial projects, major developments, highway landscaping, and urban renewal projects. These products offer significant environmental benefits, by improving plant growth; providing essential nutrients; lowering irrigation requirements; improving the drought tolerance of plants; suppressing weeds; reducing erosion; and improving soil structure, water retention, infiltration, and drainage (NSW DEC 2007). Considerable quantities of urban derived compost are sold as a component in blended soil products, such as garden soil mix, turf underlay, or turf top dressing.

7.1 Turf establishment and maintenance

Compost is well-suited for the establishment, maintenance, and renovation of turfgrass on sports fields, golf courses, lawns, and council parks. Turfgrass usually responds well to compost, whose benefits can include disease suppression. Compost is incorporated into soils prior to seeding or sodding for the establishment of turfgrass and is also used for top-dressing turfgrass to fertilize and maintain established fields and lawns. Alternatively, ready made soil mixes containing compost and various other ingredients are supplied for the establishment and maintenance of turf.

Compost used for turfgrass should be mature and must be sanitized of pathogens and weed seeds. Composts having relatively high N contents and low C/N ratios are more favorable for high N-requiring turfgrass. Turfgrass benefits from compost with appreciable levels of micronutrients, especially iron. For topdressing, fine-textured and compost should be used, screened to 10 mm (3/8 in.) or less. Additional information about compost use on turfgrass is available in "*Using Composts to Improve Turf Performance*" (Landschoot, 2009).

7.1.1 Turf establishment

Compost use supports rooting and growth of newly established turfgrass on clay soils by increasing air-filled porosity and oxygen exchange, and by reducing compaction. On sandy and sandy loam soils, incorporation of compost helps to retain moisture for plant growth, reduce irrigation requirements, and improve the efficiency of irrigation programs. Turfgrass grown with the application of compost can require up to 30% less water. Regardless of soil type, compost supplies macro- and micronutrients that are beneficial for turfgrass growth and enable reduced inorganic fertilizer usage.

To establish or renovate sod fields, lawns, and sports fields, compost is either broadcast over the entire area or piled and spread with a tractor- or grader-mounted blade. Compost is typically applied evenly to depths of 25−50 mm (1−2 in.), which is equivalent to 250−500 m^3/ha (about 130−260 yd^3/acre). After application, the compost is incorporated into the soil to a depth of 100−150 mm (4−6 in.). On small plots, the application is often done manually with wheelbarrows and rakes.

For sod crops, compost is usually applied after the previous sod harvest to improve the soil before establishing the next crop. Some of the soil is removed when sod is harvested; therefore, applying compost before the next sodding replaces removed organic matter and nutrients. However, compost also can be topdressed early in the growth of a sod crop.

7.1.2 Topdressing sports turfgrass, golf courses, and lawns

Most turfgrasses in heavily used sports fields suffer from compaction, contributing to stunting or death of turfgrasses, increased turfgrass diseases, lack of water infiltration, and undulations in the surface. Hard-compacted surfaces that have low turfgrass coverage can even increase the risk of injuries to players. Topdressings are used to ameliorate these problems, specifically to improve soil structure and drainage, reduce lime requirements by moderating soil pH, and leveling the playing surface. It has been shown that the use of compost improves the availability and playability of sports fields, a key performance indicator for municipally owned sporting facilities. Topdressing with finely screened compost is a very effective and low-cost substitute for conventional topdressings used in maintaining the quality of sporting field surfaces, such as sand and soil (NSW DEC, 2006). If a local government maintains 10 sporting fields (about 15 ha or 37 acres) with compost-based topdressing, savings in fertilizer costs could amount to between US$8,625 and US$21,000 per year (NSW DEC, 2007).

Application depths for topdressing sports turfgrass typically range from 3 to 6 mm (1/8 to 1/4 in.), which equals 30–60 m³/ha (430–860 ft³/acre). Topdressing is accomplished with a straight-drop brush or wide-spread beater spreaders. Conventional spinner-type spreaders can be used if the compost has consistent texture and less than 50% moisture content.

Topdressing is often performed after aerification (punching holes in the soil of turfed surfaces), which enables some of the topdressed material to enter the soil profile and improve soil conditions more effectively. This is often helped by dragging or raking turfgrass fields after topdressing. Regular topdressing through the growing season or annual application in the off-season can assist in turfgrass growth and maintaining the quality of the turfgrass surface.

On golf courses, compost is often used for topdressing tees, fairways, and "roughs." It also can be topdressed onto greens when finely screened (less than 4 mm or 1/8 in.). Golf courses commonly topdress with a mix of compost and washed sand. A mix of 33% compost and 67% sand is a common topdressing blend. Grass should be growing vigorously when the compost is applied and watered into the turfgrass if conditions are dry.

Residential and commercial lawns are typically top-dressed at the beginning or end of the growing season, and especially after core aeration. Straight-drop brush and wide beater spreaders are used by professional landscape contractors. Homeowners often have to tackle the job manually because most composts do not fit through home fertilizer spreaders. However, some compost suppliers offer spreading services or make residential-scale spreading equipment available for customers to borrow or rent (Fig. 16.26).

7.2 Landscaping

If there is one green industry in which compost can be considered a staple product, it is landscaping. Compost is primarily used to improve landscape and garden soils, but it also offers the bonus of nutrients, released gradually and over a long term. For landscape service providers, compost provides a measure of insurance as plants remain healthier in between maintenance visits.

Compost is used when garden beds are first established and when gardens are renewed. The garden in Fig. 16.27 was installed on a section of lawn underlain by compacted soil with low organic matter. A single application of spent mushroom compost, approximately 40 mm (1.5 in.) thick was tilled into the existing soil to a depth of approximately 10 cm (4 in.). Afterward, the garden was planted and mulched. No synthetic inorganic fertilizers have been used.

For creating new beds, typically a 50 mm (2 in.) layer of compost is mixed into the soil to a depth of 150–300 mm (6–12 in.). This requires about 50 m³ (7,15 ft³) of compost for an area of 1000 m² (quarter acre). However, the amount of compost needed depends on the quality of the existing soil. For soils with low organic matter, dominated by either sand or clay, up to 100 mm (4 in.) of compost may be justifiable. Where soils already have adequate organic matter content and are of good quality, a 25 mm (1-in.) application is reasonable.

Once a garden and landscaped area is established, compost is used regularly for maintenance, and also for new and replacement plantings. For existing plantings,

FIGURE 16.26

Topdressing turfgrass with compost: (A) using a straight-drop brush spreader to apply compost to a sports field; and (B) a small-scale walk behind spinner spreader commonly used for small area applications, including lawns and turf.

Sources: (A) R. Alexander; (B) Dirt Hugger.

compost is surface applied in a 25–50 mm (1–2 in.) layer and either gently worked into the soil or left on the surface as a mulch.

An important market for compost is soil remediation of disturbed soils due to development. In many land-development situations, the topsoil is removed, leaving only subsoils, which often result in poor growth or even death of turf and landscape plants. To prevent such problems, a layer of 25–75 mm (1–3 in.) of compost is applied and tilled into the soil before laying sod or planting trees and shrubs. This practice, plus the use of composted mulch around shrubs and trees, can increase success in establishing new plantings, especially in areas and seasons when rainfall is scarce. Some progressive communities have stipulated minimum soil organic matter levels in their development regulations. Application rates required to meet soil organic matter targets can be calculated as described in Section 2.

FIGURE 16.27

Compost used for creating garden landscapes.

Source: J. Puddester.

The state of Washington strongly promotes the use of compost to revitalize disturbed soils as a means of protecting water quality, particularly in the Puget Sound region. For example, the "Soils for Salmon" program advocates for healthy soils in urban environments to protect salmon habitats (Soils for Salmon, 2021). This and other programs in the state provide guidance on how to enrich soils through the use of compost and other means, including a manual about low-impact development (Hinman, 2012) and another manual about building healthy soils (Stenn, 2018).

7.3 Surface mulching

Surface application of composted mulch to garden beds, landscaped areas, and around trees (Fig. 16.28) is widely practiced as important erosion control and water conservation measure. The application of mulch can reduce moisture evaporation by up to 70% compared to a bare soil surface, also reducing irrigation requirements by up to 70% (NSW DEC, 2007). Furthermore, mulches help hold water and reduce leaching losses, and therefore improve the drought resistance of plants. Another key benefit of mulching is the physical protection of soils from the erosive forces of wind and rain, resulting in a reduction of runoff by more than 70% and soil erosion by more than 90% on slopes of up to 15%. A thick mulch layer inhibits the establishment of weeds and can suppress existing weeds, thereby reducing herbicide usage. In regions that are subject to freezing and thawing in winter, organic mulches may prevent winter damage to trees and shrubs by moderating temperature changes to the root and crown areas.

Fully composted mulch is preferred as it is stable and does not degrade as rapidly after application compared to mulches that have been composted only to the point of sanitization (i.e., pasteurized, see Fig. 16.3). Composted mulches should last a

FIGURE 16.28

Composted garden trimmings and bark used for mulching around trees in a public park.

Source: J. Biala.

minimum of 3 years. The rate of application of composted mulch depends on the coarseness of the product, although a mulch layer of up to 100 mm (4 in.) should be fine in most cases. Gas exchange and oxygen movement through the mulch layer into soil diminishes as the thickness of the mulch layer and the proportion of small particles increases. Lack of oxygen in the root zone can impair plant growth.

7.4 Manufactured soils for rooftop gardens

Green roofs or rooftop gardens rely on manufactured growing medium to grow perennial and annual vegetation on roofs (Fig. 16.29). Green roofs have recently gained popularity in urban areas for their aesthetics and environmental benefits (i.e., moderating building temperatures to lower heating and cooling costs, reducing GHG emissions, absorbing airborne pollutants, and mitigating stormwater runoff). Green roofs have been identified as a best management practice (BMP) for improving stormwater quality and reducing runoff volume.

Composts can be a good addition to growing media for rooftop gardens since they have lower bulk densities than mineral soils or sand. Composts for rooftops must be consistent in texture and quality. Also, it should be low in available N, P, and K to minimize nutrients in the stormwater that does run off. Hence, compost made from yard trimmings is better suited to green roof media than composts made from animal manure.

Green roof media typically use 10%–20% organic ingredients. A typical composition of green roof medium might be 15% compost, 30% sand, and 55% expanded minerals (e.g., Perlite or expanded clay). A higher proportion of compost might improve plant growth but may also increase N and P enrichment of stormwater. The addition of feedstocks high in minerals such as Fe and Al help to reduce the

FIGURE 16.29

Plants on a green roof soil amended with 10% compost by volume.

solubility of P. Nitrogen runoff problems can be reduced by growing plants that assimilate abundant N.

Compost producers can take advantage of green roof markets by producing compost with characteristics tailored to green roof media or by manufacturing the green roof media itself. The latter enterprise involves obtaining the other desired media ingredients and blending them with the appropriate compost products.

8. Erosion control and stormwater management

The US EPA has designated over 50,000 impaired water body segments in the United States, identifying sediment as the leading source of surface water pollution. In recent years, several US federal agencies and at least one transportation or environmental regulatory agency in every state has published a specification or guidance document for using compost to protect water quality, notably applications for erosion control, sediment control, and postconstruction stormwater management (Table 16.10).

Soil erosion is the process in which water or wind loosens soil particles and, in some cases, transports these soil particles to a new location in the form of sediment. Sedimentation is the process in which the sediment is deposited in a new location. While wind erosion can be a significant issue in arid climates, where dense vegetation establishment is difficult, erosion from water sources is by far the biggest contributor to water quality impairment and therefore receives the most attention in the form of BMPs, site design, and regulations.

Erosion control is a practice taken to prevent or slow the soil erosion process. A *compost blanket* has become a common, high-performance method for controlling erosion (Fig. 16.30). A compost blanket is a high-quality compost with coarse

Table 16.10 Compost quality specifications for compost filter socks, berms, and blankets (USCC, 2020b).

Parameter	Filter sock	Filter berm	Blanket
Stability[a]	<4	<4	<4
Maturity[b]	>80	>80	>80
Moisture content[c]	0–60	30–60	30–60
Organic matter content[d]	25–100	25–65	25–65
Particle size	99% < 75 mm (3 in.) 50% > 10 mm ($^3/_8$ in.)	100% < 75 mm (3 in.) 90% < 25 mm (1 in.) 70% < 19 mm ($^3/_4$ in.) 30% < 6 mm (¼ in.)	99% < 75 mm (3 in.) 65%–99% < 19 mm ($^3/_4$ in.) 0%–75% < 6 mm (¼ in.)
pH	5.0–8.5	6.0–8.5	6.0–8.5
Soluble salts	<10	<5	<5
Physical contaminants[d]	<1	<0.25	<0.25

Testing parameters above are based on the Test Methods for the Examination of Composting and Compost (TMECC). https://www.compostingcouncil.org/page/TMECC.
[a] mg CO2-C per gram of organic matter per day
[b] % seed emergence and vigor
[c] % of wet weight
[d] % of dry weight

FIGURE 16.30

Compost blanket applied to the steep slope above a stream at a construction site.

Source: B. Faucette.

particles applied evenly to cover 100% of the soil surface at a thickness of 25−50 mm (1−2 in.) on slopes up to 2:1 (horizontal:vertical). It is typically applied using a pneumatic blower truck. Seed is commonly mixed into the compost prior to application or injected during application by the pneumatic blower system. San Diego State University conducted a study to evaluate the performance of 14 commonly used erosion control BMPs, including compost blankets and reported that compost blankets reduced soil erosion by 67%−99% on slopes up to 2:1 with rainfall intensities up to 150 mm (6 in.) per hour, outperforming all other technologies evaluated in the study (Faucette et al., 2009c). A pair of University of Georgia studies also showed over 80% reduction in soil loss on construction sites relative to blown straw or hydroseeding (Faucette et al., 2005, 2007).

Sediment "control" captures sediment after soil erosion has occurred and prevents it from leaving a site or reaching surface water. Compost *filter socks* and *filter berms* have been used widely as sediment control BMP, including concentrated water flow environments. In addition, compost filter socks have also been used as check dams, inlet protection, and slope interruption BMPs. Since they are contained in a biodegradable or photodegradable mesh, compost filter socks are easy to handle and transport.

The performance of compost filter socks and berms has been widely reported in the research literature. A USDA Agricultural Research Service (ARS) study reported that compost filter socks remove 90% of total suspended solids (TSSs) and reduce 78% of turbidity in sediment-laden runoff (Faucette et al., 2008). A study conducted at the University of Georgia compared the performance of compost filter socks, straw bales, and mulch berms on field test plots and found compost filter socks reduced runoff TSS by 76%, while straw bales and mulch berms reduced TSS by 54% and 51%, respectively (Faucette et al., 2009a). Finally, an Ohio State University study evaluated the hydraulic flow-through rate for compost filter socks and silt fences and reported that compost filter socks have a 50% greater flow-through rate than silt fences without a reduction in sediment removal performance. This research has led to increased design capacity and spacing allowances for compost filter socks relative to silt fences in specifications across the United States (Keener et al., 2007).

Stormwater is the generation of runoff when precipitation exceeds the absorption capacity of ground surfaces. Stormwater typically transports a number of pollutants, directly leading to surface water pollution and impairment. Reducing stormwater volume and treating stormwater using natural processes is the foundation of green infrastructure stormwater management. Compost is widely used for green infrastructure stormwater management: (1) as a stormwater biofilter to treat stormwater; (2) to absorb rainfall and runoff to reduce stormwater volume; and (3) to establish and sustain permanent vegetation. As a biofilter, compost is typically used in bioretention and rain garden systems, bioswales, and compost filter socks. Compost used for the volumetric reduction of stormwater is typically applied as a compost blanket, mixed with native soil, or as a component in an engineered soil media, such as green roof applications. Often, these practices are applied together in a treatment train to minimize stormwater generation and surface water pollution (Fig. 16.31).

FIGURE 16.31

Compost-filled socks used in combination with compost blankets (yet to be applied) for minimizing stormwater runoff.

Source: B. Faucette.

Research from the USDA ARS has shown that compost filter socks can remove 74% of total coliform bacteria and 75% of *E. coli*; 37%−72% of cadmium (Cd), chromium (Cr), copper (Cu), nickel (Ni), lead (Pb), and zinc (Zn); 99% of diesel fuel; 84% of motor oil; 43% of gasoline; 65% of P; and 17% of N from stormwater (Faucette et al. 2008, 2009b). A University of Georgia study evaluated stormwater volume from compost blankets and conventional seeding applications and concluded compost blankets reduced runoff volume by 50% and peak runoff rate by 36% (Faucette et al., 2005). A follow-up study conducted on construction site soils reported compost blankets absorbed 80% of a 100 mm (4 in.) rainfall event and reduced runoff stormwater volume by 60% (Faucette et al., 2007). Researchers also evaluated vegetation growth (percent cover and biomass of weeds and seeded grasses) of seeded compost blankets and hydroseed over 18 months and reported compost blankets provided nearly three times greater vegetative cover than conventional seeding applications with approximately five times less weed biomass (Faucette et al., 2006).

More information about the use of compost for erosion, sediment, and stormwater management can be found in *A Watershed Manager's Guide to Organics* (Faucette and Cannon, 2014) and *The Sustainable Site: Design Manual for Green Infrastructure and Low Impact Development* (Faucette et al., 2010).

References

Cited references

Alexander, R., 2016. Phosphorous and compost use dynamics. BioCycle. https://www. biocycle.net/phosphorus-compost-use-dynamics/.

Alexander, R., Wagner, C., Date not identified. Improving compost use through application methods, Texas A&M University Cooperative Extension. http://compost.tamu.edu/docs/compost/pubs/applicationmethods.pdf.

Alvarez-Campos, O., Evanylo, J.K., 2019. Plant available nitrogen estimation tools for a biosolids-amended, clayey urban soils. Soil Sci. Soc. Am. J. https://doi.org/10.2136/sssaj2018.11.0441.

Amlinger, F., Götz, B., Dreher, P., Geszti, J., Weissteiner, C., 2003. Nitrogen in biowaste and yard waste compost: dynamics of mobilisation and availability—a review. Eur. J. Soil Biol. 39, 107—116.

ANL, 2021. Viticulture and Orchards Supplies. Australian Native Landscapes. https://anlscape.com.au/Viticulture-orchards-supplies.

Badzmierowski, M.J., Evanylo, G.K., Ervin, E.H., 2015. Biosolids amendments improve an anthropogenically disturbed urban turfgrass system. Crop Sci. 2020 (60), 1666—1681. https://doi.org/10.1002/csc2.20151.

Baldi, E., Cavani, L., Mazzon, M., Marzadori, C., Quartieri, M., Toselli, M., 2021. Fourteen years of compost application in a commercial nectarine orchard: effect on microelements and potential harmful elements in soil and plants. Sci. Total Environ. 752, 918—925.

Biala, J., 2000. The use of compost in viticulture - a review of the international literature and experience. In: Willer, H., Meier, U. (Eds.), Proceedings of the 6th International Congress on Organic Viticulture, 25 — 26 August 2000, Basel, Switzerland, pp. 130—134.

Biala, J., 2011. The Benefits of Using Compost for Mitigating Climate Change. Office of Environment and Heritage, Department of Premier and Cabinet, Sydney South, Australia.

Biala, J., Prange, R., Raviv, M., 2014. Proceedings of the First International Symposium on Organic Matter Management and Compost Use in Horticulture. Acta Horticulturae 1018.

Biala, J., 2017. Benefits of Accounting for Nutrients in Organic Soil Amendments. Vegetables Australia, pp. 24—25.

Biala, J., Chapman, S., 2017. Can the use of compost and Twin N reduce synthetic fertiliser use in macadamia? Aust. Nutgrower 36—39.

Biala, J., Wilkinson, K., Henry, B., Singh, S., Bennett-Jones, J., De Rosa, D., 2021. The potential for enhancing soil carbon levels through the use of organic soil amendments in Queensland, Australia. Region. Environ. Change 21, 95. https://doi.org/10.1007/s10113-021-01813-y.

BioCycle, 2020. Compost and Mulch Utilization on California Almond Farm. https://www.biocycle.net/compost-and-mulch-utilization-on-california-almond-farm/.

Brown, M.W., Tworkowski, T., 2004. Pest management benefits of compost mulch in apple orchards. Agric. Ecosyst. Environ. 103, 465—472.

Brown, S., Cotton, M., 2011. Changes in soil properties and carbon content following compost application: results of on-farm sampling. Compost Sci. Util. 19, 88—97.

Brown, S., Beecher, N., 2019. Carbon accounting for compost use in urban areas. Compost Sci. Util. 27 (4), 227—239.

Bundesamt für Justiz, 2017. http://www.gesetze-im-internet.de/d_v_2017/.

Clark, S., Cavigelli, M., 2005. Suitability of composts as potting media for production of organic vegetable transplants. Compost Sci. Util. 13 (2), 150−155.

Clean Energy Regulator, 2021a. Emissions Reduction Fund. Australian Government. http://www.cleanenergyregulator.gov.au/ERF.

Crisp, P., Wheeler, S., Baker, G., 2013. A Benefit/cost Assessment in Citrus IPM Following the Application of Soil Amendments . Horticulture Australia Limited Final Report No CT10022. South Australian Research and Development Institute (Sustainable Systems) Adelaide, pp. 1−24.

De Rosa, D., Basso, B., Rowlings, D.W., Scheer, C., Biala, J., Grace, P.R., 2017. Can organic amendments support sustainable vegetable production? Agron. J. 109, 1−14.

DLG, 2012. Spreading Quality − Farm Manure and Compost, Universal Spreader BERG-MANN TSW2120 Tandem, DLG Test Report 6031F. Gross-Umstadt, Germany.

Dupont, T., Granatstein, D., 2018. Compost Use for Tree Fruits. Washington State University Extension. http://treefruit.wsu.edu/publications/compost-use-for-tree-fruit/.

Ebertseder, T., Gutser, R., 2003. Nutrition potential of biowaste composts. In: Applying compost benefits and needs, Seminar Proceedings, Jointly Published by the Federal Ministry of Agriculture, Forestry, Environment and Water Management, Austria, and the European Communities, pp. 117−128. In: https://ec.europa.eu/environment/archives/waste/pdf_comments/040119_proceedings.pdf.

Eghball, B., 2000. Nitrogen mineralization from field-applied beef cattle feedlot manure or compost. Soil Sci. Soc. Am. J. 64, 2024−2030.

Eleonora, N., Dobrei, A., Alina, D., Erzsebet, K., Valeria, C., 2014. Grape pomace as fertilizer. J. Hortic. Forestry Biotechnol. 18 (2), 141−145.

Faucette, L.B., Jordan, C.F., Risse, L.M., Cabrera, M., Coleman, D.C., West, L.T., 2005. Evaluation of stormwater from compost and conventional erosion control practices in construction activities. J. Soil Water Conserv. 60 (6), 288−297.

Faucette, L.B., Jordan, C.F., Risse, L.M., Cabrera, M.L., Coleman, D.C., West, L.T., 2006. Vegetation and soil quality effects from hydroseed and compost blankets used for erosion control in construction activities. J. Soil Water Conserv. 61 (6), 355−362.

Faucette, L.B., Governo, J., Jordan, C.F., Lockaby, B.G., Carino, H.F., Governo, R., 2007. Erosion control and storm water quality from straw with PAM, mulch, and compost blankets of varying particle sizes. J. Soil Water Conserv. 62 (6), 404−413.

Faucette, L.B., Sefton, K.A., Sadeghi, A.M., Rowland, R.A., 2008. Sediment and phosphorus removal from simulated storm runoff with compost filter socks and silt fence. J. Soil Water Conserv. 63 (4), 257−264.

Faucette, B., Governo, J., Tyler, R., Gigley, G., Jordan, C.F., Lockaby, B.G., 2009a. Performance of compost filter socks conventional sediment control barriers used for perimeter control on construction sites. J. Soil Water Conserv. 64 (1), 81−88.

Faucette, B., Cardoso-Gendreau, F., Codling, E., Sadeghi, A., Pachepsky, Y., Shelton, D., 2009b. Stormwater pollutant removal performance of compost filter socks. J. Environ. Qual. 38, 1233−1239.

Faucette, L.B., Scholl, B., Beighley, R.E., Governo, J., 2009c. Large-scale performance. and design for construction activity erosion control best management practices. J. Environ. Qual. 38, 1248−1254.

Faucette, B., Tyler, R., Marks, A., 2010. The Sustainable Site: Design Manual for Green Infrastructure and Low Impact Development, first ed. Forester Press, Santa Barbara, CA.

Faucette, B., Cannon, C., 2014. The Soil and Water Connection: A Watershed Manager's Guide to Organics, second ed. US Composting Council and Composting Council Research and Education Foundation.

Follet, R.F., Reed, D.A., 2010. Soil carbon sequestration in grazing lands: societal benefits and policy implications. Rangel. Ecol. Manag. 63 (1), 4–15.

Gagnon, B., Demers, I., Ziadi, N., Chantigny, M.H., Parent, L.-É., Forge, T.A., Larney, F.J., Buckley, K.E., 2012. Forms of phosphorus in composts and in compost-amended soils following incubation. Can. J. Soil Sci. 92, 711721.

Gilbert, J., Ricci-Jürgensen, M., Ramola, A., 2020. Quantifying the Benefits of Applying Quality Compost to Soil. International Solid Waste Association (ISWA), Rotterdam, The Netherlands.

Gov, U.K., 2018a. Nitrate Vulnerable Zones (last updated). www.gov.uk/government/collections/nitrate-vulnerable-zones#:~:text=Nitrate%20Vulnerable%20Zones%20(NVZs)%20are,for%20changes%20in%20nitrate%20concentrations.

Gov, U.K., 2018b. Using Nitrogen Fertilisers in Nitrate Vulnerable Zones (last updated). https://www.gov.uk/guidance/using-nitrogen-fertilisers-in-nitrate-vulnerable-zones#how-much-nitrogen-you-can-apply-to-your-crops.

Granatstein, D., Dauer, P., 1999. Compost effects on apple tree growth, parts 1 and 2. In: Compost Connection Newletters. Washington State University, Wenatchee, WA.

Haber, N., Kluge, R., Deller, B., Flaig, H., Schulz, E., Reinhold, J., 2008. Nachhaltige Kompostanwendung in der Landwirtschaft (Sustainable use of compost in agriculture). Landwirtschaftliches Technologiezentrum Augustenberg, Karlsruhe, Germany.

Handreck, K.A., 2014. Composts in the production and performance of growing media for containers. In: Biala, J., Prange, R., Raviv, M. (Eds.), Proceedings of the First International Symposium on Organic Matter Management and Compost Use in Horticulture, Acta Horticulturae, 1018, pp. 505–512.

Handreck, K., Black, N., 2010. Growing media for ornamental plants and turf. University of NSW Press, Sydney, Australia.

Hansen, B., Thorling, L., Schullehner, J., Termansen, M., Dalgaard, T., 2017. Groundwater nitrate response to sustainable nitrogen management. Sci. Rep. 7, 8566.

Hartin, J., Crohn, D., 2007. Compost Use for Landscape and Environmental Enhancement. California Integraded Waste Management Board, Sacramento, CA.

Henry, C., Bergeron, K., 2005. Compost Use in Forest Land Restoration. U.S. EPA's Office of Wastewater Management, Washington, DC.

Hinman, C., 2012. Low Impact Development. Technical Guidance Manual for Puget Sound. Washington State University Extension, Puyallup, WA.

Holechek, J.L., Pieper, R.D., Herbel, C.H., 1998. Range Management Principles and Practices, third ed. Prentice Hall, NJ.

IPPC, 2019. Intergovernmental Panel on Climate Change Special Report on Climate Change and Land. https://www.ipcc.ch/srccl/.

Keener, H., Faucette, B., Klingman, M., 2007. Flow- through rates and evaluation of solids separation of compost filter socks vs. silt fence in sediment control applications. J. Environ. Qual. 36 (3), 742–752.

Lal, R., 2007. Soil science and the carbon civilization. SSSAJ 71 (5), 1425–1437.

Lal, R., Augustin, B. (Eds.), 2012. Carbon Sequestration in Urban Ecosystem. Springer Science+Business Media B.V.

Lal, R., Negassa, W., Lorenz, K., 2015. Carbon sequestration in soil. Curr. Opin. Environ. Sustain. 15, 79–86. https://doi.org/10.1016/j.cosust.2015.09.002.

Landschoot, P., 2009. Using Composts to Improve Turf Performance. The Pennsylvania State University. https://extension.psu.edu/using-composts-to-improve-turf-performance.

Lee, A., 2016. Nitrogen and Phosphorus Availability from Various Composted Wastes for Use in Irish Agriculture and Horticulture (Doctoral thesis). Technological University Dublin. https://doi.org/10.21427/D71S3K.

Ludwig Bergmann GmbH. Media Center, 2015. https://www.bergmann-goldenstedt.de/en/media-center/pictures/.

Mayer, A., Silver, W.L., 2018. 2: The potential for carbon sequestration in California from compost amendments to rangelands in: L Flint et al, 2018. In: Increasing Soil Organic Carbon to Mitigate Greenhouse Gases and Increase Climate Resiliency for California. https://www.energy.ca.gov/sites/default/files/2019-11/Agriculture_CCCA4-CNRA-2018-006_ADA.pdf.

MCP, 2018. Carbon Farming in Marin. Marin Carbon Project. https://www.marincarbonproject.org/home.

Millner, P.D., 2014. Pathogen disinfection technologies, metrics and regulations for recycled organics used in horticulture. In: Biala, J., Prange, R., Raviv, M. (Eds.), Proceedings of the First International Symposium on Organic Matter Management and Compost Use in Horticulture, Acta Horticulturae 1018, pp. 621–630.

NSW DEC, 2006. Cost/benefit of Using Recycled Organics in Council Parks and Gardens Operations in NSW. New South Wales Department of Environment and Conservation Paramatta, NSW, Australia.

NSW DEC, 2007. Sustainable Landscaping Using Compost. Fact Sheet. New South Wales Department of Environment and Conservation Paramatta, NSW, Australia.

Ozores-Hampton, M., Obreza, P., Stoffella, G., 2001. Mulching with composted MSW for biological control of weeds in vegetable crops. Compost Sci. Util. 9 (4), 105–113.

Ozores-Hampton, M., Obreza, T., Stoffella, P., Fitzpatrick, G., 2002. Immature compost suppresses weed growth under greenhouse conditions. Compost Sci. Util. 10, 105–113.

Ozores-Hampton, M.P., Stansly, P.A., Salame, T.P., 2011. Soil chemical, biological and physical properties of a sandy soil subjected to long-term organic amendments. J. Sustain. Agric. 353, 243–259.

Ozores-Hampton, M., 2012. Developing a vegetable fertility program using organic amendments and inorganic fertilizers. HortTechnology 22, 743–750.

Ozores-Hampton, M., Adair, R., Stansly, P., 2015. Using compost in citrus. Citrus Ind 96, 8–11.

Ozores-Hampton, M., 2017a. Guidelines for assessing compost quality for safe and effective utilization in vegetable production. HortTechnology 27, 150.

Ozores-Hampton, M., 2017b. Impact of soil health and organic nutrient management on vegetable yield and quality. HortTechnology 27, 162–165.

Ozores-Hampton, M. (Ed.), 2021. Compost Utilization in Production of Horticultural Crops. CRC Press, Boca Raton, FL.

Patti, A.F., Issa, G., Smernik, R., Wilkinson, K., 2009. Chemical composition of composted grape marc. Water Sci. Technol. 60 (5), 1265–1271.

Paulin, R.,B., Wilkinson, K., O'Malley, P., Flavel, T., 2005. Identifying the Benefits of Composted Soil Amendments to Vegetable Production. Final Report, Department of Agriculture Western Australia, Perth.

Paungfoo-Lonhienne, C., Visser, J., Lonhienne, T.G.A., Schmidt, S., 2012. Past, present and future of organic nutrients. Plant Soil 359, 1–18.

Peries, R., 2013. Subsoil Manuring: An Innovative Approach to Addressing Subsoil Problems Targeting Higher Water Use Efficiency in Southern Australia. Southern Farming Systems, Inverleigh, Victoria, Australia. http://www.sfs.org.au/trial-result-pdfs/Trial_Results_2013_TAS/2013_SubsoilManuring_TAS.pdf.

Pittaway, P., 2004. Know Your Product before Marketing Compost to Agriculture. Published in towards Zero Waste Adelaide. Waste Management Association of Australia, South Australian Branch, Adelaide, Australia.

Port Phillip and Westernport CMA, 2020. Guide to Compost Use in Vineyards. https://www.ppwcma.vic.gov.au/what-we-do/past-projects/compostundervinemornington/.

Rahman, S., Widerholt, R., 2012. Options for Land Application of Solid Manure. North Dakota State University Extension Service, Fargo, ND.

Raun, W.R., Johnson, G.V., 1999. Improving nitrogen use efficiency for cereal production. Agron. J. 91, 357−363. https://doi.org/10.2134/agronj1999.00021962009100030001x.

Rockwood, D.L., Becker, B., Ozores-Hampton, M., 2012. Municipal solid waste compost benefits on short rotation woody crops. Compost Sci. Utili. 20, 67−72.

Ryals, R., Silver, W.L., 2013. Effects of organic matter amendments on net primary productivity and greenhouse gas emissions in annual grasslands. Ecol. Appl. 23 (1), 46−59.

Ryals, R., Kaiser, M., Torn, M.S., Berhe, A.A., Silver, W.L., 2014. Impacts of organic matter amendments on carbon and nitrogen dynamics in grassland soils. Soil Biol. Biochem. 52e61. https://doi.org/10.1016/j68, 2014.

SA Centre for Economic Studies, 1999. Benefit Cost Analysis of Composted Organic Mulch in Horticultural Industries: Vines, Cherries, Almonds, Pears, Citrus and Potatoes, Capsicums, Carrots, Flowers. Adelaide, South Australia.

Schneider, M., 2017. Kompost in der Düngeverordnung, Sonderdruck aus Getreidemagazin 6/2017.

Siebert, S., Gilbert, J., 2018. Specifications for Use of Quality Compost in Growing Media. European Compost Network. https://www.compostnetwork.info/.

Snohomish County Extension, 2020. Compost Spreading Equipment and Techniques. Washington State University. www.snohomish.wsu.edu/Compost.

Soils for Salmon, 2021. Washington Organic Recycling Council. Gig Harbor, WA. https://www.soilsforsalmon.org/.

Stenn, H., 2018. Building Soil: Guidelines and Resources for Implementing Soil Quality and Depth BMP T5.13 in Stormwater Management Manual for Western Washington. Washington Department of Ecology, Olympia, WA.

Stofella, P.J., Kahn, B.A. (Eds.), 2001. Compost Utilization in Horticultural Cropping Systems. Lewis Publishers, Boca Raton, FL.

Strik, B.C., Vance, A.J., Bryla, D.R., Sullivan, D.M., 2017. Organic production systems in northern highbush blueberry: I. Impact of planting method, cultivar, fertilizer, and mulch on yield and fruit quality from planting through maturity. HortScience 52, 1201−1213.

Strik, B.C., Vance, A.J., Bryla, D.R., Sullivan, D.M., 2019. Organic production systems in northern highbush blueberry: II. Impact of planting method, cultivar, fertilizer, and mulch on leaf and soil nutrient concentrations and relationships with yield from planting through maturity. HortScience 54, 1777−1794.

Sullivan, D., 2015. Compost for Blueberry Plants: Testing and Tips. Oregon State University Extension Service. https://extension.oregonstate.edu/crop-production/berries/compost-blueberry-plants-testing-tips.

Teske, S., 2019. Achieving the Paris Climate Agreement Goals: Global and Regional 100% Renewable Energy Scenarios to Achieve the Paris Agreement Goals with Non-energy GHG Pathways For+ 1.5° C And+ 2° C. Springer Nature Switzerland.

Travis, J.W., Halbrendt, N., Hed, B., Rytter, J., Anderson, E., Jarjour, B., Griggs, J., Bates, T., Butler, S., Levengood, J., Roth, P., 2003. A Practical Guide to the Application of Compost in Vineyards. Penn State University in cooperation with Cornell University. http://lergp. cce.cornell.edu/Bates/NonRefereed/2003TravisCompostGuide.pdf.

Uckoo, R.M., Enciso, J.M., Wesselmann, I., Jones, K., Nelson, S.D., 2009. Impact of Compost Application on Citrus Production under Drip and Microjet Spray Irrigation Systems. In: Tree and Forestry Science and Biotechnology, vol. 3. Global Science Books, pp. 59–65 (special issue 1).

USCC [US Composting Council], 2020a. What Equipment Do I Need For My Project. https:// www.compostingcouncil.org/general/custom.asp?page=CompostUseEquipment.

USCC [US Composting Council], 2020b. Compost Erosion Control Uses Specific Guides. https://www.compostingcouncil.org/page/CompostErosionControlUses.

Van der Wurff, A.W.G., Fuchs, J.G., Raviv, M., Termorshuizen, A.J. (Eds.), 2016. Handbook for Composting and Compost Use in Organic Horticulture. BioGreenhouse COST Action FA 1105. www.biogreenhouse.org.

Walliser, J., 2019. DFIY Potting Soil. Savvy Gardening. https://savvygardening.com/diy-potting-soil/.

Westover, F., 2019. Compost Use in Vineyard. https://grapes.extension.org/compost-use-in-vineyards/.

Wilkinson, K., Paulin, R., Tee, E., O'Malley, P., 2002. Grappling With Compost Quality Down Under. In: Michel Jr, S.C., Rynk, R.F., Hoitink, H.A.J. (Eds.), 'Proceedings of the 2002 International Symposium on Composting and Compost Utilization. 6–8 May 2002, Columbus, Ohio'. Ohio State University, Columbus, pp. 527–539.

WRAP, 2014a. Compost Production for Use in Growing Media — a Good Practice Guide. WRAP, Banbury, UK. www.wrap.org.uk.

WRAP, 2014b. Guidelines for the Specification of Quality Compost for Use in Growing Media. WRAP, Banbury, UK. www.wrap.org.uk.

Zentralverband, G. (Ed.), 2002. Handbuch Kompost im Gartenbau ([Handbook Compost in Horticulture). FGG Förderungsgesellschaft Gartenbau, Bonn, Germany.

Consulted and suggested references

Albrecht, W.A., 1938. Loss of soil organic matter and its restoration. Part II. Soils and Men. In: Yearbook of Agriculture. USDA., pp. 347–360 (US GPO).

Alexander, R., 2001. Field Guide to Compost Use. U.S. Composting Council. http://www. compostingcouncil.org.

Biala, J., Wilkinson, K., 2020. International Comparison of the Australian Standard for Composts, Soil Conditioners and Mulches (AS4454-2012), Report for the Australian Organics Recycling Association, Hove, SA, Australia. https://www.aora.org.au/sites/default/files/uploaded-content/website-content/International_Comparison_AS4454_Final.pdf.

Baldi, E., Cavani, L., Margon, A., Quartieri, M., Sorrenti, G., Marzadori, C., Toselli, M., 2018. Effect of compost application on the dynamics of carbon in a nectarine orchard ecosystem. Sci. Total Environ. 637–638, 918–925.

Bhogal, A., Nicholson, F.A., Young, I., Sturrock, C., Whitmore, A.P., Chambers, B.J., 2011. Effects of recent and accumulated livestock manure carbon additions on soil fertility and quality. Eur. J. Soil Sci. 62, 174–181.

Buchanan, M., 2002. Compost Maturity and Nitrogen Release Characteristics in Central Coast Vegetable Production. California Integrated Waste Management Board, Sacramento, CA.

Chenu, C., Angers, D.A., Barré, P., Derrien, D., Arrouays, D., Balesdentf, J., 2019. Increasing organic stocks in agricultural soils: knowledge gaps and potential innovations. Soil Tillage Res. 188, 41–52. https://doi.org/10.1016/j.still.2018.04.011.

Chivenge, P., Vanlauwe, B., Six, J., 2011. Does the combined application of organic and mineral nutrient sources influence maize productivity? A meta-analysis. Plant Soil 342, 1–30. https://doi.org/10.1007/s11104-010-0626-5.

Dixon, J., Gulliver, A., Gibbon, D., 2001. Farming Systems and Poverty: Improving Farmers' Livelihoods in a Changing World. FAO & World Bank, Rome, Italy & Washington, DC, USA. http://www.fao.org/3/ac349e/ac349e03.htm.

Eksi, M., Roweb, D.B., Fernández-Cañeroc, R., Cregg, B.M., 2015. Effect of substrate compost percentage on green roof vegetable production. Urban For. Urban Green. 14 (2), 315–322.

Emerson, D., 2013. Compost use on athletic fields and golf courses. BioCycle 54 (10), 30.

Epstein, E., 1997. The Science of Composting. Technomic Publishing, Basel, Switzerland.

Fan, J., Ding, W., Xiang, J., Qin, S., Zhang, J., Ziadi, N., 2014. Carbon sequestration in an intensively cultivated sandy loam soil in the North China Plain as affected by compost and inorganic fertilizer application. Geoderma 230–231, 22–28. https://doi.org/10.1016/j.geoderma.2014.03.027.

Farrell, M., 2015. An Assessment of the Carbon Sequestration Potential of Organic Soil Amendments. Report for the Commonwealth Department of Agriculture, CSIRO, Urrbrae, Australia.

Gerzabek, M., Pichlmayer, F., Kirchmann, H., Haberhauer, G., 1997. The response of soil organic matter to manure amendments in a long-term experiment at Ultuna, Sweden. Eur. J. Soil Sci. 48, 273–282. https://doi-org.ezproxy.library.uq.edu.au/10.1111/j.1365-2389.1997.tb00547.x.

Godde, C.M., Thorburn, P.J., Biggs, J.S., Meier, E.A., 2016. Understanding the impacts of soil, climate and farming practices on soil organic carbon sequestration: a simulation study in Australia. Front. Plant Sci. 7, 661. https://doi.org/10.3389/fpls.2016.00661.

Granatstein, D., Mullinix, K., 2008. Mulching options for northwest organic and conventional orchards. Hortscience 43, 45–50.

Hao, X., Chang, C., Travis, G.R., Zhang, F., 2003. Soil carbon and nitrogen response to 25 annual cattle manure applications. Z. Pflanzenernähr. Bodenk. 166, 239–245. https://doi-org.ezproxy.library.uq.edu.au/10.1002/jpln.200390035.

Hassink, J., 1997. The capacity of soils to preserve organic C and N by their association with clay and silt particles. Plant Soil 191, 77–87. https://doi.org/10.1023/A:1004213929699.

Helgason, B.L., Larney, F.J., Janzen, H.H., 2005. Estimating carbon retention in soils amended with composted beef cattle manure. Can. J. Soil Sci. 85, 39–46.

Jiang, G., Zhang, W., Xu, M., Kuzyakov, Y., Zhang, X., Wang, J., et al., 2018. Manure and mineral fertilizer effects on crop yield and soil carbon sequestration: a meta-analysis and modeling across China. Global Biogeochem. Cycles 32, 1659–1672. https://doi.org/10.1029/2018GB005960.

Johnston, M., 2017. Compost use on turf. BioCycle 57 (10).

Lim, S.S., Lee, K.S., Lee, S.I., Lee, D.S., Kwak, J.H., Hao, X., Ro, H.M., Choi, W.J., 2012. Carbon mineralization and retention of livestock manure composts with different substrate qualities in three soils. J. Soils Sediments 12, 312–322. https://doi-org.ezproxy.library.uq.edu.au/10.1007/s11368-011-0458-9.

Liu, S., Wang, J., Pu, S., Blagodatskaya, E., Kuzyakov, Y., Razavi, B.S., 2020. Impact of manure on soil biochemical properties: a global synthesis. Sci. Total Environ. 745. https://doi.org/10.1016/j.scitotenv.2020.141003.

Luo, Z., Wang, E., Bryan, B., King, D., Zhao, G., Pan, X., Bende-Michl, U., 2013. Meta-modelling soil organic carbon sequestration potential and its application at regional scale. Ecol. Appl. 23, 408−420. http://www.jstor.org/stable/23441005.

Maillard, É., Angers, D.A., 2014. Animal manure application and soil organic carbon stocks: a meta-analysis. Global Change Biol. 20, 666−679. https://doi-org.ezproxy.library.uq.edu.au/10.1111/gcb.12438.

Miller, J.J., Beasley, B.W., Drury, C.F., Larney, F.J., Hao, X., 2015. Influence of Long-Term (9 yr) Composted and stockpiled feedlot manure application on selected soil physical properties of a clay loam soil in southern Alberta. Compost Sci. Util. 23, 1−10. https://doi.org/10.1080/1065657X.2014.963741.

OMAFR, No date specified. Adding compost to grapes, Ontario Ministry of Agriculture, Food and Rural Affairs.

Ozores-Hampton, M., 2006. Soil and nutrient management: compost and manure. In: Gillett, J.L., Petersen, H.N., Leppla, N.C., Thomas, D.D. (Eds.), Grower's IPM Guide for Florida Tomato and Pepper Production. Univ. Florida, Gainesville, FL, pp. 36−40.

Page, K.L., Dalal, R., Pringle, M.J., Bell, M., Dang, Y.P., Radford, B., Bailey, K., 2013. Organic carbon stocks in cropping soils of Queensland, Australia, as affected by tillage management, climate, and soil characteristics. Soil Res. 51, 596−607. https://doi-org.ezproxy.library.uq.edu.au/10.1071/SR12225.

Paustian, K., Larson, E., Kent, J., Marx, E., Swan, A., 2019. Soil carbon sequestration as a biological negative emission strategy. Front. Climate 1, 8. https://doi.org/10.3389/fclim.2019.00008.

Post, W.M., Kwon, K.C., 2000. Soil carbon sequestration and land-use change: processes and potential. Global Change Biol. 6, 317−327.

Potter, P., Ramankutty, N., Bennett, E.M., Donner, S.D., 2010. Characterizing the spatial patterns of global fertilizer application and manure production. Earth Interact. 14 (2), 1−22.

Pribyl, D.W., 2010. A Critical Review of the Conventional SOC to SOM Conversion Factor.

Roe, N.E., 2003. Why vegetable growers don't use compost. In: Ozores-Hampton, M., Roe, N.E., Hanlon, E. (Eds.), Training in the Production and Utilization of Compost in Florida. Florida Cooperative Extension Service, Naples, FL.

Sanderman, J., Henglb, T., Fiskea, G.J., 2017. Soil carbon debt of 12,000 years of human land use. Proc. Natl. Acad. Sci. USA 114, 369575−369580. www.pnas.org/cgi/doi/10.1073/pnas.1706103114.

Shukla, S., Saxena, A., 2018. Global Status of Nitrate Contamination in Groundwater: Its Occurrence, Health Impacts, and Mitigation Measures.

Sela, R., Goldrat, T., Avnimelech, Y., 1998. Determining optimal maturity of compost used for land application. Compost Sci. Util. 6, 83−88. https://doi.org/10.1080/1065657X.1998.10701913.

Srivastava, A.K., 2009. Integrated nutrient management: concept and application in citrus. In: Tree and Forestry Science and Biotechnology, vol. 3. Global Science Books, pp. 32−58 (special issue 1).

Standards Australia, 2003. Australian Standard Composts, Soil Conditioners and Mulches AS 4454 - 2003. Homebush, NSW, Australia.

Stoffella, P.J., He, Z.L., Wilson, S.B., Ozores-Hampton, M., Roe, N.E., 2014. Utilization of Composted Organic Wastes in Vegetable Production Systems. http://www.agnet.org/htmlarea_file/library/20110808105418/tb147.pdf.

Sullivan, D.M., Bary, A.I., Nartea, T.J., Myrhe, E.A., Cogger, C.G., Fransen, S.C., 2003. Nitrogen availability seven years after a high-rate food waste compost application. Compost Sci. Util. 11, 265−275.

Sullivan, D.M., Bell, N., 2015. Preplant compost application improves landscape plant establishment and sequesters carbon in compacted soil. In: 16th International Conference Rural-Urban Symbiosis, 8th − 10th September 2015, Hamburg, Germany.

Tolland, L., 2010. Design Guidelines and Maintenance Manual for Green Roofs in the Semi-arid and Arid West. https://www.epa.gov/sites/production/files/documents/GreenRoofsSemiAridAridWest.pdf.

US EPA, 2007. Total Maximum Daily Loads National Section 303(d) List Fact Sheet. U.S. Environmental Protection Agency.

USCC, 2008. Compost Use Guidelines: Planting Bed Establishment with Compost. U.S. Composting Council.

Westover, F., 2006. Notes on Composting Grape Pomace. Virginia Tech. University. https://www.arec.vaes.vt.edu/content/dam/arec_vaes_vt_edu/alson-h-smith/grapes/viticulture/extension/growers/documents/composting-grape-pomace.pdf.

WORC, 2009. A User's Guide to Compost. Washington Organic Recycling Council, Gig Harbor, WA. https://www.compostwashington.org/resources-1.

WRAP, 2006. Uses of Compost in Regeneration and Remediation of Brownfield Sites in the UK. WRAP, Banbury, UK. www.wrap.org.uk.

(assumed) WRAP, 2007. Using Quality Compost to Benefit Crops. WRAP, Banbury, UK. www.wrap.org.uk.

(assumed) WRAP, 2008. Using Quality Compost to Benefit Potato Crop. WRAP, Banbury, UK. www.wrap.org.uk.

(assumed) WRAP, 2009. Quality Compost for Use in Green Roof Construction. WRAP, Banbury, UK. www.wrap.org.uk.

Xia, L., Lam, S.K., Yan, X., Chen, D., 2017. How does recycling of livestock manure in agro-ecosystems affect crop productivity, reactive nitrogen losses, and soil carbon balance? Environ. Sci. Technol. 51, 7450−7457.

Compost use for plant disease suppression

Authors: Deborah A. Neher[1], Harry A. Hoitink[2]

[1]*University of Vermont, Burlington, VT, United States;* [2]*Professor Emeritus, Ohio State University, Wooster, OH, United States*

Contributors: Johannes Biala[3], Robert Rynk[4], Ginny Black[5]

[3]*Centre for Recycling of Organic Waste & Nutrients (CROWN), The University of Queensland, Gatton, QLD, Australia;* [4]*SUNY Cobleskill, Cobleskill, NY, United States;* [5]*Compost Research and Education Foundation (CREF), Raleigh, NC, United States*

1. Introduction

1.1 History of organic amendments used for plant disease suppression

Farmers have used manures and composts for thousands of years to maintain plant health and yield. Even so, the first experimental data which proved that "barnyard" cow manure applications could successfully control diseases caused by a soilborne plant pathogen was not published until the 1930s. Application of manure increased cotton yields but also controlled an epidemic of root rot caused by *Phymatotrichum*, although disease control was incomplete because some roots still were affected by the disease. Scientists speculated that beneficial microorganisms on plant roots competed with the plant pathogen in the soil and reduced its activity through production of antibiotics. They even suggested that roots on plants in the manured plots were more resistant to disease based on the presence of many infected roots, but the plants did not die from root rot, as was the case in controls. It was recognized even then that there was an interaction between the crop, beneficial microorganisms, and the organic amendment that might play a role in this "muck and magic" type of biological control of plant diseases. Only much later, have these ideas been supported scientifically through published research, starting in the late 1950s for example in East Germany (Bochow and Seidel, 1961) (Fig. 17.1).

Large scale disease control with compost applications began during the 1950s when the nursery industry in the US and Australia developed lower-cost potting mixes and soil amendments for woody ornamentals that were bark-based rather than peat-based products (Hoitink and Ramos, 2004). Several growers found that composted bark could suppress root rots caused by *Phytophthora cinnamomi* for which effective resistant varieties or chemical control procedures other than

FIGURE 17.1

Taxus plants produced in a bark-based potting mix naturally suppressive to *Phytophthora* root rot. Note the uniform growth of the plants and the absence of root rot symptoms. This photo was taken before systemic fungicides for control of this disease were available. It was impossible to grow this crop on a commercial scale in peat-based mixes until after such fungicides became available.

Source: H.A. Hoitink.

methyl bromide were unavailable. The nursery industry discovered that the same composts could also control this disease in field soil but only if ideal drainage was provided. Without compost, Phytophthora root rot was particularly severe in heavy soils, in peat-based potting mixes, and in soils amended with fresh sawdust (Fig. 17.2). Thus, it was recognized early that not all sources of organic matter were effective for disease control.

During the 1970s, plant pathologists performed the first controlled experiments in the US and Australia that confirmed growers' experiences (Hoitink and Ramos, 2004). Phytophthora root rot of rhododendron in potting mixes and Phytophthora root rot of avocado in field soil were early examples. Peat-based mix stimulated release of zoospores by sporangia of *P. cinnamomi* but the zoospores died in potting mixes amended with composted bark. It was realized early that composts did not kill all pathogens. Researchers did not know whether the pathogen was inactivated by a toxin or other means. Only later, was it understood that pathogens are suppressed primarily by microorganisms in compost-amended soils.

FIGURE 17.2

Typical crop losses in azaleas caused by *Phytophthora cinnamomi* in 1971, before composted bark was used.

Source: H.A. Hoitink.

Since the early 1990s, animal manures have reclaimed their value with farmers for a variety of reasons, including that livestock farmers began to add value to raw manures through composting. Compost could solve problems with nutrient management and improve soil health. The types of experiments performed on Phymatotrichum root rot of cotton were repeated with composted manures across the globe in hundreds of tests for numerous crops and for many soilborne diseases. In general, these studies show that composted manures reduced the severity of diseases caused by essentially all types of soilborne plant pathogens including bacteria, fungi, Oomycota, and some nematodes. However, diseases also can be more severe after compost application if immature compost is used or if the timing of compost application is not synchronized with crop needs (Termorshuizen et al., 2006). Therefore, several factors must be addressed to obtain disease control consistently.

Research conducted in the 1990s demonstrated that composts applied to soil could also suppress foliar diseases, but the effect was often minor (Hoitink and Ramos, 2004). It was discovered that specific microorganisms in the rhizosphere of plants (the interface between plant roots and surrounding soil) can reduce the severity of diseases on the entire plant. In a study from Germany, composted cow manure applied to soil with small grains and grapes suppressed powdery and downy mildew, respectively. Other early reports from Florida and Ohio showed that application of composted municipal waste or composted yard wastes to field soil reduced the severity of bacterial spot (*Xanthamonas*) and of early blight (*Alternaria*) of tomato. Several foliar diseases of beans and cucumber were reduced by incorporating composted paper mill sludge into a sandy Wisconsin soil. Although composted sludge was effective, fresh paper mill sludge did not have this effect.

What the early research has shown is that biological elements in composts, at least some composts, can suppress plant disease. The missing piece is understanding the mechanism(s) by which compost microorganisms suppress plant pathogens and disease. Subsequent studies confirmed that different compost types can stimulate plant defense responses. Although scientists are starting to reveal the diversity of compost organisms, much remains to be known about the ecological function of these microorganisms (Vacheron et al., 2013; Lugtenberg and Kamilova, 2009; Andrews, 2018). Furthermore, our current understanding of the relationships between compost chemistry and the induction of systemic resistance in the plant is limited.

1.2 Plant diseases prone to suppression by compost

Compost naturally suppresses Oomycota pathogens, *Pythium* and *Phytophthora*, which cause root rots on vegetables, fruit, woody ornamentals, and forest trees (Table 17.1). Wilt diseases on cereals and grasses caused by *Fusarium oxysporum* and *Verticillium dahliae* are manageable with composts. *Rhizoctonia solani*, which causes damping-off disease in most crops, has a more checkered history in relationship to control by compost. It is the most studied, but least consistently managed by compost of the major soilborne fungal pathogens. The taxonomy and genetics of *R. solani* reveal immense variation within the genus resulting in the original species being divided into multiple species.

Table 17.1 Soilborne pathogens demonstrated to be suppressed by compost.

Pathogen	Crop	Compost feedstock(s)	References
Aphanomyces euteiches	Snap bean	Paper mill waste, bark	Noble and Coventry (2005)
Aphanomyces euteiches	Root rot on pea	Sewage sludge, wood chips	Litterick et al. (2004)
Fusarium oxysporum f. sp. *lini*	Flax	Horse manure (20%) and green waste (80%) (wheat straw, corn straw, conifer bark)	Termorshuizen et al. (2006)
Fusarium oxysporum f. sp. *lini*	Flax	Yard waste (without grass)	Termorshuizen et al. (2006)
Fusarium oxysporum f.sp. *lycopersici*	Tomato	Municipal solid waste	Vida et al. (2020)
Fusarium spp.	Several hosts	Vegetal	Bonilla et al. (2012)
Fusarium wilt and stem rot	Cucumber	Greenhouse compost	Vida et al. (2020)
Microdochium nivale	Turf	Bark, poultry manure	Noble and Coventry (2005)
Phytophthora capsici	Pepper	Chitin in crab shell	Bonilla et al. (2012)
Phytophthora cinnamomi	Lupin	Spent mushroom compost (wheat straw 56%, chicken manure 39%, gypsum 5%)	Termorshuizen et al. (2006)
Phytophthora cinnamomi	White lupin	Chicken manure	Vida et al. (2020)

Table 17.1 Soilborne pathogens demonstrated to be suppressed by compost.—*cont'd*

Pathogen	Crop	Compost feedstock(s)	References
Phytophthora cinnamomi	Avocado	*Eucalyptus* trimmings	Bonilla et al. (2012)
Phytophthora cinnamomi	Avocado	Vegetal	Bonilla et al. (2012)
Phytophthora nicotianae	Tomato	Organic residue of wine grapes, green waste	Termorshuizen et al. (2006)
Phytophthora nicotianae	Tomato	Woodcut, plants, horse manure	Termorshuizen et al. (2006)
Phytophthora nicotianae	Tomato	Woody waste, poultry manure	Termorshuizen et al. (2006)
Phytophthora nicotianae	Tomato	Yard waste (without grass)	Termorshuizen et al. (2006)
Phytophthora nicotianae	Tomato	Yard waste (without grass)	Termorshuizen et al. (2006)
Phytophthora nicotianae	Tomato	Yard waste (woody materials, grass clippings)	Termorshuizen et al. (2006)
Pythium graminicola	Turf	Brewery and sewage sludges	Noble and Coventry (2005)
Pythium ultimum	Cucumber	Green vegetable waste, horse manure	Noble and Coventry (2005)
Pythium ultimum	Garden cress	Animal and vegetal	Bonilla et al. (2012)
Pythium ultimum	Garden cress	Bark	Bonilla et al. (2012)
Ralstonia solanacearum	Potato	Organic waste	Vida et al. (2020)
Rhizoctonia solani	Basil	Cow manure	Bonilla et al. (2012)
Rhizoctonia solani	Cauliflower	Wood chips, horse manure	Termorshuizen et al. (2006)
Rhizoctonia solani	Cauliflower	Wood chips 88%, manure 2.5%, clay 10%	Termorshuizen et al. (2006)
Rhizoctonia solani	Cauliflower	Yard waste (with grass)	Termorshuizen et al. (2006)
Rhizoctonia solani	Cauliflower	Yard waste (without grass)	Termorshuizen et al. (2006)
Rhizoctonia solani	Cauliflower	Yard waste (woody materials, grass clippings)	Termorshuizen et al. (2006)
Rhizoctonia solani	Garden cress	Viticulture waste	Bonilla et al. (2012)
Rhizoctonia solani	Pine	Urban biowaste	Termorshuizen et al. (2006)
Rhizoctonia solani	Pine	Wood chips, horse manure	Termorshuizen et al. (2006)

Continued

Table 17.1 Soilborne pathogens demonstrated to be suppressed by compost.—*cont'd*

Pathogen	Crop	Compost feedstock(s)	References
Rhizoctonia solani	Radish	5:5:3 ratio of manure/silage: hardwood bark: softwood shavings resulting in a C:N ratio of 34:1	Neher et al. (2017)
Rhizoctonia solani	Radish	20% food residuals, 10%—15% hardwood bark/mixed wood chips, 10% hay, ≤ 5% shredded paper, ≤ 2% dry sawdust/shavings, 50%—60% mixed horse/cattle manure with bedding	Neher et al. (2017)
Rosellinia necatrix	Avocado	Vegetal	Bonilla et al. (2012)
Sclerotium minor	Garden cress	Municipal biowaste, cow manure	Bonilla et al. (2012)
Typhula incarnata	Tomato	Cotton gin trash	Vida et al. (2020)
Typhula incarnata	Turf	Sewage sludge	Noble and Coventry (2005)
Verticillium dahliae	Turf	Bark, poultry manure	Noble and Coventry (2005)
Verticillium dahliae	Eggplant	Horse manure (20%) and green waste (80%) (wheat straw, corn straw, conifer bark)	Termorshuizen et al. (2006)

All of the above soilborne pathogens have many host plants and are distributed globally. Pathogenic and nonpathogenic strains of *Pythium*, *F. oxysporum*, and *R. solani* can coinhabit a given soil. This coexistence complicates diagnostics given similar fungal morphology and disease symptoms. Environmental properties likely affect relative pathogenicity. For example, pH influences suppression of *Phytophthora nicotianae* on tomato and Fusarium wilt on carnations.

Soilborne pathogens are masters of survival, making them difficult to kill. They exist primarily in dormant forms (e.g., spores), which allows them to survive for a decade or more. They are stimulated to germinate, grow, and infect roots when they sense nutrition available through root and seed exudates or added nutrients. Soilborne pathogens produce different enzymes to obtain nutrition. The biochemical properties of the enzymes determine their ability to compete with other species. Poor competitors, such as *Pythium* and *Phytophthora* species, thrive in high nutrient conditions and easily degradable simple carbohydrates (e.g., sugars). In contrast, *R. solani* is more competitive than *Pythium* and *Phytophthora* because it can also metabolize starches and cellulose, both of which are abundant in compost (Scotti et al., 2020).

Between growing seasons, many pathogens live as saprophytes to varying abilities. Saprophytes survive on plant debris and detritus. Mature composts, especially containing wood chips and/or bark, contain microorganisms with strong saprophytic ability that can outcompete soil pathogens with weak saprophytic ability (e.g., *Verticillium dahliae*, *Thielaviopsis basicola*). The composition of carbon sources in compost differentially attracts specific species of bacteria and fungi that naturally colonize the compost during the cooling phase of the process and are antagonistic to pathogens (Hadar and Papadopoulou, 2012; Neher et al., 2013). These saprophytic microorganisms in compost may also suppress foodborne pathogens such as coliforms, *Listeria*, or *Salmonella* species (Limoges et al., 2021). Similar antagonism against animal pathogens has also been observed in animal bedding that has been composted or allowed to compost in place (Box 17.1).

Box 17.1 Biological control of animal pathogens in bedding

Compost-bedded pack barns (CBPs) are receiving increasing attention as a housing system for dairy cows that has potential to improve animal welfare. Bedded-pack barns not only provide comfort and better foot and leg health but also microorganisms in the bedded pack have potential to decrease animal pathogens (Leso et al., 2020). Dairy farmers identify mastitis as a top animal health challenge area. Prevention is critical to limiting mastitis, particularly on organic dairy farms, where efficacy of products approved to treat infections is limited. Organic dairy farmers report less incidence and severity of mastitis on cows bedded on compost-bedded pack. Traditionally, bedded packs have been thought to increase risk of mastitis due to the presence of pathogenic bacteria and the favorable moisture and temperature for the growth of these pathogens. However, there is empirical evidence that bedded-pack systems do not increase the prevalence of mastitis, and potentially change the ability of these communities to buffer against disease (Andrews, 2018). These packs are also home to large populations of predaceous mites that prey upon fly larvae.

This system requires excellent pack and ventilation management for barns to perform well. Because of the high bacterial concentrations in bedding, regular additions of ample bedding and excellent teat preparation procedures in milking are recommended. Repeatedly adding bedding materials generates a layering of bedding and animal excrement. CBP use wood chips or sawdust as bedding instead of straw. The wood residue binds the excrement and daily aerating incorporates the manure and starts the composting process. Researchers and dairy producers from Minnesota suggest that dry, fine wood shavings, or sawdust, preferably from pine or other softwoods, are the choice bedding materials in CBP. The size of bedding particles is particularly important for regulating microbial access to the food source. Additionally, shavings or sawdust provide structure that can be easily stirred and remain fluffy enough to assure oxygen transfer within the bedding material. Especially under cold and humid weather conditions, large amounts of bedding may be necessary to keep the pack adequately dry and comfortable for the cows. Published estimates range from 8.2 to 25.6 m^3/cow per year (Leso et al., 2020). Dairy cow feces have a low C:N ratio, ranging from 15:1 to 19:1 and the most commonly used bedding materials are dry and have a very high C:N ratio. In CBP, adding fresh bedding may be necessary to absorb excessive pack moisture and to keep the pack C:N ratio within the optimal range. Otherwise, composting is inhibited in CBP if the C:N ratio decreases to 15:1 or below.

1.3 Compost organisms that suppress plant diseases

Every compost owes its disease suppressiveness to the microorganisms that inhabit it. These microorganisms naturally colonize compost during maturation and curing phases. Much scientific research has focused on identifying specific strains or species that control specific pathogens or diseases. Although some diseases can be suppressed by a single strain or species, inconsistencies may be attributed to the concept that mixtures of strains, organisms, or mechanisms are involved. Composts support a spectrum of microbial groups that offer multiple modes of action against a target pathogen or disease. Species of many biocontrol organisms that have been cultured and tested in bioassays are listed in Table 17.2. The list is likely to expand exponentially in the next decade with the use of molecular genetic tools that allow us to identify organisms from compost that are not culturable, yet prevalent and pivotal in disease suppression. Ideally, it would be most practical to have composts designed to suppress multiple pathogens and/or crop diseases.

1.4 Specific versus general compost-mediated disease suppression

Compost-mediated disease suppression ranges from specific to general. Specific suppression is provided by activities of a narrow spectrum of one or a few specific populations of beneficial microorganisms of which some do not colonize composts. General suppression results from the collective activity of many species of microorganisms in field or potting soils.

With specific suppression, the beneficial organisms deter pathogen growth through particular biological control mechanisms such as competition, parasitism, antagonism, and/or induced plant resistance. Suppression of damping-off caused by *R. solani* is an example. Other examples of specific suppression include *Streptomyces* A1RT on potato scab (*Streptomyces scabies*) and *Brachyphoris oviparasitica* (syn. *Dactylella oviparasitica*) on sugarbeet cyst nematodes. *Trichoderma* and other inoculants to control *Rhizoctonia* and *Fusarium* diseases are a proven practice for potted greenhouse crops. Strains of some *Trichoderma* spp. can kill sclerotia (resting structures) of *R. solani* (Coventry et al., 2006).

Companies that formulate, produce, and market specific antagonistic strains of microorganisms take advantage of specific suppression. Commercial products are limited to microbial species that can be cultured and have stable spores to extend shelf life (Grosch et al., 2004).

In contrast, no single species by itself is responsible for general suppression (Bonanomi et al., 2010). This type of suppression best explains biological control of root rots caused by *Pythium* spp. and *Phytophthora* spp. and some nematodes (e.g., lesion, root knot). General suppression relies on the activity and interaction among bacterial and fungal communities, and their chemical communication with the plant.

Table 17.2 Compost microorganisms identified as beneficial to biological control.

Microorganism	Disease or pathogen	Crop	Reference
Bacillus amyloliquefaciens Bg-C31	Capsicum bacterial wilt (*Ralstonia solanacearum*)	Pepper	Eljounaidi et al. (2016)
Bacillus amyloliquefaciens BZ6-1	Peanut bacterial wilt (*Ralstonia solanacearum*)	Peanut	Eljounaidi et al. (2016)
Bacillus subtilis	Large patch (*Rhizoctonia solani*)	Turf	Noble and Coventry (2005)
Bacillus subtilis	Rhizoctonia bottom rot (*Rhizoctonia solani*)	Lettuce	Grosch et al. (2004)
Bacillus subtilis Jaas ed1	Verticillium wilt (*Verticillium dahliae*)	Eggplant	Eljounaidi et al. (2016)
Bacillus subtilis strains	Damping-off (*Rhizoctonia solani*)	Carrot, cucumber, tomato	Grosch et al. (2004)
Brachyphoris oviparasitica (syn. *Dactylella oviparasitica*)	Southern root knot nematode (*Meloidogyne incognita*)	Peach	Timper (2014)
Brachyphoris oviparasitica (syn. *Dactylella oviparasitica*)	Sugarbeet cyst nematode (*Heterodera schachtii*)	Sugar beet	Timper (2014)
Burkholderia cepacia (syn. *Pseudomonas cepacia*)	Rhizoctonia bottom rot (*Rhizoctonia solani*)	Lettuce	Grosch et al., (2004)
Enterobacter HA02	Verticillium wilt (*Verticillium dahliae*)	Cotton	Eljounaidi et al. (2016)
Fusarium oxysporum F2	Verticillium wilt (*Verticillium dahliae*)	Eggplant	Hadar and Papadopoulou (2012)
Fusarium oxysporum Fo162	Burrowing nematode (*Radopholus similis*)	Banana	Timper (2014)
Fusarium oxysporum Fo162	Southern root knot nematode (*Meloidogyne incognita*)	Tomato	Timper (2014)
Gliocladium virens G-21	Bottom rot (*Rhizoctonia solani*)	Lettuce	Grosch et al. (2004)
Hirsutella minnesotensis	Soybean cyst nematode (*Heterodera glycines*)	Soybean	Timper (2014)
Hirsutella rhossiliensis	Soybean cyst nematode (*Heterodera glycines*)	Soybean	Timper (2014)
Paecilomyces variotii MSW312	Fusarium wilt (*Fusarium oxysporum* f.sp. *melonis*)	Melon	Suárez-Estrella et al. (2013)
Paenibacillus K165	Verticillium wilt (*Verticillium dahliae*)	Eggplant, potato	Eljounaidi et al. (2016)
Pseudomonas fluorescens CHA0	Root knot nematode (*Meloidogyne javonica*)	Tomato	Timper (2014)
Pseudomonas fluorescens CHA0	Southern root knot (*Meloidogyne incognita*)	Soybean, mung bean, tomato	Timper (2014)
Pseudomonas fluorescens EB69	Eggplant wilt (*Ralstonia solanacearum*)	Eggplant	Eljounaidi et al. (2016)

Continued

Table 17.2 Compost microorganisms identified as beneficial to biological control.—cont'd

Microorganism	Disease or pathogen	Crop	Reference
Pseudomonas fluorescens PICF7	Verticillium wilt (Verticillium dahliae)	Olive	Eljounaidi et al. (2016)
Pseudomonas putida	Apple replant disease	Apple	Weller et al. (2002)
Pseudomonas putida B10	Take-all (Gaeumannomyces graminis var. tritici)	Wheat	Haas and Défago (2005)
Serratia marcescens UPM39B3	Fusarium wilt (Fusarium oxysporum)	Banana	Eljounaidi et al. (2016)
Serratia plymuthica HRO-C48	Verticillium wilt (Verticillium dahliae)	Oilseed rape	Eljounaidi et al. (2016)
Trichoderma hamatum	Damping-off (Rhizoctonia solani)	Radish	Chung et al. (1988)
Trichoderma hamatum	Rhizoctonia stem canker and black scurf (Rhizoctonia solani)	Potato	Beagle-Ristaino et al. (1985)
Trichoderma harzianum	Damping-off (Pythium ultimum)	Cucumber	Pugliese et al. (2011)
Trichoderma harzianum	Rhizoctonia bottom rot (Rhizoctonia solani)	Lettuce	Grosch et al. (2004)
Microorganisms that control multiple diseases			
Bacillus subtilis GBO3	Rhizoctonia, Fusarium, Aspergillus, and others	Seed treatment for cotton, peanuts, soybeans, wheat, barley, peas, beans	Fravel (2005)
Bacillus subtilis MBI 600	Fusarium, Rhizoctonia, Alternaria, and Aspergillus	Cotton, beans, barley, wheat, corn, peas, peanuts, soybeans	Fravel (2005)
Bacillus pumilus GB 34	Fusarium wilt and Rhizoctonia damping-off	Soybean	Fravel (2005)
Bacillus subtilis var. amyloliquefaciens FZB24	Fusarium wilt and Rhizoctonia damping-off	Shade and forest tree seedlings, ornamentals, shrubs	Fravel (2005)
Pseudomonas chlororaphis 63-28	Pythium sp., Rhizoctonia solani, Fusarium oxysporum	Vegetables, ornamentals	Fravel (2005)

Combinations of microorganisms that perform better than individual species or strains

Trichoderma strains + bacteria	Fusarium wilt	Radish	Hoitink and Boehm (1999)
Trichoderma viride and/or Trichoderma harzianum	Root rot (Phytophthora nicotianae)	Tomato	Pugliese et al. (2011)
Trichoderma harzianum, Verticillium chlamydosporium + Glomus mosseae	Heterodera cajani-Fusarium udum wilt disease complex	Pigeonpea	Meyer and Roberts (2002)
Arthrobotrys oligospora and different unidentified bacteria	Root knot nematode (Meloidogyne mayaguensis)	Tomato	Meyer and Roberts (2002)
Anabaena oscillarioides C12 and Bacillus subtilis B5	Damping-off by combination of Fusarium sp., Pythium sp. and Rhizoctonia solani	Tomato	Dukare et al. (2011)
Trichoderma virens G1-3 + B. cepacia Bc-F	Rhizoctonia solani and Pythium ultimum, alone or in combination with Sclerotium rolfsii and Fusarium oxysporum f. sp. lycopersici.	Tomato	Meyer and Roberts (2002)
Verticillium chlamydosporium + Pasteuria penetrans	Southern root knot nematode (Meloidogyne incognita)	Tomato	Meyer and Roberts (2002)
Escherichia coli S17R1 + Burkholderia cepacia Bc-B	Pythium and Fusarium spp.	Cucumber	Meyer and Roberts (2002)
Embellisia chlamydospora, Verticillium chlamydosporium, and a sterile fungus	Sugarbeet cyst nematode (Heterodera schachtii)	Sugar beet	Meyer and Roberts (2002)

The challenge with general suppression is that it establishes in place; it cannot be transferred, and it can be disrupted by changes in management. Therefore, the degree and longevity of disease suppression can vary greatly. The efficacy and duration of suppressiveness depend on a number of compost and soil factors, including feedstocks from which compost is prepared, the thermophilic and curing process, maturity and phytotoxicity, salinity and nutrient content, and the microorganisms that colonize composts after peak heating and before planting in soil (Box 17.2).

Box 17.2 Biological Character of Organic Matter

In the 1950s and 1960s, organic amendment chemistry was defined using parameters such as cellulose and lignin content, and the ratio of total carbon to total nitrogen (C:N). These measures provided estimates of decomposition rate and are a component of compost recipe development but only have limited usefulness to predict the impact of compost on disease suppression (Neher et al., 2015). One explanation is that not all carbon is alike. Carbon substrates differ in water repellency, hydrocarbon content, and biochemical composition (simple sugars to highly aromatic materials recalcitrant to decomposition). There are now several high-throughput methods available to characterize organic matter, including pyrolysis—gas chromatography/mass spectrometry, near-infrared reflectance, and Fourier transform infrared spectroscopy. These methods provide a detailed view of organic matter composition that changes our understanding of the mechanisms of decomposition. They also help identify what properties of organic matter offer disease suppression more than others. Organic matter acquires beneficial characteristics soon after it begins to decay. Beneficial saprophytic microorganisms derive their nutrition and energy from this decaying material. The particle size, particle density, and age (degree of decomposition) of soil organic matter seem to set limits on disease suppression. The largest, least decomposed particles of organic matter do not seem to contribute directly to disease control, but as they decrease in particle size through decomposition, their effectiveness increases. However, there is a point of diminishing returns on particle size. Clumps that resemble soil aggregates (roughly 6 mm or ¼-inch in diameter) contain beneficial biocontrol organisms that can be lost by very fine screening (Neher et al., 2019a). Several reports suggest that the finest, most stable fraction (humus) also does not contribute to long-term biological control. This fraction is so biologically stable that it cannot nutritionally support populations of beneficial microorganisms. However, stable materials like biochar provide a porous structure that can physically sustain colonies of biocontrol organisms such as *Pseudomonas chlororaphis*, *Bacillus pumilus*, and *Streptomyces pseudovenezuelae* to suppress diseases caused by *Pythium aphanidermatum* and *F. oxysporum* f. sp. *lycopersici* in tomato (Bonanomi et al., 2018).

The composition of carbon compounds in the final product differentially selects a suite of microorganism species that colonize and are antagonistic to pathogens (Hadar and Papadopoulou, 2012). Wood-based composts have higher lignin: cellulose ratios than hay or straw carbon-based composts. Tree bark and other woody materials also contain tannins and waxes that resist decomposition. Sophisticated ^{13}C CPMAS-NMR spectroscopy methods have identified phenolic carbon and methoxyl carbon molecules associated with suppressive mechanisms (Pane et al., 2013). Both phenolic and methoxyl carbon are products of lignin degradation in woodchips. Intermediate products and residues from lignin degradation contribute to the humified matter in composts. Phenol and methoxyl carbons are unique soil carbons, requiring specialty enzymes that only a subset of saprophytic microorganisms can produce under conditions of carbon or nutrient limitations. Lignin is a complex molecule that requires a suite of 14 different enzymes to completely degrade (Chapter 4). The white-rot fungi of the Basidiomycota are among the main group that can fully degrade lignin.

Maintaining high levels of organic matter (e.g., greater than 6%) using mature compost is perhaps the single most reliable way to establish and preserve general suppression (Fig. 17.3). There are interactions between the level of organic matter decomposition and soil physical conditions, and those interactions affect disease incidence and severity in some crops. Partially decomposed organic matter improves soil structure. This transformation results in better water retention under dry conditions and improved drainage during periods of high precipitation. The improved soil conditions, in turn, lead to natural root rot suppression in wet soils in ridge tillage systems and some degree of suppression of wilt diseases in dry soils. Examples are Phytophthora root rot, which is prominent in wet soils, and the early dying disease of potato, caused by a complex involving nematodes and *Verticillium* in dry soils. These levels of organic matter require a combination of compost amendments, minimal to no tillage, and maintaining continuous vegetation cover to promote a more complex food web of abundant and active microorganisms that can suppress the incidence and severity of root diseases.

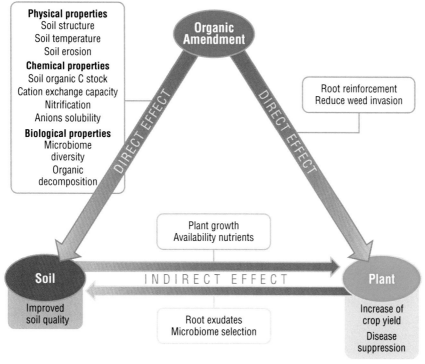

FIGURE 17.3

Interactions of compost, soil, and plants that support general suppression of plant diseases.

Adapted from Vida, C., De Vicente, A., Cazorla, F.M., 2020.

Rotating from a good host for a particular plant pathogen with crops that are nonhosts is an effective management strategy. In the years when a nonhost is planted, pathogen populations progressivly decline to low population densities. This decline, however, can be detrimental to microorganisms that are naturally antagonistic toward pathogens. A historical example of long-standing suppression is avocado root rot caused by *P. cinnamomi*, established in early 1940s in Queensland, Australia. The crop remained healthy after more than 40 years in soil infested with *P. cinnamomi* in an environment highly favorable for disease development, which was correlated with a diverse and abundant bacterial community that could antagonize the pathogen. A second example is "take-all" disease of wheat and barley caused by the fungus, *Gaeumannomyces graminis*. The suppressiveness was attributed to a build-up of populations of a specific *Pseudomonas fluorescence* strain that produced a broad-spectrum antibiotic that was especially active against the take-all pathogen. There are other examples of long-term, no-till monoculture with high organic matter content that suppress pathogens and pests, including soybean cyst nematodes (Neher et al., 2019b).

1.5 Compost factors that affect disease suppression

The degree of disease suppression experienced when soils are amended with organic amendments can vary greatly. Furthermore, organic amendments suppress diseases only for a limited period of time. The duration of suppressiveness and degree of efficacy, depend on a number of compost and soil factors, including:

- Feedstocks and characteristics,
- Composting process and curing,
- Compost decomposition level/maturity,
- Compost microorganisms and plant protection,
- Compost nutrient content,
- Compost salinity, and,
- Timing of compost application.

Extensive planning is required to implement strategies that maximize compost-induced disease suppression. This includes considering interactions among organic amendments, soils, and crops. Each of the factors will be reviewed here. Examples of disease control on several crops are used to illustrate reasons for success and failure in disease control.

1.5.1 Feedstocks and compost characteristics

There is no doubt that feedstocks determine the character of the compost and, thus, the compost's potential to suppress plant diseases and duration of the suppressive effect (Hoitink and Boehm, 1999). It is, however, difficult to prescribe or predict the effect of specific feedstocks on the disease-suppressive qualities of the resulting compost. With the exception of *Pythium* and *Phytophthora*, different compost recipes and maturity affect pathogen(s) and host crop(s) differently. Scientific results

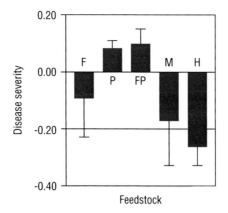

FIGURE 17.4

Disease severity of damping-off caused by *Rhizoctonia solani* in greenhouse studies, as affected by composting feedstock ($F = 0.59$, $P = 0.674$). Both controls and treatments were inoculated with virulent *Rhizoctonia solani*. Illustrated are means $\pm$ 1 standard error of percent change from noncompost control. *F*, Food waste; *FP*, Food waste + Poultry manure; *H*, Hardwood bark; *M*, dairy Manure; *P*, Poultry manure.

Source: Modified from Neher et al. (2017)

have provided some guidance but have so far fallen short of providing general recommendations regarding feedstock and disease suppression (Fig. 17.4). For example, composts made from green waste (yard trimmings, food scraps, animal manures, paper, and wood wastes) reduced damping-off caused by *R. solani* on tomato but had the opposite effect on lettuce (Noble, 2011). Another study compared two composts containing different vegetable residues amended with 10% wood chips. The compost containing rocket/arugula residues suppressed *R. solani* and *Sclerotinia minor* while another containing endive was conducive to disease (Scotti et al., 2020). Other studies examining the disease-suppressive effects of animal manure composts have generally shown positive but inconsistent results. Composts derived from poultry manure are particularly difficult to characterize because of the typically high ammonia content and salt concentrations.

One generalization is that the most reliable disease-suppressive composts are those made with woody materials, like tree bark, wood chips, and woody yard trimmings. Tree bark and other woody materials consist mostly of lignin, cellulose, tannins, and waxes, a mixture that resists decomposition. After composting, the disease-suppressive effects of composted barks last for several years in soil, depending on the tree species from which the bark was removed and how much compost was added to the soil. Decomposition of wood waste releases nutrients very slowly and produces humic acids (large molecular weight organic acids that are very complex and difficult to degrade). In contrast, food and feed wastes, animal manures and biosolids mostly consist of readily decomposable compounds and nutrients and produce fulvic acids (low molecular weight organic acids). Both fulvic and humic acids grab and bind (chelate) essential micronutrients and keep them available for uptake by plants. Chelates can strongly mediate the severity of diseases caused by soilborne plant pathogens. However, these beneficial effects usually do not last more than one or 2 years in soils.

1.5.2 Composting process and curing

Many composting feedstocks carry microorganisms that are pathogenic to plants and/or humans. Fortunately, pathogens and weed seeds are destroyed by the high temperatures achieved during the sanitization stage (e.g., pasteurization, PFRP) of composting (Neher et al., 2015). Therefore, properly prepared compost not only delivers the potential to suppress plant pathogens in the soil but also delivers few to no new pathogens to the plant environment.

Because numerous beneficial microorganisms contribute to biological control of plant diseases, the question becomes whether such organisms consistently colonize composts after peak heating. Chances of this happening are poor in large windrow or pile composting systems in which temperatures typically persist above 40°C (104°F) for prolonged periods after sanitation. Most biocontrol agents cannot grow or survive long-term in these temperatures, except for spores of *Bacillus* spp. Conversely, colonization is rapid when postsanitation temperatures are maintained below 35°C (95°F), especially at soil temperatures of 25°C (77°F) or lower. Bacterial biocontrol agents such as *Pseudomonas* spp. colonize the substrate fully in one to 2 days to establish general suppression. This does not occur, however, when the moisture content of the compost is below 30% on a weight basis. Dry composts become dusty and fungi become the principal colonizers. Some of these are nuisance fungi that delay or even inhibit plant growth. Therefore, it is important to manage moisture content during peak heating as well as during curing of compost to enhance the potential for natural colonization by the beneficial microflora during the process. Although most compost that is used in container media (predominantly made from bark, sawdust, etc.) is made in tall windrows/piles, these products still offer ideal opportunities for inoculation with specific biocontrol strains for use in greenhouse and nursery crops, as long as moisture and temperature regimes are managed appropriately.

The situation can be quite different for small windrow composting systems that are turned frequently, especially when the moisture content of the compost is maintained above 45%. These composts, especially those high in microbial activity such as manure composts, are much more likely to be colonized by a great diversity of biocontrol agents as the compost matures.

1.5.3 Level of organic matter decomposition

Fresh organic matter often has negative effects on plant health for some time after their application to soils. Fresh residues typically stimulate the growth of pathogens and increase disease incidence and severity for some time after their incorporation.

For example, the pathogens *R. solani* (causes damping-off on almost all crops) and *Armillaria mellea* (can kill mature trees including oaks, kiwi) can grow on fresh straw and wood. These fungi cannot grow on partially decayed or composted products. Both *Pythium* and *Phytophthora* cause root rot on many plants, particularly in wet soils, and thrive on fresh green manures. For example, fresh straw applied in the fall as mulch under apple trees or red raspberry bushes increases water retention in soil and immobilizes nitrogen if it has not decayed adequately. As a result, Phytophthora collar rot is aggravated in the wet soil when trees break dormancy in the spring when the Phytophthora collar rot pathogen becomes active. Fresh ground wood can have similar negative effects in the landscape, but the effect lasts much longer because wood breaks down much more slowly than straw. In contrast, composted wood, which is more like forest litter, improves soil drainage and aeration while it also improves water retention and supports the growth of mycorrhizal fungi, all of which leads to suppression of Phytophthora root rots. Other diseases are also aggravated by shredded raw wood mulches but are controlled by composted wood. Examples include diseases caused by *Armillaria*, *Pythium*, and *Rhizoctonia*. This principle applies to many crops!

Green manures need to decompose for 10 to 14 days after they are plowed into the soil prior to crop planting to prevent a drastic increase in Pythium damping-off on many crops. A California study showed that lettuce, planted in soil one day after vetch was incorporated, suffered severe preemergence damping-off due to increased *Pythium* activity. In contrast, planting one week after incorporation provided control. Allowing green manures to degrade before planting a crop provides time for beneficial microorganisms to colonize the decaying vegetation. Colonization may take several weeks, depending upon crop species and maturity, soil temperature and moisture content.

Fresh residues generally cause problems unless they decompose to some degree before planting of the next crop. There are strategies to encourage breakdown of crop residues using nutrient-rich composts, like poultry manure-based composts. For example, application of 2.5 to 5 tonne per hectare (1.1 to 2.2 tons per acre) of fresh or composted poultry manure immediately after the harvest of corn accelerates the decomposition rate of corn stover in the field. The added nitrogen combined with minimum tillage decreases survival of plant pathogens. Thus, seed rot caused by *Pythium* and seed, stalk and ear rot of corn caused by *Fusarium graminearum* can be reduced in severity by such applications in this tillage practice. The ear rot pathogen produces mycotoxins (i.e., vomitoxin) which have serious detrimental effects on livestock.

1.5.4 Compost maturity

Plant disease suppression is the result of the activity of antagonistic microorganisms that naturally recolonize the compost during the cooling phase of the process. Composition of microbial communities starts similarly after the sanitation phase and the composition of the community changes as the temperature declines and the chemistry of the compost changes (Neher et al., 2013). These natural patterns resemble ecological succession of decomposer communities in forest litter and wood decay reported by soil ecologists. Microorganisms secrete enzymes that target portions of the decaying organic matter that provide the nutrients or energy they need. Microbial feeding on organic matter alters the chemistry of compost which, in turn, promotes a microbial turnover that further changes organic carbon chemistry. This curing phase offers a substrate and environmental conditions conducive for microbial recolonization that can be expedited by inoculating post-thermophilic compost or preparing a palatable substrate that provides a competitive advantage for colonization by bacteria and fungi that offer biological control.

Immature compost corresponds with early stages in succession that favor microbial species that are most competitive when simple carbohydrates are abundant, earning them the ecological title, copiotrophs. Mature compost corresponds with later succession that favor microbial species that are most competitive with complex carbohydrates (e.g., lignin, tannins) earning them the ecological title, oligotrophs. The ratio of oligotrophic to copiotrophic organisms increases through maturation, which corresponds to enhanced disease suppression of mature compared with immature composts (Fig. 17.5).

Biological control organisms can grow effectively on both immature and mature products but shift to become relatively more or less competitive against pathogens depending on the relative competitiveness on particular substrates. A classic example is for pathogens like *R. solani* that are favored in early stages of composting when concentrations of water-soluble carbon compounds are high (Chung et al., 1988). Once this carbon is depleted, as in mature compost, the efficacy of biological control fungi such as *Trichoderma harzianum* increases because it produces

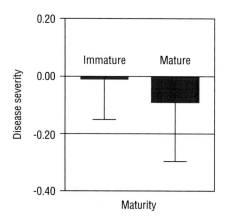

FIGURE 17.5

Relative disease severity of damping-off caused by *Rhizoctonia solani* affected by compost maturity ($F = 4.59$, $P = 0.041$). Both controls and treatments were inoculated with virulent *Rhizoctonia solani* (Illustrated are means ± 1 standard error of percent change from noncompost control.)

Source: Modified from Neher et al. (2017).

enzymes that degrade cellulose. Managing carbon quality and compost maturity are the tools that one has to manage the colonization of compost with microorganisms that will favor disease suppression.

At the other end of the spectrum, composts that are excessively stable or fully decomposed after months or years of decomposition in soil no longer have the ability to support populations of beneficial organisms. As beneficial organism populations decline, plant pathogens increase in numbers and activity and diseases increase in severity in the now "worn out" soil, even though humic acids produced from the compost are still present in the soil. For this reason, highly stabilized organic matter, such as peat or geologically old soil organic matter (as found in soils derived from prairies or in organic soils after they have been farmed for many years) are not effective in controlling plant pathogens unless new sources of stable organic matter are added.

1.5.5 Compost microorganisms and plant protection

Microorganism communities are abundant and diverse in compost. The assembly of microbial communities (consortium) are organized and influenced by recipe, choice of post-thermophilic process, and duration of curing (maturation) of composts (Neher et al., 2013). When added to soil, the nature and behavior of the consortium are modified by plant (crop) and soil type. It is only recently that scientists have the tools to solve the mysteries of how these highly coevolved relationships work so we can incorporate those insights into management practices.

Until the 1990s, the knowledge of compost microbiology was limited to organisms that would grow in Petri dish culture. Suppression was tested by exposing the pathogen to cultured organisms *in vitro* or measuring reduction of disease symptoms when they were inoculated in a conductive soil. This era identified the strains commonly seen on the market including species of *Pseudomonas* (γ-Proteobacteria), *Bacillus* (Firmicutes), *Streptomyces* (Actinobacteria), and *Trichoderma* (Ascomycota). Unfortunately, this approach missed 99% of the species that are now detectable by modern molecular techniques that detect microorganisms independent of their ability to grow in a Petri dish.

Molecular technology provides a new perspective on the microbial community or microbiome. A microbiome contains the genes, metabolites, proteins, and species associated with various habitats of a plant host whether it be whole plants, specific organs such as roots, or the rhizosphere (root-soil interface). Now, studies of suppressive soils can use microbiome analyses and examine the structure and complexity of interactions among microorganisms themselves and in interaction with roots and/or soil. Some generalizable patterns are emerging as the use of molecular techniques increases. For example, the assembly of bacterial and fungal communities that colonize compost during curing and maturation phases depend on whether the carbon source was hay, straw, softwood (e.g., pine), or hardwood (e.g., birch) (Neher et al., 2013). Different microbial species produce different types and diversity of enzymes that generally or very specifically target particular types of decaying organic matter. The consortium of microorganisms that can suppress disease is able to metabolize and degrade complex matrices better than the community found in conducive compost (Scotti et al., 2020). Members of this "suppression" consortium include an abundance of bacteria in the phyla Proteobacteria, Bacteroidetes, Actinobacteria, and Deinococcus-Thermus. In contrast, members of the conductive microbial consortium include an abundance of bacteria in the phyla Verrucomicrobia, Gemmatimonadetes, Acidobacteria, and Planctomycetes. Suppressive composts contain more fungi in the phylum Basidiomycota and fewer Ascomycota than the conducive compost. The Basidiomycota contain fungi that can decompose lignin associated with wood, such as the brown-rot and white-rot fungi.

The rhizosphere provides the frontline defense for plant roots against attack by soilborne pathogens. As Fig. 17.6 illustrates, plants are able to influence the composition and activation of their rhizosphere microbiome through exudation of compounds that stimulate (green arrows) or inhibit (red blocked arrows). Vice versa, a wide range of soilborne pathogens is able to affect plant health. Prior to infection, these deleterious microbes are in competition with many other microbes in the rhizosphere for nutrients and space. In this battle for resources, beneficial microbes limit the success of the pathogen through production of biostatic compounds, consumption of (micro)nutrients, or by stimulating the immune system of the plant. Most microbes neither affect the plant nor the pathogen directly

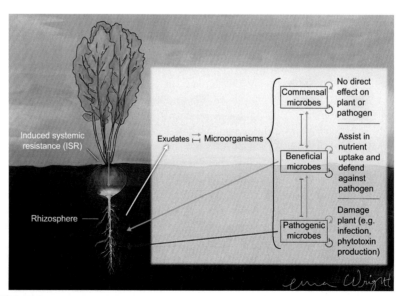

FIGURE 17.6

Interactions in the rhizosphere.

Source: Emma Wright (Research Technician, University of Vermont).

because they occupy different ecological niches (commensal microbes) but are likely to affect every other organism to a certain extent through a complex network of interactions.

The action within the rhizosphere takes place at the scale of micrometers (1 μm = 0.000039370 inch), which is why molecular scale tools are necessary to study the phenomenon. Plants naturally leak sugars and a wide variety of small molecules (e.g., amino acids, organic acids, phenolics, alkaloids) through their roots as lubrication for roots to grow by extension into soil and for root defense. This leakage (exudates) is food and energy for soil organisms, attracting them to root surfaces and the volume of soil adhering to the root. Plants actively recruit beneficial soil microorganisms in their rhizospheres to counteract pathogen assault (Lakshmanan, 2015). In return, microorganisms can promote plant growth by mineralizing nutrients, producing plant hormone imitations, or secreting antibiotics to defend against other microorganisms (Bais et al., 2006).

Rhizosphere microorganisms can differentiate plant species by exudate "flavors." For example, *R. solani* is an invading pathogenic fungus that induces, directly or via the plant, a stress signal detected by the rhizosphere bacterial community (Chapelle et al., 2015). In response, the microbiome shifts in community composition and activates traits that protect roots by restricting the ability of the pathogen to infect and cause disease. For example, when *R. solani* attacks the rhizosphere of sugar beet, a consortium of bacteria response by antagonizing the pathogen, e.g., *Paraburkholderia* (β-Proteobacteria), *Pseudomonas* (γ-Proteobacteria), and *Streptomyces* (Actinobacteria) (Mendes et al., 2011). Additional known pathogen antagonist groups include bacteria (β-Proteobacteria: *Burkholderia*; γ-Proteobacteria: *Serratia*; Firmicutes *Bacillus*), and fungi (*Trichoderma*, *Penicillium*, *Gliocladium*, *Sporidesmium*, nonpathogenic *Fusarium* spp.). The means or mechanism of antagonism can include production of antibiotics (against bacteria or fungi) or competition (for nutrients, trace elements, or colonization sites). For example, *Streptomyces* (Actinobacteria) and *Trichoderma* (Ascomycota) species are prolific producers of antibiotics. Antibiotics secreted in low concentrations can mediate intercellular signaling (communication), and in high concentrations can inhibit pathogen growth. Sometimes, it takes multiple species of microorganisms working together collectively to antagonize pathogens (syntropy). Root-associated bacteria can distinguish among their neighbors and fine-tune the biosynthesis of antimicrobial metabolites. These types of natural suppression are highly coordinated events influenced by the plant host and soil (Bais et al., 2006).

If the pathogen breaks through this first line of defense, it encounters the basal and induced defense mechanisms of the plant. Plants communicate with microorganisms in the rhizosphere through root exudates. A direct attack by pathogens stimulates the plant to release chemical signals consisting of phenolic compounds (e.g., coumaric, cinnamic, salicylic acids) or saponins (glycosides with triterpene or steroid backbones). Phenolic compounds stimulate germination of fungal conidia in low concentrations and inhibit fungal growth in high concentrations.

Mal-timed chemical signals can trick fungal pathogens into germination in unfavorable conditions disarming them from a successful infection. Saponins form complexes with sterols and damage the cell membranes of plant pathogens. Indirectly, root exudates stimulate microorganisms to produce small water-soluble molecules, called volatile organic compounds (VOCs). The type and temporal dynamics of VOC production are extremely species-specific, at least for *Trichoderma* that produces a diversity of sesquiterpene emission patterns (Guo et al., 2020). VOCs evaporate easily at room temperature and distribute into the surrounding air, enabling them to act as communication signals within and among organisms. When the density of these molecules exceeds a certain threshold (measured as parts per trillion with modern instrumentation), it triggers a coordinated community response (quorum sensing) that activate various plant defense-related genes that either suppresses disease symptoms or stimulate plant growth. VOC-producing organisms in the rhizosphere suppress symptoms include *Pseudomonas trivialis* (γ-Proteobacteria), *Pseudomonas fluorescens* (γ-Proteobacteria), *Bacillus subtilis* (Firmicutes), *Burkholderia cepacia* (γ-Proteobacteria), and *Trichoderma* (Ascomycota). Those that promote plant growth are produced by *Flavobacterium* (Bacteroidetes), *Streptomyces* (Actinobacteria), and *Trichoderma*. Microbial species promote plant growth by production of antibiotics to combat pathogens, manufacturing plant growth mimics, and/or induced systemic resistance (ISR) that protects noninfected tissues throughout the plant. ISR in plants by compost occurs in cucumber Pythium root rot (i.e., *Pythium ultimum* and *Pythium aphanidermatum*). Traditionally, the role of VOCs was overlooked partly due to analytical limitations. With modern tools, we are likely to gain knowledge about how they operate ecologically and are modified by soil type, neighboring species, and plants.

By intentionally designing recipes and curing methods, compost can become a tool to manipulate or deliver a natural consortium of microorganisms in soil, onto seeds, and planting materials. The advantage of assembling microorganisms with complementary or synergistic traits provides a more effective and consistent effect. For example, a consortium containing both *Flavobacterium* and *Chitinophaga* conferred significant and more consistent protection against fungal root infection than individual consortium members (Scotti et al., 2020). However, the next scientific challenge is to find or select the right players of a consortium. Rather than single species, consortia are communities that mimic general disease suppression in soil. One may argue that control of different pathogens on different crops requires a different combination of microorganisms and/or mechanisms (Termorshuizen et al., 2006; Bonanomi et al., 2010). This is probably true for different groups of soilborne plant pathogens, that is, bacteria, fungi, oomycetes, and nematodes. However, studies on natural disease-suppressive soils have pointed to common players and identical mechanisms and genes in the suppressiveness of soils to different fungal pathogens. Furthermore, the onset of natural disease suppressiveness of soils follows a similar pattern for various fungal pathogens suggesting that similar processes, mechanisms, and microorganisms may be required for the transition of a soil from a conducive to a suppressive state.

1.5.6 Compost salinity

Many plants experience stress by excessively high concentrations of salts in the root environment (EC readings >10 dS/m). In turn, stressed plants are more susceptible to diseases, particularly root diseases caused by *Pythium* and *Phytophthora*. Even when the salinity levels are not stressful to plants, elevated concentrations of salts can negate the disease suppression benefits supplied by the organic and biological components of the compost. Composted livestock manure tends to have relatively high levels of soluble salts and may not produce the expected disease benefits in container-grown plants.

It is often impractical to adjust feedstock recipes to lower compost salinity levels. Instead, one can blend high-salinity composts with low-salinity compost (e.g., most yard trimmings composts). It is advisable to apply compost to fields well ahead of planting to allow salts to leach below the root zone if there is adequate rain (or snow). However, an unintended side effect could be the loss of mineral nutrients to leaching too. For mulches, the best approach is to blend composted materials high in salinity with woody or bark mulches which typically are low in salinity to dilute the negative factors and provide long-lasting beneficial effects for value added markets. These blends can be applied as mulches at any time of year and provide beneficial effects more consistently.

1.5.7 Compost nutrient content

A growing body of research shows a link between soil fertility and plant disease incidence/severity. Generally, plants are more susceptible to disease if they are either nutrient stressed (limited) or their roots are surrounded by a nutrient surplus. Nitrogen (N) is the primary nutrient to consider for two reasons. First, availability of mineral N (ammonium and nitrate) from composts or mulches varies more than that of any other nutrient. Composts made from biosolids or animal manures typically contain between about 1.5% and 2.5% total N (dry weight) but values may exceed 4%, particularly for composted poultry manure. Most N, usually above 90%, in compost is organically bound, with only a small proportion being immediately plant available (i.e., ammonium, nitrate). Typically, only a small part (0% to 10%) of the total N in these composts is converted to mineral N forms (ammonium and nitrate) within the first three months after their amendment to soil in spring/early summer. A much higher proportion would be mineralized from raw, high N manures. The remaining organically bound N may be released in as little as two to five years or as long as 30 years or more. If very mature composts with high nitrate concentrations are used, or if mineralization supplies high concentrations of nitrate during the growing season, certain plant diseases may be exacerbated unless care is taken to avoid N overloading of the soil. Nutrient budgets that account for mineral soil N levels and inputs from both organic soil amendments and mineral fertilizers can prevent this. Composted cow manure contains mostly nitrate whereas poultry manure composts are high in ammonia.

Mineral N availability has a major effect on plant disease. Bacterial leaf spots (e.g., *Pseudomonas syringae*, *Xanthomonas campestris*), fire blight on apples and pears (*Erwinia amylovora*), and Pythium root rots and Fusarium wilts are more severe with immature composts that are high in ammonium and low in nitrate, even in hydroponic systems. Thus, great care should be taken in selecting composts for use on crops susceptible to these *Fusarium* diseases (e.g., celery, basil, cyclamen). Low N composts such as those prepared from bark or yard trimmings are best for control of these diseases and then particularly so in greenhouse crops if inoculated with *Trichoderma* strains capable of suppressing Fusarium wilts.

A lack of available N can also aggravate plant disease incidence/severity, especially when immature compost with a high C:N ratio is used. N-poor composts that are relatively young with a high degree of biological activity can tie up and immobilize N when applied to the soil. Because N release or mineralization from these composts is dependent on soil microorganisms, they may create a temporary N deficiency for plants and favorable conditions for plant pathogens to establish.

1.5.8 Timing of compost applications

Compost analysis, soil test results, and crop requirements together should form the basis for determining compost application rates. Nutrient release from composts in the field, particularly that of N must be balanced against what is in the soil, the requirements of the crop, and what is applied with mineral fertilizers. Once growers have applied compost to soil, it becomes more important to properly estimate the quantity of nutrients released from compost before any additional nutrients (organic or mineral) are applied. Fruit growers who do not address this issue will increase fire blight on apple due to excessive N supply even though they will maintain control of Phytophthora collar rot (Fig. 17.7). Grape producers would decrease wine quality. This problem was identified in Italian field studies during the 1990s as the only possible negative aspect associated with 30 years of compost use in vineyards when soil fertility and nutrient supply were not addressed adequately.

FIGURE 17.7

Suppression of Phytophthora collar rot on apple in a bark mix versus a conducive peat mix with three *Phytophthora cactorum* inoculum density levels in each (Spring et al., 1980).

Source: Courtesy of Harry A. Hoitink.

Application of bark or leaf-derived composts with boosted N content (up to between 1.7% and 2.2%) at high rates (5.0−7.5 cm; 2−3 inches; equivalent to approximately 500−750 m^3/ha) followed by incorporation into the topsoil (15−20 cm; 6−8 inches) has been used to control soilborne diseases in ornamental nursery crops. This approach eliminated the need for methyl bromide fumigation in these crops since the 1970s. Moderate to high application rates of composts can replace methyl bromide for control of the strawberry black root rot complex caused by several pathogens.

Application of N-rich composted biosolids or manure to a depth of 2.5 cm (1 inch; equivalent to approximately 250 m^3/ha) and incorporation into the topsoil (10 cm; 4 inches) prepares an ideal seed bed for most woody plants and for seeding new lawns. As mentioned above, only highly stabilized composts in which mineral N is present primarily as nitrate-N should be used to amend soils growing crops that are sensitive to ammonium-N and Fusarium wilt. Especially on sandy soils and in potting mixes, care must be taken to avoid ammonium toxicity.

Composts with high ammonia or salinity are best applied several weeks before planting so that salts disperse and much of the ammonium is adsorbed and converted to nitrate. This is especially critical for manure-based composts applied to crops highly susceptible to Phytophthora root rots and salinity (e.g., soybeans). Crops such as small grains are much less sensitive to these diseases than many vegetables, but they may suffer from Rhizoctonia damping-off. For these crops also, it is better to apply the compost a month or more ahead of planting to minimize this problem. This also avoids ammonium toxicity induced by composted manures that still are high in ammonium content. In regions where crops are planted immediately after another has been harvested, it would be best to apply compost to a disease resistant crop (e.g., corn) that is grown before a susceptible crop is planted.

1.6 Indicators of suppression

Based on a simple understanding of ecological succession and compost maturity, one might anticipate that a simple fungal to bacterial ratio would suffice as a measure of mature and suppressive compost. However, this is oversimplified as illustrated above. The resolution of identification of both bacteria and fungi must include information about their ecological role and lifestyle, e.g., parasitic, saprophytic, oligotrophic, copiotrophic. That said, it is neither practical nor affordable for farmers to run DNA tests to look at specific bacterial and fungal species. Nonetheless, there is an unfulfilled need for reliable indicators to detect composts that suppress soilborne pathogens. Furthermore, these tools are likely to have tweaks and modifications tailored to specific diseases or pathogens.

Disease suppression is best tested by plant bioassays (Wichuk and McCartney, 2010). Effective plant bioassays are standardized by plant cultivar and environmental conditions but are time-consuming (2−4 weeks) to complete which may be longer than desired. Comparably robust, but quicker (1−2 days) assays would be ideal for quality control and quarantine programs.

Microbial biomass and activity: Simple measures of microbial activity or biomass predict *Pythium ultimum* and *Pythium irregulare* but not *R. solani* (Scheuerell et al., 2005). Compost analytical labs have a variety of measures to reflect microbial activity by respiration (CO_2 evolution) by dehydrogenase, Solvita test, and/or hydrolysis of fluorescein diacetate (FDA) (Green et al., 2006). Advantages of these methods are their simplicity and rapid response. However, the methods are criticized for imprecision and weak associations with populations of known biological control agents such as fluorescent *Pseudomonas* and *Trichoderma* spp. (Pane et al., 2013; Scotti et al., 2020). FDA has been a tool that works to predict suppression of Oomycota pathogens (e.g., *Pythium* and *Phytophthora*) but not necessarily fungal soilborne pathogens (Hadar and Papadopoulou, 2012).

Compost maturity: Mature composts have greater C:N and lignin, cellulose ratios, and slow-release of nutrients than immature composts. Mature composts are promoted as suppressive to *R. solani* (Scheuerell et al., 2005; Coventry et al., 2006).

Ideally, methods should reflect a composite of species and mechanisms and do not require a specialist and expensive analytical equipment. Promising candidates are (1) competition plate assays (Pane et al., 2013; Neher and Weicht, 2018), (2) ecoenzymes (Neher et al., 2017), and/or (3) physiological profiles using Biolog EcoPlates that screen for utilization of 31 carbon types (Scotti et al., 2020; Neher, 2021). Antibiosis activity on plate assays are effective tests for *R. solani* (Neher et al., 2017), *Streptomyces scabies* (Bakker et al., 2013), *Sclerotinia minor* (Pane et al., 2013), and *Fusarium* (Borrero et al., 2006). Suppressive colonies create a visible zone of inhibition around the pathogen colony. Microbial ecoenzymes active on chitin and cellulose are better predictors of disease suppressiveness by fungal pathogens than microbial respiration (Neher et al., 2017). The enzymes might damage cell walls of fungal pathogens such as *Rhizoctonia*, *Fusarium*, and *Verticillium*. However, Oomycota *Phytophthora* and *Pythium* have cellulose in their cell walls instead of chitin.

1.7 Conclusions

Everyone wants shelf-ready products that are inexpensive and easy to use. Unfortunately, the science is lagging to provide these immediately for composts, especially composts that allege disease-suppression benefits. As living entities, composts require more care than synthetic fertilizers and pesticides. For example, the effectiveness changes with age. Disease suppression may be negligible or even harmful in young composts; it may diminish in old material. Compost biology may be altered by high moisture, extreme heat, direct sunlight or, less likely, freezing temperatures.

Still, composts applied to soils can provide biological control of root diseases and occasionally also of foliar diseases of plants. Many factors must be considered to obtain consistent disease-suppressive effects with composts. First, the compost must have met temperature and time requirements for sanitation. In addition, it must have been adequately cured and matured, sufficiently enough for beneficial biocontrol organisms to colonize and proliferate. Compost that performs consistently must be prepared by a consistent process and from a relatively consistent feedstock recipe. Compost is at a point in history where organic farming was 20 to 25 years ago. The landscaping and road construction industries have widely adopted compost standards, but similar standards are scarce for field crops production.

Compost must be applied at a time of year and an application rate that meets, but does not overwhelm, the fertility needs of the crop. Soil fertility and nutrient supply must be included in these decisions. The quantity of essential plant nutrients such as N and phosphorus in the soil accumulates with each compost application. Soil fertility and mineral fertilizer application must be considered when subsequent compost application rates are determined to avoid increasing the severity of disease or cause other negative side effects due to excessive soil nutrient levels. Fall application is required for crops sensitive to Phytophthora root rot if composts high in salinity are used. In general, it is safest to apply compost weeks to months in advance of planting to avoid possible problems with ammonia, salts and immaturity.

Concerning disease-suppressive composts, the curing stage remains underappreciated. Commercial composting guidelines require a high-temperature phase designed to facilitate the removal of human and plant pathogens. However, these requirements stop short of guidelines for compost curing (cooling) in the post-thermophilic phase. The curing phase offers favorable conditions for microbial recolonization, accomplished by either inoculating post-thermophilic compost or creating a palatable substrate that offers a competitive advantage for colonization by bacteria and fungi capable of suppressing soilborne pathogens. With a better understanding of the microbiology of composting, the pivotal conditions can be managed to enhance disease suppressiveness either by regulating the microbe-to-microbe interactions or microbe-to-plant interactions in soil. This knowledge will elucidate which recipe and post-thermophilic practices are best to develop compost for more reliable strategies to manage ubiquitous and difficult to manage soil pathogens.

References

Cited references

Andrews, T., 2018. Ecology of Composted Bedded Pack and its Impact on the Udder Microbiome With an Emphasis on Mastitis Epidemiology (M.S. thesis). University of Vermont.

Bakker, M.G., Otto-Hanson, L., Lange, A.J., Bradeen, J.M., Kinkel, L.L., 2013. Plant monocultures produce more antagonistic soil *Streptomyces* communities than high-diversity plant communities. Soil Biol. Biochem. 65, 304–312.

Bais, H.P., Weir, T.L., Perry, L.G., Gilroy, S., Vivanco, J.M., 2006. The role of root exudates in rhizosphere interactions with plants and other organisms. Annu. Rev. Plant Biol. 57, 233–266.

Beagle-Ristaino, J.E., Papavizas, G.C., 1985. Biological control of Rhizoctonia stem canker and black scurf of potato. Phytopathology 75, 560–564.

Bochow, H., Seidel, D., 1961. Die Wirkung einer organischen Duengung auf den Befall durch *Plasmodiophora brassicae* und *Pythicum debarianum* D. Akad. d. Landbauwiss. Z. Berlin, Tagungsberichte Nr 41, 69−82.

Bonanomi, G., Antignani, V., Capodilupo, M., Scala, F., 2010. Identifying the characteristics of organic soil amendments that suppress soilborne plant diseases. Soil Biol. Biochem. 42, 136−144.

Bonanomi, G., Lorito, M., Vinale, F., Woo, S.L., 2018. Organic amendments, beneficial microbes, and soil microbiota: toward a unified framework for disease suppression. Annu. Rev. Phytopathol. 56, 1−20.

Bonilla, N., Gutiérrez-Barranquero, J.A., de Vicente, A., Cazorla, F.M., 2012. Enhancing soil quality and plant health through suppressive organic amendments. Diversity 4, 475−491.

Borrero, C., Ordovás, J., Trillas, M.I., Avilés, M., 2006. Tomato Fusarium wilt suppressiveness. The relationship between the organic plant growth media and their microbial communities as characterised by Biolog®. Soil Biol. Biochem. 38, 1631−1637.

Chapelle, E., Mendes, R., Bakker, P.A.H.M., Raaijmakers, J.M., 2015. Fungal invasion of the rhizosphere microbiome. ISME J. 10, 265−268.

Chung, Y.R., Hoitink, H.A.J., Lipps, P.E., 1988. Interactions between organic-matter decomposition level and soilborne disease severity. Agric. Ecosyst. Environ. 24, 183−193.

Coventry, E., Noble, R., Mead, A., Marin, F.R., Perez, J.A., Whipps, J.M., 2006. Allium white rot suppression with composts and *Trichoderma viride* in relation to sclerotia viability. Phytopathology 96, 1009−1020.

Dukare, A.S., Chaudhary, V., Singh, R.D., Saxena, A.K., Prasanna, R., et al., 2011. Evaluating microbe-amended composts as biocontrol agents in tomato. Crop Protect. 30, 436−442.

Eljounaidi, K., Lee, S.K., Bae, H., 2016. Bacterial endophytes as potential biocontrol agents of vascular wilt diseases − review and future prospects. Biol. Contr. 103, 62−68.

Fravel, D.R., 2005. Commercialization and implementation of biocontrol. Annu. Rev. Phytopathol. 43, 337−359.

Green, V.S., Stott, D.E., Diack, M., 2006. Assay for fluorescein diacetate hydrolytic activity: optimization for soil samples. Soil Biol. Biochem. 38, 693−701.

Grosch, R., Koch, T., Kofoet, A., 2004. Control of bottom rot on lettuce caused by *Rhizoctonia solani* with commercial biocontrol agents and a novel fungicide. Z. für Pflanzenkrankh. Pflanzenschutz 111, 572−582.

Guo, Y., Jud, W., Ghirarodo, A., Antritter, F., Benz, J.P., Schnitzler, J.-P., Rosenkranz, M., 2020. Sniffing fungi − phenotyping of volatile chemical diversity in *Trichoderma* species. New Phytol. 227, 244−259.

Haas, D., Défago, G., 2005. Biological control of soil-borne pathogens by fluorescent pseudomonads. Nat. Rev. Microbiol. 3, 307−319.

Hadar, Y., Papadopoulou, K.K., 2012. Suppressive composts: microbial ecology links between abiotic environments and healthy plants. Annu. Rev. Phytopathol. 50, 133−153.

Hoitink, H.A.J., Boehm, M.J., 1999. Biological control within the context of soil microbial communities: a substrate-dependent phenomenon. Annu. Rev. Phytopathol. 37, 427−446.

Hoitink, H., Ramos, L., 2004. Disease suppression with compost: history, principles and future. In: International Conference Soil and Compost Eco-Biology. September 15th − 17th 2004, León − Spain.

Lakshmanan, V., 2015. Root microbiome assemblage is modulated by plant host factors. Plant Microbe Interact. 75, 57−79.

Leso, L., Barbari, M., Lopes, M.A., Damasceno, F.A., Galama, P., Taraba, J.L., Kuipers, A., 2020. Invited review: compost-bedded pack barns for dairy cows. J. Dairy Sci. 103, 1072−1099.

Limoges, M.A., Neher, D.A., Weicht, T.R., Millner, P.D., Sharma, M., Donnelly, C., 2021. Differential Survival of Generic E. coli and Listeria spp. in Northeastern U.S. Soils Amended with Dairy Manure Compost, Poultry Litter Compost, and Heat-Treated Poultry Pellets and Fate in Raw Edible Radish Crops. Journal of Food Protection. https://doi.org/10.4315/JFP-21-261.

Litterick, A.M., Harrier, L., Wallace, P., Watson, C.A., Wood, M., 2004. The role of uncomposted materials, composts, manures, and compost extracts in reducing pest and disease incidence and severity in sustainable temperate agricultural and horticultural crop production—a review. Crit. Rev. Plant Sci. 23, 453−479.

Lugtenberg, B., Kamilova, F., 2009. Plant growth promoting rhizobacteria. Annu. Rev. Microbiol. 63, 541−556.

Mendes, R., Kruijt, M., De Bruijn, I., Dekkers, E., Van Der Voort, M., Schneider, J.H.M., Piceno, Y.M., Desantis, T.Z., Andersen, G.L., Bakker, P.A.H.M., Raaijmakers, J.M., 2011. Deciphering the rhizosphere microbiome for disease-suppressive bacteria. Science 332, 1097−1100.

Meyer, S.L., Roberts, D.P., 2002. Combinations of biocontrol agents for management of plant-parasitic nematodes and soilborne plant-pathogenic fungi. J. Nematol. 34, 1−8.

Neher, D.A., Weicht, T.R., Bates, S.T., Leff, J.W., Fierer, N., 2013. Changes in bacterial and fungal communities across compost recipes, preparation methods, and composting times. PLoS One 8 (11), e79512.

Neher, D.A., Weicht, T.R., Dunseith, P., 2015. Compost for management of weed seeds, pathogen, and early blight on brassicas in organic farmer fields. Agroecol. Sustain. Food Syst. 29, 3−18.

Neher, D.A., Fang, L., Weicht, T.R., 2017. Eco- enzymes as indicators of compost to suppress *Rhizoctonia solani*. Compost Sci. Util. 25, 251−261.

Neher, D.A., Weicht, T.R., 2018. A plate competition assay as a quick preliminary assessment of disease suppression. JoVE e58767.

Neher, D.A., Cutler, A.J., Weicht, T.R., Sharma, M., Millner, P.D., 2019a. Composts of poultry litter or dairy manure differentially affect survival of enteric bacteria in fields with spinach. J. Appl. Microbiol. 126, 1910−1922.

Neher, D.A., Nishanthan, T., Grabau, Z.J., Chen, S.Y., 2019b. Crop rotation and tillage affect nematode communities more than biocides in monoculture soybean. Appl. Soil Ecol. 140, 89−97.

Neher, D.A., 2021. Biological indicators and compost for managing plant disease. Acta Hortic 1317, 33−46. https://doi.org/10.17660/ActaHortic.2021.1317.5.

Noble, R., Coventry, E., 2005. Suppression of soil-borne plant diseases with composts: a review. Biocontrol Sci. Technol. 15, 3−20.

Noble, R., 2011. Risks and benefits of soil amendment with composts in relation to plant pathogens. Australas. Plant Pathol. 40, 157−167.

Pane, C., Piccolo, A., Spaccini, R., Celano, G., Villecco, D., Zaccardelli, M., 2013. Agricultural waste-based composts exhibiting suppressivity to diseases caused by the phytopathogenic soil-borne fungi *Rhizoctonia solani* and *Sclerotinia minor*. Appl. Soil Ecol. 65, 43−51.

Pugliese, M., Liu, B., Gullino, M.L., Garabaldi, A., 2011. Microbial enrichment of compost with biological control agents to enhance suppressiveness to four soil-borne diseases in greenhouse. J. Plant Dis. Prot. 118, 45−50.

Scheuerell, S.L., Sullivan, D.M., Mahafee, W.F., 2005. Suppression of seedling damping-off caused by *Pythium ultimum, P. irregulare*, and *Rhizoctonia solani* in container media amended with a diverse range of Pacific Northwest compost sources. Phytopathology 95, 306–315.

Scotti, R., Mitchell, A.L., Pane, C., Finn, R.D., Zaccardelli, M., 2020. Microbiota characterization of agricultural green waste-based suppressive composts using omics and classic approaches. Agriculture (Basel) 10, 61.

Spring, D.E., Ellis, M.A., Spotts, R.A., Hoitink, H.A.J., Schmitthenner, A.F., 1980. Suppression of apple collar rot pathogen in composted hardwood bark. Phytopathology 70, 1209–1212.

Suárez-Estrella, F., Arcos-Nievas, M.A., López, M.J., Vargas-García, M.C., Moreno, J., 2013. Biological control of plant pathogens by microorganisms isolated from agro-industrial composts. Biol. Contr. 67, 509–515.

Termorshuizen, A.J., van Rijn, E., van der Gaag, D.J., Alabouvette, C., Chen, Y., Lagerlöf, J., Malandrakis, A.A., Paplomatas, E.J., Rämert, B., Ryckeboer, J., Steinberg, C., Zmora-Nahum, S., 2006. Suppressiveness of 18 composts against 7 pathosystems: variability in pathogen response. Soil Biol. Biochem. 38, 2461–2477.

Timper, P., 2014. Conserving and enhancing biological control of nematodes. J. Nematol. 46, 75–89.

Vacheron, J., Desbrosses, G., Bouffaud, M.L., Touraine, B., Moenne-Loccoz, Y., Muller, D., Legendre, L., Wisniewski-Dye, F., Prigent-Combaret, C., 2013. Plant growth-promoting rhizobacteria and root system functioning. Front. Plant Sci. 4. https://doi.org/10.3389/fpls.2013.00356.

Vida, C., De Vicente, A., Cazorla, F.M., 2020. The role of organic amendments to soil for crop protection: induction of suppression of soilborne pathogens. Ann. Appl. Biol. 176, 1–15.

Weller, D.M., Raaijmakers, J.M., Mcspadden Gardener, B.B., Thomashow, L.S., 2002. Microbial populations responsible for specific soil suppressiveness to plant pathogens. Annu. Rev. Phytopathol. 40, 309–348.

Wichuk, K.M., McCartney, D., 2010. Compost stability and maturity evaluation - a literature review. Can. J. Civ. Eng. 37, 1505–1523.

Consulted and suggested references

Antoniou, A., Tsolakidou, M.-D., Stringlis, I., Pantelides, I., 2017. Rhizosphere microbiome recruited from a suppressive compost improves plant fitness and increases protection against vascular wilt pathogens of tomato. Front. Plant Sci. 8. https://doi.org/10.3389/fpls.2017.02022.

Chalker-Scott, L., 2005. Literature on Compost Tea and Disease Suppression. Washington State University Extension. https://puyallup.wsu.edu/lcs/reference-compost-tea/.

Compost for Soils, 2009. Disease Suppression. Compost Australia. https://www.aora.org.au/resources/disease-suppression.

De Corato, U., 2020. Disease-suppressive compost enhances natural soil suppressiveness against soil-borne plant pathogens: a critical review. Rhizosphere 13, 00192. https://doi.org/10.1016/j.rhisph.2020.100192.

Doungous, O., Minyaka, E., Longue, E.A.M., Nkengafac, N.J., 2018. Potentials of cocoa pod husk-based compost on Phytophthora pod rot disease suppression, soil fertility, and *Theobroma cacao* L. growth. Environ. Sci. Pollut. Control Ser. 25, 25327–25335. https://doi.org/10.1007/s11356-018-2591-0.

Ersahin, Y.S., Haktanir, K., Yanar, Y., 2009. Vermicompost suppresses *Rhizoctonia solani* Kühn in cucumber seedlings. J. Plant Dis. Prot. 116, 182—188.

EUCLID, 2018. Guidelines on Use of Compost for Controlling Plant Diseases. EU-CHINA Lever for IPM Demonstration. http://www.euclidipm.org/images/documents/deliverables/D1_1_Guidelines-on-use-compost.pdf.

Evans, K., Percy, A., 2014. Integrating compost teas in the management of fruit and foliar diseases for sustainable crop yield and quality. In: Maheshwari, D. (Ed.), Composting for Sustainable Agriculture, Sustainable Development and Biodiversity, vol. 3. Springer, Cham. https://doi.org/10.1007/978-3-319-08004-8_9.

Hadar, Y., Mandelbaum, R., Gorodecki, B., 1992. Biological control of soilborne plant pathogens by suppressive compost. In: Tjamos, E.C., Papavizas, G.C., Cook, R.J. (Eds.), Biological Control of Plant Diseases, NATO ASI Series (Series A: Life Sciences), vol. 230. Springer, Boston, MA. https://doi.org/10.1007/978-1-4757-9468-7_1.

Hoitink, H.A.J., Boehm, M.J., Hadar, Y., 1993. Mechanisms of suppression of soilborne plant pathogens in compost-amended substrates. In: Hoitink, H.A.J., Keener, H.M. (Eds.), Science and Engineering of Composting: Design, Environmental, Microbiological and Utilization Aspects. Renaissance Publications, Worthington, OH, pp. 601—621.

Koné, S.B., Dionne, A., Tweddell, R.J., Antoun, H., Avis, T.J., 2010. Suppressive effect of non-aerated compost teas on foliar fungal pathogens of tomato. Biol. Control 52, 167—173.

Mehta, C.M., Palni, U., Franke-Whittle, I.H., Sharma, A.K., 2014. Compost: its role, mechanism and impact on reducing soil-borne plant diseases. Waste Manag. 34, 607—622. https://doi.org/10.1016/j.wasman.2013.11.012.

Millner, P.D., Ringer, C.E., Maas, J.L., 2004. Suppression of strawberry root disease with animal manure composts. Compost Sci. Util. 12 (4), 298—307. https://doi.org/10.1080/1065657X.2004.10702198.

Neher, D.A., 2019. Compost and plant disease suppression. BioCycle 60 (8), 22—25.

Nelson, E.B., Boehm, M.J., 2002. Microbial mechanics of compost-induced disease suppression. Biocycle 43, 45—47.

Pittaway, P., 2014. Interpreting results for plant growth promotion and disease suppression bioassays using compost. Acta Hortic. 1018, 181—186. https://doi.org/10.17660/ActaHortic.2014.1018.17.

Postma, J., Schilder, M.T., 2015. Enhancement of soil suppressiveness against Rhizoctonia solani in sugar beet by organic amendments. Appl. Soil Ecol. 94, 72—79.

Scheuerell, S., Mahaffee, W., 2002. Compost tea: principles and prospects for plant disease control. Compost Sci. Util. 10 (4), 313—338. https://doi.org/10.1080/1065657X.2002.10702095.

Scheuerell, S.J., 2004. Compost tea production practices, microbial properties, and plant disease suppression. In: International Conference Soil and Compost Eco-Biology, September 15th — 17th 2004, León — Spain.

Swain, S., Garbelotto, M., 2016. Presentation: Disease Suppression in the Composting Process. University of California Agriculture and Natural Resources, Marin County Cooperative Extension. http://cemarin.ucanr.edu/Programs/Custom_Program97/Integrated_Pest_Management_IPM/.

Suárez-Estrella, F., Arcos-Nievas, M.A., López, M.J., Vargas-García, M.C., Moreno, J., 2013. Biological control of plant pathogens by microorganisms isolated from agro-industrial composts. Biol. Control 67 (3), 509–515.

Tjamos, E.C., Papavizas, G.C., Cook, R.J. (Eds.), 2013. Biological Control of Plant Diseases: Progress and Challenges for the Future, Vol. 230. Springer Science & Business Media.

Tuitert, G., Bollen, G.J., 1996. The effect of composted vegetable, fruit and garden waste on the incidence of soilborne plant diseases. In: de Bertoldi, M., Sequi, P., Lemmes, B., Papi, T. (Eds.), The Science of Composting. Springer, Dordrecht. https://doi.org/10.1007/978-94-009-1569-5_178.

Van der Wurff, A.W.G., Fuchs, J.G., Raviv, M., Termorshuizen, A.J. (Eds.), 2016. Handbook for composting and compost use in organic horticulture. BioGreenhouse COST Action FA 1105. www.biogreenhouse.org.

Wright, E., 2020. Compost Suppression of a Fungal Pathogen, Rhizoctonia solani, and its Impact on Root Microbiomes (Biological Sciences Honors thesis). University of Vermont.

Compost marketing and sales

18

Author: Ron Alexander[1],*

[1]*R. Alexander Associates, Inc., Apex, NC, United States*

Contributors: Rod Tyler[2], Mary Schwarz[3], Jeff Ziegenbein[4]

[2]*Green Horizons Environmental, Medina, OH, United States;* [3]*Cornell Waste Management
Institute, Cornell University, Ithaca, NY, United States;* [4]*Regional Compost Operations, Inland
Empire Utilities Agency, Chino, CA, United States*

1. Introduction

Over the past 30 years, the emphasis of composting has shifted from "just" treating unwanted waste materials, to manufacturing "products"—compost products. This trend holds for composting facilities of all sizes, whether on-farm, operated commercially, or by a municipality. With the increasing emphasis on compost manufacturing, marketing is essential to nearly all composting operations, and it should begin before the first truckload of compost is produced. A successful marketing program helps both the compost producer and compost end users operate more profitably.

Marketing encompasses a combination of steps that successfully brings compost products to those end users who are willing to pay a fair market value for them. Marketing involves many things, including:

- matching product characteristics to end users' needs,
- researching applications and customers that might benefit from your products,
- developing strategies to reach potential customers in a cost-effective manner,
- devising systems for distributing products (possibly including packaging) to customers, and,
- giving products the proper market identity.

Marketing includes a variety of activities that go into selling compost—that is, convincing customers that the products are worth their cost. Marketing compost requires an educational component to inform customers about product benefits and how to best capture those benefits; to train sales staff about product characteristics and technical applications; and to instruct compost production staff about producing the products that customers want.

---

* www.alexassoc.net

The Composting Handbook. https://doi.org/10.1016/B978-0-323-85602-7.00014-5

To be successful over the long-term, there is "a lot to learn" about marketing composts, but this can be done pragmatically over time. A technically competent marketer and salesperson must possess knowledge in diverse subject matter, as well as skill in the area of sales. The marketing of compost products must be approached in the same way as any other commercial horticultural or agricultural product. Composts are incredibly functional products, manufactured from "recycled" feedstocks, and therefore considered to be environmentally friendly or "green" products. However, most people purchase compost because of what it can do *for them*. They want to know "how will it save or make me money or give me better results in the field." This chapter introduces the knowledge needed to answer these questions and more effectively market composts and compost-derived products. Several publications, reports, and guidelines provide additional information specific to marketing compost products including *The Practical Guide to Compost Marketing and Sales* (Alexander, 2010), *Winning the Organics Game* (Tyler, 1996), *Marketing of compost in emerging markets* (Brinkman, 2019), and *Marketing Compost: A Guide for Compost Producers in Low and Middle-Income Countries* (Rouse et al., 2008).

2. Marketplace for compost

With a few exceptions, most potential users of (and customers for) compost fall into two broad categories—agriculture and the green industry. The green industry includes businesses which are more "ornamental" in nature, such as landscapers, grounds maintenance (e.g., parks, schools, commercial properties), silviculture, nursery and greenhouse production, garden centers, other retail stores that sell landscape and garden products, sod producers, the lawn care industry, golf courses, and secondary businesses that support these enterprises, such as suppliers of topsoil, mulch, and other bulk products. Hobby and home gardeners are considered an important niche within the green industry. In some locations, home gardeners represent the largest number of potential end users for compost. Although their numbers are large, they individually use lower volumes than professional end users.

Agriculture primarily covers field crops, vegetable and fruit growers, and livestock producers, although segments of the green industry (nurseries and greenhouse) also fall within this category. Until recently, compost manufacturers had more success marketing compost to green industry markets, primarily because they are able to pay higher prices for compost. However, agriculture remains a huge untapped market that can potentially use large volumes of compost, and in certain regions, compost is quite popular with farmers. For instance, in California, compost is used to conserve water and enhance the carbon content of the soil. Within the agriculture sector, certified organic agriculture is a specialized market with enormous potential for compost use. Organic farms frequently use composts to augment soil fertility and build soil organic matter. Serving the organic agriculture market requires some understanding of organic farming and certification regulations as well the rules pertaining to "input materials," such as compost.

Other important market segments that do not fit neatly within the green or agricultural categories include businesses and individuals involved in road construction and maintenance, residential and commercial construction, erosion control, storm water management, mine reclamation, soil remediation, and landfill maintenance. In all cases, the intended functions of compost products are to improve soils (or soil-like media) and assist the plants and/or organisms that inhabit those soils. With a changing climate, compost-based systems have proven to be highly efficacious in erosion control and storm water management.

The pool of potential and actual buyers of compost products vary from one region to the next, and it is important to conduct some level of market research within your target market in order to develop a meaningful sales strategy. Understanding end user requirements further allow a compost manufacturer to modify compost (and compost products) to best meet users' needs.

Fig. 18.1 compares the number of potential end users in specific market segments to the actual identified market demand for compost. These data illustrate that the number of potential end users in a market segment can be an indicator for the potential compost demand, but it is not typically a reliable indicator. This discrepancy is due to a variety of factors, including market development efforts by composters, conservativeness in purchasing habits, size of the business, etc. Two striking pieces of data that can be gleaned through this comparison are that:

- Organic farms and government entities were early adopters of compost usage, while
- Nurseries and golf courses proved to be very conservative buyers (slow to adopt new products).

Fig. 18.2 illustrates compost and mulch demand in the State of California, and compares and contrasts similar data broken down into four regions of the State

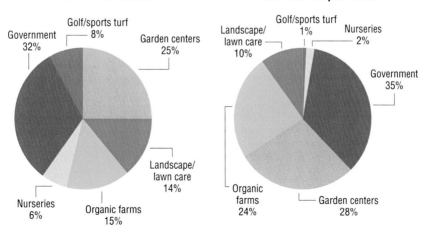

Actual end user demand   **Estimated compost demand**

Actual end user demand: Government 32%, Golf/sports turf 8%, Garden centers 25%, Landscape/lawn care 14%, Organic farms 15%, Nurseries 6%

Estimated compost demand: Landscape/lawn care 10%, Golf/sports turf 1%, Nurseries 2%, Government 35%, Garden centers 28%, Organic farms 24%

FIGURE 18.1

Example analysis of potential end users and actual market demand in the State of Iowa in 1998 (E&A Environmental Consultants, 1998).

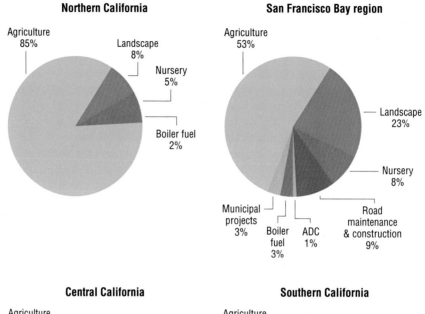

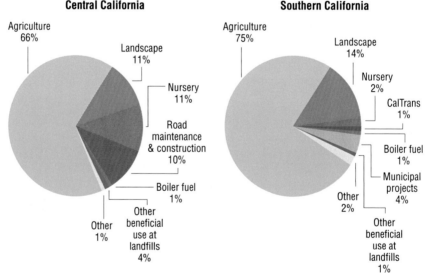

FIGURE 18.2

Example analysis of state versus regional demand for compost in the State of California in 2017.

Adapted from IWMC (Integrated Waste Management Consulting), 2019.

(IWMC, 2019). It should be noted that California is the only state in the United States where agriculture is the largest purchaser of compost. There are many reasons for this; including the climate, soil conditions, lack of irrigation water, the overall size of the state agricultural industry, acreage of high value crops (fruits, vegetable, nuts, etc.), and the completion of relevant agricultural field research with compost. The regional comparisons illustrate that in the San Francisco Bay Area and southern California, where population is the greatest, landscape usage of compost and mulch are greater than in less populated areas of the State. It also illustrates how the Department of Transportation engineers in the San Francisco Bay Area and Central California are specifying the use of compost and mulch in larger volumes than in other regions of the State.

In contrast to most of North America, agriculture is the largest user of compost in much of Europe. However, the share of compost used by agriculture differs among countries. Fig. 18.3 shows this difference by the examples of France, Germany, and the UK. In addition to market demand, the use of compost is influenced by agricultural nutrient regulations and compost quality standards within various countries (Saveyn and Eder, 2014).

This type of comparative data illustrates that compost markets are geographically dependent and should be understood as such when market planning. However, the market opportunities for a given compost producer are local, generally within 80−160 km of the facility (50−100-miles). Therefore, it is a mistake to assume that national, state, provincial, or even regional market data relates specifically to small geographical market of a particular composting facility.

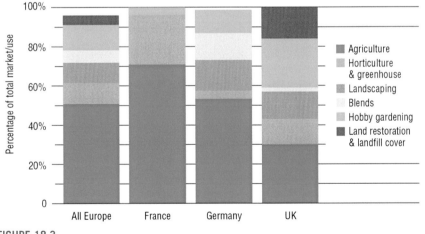

FIGURE 18.3

Compost use in selected European countries and Europe-wide, circa 2008.

Developed from data presented by Saveyn and Eder (2014).

3. The Product(s)—compost(s)

Some would say that compost is not a product but rather a class of products. Although versatile, different composts have different characteristics and intended uses. The characteristics of compost products vary with the ingredients and methods used to manufacture and refine them (see Chapter 15). Thus, manufacturers have some latitude in deciding what product(s) they produce.

As compost production has increased, compost manufacturers have endeavored to convince the "market" to accept composts as mainstream horticultural and agricultural products. Compost is often compared to fertilizer, topsoil, peat moss and, more recently, biochar and anaerobic digestate (Box 18.1). However, it is not typically used in the same ways as those products. Compost is a different type of product. It possesses benefits that, in many ways, make it superior to other "traditional"

Box 18.1 Comparing compost, anaerobic digestate, and biochar

Compost

Compost is the product manufactured through the controlled aerobic, biological decomposition of biodegradable materials. The product has undergone mesophilic and thermophilic temperatures, which significantly reduces the viability of pathogens and weed seeds, and stabilizes the carbon, such that it is beneficial to plant growth. Compost is typically used as a soil amendment, but may also contribute plant nutrients.

Lately, compost has been sharing the soil amendment stage with two other derivatives of organic feedstocks — anaerobic digestate and biochar (Fig. 8.4). Despite some similarities, these products are different from compost in their qualities and use.

Anaerobic digestate

Definition: Anaerobic digestate is the liquid or solid material processed through anaerobic digestion. Labeling digestate materials shall be designated by prefixing the name of the feedstock from which it is produced, i.e., cow manure digestate, biosolids digestate, etc.

How it differs from compost: Digestate can be a liquid or solid depending on the digestion technology employed and postprocessing (e.g., dewatering, drying, pelletizing). Digestate is not typically considered a "finished" product for horticulture, but it is suitable for application to agricultural land. Digestate is not biologically stable. It can decompose and generate odors and organic acids if not handled promptly or appropriately.

Biochar

Definition: Biochar is a solid material obtained from thermochemical conversion of biomass in an oxygen-limited environment (pyrolysis) containing at least 60% carbon. Feedstocks may be composed of crop residue, wood or other forest waste, and animal manures. Materials transported in salt water, painted, or treated with preservatives are not permitted. When listing biochar in an ingredient statement, the feedstock shall be designated by prefixing the term biochar with the feedstock from which it was produced, i.e., poultry litter biochar, green waste biochar, papermill biochar, etc. When more than one feedstock is involved, all feedstocks greater than 10% of the total volume are to be listed by decreasing volume. Their uses include soil amendments.

How it differs from compost: Biochar is a dry product that is rich in biologically stable carbon (often over 60% carbon). It possesses a high surface area and cation exchange capacity. The nutrient content of biochar is directly related to the feedstock for which it was derived. Wood-based biochars lack nitrogen, but they conserve nitrogen in the receiving soils. Biochar is commonly available in packaged form, in 1 to 2 cubic yard totes, and sometimes in bulk form. It is more expensive than compost on a volume and weight basis.

Box 18.1 Comparing compost, anaerobic digestate, and biochar—cont'd

FIGURE 18.4

Contrasting images of (A) compost, (B) anaerobic digestate, and (C) biochar.

Source: Ref: R.Alexander Associates, Inc.

Based on Association of American Plant Food Control Officials Product Definitions as of 2020. (AAPFCO, 2020).

horticultural and agricultural products, but its complex and diverse nature makes it unfamiliar to many potential users. Of course, familiarity with the product increases each year. Selling compost against these traditional products requires a marketing approach that recognizes the differences and emphasizes the advantages of compost.

Although most composts contain essential plant macro and micronutrients, their concentrations vary greatly depending on the feedstocks used and composting management practices employed. As such, they are not normally used as fertilizer substitutes. In most applications, composts are used for their soil amending properties. Typically, marketing composts with explicit nutrient claims requires the product to be specifically registered as a fertilizer, not a soil amendment. The regulations in each potential market jurisdiction (state, province, or country) should be consulted to determine what can actually be stated or claimed about compost products. In North America, such regulations are influenced by Association of American Plant Food Control Officials (AAPFCO), which is an organization of fertilizer control officials from each state in the United States, from Canada and from Puerto Rico. These same "control officials" are engaged in the administration of soil amendment laws and regulations. Recommendations made by AAPFCO and similar organizations lean heavily on scientific research, not users' experiences (Alexander, 2005).

Regardless of how it is labeled, it should be understood that many composts, particularly those derived from nutrient rich feedstocks, such as biosolids, animal manure, or livestock processing by-products, can supply a significant quantity of nutrients, and these nutrients should be accounted for when these types of compost are used. Indeed, many nutrient-rich composts have been used successfully as primary nutrient sources in specific agricultural applications, such as organic vegetable or sod production.

Composts, by themselves, should not be used as topsoil because they do not possess the long-term structural stability of topsoil, or its higher bulk density. Composts, although biologically stable compared to the raw feedstocks used in its production, continue to decompose slowly following incorporation into soil. Improperly produced or utilized composts can lose significant volume (shrinkage) over time. However, compost is a very popular component to "blended soil" products and is a popular soil amendment or a topsoil additive (Table 18.1).

Compared to peat moss (and some peat or muck soils), whose organic matter is hundreds to thousands of years old, composts represent much younger forms of organic matter. Because composts are manufactured, not mined, they should be considered as renewable resources, which can be an important marketing advantage compared to peat. Although peat is more biologically stable, composts can provide many of the same benefits of peat (porosity, water holding capacity). In addition, composts are more biologically active and supply and retain significant quantities of both micro- and macronutrients. The benefits of compost biological activity and diversity should never be underestimated. Composts can both support the growth of beneficial microorganisms (carbon is a food source) and be a source of them to soil or media.

Table 18.1 Topsoil modification with yard trimmings compost (Alexander, 2010).

Soil characteristics	Units	100% Native soil	75%/25%* Soil/Compost	50%/50%* Soil/Compost
Chemistry				
Soil pH		6.7	6.9	7
Soluble salts	dS/m	0.51	0.38	0.45
Organic matter	%	1.7	4.6	8.8
CEC	meq/100g	5.4	7.4	10.6
Nutrients				
Phosphorus (P)	ppm	26	60	85
Potassium (K)	ppm	80	345	542
Calcium (Ca)	ppm	788	977	1402
Magnesium (Mg)	ppm	110	169	228
Sulfur (S)	ppm	29	19	23
Boron (B)	ppm	0.2	0.5	0.7
Copper (Cu)	ppm	1.5	1.1	1.2
Iron (Fe)	ppm	185	266	252
Manganese (Mn)	ppm	28	43	55
Zinc (Zn)	ppm	3.2	5.7	8.6
Sodium (Na)	ppm	34	32	38
		*v/v basis		Test 10.31.2019

Over the past several decades, research and demonstration studies have illustrated the many benefits of using composts. Composts have further proven their usefulness "in the field." For example, there are numerous commercial horticulture businesses that now use composts routinely in container growing media mixes, in landscape installations, and for turf establishment and topdressing. Because of these successes, high quality composts can compete with traditional soil amendment products, based on their cadre of superior traits. To expand compost markets, however, it is important not only to replace existing products but also to expand overall usage of soil amendments and to create new applications for composts. The customer already identifies value in the product being replaced by compost (e.g., topsoil or peat). The production of consistently high-quality products must be assured for the benefits to be realized in our market development efforts.

Keeping in mind that composting is a manufacturing process, *composts are what they are processed to be*. Composts can be processed for use as soil amendments, turf topdressing, mulch, erosion control media, etc. They can also be further refined for optimal use as growing media components (e.g., well matured/stabilized), nutrient sources, or blended for its use in a variety of creative horticultural and environmental applications. The inherent variability of compost attributes, derived from a combination of feedstock characteristics and management process differences, can be used as an asset to develop diverse market outlets for composts and compost-derived products. A particular compost product has particular characteristics. It cannot be everything to all people. Recognizing that fact, each compost product should be sold to its strengths (characteristics). Likewise, if compost is produced for a specific market, manufacturers and marketers need to understand the requirements of those markets and modify the feedstocks and composting management practices to meet those market needs.

4. Marketing concepts

Before moving ahead with a compost marketing program, it is useful to understand common marketing concepts and how these concepts relate to selling compost. Concepts such as marketability, positioning, branding, and market development describe how marketers deliver information about products to the pool of potential end users.

4.1 Marketability

The "marketability" of a product refers to its sales *potential*. Is the compost produced what the target customers need? Are they willing to pay the asking price for the product? Are there enough customers in the marketplace to purchase the full volume of compost available? There are many factors that affect the marketability of compost products. Producing a compost product that possesses the characteristics required for a specific application(s), or end user group, is key. Providing a competitive pricing structure and good customer service is also essential. However, these factors alone do not determine a product's marketability. Appropriate sales activity improves marketability and increases *market share*. Luckily, the compost manufacturer can influence all of these factors. Yet there are market factors that a composter has little ability to impact. The marketability of specific compost products also depends on the innate regional demand for compost (and other soil amendments), which is based on factors such as population, business demographics, available green space, and the degree of competition. Sales staff cannot alter these factors, so the location of the composting facility should be thoroughly investigated before committing to a site.

The marketability of compost can be influenced by specific barriers that exist in specific locations. Table 18.2 presents the results of a marketing survey conducted in Iowa in 1998. The first column lists the barriers that stand in the way of increasing sales, as perceived by compost manufacturers. The second column gives factors that potential users said would encourage them to purchase more compost. Such barriers can be identified and overcome through strong messaging and market development efforts.

Table 18.2 Perceived compost market barriers.

Compost marketing barriers as perceived by compost manufacturers (percent of survey respondents identifying the specified barrier)	Factors to encourage compost use as perceived by potential users (percent of survey respondents identifying the specified factor)
Transportation costs (23%)	Improved local availability (42%)
Quality of product (20%)	Lower cost (39%)
Markets undeveloped (20%)	Availability of consistent products (30%)
Lack of marketing experience/staff (17%)	Availability of higher quality product (29%)
Volume of marketable product (14%)	List of local compost sources (27%)

Adapted from Iowa Statewide Compost Market Assessment; Iowa Department of Natural Resources, 1998. Iowa Statewide Compost Market Assessment Prepared by: Resource Conservation and Development of Northeast Iowa, Inc., Postville, IA.

4.2 Positioning

The "positioning" of a product refers to where and to whom you are selling the product, and the specific application. Positioning your compost product for distribution within an existing market takes an understanding of market demographics, competition, as well as end user requirements. Market positioning considerations include:

- Geography—Where will marketing efforts be concentrated geographically, and in how large of an area?
- Market segments—In which market segments will sales efforts concentrate (e.g., nurserymen, landscapers, etc.)?
- Product/application (uses)—What are the typical characteristics of the product(s), and what specific type(s) of product will be offered (e.g., soil amendment, mulch, etc.)?
- Competition—What are the competing products? How do they compare on a quality and economic basis? What differentiates your product(s) from theirs? How familiar are potential customers with compost use for their specific applications?
- Product or feedstock characteristics—Does the finished compost product possess any characteristics which make it unique, allowing for niche markets to be developed or limiting the product's use within specific markets? Along with the innate characteristics of the feedstock and the qualities of the finished product, the composter's ability to produce a consistent product will affect the markets that can be targeted, as well as the product's value (Fig. 18.5). *It is better to market a consistently mediocre product, than a product whose quality is inconsistent. With a consistent product, the customer can be advised how best to use the product!* Producing a consistent compost product requires frequent testing of the product, and possibly the feedstocks.

FIGURE 18.5

Compost products can be differentiated by the niche markets that they best serve and/or the feedstocks from which they were derived; whatever best serves the compost users and the marketing effort.

- Sales price—Is there a specific price that must be obtained for the product in order to meet necessary revenue requirements of your company?
- Transportation—Are transportation costs low enough to provide an economic advantage, or allow product to be transported farther (expand geographical market area)?
- Infrastructure—Is the necessary equipment economically available to produce the desired finished product (e.g., screens)? Is there enough space to properly produce and stockpile the product for peak market demand?
- Technical expertise—Does the company possess (or can it access) the technical expertise necessary to sell to highly technical market segments (e.g., golf courses, nurseries)? Is the staff adequately prepared to educate potential customers about the benefits of using composts (help them set up demonstration trials, for example, explain the particular advantages of compost over the competing soil amendment, etc.)?

All of these factors must be considered in product positioning. Market research plus knowledge of the industry assists in making these product positioning decisions. Marketing efforts can be more focused, and more effective, after evaluating these factors and making these positioning decisions during the upfront stages of planning and market development.

4.3 Product branding

Branding is a marketing technique that gives a specific product an identity and image distinct from other products in the same category. Adding a trademarked name strengthens the brand's identity. A product that does not have a brand is generic or often considered a "commodity." It may or may not be a good idea to create one's own brand of compost. The benefits of branding results from several years of coordinated effort, sales, and product positioning.

What is the nation's leading brand of compost? Mulch? Topsoil? Most products in these categories have a generic identity because marketers early on did not focus on branding and establishing a good, trademarked product name. Bulk composts are more difficult to brand on a national basis because their distribution is highly impacted by transportation costs. A compost product is often perceived to have the same value as the product it replaces, and when this value is low, the compost product is not perceived to be special. For example, when compost is considered as a topsoil replacement, the compost is valued at roughly the same price as topsoil in that marketplace, which can be as low as $10 to $15 per cubic meter or cubic yard, delivered. Branding differentiates a particular product from other products in the local marketplace and differentiates its value as well. Also, a brand represents intellectual capital that creates value for the company and anyone who might purchase the company in the future, especially if the brand is accompanied by a trademark and logo. Table 18.3 presents example brand names for a wide variety of compost products. Fig. 18.6 shows the product logos associated with these brands.

The brand should portray the values of the company and/or the product, as demonstrated by the products in Table 18.3. If the company stands for quality and service, the brand should shout this message in its name and the artwork associated with the logo. Too many cases occur where a compost producer becomes successful only to find later that their "catchy name" really does not match the company's values or ideals. When this happens, the product name, logo, or company name change, and customer confusion results. Most people over think product and brand names. Simplicity is the key. Once it has been decided to work with a particular name, the name should be registered with the appropriate patent and trademark authority (e.g., US patent and trademark office) to protect the embedded marketing intellectual capital.

In some cases, it may be better to support a national or regional brand than to create one's own, especially if marketing expertise is lacking. This option is primarily available with packaged products, whereas a composter may package under another company name (brand), or simply sell them the compost they require to produce their own branded product. With that stated, some private composters purchase compost from a municipal composting facility and sell it under the private company's name. This type of decision also depends on the compost manufacturer's ability to invest in market development, the degree of competition, recognition of national or regional brands established in the target geographical market. Market research can reveal "who's who" in the compost product arena and whether or not supporting another brand is worth considering.

4.4 Market development

Market development encompasses those activities that help to build a large and stable customer base. Market development includes a variety of activities that improve

Table 18.3 Examples of compost product brands.[a-c]

Product brand name[c]	Key feedstocks	Region	Product packaging	Prime uses	Prime market segments
All Treat Farms	Leaves and yard trimmings Cattle manure Source-separated organics	Ontario (Canada) with some sales outside the province	Packaged in 18 kg bags (40 lbs) and sold to retailers by the pallet load (70 bags). Bulk—accommodate all truck sizes.	Primarily as a soil amendment for gardens and lawns	National retail chains, golf course construction and maintenance, landscaping and gardening markets, roadways, and various uses within parks and recreational areas.
Dr. Gobbler's Soil RX	Turkey manure, rice hulls, wood shavings, and chips	Central and South Texas	Bags—1.5 cubic ft bags (43-L); Bulk—50 cubic yards bulk trucks, delivery within 100-mile radius of Waco TX;	Soil enrichment for landscapes and nurseries	Landscaping, wholesale nurseries, highway reseeding after construction
Eco Mix Compost	Green waste	Southern England	Bulk truckloads; Bulk bags Retail bags (50 lt)	Soil amendment for landscapes, turf, soil blending, and agriculture	Landscapers, homeowners, garden centers, landscape supply yards, farmers
Greenhouse Gold Seafood Compost	Tree bark, lobster, and crab shells	Atlantic Canada, Ontario, and New England	Bags—12, 5 and 15 kg bags (28 and 33 lbs; or bulk	Soil amendment for lawns, gardens, and landscapes	Landscaping companies, retail to gardeners
McGill Soil Builder Premium Compost	Biosolids, yard trimmings, food by-products, and wood chips	North Carolina and Virginia	Bulk truckloads, some bagged	Soil amendment for landscapes, turf, soil blending, and agriculture	Landscape, turf management, erosion control, resale, farmers
Moo Doo	Cow manure and sawdust	Northeastern US	14 and 28-L bags (0.5 and 1 ft³); Bulk to local landscapers	Soil amendment for retail and landscape use	Retail, landscape

Table 18.3 Examples of compost product brands.[a–c]—cont'd

Product brand name[c]	Key feedstocks	Region	Product packaging	Prime uses	Prime market segments
Queen City Compost	Yard trimmings	Mecklenburg County (NC) metropolitan area	Bulk truckloads, some bagged	Soil amendment for landscapes, turf, soil blending, and bulk resale	Landscapers, homeowners, garden centers, landscape supply yards
RICHGRO *Black Marvel* Garden Compost	Wood fiber, pine bark, manure plus *Black Marvel* supplemental plant nutrients	Australia (national)	25 L bags (1 ft³)	Soil enrichment for gardens	Retail to gardeners
RETERRA ActivCompost With RAL-Quality Assurance Label	Biowaste from separate collection of household waste (biobin) and greenwaste	Germany—from all operating RETERRA/ REMONDIS composting/ digestion plants	Bulk material/truck loads	Use as organic fertilizer in conventional and in organic farming)	Farmers—conventional and organic farms
SoilPro Premium Compost	Biosolids, green waste, animal bedding, and wood chips	Southern California	Bulk truckloads	Soil amendment for landscapes, turf, soil blending, and agriculture	Landscapers, turf managers, soil blenders, and farmers
Thoroughbred Compost	Thoroughbred horse farm manure with straw or wood shavings bedding	Kentucky and surrounding area (100-mile radius from Lexington)	Bulk—truck load pickup and delivery	Soil amendment for landscapes; top dressing for turf; erosion control	Landscaping, highway construction, retail garden centers, and golf courses

[a] All brand names are trademarks or the respective companies and used here with their permission.
[b] The information in this table was supplied by the compost manufacturers. Examples were selected to display a diversity of products, feedstocks, markets, and geography.
[c] Most of the manufacturers represented here offer a multiple of compost-based products sold under brand concept identified, in addition to the specific product listed in the table.

FIGURE 18.6

Compost brand example logos and product images. All images are trademarks of the respective companies and used here with their permission. Note: not all products in Table 18.3 are represented.

market penetration and overall sales volumes. They may include various types of promotions (e.g., hosting end use training sessions, industry magazine advertising, and Website and internet promotion) and sales calls. Often, promotions provide customers with information about product quality and use recommendations. For new clients unfamiliar with the potential benefits of compost use, it even might be worthwhile to set up a demonstration trial comparing the compost products with their current growing media or cultural practice. Creating a brand name can also serve as a cornerstone for an organization's market development efforts.

Facilities that do not invest resources (time, effort, money) into market development and those which simply do not take it seriously have a much greater frequency of failure. Successful market development requires more than obtaining a few advertisements in newspapers or industry journals. Nothing replaces face-to-face sales activities, such as "sales calls" and speaking with clients at trade conferences for professional end users. Similar techniques may be used on a retail level. As an example, recognizing the need for active market development efforts, a compost producer in the Seattle area hired a full-time staff member to cultivate relationships with the horticulture and home gardener markets in the region. The person hired joined several garden clubs, was a frequent guest on a popular radio gardening program and attended all regional horticultural and home gardening tradeshows. As a result of these efforts, the compost products have acquired strong brand recognition and are widely considered as the compost to buy in the region.

A general rule of thumb is that it takes 3 to 5 years of intense market development effort to establish a market that is sustainable. This time estimate is very dependent on the volume of compost generated, the local and regional market demographics, and the goals for producing compost, as well as other factors. As the volume of compost produced increases, market development efforts need to begin earlier, *and* more resources need to be committed toward the effort. It is often worthwhile to start marketing before compost is produced (premarketing) to introduce and prepare the market for the new product. Remember, the investment in time, effort, and money is just that—*an investment*. If done properly, it will pay back great *returns* for many years. The overall market development investment should decrease with time, as the marketing program and customer base become established.

5. Market options

"Markets" for compost products are typically defined by the end-user market segment targeted, such as landscapers, nurseries, certified organic farmers, state departments of transportation, and gardeners. End users generally require composts with distinct properties, expect composts to perform particular functions, or have a specific price range that they are able or willing to pay. Thus, the compost marketing program must consider the character of specific user groups and the characteristics that they require of compost.

5.1 Market versus self-use

One market option is whether to market compost at all. A major decision for any composting operation is how to recover costs and generate additional income from composting. Revenues are generated through tipping fees and/or aggressive marketing of compost products. Cost savings can be gained through improving efficiencies in waste collection and processing, and/or by the internal usage of the product(s). For farm operations especially, on-farm compost use can offset the costs of purchasing compost from other sources or costs of other inputs, such as fertilizers or pesticides. This opportunity is also available to many public and green industry operations (e.g., municipalities, parks, golf courses). In addition, public entities have the option of distributing compost among its citizens at a subsidized cost. The major benefit of selling compost is that, potentially, a better economic return can be realized, because customers may place a higher value on the compost, compared to what can be recovered by using the compost internally. Many compost manufacturers believe that the greatest benefit of a planned and aggressive compost-marketing program is that it creates (or raises) the inherent value of the product, whereas self-uses typically do not. On the other hand, marketing can be a somewhat costly undertaking, depending on the volume of compost produced. The decision is not an "either/or" situation. Composts can be both sold and used internally depending on the market conditions and the internal need for improving soil conditions and crop production. This option is particularly well suited to on-farm operations where land is available to apply, and benefit from, compost.

5.2 Feedstocks versus market characteristics

Experience has shown that attributes of finished composts are linked strongly to characteristics of their initial feedstocks. For instance, biosolids and manure-based composts typically possess higher nutrient contents than composts derived from yard trimmings. Manure-based composts tend to have the highest soluble salt contents (electrical conductivity), as well as a high bulk density, while yard trimmings composts have lower soluble salt contents and high water-holding capacity properties. It is also important to understand that composts produced from specific feedstocks (e.g., biosolids, municipal/mixed solid waste, manure) may have certain stigmas attached to them that warrant special consideration during your market development activities. When considering which market segments to direct sales efforts, the chemical, physical, and biological attributes of the compost must first be considered to determine if the product is suitable for that use. Further, certain feedstocks (e.g., curbside collected food waste) can contain more inert contamination (trash) than others.

Sometimes connections can be made between feedstock-based characteristics of composts and suitability for specific target markets. However, the value and marketability of a compost product is not determined by the feedstocks from which it is derived. A 1995 survey of composting facilities handling biosolids, yard trimmings, manure, or municipal solid waste (MSW) demonstrated similar product values and marketing success across all compost feedstocks, except MSW (E & A Environmental Consulting, 1995). The difficulties associated with the MSW compost were due to finished product qualities (e.g., immaturity, contamination) rather than the user's perceptions about the feedstock. This study suggests that customers are more concerned with the product's overall quality and efficacy rather than the ingredients used to make it.

5.3 Volume versus value markets—market hierarchies

"Product positioning" refers specifically to the market segments being targeted for sale. The idea is to position your product among the markets that bring the greatest economic return (where resources make it possible), and/or can utilize the greatest volume. Based on the economic return, a hierarchy can be established among the market segments for compost (Fig. 18.7). The markets at the top of the hierarchy generally pay higher prices for compost. However, they also have stricter product requirements and may be more difficult to capture.

The hierarchy can be divided into two general market categories: "volume" and "value" markets. Volume market customers are at the bottom of the hierarchy. They tend to use large volumes of compost but usually are not willing to pay a high unit price. Value market customers tend to pay higher prices but use smaller quantities. Typical examples of volume markets are agriculture and land reclamation. Value markets are typified by wholesale nurseries (potted plant growers), garden centers,

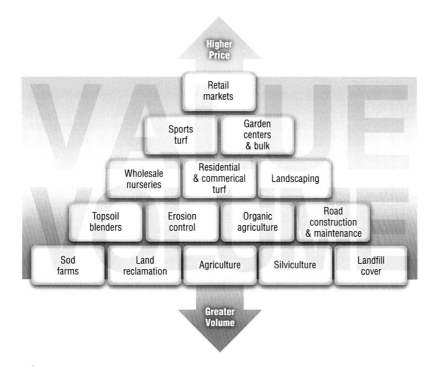

FIGURE 18.7

Relative categorization of volume and value markets for compost.

Adapted from Winning the Organics Game *(Tyler, R.W., 1996. Winning the Organics Game, ASHS Press, Alexandria, VA).*

and landscapers. However, specific markets arguably can be categorized as either value or volume.

The concept is to position the marketing program to sell to either "value" markets or to "volume" markets. By doing this, you reduce confusion within the market place. Otherwise, customers may question whether your product is high quality or low quality and what is its fair price. However, it is better to market to volume and value markets concurrently rather than dumping excess compost into a value market at an artificially low price. If you choose to market to both value and volume markets at the same time, separate branding and/or product naming may be necessary to differentiate the products (as well as their characteristics and pricing).

Historically, there are markets that have continued to pay a higher overall price for the compost that they purchase (e.g., wholesale nurserymen, landscapers; sports and residential turf managers) compared to most farm or highway applications. These value markets often take top priority in many market development programs, especially with those compost manufacturers that concentrate on compost sales for revenue generation. As a group, the largest value market is probably the home gardener. However, it is often problematic selling directly to home gardeners because of their need for individual technical assistance and the small volume that each customer purchases. Value markets tend to have stricter product quality demands and there is more competition for their business. Value markets usually require a greater investment in market development.

From a marketing standpoint, volume markets should be approached after value markets are exhausted. However, there are circumstances where it is advantageous, or necessary, to focus marketing efforts primarily on volume markets. Such circumstances include: when there is excessive inventory; where volume users are near the production facility; if product qualities are inherently ill-suited to value markets; or where large volumes are produced because feedstocks must be processed. Volume markets possess great potential for compost use—mainly because of the sheer volume that they represent. However, this market potential may never be met, unless the compost producer and the buyer can agree upon mutually acceptable price.

5.4 Bulk versus packaged products sales

Composts can be sold in both bulk and packaged forms. Marketing composts in bulk form allows for easier processing and delivery, but the product possesses a lower "per unit" value. Many composters are lured to the concept of packaging compost, even though the vast majority of compost products are distributed in bulk form. The lure is related to the potential for increased revenue and better overall name recognition (branding). Although compost is sold in bulk form to both professional and retail customers, retail applications for compost tend to be in small volumes. Therefore, the convenience of packaged products is a great advantage if retail sales are to be pursued. Packaging compost has also been shown to improve end use by landscapers in certain regions of the US.

Transportation issues are unique with packaged products. For shipping, bags are usually placed on pallets and then the pallet is wrapped in plastic to keep the bags in place and protected them from the elements. Different types of delivery trucks are required for packaged products, and most buyers cannot purchase a full tractor-trailer load of a single type of packaged product (e.g., 800 to 1200 bags at 20 kg or 40 pounds). Therefore, a compost producer may need to produce a series of packaged products (e.g., soil amendment, potting mix, mulch, etc.) to optimize marketability. By offering customers a variety of products (a product line), a compost producer can more easily fill an order that is large enough to competitively deliver.

Before committing to packaging compost products, market research should be completed and a detailed understanding of the market should be acquired. Obviously, the ability to sell packaged compost can improve a composter's ability to market bulk product, as well as increase the bulk product's value. However, several compost manufacturers have gone into packaging, only to fail because they did not understand the complexities of the packaged product industry and underestimated the additional costs necessary for packaging. It typically costs more for bagging (packaging) and transportation, than it does for the contents of the bag. Some composters have determined that they generate greater profit through the sale of bulk product, than they do selling packaged product because of the additional costs associated with packaging, storage, and distribution. When evaluating the economics of packaging, consider the overall return on investment, as well as costs. For example, it does not make economic sense to sell compost in bags for $50 per cubic meter (or cubic yard) if it costs $45 to package, while bulk sells at $15 per cubic meter (or cubic yard). It should be mentioned that large regional and national "soils" packaging companies purchase large volumes of compost to be used in the production of their own branded product lines. These firms can purchase large volumes of compost, and as such, expect significant product discounts.

5.5 Derivative products

Another means to expand compost markets, and increase marketability, is to create derivative products by blending compost with other components, or by screening to adjust the particle size (Figs. 18.8 and 18.9). This approach has allowed compost manufacturers to better service horticultural and agricultural supply companies. While compost manufacturers commonly screen compost to remove large particles, refined screening produces specialty products like turf topdressing, mulches, and erosion control substrates. New products also have resulted from blending compost with other materials such as sand, soil, organic fertilizers, peat moss, coconut coir, and wood chips. This strategy has allowed compost manufacturers to expand their sales by diversifying their product lines and marketing products that are more familiar to end users (e.g., topsoils, growing media). In addition, blending can correct quality problems including elevated salt levels, pH, and marginal immaturity.

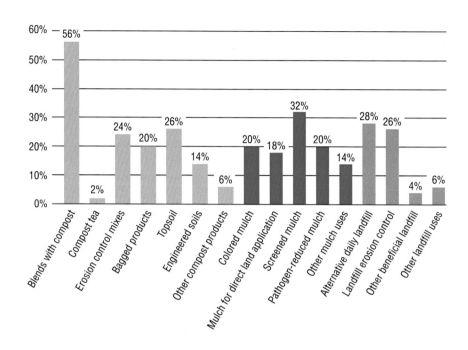

FIGURE 18.8

Examples and types of derivative products produced by composters in California.

Adapted from IWMC (Integrated Waste Management Consulting), 2019.

FIGURE 18.9

Offering a range of compost and complementary products offers producers a means to expand sales and penetrate new markets.

5.6 Distribution options

Several options exist for distributing compost to the marketplace that require a variety of key decisions. Options include:

- operating in-house marketing program (sales done with in-house staff),
- contracting with a product broker, which may include another composter that requires additional volumes of product to market,
- creating a distribution network, and,
- selling directly from the production site.

A compost manufacturer can employ combinations of these options (Fig. 18.10). The distribution strategy that best suits a particular composting facility depends on many factors, including the quantity of compost produced, current/future staffing, sales and technical skills, financial constraints, and the goals of the facility.

An early decision in the development of a composting program relates to who will be directly responsible for the distribution of the products. Will internal staff be responsible, or will a specialized company (broker) be hired to do the marketing and distribution of the products? With an in-house marketing program, the compost producer is in ultimate control and can better react to the on-going requirements of their facility and their customers. In addition, the composter receives the highest revenue for the product because that revenue does not have to be shared with an outside company. The material can be sold directly from your site in larger "wholesale" volumes or smaller "retail" volumes. Hiring a broker, on the other hand, may provide a more immediate and efficient distribution of product, as this company *should* possess industry understanding, *contacts* and the technical product background.

No end user within the market should be completely ignored, if possible. Even if a composter plans to concentrate its sales efforts on customers purchasing large truckloads, bulk sales of product to smaller professionals and homeowners can still be made if a chain of distributors or resellers is established. Marketing packaged compost through these *middlemen is common*, but it can also be practical with bulk composts. Using an established distribution channel is an excellent method to market smaller volumes of compost and also a means to approach the homeowner market without having to bag compost. For example, garden centers can redistribute small bulk loads (e.g., less than 10 cubic yards) more efficiently than a compost producer; especially if most of their loading equipment is large in size. As such, it may make sense to set up distributors within your geographical market to assist in distributing to smaller professional, as well as retail, customers.

6. Market planning

One of the biggest mistakes made by compost manufacturers is a lack of proper market planning due to the belief that market planning is not necessary or important, or

(A)

(B)

(C)

FIGURE 18.10

Compost manufacturers distribution options: (A) bulk sales delivered directly to customers and/or brokers; (B) bulk sales to customers from the composting facility; and (C) bagged product sales from the facility and/or retail outlets.

from over eagerness to get out in the "field" and sell. Often, the lack of attention to marketing stems from the focus on compost production and the gate fee side of the business. The planning, engineering design, construction, and budgeting aspects of developing a composting facility are often extensive. Rarely, however, is that the case when it comes to market planning. By developing a marketing plan, manufacturers can approach the market pragmatically, allowing staff to better understand the demographic nuances of the geographic market area. It also assists the development of a successful program in a more efficient manner—saving time and money.

The marketing plan is simply the blueprint or guide for market development and then realizing the market potential via the sales program. It should be modified as information and experience are gained and competitive forces change. The best marketing plan starts with a mission statement and company goals (Fig. 18.11). The plan should be specific in its own marketing goals, objectives, and activities. It should be specific enough to identify daily activities in all of its elements.

The components and structure of a marketing plan vary depending on the resources at hand and how a company positions its products. For example, a marketing mix may be considered a combination of four Ps—product, place, price, and promotion (Box 18.2). As shown in Fig. 18.11, the major components considered and discussed here are: compost production, market research, product research, promotion, education, and sales and distribution plan.

- *Compost production/facility management*—Compost manufacturers must produce products that meet the requirements and specifications of end users. As the manufacturing approach to composting has gained prominence, the composting industry has learned to better determine and produce the products that the various market segments require. Developing internal product specifications, which describe the characteristics the products must possess for sale to particular target markets, facilitates long-term market development. A quality

FIGURE 18.11

Market planning starts with the company's mission and goals, which are served by the marketing plan and its components.

Box 18.2 The four Ps marketing model

The publication, *Marketing of compost in emerging markets* (Brinkman, 2019), envisions the structure of a marketing mix as a combination of 4 "Ps"—product, place, price, and promotion. As this guidance document explains, these four elements determine a product's market position and how customers understand product and its benefits.

Product: This P refers to the compost product, or more specifically the product's quality and characteristics (Chapter 12). The quality and characteristics of a compost product determine its benefits for a given use. Such characteristics range from nutrients and pH to color and degree of contamination. From a marketing standpoint, compost users have expectations of the product, and perhaps even specifications, the compost manufacturer, and marketer, must meet. Any product claims made must be supported with evidence—analysis, demonstrations of performance, etc. The requirements for compost quality and characteristics can differ greatly among market segments. If a product cannot meet the needs of a given market segment, then it must be marketed to another segment that can use the product beneficially.

Place: Place refers to the location or the source of the compost product, which can be the production facility or a distribution point. Generally, compost is a local product. The market for bulk compost is typically within 50 km (30 miles) of the facility and/or distribution point. Therefore, the facility location greatly influences the potential markets. If the facility location is unsuitable, other distribution options can be established, although remain somewhat limited by distance due to transportation costs. A useful tool for considering potential markets, and/or facility locations, is a marketing map that pin-points potential customers and sources of feedstocks.

Price: Price is an important marketing factor. It must be set at point that at least covers financial, production, transportation, and marketing costs and, hopefully, generates profit. The steps for determining price are: (1). Make a market assessment for the compost product; (2). Determine the costs of production (including overhead and marketing) in order to determine a unit cost (e.g., $ per ton); (3). Estimate the demand curve, that is, how demand varies with price; (4). Determine the pricing objective; and (5). Decide upon the price.

The options for pricing objective include cost-plus, penetration, competition, premium, and product-line pricing. In cost-plus pricing, the producer adds a profit margin to the unit production cost. The margin is typically a percentage of the unit cost (e.g., 50%). Penetration pricing sets the price at an intentionally low level in order to grow sales and establish a market share. Over time, the price may be increased to meet company goals. Competition pricing bases the price on the price's competitors charge. The company may choose to set a price that is slightly higher to signify higher quality or lower to gain a market price advantage. Alternatively, the company can match competitors' prices and compete on other terms. Premium pricing is used to set apart a product or service that is special or high quality. The qualities of the product should truly merit the distinction or customers will abandon it. Similarly, product-line pricing fixes different prices for different market segments based on their respective expectations, compost benefits, and ability to accept a given price. The higher prices may be justified by greater quality control and/or additional service.

Promotion: The purpose of promotion is to stimulate demand for the product. It tells potential customers about the company and its products; it tells customers how the product benefits; and it convinces customers to purchase the company's product instead of those of competitors. Furthermore, compost can also be promoted as an environmental benefit, both because it is a recycled product but also because its use benefits the environment. There are numerous promotional techniques including word-of-publicity, face-to-face customer contact, advertising, sales discounts (e.g., higher volume/lower price), direct marketing (e.g., mail brochures), general publicity (e.g., news articles), sponsorships, displays, demonstrations, branding, and packaging (Fig. 18.12).

Box 18.2 The four Ps marketing model—cont'd

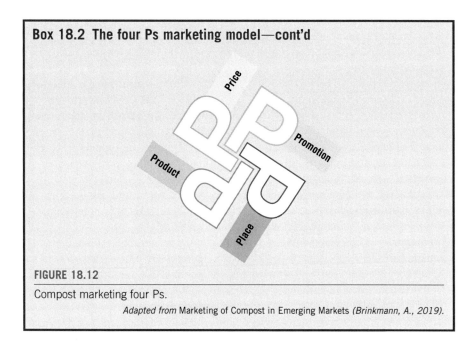

FIGURE 18.12

Compost marketing four Ps.

Adapted from Marketing of Compost in Emerging Markets *(Brinkmann, A., 2019).*

control program should be implemented based on end-use, while continuing to meet regulatory requirements. A facility should have adequate storage, an inventory control and tracking system, and procedures that ensure clean products free of physical and biological contaminants including litter, stones, weed seeds, and pathogens. For example, if customers insist on weed free compost, the facility must manage the process to achieve temperatures high enough to thoroughly kill weed seeds and then the finished compost should be stored in way that minimizes the introduction of new seeds.

- *Market research*—Market research gathers data about the current market conditions and trends. It typically entails surveying a percentage of the pool of potential customers and then analyzing the survey data. Information gathered within the target geographic area should include demographic data, competing product data, identify any market constraints, end user preferences, plus services, and tools that can expand compost use. This information is useful for developing product specifications and pricing strategies. One of the most important functions of market research is identifying the target markets—those markets with greatest revenue potential or that can be penetrated in a short period of time. Extensive market research can be time consuming and expensive and thus more common among larger facilities. The facility staff can perform market research, though there is a risk that their point of view may be biased. An experienced market research company can be hired to provide a more objective assessment. However, the company should be knowledgeable about agriculture, the green industry, and other compost markets and preferably possess sales and marketing experience.

- *Product research*—The first step in product research is performing a literature search related to compost qualities and applications. This step will provide access to information that has already been generated. Afterward, any additional product research should focus on end-user applications that represent the largest potential markets or on niche applications for your products. Product research can be conducted in partnership with universities (including local extension programs), private research firms, potential customers (e.g., nurseries or greenhouses or state departments of transportation), and testing laboratories. Participation in end user and industry trade associations leads to opportunities for field research studies, demonstrations, and funding.
- *Promotion*—When developing a marketing program, compost manufacturers should consider establishing a product name and logo (i.e., brand) that serve as a focus for its promotional efforts. Promotion encompasses a multitude of activities that include newsletters, cooperative advertising programs with distributors, billboards, radio ads, and the internet promotion. Magazine advertisements usually reach a regional and national audience and therefore are suited to compost companies that distribute beyond the local area. Many promotional efforts begin by having sales staff exhibit at trade conferences. Other methods of promotion include demonstration plots and field trials, especially with large potential end users and innovators. When considering promotional activities, it is important to determine the audience that the message needs to reach, and how to measure the success of the promotional campaign.
- *Education*—Education is an essential part of marketing that covers many facets of the marketing program. Early in the program's development, it is important to educate the composting facility staff about the required attributes of compost products. Over the long term, the end users in the target market become the primary focus of educational activities. These activities can include research articles, product literature, presentations, field research, videos, and seminars. Providing visual documentation of a products performance is especially effective and providing this documentation has been simplified through the development of visually oriented websites. Education efforts often include the general public in the geographic market area and in the community where the facility is located. Opportunities for reaching the public include garden clubs, public interest groups, and schools. The public can be informed through retail distribution centers (e.g., garden centers, nurseries, home centers). Staff at these outlets can be trained through personal training sessions, videos, and other educational tools. Each year, composting organizations from several countries

jointly sponsor International Compost Awareness Week (ICAW), which offers an excellent opportunity to conduct educational programs to raise the public's awareness about composting and compost.

- *Sales/distribution*—This important market development task is dictated by the results of market research, including identification of target markets and the geographic market areas. Sales/distribution planning determines pricing policies, how target markets are approached, and whether in-house marketing staff, product brokers, and/or distributors are used to reach target markets. Within the sales/distribution program, it is important to devise ways to provide technical assistance to end users. This marketing component also includes developing a product delivery infrastructure and customer services, such as access to application equipment.

These six facets of market planning comprise the market development program. They are intertwined, in that activities completed in one area often complement and overlap with another. Efforts completed for one facet will often overlap into another facet. Market research data and a good understanding of the overall market will assist in the development of the most appropriate course of action.

---

7. Compost sales

Many people consider "sales activities" to be the major activity within a marketing program. The sales function enables a supplier or manufacturer to physically introduce a product, service, or concept to a potential customer. Selling composts requires a knowledgeable and skillful salesperson, an able supplier and can be aided by a variety of sales techniques and tools.

7.1 The compost supplier and salesperson

A compost supplier must provide good customer service, reliability, and a consistent compost product. A good supplier is always open to suggestions and works to satisfy their customers' particular needs. It is usually less expensive to make a current customer happy, than it is to find a new customer. The supplier must be able to provide compost when and where the customer needs it, whether you are a compost producer or broker. This requirement is accentuated by the seasonality of the green industry and other users of compost. To assure product availability, it is helpful to know the production cycles and storage capacity of your facility, and to improve efficiencies in both areas (as long as product quality is not negatively affected). Other issues that must be considered include how the compost will be delivered (transportation issues), site hours, payment terms, quality control, instructional literature, etc.

Most good compost salespeople are passionate about selling compost and a believer in the industry cause—recycling organic residuals back to the soil and protecting the environment. In addition, the compost salesperson must possess sales skills and a working knowledge of the green industry (e.g., agriculture, horticulture) and composting. The salesperson's compost knowledge should encompass the functions and benefits of compost and competing products, relevant regulations, health/safety/environmental issues, and the basics of the composting process. Questions about how compost products are manufactured frequently arise during sales calls. A salesperson who understands the composting process can help facility operators manufacture products that meet end users needs and rectify compost quality problems that customers may experience. Sales staff are typically hired with experience in *either* sales, the green industry, *or* compost production. The salesperson then is trained to fill in the gaps in his or her background.

7.2 Prospecting

A key element of compost sales and market development is prospecting—identifying and rating potential customers before making the sales call. Prospecting starts with creating an inventory of potential customers that you already know, then expanding the list by obtaining or developing a database. Management of databases has become much simpler with the use of computers, and the ability to purchase electronic database, as well as customer relationship management programs. Excellent computer-based programs exist which were developed explicitly for prospecting and customer contact. In the past, composters often started databases or prospect lists by reviewing telephone directories in their specific geographic target market. Today, these lists can be purchased from list companies or harvested from "on-line" resources. Joining end user trade organizations, to access their membership and membership directory, is another common and effective strategy. One of the best methods to build upon this initial prospect list is by assertively obtaining referrals from customers and other prospects. *Note that prospects who don't want to buy your compost may still provide you with viable referrals—so ask for them.*

Once the initial prospect list is developed, it must be refined and the prospects should be rated. Refining and rating prospects help the sales staff to set priorities for leads and allows them to be more productive. For example, if the decision is made to only sell compost in large tractor-trailer volumes, then making sales calls on small customers does not make sense (at least in the short-term). Prospects may be rated by market segment, by size (number of employees, annual income), by estimated current or potential compost use volume, etc. While some composters may want to place a priority on the largest companies, others may focus on a particular market segment (e.g., landscapers, resellers) or product application. The marketing plan should guide the prospect rating system. Once rating is completed, the prospect list is more readily useable in lead generation and sales activities. *Note that rating prospects will likely be an ongoing activity.*

Two things must be done to refine a prospect list: create a rating system, and then begin contacting the prospects. A fear of contacting new potential customers, whether by the telephone or in person, is the downfall of many potential sales people. This reluctance may be a personality trait of an individual, or it may be brought on by on-going difficulties "in the field" (e.g., poor sales results). Regardless of the reason, both situations can be overcome. *Sales is a* "numbers game"—*when more calls are made (to try to get sales appointments), more sales appointments are obtained. When more sales appointments are obtained, more customers are obtained.*

7.3 Lead generation

The term "leads" refers to prospects that have been qualified in some manner; that is, there is some evidence of their interest in your products (e.g., response to an advertisement). Like all effective product manufacturers, compost manufacturers have systems in place to generate leads for their sales staff. Leads may be generated by normal sales activities (referrals), or by a variety of promotional and educational activities. Typical promotional activities used for lead generation include mailings, use of social media (e-blasts, websites), staffing trade shows and advertisements in trade journals, newspapers, and on the web. The promotional activities used to generate leads, and expand sales, vary greatly based on whether you are focusing your sales efforts on professional or nonprofessional end users. When focusing sales activities on professional end users, developing and managing a database of prospects is important. Since transportation costs limit compost distribution to narrow geographic regions, focused promotional activities should be used to generate leads. Experience suggests that directed mailings and e-blasts, with telephone follow up, can be extremely effective, as are activities through end user trade conferences and social media.

7.4 The sales call

Many sales call techniques have developed over the years to sell everything from shovels to automobiles. Regardless of the technique, it is imperative that the salesperson be properly prepared for the sales call and present herself/himself as being a "problem solver." Remember that the objective is not selling compost, but selling a means to solve a customer's particular problem. The salesperson can be more effective if armed with technical knowledge about the market in which he/she is selling. The *bottom line* is that the potential customer wants to be shown how compost is going to save or make them money; provide a superior result, or both! Although many end users believe in recycling, this is not the primary reason that most people buy compost. In some situations, preparing for a sales call can be time-consuming, involving research about the prospect's business approach and services. In other situations, it may simply require gathering the appropriate sales literature necessary to make an organized presentation. Either way, it is helpful to be pragmatic and time-efficient during the sales call but also thorough.

Starting a sales call can be uncomfortable. Asking open ended questions like, "What have you heard about our product?" rather than, "Are you interested?" helps get the buyer involved in an interactive discussion. Asking the customer about his/her needs through this process helps them feel as though the salesperson is not selling, but professionally advising them as to what product to purchase. Asking, "If you were our company, what types of end users would you be trying to sell to?" is an unassuming question that can start the bonding process with the potential customer.

Different market segments use compost for different purposes and have different priorities. Therefore, compost salespeople should learn about the technical requirements of a specific end user group before approaching it. *Always keep in mind the needs of the end user that were identified during market research and prospecting.* With this in mind, a salesperson has a better chance of getting the prospect's attention if the sales presentation concentrates on the issues that are most relevant to that particular end user. Product benefits that are most important to each specific end user should be presented before all others. For this purpose, sales staff can make use of pitch pages, which list common applications of compost within each market segment and market-specific sales call suggestions and marketing tips.

7.5 Compost sales tools and services

In order to assist sales staff to be more successful *in the field*, specific tools and services can be made available (Table 18.4). Sales tools not only improve sales success but also demonstrate the professionalism of your company and provide the customer

Table 18.4 Compost sales tools and services.

Compost sales tools	Compost sales–related services
Sales literature	Product delivery
Before and after pictures, end use videos	Application equipment
Product quality analyses and test results	Technical assistance from staff and specialized consultants
Product samples	Training presentations
National publications and pertinent trade publication articles	Informational library
Websites containing appropriate content	
Case studies	
Product research	
Computer programs (e.g., customer relationship management programs) and databases	
Computer models	

with useful technical information. For instance, compost samples, before and after photographs and letters of recommend (references) allow the customer to make her or his own judgment about the sales claims.

Specific sales related services also help to the marketing program. The need for these services increases with the scale of compost production (i.e., greater compost volumes). While all market segments have a basic need for product delivery, specific market segments, such as wholesale nurseries and golf courses, require a greater degree of technical assistance. Approaching some markets without the proper technical "back-up" increases the potential liability of the compost producer if the customer's crops or property are damaged. Additional technical assistance may be available through horticultural/agronomic consultants, university researchers, agricultural extension agents, and other technical experts. It is important, however, that these experts understand compost. Without this context, it is difficult to get the proper assistance. Product liability insurance further protects the compost producer, and also provides comfort to potential end-users.

References

Cited references

AAPFCO, 2020. AAPFCO Product Definitions. American Association of Plant Food Control Officials. https://www.aapfco.org/.

Alexander, R., 2005. AAPFCO Soil Amendment/Compost Uniform Product Claims. R. Alexander Associates, Inc, Apex, NC.

Alexander, R.A., 2010. Practical Guide to Compost Marketing and Sales, first & second ed. R. Alexander Associates, Inc, Apex, NC.

Brinkmann, A., 2019. Marketing of Compost in Emerging Markets. Netherlands Enterprise Agency. https://bvor.nl/.

IWMC (Integrated Waste Management Consulting), 2019. SB 1383 Infrastructure and Market Analysis. California Department of Resources Recycling and Recovery (CalRecycle), Sacramento, CA. https://www2.calrecycle.ca.gov/Publications/Details/1652.

E & A Environmental Consultants, Inc., 1995. Unpublished Study Conducted for New York City. Editor's note: The Chapter's author, R. Alexander, Was Principally Involved in this Study. E & A Consultants has since been purchases by Tetra Tech, Inc.

E & A Environmental Consultants, Inc., 1998. Unpublished Study Conducted for the State of Iowa. Editor's note: The Chapter's author, R. Alexander, Was Principally Involved in this Study. E & A Consultants has since been purchases by Tetra Tech, Inc.

Iowa Department of Natural Resources, 1998. Iowa Statewide Compost Market Assessment Prepared by: Resource Conservation and Development of Northeast Iowa, Inc. (Postville, IA).

ORBIT/ECN, 2008. Compost Production and Use in the EU, Final Report of ORBIT e.V. European Compost Network ECN to European Commission, Joint Research Centre.

Rouse, J., Rothenberger, S., Zurbrugg, C., 2008. Marketing Compost: A Guide for Compost Producers in Low and Middle-Income Countries. Eawag, Dübendorf, Switzerland.

Saveyn, H., Eder, P., 2014. End-of-Waste Criteria for Biodegradable Waste Subjected to Biological Treatment (Compost & Digestate): Technical Proposals. European Commission, EU Science Hub. https://ec.europa.eu/jrc/en/publication/eur-scientific-and-technical-research-reports/end-waste-criteria-biodegradable-waste-subjected-biological-treatment-compost-digestate.

Tyler, R.W., 1996. Winning the Organics Game. ASHS Press, Alexandria, VA.

Consulted and suggested references

Alexander, R., 2014. 10 Trends in the compost marketplace. BioCycle. https://www.biocycle.net/2014/09/18/10-trends-in-the-compost-marketplace/.

Eggerth, L.L., Díaz, L., Chang, M., Iseppi, L., 2007. Chapter 12 marketing of composts. Waste Manag. In: Diaz, et al. (Eds.), Compost Science and Technology, vol. 8. Elsevier, Boston, MA, pp. 325−355.

European Commission, 2000. Success Stories on Composting and Separate Collection. European Communities. Directorate-General for the Environment. https://ec.europa.eu/environment/waste/publications/pdf/compost_en.pdf.

Freeman, J., Skumatz, L., D'Souza, D., SERA, 2012. Boulder County Compost Market Study. Skumatz Economic Research Associates Inc., Superior, CO. www.serainc.com.

NERC, 2010. Making Your Compost Product Work for You. Northeast Recycling Council, Inc. https://nerc.org/.

R. Alexander & Associates, Inc., 2017. Unpublished Study Conducted for the University of Wisconsin-Oshkosh. https://www.alexassoc.net/.

Texas Cooperative Extension, 2006. Marketing Composted Manure to Public Entities. Final Report. The Dairy Compost Utilization Project. Texas A&M University. http://compost.tamu.edu/reports/.

Composting economics

Authors: Craig S. Coker[1], Mark King[2]

[1]*Coker Composting and Consulting, Troutville, VA, United States;* [2]*Maine Dept. of Environmental Protection, Bangor ME, United States*

Contributors: Jane Gilbert[3], Jonathan M. Rivin[4], Rudy Wentz[5], Mary Schwarz[6], Britt Faucette[7], Rod Tyler[8]

[3]*Carbon Clarity, Rushden, Northamptonshire, United Kingdom;* [4]*Oregon Department of Environmental Quality, Portland, OR, United States;* [5]*Formerly with the State University of New York, Cobleskill, NY, United States;* [6]*Cornell Waste Management Institute, Cornell University, Ithaca, NY, United States;* [7]*Filtrexx International, Decatur, GA, United States;* [8]*Green Horizons Environmental, Medina, OH, United States*

1. Introduction

In essence, composting economics refers to the *business* of composting—how to manage feedstocks and manufacture a marketable product in such a way as to produce a profit. Compost production is about manufacturing products. It involves manufacturing-type expenses—processing, product quality control, labor, health and safety, and environmental protection. It is a manufacturing process that produces a bulky commodity with a modest price point that is costly to transport long distances (Fig. 19.1).

Business economics are about making sure revenues exceed the cost of production. However, the economics are more complex with some composting enterprises such as municipal, farm, and institution operations (e.g., universities). There are a number of constraints to the traditional "economics first" approach. Municipalities need to consider factors like the avoided cost of transportation and other means of discarding yard trimmings, biosolids, source-separated organics (SSOs), and other municipal solid waste (MSW). For example, composting of SSOs or MSW might be economically justified because it extends the useful life of local landfill, delaying the cost of a new landfill, or reducing the cost of transportation to a distant regional landfill. Commercial compost operations, on the other hand, need to balance the overall costs of compost production against the income derived from raw materials receipt and overall sales of finished compost products. Finally, on-farm composting operations might realize difficult-to-quantify yet real economic benefits from more efficient manure management, nutrient retention, and healthier soils. There are also external benefits associated with environmental protection and public relations.

The Composting Handbook. https://doi.org/10.1016/B978-0-323-85602-7.00016-9

FIGURE 19.1

Composting economics is tied to the fact that compost is a manufactured bulk product of low density and a relatively low price point.

This chapter discusses the major elements of composting economics. It presents actual economic revenue and costs figures in US dollars circa 2020. In some cases, cost estimates from earlier years are adjusted to 2020 values by applying appropriate inflation factors. Readers should understand that the economic values presented are typical or example values. They are neither exact nor universal. Readers are advised to seek local and current economic estimates. Print and on-line publications like *BioCycle* and *Composting News* frequently offer current information about the economics of composting (BioCycle, 2021; Composting News, 2021). Specifically, *BioCycle* publishes a regular series of articles that concern composting economics under the category of "Business and Finance" (Coker, 2020). *Composting News* updates US regional market prices for compost annually.

2. Economics overview

Compost production and/or compost utilization, in and of themselves, can only be "profitable" if incoming revenues plus operational savings plus other benefits exceed the combination of operating expenses and other detriments (Table 19.1).

Typical revenue sources are gate fees (also called tip or tipping fees), sales of compost and/or compost-based soil products, and services (e.g., transportation, compost application). Revenues from compost use typically derive from direct compost sales to other users. For on-farm operations that use some or all of the compost produced, revenues may arise from increased crop yield from compost

Table 19.1 Example income and expenditure items at a typical medium-scale composting operation.

Revenue	Expenditure	Savings
Gate fees	Labor	Reducing storage, handling, transportation, and disposal costs for manure and other residuals
Sales of compost	Fuel	
Sales of related services (e.g., compost application)	Debt service	Reduced fertilizer, herbicide, and pesticide use on crops
	Feedstock amendments	
Increased crop yields for on-farm composters	Management and administrative costs	Reduced irrigation of crop for on-farm composters
	Compost transportation and application costs	

use and possibly higher prices for healthier produce. The associated savings include reduced fertilizer, herbicide and pesticide use, and avoidance of crop losses to disease. The avoided cost of purchasing compost, fertilizer, and other products is a potential savings for farms, nurseries, landscaping companies, municipalities, universities, and other institutions that produce and use their own compost.

Other benefits occur for all composting applications, but they are difficult to assign accurate revenue. Some examples include improved environmental impacts, avoided costs of permitting and improved soil health due to improved soil quality. Additionally, as the world's response to climate change intensifies, traded carbon credits, CO_2 equivalents, or methane reduction equivalents may become a more reliable and predictable revenue stream. Both composting and compost use may yield carbon credits.

Compost production expenses are labor, fuel, debt service on improvements or equipment, feedstock amendments (if purchased), business development (including securing feedstocks, marketing and selling products), and management and administrative costs (Box 19.1). Some of these expenses are direct expenses (e.g., costs of

---

Box 19.1 Understanding cost rates

Understanding and evaluating costs in planning a composting enterprise requires more than just knowing the cost of a piece of equipment. It requires knowledge of cost rates. A cost rate is a special ratio in which the two terms are in different units, for example: dollars per hour, dollars per square meter or cubic foot, dollars per ton, and dollars per cubic meter or cubic yard.

Cost guides for estimating construction costs are often expressed in dollars per unit area (i.e., asphalt paving at $10 per square meter or square yard). Costs for operating machinery are known as machine rates (i.e., dollars per ton of material handled) whereas labor rates are dollars per unit time (usually hours). Machine rates reflect the amortized cost of the equipment (roughly capital cost plus loan interest costs divided by the number of tons of material the machine is expected to handle over its "economic" life, which is usually 5 to 7 years) plus the operating costs (fuel, maintenance, and replacements) plus the labor rate of the operator. In calculating the tonnage a piece of equipment will handle over its life, it is best to use realistic estimates of the quantity handled per hour, the number of hours per day (usually no more than 6 h per day) and days per week (5–5.5, depending on Saturday work schedules) the machine will be used. The labor rate (for all personnel) should be calculated based on actual pay plus the hourly value of fringe benefits. Fringe benefits include paid vacation, retirement, sick leave, insurance, etc. This is also known as the "loaded" labor rate.

goods sold) and some are indirect expenses (e.g., overhead). Compost utilization expenses are the costs to apply the compost and costs to get compost to customers.

2.1 Revenue/savings items

The composting business is a unique industry that bridges both manufacturing and waste management. It produces a salable product and can also earn revenue from processing feedstocks that others view as waste materials. While some composting enterprises rely on *either* gate fees *or* compost sales, most composters gain revenue from both items. The most successful private-sector composting business models are set up as stand-alone enterprises, with careful cost accounting, and derive approximately 75% of the composting enterprises' revenue from processing gate fees, with approximately 25% of the revenue coming from compost sales (Coker, 2020). In general, every one dollar of gate fee requires a compost price of five dollars to earn the same revenue because of the shrinkage in volume from feedstock to compost (Tyler, 1996). However, there are a number of successful marketing focused companies that have the inverse ratios due to excellent marketing and higher revenues from product sales (Bradley, 2018). In addition to sources of incoming revenue, composting offers opportunities for savings through reduced or avoided costs and through improved operational efficiencies at the point of waste generation.

2.1.1 Revenue from gate fees

Gate fees are payments made by the waste generator to have waste composted (or otherwise disposed). Gate fees are normally calculated on a per ton,[1] per cubic meter (m^3), or per cubic yard basis (yd^3), as received at the composting facility. Some composting operations have truck scales or weighbridges for weighing and invoicing trucks, others calculate the volume of material arriving in the delivery vehicle (Fig. 19.2). Trucking costs for hauling the feedstocks to the facility are usually a separate charge.

Feedstocks that typically earn gate fees include biosolids, municipal yard trimmings, SSOs, vegetative and land clearing debris, by-products from food manufacturing (which may also have enterprise value as animal feed in the winter), food scraps from markets and food service facilities, industrial residuals like short-fiber paper sludge from papermaking, seafood plant processing wastes, dissolved air flotation treatment residuals from poultry processing, commercial grease trap wastes, and occasionally animal manures. Fees received for these wastes vary with the difficulty of managing and composting the waste, the moisture content, and alternative disposal costs in the region (either landfilling or direct land application). Biosolids are one of the higher gate fee materials available because process and environmental control costs are higher. Wet materials tend to command a higher

---

[1] In this chapter, "ton" refers to *either* a metric ton or US ton, unless explicitly identified in the text. The two units are nearly equal. 1 metric ton = 1.1 US ton.

FIGURE 19.2

Small truck crossing a scale (weighbridge) at a farm-based composting facility that collects gate fees for community yard trimmings and commercial food wastes.

fee because of higher transportation and handling costs. Opportunities for gaining a gate fee for composting food-based feedstocks are increasing due to regulations and policies discouraging landfilling.

In most cases, the maximum gate fee is determined by the disposal fees charged by the local or regional landfill or waste combustion facility. Fig. 19.3 presents trends in landfill fees up to 2020 for various regions of the United States as reported by the Environmental Research and Education Foundation, which annually produces a report on landfill gate fees and other waste statistics (EREF, 2021). The report further breaks down gate fees by individual US states. As Fig. 19.3 suggests, disposal fees differ considerably among regions. They can also vary greatly among localities within states and counties. Gate fees in other countries also range widely and are greatly influenced by waste disposal taxes or levies imposed by various countries, states, or provinces (Pickin et al., 2018; EEA, 2013, 2016).

The fee that a composter can capture tends to be less than the local cost of disposal at a landfill or waste-to-energy facility because feedstock suppliers must be more discriminating about the material delivered to a composting operation compared to a landfill. However, a composter might be able to charge a higher fee than a landfill if the composting site is closer to the source of the feedstock than the landfill, or if alternative treatment methods are required by law. Table 19.2 provides examples of the range of gate fees charged by composters in the Southeastern US in 2020 dollars.

Competition for feedstock and gate fees comes not only from landfilling but also from diversion for other beneficial uses. For example, paper mill sludges have been used to generate alternative energy thereby reducing the amount available for composting. Similarly, when wood is used for heating fuel, wood chips, shavings, and even sawdust are less available as amendments for composting. As indicated in Table 19.2, it is not always possible to get a gate fee for these wood residuals;

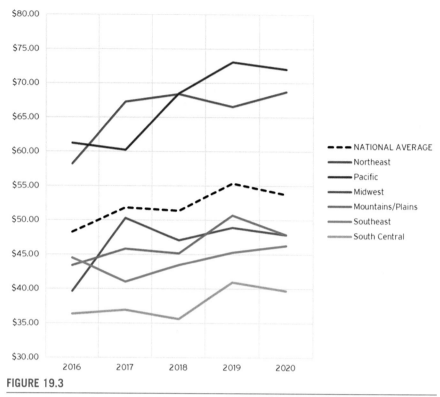

FIGURE 19.3

Average landfill gate fees in US $/ton and trends in regions of the US as reported by the Environmental Research and Education Foundation (EREF, 2021).

in many cases, woody amendments must be purchased and/or transported by the composter. Wet feedstocks, including food residuals, are also target feedstocks for anaerobic digestion (AD) facilities, and composters may have to compete with AD enterprises for them. However, composters also have an opportunity to partner with AD facilities by composting the solid residues left over from AD treatment.

In limited cases, there may be willingness to pursue composting rather than other alternatives due to a generator's interest in environmental sustainability, even in the face of higher costs. The bottom line is that the cheapest allowable disposal cost option generally wins.

Table 19.2 Gate fees for compostable feedstocks in US Dollars per US ton (Southeastern US).

Feedstock	Gate fee range
Biosolids	$40.00–60.00+
Manures	$0.00–20.00
Wood wastes	$0.00–25.00
Municipal solid wastes	$25.00–45.00
Yard wastes	$15.00–30.00
Liquids	$45.00–65.00

Source: Craig Coker, professional data.

2.1.2 Revenue from compost sales

Revenue can be realized from sale of composts and compost-based soil products like topsoil and engineered / manufactured soils. Composted manure has higher demand and commands a higher price than raw manure. It is richer in organic matter and has more nutrients than topsoil; and is lighter weight than either manure or topsoil. The higher value of the lighter compost means it can be cost-effectively transported over greater distances. Even so, the cost of transport limits viable compost market areas to less than 160 km (100 miles) from the composting facility.

Some compost producers offer several different product mixes, each being tailored to particular markets (Fig. 19.4). Different compost products are frequently distinguished by the level of screening, with more-finely screened compost earning a higher price. Some higher-value products are created by blending. For example, compost can be blended with acidic materials (e.g., pine bark) to make a soil amendment suitable for acid-loving (ericaceous) plants like azaleas. Similarly, composters taking in iron salt residuals from drinking water treatment plants can market their higher iron-content compost as more suitable for turf topdressing. These blends, and other value-added products, cost more to produce but also command a higher price and expand the market. Blending with other products also helps to stretch compost supply. For example, 100 cubic meters of compost can be sold alone or it could be sold within a 200 cubic meter compost-pine bark container-mix.

FIGURE 19.4

Multiple bulk compost-based products, each targeted to a different market and set of users.

Where the market supports only low prices (e.g., volume markets), the composter can offer a product that not been processed as much, such as an unscreened compost, or a compost that is less mature. Municipally owned and operated composting facilities tend to sell their products at below-market pricing, as they do not have a strong profit motive. On-farm composters may similarly discount compost prices if the composting operation is reducing costs in other areas of the farm (Section 2).

Compost and soil products are sold in either bulk (i.e., loose) or bagged form. Bulk products are normally sold on a per unit volume basis (per cubic meter or cubic yard). Delivery charges are typically added to sales prices. Bagged products command a higher price; however, the cost of producing bagged product are also higher and the competition is greater. Therefore, the economics of bagged production do not normally work well at the smaller scale of composting operations, unless a high-value niche can be created, such as sales through a farm market.

Table 19.3 contains illustrative pricing for both bulk and bagged manure-based compost for selected US market areas in 2020, as compiled by Composting

Table 19.3 Prices in $US for composted cow manure in selected regions of the US as of August 2020 (Composting News, 2020).

Bulk retail prices per yd³ (0.76 m³)			
Region	**Average**	**High**	**Low**
Northeast	$45.50	$46.00	$45.00
Southeast	$31.33	$45.00	$12.00
Cleveland	$24.00	$24.00	$24.00
Iowa	$42.00	$60.00	$16.00
Minneapolis	$29.33	$32.00	$26.00
Texas	$29.33	$47.50	$15.00
Denver	$21.71	$30.00	$12.00
Phoenix	$20.00	$20.00	$20.00
Northwest	$23.85	$33.75	$17.45
San Francisco Bay	$26.00	$26.00	$26.00
Southern California	$17.25	$24.00	$12.00

Bagged retail prices for a 40-lb. (1-kg) bag			
Region	**Average**	**High**	**Low**
Northeast	$6.53	$9.87	$3.85
Chesapeake	$5.97	$7.95	$3.99
Southeast	$5.16	$6.97	$3.34
Florida	$1.98	$1.99	$1.97
Cleveland	$3.27	$3.99	$2.49
Iowa	$2.18	$2.49	$1.75
Minneapolis	$3.70	$7.00	$1.99
Texas	$4.59	$6.99	$1.99
Phoenix	$2.99	$3.99	$2.49
Northwest	$3.99	$3.99	$3.99
San Francisco Bay	$7.00	$7.00	$7.00

News (2020).[2] Manure-derived compost is familiar to compost users, and is relatively rich in nutrients, so it tends to bring a slightly higher price than composts derived from most other feedstocks (e.g., green waste, biosolids). Compost prices in other countries are comparable but likewise depend on the feedstocks, level of processing, and local availability and demand. Local sources of compost can be found through local public agencies and web searches and direct pricing inquiries can be made of nearby sources. In localities where compost prices are not well established, the local price of topsoil can serve as an approximate indicator of the minimum price customers are willing to pay for high-quality compost.

2.1.3 Savings/avoided costs

For some composting operations, the economic benefits come from savings to other parts of the enterprise, rather than, or in addition to, direct revenues. For public agencies and institutions (e.g., municipalities, parks, universities), savings can result from the avoided cost of feedstock disposal or savings in the purchase of compost, topsoil, mulch, and similar inputs. Savings can also result from reduced cost of labor and/or transportation relative to the handling of uncomposted feedstocks. For example, composting deciduous leaves at a local composting site reduces hauling costs if the leaves are otherwise delivered directly to remote fields.

The potential for operational savings resulting from composting are most evident and numerous for on-farm enterprises, particularly in manure management. Environmental regulations governing application rates and neighbor concerns about odors and flies have increased the cost of direct land application of manures in some areas. In particular, the need for more and more distant land areas to spread manure adds to the expense (Box 19.2).

Box 19.2 Examples of manure handling cost comparisons

In a study of bedded hog manure by Practical Farmers of Iowa (2000), one farmer participant found that composting reduced the volume of the manure bedding by 18%—54%. At the average hauling distance of 0.6 km (0.4 mi.) and spreading field length of 0.6 km, composting saved about as much labor time as it required to produce the compost. Since no special equipment was used by the farmer for composting, composting proved to be a "break even" enterprise on the study farm. Another participant at a 423-head hog farm evaluated composting versus direct land application strictly from a time management perspective and concluded it required approximately 53 man-hours of labor to clean, load, haul, and apply 100 loads of manure to fields 3 km away (1.8 miles), but that it required the same amount of time (53 h) to process, haul, and apply the corresponding 56 loads of compost to that same field. With larger quantities of manure and/or longer travel distances to land application fields, the lesser volume produced by composting can be economically favorable to direct land application.

---

[2] Composting News annually compiles and publishes prices of bulk and bag compost for selected locations within the US.

In addition to manure handling costs, other tradeoffs to consider include odor management and crop nutrients. Land application of compost is nearly odor-free. Nutrients are less subject to losses in the field if applied as compost because the composting process converts available forms of nitrogen to organic nitrogen, which is resistant to leaching. In addition, nutrients from subsequent fertilizer applications remain available for plant uptake longer due to the higher cation exchange capacity of compost and proven nutrient-holding capacity of compost amended soils. These and other factors (e.g., moisture content, product uniformity) give the farmer greater flexibility as to when to land apply the material.

Applying composted versus raw manure can potentially return value in the form of nitrogen (N) fertility. Much depends on how manure is otherwise handled and how it is composted (the opposite may also occur—composting can lead to greater N losses). To avoid nitrogen loss due to ammonia volatilization, raw manure must be promptly incorporated into the soil. If incorporation is delayed for four days, as much as 40% of the nitrogen is lost in hot, dry periods (Fulhage et al., 2002). For manure with a 3% N content, and a value of N at US $1.14/kg ($0.52/lb), this means a potential loss of $13.68 per *dry* metric ton ($12.48 per *dry* US ton). In comparison, losses of N in composting have been measured on the order of 7% if the C:N ratio of the material being composted is above 30:1 (Tubail et al., 2008). This potential loss of N would be valued at $2.39 per dry metric ton ($2.18 per dry US ton).

In addition to nutrient management, application of raw manure spreads weed seeds while composting it first kills weed seeds, insect larvae and disease from the heat cycle of composting. If animals feed contains contaminants like weed seeds, they will easily spread in raw manure increasing problems, especially with noxious weeds.

2.2 Expense items

The expenses of a composting enterprise include fixed costs, like loan payments, and variable costs, like diesel fuel. The direct costs of composting are those associated with making the product (cost of bulking agent/amendment, loader maintenance, laboratory testing) and/or obtaining feedstocks (collection container rental, collection route operating costs), while the indirect costs are overhead items like costs for starting the operation, marketing and sales expenses, human resource expenses, insurance costs, and professional service fees.

Table 19.4 summarizes expenses that are common to composting operations. These expense items can be further categorized by their function such as initial costs, operational costs, product distribution, and management and administrative costs. Government entities require some composting facilities to provide financial assurance in the event that the facility closes abruptly, although these "closure costs" are rare for municipal and on-farm composting situations.

2.2.1 Initial cost

The cost of developing a composting operation is considered "front-end loaded." The entire facility has to be planned, permitted, designed, built, and in operation for approximately 6 months before the first sale of finished product is possible.

Table 19.4 Common composting expense items.

Initial (start-up)	Operational	Product distribution	Administrative and management
Business research and business plan development	Labor	Product promotion and advertising	Salaries for managers and office staff
Attorney fees	Fuel and lubricants	Sales	Office supplies
Accountant fees	Equipment amortization	Broker commissions	Office utilities
Engineering fees Paving (e.g. hard standing) Civil engineering works	Amendment purchase	Travel	Rent
Financing costs	Electricity	Product samples	Business insurance
Permits	Water Leachate management	Demonstration projects	Health insurance
Feedstock procurement costs	Equipment wear parts	Sponsorships	Payroll taxes
Coverage for 6–12 months of operational expenses	Equipment replacement	Shipping/ delivery	Business taxes
Site acquisition and/or development	Trucking	Application	Professional services (attorneys, accountants, engineering)
Equipment purchases	Soil blending ingredients	Trucking	Training
Financial assurance for closure	Compliance testing		

Gate fee revenues can help offset some of the initial cost of operations, but only after the facility starts operation.

The initial cost of enterprise development can include business plan research and development, professional service fees for attorneys, accountants, and consultants, financing fees, permit application preparation, and government permit and approval fees. If the enterprise is planning on accepting gate fee materials, the time and effort needed to secure firm contracts for those wastes is considerable. The extent of permitting documentation and regulatory approvals needed is often directly correlated to the types and sources of materials to be composted. Farm-generated wastes have the least regulatory oversight. Biosolids, food and solid wastes have more oversight. A larger-scale composting operation, taking off-site waste, may have to invest

FIGURE 19.5

Site development and construction costs are a major expense for most composting facilities.

$125,000 to $250,000 in engineering and permitting costs alone (Coker, 2020). The actual expenses depend on the prevailing rates and permitting requirements of the specific location. In some regions, these costs can be appreciably higher than $250,000.

Facilities subject to regulatory permits normally have to be built, then inspected by regulators for adherence to the approved plans, after which they receive permission to begin operations. Construction activities include all the work and cost associated with site development: clearing and grading, utilities, storm water management systems, paving, constructing buildings and roads, and installing scales, fences, signage, and landscaping (Fig. 19.5). Before the facility begins operations, the employees who will work at the composting facility should attend one of the many compost operator training courses offered by professional organizations, public agencies, or universities. The facility may also need to have an approved operations plan and closure plan financial assurance before it can begin operations.

2.2.2 Operational cost

Operational cost includes both direct and indirect costs. In addition to capital for land, buildings, and development, making compost requires labor, fuel, equipment usage, electrical power, and often a carbon-based amendment/bulking agent. The financial statements for the composting enterprise account for these costs as "cost of goods sold" (COGS), which affects the gross profit of the enterprise.

Labor costs are the largest component of COGS. Labor is needed to prepare and mix feedstocks, build and monitor compost piles, move compost to curing, and to screen and prepare the final product for market. Sometimes labor costs attributable to the composting enterprise can be reduced by sharing those resources with other functions of the organization, such as road maintenance in a municipality, landscape services at a nursery or college, and manure management on a farm. If employees can be cross trained to perform other functions (e.g., training a loader operator to monitor the process or training a truck driver to perform routine diesel equipment maintenance, etc.), then overall labor costs can be optimized. Cost of labor can range from $4.00 to $12.00 per ton of feedstock handled, depending on composting method, equipment available, and regional labor rates (Coker, 2020).

Fuel costs are often the second-largest cost, as most equipment used to handle the materials of composting is diesel fuel-driven, and transport is often needed to bring feedstocks to the composting facility. Transportation is almost always needed to take product to market.

Transport costs in mid-2020 approached $4.50 per loaded mile in the United States (i.e., a loaded truck traveling one way from point to point). Trucking costs to bring feedstocks to a composting facility are often recovered from the generator as a separate charge from the material gate fee. Because the cost to take product to market has to be absorbed in the cost of the product, there is usually a maximum, cost-effective product delivery distance from the point of generation, usually between 80 and 160 km (50−100 miles). This optimal "trucking" distance may also be reduced as fuel surcharges become necessary due to fluctuating fuel costs. In general, fuel surcharges are calculated by subtracting the fuel threshold amount (price at which you need to begin enacting a surcharge) from the actual price per gallon and divide that amount by vehicle's miles per gallon. For example, assume that a surcharge is enacted when diesel fuel tops $0.90 per liter ($3.40 per gallon) and it is now $1.35 per liter ($5.10/gallon), a difference of $0.45. Further assume that the transport truck gets approximately 1.8 km per liter when loaded (4.2 miles/gallon). Therefore, the surcharge would be: $0.45 per liter/1.8 km per liter = $0.25 surcharge per km ($.402 per mile). This would result in a surcharge of approximately $25.00 per 100 km.

Equipment used in composting needs regular and thorough maintenance. Types of equipment used in composting are trucks, loaders, conveyors, turners, grinders, shredders, fans, screens, etc (Fig. 19.6). In particular, hammer mills on grinders and teeth on shredders can wear down rapidly, resulting in high maintenance costs. Large scale equipment capable of processing thousands of cubic meters or yards per year may be over $300,000 to $600,000 per item depending on processing throughput ability.

Electrical power can be a significant cost, particularly if aerated static pile or aerated bin composting is used. An aerated biofilter also adds greatly to the electrical use. Electrical power costs can be on the order of $1.05 to $1.50 per ton of feedstock handled in the United States, depending on availability of three-phase power and electric utility rates.

Dry and carbon-based amendments can be a significant operating expense. If the facility is located in an area where there is a competing demand for wood-based industrial boiler fuel, cost for materials such as sawdust can reach $25.00 per ton plus transport. Some on-farm operations look to vegetative waste materials to serve as both carbon source and structural bulking agent. These materials are often available for minimal cost (sometimes free with only the cost of hauling). These materials include ground or unground municipal yard trimmings (leaves, grass, and brush),

FIGURE 19.6

Labor, fuel, and equipment comprise the largest share of operating costs for most composting enterprises.

chipped woody debris from clearing operations along utility rights-of-way, and ground vegetative debris from land clearing operations. These types of woody wastes are not as sought after by boiler fuel markets (due to higher moisture and dirt content) than clean sawdust, wood chips, and wood shavings.

2.2.3 Product distribution cost

The cost of product distribution and/or product use includes marketing expenses (e.g., brochure printing, newsletter mailings, demonstration projects, etc.), sales expenses (e.g., advertising, travel costs, broker commissions, etc.), delivery expenses (usually reimbursed by the buyer), and, if offered, application expenses (e.g., soils testing costs, application equipment costs, etc.).

These costs are highly variable and depend on the extent and nature of the compost facility's product distribution program. Marketing and sales costs, for example, vary from the relatively low cost of an informative website to the relatively high cost of customized 30-second television advertisements. Brokerage commissions are usually about 10%, which can often be factored into the product price. Delivery charges and application services fees are almost always billed to the customer. Compost application services can be a separate profitable enterprise on its own, in some cases.

2.2.4 Management and administrative cost

The routine cost of managing any business enterprise includes salaries for managers, secretaries, and salespeople, office supplies, insurance, payroll taxes, and office rent. Also included are professional service fees for legal, accounting, and engineering expenses. Any labor cost associated directly with making or selling products should be accounted for as a direct expense, not an indirect (overhead) expense. This approach gives the most accurate accounting of COGS, and thus, gross profit. For start-up composting enterprises, budget 25%−30% of direct operating costs to cover management and administration.

2.2.5 Financial assurance for closure

Facilities that become permitted and regulated by environmental agencies are sometimes required to maintain financial assurance to cover closure costs in the event the facility is abandoned. This requirement is rare for municipal and on-farm composting situations, but much depends on the policies of the regulatory agency.

Financial assurance is a regulatory program designed to assure regulators that a composting facility has the means to finance the closure of the facility, should that be needed. This type of program originated in the solid waste landfill industry so that public tax dollars would not have to be spent to clean up privately owned, but abandoned, landfills. Financial assurance mechanisms acceptable to most government agencies include certificates of deposit, irrevocable letters of credit, trust funds, surety bonds, and insurance policies. Financial assurance mechanisms can be a significant cost to a composting facility. As few have the resources to self-finance closure costs (which can easily exceed US $250,000), pollution liability insurance policies are often used. The annual premium for a $1,000,000 insurance policy can exceed $50,000.

Closure costs are usually calculated by assuming all compostable materials on the composting pad is a solid waste that must be collected and hauled to disposal by an outside contractor. Many US states require these estimates be prepared by a Professional Engineer. Closure costs often also include the cost to pump out storm water management ponds and truck the contents to a wastewater treatment plant. In some states, operators are given a finite period of time (e.g., 6 months) after cessation of waste acceptance to convert all wastes to compost and move the product to market. Following this, and an inspection by the regulatory agency for satisfactory closure, the owner is released from the financial assurance requirement .

2.3 Typical costs

As noted previously, costs can be categorized as initial costs, operational costs, and product distribution costs. Tables 19.5 and 19.6 supply general cost guidance information. However, it is important to recognize that most types of costs are very specific to a particular site, to a particular enterprise and to local market conditions. In Table 19.5, the initial equipment costs listed are for new equipment. However, many composting operations start out with low-hour used equipment, which is available at substantial discount to new equipment prices. Overall costs and amount of time required to turn windrows is dependent on the character, volume and bulk density of the material being turned, on the type of equipment used, its power and size, volume capacity, number of times windrows are turned, and on

Table 19.5 Typical initial costs (2020 United States dollars).

Initial (start-up)	Administrative and management
Land acquisition	$3,000 to $20,000+/acre $7,400 to $50,000+/hectare
Design drawings	$5,000 to $8,000/sheet[a]
Permitting	$20,000 to $250,000+
Clearing and grading	$10,000 to $15,000/acre $25,000 to $37,000/hectare
Paving (installed)	
Lined gravel pad[b]	$40.50/yd^2 ($48.44/m^2)
Asphalt[c]	$42.23/yd^2 ($50.50/m^2)
Roller-compacted concrete base	$110.00/yd^3 of concrete ($144/m^3)
Lime-stabilized soil-cement base	$8.00 to $10.00/yd^2 ($9.57 to $11.97/m^2)
Concrete slab[d]	$29.52/yd^2 ($35.31/m^2)
Concrete mixing pits	$250/yd^3 of concrete installed ($328/m^3)
Standpipe windrow irrigation ponds	$40.00 to $60.00/linear ft ($131 to $197/m)
Storm water management system	$0.92/ft^2 of watershed area ($9.90/m^2)
Equipment (new)	
Grinders	$80,000 to $500,000+
Mixers	$40,000 to $200,000
Loaders	$75,000 to $150,000
Turners	$30,000 to $500,000+
Moisture addition	$20,000 to $90,000
Screens	$75,000 to $250,000+
Baggers	$65,000 to $300,000+

[a] A sheet is a one page of a typical engineering drawing, roughly 24 in. x 36 in. (or 600 mm x 900 mm).
[b] Estimate based on 60 mm. HDPE liner, 6 in. (15 cm) of #67 stone, 6 in (15 cm). ABC stone with two layers of geosynthetic fabric.
[c] Estimate based on 4 in (10 cm). intermediate course plus 2 in (5 cm). surface course over 12 in (30 cm). stone base.
[d] Estimate based on 6 in (15 cm). unreinforced fibermesh slab over 6 in (15 cm). gravel base.
Source: Craig Coker, professional data.

Table 19.6 Typical operating costs (US dollars, 2020).

Element	Estimated hourly operating cost (US $/hour)
Mobile self-propelled windrow turner, 3-meter width (16 ft)	$163.00
Wheel loader, 6.5 cubic yard bucket (new)	$60.00
Wheel loader, 5 cubic yard bucket (used)	$82.00
Deck screen	$185.00
Compost field applicator	$66.00
Live-bottom tractor trailers	$104.00
Low-boy trailer (equipment transport)	$23.00

Based on data presented by Rynk, R., 2001.

operator skill. Composting facilities required to meet regulatory requirements for pathogen destruction will need to turn windrows more frequently than others. Turning becomes less costly on a per-unit volume basis as the volume of material increases and equipment is used more efficiently. Table 19.6 contains other operating costs based on data collected at two western US composting facilities.

3. The big picture

A good first step in evaluating the feasibility of a composting operation is to prepare an "enterprise budget," a financial instrument commonly used to produce agricultural products. In financial circles, these are often called "pro formas." Enterprise budgets represent estimates of expected receipts (income), costs, revenues, and profits. The information contained in the enterprise budgets can be used by financial institutions, governmental agencies, and other advisers making decisions pertaining to business issues. This type of rigorous financial planning is necessary to obtain outside debt or equity financing for a new enterprise.

Enterprise budgets provide two important types of information:

- a detailed listing of the inputs needed (i.e., composting feedstocks, labor, fuel, etc.) to produce a particular output (i.e., compost, topsoil, etc.); and
- financial data on costs, income, and profitability for a specific enterprise.

Traditional enterprise budgets consist of four separate categories: gross income, variable costs, fixed costs (overhead), and net-income matrix, which is an appropriate way to develop enterprise budgets for crop production. For a composting enterprise, however, the budget should reflect the categories, items, and format of the financial statements produced by the enterprise. This facilitates direct comparisons between budgeted and actual expenses in the future.

The budgeting process begins by creating a chart of accounts. The chart of accounts is a listing of all the accounts in the general ledger, each account accompanied by a reference number. To set up a chart of accounts, one first needs to define the various accounts to be used by the business. Each account should have a number to identify it. Setting up a budget using the same chart of accounts as is used for the enterprise's financial statements greatly simplifies comparing actual to budgeted expenses. Table 19.7 shows a typical chart of accounts for a commercial composting company. This "tree" of accounts can be subdivided down to a fine level of detail for budgeting and accounting, but extensive subcategorizing of expenses complicates routine bookkeeping. A financial advisor can help explain what should be tracked.

In Table 19.7, the chart of accounts follows traditional financial statement formatting, i.e., total income (sum of all sales) minus the COGS (how much it cost to make compost) equals gross profit. Then gross profit minus overhead expenses (the cost of running the enterprise) equals net profit.

Table 19.7 Example chart of accounts for a composting facility (based on an actual facility in the US).

Account#	Account name
Income	
2,100	Gate fees
2,200	Product sales
2,300	Delivery charges
2,400	Interest earned
2,500	Miscellaneous income
Cost of goods sold	
3,100	Amendment/bulking agent
3,200	Vehicle taxes and insurance
3,300	Payroll expenses
3,301	Payroll
3,302	Employer-paid taxes
3,400	Product supplies
3,500	Leasing costs
3,501	Windrow turner
3,502	Trommel screen
3,503	Trucks
3,600	Licenses and permits
3,700	Repairs and maintenance
3,800	Fuel
3,900	Lab expenses
Gross profit	
Overhead expenses	
4,010	Advertising
4,020	Bank charges
4,030	Charitable contributions
4,040	Dues and subscriptions
4,050	Fuel—office vehicles
4,060	Medical
4,080	Utilities
4,200	Travel expenses
4,300	Payroll—office
Net profit	

Table 19.8 illustrates an actual annual budget from an on-farm composting operation in the Eastern US that also sells landscape supplies and animal feeds.

Developing an enterprise budget is best done by estimating all costs first, then estimating revenues, and finally, adjusting the timing of expenses as much as

Table 19.8 Example on-farm enterprise budget.

Item	Amount
Revenue	
Gate fee materials	$921,482
Animal feed sales	$586,825
Mulch sales	$45,100
Compost and soils sales	$151,084
Trucking charges	$215,257
Equipment rental	
New waste contracts	$42,500
Total	**$1,969,248**
Expenses	
Accounting/lawyer	$15,000
Marketing	$15,539
Capital assets	$4,000
Compost expenses	$86,641
Feed ingredients	$200,817
Fuel	$262,758
Gifts	$1,739
Insurance	$83,247
Principal and interest	$182,481
Janitorial	$11,801
Leased and financed equipment	$506,136
Licenses and permits	$19,519
Medical	$3,391
Memberships	$4,230
Office supplies	$6,680
Repairs and maintenance	$68,431
Phone	$7,497
Sales travels/meals	$8,087
Utilities	$9,536
Wages and taxes	$354,000
Total	**$18,422,530**
Gross income/loss	**$126,717**

possible to ensure that costs are offset by revenues. Timing adjustments depend on items such as: seasonal issues, need for constant positive cash flow, availability of revolving credit line funds, etc. The budget is built, line item by line item and month by month over at least a 12-month period, preferably 36 months (3 years). If the facility operates on a cash basis of accounting, the enterprise budget should be set up on a cash basis (Box 19.3). If accrual accounting is used, the enterprise budget should be similarly structured (i.e., lagging receipts by estimated receivables aging, etc.).

Expenses include both COGS and overhead expenses. COGS includes all expenses of product manufacturing: depreciation (a way of accounting for the decline

> ### Box 19.3 Cash versus accrual accounting
>
> There are two basic accounting methods available to most small businesses: cash or accrual.
>
> Cash method: If the cash method of accounting is used, income is only recorded when cash payments are received from customers and an expense is recorded only when a check is written to a vendor. Most individuals and businesses use the cash method for their personal finances because it's simpler and less time-consuming. However, this method can distort income and expense values, especially when credit is extended to customers, or the facility buys on credit from their suppliers, or if materials are kept in inventory for prolonged periods of time.
>
> Accrual method: With the accrual method, income is recorded when the sale occurs, whether it be the delivery of a product or the rendering of a service—regardless of when payment is received. Expenses are recorded when goods or services are received, even though payment may not be made till a later date. The accrual method gives a more accurate picture of a facility's financial situation than the cash method because income is recorded on the books when it is truly earned, and expenses are recorded when they are incurred. Income earned in one period is accurately matched against the expenses that correspond to that period, so a better picture of net profits for each period is established.
>
> Pros and cons: The cash method is easier to maintain because income is not recorded until cash is received and expenses are not recorded until cash is paid out. With the accrual method, more transactions are typically recorded. With the cash method, the only transaction that is recorded is when the customer pays the bill.
>
> Another issue to be considered is the accounting method used for tax purposes. For convenience, the same method used for internal reporting should be used for tax purposes. However, the Internal Revenue Service permits use of a different method for tax purposes. Some businesses can use the cash method for tax purposes. If an inventory (i.e., finished compost in product storage) is maintained, then the accrual method would be more applicable, at least for sales and purchases of inventory for resale.

in value of an asset like a truck), amortized capital (i.e., loan or lease payments), production labor, production expenses (i.e., bulking agent or carbon source), costs for maintaining delivery vehicles, etc. Overhead costs are those costs associated with running the business, such as management and office labor, business insurance, personnel fringe benefits, etc.

Budgeted revenues (or gross income) consist of estimates of gate fees to be received, reimbursement for trucking costs, and income from sales of compost and/or compost-based products. If comparing a composting enterprise to some other functions of the organization (such as an alternative form of disposal), consider adding any known "avoided costs" to the revenue/income side of the budget. For example, if a proposed composting operation would avoid the need to truck manure 40 km (25 miles) to a land application field, then add that avoided cost of trucking to the composting enterprise revenue estimate.

3.1 Case study

This farm is a Confined Animal Feeding Operation of 7000 dairy replacement heifers located in a high desert region. The farm consists of 160 acres, of which

100 acres are feedlots and 30 acres are designated for composting. The animals are separated into 64 dry lot outdoor pens. No crops are grown, and animal feed is imported. The farm has been composting manure for approximately five years.

The on-farm waste stream consists of dairy manure and straw bedding. The manure and bedding are collected from all of the 64 pens, with each pen cleaned out about every 2–3 months. Manure and straw bedding are also imported from an adjacent dairy farm. The imported bedding is especially important as there is insufficient bedding (i.e., carbon source) on-farm to produce an acceptable feedstock mixture. This exchange is mutually beneficial. The adjacent dairy farm is disposing of its wastes and the heifer farm is using the wastes as a revenue stream. Approximately 1500 tons of bedding/manure are imported once per year when the dairy farm is cleaned. The cost of importing the manure/bedding straw is equivalent to collecting and transporting manure/bedding on-farm.

Manure is not composted during the winter months, from November through February, because of heavy rains. In the winter, the manure/bedding is piled within the pens; during the rest of the year, the manure/bedding is collected and directly hauled to the composting area. The composting area is bare ground as State regulators do not require the composting to be conducted on an impervious surface. However, as the farm intends to expand the variety of its imported feedstocks (such as yard waste) to derive gate fee revenue, an impervious storage/composting surface is planned to comply with State regulation.

Three products are sold at the farm: aged manure, compost, and topsoil. Of the total manure, about 5% is incorporated into topsoil products, 60% is sold as aged manure, and the rest is sold as compost. The aged manure is produced from manure and small amounts of residual bedding. It is piled for a year, with little managing, except for an occasional pile relocation. Water is sometimes sprayed on the piles so that a hardened surface forms, which reduces wind erosion. Once a client places an order for the aged manure, it is screened using a trommel screen. This final product is also mixed with soil to produce topsoil. The ratio of manure to soil is blended as requested. The soil is obtained from on-farm construction projects.

In the spring, starting approximately in April, manure/bedding is composted using the windrow method. Stored manure/bedding, "fresh" manure/bedding from the pens and imported manure/bedding are hauled to the composting area. Piles, approximately 6 ft high $\times$ 10 ft wide $\times$ 600 ft long (1.8 m $\times$ 3 m $\times$ 180 m) are made, with 90 to 150 piles being produced during a year. The piles are turned with a self-propelled compost turner that straddles the pile. The piles are managed so that the temperatures achieve in excess of 55°C (131°F) for at least 15 days and during that time, the piles are turned five times. Subsequently, the piles may be turned, but only to enhance cooling. The objective is to cool the piles quickly, so that the pile temperatures return to ambient, and the total composting time is approximately three to four weeks. Temperature probes are used to monitor pile temperatures with four locations per pile being tested. The farm owns three probes. As a final composting operation, the compost is run through a trommel screen to remove rocks and large chunks. The compost is certified organic.

Three people are dedicated to "manure-related" activities. One focuses on manure collection and coordinates transport and delivery, the second is involved primarily with composting operations, and the third drives the truck. Four other farm employees are used as needed.

About 25,000 tons of aged/screened manure and 15,000 tons of compost are produced annually, and both are sold in bulk. The manure is sold for $10.00/ton (there are about 1.25 yd^3 per ton of manure or 1 m^3) and the compost for $18.00/ton (1.25–1.5 yd^3 per ton of compost or 1–1.14 m^3). It costs $7.00/ton to produce the aged/screened manure, which includes loading/unloading into trucks, pile formation, and final screening. It costs $16.00/ton to make the compost. The reason for having two products is that the compost appeals to those who farm organically, while some farmers prefer a lower priced product. Approximately 20,000 yd^3 (15,300 m^3) of topsoil are produced annually and is also sold in bulk. The topsoil sells for $8.00/$yd^3$ and costs $6.00/$yd^3$ to produce (about $10.50 and $7.80 per m^3, respectively). Manure is bought primarily by local farmers, compost is primarily used by the fruit tree industry, and topsoil is primarily used by the general public, although sometimes large external construction projects, such as a golf course, have consumed most (or all) of the supply. Costs associated with this farm for equipment and labor are shown in Tables 19.9 and 19.10, revenues are shown in Table 19.11.

4. Economics of compost use

To sell compost products to end users, it is useful to understand the economics of compost use so as to educate the end user about costs and savings. The economics of compost use is particularly important to on-farm composters as many farmers produce compost primarily or entirely for use in their own crop production systems.

The economic gain from using compost is difficult to quantify as the primary benefits—improved soil health, crop yield, and crop quality—are not immediate and depend on many external factors, including soil qualities, weather, market factors, etc. Therefore, this section focuses on two contrasting facets of compost use: quantifying nutrient values and the economics of one particular use, sediment and erosion control. These facets contrast because the nutrients in compost provide little advantage to its use in controlling sediment and erosion. Additionally, the

Table 19.9 Cost of compost/manure equipment used.

Equipment (number)	Cost	Year purchased	Function	% of time[a]
Loaders, Case (6)	$150,000 each[b]	All purchased since 2003	Moves manure, bedding, and compost	100
Trommel screen, McCloskey, 516RE	$200,000	2006[b]	Screens compost	100
Compost turner, Frontier	$50,000[c]	2006	Turns compost piles[d]	100
Semi, Freightliner 475 hp	$40,000	2003	Pulls belt trailer	100
Belt trailer, R-Star Manuf. 50', 36" self-loading	$70,000	2007	Hauls manure	100
Truck, Ford, L9000	$30,000	1998[b]	Pulls end dump trailer	70
End dump trailer, 30'	$15,000	1993[b]	Manure/compost transport	70
Truck, Freightliner with manure spreader	$30,000[e]	2005	Manure spreader	100
Water truck, Kenworth	$25,000	1974	Waters compost/manure piles	30

[a] Percentage of time in manure/compost activities versus total usage.
[b] Estimated.
[c] Purchased at a bargain price. Actual worth was $125,000.
[d] Uses $25 diesel fuel/hr (2008).
[e] Cost of truck; manure spreader was made at farm shop.

Table 19.10 Labor costs.

Personnel (number)	Task	Salary per month
Truck driver (1)	Drives trucks used for hauling manure/bedding	$4,000
Manure management (1)	Manages collection/transport of manure/bedding	$4,000
Compost management (1) Part-time, as needed (4)	Manages composting operations	$4,000

Table 19.11 Revenue streams.

Product	Production	Cost of production	Revenue
Aged/screened manure	25,000 tons	$7/ton	$10/ton
Compost	15,000 tons	$16/ton	$18/ton
Topsoil	20,000 yd^3	$6/yd^3	$8/yd^3
	15,300 m^3	$7.80/m^3	$10.50/m^3

accompanying textbox discusses the potential value of compost in sequestering soil carbon (Box 19.4).

4.1 Nutrient value of compost

At a minimum, compost can be valued based on the nutrients that it supplies, knowing the corresponding cost of those nutrients in commercial fertilizers. Even though compost generally provides dilute, slow released forms of nutrients, relative to commercial chemical fertilizers, it is useful to know the fertilizer value of compost.

Box 19.4 Carbon sequestration value of compost

As Chapter 15 evinces, even a single compost application can produce long term increases in soil organic carbon (SOC), at least on noncultivated land (e.g., rangeland). Repeated applications of compost further increase the benefit when the compost is applied to soils diminished of organic matter by tillage and other factors (i.e., SOC below natural levels). The resulting sequestration of the inherent carbon is a benefit that can help reverse some of the impacts of climate change. This benefit has economic value. It also can generate revenue if compost users or compost producers can sell the resulting carbon-credits.

According to a report published by the International Solid Waste Association (ISWA), typical yard trimmings—based composts have a carbon sequestration value ranging from approximately US $5 to $9 per fresh metric ton when the compost is applied to degraded soils (Gilbert et al., 2020). Specifically, the report lists the values in Euros as €3.50 to €8.10 per fresh metric ton ($3.98 to $9.20), depending on the assumed carbon sequestration rate.

The composts have additional value due to their plant nutrients. The report calculates the value based on total nutrient content (not just plant-available) and estimates the total nutrient value in the range of approximately $20 to $23 per fresh ton, based on 2019 average fertilizer prices (specifically, €17.70 to 20.10 per fresh ton). Therefore, the estimated total values of the example composts range from $24 to 32 per fresh ton (€21.20 to 28.20) due to both nutrients and carbon sequestration. Based on these values, the report suggests that current prices for composts are only about one-third of their actual value.

The project used a model to predict the effect of annual applications of two types of compost, one made from yard trimmings compost and one made from yard trimmings plus food waste (Table 19.12). Several application rates were modeled over several different SOC-diminished soils of varying qualities. The cumulative impacts of annual compost applications were determined over periods of 10 and 20 years. The model predicts an increase of 0.40%—0.55% in SOC after 20 years of annual compost applications at 30 fresh metric tons per hectare (13 US tons/acre).

Box 19.4 Carbon sequestration value of compost—cont'd

Table 19.12 Characteristics of the compost used for the International Solid Waste Association model. Based on laboratory analysis of 21 samples for each type of compost.

	Yard trimmings compost	Yard trimmings/food
	% of dry weight	
Total N	1.37	1.79
P	0.22	0.27
P_2O_5 equivalent	0.51	0.63
K	0.83	0.82
K_2O equivalent	0.99	0.98
Organic matter	30%	32%
Moisture content	30% (wet weight)	34% (wet weight)

Gilbert, J., Ricci-Jurgensen, M., Ramola, A., 2020. Quantifying the Benefits of Applying Quality Compost to Soil, International Solid Waste Association, Rotterdam. Available at: https://www.iswa. org/uploads/media/Report_4_Quantifying_the_Benefits_to_Soil_of_Applying_Quality_Compost.pdf.

The following example illustrates the calculations needed to estimate the fertilizer value of compost *realized in the first year after application*, given assumed rates of nutrient availabilities. Considering that the nutrients from compost are largely conserved, and gradually released over several years, the actual nutrient value of the compost is much higher. The compost used in this example is based on a farm-produced compost in Virginia.

Notwithstanding the economic value of the compost as a fertilizer material, the cost of using compost is often offset by other factors such as the increased revenue from enhanced crop yields due to improvements in soil quality, better nutrient retention, and water conservation.

Example 19.1 Nutrient value of compost

A compost product, derived from paper mill sludge, food wastes, and poultry litter, has the following nutrient concentration on a dry weight basis and other characteristics:

1.2% organic N + 0.48% inorganic N = 1.68% Total N

   0.43% P

   0.99% K.

   55% solids content (45% moisture content)

   bulk density (fresh or "as is") = 534 kg/m³ (900 lb/yd³)

The compost will be used in a warmer temperate climate. It is expected that in the first year after application 10% of the organic N, 90% of the inorganic N (accounting for ammonia loss), 50% of the P, and 80% of the K will be available to plants (see Chapters 14 and 15).

Step 1: Determine the amount of nutrients in kilograms (kg) per DRY metric ton (t). (Note: to convert to lbs per dry US ton multiply the results by 2).

Organic N:	0.0120×1000 kg/t = 12.0 kg Org-N/t
Inorganic N:	0.0048×1000 kg/t = 4.8 kg Inorg-N/t
P:	0.0045×1000 kg/t = 4.5 kg P/t
K:	0.0119×1000 kg/t = 11.9 kg K/t

Step 2: Convert elemental P and K to fertilizer formulations (see Chapter 15).

P:	$P_2O_5 = P \times 2.3 = 4.5$ kg $\times 2.3 = 10.35$ kg P_2O_5
K:	$K_2O = K \times 1.2 = 11.9$ kg $\times 1.2 = 14.28$ kg K_2O

Step 3: Determine the available nutrients per dry ton.

Organic N:	12.0 kg $\times$ 0.10 = 1.2 kg/t
Inorganic N:	4.8 kg $\times$ 0.90 = 10.8 kg/t
All N:	12 kg/t
P_2O_5:	10.35 kg $\times$ 0.50 = 5.175 kg/t
K_2O:	14.28 kg $\times$ 0.80 = 11.424 kg/t

Step 4: Determine the per kilogram price of fertilizer nutrients.

Fertilizers are sold as a commodities and prices vary daily. Commodity price reports, like the World Bank's Commodity Price Data (Pink Sheet), track commodity prices for fertilizers including urea, diammonium phosphate (DAP), and potassium chloride (KCL).

Major Nutrient	Base Commodity	Commodity Price[a]	Equivalent price[b] Per dry kg
N	Urea	$245/t	$0.53/kg N
P_2O_4	DAP	306/t	$0.46/kg P_2O_5
K_2O	KCL	$2563/t	$0.42/kg K_2O

[a] US $ per dry metric ton. Source: World Bank, Commodity Price Data, 2019 Average 2020
[b] Urea is a nitrogen fertilizer that is 46% N. KCL contain K in amounts equivalent to 61% K_2O. DAP has 18% N and P in amounts equivalent to 46% P_2O_5. Converting commodity prices to individual nutrient prices is not necessarily straightforward, especially for DAP. Flynn (2014) provides an excellent explanation of the conversion.

Step 5: Determine the value of the compost based on nutrients per dry ton.

Major Nutrient	Amount per Dry metric ton	Equivalent Price/kg	Nutrient value $ per dry metric ton
N	12 kg/t	$0.53/kg N	$6.36
P_2O_5	5.175 kg/t	0.46/kg P_2O_5	$2.38
K_2O	11.424 kg/t	$0.42/kg K_2O	$4.80
Total			$13.54

Step 6: Determine the value of the compost based on nutrients per fresh ton (i.e., wet or "as is" ton) and volume.

Solids content = 55% (0.55)	Bulk density = 534 kg/m³ (0.534 t/m³)
$ per fresh ton:	$13.54/dry ton $\times$ 0.55 solids content = $7.45/fresh metric ton
$ per unit volume:	$7.45/fresh ton $\times$ 0.534 t/m³ = $3.98/m³ = $3.04/yd³

4.2 Compost use in sediment and erosion control

Compost is used in applications to prevent erosion and sediment loss from construction sites and to filter sediment out of storm-induced runoff (Chapter 16). The utility of compost in these applications stems from its durability and physical characteristics (e.g., particle size, permeability, adsorption). Nutrients are of little value in most of these applications, except to support vegetation growth. In fact, nutrients have negative value in stormwater treatment applications. Therefore, in these cases, the economic value of compost is determined by demand, supply, and the costs of competing products. The main compost products used in these instances are blankets, filter berms, and filter socks (Table 19.13).

A compost blanket is a stabilization, erosion control, and vegetation establishment practice used on slopes to stabilize bare, disturbed, or erodible soils on and adjacent to construction activities. Compost blankets are used for both temporary and permanent slope erosion control and vegetation establishment applications. Blankets are normally applied with a mulch blower truck in an even 5−cm (2-inch) layer over disturbed areas (Fig. 19.7). In addition, the compost is often preseeded with any particular seed mix as it is being discharged out of the blower hose. A 5-cm thick compost blanket consumes about 500 m^3 of compost per hectare of blanket (about 270 yd^3/acre).

Table 19.13 Cost estimates for erosion control alternatives.

Length of erosion control Duration of project (months) Would compost be removed from site? (usually not required)		150 m (500 ft) 6 No
30 cm × 60 cm (1 ft × 2 ft) compost berms	**Low-cost estimate**	**High-cost estimate**
Materials and installation cost	$1,305	$1,575
Regular inspection and sediment removal cost	$1,800	$1,800
Repair and replacement cost	$196	$236
Compost removal cost	$0	$0
Total cost	**$3,301**	**$3,611**
30 cm (12in) diameter compost filter socks	**Low-cost estimate**	**High-cost estimate**
Materials and installation cost	$1,500	$1,750
Regular inspection and sediment removal cost	$1,800	$1,800
Repair and replacement cost	$68	$68
Sock removal and disposal cost	$20	$20
Compost removal cost	$0	$0
Total cost	**$3,388**	**$3,638**
1m high (3 ft) silt fencing	**Low-cost estimate**	**High-cost estimate**
Materials and installation cost	$1,045	$1,445
Regular inspection and sediment removal cost	$1,800	$1,800
Repair and replacement cost	$523	$723
Removal and disposal cost	$180	$401
Total cost	**$3,548**	**$4,369**

FIGURE 19.7

Compost is often applied as a blanket for erosion control using a blower truck that blows compost through a hose to the application area.

Source: Cornell Waste Management Institute.

Compost blankets cost around $1.94 to $3.76 per m^2 installed ($0.18 to $0.35 per ft^2) installed. This cost typically includes seed but doesn't require additional inputs for fertilizers or lime. Costs are dependent on the cost of the source compost and distance to the application site. They compete with conventional erosion control blankets, which consist of a uniform web of interlocking fibers (e.g., straw) with net backing. Conventional erosion control blankets of low performance quality are similarly priced, but high-performance blankets cost $3.76 to $4.30 per m^2 ($0.35 to $0.40 per ft^2).

A compost filter berm is a narrow permeable mound of compost that slows the flow of runoff and traps sediment (Chapter 16). It uses a specifically sized mulch/compost filtering material installed using a berm-forming device. This technique is primarily used for temporary erosion/sediment control applications (e.g., preventing spring runoff damage along seasonal camp roads) where perimeter controls are required or necessary.

Compost filter berms are an alternative to traditional silt fencing. An advantage to compost berms is there is no need to collect used silt fencing and dispose of it. Filter berms cost range from $8.70 to 11.50 per linear meter ($2.65 to 3.50 per linear foot), where silt fencing is about $8.20 per meter ($2.50 per foot). Adding in the cost of collection and disposal for silt fence, the costs are equivalent.

A compost filter sock is a three-dimensional tubular sediment control and storm water runoff filtration device. They are typically used on and around construction activities for perimeter control of sediment and other pollutants (such as metals, nitrogen, phosphorus, petroleum hydrocarbons, pesticides, and herbicides). Compost filter socks trap sediment and other pollutants by filtering runoff water as it passes through the matrix of the sock and by allowing water to temporarily pond behind the sock, allowing deposition of suspended solids. Compost filter socks are also used to reduce runoff flow velocities on sloped surfaces (Fig. 19.8).

Like berms, filter socks are an alternative to silt fencing. However, unlike silt fence which has only sediment storage behind the fence, compost filter socks have sediment storage inside the sock between the porous coarse compost materials. In fact, 60% of the average sediment stopped is inside the socks, and only 40% behind the socks. For this reason, compost filter socks are more effective than silt fence. They are also an alternative to gravel check dams placed in drainage ways for sediment capture and filtration.

Costs for filter socks vary with the sock diameter; cost ranges from about $5.00 per linear meter ($1.50/ft) for a 20−cm (8−inch) diameter sock to $46.00 per meter ($14.00/ft) for a 60-cm (24−inch) diameter sock. Larger diameter socks are used in applications such as living walls and stream bank stabilization. Installation rates for socks also vary with sock diameter (as larger socks are slower to fill). Rates vary from 900 to 1500 linear meters per day (roughly 3000 to 5000 ft/day) for a 20-cm (8−inch) sock to 300 to 600 m per day (1000 to 2000 ft/day) for a 36-cm (14−inch) sock. compares the costs of silt fencing, compost filter berms, and compost filter socks for an example situation where 150 m (about 500 ft) of erosion and sediment

FIGURE 19.8

Compost filter sock used on a roadside construction site to keep sediment and nutrients from moving with runoff into the culvert.

Cornell Waste Management Institute.

control is required. It is based on the now archived Erosion Control Calculator from the US EPA's Greenscapes program (USEPA, 2006).

The market potential for compost use in sediment and erosion control is considerable, given the volumes involved. For instance, a 30-cm (12−inch) compost sock uses 1 m^3 for every 13 m of length, or 1 yd^3 for every 30 ft. Compost blankets require about 500 m^3 per hectare of disturbed land (270 yd^3/acre).

References

Cited references

Biocycle Magazine and BioCycle Connects, 2021. Published by J.G. Press. https://www.biocycle.net/.

Bradley, A.L., 2018. Presentation: Compost Marketing: Strategies for Success. Northeast Recycling Council. https://nerc.org/documents/compost_marketing/compost_marketing_strategies_for_success.pdf.

Coker, C., 2020. Composting business management series: Part I. Revenue forecasts for composters; Part II. Capital cost of composting facility construction. Part III. Composting facility operating cost estimate. Part IV: Net present value of composting expenditures. Part V: Tool for triple bottom line assessment. BioCycle. Available from: https://www.biocycle.net/category/businessfinance/.

Composting News, 2020. National Compost Prices. McEntee Media (Accessed 2020).

Composting News, 2021. Published by McEntee Media. http://compostingnews.com/. Cornell Waste Management Institute. http://cwmi.css.cornell.edu/composting.htm.

EEA (European Environment Agency), 2016. Municipal Waste Management across European Countries, Briefing No. 03/2016. Available at: https://www.eea.europa.eu/themes/waste/waste-management/municipal-waste-management-across-european-countries.

EEA, 2013. Typical Charge (Tip Fee and Landfill Tax) for Legal Landfilling of Non-hazardous Municipal Waste in EU Member States and Regions. Available at: https://www.eea.europa.eu/data-and-maps/figures/typical-charge-gate-fee-and.

EREF (Environmental Research and Education Foundation), 2021. Analysis of MSW Landfill Tipping Fees– 2020. Available at: www.erefdn.org https://erefdn.org/bibliography/datapolicy-projects/.

Flynn, R., 2014. Calculation Fertilizer Costs. New Mexico State University. Available at: https://aces.nmsu.edu/pubs/_a/A133/.

Fulhage, C.D., et al., 2002. Fertilizer Nutrients in Livestock and Poultry Manure. University of Missouri Extension Publication EQ, p. 351.

Gilbert, J., Ricci-Jurgensen, M., Ramola, A., 2020. Quantifying the Benefits of Applying Quality Compost to Soil. International Solid Waste Association, Rotterdam. Available at: https://www.iswa.org/uploads/media/Report_4_Quantifying_the_Benefits_to_Soil_of_Applying_Quality_Compost.pdf.

Pickin, J.P., Randell, P., Trinh, J., Grant, B., 2018. National Waste Report 2018. Australia Department of the Environment and Energy; Blue Environment Pty Ltd.

Practical Farmers of Iowa, 2000. Managing Manure and Bedding From Hoophouses. Available at: http://www.pfi.iastate.edu/ofr/Hoop_Manure_Project_Intro.htm.

Rynk, R., 2001. Exploring the economics of on-farm composting. Biocycle 42 (4), 62.

Tubail, K., et al., 2008. Gypsum additions reduce ammonia nitrogen losses during composting of dairy manure and biosolids. Compost Sci. Util. 16 (4), 285−293.

Tyler, R.W., 1996. Winning the Organics Game. ASHS Press, Alexandria, VA.

USEPA (United States Environmental Protection Agency), 2006. Erosion Control Calculator. Available at: https://archive.epa.gov/wastes/conserve/tools/greenscapes/web/pdf/erosion. pdf.

Consulted and suggested references

BioCycle, 1998. Composting for Manure Management. J.G. Press, Emmaus, PA.

CET, 2000. Building a Market-Based System of Farm Composting of Commercial Food Waste. Center for Ecological Technology, Northampton, MA. https://www. centerforecotechnology.org/.

Christian, A.H., Evanylo, G.K., Pease, J.W., 1997. On-Farm Composting: A Guide to Principles, Planning & Operations. Virginia Cooperative Extension. https://vtechworks.lib.vt. edu/handle/10919/48077.

Coker, C., 2015a. Calculating costs of organics processing. BioCycle. In: https://www. biocycle.net/calculating-costs-of-organics-processing/.

Coker, C., 2015b. Economic tool to evaluate organics recycling options. BioCycle. https:// www.biocycle.net/economic-tool-to-evaluate-organics-recycling-options/.

Coker, C., 2016. Weighting factors in organics recycling facility development. BioCycle. https://www.biocycle.net/weighting-factors-in-organics-recycling-facility-development/.

Komilis, D., Ham, R.K. Life Cycle Inventory and Cost Model for Mixed Municipal and Yard Waste Composting, U.S. Environmental Protection Agency. https://mswdst.rti.org/docs/ Compost_Model_OCR.pdf.

Massey, R., Lory, J., 2020. Calculating the Value of Manure as a Fertilizer Source. University of Missouri Extension. https://extension.missouri.edu/publications/g9330.

Miller, I., Angiel, J., 2009. Municipal yard trimmings composting Benet cost analysis. BioCycle 50 (7), 21.

Pandyaswargo, A.H., Premakumara, D., 2014. Financial sustainability of modern composting: the economically optimal scale for municipal waste composting plant in developing Asia. Int. J. Recycl. Org. Waste Agric. 3, 1−14.

Plana, R., 2015. Handbook for Compost Marketing. SCOW. https://www.acrplus.org/images/ project/SCOW/Handbook-for-compost-marketing.pdf.

Schwab, J., 2006. Erosion Control Alternatives Cost Calculator. USEPA GreenScapes Program.

Shah, E., Sambaraju, K., 1997. Technical and Economic Analysis of Composting Enterprises in Bangalore − India. WASTE Urban Waste Expertise Programme, Netherlands.

Appendices

Appendix	Title (link)
A	Resources for composting industry—trade associations, periodicals, etc. (https://www.compost foundation.org/A)
B	Typical characteristics of common feedstocks—moisture C, N, bulk density (https://www.compost foundation.org/B)
C	Compost recipe calculator example (https://www.compostfoundation.org/C)
D	Composting math tutorial (https://www.compostfoundation.org/D)
E	Profile of selected specific composting feedstocks (https://www.compostfoundation.org/E)
F	An example for sizing a basic ASP and its aeration system (https://www.compostfoundation.org/F)
G	Composting process troubleshooting guide (https://www.compostfoundation.org/G)
H	DIY compost monitoring tools (https://www.compostfoundation.org/H)
I	Common composting odorants (https://www.compostfoundation.org/I)
J	Selected health and safety resources (https://www.compostfoundation.org/J)
K	Worker safety regulations in the United States (https://www.compostfoundation.org/K)
L	Current compost training and certification opportunities in the United States (https://www.compost foundation.org/Education/COTC)
M	TMECC's section 02.01, Field Sampling of Compost Materials (https://cdn.ymaws.com/www. compostingcouncil.org/resource/resmgr/images/Guide_for_Collecting_Field_S.pdf)
N	Soil and organic matter primer (https://www.compostfoundation.org/N)
O	USCC's plant bioassay for pyridine herbicides (https://www.compostfoundation.org/O; https:// cdn.ymaws.com/www.compostingcouncil.org/resource/resmgr/images/USCC-PH-Fact-Sheet-3-for-web.pdf)
P	Example of a complete and comprehensive compost analysis report (https://www.compost foundation.org/P)
Q	STA and CCUP Primer (https://www.compostfoundation.org/Q)
R	Compilation of compost analysis data (https://www.compostfoundation.org/R)

Sources of photographs and external graphics

The photograph featured above is an artfully edited image of the West Marin composting facility in Nicasio, CA. The photo was taken and edited by Will Bakx. Will passed away days before *The Composting Handbook* was printed. In addition to being a skilled photographer, Will was a well-regarded Zero Waste leader, who shared an extraordinary legacy of compost research, practice, and teaching. A visionary in his field of composting, Will's work touched, and will yet touch, countless people. He was a friend to many of us within and beyond the composting world.

***All photographs that do not have a source listed were provided by Robert Rynk.
All artwork was created by Doug Pinkerton, unless otherwise indicated.***

Individuals

John Aber, Professor Emeritus, University of New Hampshire, Durham, NH

Ron Alexander, President, R. Alexander and Associates, Apex, NC

Will Bakx, Sonoma Compost, Nicasio, CA

Eric Bendfeldt, Extension Specialist, Virginia Tech/Virginia Cooperative Extension, Harrisonburg, VA

Johannes Biala, Director, Centre for Recycling of Organic Waste & Nutrients (CROWN), The University of Queensland, Gatton, QLD, Australia

John Biernbaum, Professor Emeritus, Michigan State University, East Lansing, MI

Jean Bonhotal, Director, Cornell Waste Management Institute, Cornell University, Ithaca, NY

Monica Bruckner, Science Education Resource Center (SERC), Carlton College, Northfield, MN

Andrew, Carpenter Principal and Soil Scientist, Northern Tilth, LLC., Belfast, ME

Craig, Coker, Principal, Coker Composting and Consulting, Troutville, VA

Leslie Cooperband, Owner and operator, Prairie Fruits Farm & Creamery, Champagne, IL

Matt Cotton, Owner, Integrated Waste Management Consulting, Richmond, CA

Jeff Creque, Director of Rangeland and Agroecosystem Management, Carbon Cycle Institute, Petaluma, CA

Bill Davis, Graphic Artist, Formerly, Cornell Waste Management Institute, Ithaca, NY

Gregory Evanylo, Professor, School of Plant and Environmental Sciences, Virginia Polytechnic Institute and State University, Blacksburg, VA

Britt Faucette, Director of Research, Technical, & Environmental Services, Filtrexx International, Decatur, GA

Scott Gamble, Organic Waste Specialist, Professional Engineer, Edmomton, AL

Greg Gelewski, Deputy Director, Madison County Solid Waste Administrative, Wampsville, NY

Jane Gilbert, Principal and Chief Scientist, Carbon Clarity, Rushden, Northamptonshire, UK

Karin Grobe, Santa Cruz, CA

James Hardin, Associate Professor, SUNY Cobleskill, Cobleskill, NY

Harry A. Hoitink, Professor Emeritus, Ohio State University, Wooster, OH

Harold Keener, Professor Emeritus, Professor Emeritus, Ohio State University, Wooster, OH

Mark King, Organics Management Specialist, Maine Dept. of Environmental Protection, Bangor, ME

Ji Li , Professor, Department of Ecology and Ecological Engineering, China Agricultural University, Beijing, China

Robert Light, Former Associate Dean, College of Food and Natural Resources, University of Massachusetts, Amherst, MA

Dan Lilkas-Rain, Recycling Coordinator, Town of Bethlehem, Bethlehem, NY

Brendon Mallia, Owner, Composting New Zealand Limited, Kapiti, New Zealand

Daryl McCartney, Professor in Environmental Engineering, University of Alberta, Edmonton, AB

James McSweeney, Owner & Technical Lead, Compost Technical Services, Arlington, MA

Frederick Michel, Professor, Ohio State University, Wooster, OH

Dustin Montey, Biomass Processing Asst Mgr, Shakopee Mdewakanton Sioux Community, Prior Lake, MN

Peter Moon, President, Principal Engineer, O2Compost, Snohomish. WA

Deborah Neher, Professor, University of Vermont, Burlington, VT

Cary Oshins, Associate Director and Director of Certification, U.S. Composting Council, Raleigh, NC

Monica Ozores-Hampton, CEO and Co-Founder, TerraNutri, LLC, Miami Beach, FL

Bob Paulin, Western Australian Department of Agriculture, South Perth, WA

Josh Payne, State Poultry Specialist, Oklahoma State University, Cooperative Extension Service, Muskogee, OK

Doug Pinkerton, Graphics Designer, Pinkerton Design, Bethlehem, PA

Jack Pos, Professor Emeritus, University of Guelph, Guelph, ON

Mary Schwarz, Extension Support Specialist, Cornell Waste Management Institute, Cornell University, Ithaca, NY

Stefanie Siebert, Executive Director, European Compost Network, Bochum

Matthew Smith, Research Ecologist, USDA National Agroforestry Center, Lincoln, NE

Bernadine Strik, Professor, Oregon State University, Corvallis, OR

Rod Tyler, President and Founder, Green Horizons Environmental, Medina, OH

Jessica Wallister, Horticulturist and author, Savvy Gardening, Sewickley, PA, https://www.jessicawalliser.com/

Eric Walters, Black Bear Composting, Crimora, VA, www.blackbearcomposting.com

John Wick, Marin Carbon Project, Point Reyes Station, CA

Kevin Wilkinson, Director, Frontier Ag & Environment, Smythes Creek, Victoria, Australia

Steven Wisbaum, President, CV Compost, Charlotte, VT

Emma Wright, Research Technician, University of Vermont, Burlington, VT

Companies

Ag-Bag Environmental, Astoria, OR, www.ag-bagfs.com

Agrilab Technologies, Enosburg Falls, VT, https://agrilabtech.com/

BDP Industries, Greenwich, NY, https://www.bdpindustries.com/

Ludwig Bergmann GmbH, Amtsgericht Oldenburg HRB, www.bergmann-goldenstedt.de

BioCycle, Emmaus, PA, https://www.biocycle.net/

Biodegradable Products Institute (BPI), New York, NY, https://bpiworld.org/

Brown Bear Corp., Corning, IA, https://brownbearcorp.com/environmental_composting.html

California State Water Resources Control Board, Sacramento, CA, https://www.waterboards.ca.gov/

Composting News, McEntee Media, Inc., Strongsville, OH, http://compostingnews.com/

Composting New Zealand Limited, Kapiti, New Zealand, https://compostingnz.co.nz/

CV Compost, CV Compost, Charlotte, VT, https://www.cvcompost.com/

C-Wise, Nambeelup, Western Australia, https://cwise.com.au/

CWMI, Cornell Waste Management Institute, Ithaca, NY, http://cwmi.css.cornell.edu/

Dirt Hugger, Dallesport, WA, https://www.dirthugger.com/

ECS, Engineered Compost Solutions, Seattle, WA, https://compostsystems.com/

EDGE Innovate, Dungannon, Co. Tyrone, N. Ireland, https://edgeinnovate.com/applications/compost/

EREF, Environmental Research and Education Foundation, Raleigh, NC, https://erefdn.org/

Global Composting Solutions, Christchurch, New Zealand, https://www.globalcomposting.solutions/

GMT, Green Mountain Technologies, NE Bainbridge Island, WA, https://www.compostingtechnology.com/

Grassdale Fertilisers, Toowoomba, QLD, Australia, https://www.grassdalefert.com.au/

IEUA, Inland Empire Utilities Agency, Inland Empire Utilities Agency, Chino, CA, https://www.ieua.org/facilities/inland-empire-regional-composting-facility/

Komptech Americas, Denver, CO, https://komptechamericas.com/application/commercial-composting/

Koster Moisture Tester, Inc., Canton, OH, https://buykoster.com/

Midwest Laboratories, Omaha, NE, https://midwestlabs.com/

Morbark, LLC, Winn, MI, https://www.morbark.com/products/recycling/

NeuStarr, Greer, SC, https://www.neustarr.com/

O2Compost, Snohomish, WA, www.o2compost.com

OCRRA, Onondaga County Resource Recovery Agency, North Syracuse, NY, www.OCRRA.org

Particle Technology Labs, Downers Grove, IL, https://www.particletechlabs.com/

Peats Soil & Garden Supplies, Willunga, South Australia, https://www.peatssoil. com.au/

Pennsylvania State University Agricultural Analytical Services Laboratory, State College, PA, https://agsci.psu.edu/aasl

Peterson Pacific (ASTEC Industries, Inc.), Eugene, OR, www.astecindustries.com

ReoTemp Instruments, San Diego, CA, https://reotemp.com/compost/

Rotochopper, Inc., St. Martin, MN, https://www.rotochopper.com/industries/ compost/

SCARAB International, LLLP, White Deer, TX, https://www.scarabmfg.com/

Soil Control Lab, Watsonville, CA, https://www.scarabmfg.com/

St. Louis Composting., Valley Park, MO, https://www.stlcompost.com/

Tel-Tru Manufacturing Co., Rochester, NY, https://www.teltru.com/

Tracy Material Recovery Facility and Solid Waste Transfer, Tracy, CA, https:// tracymaterialrecovery.com/

USCC, U.S. Composting Council, Raleigh, NC, https://www.compostingcouncil. org/

Vermeer, Pella, IA, https://www.vermeer.com/na/recycling

W. L. Gore & Associates, Inc, Newark, DE, https://www.gore.com/resources/gore-cover-for-biosolids-composting

Whatcom County Cooperative Extension, Washington State University, Bellingham, WA, https://extension.wsu.edu/whatcom/

Compost Brand Logos

All Treat Farms (All Treat Farms Compost), Arthur, ON, https://alltreat.com/

Dr. Gobbler (Dr. Gobbler's Soil RX), Clifton, TX, http://www.drgobbler.com/

eco Landscaping (Eco Mix Compost), Christchurch, Dorset, UK, https://www. thisiseco.co.uk/landscaping/compost.html

Envirem Organics Inc. (Greenhouse Gold Seafood Compost), Fredericton, NB, https://envirem.com/

McGill (McGill Soil Builder), Merry Oaks, NC, https://mcgillcompost.com/mcgill-compost-products

Mecklenburg County (Queen City Compost), Charlotte, NC, https://www.mecknc. gov/LUESA/SolidWaste/Disposal-Recycling/Pages/Compost-Mulch-Info-Prices. aspx

Reterra/Remondis Ag & Co. KG (Reterra ActivCompost), Lünen, Germany, https:// reterra.de/

RichGro, RichGro Black Marvel, Jandakot, Western Australia, https://www.richgro. com.au/

SoilPro Products (SoilPro Premium), Rancho Cucamonga, CA, https://www.
soilproproducts.com/

Thoroughbred Landscape Products (Thoroughbred Compost), Lexington, KY,
https://www.thoroughbredlandscapeproducts.com/

Vermont Natural Ag Products (Moo Doo Compost), Middlebury, VT, http://www.
vermontnaturalagproducts.com/

Index

Note: 'Page numbers followed by "*b*" indicate boxes, "*f*" indicate figures and "*t*" indicate tables.'

Printed in the United States
by Baker & Taylor Publisher Services